Climate and Weather of the Sun-Earth System (CAWSES)

Springer Atmospheric Sciences

For further volumes:
www.springer.com/series/10176

Franz-Josef Lübken
Editor

Climate and Weather of the Sun-Earth System (CAWSES)

Highlights from a Priority Program

Springer

Editor
Franz-Josef Lübken
Leibniz Institute of Atmospheric Physics
Rostock University
Kühlungsborn, Germany

ISSN 2194-5217 ISSN 2194-5225 (electronic)
Springer Atmospheric Sciences
ISBN 978-94-007-4347-2 ISBN 978-94-007-4348-9 (eBook)
DOI 10.1007/978-94-007-4348-9
Springer Dordrecht Heidelberg New York London

Library of Congress Control Number: 2012945619

Printed on acid-free paper

Springer is part of Springer Science+Business Media (www.springer.com)

Foreword

The Sun, combined with the Earth's orbit, has always governed the activities of humanity on timescales from the changing seasons to glacial cycles. Changes in climate can be driven by alterations of the total solar irradiance (TSI) but recent evidence from satellites shows only modest changes in TSI which has only a minimal impact on the current changes that are being observed in the climate system. However the solar insolation at ultra-violet and extreme ultra-violet wavelengths changes from ~7 % to more than 100 % over a solar cycle and hence can result in major changes in the middle and upper atmospheres. For example, weaker westerly winds at the Earth's surface are observed in winters at sunspot minimum, and at sunspot maximum the temperature of the thermosphere is about 400 K greater than at sunspot minimum and the winds are about twice as strong.

Energetic particles from the Sun can change the climate system too. Observations of surface air temperatures, derived from climate re-analysis data, show differences between periods when there are significant space weather events compared to those periods when events are absent [*Seppälä et al.*, 2009]. Differences can be as large as ~4 K with areas of warming and cooling approximately the same suggesting a redistribution of energy. However, the mechanism by which these changes occur is not known.

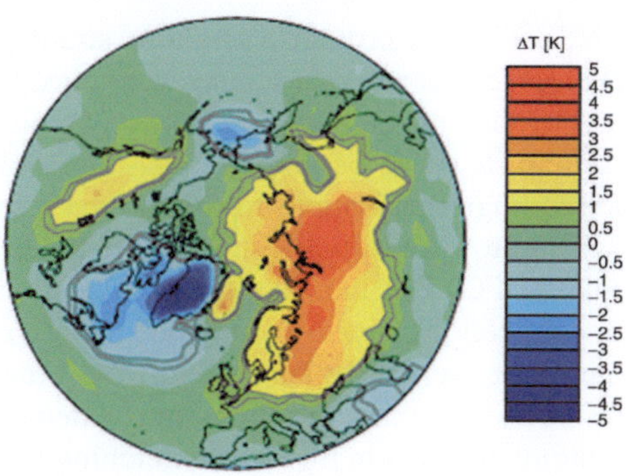

As changes in the Earth's atmosphere occur, whether due to changes of solar origin or in response to enhanced green house gas concentrations, the propagation and dissipation of gravity and planetary waves, and tides is altered. Clouds, whether they occur in the troposphere, stratosphere and mesosphere have a marked impact on the atmosphere. Thus, changes can occur on all spatial and temporal scales and both within and between different levels of the atmosphere caused by solar variations.

Addressing the Sun-Earth connections is both an essential and urgent issue. To understand both the natural variability of the climate system and those caused by humanity is essential for the prediction of future climate scenarios that have greater confidence. Therefore it is timely for the international community to have a focused scientific effort to address these critical issues.

Research over previous decades has provided solid explanations of many of the individual processes that are involved in the coupling of the Sun to the Earth environment but to make real progress at these scientific frontiers a system-level approach is now required, and indeed possible for the first time. The necessary science infrastructure has developed to a point where it is possible to address these topics. The worldwide research community now has access to international data sets from every critical region in the space environment, a highly-distributed, ground-based network of sensor, virtual observatories, advanced computational and visualisation facilities, and sophisticated Sun-to-Earth community models.

But the availability of data and models is not enough; an operating framework is also required and this has been provided by the Scientific Committee on Solar-Terrestrial Physics (SCOSTEP). It was established as an Inter-Union Commission of International Council of Science (ICSU) in 1966 with the aim to promote and organise international interdisciplinary programmes of limited duration in solar-terrestrial physics. In 2004 SCOSTEP began the Climate and Weather of the Sun-Earth System program—CAWSES—with the specific objective to enhance understanding of the space environment and its impact on space weather and the climate. CAWSES made very substantial progress since its inception not only in pushing back the frontiers of knowledge but also in building science capacity in developing countries. It held the first virtual conference which had 270 participants. After five years, it was recognised that there was much still to be achieved and hence SCOSTEP endorsed CAWSES-II to run from 2009–2013.There was re-focusing of the working groups to reflect the evolution of science questions but the fundamental aim remained the same.

CAWSES and CAWSES-II only provide a scientific strategy and framework that gives focus for the development of relevant science activities. For such international initiatives to be successful the engagement of scientists across many disciplines in many countries is essential. It needs passionate scientific leadership, and champions who can persuade funding organisations to support the research. Funding agencies also have to be willing to take initiatives, and sometimes risks. The Deutsche Forschungsgemeinschaft (DFG) German Research Foundation recognised the importance of Sun-Earth connection science and created a German CAWSES Priority Programme. The Programme, very ably led by Prof. Dr. Franz-Josef Lübken, has explored many aspects of Sun-climate links as is amply demonstrated by the comprehensive and illuminating chapters in this book stretching from fundamental solar physics through to the surface of the planet. There is an excellent blend of theory, observation and modelling. The Programme has also been very successful in educational terms giving many opportunities for early career scientists. We commend both the German CAWSES Priority Programme and the book—both are truly excellent exemplars for the research community.

References

Seppälä, A., Randall, C.E., Clilverd, M.A., Rozanov, E., & Rodger, C.J. (2009). Geomagnetic activity and polar surface level air temperature variability. *Journal of Geophysical Research, 114*, A10312. doi:10.1029/2008JA014029.

Chairs of CAWSES

Alan Rodger and Susan Avery

Preface

The Sun is the most important external driver of climate. Although the total solar irradiance (TSI) fluctuates by less than 0.1 % during a solar cycle, the solar impact on the terrestrial atmosphere can be significant, in particular in the upper atmosphere where the highly variable energetic part of the solar spectrum is absorbed. TSI variations on centennial time scales are of similar magnitude. Unfortunately, the scientific understanding of the variation of solar radiation (and its spectral components) and its impact on the atmosphere is rather limited. This concerns the direct modification of composition by solar radiation, but even more so various coupling mechanisms. For example, photochemically active gases are generated in the upper atmosphere, propagate to lower layers where they substantially alter the composition, e.g., the abundance of ozone. Planetary waves, gravity waves, and tides are excited in the atmosphere and propagate over large distances. They transport trace gases, energy, and momentum. Although these waves are most pronounced in the middle atmosphere, they may modify the background circulation in the entire atmosphere, even in the troposphere. As can be seen in various chapters in this book, major progress has been achieved in our understanding of solar radiation, its impact on the atmosphere, and coupling mechanisms within the atmosphere. Still, various uncertainties exist. For example, the observed solar signal in some parts of the middle atmosphere is larger than can be explained by models.

The purpose of this book is to summarise the scientific results related to a major research program of the international SCOSTEP organisation (Scientific Committee on Solar-Terrestrial Physics) called CAWSES (Climate And Weather of the Sun Earth System). The German Research Foundation (Deutsche Forschungsgemeinschaft, DFG) ran a priority program ('Schwerpunktprogramm', SPP) from 2005 to 2011 focusing on several important science topics from CAWSES. The aim of the CAWSES SPP is a better understanding of the influence of the Sun on the terrestrial atmosphere and the physical/chemical processes involved in various coupling mechanisms within the atmosphere. The DFG spent approximately 10 Million Euro for a total of 25–30 institutes. Most of the financial support from DFG was used for postdoctoral and PhD student positions. In 31 chapters, this book presents the scientific summaries from 28 projects supported under a total of 93 individual grants. International cooperation was strongly encouraged and supported by funds for travel, etc. At each stage of the proposals, an international team of reviewers assisted the DFG in their selections. Several hundred papers were published in international peer reviewed journals within the priority program.

The book is structured in six parts, namely i) solar radiation, heliosphere, and galactic cosmic rays (Chaps. 2 to 6), ii) solar influence on trace gases (Chaps. 7 to 10), iii) thermosphere, energetic particles, and ionisation (Chaps. 11 to 17), iv) mesospheric ice clouds (Chaps. 18–20), v) gravity waves, planetary waves, and tides (Chaps 21 to 28), and vi) large-scale coupling (Chaps. 29 to 32). The evolution of solar radiation, its effect on galactic cosmic rays (GCR), and their potential impact on cloud droplets in the troposphere are studied in detail. Some chapters concentrate on the direct effect of solar radiation on trace gases, mainly on ozone and water vapour in the stratosphere and mesosphere. Precipitating particles of solar or geomagnetic origin modify the upper atmosphere directly. They also produce photochemically active species which can be transported downward and significantly affect the mesosphere and upper stratosphere. This aspect is studied in several chapters. Results on short and long-term variations of mesospheric ice clouds, as well as related microphysical aspects are presented. Planetary waves, gravity waves, and tides play a key role in distributing and modifying a signal being imposed somewhere in the atmosphere by, for example, the solar cycle. Physical details of this process and implications of wave morphology for the entire atmosphere, from the thermosphere to the troposphere, are presented in several chapters. Finally, results on global aspects of the solar cycle, long-term variations, and comparison with anthropogenic climate change are covered.

This book addresses researchers and students who are interested in actual results on solar cycle and long term variations in the atmosphere. The chapters are written by lead scientists from nearly all major German research institutions where the terrestrial atmosphere is investigated. A brief introduction is provided at the beginning of each chapter to familiarise a broader community with the scientific background. Each chapter was subject to an international peer review process to ensure high quality.

In the name of the German CAWSES community I thank the German Research Foundation for funding the CAWSES priority program. We appreciated the constructive and stimulating support from the reviewers, some of whom accompanied our program for the entire six years. As speaker of this program I would like to express my appreciation for all the activities, excitement, and success being created in various groups. Perhaps most important, many students were involved and contributed to the enthusiasm when working on a wide range of scientific topics of solar-terrestrial physics. I thank Dr. Norbert Engler and Monika Rosenthal for their excellent assistance when compiling this book.

Kühlungsborn
February 2012

Franz-Josef Lübken

Contents

List of Contributors

Ulrich Achatz Institute for Atmospheric and Environmental Sciences, Goethe University, Frankfurt (Main), Germany

Christina Arras Dept. Geodesy and Remote Sensing, German Research Centre for Geosciences GFZ, Potsdam, Potsdam, Germany; GFZ German Research Centre for Geosciences, Potsdam, Germany

Alan D. Aylward Atmospheric Physics Laboratory, University College London, London, UK

Andreas J.G. Baumgaertner Max Planck Institute for Chemistry, Mainz, Germany; Deutsches Zentrum für Luft-und Raumfahrt (DLR), Project Management Agency, Bonn, Germany

Gerd Baumgarten Leibniz-Institute of Atmospheric Physics, Kühlungsborn, Germany

Erich Becker Institute of Atmospheric Physics, Kühlungsborn, Germany

Uwe Berger Leibniz-Institute of Atmospheric Physics, Kühlungsborn, Germany

Georg Beyerle GFZ German Research Centre for Geosciences, Potsdam, Germany

Jan Philipp Bornebusch University of Osnabrück, Osnabrück, Germany

Heinrich Bovensmann Institute of Environmental Physics, University of Bremen, Bremen, Germany

Harald Bönisch Experimental Atmospheric Research Institute for Atmospheric and Environmental Sciences, Goethe-University Frankfurt, Frankfurt am Main, Germany

Christoph Brühl Max-Planck-Institut für Chemie, Mainz, Germany

John P. Burrows Institute of Environmental Physics, University of Bremen FB1, Bremen, Germany

Ulrich Cubasch Institut für Meteorologie, Freie Universität Berlin, Berlin, Germany

Matt DeLand Science System and Applications, Inc (SSAI), Lanham, MD, USA

Matthew T. DeLand Science Systems and Applications, Inc., Lanham, MD, USA

Sebastian Dikty Institute of Environmental Physics, University of Bremen FB1, Bremen, Germany

Denis Duft Institute for Meteorology and Climate Research – Atmospheric Aerosol Research (IMK-AAF), Karlsruhe Institute of Technology (KIT), Karlsruhe, Germany

Andreas Engel Experimental Atmospheric Research Institute for Atmospheric and Environmental Sciences, Goethe-University Frankfurt, Frankfurt am Main, Germany

Norbert Engler Leibniz-Institute of Atmospheric Physics at the Rostock University, Kühlungsborn, Germany

Manfred Ern Institute of Energy and Climate Research (IEK-7), Forschungszentrum Jülich, Jülich, Germany; Institute for Energy and Climate Research – Stratosphere (IEK-7), Jülich, Germany

Antonia Faber Dept. Geodesy and Remote Sensing, German Research Centre for Geosciences GFZ, Potsdam, Potsdam, Germany

Artem G. Feofilov Laboratory of Dynamical Meteorology, École Polytechnique, Palaiseau Cedex, France

Horst Fichtner Theoretische Physik IV, Ruhr-Universität Bochum, Bochum, Germany

Jens Fiedler Leibniz-Institute of Atmospheric Physics, Kühlungsborn, Germany

Linton E. Floyd Interferometrics Inc., Herndon, Virginia, VA, USA

Kristina Fröhlich Dept. Climate and Environment, Deutscher Wetterdienst, Offenbach, Germany

Bernd Funke Instituto de Astrofisica de Andalucia (CSIC), Granada, Spain

Axel Gabriel Leibniz-Institute of Atmospheric Physics at the University Rostock, Kühlungsborn, Germany

Norbert Glatthor Karlsruhe Institute of Technology, Institute for Meteorology and Climate Research, Eggenstein-Leopoldshafen, Germany

Sven H.W. Golchert Institut für Klimaforschung, KIT Karlsruhe, Eggenstein-Leopoldshafen, Germany

Hans-F. Graf Centre for Atmospheric Science, Cambridge University, Cambridge CB2 3EN, UK

Norbert Grieger Leibniz-Institute of Atmospheric Physics, Kühlungsborn, Germany

Aleksandr N. Gruzdev A. M. Obukhov Institute of Atmospheric Physics, Russian Academy of Sciences, Moscow, Russia

Florian Haenel Karlsruhe Institute of Technology, Institute for Meteorology and Climate Research, Eggenstein-Leopoldshafen, Germany

Kristoffer Hallgren Max-Planck-Institut für Sonnensystemforschung, Katlenburg-Lindau, Germany

Jerry W. Harder Laboratory for Atmospheric and Space Physics (LASP), University of Colorado, Boulder, CO, USA

Matthew J. Harris Atmospheric Physics Laboratory, University College London, London, UK

Paul Hartogh Max-Planck-Institut für Sonnensystemforschung, Katlenburg-Lindau, Germany

Kathrin Häusler GFZ German Research Centre for Geosciences, Potsdam, Germany; High Altitude Observatory, National Center for Atmospheric Research, Boulder, CO, USA

Bernd Heber Institut für Experimentelle und Angewandte Physik, Christian-Albrechts-Universität zu Kiel, Kiel, Germany

Stefan Heise GFZ German Research Centre for Geosciences, Potsdam, Germany

Klaudia Herbst Institut für Experimentelle und Angewandte Physik, Christian-Albrechts-Universität zu Kiel, Kiel, Germany

Gerd Hochschild Institut für Klimaforschung, KIT Karlsruhe, Eggenstein-Leopoldshafen, Germany

Lars Hoffmann IEK-7, Forschungszentrum Jülich GmbH, Jülich, Germany

Peter Hoffmann Leibniz-Institute of Atmospheric Physics at the Rostock University, Kühlungsborn, Germany

Ines Höschel Institut für Meteorologie, Berlin, Germany

Christoph Jacobi Institute for Meteorology, University of Leipzig, Leipzig, Germany

Christopher Jarchow Max-Planck-Institut für Sonnensystemforschung, Katlenburg-Lindau, Germany

Patrick Jöckel Max Planck Institute for Chemistry, Mainz, Germany; Institut für Physik der Atmosphäre, Deutsches Zentrum für Luft-und Raumfahrt (DLR), Weßling, Germany

Silvio Kalisch Institute of Energy and Climate Research (IEK-7), Forschungszentrum Jülich, Jülich, Germany

May-Britt Kallenrode University of Osnabrück, Osnabrück, Germany

Martin Kaufmann Institute for Chemistry and Dynamics of Geosphere, Forschungszentrum Jülich GmbH, Jülich, Germany; Institute for Energy and Climate Research – Stratosphere (IEK-7), Jülich, Germany

Jens Kieser Max Planck Institute for Meteorology, Hamburg, Germany

Gregor Kiesewetter Institute of Environmental Physics, University of Bremen, Bremen, Germany

Johannes Kiliani Leibniz-Institute of Atmospheric Physics, Kühlungsborn, Germany

Ingo Kirchner Institut für Meteorologie, Berlin, Germany

G. Kishore Kumar Leibniz-Institute of Atmospheric Physics at the Rostock University, Kühlungsborn, Germany

Paul Konopka Institute for Energy and Climate Research – Stratosphere (IEK-7) Forschungszentrum Jülich GmbH (FZJ), Jülich, Germany

Andreas Kopp Institut für Experimentelle und Angewandte Physik, Christian-Albrechts-Universität zu Kiel, Kiel, Germany

Ralf Koppmann Faculty of Mathematics and Natural Sciences, Physics Department, University of Wuppertal, Wuppertal, Germany; Physics Department, University of Wuppertal, Wuppertal, Germany

Marc Krebsbach Physics Department, University of Wuppertal, Wuppertal, Germany

Natalie A. Krivova Max-Planck-Institut für Sonnensystemforschung, Katlenburg-Lindau, Germany

Anne Kubin Institut für Meteorologie, Freie Universität Berlin, Berlin, Germany

Markus Kunze Institut für Meteorologie, Freie Universität Berlin, Berlin, Germany

Alexander A. Kutepov Department of Physics, The Catholic University of America/NASA Goddard Space Flight Center, Greenbelt, MD, USA

Karin Labitzke Institut für Meteorologie, Freie Universität Berlin, Berlin, Germany

Ulrike Langematz Institut für Meteorologie, Freie Universität Berlin, Berlin, Germany

Catrin Lehmann Institute for Energy and Climate Research – Stratosphere (IEK-7), Jülich, Germany

Thomas Leisner Institute for Meteorology and Climate Research – Atmospheric Aerosol Research (IMK-AAF), Karlsruhe Institute of Technology (KIT), Karlsruhe, Germany

Qiang Li Leibniz-Institute of Atmospheric Physics at the Rostock University, Kühlungsborn, Germany

Andrea Linden Karlsruhe Institute of Technology, Institute for Meteorology and Climate Research, Eggenstein-Leopoldshafen, Germany

Franz-Josef Lübken Leibniz-Institute of Atmospheric Physics, Kühlungsborn, Germany

Hermann Lühr GFZ German Research Centre for Geosciences, Potsdam, Germany

Stefanie Marker ILS Kraftfahrzeuge, Technische Universität Berlin, Berlin, Germany

Vivien Matthias Leibniz-Institute of Atmospheric Physics at the Rostock University, Kühlungsborn, Germany

Alexander S. Medvedev Max Planck Institute for Solar System Research, Katlenburg-Lindau, Germany

Stergios Misios Max Planck Institute for Meteorology, Hamburg, Germany

Nicholas J. Mitchell Centre for Space, Atmospheric and Oceanic Science, Department of Electronic and Electrical Engineering, University of Bath, Bath, UK

Martin G. Mlynczak NASA Langley Research Center, Hampton, VA, USA

Tanja Möbius Experimental Atmospheric Research Institute for Atmospheric and Environmental Sciences, Goethe-University Frankfurt, Frankfurt am Main, Germany

Justus Notholt Institut für Umweltphysik, Universität Bremen, Bremen, Germany

Jens Oberheide Department of Physics and Astronomy, Clemson University, Clemson, SC, USA

Dirk Offermann Faculty of Mathematics and Natural Sciences, Physics Department, University of Wuppertal, Wuppertal, Germany

Joseph Pagaran Institute of Environmental Physics, University of Bremen FB1, Bremen, Germany

Mathias Palm Institut für Umweltphysik, Universität Bremen, Bremen, Germany

Adalbert W.A. Pauldrach University Observatory Munich/Wendelstein Observatory, Munich, Germany

Dieter H.W. Peters Leibniz-Institute of Atmospheric Physics at the University Rostock, Kühlungsborn, Germany

Peter Preusse Institute of Energy and Climate Research (IEK-7), Forschungszentrum Jülich, Jülich, Germany

Nabiz Rahpoe Institute of Environmental Physics, University of Bremen, Bremen, Germany

Markus Rapp Leibniz-Institute of Atmospheric Physics at the Rostock University, Kühlungsborn, Germany; Deutsches Zentrum für Luft- und Raumfahrt (DLR), Institute of Atmospheric Physics, Wessling, Germany

Thomas Reddmann Karlsruhe Institute of Technology (KIT), Inst. for Meteorology and Climate Research, Karlsruhe, Germany

Martin Riese Institute for Energy and Climate Research – Stratosphere (IEK-7), Jülich, Germany

Charles Robert Belgium Institute of Space Aeronomy (BIRA), Brussels, Belgium

Franz Rohrer IEK-8, Forschungszentrum Jülich GmbH, Jülich, Germany

Susanne Rohs IEK-7, Forschungszentrum Jülich GmbH, Jülich, Germany

Jürgen Röttger Max-Planck-Institute of Solar System Research, Katlenburg-Lindau, Germany

Daniel Rzesanke Institute for Meteorology and Climate Research – Atmospheric Aerosol Research (IMK-AAF), Karlsruhe Institute of Technology (KIT), Karlsruhe, Germany

Klaus Scherer Theoretische Physik IV, Ruhr-Universität Bochum, Bochum, Germany

Cornelius Schiller IEK-7, Forschungszentrum Jülich GmbH, Jülich, Germany

Hauke Schmidt Max Planck Institute for Meteorology, Hamburg, Germany

Torsten Schmidt Dept. Geodesy and Remote Sensing, German Research Centre for Geosciences GFZ, Potsdam, Potsdam, Germany

Fabian Senf Leibniz-Institute of Atmospheric Physics, Kühlungsborn, Germany

Werner Singer Leibniz-Institute of Atmospheric Physics at the Rostock University, Kühlungsborn, Germany

Björn-Martin Sinnhuber Institute of Environmental Physics, University of Bremen, Bremen, Germany; Now at Karlsruhe Institute of Technology, Karlsruhe, Germany

Miriam Sinnhuber Institut für Klimaforschung, KIT Karlsruhe, Eggenstein-Leopoldshafen, Germany; Institute of Environmental Physics, University of Bremen, Bremen, Germany; Institute for Meteorology and Climate Research, Karlsruhe Institute of Technology, Leopoldshafen, Germany

Sami K. Solanki Max-Planck-Institut für Sonnensystemforschung, Katlenburg-Lindau, Germany; School of Space Research, Kyung Hee University, Gyeonggi, Yongin, Korea

Reinhold Spang IEK-7, Forschungszentrum Jülich GmbH, Jülich, Germany

Thomas Spangehl Institut für Meteorologie, Freie Universität Berlin, Berlin, Germany; Deutscher Wetterdienst, Offenbach am Main, Germany

Gabriele Stiller Karlsruhe Institute of Technology (KIT), Inst. for Meteorology and Climate Research, Karlsruhe, Germany; Karlsruhe Institute of Technology, Institute for Meteorology and Climate Research, Eggenstein-Leopoldshafen, Germany

Irina Strelnikova Leibniz-Institute of Atmospheric Physics at the Rostock University, Kühlungsborn, Germany

Georg Teiser Leibniz-Institute of Atmospheric Physics at the Rostock University, Kühlungsborn, Germany

Stefan Versick Karlsruhe Institute of Technology (KIT), Inst. for Meteorology and Climate Research, Karlsruhe, Germany

Bärbel Vogel Institute for Energy and Climate Research – Stratosphere (IEK-7) Forschungszentrum Jülich GmbH (FZJ), Jülich, Germany

Thomas von Clarmann Karlsruhe Institute of Technology, Institute for Meteorology and Climate Research, Eggenstein-Leopoldshafen, Germany

Christian von Savigny Institute of Environmental Physics, University of Bremen FB1, Bremen, Germany; Institute of Physics, Ernst-Moritz-Arndt-University of Greifswald, Greifswald, Germany

Mark Weber Institute of Environmental Physics, University of Bremen FB1, Bremen, Germany

Jens Wickert Dept. Geodesy and Remote Sensing, German Research Centre for Geosciences GFZ, Potsdam, Potsdam, Germany

Nadine Wieters Institute of Environmental Physics, University of Bremen, Bremen, Germany

Holger Winkler Institute of Environmental Physics, University of Bremen, Bremen, Germany

Jan Maik Wissing University of Osnabrück, Osnabrück, Germany

Erdal Yiğit Department of Atmospheric Oceanic and Space Sciences, Ann Arbor, MI, USA; Space Sciences Laboratory, UC Berkeley, Berkeley, CA, USA

Chapter 1
Scientific Summary of the German CAWSES Priority Program

Franz-Josef Lübken

Abstract The German Research Foundation (Deutsche Forschungsgemeinschaft, DFG) launched a 'priority program' (Schwerpunktprogramm, SPP) focusing on science topics related to CAWSES (Climate And Weather of the Sun Earth System) which refers to a program of the international SCOSTEP organization (Scientific Committee on Solar-Terrestrial Physics). The CAWSES-SPP ran from 2005 to 2011. In 31 chapters, this book presents the scientific highlights from 28 projects supported under a total of 93 individual grants. This chapter summarizes some important results from this book and puts them into a CAWSES program perspective. Long-term trends and solar cycle variations in various parameters are studied in basically all chapters. Five chapters cover the evolution of solar radiation, its effect on galactic cosmic rays (GCR), and their potential impact on cloud droplets in the troposphere. Four chapters concentrate on the direct effect of solar radiation on trace gases, mainly on ozone and water vapor in the stratosphere and mesosphere. Precipitating particles of solar or geomagnetic origin modify the upper atmosphere directly. They also produce photochemically active species which can be transported downward and significantly affect the mesosphere and upper stratosphere. This aspect is studied in seven chapters. Microphysical aspects of mesospheric ice clouds, and their short and long-term variation, are studied in three chapters. Planetary waves and gravity waves (including tides) play a key role in distributing and modifying a signal being imposed somewhere in the atmosphere by, for example, the solar cycle. Physical details of this process and implications of wave morphology for the entire atmosphere, from the thermosphere to the troposphere, are presented in eight chapters. Finally, results on global aspects of the solar cycle and long-term variations are presented in four chapters. Basically all chapters involve observations and modeling.

F.-J. Lübken (✉)
Leibniz-Institute of Atmospheric Physics, Schlossstr. 6, 18225 Kühlungsborn, Germany
e-mail: luebken@iap-kborn.de

F.-J. Lübken (ed.), *Climate and Weather of the Sun-Earth System (CAWSES)*,
Springer Atmospheric Sciences,
DOI 10.1007/978-94-007-4348-9_1,

1.1 Introduction

The purpose of this chapter is to summarize some important scientific results from the remaining 31 chapters in this book. Considering the quantity and quality of scientific results presented in more than 600 pages, it is obvious that such a summary cannot cover all aspects. An important advantage of a concerted scientific action like this priority program of the German Research Foundation is that various groups with different expertise worked together on a common research topic. Indeed, basically all groups involved in the CAWSES-SPP had collaborations with other groups, thus creating substantial synergy effects. Some of the cross-references to other chapters in this book are explicitly mentioned in the text, while others are obvious from the context. Although CAWSES-SPP was primarily a German program, all projects had strong cooperations with several international research institutes.

The aim of CAWSES is a better understanding of the influence of the Sun on the terrestrial atmosphere on time scales from hours to centuries. The focus of the CAWSES-SPP was on scientific problems dealing with important aspects of the solar-terrestrial system. The Sun influences the atmosphere through the absorption of radiation and energetic particles, through the generation and modification of photochemically active trace gases, and through the generation of waves, including tides. The physical and chemical processes involved are coupled through various complicated mechanisms. Although the total solar irradiance fluctuates by less than 0.1 %, the solar impact on the terrestrial atmosphere can be significant, in particular in the upper atmosphere where the highly variable energetic part of the solar spectrum is absorbed. A local disturbance can be distributed over large vertical and horizontal distances through various coupling mechanisms such as transport of trace gases and propagation of waves. Figure 1.1 highlights some of the basic processes and important coupling mechanisms involved.

Medium and long-term variation of solar activity and its influence on the terrestrial atmosphere is also relevant to assess the importance of natural processes in long-term trends in comparison with anthropogenic influences. The following topics were investigated in the scope of this priority program:

- Characterization of the variability of solar forcing by electromagnetic radiation and by impact of energetic particles.
- Analysis of solar forcing on the thermal, dynamical, electro-dynamical, and compositional structure of the atmosphere in the height range from the upper troposphere to the lower thermosphere and on time scales from hours to centuries. This includes investigation of neutral gas, plasma, and aerosols.
- Investigation of the coupling mechanisms in the atmosphere, including transport of trace gases, and generation, propagation and destruction of waves (e.g., planetary waves, gravity waves, tides, turbulence).
- Identification and understanding of solar signals in atmospheric parameters which are not directly influenced by the Sun, including a study of the relevant physical and photochemical processes.
- Comparison of solar induced long-term variations with anthropogenic climate change.

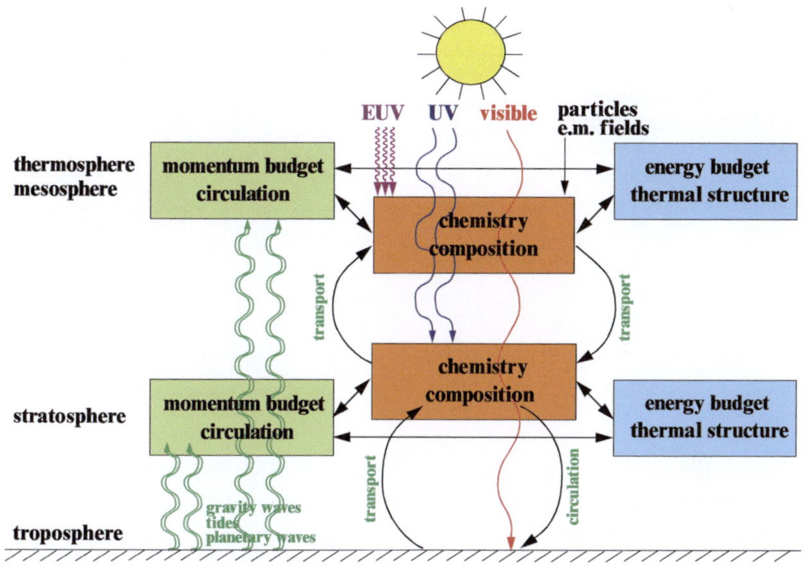

Fig. 1.1 Some fundamental physical/chemical processes of solar activity and coupling mechanisms affecting the Earth's atmosphere. Solar radiation and particles affect the terrestrial atmosphere from the thermosphere to the ground. Various coupling mechanisms exist which can distribute and modify the solar signal over large spatial scales. Several of these processes are studied in the CAWSES priority program. From *Lübken et al., J. Geophys. Res., 2010* (Copyright by American Geophysical Union)

Although space weather was not an explicit topic in the priority program, several results are highly relevant to this issue, for example tropospheric processes creating thermospheric tides which affect satellite orbits. This summary is organized according to the main topics covered in this book, namely

1. Solar radiation, heliosphere, and galactic cosmic rays (Chaps. 2 to 6)
2. Solar influence on trace gases (Chaps. 7, 8, 9, 10)
3. Thermosphere, energetic particles, and ionization (Chaps. 11 to 17)
4. Mesospheric ice clouds (Chaps. 18, 19, 20)
5. Gravity waves, planetary waves, and tides (Chaps. 21 to 28)
6. Large-scale coupling (Chaps. 29, 30, 31, 32)

For practical reasons, citations given in the chapters are not repeated here. Acronyms are listed at the end of the book.

1.2 Solar Radiation, Heliosphere, and Galactic Cosmic Rays

The first five chapters cover solar radiation and the modulation of galactic cosmic ray flux in the heliosphere and its potential impact on the troposphere. Solar variability is the most important external driver of climate. The reconstruction of total

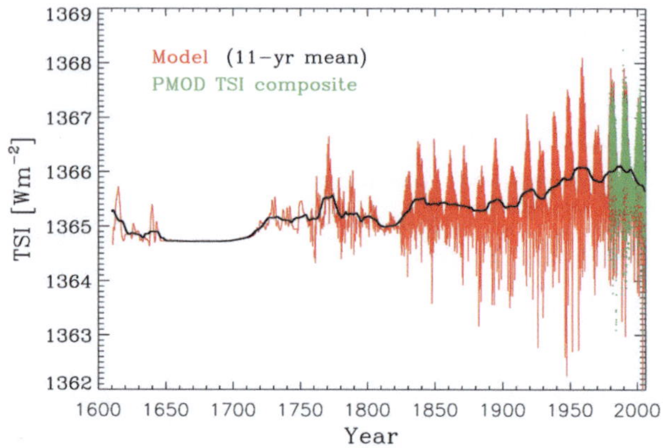

Fig. 1.2 Total solar irradiance since 1610 reconstructed using the SATIRE-T model (*red line*) and the PMOD composite of measurements since 1978 (*green*). The 11-yr smoothed TSI is shown in *black*. From *Krivova et al.* (Chap. 2)

solar irradiance (TSI) and spectral solar irradiance (SSI) is important for a reliable evaluation of the connection between solar variability and Earth's climate. Unfortunately, our information about TSI and SSI is insufficient, increasingly so when going back in time. Satellite measurements are performed only since 1978, and telescope measurements of sunspots are available since 1610. Before that, cosmogenic isotopes such as ^{14}C and ^{10}Be are used as indirect indicators of solar variability. As is shown by *Krivova et al.* (Chap. 2) the physics-based SATIRE model reproduces TSI to an accuracy of a few percent if compared to satellite observations. It is confirmed that TSI has increased by $\sim$1.25 W/m^2 since the Maunder minimum (Fig. 1.2). Surprisingly, the contribution of UV radiation to the solar cycle TSI variation is higher than previously considered. This has potential impact on the effect of solar cycle and long-term solar effects in the atmosphere.

The EUV and UV parts of the solar spectrum are absorbed in the mesosphere/lower thermosphere (MLT) and stratosphere, respectively. Therefore, a comprehensive study of the effect of solar radiation on photochemistry and composition requires measurements (or models) of spectrally resolved solar irradiance.

Weber et al. (Chap. 3) analyze SSI variations during the solar storm in October 2003 ('Halloween storm') and during solar cycles 21 to 23. Amongst others, they use first measurements of daily spectral solar irradiance from the UV to the near infrared from the SCIAMACHY satellite. During the Halloween storm, the irradiances in the near UV (above 300 nm), visible, and near IR are reduced(!) by about 0.4 %, which is in agreement with TSI reduction by the same amount observed by others. This reduction is much larger compared to solar cycle variations of TSI which is only $\sim$0.1 %. The SCIAMACHY observations are compared to the net effect of brightening by faculae and darkening by sunspots, respectively. These features on the solar surface are described by solar proxies, namely Mg-II and sunspots, respectively. It is shown that proxy based models underestimate solar cycle changes in the UV. This challenges the application of solar proxies in chemistry models and highlights the importance of long-term measurements of SSI.

The effect of solar radiation on trace gases is investigated in several chapters. Solar rotation with a period of 27 days leads to variation of UV radiation (here

205 nm) which photolysis O_2 and thereby affects ozone. From observations by SCIAMACHY *Weber et al.* find only a small 27-d signal during solar cycle 23 (0.2 % ozone change for a 1 % change in UV radiation). The effect is not persistent in time.

The Sun modulates galactic cosmic rays (GCR) entering the solar system. GCR create cosmogenic isotopes such as ^{14}C and ^{10}Be. Speculations that GCR could affect climate directly through cloud production are controversially discussed in the literature. *Fichtner et al.* (Chap. 4) investigate in detail the modulation of GCR when entering the heliosphere, magnetosphere, and the terrestrial atmosphere. They show, for example, that the production of ^{10}Be depends on various time-dependent parameters most of which are not known accurately enough. The uncertainties of various cross sections alone result in an uncertainty of the production rate of ^{10}Be by $\sim$25 %. This has potential implications for the retrieval of long-term climate records from cosmogenic isotopes.

Rohs et al. (Chap. 5) try to identify the GCR effect on clouds by studying MIPAS cloud parameters relative to neutron count data indicative for GCR. They concentrate on six events of strong solar eruptions ('coronal mass ejections') where the interaction of solar wind with GCR particles leads to a significant decrease of GCR ('Forbush decreases'). Even for these strong events they found no, or only a weak correlation between GCR and high clouds. Still, considering the entire data set they conclude that a GCR effect on clouds cannot be excluded at this point.

The impact of solar radiation via cosmic rays is believed to be introduced through particle charging affecting cloud formation. Unfortunately, many steps are involved from the charging of nuclei and droplets to cloud formation. Little is known quantitatively about the physical mechanisms involved. *Rzesanke et al.* (Chap. 6) performed laboratory studies to characterize the influence of charges on the microphysics of cloud droplets. They quantify the enhanced scavenging of aerosol particles by charged cloud droplets and measure the size dependent contact freezing probabilities. They also found a substantial decrease of saturation vapor pressure in the vicinity of charged droplets which has so far not been taken into account in cloud modeling. These laboratory results are important not only for tropospheric clouds but also for ice particles in the mesosphere.

1.3 Solar Influence on Trace Gases

Four chapters cover the direct solar effect on trace gases in the middle atmosphere. More contributions to solar modulation of trace gases are described in Sect. 1.4. Ground-based microwave radiometry allows ozone and water vapor in the stratosphere and mesosphere to be measured nearly continuously. Some observations exist for several years which allows study of the impact of solar radiation on these trace gases on time scales from days to solar cycle. *Hartogh et al.* (Chap. 7) present water vapor and ozone measurements at high and middle latitudes (69°N and 52°N). They find an anti-correlation between H_2O and solar activity in the winter mesosphere, which is expected since enhanced photo-dissociation by Ly_α during solar maximum destroys water vapor. This anti-correlation is also present in summer but is much weaker and only in the uppermost part of the height range covered by the instrument (∼70–80 km). Below that altitude the correlation is positive in summer, which is explained by increased up-welling from the 'moist' stratosphere, and/or by additional photochemical sources for H_2O (autocatalytic production). Water vapor shows a general decrease since 1996 in the mesosphere, both in summer and winter. This is surprising since an increase of H_2O is generally expected due to the increase of methane in the troposphere (source for H_2O in the mesosphere).

Palm et al. (Chap. 8) present ground based microwave observations of ozone at Spitsbergen (78°N) on time scales from days to months. As is well known, mesospheric ozone concentration is larger during night since it is destroyed by photolysis during day. The authors perform a detailed analysis by measuring ozone as a function of solar zenith angle. Even in polar summer (when the Sun is above the horizon all day) a diurnal variation is observed which is explained by the wavelength dependent attenuation of UV radiation destroying ozone. The situation is complicated by the fact that UV radiation also produces ozone by generating atomic oxygen through photo-dissociation of O_2 (atomic oxygen is a key substance in the production of ozone). During the solar storm in October/November 2003 ('Halloween storm') they measure details of a strong ozone decrease (by up to 60 %) in the mesosphere and upper stratosphere. In short, ionization by solar protons produces HO_x which destroys ozone (see next section for more details). Since HO_x is short-lived, this effect disappears shortly after the solar storm is over.

Sinnhuber et al. (Chap. 9) construct a new long-term ozone data set using various measurements from satellites and data assimilation. This data set reproduces ozone variability on intra-seasonal time scales much better compared to ECMWF ERA-40 reanalysis. The data set is analyzed, concentrating on winter at high latitudes. Systematic variations ('anomalies') related to changes in the Northern Annular Mode (NAM) are detected. NAM describes a large-scale pressure and circulation pattern in the North Atlantic. Positive NAM index is associated with stronger, colder polar vortex, and reduced ozone, whereas a negative NAM index is associated with a disturbed polar vortex and intrusion of ozone rich air from lower latitudes. Anomalies in the autumn polar vortex and related Arctic ozone variations may persist for several months and may propagate downward from the stratosphere to the troposphere (Fig. 1.3). However, this signal may be distorted by tropospheric events propagating into the stratosphere. It is therefore not surprising that an earlier observation of an unexpected (and unexplained) correlation between ozone in autumn and ozone in spring is not confirmed. The new data set is also used to study long-term and decadal ozone trends, which are of general importance for trends in the middle atmosphere (see, for example, Chap. 18). On global scales total ozone has decreased by roughly 0.1–0.3 %/year in the period 1979–1999 (much stronger at SH polar latitudes), and has increased by up to 0.3 %/year thereafter (2000–2007). Interestingly, this change of trends is not only due to anthropogenic effects in gas-phase chemistry (including ozone destroying trace gases) and polar heterogeneous chemistry, but also due to climatological changes in meteorology (transport and temperatures).

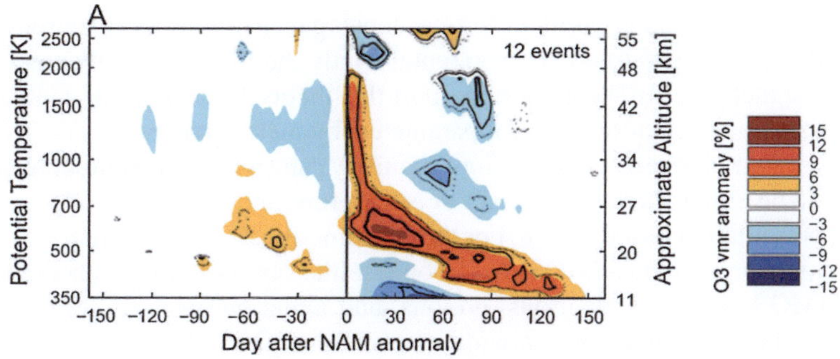

Fig. 1.3 Ozone anomalies (in percent relative to average conditions) north of 65°N relative to the beginning of a weak vortex (day 0). If the vortex is weak (negative NAM) ozone rich air from mid latitudes enters the polar region. The temporal/spatial characteristics of the anomaly is determined by a complex interaction of photochemistry and dynamics. The anomaly propagates from the stratosphere to the upper troposphere and is visible for several months. From *Sinnhuber et al.* (Chap. 9)

Kaufmann et al. (Chap. 10) deduce atomic hydrogen at ~80–95 km from measurements of hydroxyl (OH) by SCIAMACHY and ozone by GOMOS. They concentrate on equatorial latitudes and find an ~8 % increase of hydrogen from 2002 to 2008, i.e., in the declining phase of the solar cycle. They confirm that chemical heating rates are on the order of 4–10 K/day which is important for the energy budget at these altitudes. Heating rates have decreased from 2002 to 2008, which needs to be taken into account when studying solar induced long-term temperature variations in the MLT region.

1.4 Thermosphere, Energetic Particles, Ionization, and Impact on Trace Gases

From CHAMP measurements *Lühr and Marker* (Chap. 11) find significant local density enhancements (up to nearly doubling) in the upper thermosphere (~400 km) at high latitudes. Combining these observations with EISCAT measurements they provide an explanation of these enhancements which involves a chain of processes originating from soft-energy particle precipitation, Joule heating caused by small-scale field-aligned currents, increasing temperatures, and finally an upwelling of 'molecular-rich' air from below. The amplitude of the enhancement increases with solar flux. This demonstrates that solar activity can influence thermospheric background not only by absorption of EUV radiation but also by rather indirect effects.

Satellite observations now allow to study ionospheric effects on planetary scales. *Arras et al.* (Chap. 12) use GPS measurements by CHAMP and other satellites to derive, for the first time, a global and comprehensive coverage of sporadic E-layers (E_s), i.e., an enhancement of ionization at ~110 km altitude. They find maximum occurrence during daytime at mid latitudes of the summer hemisphere. The current theory of E_s formation is based on vertical convergence of ions and electrons caused by $\mathbf{v} \times \mathbf{B}$-forces originating from horizontal winds and the horizontal component of Earth's magnetic field. The authors support this theory by combining the morphology of E_s with observations of tidal parameters by meteor radars.

As is shown in several chapters, ion chemistry can significantly affect trace gases in the middle atmosphere (see below). It is therefore important for models to have a good representation of ionization. For the first time *Wissing et al.* (Chap. 13) present a 3D ionization model (AIMOS) based on satellite observations of energetic particles. AIMOS considers electrons, protons, and alpha particles. It combines solar and geomagnetic sources over a wide range of energies. The model is tested and improved by using observations from EISCAT. It turns out that ignoring some ionization processes (e.g., particles during the night) may lead to a substantial underestimate of electron densities, which may have severe consequences for ion chemistry modeling.

As has been mentioned above, high-quality radar measurements by EISCAT strongly supported the development of models and the interpretation of satellite observations. *Röttger et al.* (Chap. 14) summarize the operation of EISCAT and present the main scientific highlights where EISCAT contributed. There are several examples where the combination of EISCAT observations and models led to a better understanding of the physical/chemical processes involved.

It is known since several years that photochemically active species, being produced in the lower thermosphere and mesosphere, may be transported into the stratosphere where they significantly alter the concentration of trace gases such as ozone. In Chaps. 15 and 16 by *Reddmann et al.* and *Sinnhuber et al.*, respectively, new measurements from MIPAS and simulations by various models are used to study this important coupling mechanism. Some new and crucial details are identified. It is important to realize that particles with different energies ionize the polar atmosphere at different altitudes. For example, regular low energy electrons leading to auroras are absorbed in the MLT region, whereas high energy protons during solar proton events (SPEs) penetrate into the stratosphere. Both produce ions, which lead to photochemically active gases such as NO_x ($= N + NO + NO_2$) and HO_x ($= OH + H + HO_2$) destroying ozone. The relative importance and the altitude dependence of photochemical effects and their influence on trace gases are studied in these chapters.

Auroral electrons are much more frequent than SPEs but are commonly not considered to be important for stratospheric chemistry because they are absorbed at higher altitudes. However, during polar winter NO_x is transported downward to stratospheric altitudes. The photochemical lifetime of NO_x is large (weeks to months) because sunlight destroying NO_x is absent. In total, NO_x intrusions from the MLT region are likely to be more important for the chemical state of the middle atmosphere than SPEs (Fig. 1.4). NO_x intrusion can add 5 %–10 % of the total global NOy (NOy = all active nitrogen species including NO_x). At high latitudes, ozone is reduced by up to 50 % in the middle atmosphere, and total ozone is reduced by about 20 Dobson units. The effect lasts for several months. More important details are described in these chapters. For example, ion chemistry plays a crucial role

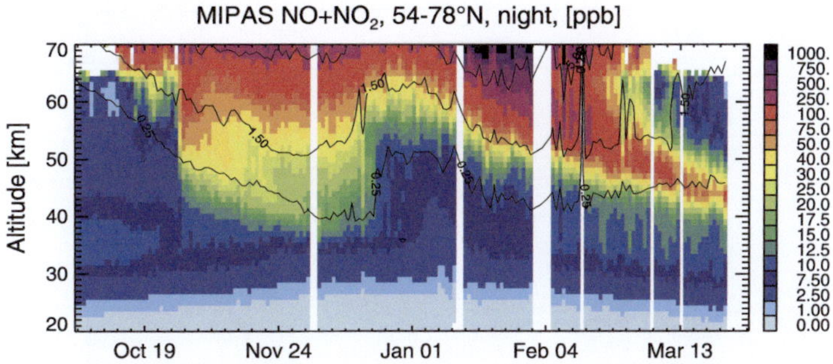

Fig. 1.4 MIPAS observations of NO_x ($NO + NO_2$) in the polar night. The color bar gives NO_x mixing ratios in ppb. The *black lines* are isolines of the CO mixing ratio for 0.25, 1.5 and 5 ppm, indicating transport. A large solar proton event in late October 2003 caused a significant enhancement of NO_x. However, even more NO_x is transported from the thermosphere to the stratosphere in January and February 2004 due to high geomagnetic activity which is not directly related to the SPE

in explaining the conversion of HCl to other chlorine species affecting ozone. Other photochemically active gases, such as HNO_3, are formed during SPEs. For the first time, an enhancement of H_2O_2 during a SPE was observed by MIPAS.

The downward transport of species from the lower thermosphere to the stratosphere depends on the general circulation of the atmosphere and thereby on gravity waves (see Sect. 1.6). This allows use of NO_x observations in the middle atmosphere to validate models which use different parameterizations for gravity waves. This topic was also covered in the HEPPA model comparison initiative. In summary, the transport of photochemically active gases from the thermosphere to the stratosphere is an important coupling process in the atmosphere which, amongst others, depends on dynamical coupling.

Baumgärtner et al. (Chap. 17) applied a whole-atmosphere model (MESSy), including various extensions, to study in detail particle precipitation effects in the entire atmosphere. Surprisingly, they find (in the NH winter) localized effects of geomagnetic activity from the MLT region all the way down to the troposphere(!), both in observations (ERA-40) and in the model. Sensitivity studies were performed to identify the physical mechanism which involves the production of NO_x, ozone destruction, heating of the stratosphere due to the reduction of IR cooling, and finally a modification of the polar vortex related to a stronger NAM index (stronger vortex) at high geomagnetic activity. In the troposphere NAM anomalies are related to weather anomalies. Further studies are required to confirm the effect of geomagnetic activity in the entire atmosphere and to better understand details of the proposed mechanism.

1.5 Mesospheric Ice Clouds

Mesospheric ice clouds are very sensitive to background conditions, in particular to temperatures, and are therefore suitable to improve our understanding of the middle

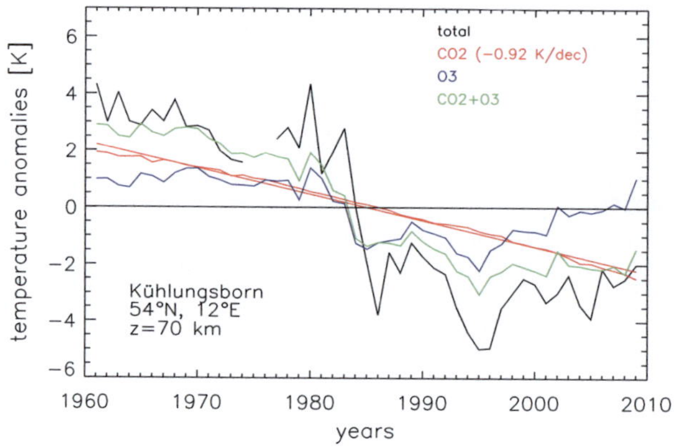

Fig. 1.5 Temperature trends from the LIMA model in the mesosphere (70 km) over Kühlungsborn (54°N, 12°E) in summer. The trend is not uniform in time. Various long-term variations contribute to the total trend, namely CO_2 (*red*), ozone in the upper stratosphere/lower mesosphere (*blue*), and ozone in the rest of the stratosphere (difference between *black* and *green* curve). LIMA trends nicely agree with various observations (SSU satellites, lidars, phase heights etc.). More details and trends in mesospheric ice clouds are presented by *Lübken et al.* (Chap. 18)

atmosphere, to test models, and to study potential long-term changes. This topic is covered in three chapters. Long-term variations of noctilucent clouds (NLCs) and the background atmosphere are analyzed by *Lübken et al.* (Chap. 18). They present LIMA model studies, which include the effects of the stratosphere and of greenhouse gas trends (CO_2 and O_3) on the mesosphere. At lower altitudes, i.e. from the ground to $\sim$35 km, LIMA adapts to the real atmosphere from ECMWF analysis which introduces year-to-year variations. It turns out that atmospheric shrinking in the stratosphere alone (i.e., no CO_2 and O_3 trends in the mesosphere) causes a significant trend in the mesosphere. In general, temperature trends are negative in the stratosphere and mesosphere, but they are positive in the summer polar mesopause region which includes the upper part of the ice particle regime. This complicates a simple prediction of long-term temperature trends on NLCs. LIMA is able to reproduce long-term variations of polar mesospheric clouds (PMCs) observed by SBUV. Temperature trends are not uniform in time but vary on decadal time scales (Fig. 1.5). Stratospheric ozone contributes significantly to this time variation of trends. The background atmosphere (temperature and water vapor) is modulated by the solar cycle which also influences NLC. However, natural year-to-year variability can destroy a simple (anti-)correlation. This is consistent with lidar observations at ALOMAR which do not show a clear anti-correlation between solar cycle and NLC.

Rapp et al. (Chap. 19) study in detail the effects of charged aerosols on radar returns coming from the D region. They consider mesospheric smoke particles (MSP) which are believed to act as condensation nuclei for ice clouds. For the first time MSP particles were identified in the spectral shape of the backscattered signal of EISCAT radars in Northern Scandinavia and from the Arecibo radar in Puerto Rico. The S/N ratio for the latter is so strong that it allowed height profiles of particle radii (typically $r = 0.8$–1 nm) and number densities ($N = 10$–1000/ccm) to be deduced. These numbers are consistent with the theory of meteor ablation. Regarding mesospheric ice clouds, they derive particle properties by analyzing the signal strength from EISCAT radars at three frequencies applying an established theory for ice particle affecting radar echoes known as 'polar mesosphere summer echoes' (PMSE). The observations confirm the PMSE theory and give ice particle radii of 10–60 nm which nicely agrees with lidar and satellite measurements of PMCs. These results promise a new direction in MLT research, namely aerosol particle studies by radars.

Von Savigny et al. (Chap. 20) investigate a significant reduction of satellite based detection of PMCs during a solar proton event (SPE). Various processes were considered, for example Joule heating, but cannot explain this event. Instead, a potential candidate has been studied in detail by KMCM modeling, namely dynamical heating caused by gravity wave filtering at lower heights, induced by reduced zonal winds caused by SPE destruction of ozone. Indeed, this process can explain most, but not all, of the observed features of PMC reduction. The authors also discuss the recently discovered 27-day solar cycle signature in PMCs, which is generated similarly to the solar cycle effect on NLCs: larger solar activity implies less favorable conditions for ice particles because it is warmer and water vapor is less abundant because it is destroyed by photo-dissociation.

1.6 Gravity Waves, Planetary Waves, and Tides

Hydrodynamical waves play a key role in understanding the thermal and dynamical state of the terrestrial atmosphere. They can transport momentum, energy, and trace gases over large distances and may cause drastic deviations from the state without waves. For example, wave driven circulation pushes the summer mesopause at high latitudes away from radiative equilibrium by approximately 100 K(!) and makes it the coldest place in the Earth's atmosphere, despite permanent sunshine. Unfortunately, measurements of waves are complicated and sparse. Global scale models use crude parameterizations to describe the impact of these waves. Several chapters cover the role of gravity waves, planetary waves, and tides for the mean state and also for the variability of the atmosphere from the ground to the thermosphere.

Ern et al. (Chap. 21) present the first multi-year global data set of gravity wave parameters from satellite observations complemented by long-term local measurements by radars. In the lower stratosphere, GW induced temperature fluctuations are largest in the high latitude winter season, and in the sub-tropics of the summer hemisphere (Fig. 1.6). Some GW sources are identified, e.g., winds over mountains or deep convection. The quasi-biennial oscillation (QBO) modifies GW activity not only in the stratosphere but also at higher altitudes. There is some weak and non-conclusive evidence for solar cycle variations of GW activity in the stratosphere. These GW observations have been implemented in a GCM model to study the impact of the associated drag on the basic state of the atmosphere. Furthermore, ray tracing studies have been performed to investigate the propagation characteristics

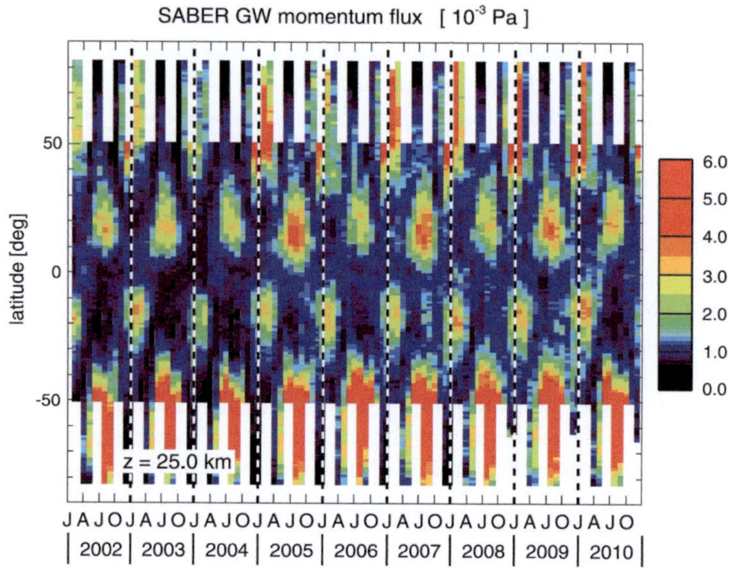

Fig. 1.6 Time series of zonal average gravity momentum flux at 25 km from SABER for gravity waves with $\lambda_z = 4$–25 km. Activity is enhanced in the summer subtropics and at the locations of the winter polar night jets. Note that the subtropical maxima are larger every other year, especially in the northern hemisphere. More precisely they are modulated by the quasi-biennial oscillation (QBO). From *Ern et al.* (Chap. 21)

of GW. It turns out that GW may propagate large horizontal distances when traveling from the troposphere to the mesosphere. This study shows that many assumptions made in models using standard GW parameterizations are too simple and need to be improved. Gravity waves can be important even for the troposphere. That is why some constraints regarding non-orographic GW drag schemes have been implemented in the ECMWF weather forecast model.

Compared to satellites, radars measure waves with higher vertical and temporal resolution, but are constrained to few locations. *Singer et al.* (Chap. 22) present a study on long-term wind variations in the MLT region at 54°N and 69°N. They find strong maximum wave activity in winter and a smaller maximum in summer which is explained by filtering of GW at lower heights. In summer, GW activity is larger at middle latitudes compared to high latitudes. Applying the KMCM model which includes GW generation, filtering, and damping, this is explained by the latitudinal variation of zonal winds causing differences in GW filtering. Winds in the MLT region at 69°N increase with solar activity and vary by as much as 6–8 m/s during the solar cycle. Using 20 years of wind measurements at 54°N a long-term enhancement of westward (eastward) directed winds in the summer MLT at ~70–80 km (80–90 km) by up to 1 m/s/year is detected. The wind trend at ~75 km causes some GW trends at higher altitudes. It is also shown that zonal winds at ~85 km regularly reverse during sudden stratospheric warmings, a feature which is not well captured

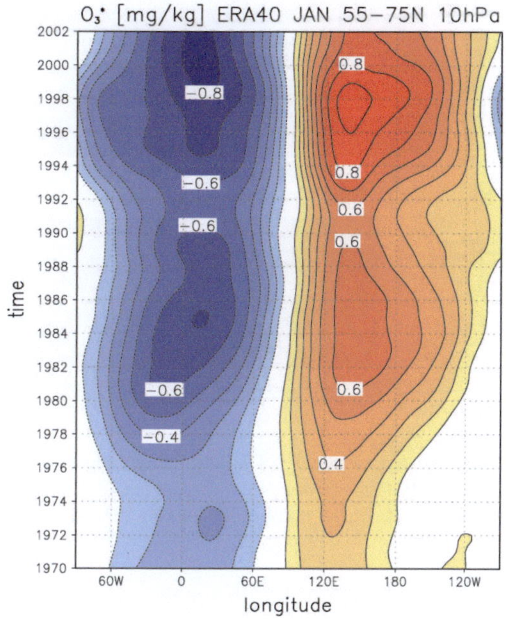

Fig. 1.7 Long-term trend of the longitudinal distribution of ozone. More precisely, the deviation of ozone (in ppm) from a zonal mean is shown for the northern polar region (55°–75°N) in January at a pressure level of 10 hPa (∼32 km). This zonally asymmetric ozone component has more than doubled in the last 32 years. More details are presented by *Gabriel et al.* (Chap. 24)

in GCM models. Wind reversals can also be induced by strong solar proton events, in particular regarding meridional winds.

Gravity waves lead to temperature variations of the background atmosphere. This alters radiative cooling/heating rates at infrared wavelengths. By considering non-LTE effects *Kutepov et al.* (Chap. 23) show that gravity waves can thereby enhance cooling around the mesopause. The effect is small in summer (<-1 K/day) but may be substantial in the winter hemisphere (up to -4 K/day).

Ozone in the winter stratosphere is not completely zonally symmetric because of large-scale disturbances and heterogeneous chemistry. This zonally asymmetric component in ozone (called $O_3^{\star}$) induces asymmetries in heating and cooling, which in turn excites planetary waves. *Gabriel et al.* (Chap. 24) analyze in detail physical details of this process and show that these waves may influence not only the entire stratosphere but also (by propagation and non-linear interaction) the mesosphere and the troposphere. For example, the geopotential height of the 1000 hPa pressure level (i.e., near the Earth's surface) may vary by up to ± 30 m which is ± 20–30 % of the undisturbed values. Interestingly the zonally asymmetric component in ozone has increased substantially in the last 40 years (Fig. 1.7) which (through dynamical coupling involving planetary waves) introduces long-term variations in the stratosphere, mesosphere, and even in the troposphere. This is another example of a coupling mechanism from above to below.

The effect of gravity waves in the highly dissipative regime above the turbopause has largely been ignored in the past because the effects were believed to be small and parameterizations were not yet available. As is shown by *Yiğit and Medvedev* (Chap. 25) dissipating gravity waves contribute significantly to the thermal and

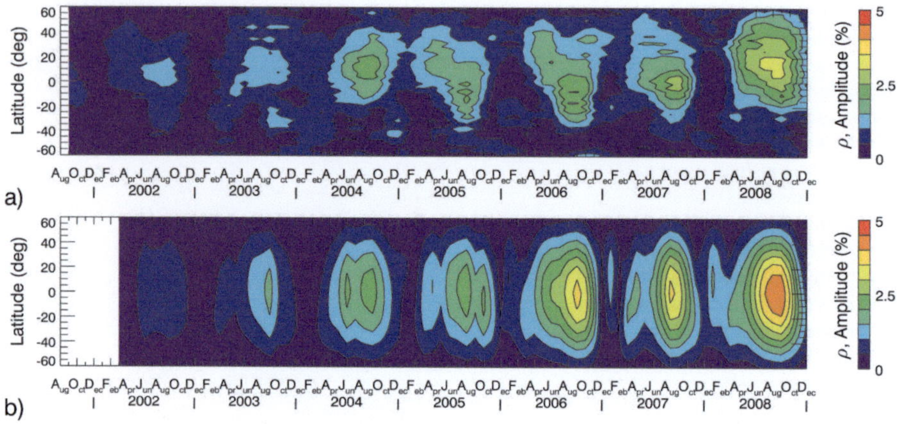

Fig. 1.8 *Upper panel*: DE3 tidal component of relative density modulation observed by CHAMP close to 400 km. *Lower panel*: model calculations (Hough mode extensions) of the DE3 tide at the same altitude based on observations by TIMED in the MLT. From *Häusler et al.* (Chap. 26)

dynamical state of the lower thermosphere, in particular at high latitudes. Here the momentum deposition due to gravity waves is comparable to ion drag. The net thermal effect is cooling by downward heat conduction. Future GCM modeling studies of the thermosphere will have to consider this substantial contribution from gravity waves originating at lower heights.

Tides are global scale oscillations with main periods of 24 h and 12 h. They are called 'non-migrating' if the disturbance does not follow the apparent motion of the Sun around the Earth. Atmospheric tides have been known for a long time, but several exciting and unexpected results have been discovered in recent years. As is shown by *Häusler et al.* (Chap. 26) non-migrating tides (here the eastward propagating diurnal tide with zonal wavenumber 3, labeled DE3) being excited in the troposphere by strong regional latent heat release can propagate all the way into the mesosphere and even into the upper thermosphere. Such tides have not been expected in the upper thermosphere since the main sources are in the troposphere/stratosphere and dissipation in the lower thermosphere hinders propagation. However, they are observed by satellites. Upper atmosphere tides are identified in various observations such as winds up to 105 km by TIDI, temperatures at ~100 km by SABER, nitric oxide concentration at 100–150 km by SNOE, and mass density and winds at ~400 km from CHAMP (Fig. 1.8). Tidal signatures are also identified in several ionospheric components. The physical explanation is supported by applying Hough mode extensions (HME) which is a theoretical representation of tides obtained through assimilation of measurements. HME range way into the upper thermosphere and thereby allow interpretation of CHAMP observations. This chapter demonstrates another important and yet unexplored coupling process from the troposphere through the mesosphere all the way into the upper thermosphere.

More details of the propagation of tides and their interaction with planetary and gravity waves is presented in a theoretical study by *Achatz et al.* (Chap. 27).

Different from classic GW parameterizations, they include time-dependent background conditions and horizontal refraction of waves. The former leads to a disappearance of critical layers and thereby drastically decreases damping of tides by GWs. Planetary waves can lead to a non-migrating tidal component in the mesosphere caused by migrating forcing in the troposphere.

Trends and solar cycle effects in mesospheric waves, winds, and temperatures in the upper mesosphere are studied by *Offermann and Koppmann* (Chap. 28) applying mainly ground based measurements of OH temperatures and satellite observations of winds and temperatures. In their 1987–2008 data set of OH temperatures (altitude: $\sim$87 km) they find a significant solar cycle signal of $\sim$3.5 K/100 SFU (i.e., approximately 5 K during a solar cycle) and an annual mean temperature decrease of -0.23 K/y with rather small trends in summer. Note, that the total variation of solar radiation throughout a solar cycle is approximately 150 SFU. The length of the summer season increases by approximately 1.2 days/year. It is suggested that these changes are related to trends in dynamic parameters (gravity wave activity, winds, quasi-two day waves) which points to a long-term trend in the large-scale circulation.

1.7 Large-Scale Coupling

Several papers study the solar impact in the atmosphere on global dimensions and on time scales from decades to thousands of years. This includes the direct impact of solar irradiance variations from the troposphere to the MLT region, but also dynamical feedback mechanisms, for example by planetary and gravity waves affecting the mean circulation.

Langematz et al. (Chap. 29) study the solar impact (including precipitating particles) on the Earth's atmosphere from decadal to millennium time scales applying sophisticated models and observations. They concentrate on stratospheric ozone, temperatures, and winds and their impact on tropospheric weather and climate. From solar minimum to maximum, ozone in the stratosphere increases by approximately 1.5–3 % (similar to the effect of man-made halo-carbons) and temperatures increase by 0.6–1.2 K (note that Fig. 1.9 shows the ozone changes per 100 SFU). It is confirmed that these stratospheric changes can cause effects in the troposphere purely by dynamical feedback processes (Fig. 1.9). From solar minimum to maximum, the geopotential height of the 700 hPa level increases by as much as 25 m, which is associated with a poleward shift of the tropospheric jet stream. Local temperatures in the lower troposphere (850 hPa) increase by as much as 1.5–2 K. Regarding centennial time scales, it is well known that TSI, SSI, and greenhouse gas concentrations have increased since the Maunder Minimum (1645–1715). During this phase of extremely low solar activity, annual global mean temperatures were lower in the upper troposphere and stratosphere by $\sim$0.4 K compared to pre-industrial times. The polar vortex was weaker. At the Earth's surface the TSI increase and natural variability (volcanoes) cause some effects, but the strong temperature increase from industrialization onwards is mainly caused by the increase of GHG concentrations.

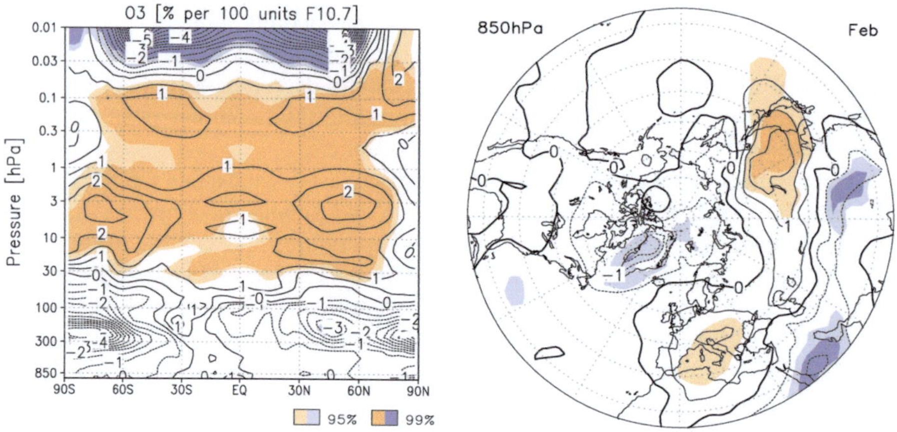

Fig. 1.9 *Left panel*: Variation of annual mean ozone per 100 units of solar F10.7 cm flux (= Sun flux units = SFU). *Right panel*: Temperature variation in the lower troposphere (at 850 hPa) per 100 SFU. The distribution of the signal is indicative of an enhancement of the AO index (stronger vortex), associated with a cooling of the polar areas and a warming of middle/southern Europe and eastern Eurasia during solar maximum. From *Langematz et al.* (Chap. 29)

Several years ago large solar influences in some parts of the stratosphere were identified if the years were grouped according to the phase of the QBO (east or west). In the meantime the data base has been extended to 70 years and the solar impact is fully confirmed (see *Labitzke and Kunze*, Chap. 30). During summer and for QBO-east the solar signal in the tropics/sub-tropics at 30 hPa is highly significant and large, namely an increase of up to 2.4 K from solar minimum to solar maximum. During Arctic winter the situation is more complicated because of much larger natural variability. It is therefore difficult to derive trends, even more so if the trend changes in time. Still, after grouping into QBO-east/west, a clear and significant correlation between the geopotential height of the 30 hPa level and solar flux is observed also in winter. Based on 70 years of observations and after selecting for QBO-east, this height is found to increase by as much as $\sim$1000 m from solar minimum to maximum. Significant progress has been made in recent years in identifying a physical mechanism for the QBO/solar-cycle relationship and its effect on stratospheric temperatures. Still, the effect cannot be reproduced to its full extent in models (see below).

A comprehensive analysis of solar effects in the entire atmosphere is performed by *Schmidt et al.* (Chap. 31) applying AOGCM models, even considering the ocean. A 'secondary maximum' of solar response is found in the equatorial lower stratospheric ozone and temperatures which, however, may easily be obscured by variability stemming from ENSO. The 27-d variation of solar radiation leaves a clear signal in the stratosphere and mesosphere, but is intermittent for not identified reasons. From solar cycle minimum to maximum a 3 K temperature and 7 % ozone increase at the mesopause is detected. Values in the stratosphere are generally smaller ($\sim$0.2–0.3 K between 100 hPa and 10 hPa). Attempts to identify the physical mechanisms relating QBO and solar cycle influence on the stratosphere and troposphere (via planetary waves and their effects on circulation, see Chap. 30) have been made but were not conclusive. In general, the signal in the real atmosphere is significantly larger than in models. Recent satellite observations suggest that solar UV variability throughout the solar cycle may be much stronger than previously assumed. This would presumably have severe impacts on the GCM modulations presented in this chapter.

A long-term increase of the Brewer-Dobson circulation is expected which reduces the transport time of man-made inert trace gases from the troposphere into the stratosphere, i.e., the 'age-of-air' in the stratosphere is expected to decrease. Based on 30 years of balloon borne measurements of SF_6 and CO_2 *Stiller et al.* (Chap. 32) show that the age of stratospheric air is approximately 5 years and has increased(!) with a rate of $+0.24 \pm 0.22$ years/decade. This is contrary to most model results which predict a reduction(!) by to -0.25 years/decade. The authors present for the first time global measurements of age-of-air based on satellite observations of SF_6 from 2002 to 2010. The values for the age-of-air and its trend are in agreement with other observations, including the balloon borne observations mentioned above. It is not clear at the moment how this apparent discrepancy between observations and GCM models can be solved.

1.8 Concluding Remarks

Some of the more general aspects of the numerous scientific results presented in this book may be summarized as follows.

- Long-term variations of solar radiation (from solar cycle to centennial time scales) directly influence the thermosphere, mesosphere, and stratosphere. Even the troposphere is affected through coupling mechanisms which involve gravity/planetary waves and their influence on the mean circulation. Since the mid of the 20th century solar activity is larger compared to several hundred years before. This has contributed to long-term variations in the atmosphere and created, for example, cold conditions during the Maunder Minimum in the 17th century. The long-term variation of solar activity also plays a role in more recent climate issues, although anthropogenic greenhouse gas increase is the most important driver of current climate. Not only the total solar irradiance but also the variation in its spectral decomposition have shown unexpected features. In some parts of the atmosphere, the observed solar modulation is still larger than explained in models. Some basic features of the physical mechanisms explaining the role of solar-cycle and QBO interaction in creating relatively large temperature variations are meanwhile established. Still, there is a mismatch between models and observations which points to some deficiencies in our understanding.
- Several coupling mechanisms exist and are crucial to understand the reaction of the atmosphere to external and internal drivers including their temporal/spatial variability. Coupling exists from above to below and vice versa. For example,

photochemically active species being created by solar particles and geomagnetic activity in the thermosphere can be transported down to the stratosphere where they significantly reduce ozone for an extended period of time. The stratosphere couples downward by a combination of mechanisms involving trace gases, solar heating, planetary/gravity wave generation, propagation, breaking, and finally their effect on the general circulation. The transport of trace gases may also be used to validate circulation models. There are various coupling processes from below to above. For example, trace gases emitted from anthropogenic activity propagate to higher altitudes. Planetary waves and gravity waves are typically excited in the troposphere and propagate through the entire middle atmosphere. Even thermospheric tides can be generated by tropospheric processes. It is shown that the parametrization of these waves in models needs to be improved.
- Large trends exist in some parts of the middle atmosphere, much larger compared to the troposphere. But trends may not be uniform in time. For example, temperature trends on decadal scales in the middle atmosphere vary in time since several processes with different temporal characteristics are involved. Care must be taken when deducing trends from too short time series since these 'short term' variations (including solar cycle) may play a role.
- It is important to distinguish local from global effects. Some trends and solar variability effects are important in certain regions only, whereas they may be negligible on global scales. The effect may propagate from one region to another by coupling processes.

Many important and new scientific results have been achieved in the CAWSES priority program. Some old puzzles were solved and some new and surprising results were discovered. This program has brought together a rather diverse community, with expertise ranging from the Sun through the thermosphere, ionosphere, mesosphere, and stratosphere, down to the troposphere. This concentration of specialists has created several synergy effects, for example when considering the consequences of solar spectral irradiance variations for atmospheric photochemistry and dynamics. The CAWSES community in Germany has greatly benefited from the DFG priority program. We have developed plans to continue this successful cooperation within Germany but also with international groups in the future.

Acknowledgements In the name of all participants I would like to thank the German Research Foundation DFG for funding this substantial program. Furthermore, the German CAWSES community appreciates the constructive and stimulating support from the reviewers, some of whom accompanied our program for the entire six years. As speaker of this program I would like to express my appreciation for all the activities, excitement, and success being created in various groups. Perhaps most important, many students were involved and contributed to the enthusiasm when working on a wide range of scientific topics of solar-terrestrial physics. I thank Dr. Norbert Engler and Monika Rosenthal for their work in compiling this book.

References

Lübken, F.-J., Austin, J., Langematz, U., & Oberheide, J. (2010). Introduction to special section on Climate and Weather of the Sun Earth System. *Journal of Geophysical Research, 115*, D00I19. doi:10.1029/2009JD013784.

Chapter 2
Models of Solar Total and Spectral Irradiance Variability of Relevance for Climate Studies

Natalie A. Krivova and Sami K. Solanki

Abstract The variable radiative output of the Sun is a prime external driver of the Earth's climate system. Just how effective this driver is has remained relatively uncertain, however, partly due to missing knowledge on the exact variation of the Sun's irradiance over time in different parts of the solar spectrum. Due to the limited length of the time series of measured irradiance and inconsistencies between different measurements, models of solar irradiance variation are particularly important. Here we provide an overview of progress over the last half decade in the development and application of the SATIRE family of models. For the period after 1974, the model makes use of the full-disc magnetograms of the Sun and reproduces up to 97 % of the measured irradiance variation. Over this time frame, there is no evidence for any non-magnetic change in the solar irradiance on time scales longer than about a day. We have also been able to compute total solar irradiance since the Maunder minimum and further into the past throughout the whole Holocene. The Sun's spectral irradiance from the Lyman α line in the UV to the far infrared has also been reconstructed throughout the telescopic era.

2.1 Introduction

The Earth's global surface temperature has being growing rapidly over the last decades [see, e.g., *Solomon et al.*, 2007]. This has in large measure been attributed to human activity. However, a quantitative assessment of the anthropogenic contribution to the change in climate is still hampered by inadequate understanding of the relative roles of different climate drivers, both internal and external [e.g., *Hansen et al.*, 2002, 2005; *Jungclaus et al.*, 2010; *Schmidt et al.*, 2011].

The main external driver of the Earth's climate system is the solar radiative output [e.g., *Eddy*, 1976; *Reid*, 1987; *Hansen*, 2000; *Bond et al.*, 2001; *Neff et al.*, 2001;

N.A. Krivova (✉) · S.K. Solanki
Max-Planck-Institut für Sonnensystemforschung, 37191 Katlenburg-Lindau, Germany
e-mail: natalie@mps.mpg.de

S.K. Solanki
School of Space Research, Kyung Hee University, Yongin, Gyeonggi 446-701, Korea

F.-J. Lübken (ed.), *Climate and Weather of the Sun-Earth System (CAWSES)*, Springer Atmospheric Sciences,
DOI 10.1007/978-94-007-4348-9_2,

Hansen et al., 2002; *Camp and Tung*, 2007; *Gray et al.*, 2010]. The strength of Sun's influence and which process plays the main role remain, however, unclear. Variations in solar total and/or spectral irradiance are the prime suspects. Solar total (i.e. integrated over all wavelengths) irradiance (TSI) is the total solar energy flux at the top of the Earth's atmosphere, and thus any changes in the TSI affect the overall energy balance of the climate system. Variations in the spectral distribution of the irradiance, in particular in the UV but also in the visible and IR, have a pronounced effect on the chemistry and dynamics of the Earth's atmosphere [e.g., *Haigh*, 1994, 2007; *Haigh et al.*, 2010; *Rozanov et al.*, 2004; *Kodera and Kuroda*, 2002, 2005; *Langematz et al.*, 2005; *Matthes et al.*, 2006; *Gray et al.*, 2010].

Space-based instruments have being monitoring solar total and spectral (SSI) irradiance since 1978 [e.g., *Willson et al.*, 1981; *Fröhlich et al.*, 1997; *Floyd et al.*, 2003; *Willson and Mordvinov*, 2003; *Skupin et al.*, 2005; *Kopp et al.*, 2005; *Fröhlich*, 2006; *Harder et al.*, 2009]. Different mechanisms have been proposed to explain the observed changes in the irradiance [see review by *Domingo et al.*, 2009], of which most successful was the modulation by the solar surface magnetic field. Models assuming that solar brightness changes due to the varying relative contributions of dark sunspots, bright faculae and the bright network explain over 90 % of the measured TSI variation on time scales of days up to the solar cycle [*Fröhlich and Lean*, 1997; *Fligge et al.*, 2000; *Preminger et al.*, 2002; *Ermolli et al.*, 2003; *Krivova et al.*, 2003; *Wenzler et al.*, 2006, 2009; *Ball et al.*, 2011].

Despite the great success of the models in reproducing TSI measurements, a number of open questions remain, including the presence and the magnitude of the secular trend in the irradiance during the last 3 cycles [*Fröhlich*, 2009; *Scafetta and Willson*, 2009; *Krivova et al.*, 2009a, 2011a], the absolute level of the TSI [*Kopp et al.*, 2007; *Kopp and Lean*, 2011] or the contribution of different spectral ranges to the irradiance variation [*Krivova et al.*, 2006; *Harder et al.*, 2009; *Pagaran et al.*, 2009; *Morrill et al.*, 2011].

Understanding the mechanisms of the irradiance variability is not of purely theoretical interest. Only when the physical origin of the variation is recognised and the measured changes are reproduced with high accuracy, can a reconstruction of the past solar irradiance be trustworthy. Such longer-term reconstructions are prerequisites for a reliable evaluation of the connection between solar variability and the Earth's climate change, since the time series of direct measurements is just over 30 years long and is too short for this.

Extension of the models back in time poses additional challenges. Available historical data featuring solar activity become sparser and of lower quality when going further back in time. The magnitude of the secular trend in the irradiance has remained by far the most speculative aspect of long-term reconstructions [cf. *Lean et al.*, 1992; *Hoyt and Schatten*, 1993; *Zhang et al.*, 1994; *Soon et al.*, 1996; *Mendoza*, 1997; *Foster*, 2004; *Wang et al.*, 2005; *Krivova et al.*, 2007, 2010; *Schrijver et al.*, 2011; *Shapiro et al.*, 2011].

Here we describe our recent progress in modelling solar irradiance on time scales of days to millennia using the SATIRE set of models. We refer to *Domingo et al.* [2009] for a recent review of different irradiance models on time scales of days to the solar cycle. The present paper is structured as follows. Section 2.2 sketches out essentials of the SATIRE model. Section 2.3 describes our efforts in modelling solar total and spectral irradiance over the satellite era, whereas Sects. 2.4 and 2.5 deal with reconstructions of solar irradiance for the telescope and pretelescopic eras, respectively. Section 2.6 summarises the results.

2.2 SATIRE

SATIRE (Spectral And Total IRradiance REconstructions) is a set of routines developed for calculations of solar irradiance from other magnetic activity proxies [*Solanki et al.*, 2005; *Krivova and Solanki*, 2008; *Krivova et al.*, 2011b]. SATIRE is based on the assumption that all irradiance variation on time scales longer than a day is entirely due to the changes in the number and distribution of magnetic features (such as sunspots or faculae) on the solar surface.

Irradiance changes on shorter time scales are dominated by other sources, such as the p-mode oscillations peaking at a period of around 5 minutes [see review by *Christensen-Dalsgaard*, 2002] or solar granulation, i.e. the convective cells on the surface of the Sun [e.g., *Seleznyov et al.*, 2011] and are not described by the SATIRE. These variations are, however, of no importance for climate studies. On time scales related to climate variability, no evidence for other mechanisms of intrinsic (i.e. not related to changes in the Earth's orbit) solar irradiance modulation has yet been provided [see, e.g., *Domingo et al.*, 2009, as well as Sect. 2.3].

Magnetic features that can be observed on the solar surface are divided in SATIRE into various classes. Sunspot umbrae and sunspot penumbrae, that are cooler than the surrounding quiet photosphere, appear dark and reduce solar brightness. Faculae and the network (represented by a single component in SATIRE) are bright. Solar surface essentially free (above the noise and detectability thresholds) of magnetic field is called 'the quiet Sun'. Thus, SATIRE currently distinguishes four different photospheric components.

The brightness of each component is assumed to be time-invariant but depends (as attested by observations) on the wavelength and the position on the solar disc. These brightnesses were calculated [*Unruh et al.*, 1999] in the LTE approximation using the ATLAS9 code by *Kurucz* [1993].

The area covered by different features on the solar surface changes with time, which leads to the irradiance modulation. To describe the surface coverage by each component and its evolution in time, we employ different observational data. Most detailed information is provided by the direct measurements of the solar photospheric magnetic field, i.e. by the full-disc magnetograms. The version of SATIRE which makes use of the magnetograms is called SATIRE-S (S stands for Satellite era; *Krivova et al.*, 2003; *Wenzler et al.*, 2005, 2006) and is discussed in Sect. 2.3. Magnetograms with sufficient quality and cadence have been recorded for less than four decades. Thus reconstructions of solar irradiance over longer periods have to content themselves with data having lower quality and resolution (both spatial and temporal). For the period after 1610, the sunspot number can be employed to reconstruct the evolution of the solar surface magnetic field and thus irradiance. This is done in the SATIRE-T (for the Telescope era; *Solanki et al.*, 2002; *Balmaceda et al.*, 2007; *Krivova et al.*, 2007, 2010) model described in Sect. 2.4. Finally, before 1610 even the sunspot number is not available. Then, concentrations of cosmogenic isotopes in terrestrial archives can be employed as a proxy of solar magnetic activity if the effect of the Earth's magnetic field is taken into account. This is possible for the Holocene [*Solanki et al.*, 2004; *Usoskin et al.*, 2006a; *Vieira and Solanki*, 2010]. The SATIRE-M model [*Vieira et al.*, 2011] deals with solar irradiance on millennial time scales (Sect. 2.5). Knowing brightnesses of individual features and their surface coverage, it is then possible to calculate solar irradiance as the sum of the contributions of all components.

Since brightnesses of the photospheric components depend on the wavelength, calculations are done on a grid of wavelengths from 10 to 160 000 nm. An integral over all wavelengths gives the total solar irradiance (TSI). The LTE approximation involved in calculations of the brightness spectra of photospheric components is not valid in some spectral lines, mainly in the UV (see also *Danilovic et al.*, 2007, 2011 for an example of SATIRE's performance in spectral lines in the visible) and below roughly 200–250 nm. The contribution of these short wavelengths to the TSI is less than 1 % [*Krivova et al.*, 2006] and the computed TSI is still quite accurate. However, the calculated UV fluxes are not reliable. We therefore correct the model at shorter wavelength in the following way. *Krivova and Solanki* [2005] and *Krivova et al.* [2006] found that SATIRE reproduces quite accurately variations of solar spectral irradiance in the range between 220 and 240 nm as observed by the UARS/SUSIM instrument. Thus we have derived empirical relationships between the measured irradiance in this range and irradiance at other wavelengths covered by SUSIM (115–410 nm). Applying these relationships to the modelled irradiance at 220–240 nm, SATIRE is extended down to 115 nm.

2.3 Satellite Era

2.3.1 TSI

Employment of the solar full-disc magnetograms for irradiance modelling turned out to be very successful in reproducing the measured irradiance variations. *Krivova et al.* [2003] and *Ball et al.* [2011] employed the SoHO/MDI magnetograms and continuum images to model solar irradiance over the ascending (1996–2002) and descending (2003–2009) phases of cycle 23, respectively. The model captures 92 to 97 % of the TSI variation (given by the squared correlation coefficient r_c^2, with maximum r_c values reaching 0.984) measured by the SoHO/VIRGO and SORCE/TIM instruments (see Fig. 2.1). Employment of the ground-based NSO KP magnetograms and continuum images [*Wenzler et al.*, 2004, 2005, 2006] results in a somewhat lower correlation of $r_c = 0.91$ ($r_c^2 = 0.83$) over the period 1974–2003. Taking

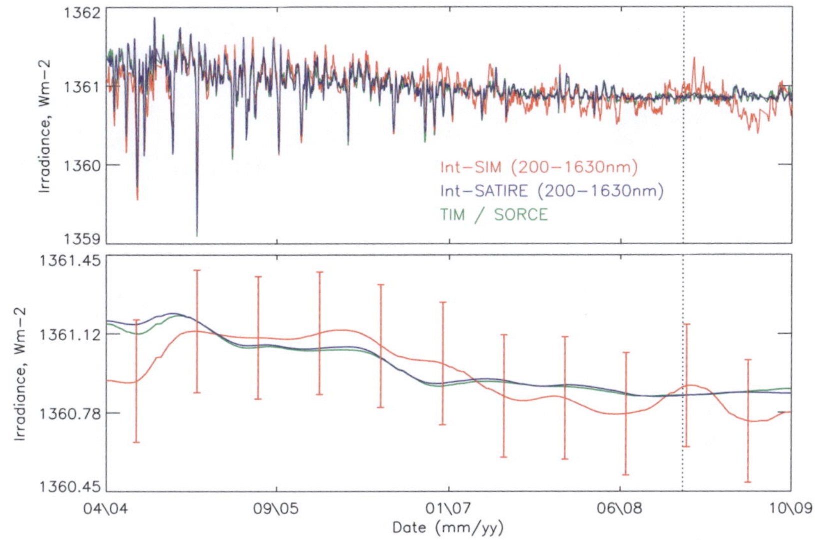

Fig. 2.1 TSI measured by SORCE/TIM (*green line*) and irradiance integrated (Int) over the range 200–1630 nm (i.e. the spectral range covered by SORCE/SIM) as measured by SORCE/SIM (*red*) and modelled with SATIRE-S (*blue*). The Int-SIM and Int-SATIRE curves are shifted in absolute levels to compensate for the missing contributions from the spectral ranges below 200 nm and above 1630 nm. *Top*: daily values; *bottom*: smoothed to remove short-term fluctuations. Error bars represent one standard deviation in the long-term stability of Int-SIM and are 0.259 $\mathrm{W\,m^{-2}}$ or 212 ppm. This figure is taken from *Ball et al.* [2011]

into account the significantly lower quality of the ground-based data, which suffer from numerous artefacts, such a good agreement of the model with the measurements provides strong support for surface magnetism as the dominant driver of irradiance variations on time scales from about a day to the solar cycle.

The space-based TSI measurements started in 1978 and were carried out by a number of different experiments, whose observing time span partly overlapped. Each of the instruments suffered from its individual degradation, calibration or other problems, making a construction of a composite time series nontrivial. The three currently available composites, PMOD [*Fröhlich*, 2009], ACRIM [*Willson and Mordvinov*, 2003] and IRMB [*Dewitte et al.*, 2004] show significantly different long-term trends (i.e. differences in the relative irradiance levels during different activity minima). This initiated the debate on the presence and magnitude of a secular trend in the TSI during the satellite era [e.g., *Fröhlich*, 2006, 2009; *Scafetta and Willson*, 2009].

Whereas the ACRIM and IRMB composites show an increase in the TSI between the minima preceding cycles 22 and 23 (though the increase is not statistically significant in the IRMB composite), the PMOD composite suggests a slight decrease (by 0.053 $\mathrm{W/m^2}$, given by *Fröhlich*, 2011, to 0.123 $\mathrm{W/m^2}$, from *Fröhlich*, 2009). From the minimum in 1996 preceding cycle 22 to the most recent minimum in 2008, TSI has been found to decrease by a further 0.22 to 0.273 $\mathrm{W/m^2}$ [*Fröhlich*,

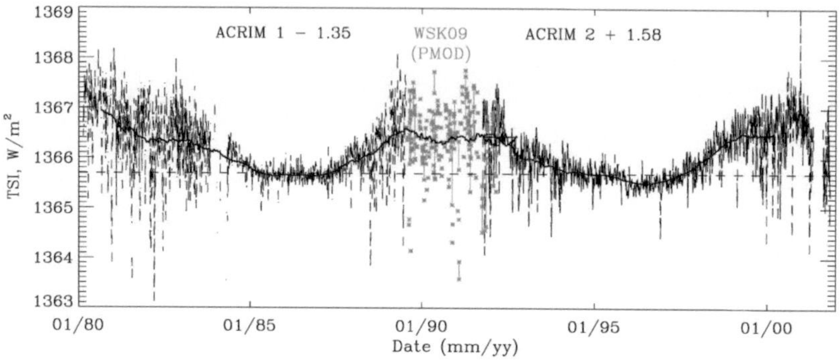

Fig. 2.2 'Mixed' TSI composite constructed from ACRIM-1 and ACRIM-2 data (*black dashed line*), with the gap bridged using the SATIRE-S (PMOD-optimised) model (asterisks connected by *grey solid line* when there are no gaps). The *heavy solid line* is the 1-year smoothed TSI, and the *horizontal dashed line* shows the level of the minimum preceding cycle 22. From *Krivova et al.* [2009a]

2009, 2011]. Note that SORCE/TIM measures a shallower decrease [*Kopp and Lean*, 2011]. This most recent decrease appears unusual when compared to other solar magnetic activity proxies according to *Fröhlich* [2009], which has been interpreted by him as a sign of non-magnetic origin in the TSI secular change. We have therefore tested whether the SATIRE-S model is able to reproduce all the observed TSI changes within the surface magnetism concept.

First, *Wenzler et al.* [2009] have compared the TSI reconstructed from the NSO KP magnetograms with all three available composites. They found a significantly better agreement with the PMOD composite than with the other two records. Both the correlation coefficients were significantly higher (over 0.91 for PMOD compared to 0.79–0.82 for ACRIM and IRMB) and the slopes of the linear regressions between the data and the model were closer to 1 (0.98 compared to 0.81–0.82). In particular, the upward trend between the minima in 1986 and 1996 in the ACRIM and IRMB composites could not be reproduced.

The main moot point in the discussion on the trend between the minima 21/22 and 22/23 is the question on the 'cross-calibration' of the ACRIM-1 and ACRIM-2 instruments caused by an unplanned 2-year long gap between their operations. Therefore *Krivova et al.* [2009a] have used the TSI reconstructed by SATIRE-S to bridge this gap, as proposed by *Scafetta and Willson* [2009]. Such a 'mixed' (observations—model—observations) composite is shown in Fig. 2.2 and suggests that the TSI has decreased between the two minima by 0.15–0.38 W/m^2. The exact value of the change depends somewhat on the SATIRE-S optimisation (i.e. whether it is optimised to best fit the PMOD, ACRIM or IRMB composite). But independently of the optimisation (1) an increase in the minima levels could not be achieved, and (2) the decrease of 0.15–0.38 W/m^2 in the TSI from 1986 to 1996 is slightly larger than is shown by the PMOD composite (0.053–0.12 W/m^2; the 1-σ uncertainty of the PMOD composite for the minimum in 1996 is listed as 0.1 W/m^2; *Fröhlich*, 2009).

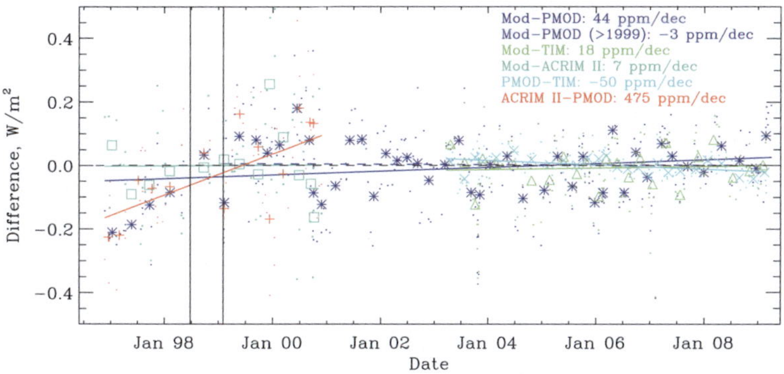

Fig. 2.3 The difference, in W/m^2, between the model and the data: PMOD (*blue*), ACRIM II (*pine green*), TIM (*light green*), and between the PMOD and the two other data sets: ACRIM II (*orange*) and TIM (*cyan*). *Dots* represent values for individual days, whereas *bigger symbols* represent bins of five individual points. The corresponding trends with time are indicated by the *straight lines* and their slopes are listed in the *right top corner*. The *vertical black lines* show the beginning and the end dates of the period without regular contact with SoHO. From *Krivova et al.* [2011a]

Finally, *Krivova et al.* [2011a] have used 60-min averaged MDI magnetograms sampled roughly every two weeks to reconstruct the TSI over the period November 1996 to April 2009. They have compared the modelled TSI with the PMOD composite and with the measurements by two individual instruments, UARS/ACRIM-2 and SORCE/TIM that monitored the TSI over the minimum in 1996 and during the declining phase of cycle 23, respectively. Excellent agreement has been found between the model and all sets of measurements with the exception of the early (1996–1998) PMOD data. The difference between the model and all sets of measurements as well as between the PMOD and the other two records is plotted in Fig. 2.3.

Whereas the agreement between the SATIRE-S and the PMOD composite between 1999 and 2009 is essentially perfect, the modelled TSI increases faster from the end of 1996 to 1999 than implied by the PMOD composite. On the other hand, model's steeper trend agrees remarkably well with the ACRIM-2 data. This is further supported by the fact that the shallower trend of the PMOD data is only seen in the measurements of one VIRGO radiometer, the PMO6V, whose data are used in the composite, whereas the other VIRGO radiometer, the DIARAD, shows a steeper trend in agreement with ACRIM-2 and SATIRE-S results. This all suggests that the TSI level during 1997–1998 might be overestimated in the PMOD data set by roughly 0.1–0.15 W/m^2. This inconsistency of the PMOD composite with other measurements and the model around the minimum preceding cycle 23 explains why the earlier study by *Steinhilber* [2010] found a discrepancy between the observed and modelled TSI in the minima preceding cycles 23 and 24. At the same time, the study of *Krivova et al.* [2011a] clearly demonstrates that there is no evidence for any non-magnetic long-term change in the TSI over the period of satellite measurements.

2.3.2 SSI

Although regular monitoring of solar spectral irradiance (SSI) in the UV also started in 1978, a consistent time series does not exist. The reason is that, in addition to different absolute levels, degradation trends and other problems in the data from various instruments depend strongly on the wavelength, which makes a proper self-consistent cross-calibration of measurements by different instruments essentially impossible [cf. *DeLand and Cebula*, 2008]. Regular observations in a broader spectral range covering the visible and the near-IR began only in 2002/2003 with ENVISAT/SCIAMACHY and SORCE/SIM.

Poorer spectral data is one of the factors hindering SSI models in their development. Another one is the failure of the LTE approximation in the UV part of the spectrum. Non-LTE models still need to mature to provide SSI reconstructions over a broad spectral range [e.g., *Fontenla et al.*, 2004, 2006] although significant progress has been made recently [*Shapiro et al.*, 2010].

As described earlier in Sect. 2.2, SATIRE makes use of the empirical relationships derived from SUSIM observations at wavelength below 270 nm and thus allows a reconstruction of the solar spectral irradiance over essentially the whole range that is of interest for climate models (115–160 000 nm). Detailed comparisons of the SATIRE-S results with different spectral data can be found in *Krivova et al.* [2003, 2006, 2009b]; *Krivova and Solanki* [2005]; *Unruh et al.* [2008]; *Danilovic et al.* [2007, 2011]; *Ball et al.* [2011]. Reconstructions of the SSI from magnetograms over cycles 21–23 were presented by *Krivova et al.* [2006, 2009b, 2011b]. They have shown that the contribution of the UV radiation below 400 nm to the TSI variation might be significantly higher than was previously estimated from UARS/SOLSTICE and UARS/SUSIM data [see, e.g., *Lean*, 1989; *Lean et al.*, 1997; *Krivova et al.*, 2006]. The reason is that the long-term stability of these (and earlier) instruments above 250–300 nm was comparable to or even lower than the solar cycle changes in this range.

The relative contributions of different wavelength ranges to the TSI and its solar cycle variation obtained by *Krivova et al.* [2006] are listed in Table 2.1 together with the values obtained by other authors from both observations and models. All recent estimates suggest a relatively high contribution of the UV wavelengths to the TSI changes. At the same time, considerable uncertainty remains.

In particular, Table 2.1 makes it apparent that, of the recent estimates of the contribution of the UV wavelengths below 400 nm to the total change in the TSI, all but SIM are lying in the range 47–63 %. At the same time, the SIM estimate of roughly 180 % is by a factor of 3 higher. The results based on the SORCE/SIM measurements [*Harder et al.*, 2009] are particularly surprising, since they imply that the total change in the irradiance at 200–400 nm is roughly a factor of 2 higher than the TSI change over the same period, which is almost compensated by the negative trends in the visible and the IR. If confirmed, this may have a significant effect on climate models [*Haigh et al.*, 2010; *Garcia*, 2010]. Note that values from *Harder et al.* [2009] listed in the table are rough estimates from their plot and depend strongly

Table 2.1 The relative contribution of different spectral ranges to the TSI variation ($\Delta F_\lambda / \Delta F_{\mathrm{TSI}}$) as measured or modelled by different authors (listed in the 1st column). The solar cycle upon which the results are based is identified in the 2nd column. The last line lists the contribution of the irradiance in the corresponding intervals, F_λ, to the total solar irradiance, F_{TSI}, according to *Krivova et al.* [2006]. If only a part of the corresponding wavelength interval or a larger range was considered, then the exact range is given in brackets (for PWB-09 and MFM-11)

Ref	Cycle	Wavelength range, nm				
		200–400	400–700	700–1000	1000–3000	> 3000
Lea-97	22	30.8				
KSF-06	23	61.8	26.1	15.4	−5.5	1.4
Hea-09	23	≈ 180	≈ −90	≈ 10	≈ −50	
PWB-09	23	47 (> 240)	20.6	5.6 (< 900)	3.2 (1100–1600)	
MFM-11	21–23	63.3 (> 150)				
$F_\lambda / F_{\mathrm{TSI}}$, %		7.7	38.7	22.7	28.8	2.0

Lea-97 *Lean et al.* [1997]
KSF-06 *Krivova et al.* [2006]
Hea-09 *Harder et al.* [2009]
PWB-09 *Pagaran et al.* [2009]
MFM-11 *Morrill et al.* [2011]

on whether the SORCE/TIM data are used as a measure for the TSI change or an integral over the SORCE/SIM data (corrected for the missing wavelengths).

The unusual behaviour displayed by the SORCE/SIM data can also be seen in Fig. 2.1. In this figure, the red line shows the data integrated over the entire SORCE/SIM spectral range (200–1630 nm) shifted to account for the missing wavelengths, the blue line shows the same quantity, but now provided by the SATIRE-S model [*Ball et al.*, 2011], and the red line represents the SORCE/TIM TSI measurements. In the bottom panel, all data are smoothed to emphasise the long-term trends. Whereas SATIRE-S results agree amazingly well with the TIM data, the long-term behaviour of the integrated SIM data is rather different. A comparison of the SATIRE-S with the SIM data in individual spectral bands reveals that most significant differences between the modelled and measured trends appear at the beginning of the considered period, i.e. before 2006–2007 [*Ball et al.*, 2011]. Since SATIRE-S also reproduces UV measurements by UARS/SUSIM between 1991 and 2002 [*Krivova and Solanki*, 2005; *Krivova et al.*, 2006, 2009b; *Ball et al.*, 2011; *Morrill et al.*, 2011] this implies that either the mechanism of SSI variation fundamentally changed around the peak of cycle 23 or there is an inconsistency between SUSIM and SIM measurements. Thus the origin of the very different trends measured by SIM needs to be further investigated.

We note that the estimate of the relative changes in different spectral ranges by *Pagaran et al.* [2009] based on SCIAMACHY measurements should be considered as a lower limit. The reason is that their normalisation to the TSI change is not self-consistent (the total change in the irradiance used for the normalisation is taken from the PMOD composite and is not estimated from SCIAMACHY data). As a result, the sum of the contributions of all the wavelength ranges they list is only 76 % of the TSI. The missing wavelengths (below 240 nm and above 1600 nm) are unlikely to contribute the remaining 24 % if other studies are to be believed (see Table 2.1 and *Krivova et al.*, 2006). We therefore believe that the real numbers should be roughly 20 % higher, i.e. the contribution of the 240–400 nm range to the TSI variation, as derived from SCIAMACHY data, is most probably lying around 60 %.

2.4 Telescope Era

Since high-quality full-disc magnetograms of the Sun are not available before the 1970s, reconstructions on longer time scales have to rely on disc-integrated quantities. To describe the evolution of sunspots, the records of their areas and numbers have widely been employed. Sunspot areas and positions have regularly been measured by the Royal Greenwich Observatory between 1874 and 1976. *Balmaceda et al.* [2009] have combined these data with more recent observations from a number of other observatories taking into account their systematic differences to provide a homogeneous record covering more than 130 years. The historic sunspot number record goes back to 1610. A long, reliable proxy of facular areas is not yet available. However their evolution is related to that of sunspots, since both are found in active regions. Hence, a given relationship of the magnetic flux emerging in plage to that in sunspots (such as that found by *Chapman et al.*, 1997) can be employed. The changes of the weak magnetic field on the solar surface, however, are not well represented by the sunspot proxies, which makes an estimate of the secular trend in the irradiance a particularly challenging task. It is this secular trend that is of particular interest for climate studies.

Most of the early reconstructions of TSI back to the Maunder minimum [e.g., *Lean et al.*, 1992; *Zhang et al.*, 1994] relied on indirect estimates of the secular change in irradiance derived from a comparison with other Sun-like stars. The basis for these estimates turned out to be flawed [e.g., *Hall and Lockwood*, 2004; *Wright*, 2004] and they have later been strongly criticised. A physical mechanism that produces a secular change in the irradiance has been identified by *Solanki et al.* [2000, 2002]. *Harvey* [1992] has shown that a significant amount of fresh flux in ephemeral active regions (smaller than normal active regions and spread over the whole solar surface) appears at the surface already during the decay phase of the previous cycle, so that consecutive cycles of ephemeral region emergence overlap. Significant background magnetic flux is thus present on the solar surface even at activity minima. This flux is actually comparable to the flux in active regions at activity maximum [*Harvey*, 1994; *Krivova and Solanki*, 2004]. Since the length and the strength of the solar cycle vary with time, so does the overlap between the ephemeral region flux from different activity cycles. This should lead to a secular change in the background field.

This idea underlies our reconstructions of the open and total solar magnetic flux back to the Maunder minimum [*Solanki et al.*, 2000, 2002; *Krivova et al.*, 2007;

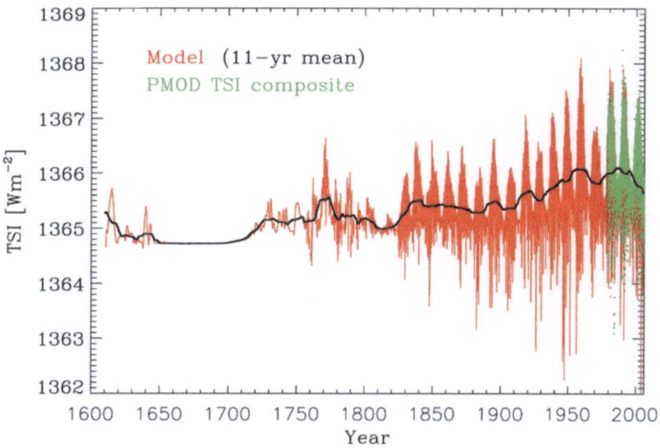

Fig. 2.4 Solar total irradiance since 1610 reconstructed using the SATIRE-T model (*red line*). Also shown are the 11-yr smoothed TSI (*thick black line*) and PMOD composite of measurements since 1978 (*green*). From *Krivova et al.* [2010]

Vieira and Solanki, 2010] from the sunspot number. The total flux agrees well with observations available for the last 4 cycles, while the open flux closely follows the empirical reconstruction by *Lockwood et al.* [2009]. Moreover, if the modelled open flux is used to calculate ^{44}Ti activity following *Usoskin et al.* [2006b], the latter agrees well with ^{44}Ti activity measured in stony meteorites [*Vieira et al.*, 2011].

Using the modelled solar surface magnetic flux as input to SATIRE-T (see Sect. 2.2), *Balmaceda et al.* [2007] and *Krivova et al.* [2007, 2010] have reconstructed solar total and spectral irradiance since the Maunder minimum. The reconstructed TSI is shown in Fig. 2.4. We find that between the end of the Maunder minimum and the end of the 20th century, the cycle-averaged TSI has increased by 1.25 W/m^2 or about 0.9 %. *Krivova et al.* [2007] have estimated the possible range of the secular increase as 0.9–1.5 W/m^2. This range is consistent with most other recent estimates derived under various assumptions [*Foster*, 2004; *Wang et al.*, 2005; *Crouch et al.*, 2008; *Steinhilber et al.*, 2009], although two different, fairly controversial estimates have recently been published by *Schrijver et al.* [2011] and *Shapiro et al.* [2011]. This is because they considered rather extreme assumptions about the solar activity state during the Maunder minimum. *Schrijver et al.* [2011] assumed that the minimum solar activity state, similar to that reached during the Maunder minimum, was globally approached during the last minimum in 2008, which suggests a decrease of only about 0.014–0.036 % compared to the minimum in 1996. *Shapiro et al.* [2011], in contrast, assumed that during the Maunder minimum the entire solar surface had the intensity that is currently observed only in the darkest parts of supergranule cells, which results in a value of 0.4 % (6 ± 3 W/m^2) for the TSI change between the Maunder minimum and the present. Therefore these two extreme estimates may be considered as lower and upper limits.

Figures 2.5 and 2.6 show solar irradiance in Ly-α and several other spectral intervals of special interest for climate models reconstructed by *Krivova et al.* [2010]. Interestingly, the irradiance variation in the IR, between 1500 and 2500 nm, is reversed compared to other spectral ranges, as also seen in SORCE/SIM data [*Harder et al.*, 2009], which, however, also displays such a reversed behaviour in the visible.

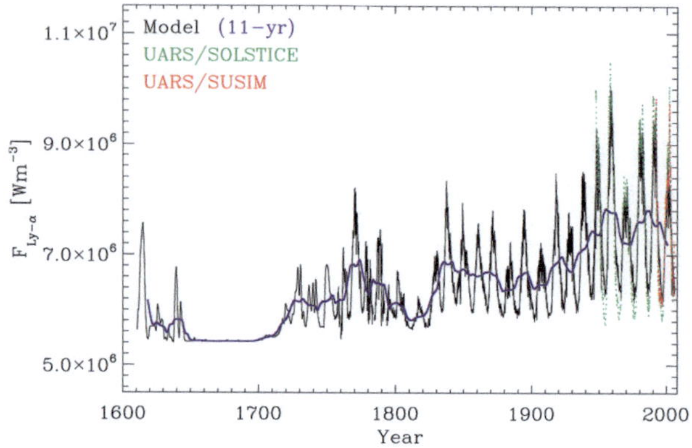

Fig. 2.5 Solar irradiance in Ly-α reconstructed using the SATIRE-T model (*black solid line*). Also shown are the 11-yr smoothed Ly-α irradiance (*dark blue*), measurements by the UARS/SUSIM instrument (*red*) as well as the composite (*green*) of UARS/SOLSTICE measurements and proxy models by *Woods et al.* [2000]

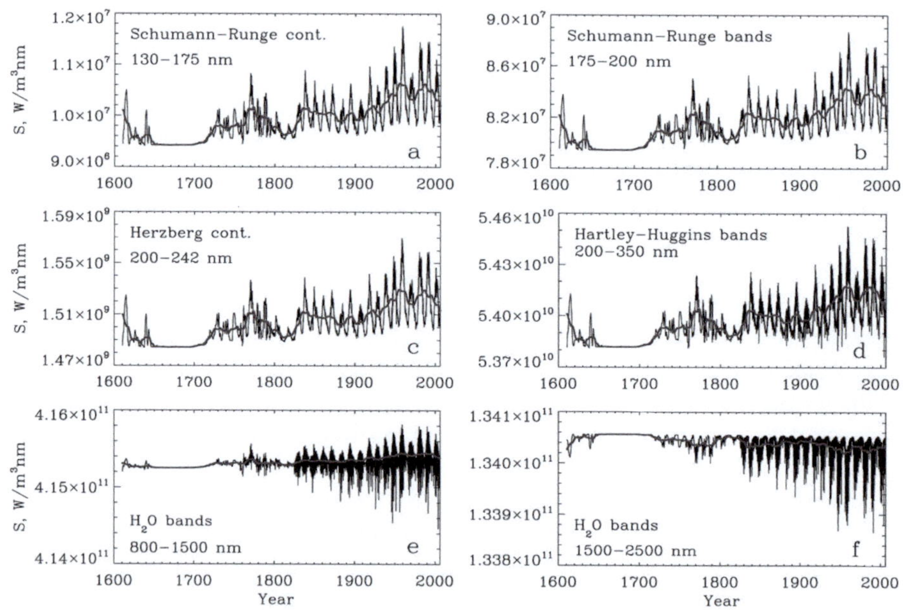

Fig. 2.6 Reconstructed solar irradiance in selected spectral intervals of special interest for climate models: daily (*thin lines*) and smoothed over 11 years (*thick lines*). (**a**) Shumann-Runge oxygen continuum; (**b**) Schumann-Runge oxygen bands; (**c**) Herzberg oxygen continuum; (**d**) Hartley-Huggins ozone bands; (**e**) and (**f**) water vapour infrared bands. The exact wavelength ranges are indicated in each panel. From *Krivova et al.* [2010]

Potentially, information on the magnitude of the secular change in irradiance can be provided by full-disc solar images in the Ca II line. A number of observatories around the globe carried out such observations since the beginning of the 20th century, and some of these archives have recently been digitised. Unfortunately, these images suffer from numerous problems and artefacts (e.g., plate defects and aging, geometrical distortions, degradation and changes of the instrumentation etc.), some of which, such as photometric uncertainties, are difficult to correct [*Ermolli et al.*, 2009]. Thus more work is needed before the Ca II archives can be employed for reliable irradiance reconstructions, although their potential remains very large. A simultaneous analysis of different archives would be advantageous.

2.5 Pretelescopic Era

The sunspot number record goes back to 1610. Information on the solar magnetic activity prior to 1610 is provided by concentrations of cosmogenic isotopes, such as ^{14}C or ^{10}Be, in natural archives. These can be used to calculate the Sun's open and total magnetic flux and the sunspot number [*Solanki et al.*, 2004; *Usoskin et al.*, 2006a; *Steinhilber et al.*, 2009; *Vieira and Solanki*, 2010].

The main complication for the irradiance reconstructions is that only cycle-averaged values can be derived. Thus the SATIRE-T model cannot be applied directly. *Vieira and Solanki* [2010] and *Vieira et al.* [2011] have reconstructed the evolution of the decadally averaged magnetic flux from decadal values of ^{14}C concentrations employing a series of physics-based models connecting the processes from the modulation of the cosmic ray flux in the heliosphere to the isotope records in natural archives. They have also found that the variation in the TSI is produced by contributions of the magnetic flux from two cycles. This result suggests that reconstructions of TSI based on linear relationships between the open flux and TSI [e.g., *Steinhilber et al.*, 2009] are not justified physically, although the practical consequences are expected to be significant mainly for time scales shorter than 40–50 years. *Vieira et al.* [2011] thus compute the TSI (Fig. 2.7) as a linear combination of two consecutive decadal values of the open magnetic flux and employ different paleomagnetic models to evaluate the uncertainties.

Note that the TSI reconstructed in this way shows a stronger increase since the Maunder minimum than the reconstruction by *Steinhilber et al.* [2009] from the ^{10}Be data. This is expected to be largely due to the difference in the methods employed by the two groups, although differences in the input data, ^{10}Be vs. ^{14}C, may also contribute. In particular, the linear relationship between the TSI and the open flux employed by *Steinhilber et al.* [2009] is based on the measurements for the last 3 minima only [*Fröhlich*, 2009]. Of these, the levels of the last two in the TSI are rather uncertain (see Sect. 2.3.1).

2.6 Summary

The SATIRE models have been used to reconstruct solar total and spectral irradiance on time scales ranging from a day up to millennia from different available proxies of solar magnetic activity. Most accurate are reconstructions from the full-disc magnetograms and continuum images (SATIRE-S) covering the period after 1974. Reconstructions from the sunspot number (SATIRE-T) go back to 1610, while the

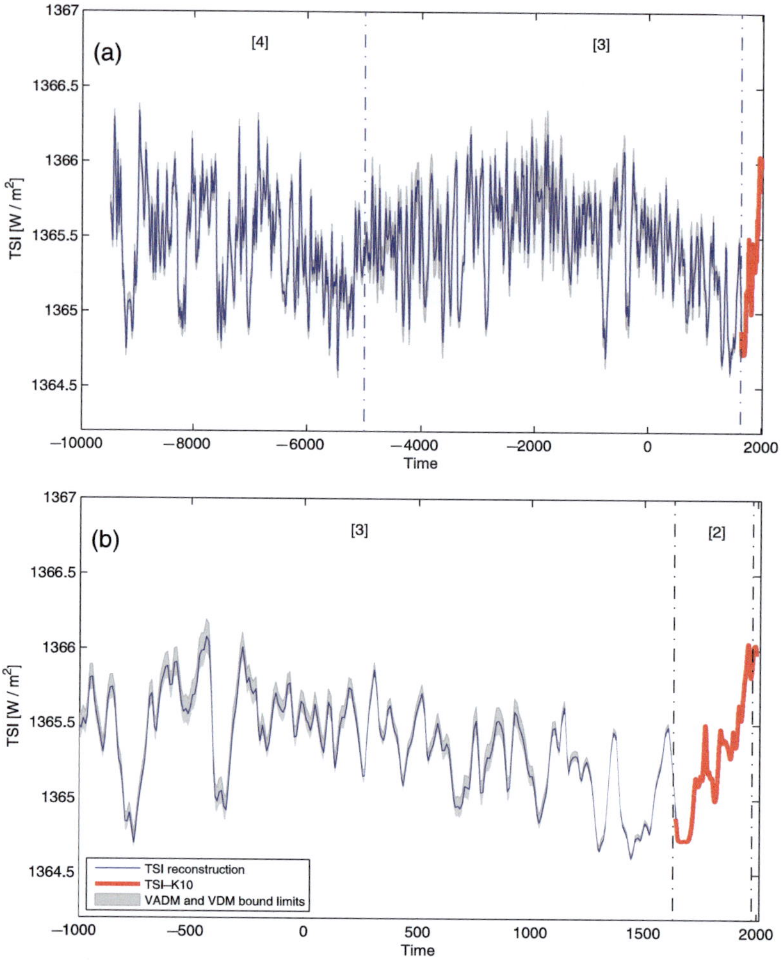

Fig. 2.7 (**a**) TSI reconstruction since 9500 BC from the SATIRE-M (*blue*) and SATIRE-T (*red*) models. (**b**) Enlargement of panel a for the last 3000 years. From *Vieira et al.* [2011]

SATIRE-M version, relying on the cosmogenic isotope concentrations, has provided TSI over the whole Holocene. The following data sets produced in the framework of the CAWSES/SOLIVAR project described here are available from the MPS Sun-Climate web page http://www.mps.mpg.de/projects/sun-climate/data.html:

- the solar spectral UV irradiance at 115–400 nm between 1 Jan 1974 and 31 Dec 2007 reconstructed from magnetograms [*Krivova et al.*, 2006, 2009b];
- TSI and SSI since 1610 reconstructed from the sunspot number [*Balmaceda et al.*, 2007; *Krivova et al.*, 2007, 2010];
- TSI for the Holocene reconstructed from the ^{14}C data [*Vieira et al.*, 2011];
- the composite of daily sunspot areas and the PSI index calculated after cross-calibration of measurements by different observatories [*Balmaceda et al.*, 2009].

Acknowledgements This work was supported by the Deutsche Forschungsgemeinschaft, DFG project number SO 711/1 and by the WCU grant No. R31-10016 funded by the Korean Ministry of Education, Science and Technology.

References

Ball, W. T., Unruh, Y. C., Krivova, N. A., Solanki, S. K., & Harder, J. W. (2011). Solar irradiance: a six-year comparison between SORCE observations and the SATIRE model. *Astronomy & Astrophysics*, *530*, A71.

Balmaceda, L., Krivova, N. A., & Solanki, S. K. (2007). Reconstruction of solar irradiance using the Group sunspot number. *Advances in Space Research*, *40*, 986–989.

Balmaceda, L. A., Solanki, S. K., Krivova, N. A., & Foster, S. (2009). A homogeneous sunspot areas database covering more than 130 years. *Journal of Geophysical Research*, *114*, A07104. doi:10.1029/2009JA014299.

Bond, G., Kromer, B., Beer, J., Muscheler, R., Evans, M. N., Showers, W., Hoffmann, S., Lotti-Bond, R., Hajdas, I., & Bonani, G. (2001). Persistent solar influence on North Atlantic climate during the Holocene. *Science*, *294*, 2130–2136.

Camp, C. D., & Tung, K. K. (2007). Surface warming by the solar cycle as revealed by the composite mean difference projection. *Geophysical Research Letters*, *34*, L14703. doi:10.1029/2007GL030207.

Chapman, G. A., Cookson, A. M., & Dobias, J. J. (1997). Solar variability and the relation of facular to sunspot areas during solar cycle 22. *The Astrophysical Journal*, *482*, 541–545.

Christensen-Dalsgaard, J. (2002). Helioseismology. *Reviews of Modern Physics*, *74*, 1073–1129. doi:10.1103/RevModPhys.74.1073.

Crouch, A. D., Charbonneau, P., Beaubien, G., & Paquin-Ricard, D. (2008). A model for the total solar irradiance based on active region decay. *The Astrophysical Journal*, *677*, 723–741. doi:10.1086/527433.

Danilovic, S., Solanki, S. K., Livingston, W., Krivova, N. A., & Vince, I. (2007). Magnetic source of the solar cycle variation of the Mn I 539.4 nm line. In *Modern solar facilities—advanced solar sciences* (pp. 189–192). Göttingen: Universtitätsverlag.

Danilovic, S., Solanki, S. K., Livingston, W., Krivova, N. A., & Vince, I. (2011). The solar cycle variation of the Mn I 539.4 nm line. *Astronomy & Astrophysics*. doi:10.1051/0004-6361/201116565. Submitted.

DeLand, M. T., & Cebula, R. P. (2008). Creation of a composite solar ultraviolet irradiance data set. *Journal of Geophysical Research*, *113*(A12), A11103. doi:10.1029/2008JA013401.

Dewitte, S., Crommelynck, D., Mekaoui, S., & Joukoff, A. (2004). Measurement and uncertainty of the long term total solar irradiance trend. *Solar Physics*, *224*, 209–216.

Domingo, V., Ermolli, I., Fox, P., Fröhlich, C., Haberreiter, M., Krivova, N., Kopp, G., Schmutz, W., Solanki, S. K., Spruit, H. C., Unruh, Y., & Vögler, A. (2009). Solar surface magnetism and irradiance on time scales from days to the 11-year cycle. *Space Science Reviews*, *145*, 337–380. doi:10.1007/s11214-009-9562-1.

Eddy, J. A. (1976). The Maunder minimum. *Science*, *192*, 1189–1202.

Ermolli, I., Berrilli, F., & Florio, A. (2003). A measure of the network radiative properties over the solar activity cycle. *Astronomy & Astrophysics*, *412*, 857–864.

Ermolli, I., Solanki, S. K., Tlatov, A. G., Krivova, N. A., Ulrich, R. K., & Singh, J. (2009). Comparison among Ca II K spectroheliogram time series with an application to solar activity studies. *The Astrophysical Journal*, *698*, 1000–1009.

Fligge, M., Solanki, S. K., & Unruh, Y. C. (2000). Modelling irradiance variations from the surface distribution of the solar magnetic field. *Astronomy & Astrophysics*, *353*, 380–388.

Floyd, L., Rottman, G., DeLand, M., & Pap, J. (2003). 11 years of solar UV irradiance measurements from UARS. *ESA SP*, *535*, 195–203.

Fontenla, J. M., Harder, J., Rottman, G., Woods, T. N., Lawrence, G. M., & Davis, S. (2004). The signature of solar activity in the infrared spectral irradiance. *The Astrophysical Journal Letters, 605*, L85–L88.

Fontenla, J. M., Avrett, E., Thuillier, G., & Harder, J. (2006). Semiempirical models of the solar atmosphere. I. The quiet- and active sun photosphere at moderate resolution. *The Astrophysical Journal, 639*, 441–458.

Foster, S. (2004). *Reconstruction of solar irradiance variations for use in studies of global climate change: application of recent SOHO observations with historic data from the Greenwich observatory*. Ph.D. thesis, University of Southhampton, School of Physics and Astronomy.

Fröhlich, C., & Lean, J. (1997). Total solar irradiance variations: the construction of a composite and its comparison with models. *ESA SP, 415*, 227–233.

Fröhlich, C. (2006). Solar irradiance variability since 1978: revision of the PMOD composite during solar cycle 21. *Space Science Reviews, 125*, 53–65.

Fröhlich, C. (2009). Evidence of a long-term trend in total solar irradiance. *Astronomy & Astrophysics, 501*, L27–L30. doi:10.1051/0004-6361/200912318.

Fröhlich, C. (2011). Solar constant. Construction of a composite total solar irradiance (TSI) time series from 1978 to present. http://www.pmodwrc.ch/pmod.php?topic=tsi/composite/SolarConstant.

Fröhlich, C., Andersen, B., Appourchaux, T., Berthomieu, G., Crommelynck, D. A., Domingo, V., Fichot, A., Finsterle, W., Gomez, M. F., Gough, D., Jimenez, A., Leifsen, T., Lombaerts, M., Pap, J. M., Provost, J., Cortes, T. R., Romero, J., Roth, H., Sekii, T., Telljohann, U., Toutain, T., & Wehrli, C. (1997). First results from VIRGO, the experiment for helioseismology and solar irradiance monitoring on SOHO. *Solar Physics, 170*, 1–25.

Garcia, R. R. (2010). Atmospheric physics: solar surprise? *Nature, 467*, 668–669. doi:10.1038/467668a.

Gray, L. J., Beer, J., Geller, M., Haigh, J. D., Lockwood, M., Matthes, K., Cubasch, U., Fleitmann, D., Harrison, G., Hood, L., Luterbacher, J., Meehl, G. A., Shindell, D., van Geel, B., & White, W. (2010). Solar influences on climate. *Reviews of Geophysics, 48*, RG4001. doi:10.1029/2009RG000282.

Haigh, J. D. (1994). The role of stratospheric ozone in modulating the solar radiative forcing of climate. *Nature, 370*, 544–546.

Haigh, J. D. (2007). The sun and the earth's climate. *Living Reviews in Solar Physics*. http://solarphysics.livingreviews.org/Articles/lrsp-2007-2/.

Haigh, J. D., Winning, A. R., Toumi, R., & Harder, J. W. (2010). An influence of solar spectral variations on radiative forcing of climate. *Nature, 467*, 696–699. doi:10.1038/nature09426.

Hall, J. C., & Lockwood, G. W. (2004). The chromospheric activity and variability of cycling and flat activity solar-analog stars. *The Astrophysical Journal, 614*, 942–946.

Hansen, J., Sato, M., Nazarenko, L., Ruedy, R., Lacis, A., Koch, D., Tegen, I., Hall, T., Shindell, D., Santer, B., Stone, P., Novakov, T., Thomason, L., Wang, R., Wang, Y., Jacob, D., Hollandsworth, S., Bishop, L., Logan, J., Thompson, A., Stolarski, R., Lean, J., Willson, R., Levitus, S., Antonov, J., Rayner, N., Parker, D., & Christy, J. (2002). Climate forcings in Goddard Institute for space studies SI2000 simulations. *Journal of Geophysical Research, 107*. doi:10.1029/2001JD001143.

Hansen, J., Sato, M., Ruedy, R., Nazarenko, L., Lacis, A., Schmidt, G. A., Russell, G., Aleinov, I., Bauer, M., Bauer, S., Bell, N., Cairns, B., Canuto, V., Chandler, M., Cheng, Y., Del Genio, A., Faluvegi, G., Fleming, E., Friend, A., Hall, T., Jackman, C., Kelley, M., Kiang, N., Koch, D., Lean, J., Lerner, J., Lo, K., Menon, S., Miller, R., Minnis, P., Novakov, T., Oinas, V., Perlwitz, J., Perlwitz, J., Rind, D., Romanou, A., Shindell, D., Stone, P., Sun, S., Tausnev, N., Thresher, D., Wielicki, B., Wong, T., Yao, M., & Zhang, S. (2005). Efficacy of climate forcings. *Journal of Geophysical Research, 110*(D9), D18104. doi:10.1029/2005JD005776.

Hansen, J. E. (2000). The Sun's role in long-term climate change. *Space Science Reviews, 94*, 349–356.

Harder, J. W., Fontenla, J. M., Pilewskie, P., Richard, E. C., & Woods, T. N. (2009). Trends in solar spectral irradiance variability in the visible and infrared. *Geophysical Research Letters,*

36. doi:10.1029/2008GL036797.
Harvey, K. L. (1992). The cyclic behavior of solar activity. In *ASP conf. ser.: Vol. 27. The solar cycle* (pp. 335–367).
Harvey, K. L. (1994). The solar magnetic cycle. In R. J. Rutten & C. J. Schrijver (Eds.), *Solar surface magnetism* (p. 347). Dordrecht: Kluwer.
Hoyt, D. V., & Schatten, K. H. (1993). A discussion of plausible solar irradiance variations, 1700–1992. *Journal of Geophysical Research*, *98*, 18895–18906.
Jungclaus, J. H., Lorenz, S. J., Timmreck, C., Reick, C. H., Brovkin, V., Six, K., Segschneider, J., Giorgetta, M. A., Crowley, T. J., Pongratz, J., Krivova, N. A., Vieira, L. E., Solanki, S. K., Klocke, D., Botzet, M., Esch, M., Gayler, V., Haak, H., Raddatz, T. J., Roeckner, E., Schnur, R., Widmann, H., Claussen, M., Stevens, B., & Marotzke, J. (2010). Climate and carbon-cycle variability over the last millennium. *Climate of the Past*, *6*, 723–737. doi:10.5194/cp-6-723-2010.
Kodera, K., & Kuroda, Y. (2002). Dynamical response to the solar cycle. *Journal of Geophysical Research*, *107*, D24. doi:10.1029/2002JD002224.
Kodera, K., & Kuroda, Y. (2005). A possible mechanism of solar modulation of the spatial structure of the North Atlantic Oscillation. *Journal of Geophysical Research*, *110*, D02111. doi:10.1029/2004JD005258.
Kopp, G., & Lean, J. L. (2011). A new, lower value of total solar irradiance: evidence and climate significance. *Geophysical Research Letters*, *38*, L01706. doi:10.1029/2010GL045777.
Kopp, G., Lawrence, G., & Rottman, G. (2005). The total irradiance monitor (TIM): science results. *Solar Physics*, *230*, 129–139. doi:10.1007/s11207-005-7433-9.
Kopp, G., Heuerman, K., Harber, D., & Drake, G. (2007). *The TSI radiometer facility: absolute calibrations for total solar irradiance instruments*. In *Society of photo-optical instrumentation engineers (SPIE) conference series: Vol. 6677*. doi:10.1117/12.734553.
Krivova, N. A., & Solanki, S. K. (2004). Effect of spatial resolution on estimating the Sun's magnetic flux. *Astronomy & Astrophysics*, *417*, 1125–1132.
Krivova, N. A., & Solanki, S. K. (2005). Reconstruction of solar UV irradiance. *Advances in Space Research*, *35*, 361–364.
Krivova, N. A., & Solanki, S. K. (2008). Models of solar irradiance variations: current status. *Journal of Astrophysics and Astronomy*, *29*, 151–158.
Krivova, N. A., Solanki, S. K., Fligge, M., & Unruh, Y. C. (2003). Reconstruction of solar total and spectral irradiance variations in cycle 23: is solar surface magnetism the cause? *Astronomy & Astrophysics*, *399*, L1–L4.
Krivova, N. A., Solanki, S. K., & Floyd, L. (2006). Reconstruction of solar UV irradiance in cycle 23. *Astronomy & Astrophysics*, *452*, 631–639.
Krivova, N. A., Balmaceda, L., & Solanki, S. K. (2007). Reconstruction of solar total irradiance since 1700 from the surface magnetic flux. *Astronomy & Astrophysics*, *467*, 335–346.
Krivova, N. A., Solanki, S. K., & Wenzler, T. (2009a). ACRIM-gap and total solar irradiance revisited: is there a secular trend between 1986 and 1996? *Geophysical Research Letters*, *36*, L20101. doi:10.1029/2009GL040707.
Krivova, N. A., Solanki, S. K., Wenzler, T., & Podlipnik, B. (2009b). Reconstruction of solar UV irradiance since 1974. *Journal of Geophysical Research*, *114*, D00I04. doi:10.1029/2009JD012375.
Krivova, N. A., Vieira, L. E. A., & Solanki, S. K. (2010). Reconstruction of solar spectral irradiance since the Maunder minimum. *Journal of Geophysical Research*, *115*, A12112. doi:10.1029/2010JA015431.
Krivova, N. A., Solanki, S. K., & Schmutz, W. (2011a). Solar total irradiance in cycle 23. *Astronomy & Astrophysics*, *529*, A81. doi:10.1051/0004-6361/201016234.
Krivova, N. A., Solanki, S. K., & Unruh, Y. C. (2011b). Towards a long-term record of solar total and spectral irradiance. *Journal of Atmospheric and Solar-Terrestrial Physics*, *73*, 223–234. doi:10.1016/j.jastp.2009.11.013.
Kurucz, R. (1993). ATLAS9 stellar atmosphere programs and 2 km/s grid. In *ATLAS9 stellar atmosphere programs and 2 km/s grid*. Cambridge: Smithsonian Astrophysical Observatory. Kurucz CD-ROM No. 13.

Langematz, U., Matthes, K., & Grenfell, J. L. (2005). Solar impact on climate: modeling the coupling between the middle and the lower atmosphere. *Memorie Della Società Astronomica Italiana*, *76*, 868–875.

Lean, J. (1989). Contribution of ultraviolet irradiance variations to changes in the sun's total irradiance. *Science*, *244*, 197–200.

Lean, J., Skumanich, A., & White, O. (1992). Estimating the sun's radiative output during the Maunder minimum. *Geophysical Research Letters*, *19*, 1595–1598.

Lean, J. L., Rottman, G. J., Kyle, H. L., Woods, T. N., Hickey, J. R., & Puga, L. C. (1997). Detection and parameterization of variations in solar mid- and near-ultraviolet radiation (200–400 nm). *Journal of Geophysical Research*, *102*, 29939–29956.

Lockwood, M., Rouillard, A. P., & Finch, I. D. (2009). The rise and fall of open solar flux during the current grand solar maximum. *The Astrophysical Journal*, *700*, 937–944. doi:10.1088/0004-637X/700/2/937.

Matthes, K., Kuroda, Y., Kodera, K., & Langematz, U. (2006). Transfer of the solar signal from the stratosphere to the troposphere: northern winter. *Journal of Geophysical Research*, *111*, D06108. doi:10.1029/2005JD006283.

Mendoza, B. (1997). Estimations of Maunder minimum solar irradiance and Ca II H and K fluxes using rotation rates and diameters. *The Astrophysical Journal*, *483*, 523–526.

Morrill, J. S., Floyd, L., & McMullin, D. (2011). The solar ultraviolet spectrum estimated using the Mg II index and Ca II K disk activity. *Solar Physics*, *269*, 253–267. doi:10.1007/s11207-011-9708-7.

Neff, U., Burns, S. J., Mangini, A., Mudelsee, M., Fleitmann, D., & Matter, A. (2001). Strong coherence between solar variability and the monsoon in Oman between 9 and 6 kyr ago. *Nature*, *411*, 290–293.

Pagaran, J., Weber, M., & Burrows, J. (2009). Solar variability from 240 to 1750 nm in terms of faculae brightening and sunspot darkening from SCIAMACHY. *The Astrophysical Journal*, *700*, 1884–1895. doi:10.1088/0004-637X/700/2/1884.

Preminger, D. G., Walton, S. R., & Chapman, G. A. (2002). Photometric quantities for solar irradiance modeling. *Journal of Geophysical Research*, *107*(A11), 1354. doi:10.1029/2001JA009169.

Reid, G. C. (1987). Influence of solar variability on global sea surface temperatures. *Nature*, *329*, 142–143.

Rozanov, E. V., Schlesinger, M. E., Egorova, T. A., Li, B., Andronova, N., & Zubov, V. A. (2004). Atmospheric response to the observed increase of solar UV radiation from solar minimum to solar maximum simulated by the University of Illinois at Urbana-Champaign climate—chemistry model. *Journal of Geophysical Research*, *109*, D1. doi:10.1029/2003JD003796.

Scafetta, N., & Willson, R. C. (2009). ACRIM-gap and TSI trend issue resolved using a surface magnetic flux TSI proxy model. *Geophysical Research Letters*, *36*, L05701. doi:10.1029/2008GL036307.

Schmidt, G. A., Jungclaus, J. H., Ammann, C. M., Bard, E., Braconnot, P., Crowley, T. J., Delaygue, G., Joos, F., Krivova, N. A., Muscheler, R., Otto-Bliesner, B. L., Pongratz, J., Shindell, D. T., Solanki, S. K., Steinhilber, F., & Vieira, L. E. A. (2011). Climate forcing reconstructions for use in PMIP simulations of the last millennium (v1.0). *Geoscientific Model Development*, *4*, 33–45. doi:10.5194/gmd-4-33-2011.

Schrijver, C. J., Livingston, W. C., Woods, T. N., & Mewaldt, R. A. (2011). The minimal solar activity in 2008–2009 and its implications for long-term climate modeling. *Geophysical Research Letters*, *38*, L06701. doi:10.1029/2011GL046658.

Seleznyov, A. D., Solanki, S. K., & Krivova, N. A. (2011). Modelling solar irradiance variability on time scales from minutes to months. *Astronomy & Astrophysics*, *532*, A108.

Shapiro, A. I., Schmutz, W., Schoell, M., Haberreiter, M., & Rozanov, E. (2010). NLTE solar irradiance modeling with the COSI code. *Astronomy & Astrophysics*, *517*, A48. doi:10.1051/0004-6361/200913987.

Shapiro, A. I., Schmutz, W., Rozanov, E., Schoell, M., Haberreiter, M., Shapiro, A. V., & Nyeki, S. (2011). A new approach to the long-term reconstruction of the solar irradiance leads to large his-

torical solar forcing. *Astronomy & Astrophysics*, *529*, A67. doi:10.1051/0004-6361/201016173.
Skupin, J., Noël, S., Wuttke, M. W., Gottwald, M., Bovensmann, H., Weber, M., & Burrows, J. P. (2005). SCIAMACHY solar irradiance observation in the spectral range from 240 to 2380 nm. *Advances in Space Research*, *35*, 370–375.
Solanki, S. K., Schüssler, M., & Fligge, M. (2000). Evolution of the Sun's large-scale magnetic field since the Maunder minimum. *Nature*, *408*, 445–447.
Solanki, S. K., Schüssler, M., & Fligge, M. (2002). Secular variation of the Sun's magnetic flux. *Astronomy & Astrophysics*, *383*, 706–712.
Solanki, S. K., Usoskin, I. G., Kromer, B., Schüssler, M., & Beer, J. (2004). Unusual activity of the Sun during recent decades compared to the previous 11,000 years. *Nature*, *431*, 1084–1087.
Solanki, S. K., Krivova, N. A., & Wenzler, T. (2005). Irradiance models. *Advances in Space Research*, *35*, 376–383.
Solomon, S., Qin, D., Manning, M., Chen, Z., Marquis, M., Averyt, K. B., Tignor, M., & Miller, H. L. (Eds.) (2007). *Climate change 2007: the physical science basis. Contribution of working group I to the fourth assessment report of the intergovernmental panel on climate change*. Cambridge: Cambridge University Press.
Soon, W. H., Posmentier, E. S., & Baliunas, S. L. (1996). Inference of solar irradiance variability from terrestrial temperature changes, 1880–1993: an astrophysical application of the sun-climate connection. *The Astrophysical Journal*, *472*, 891–902. doi:10.1086/178119.
Steinhilber, F. (2010). Total solar irradiance since 1996: is there a long-term variation unrelated to solar surface magnetic phenomena? *Astronomy & Astrophysics*, *523*, A39.
Steinhilber, F., Beer, J., & Fröhlich, C. (2009). Total solar irradiance during the Holocene. *Geophysical Research Letters*, *36*. doi:10.1029/2009GL040142.
Unruh, Y. C., Solanki, S. K., & Fligge, M. (1999). The spectral dependence of facular contrast and solar irradiance variations. *Astronomy & Astrophysics*, *345*, 635–642.
Unruh, Y. C., Krivova, N. A., Solanki, S. K., Harder, J. W., & Kopp, G. (2008). Spectral irradiance variations: comparison between observations and the SATIRE model on solar rotation time scales. *Astronomy & Astrophysics*, *486*, 311–323.
Usoskin, I. G., Solanki, S. K., & Korte, M. (2006a). Solar activity reconstructed over the last 7000 years: the influence of geomagnetic field changes. *Geophysical Research Letters*, *33*, L08103. doi:10.1029/2006GL025921.
Usoskin, I. G., Solanki, S. K., Taricco, C., Bhandari, N., & Kovaltsov, G. A. (2006b). Long-term solar activity reconstructions: direct test by cosmogenic ^{44}Ti in meteorites. *Astronomy & Astrophysics*, *457*, L25–L28.
Vieira, L. E. A., & Solanki, S. K. (2010). Evolution of the solar magnetic flux on time scales of years to millennia. *Astronomy & Astrophysics*, *509*, A100. doi:10.1051/0004-6361/200913276.
Vieira, L. E. A., Solanki, S. K., Krivova, N. A., & Usoskin, I. (2011). Evolution of the solar irradiance during the Holocene. *Astronomy & Astrophysics*, *531*, A6. doi:10.1051/0004-6361/201015843.
Wang, Y.-M., Lean, J. L., & Sheeley, N. R. (2005). Modeling the Sun's magnetic field and irradiance since 1713. *The Astrophysical Journal*, *625*, 522–538. doi:10.1086/429689.
Wenzler, T., Solanki, S. K., Krivova, N. A., & Fluri, D. M. (2004). Comparison between KPVT/SPM and SoHO/MDI magnetograms with an application to solar irradiance reconstructions. *Astronomy & Astrophysics*, *427*, 1031–1043.
Wenzler, T., Solanki, S. K., & Krivova, N. A. (2005). Can surface magnetic fields reproduce solar irradiance variations in cycles 22 and 23? *Astronomy & Astrophysics*, *432*, 1057–1061.
Wenzler, T., Solanki, S. K., Krivova, N. A., & Fröhlich, C. (2006). Reconstruction of solar irradiance variations in cycles 21–23 based on surface magnetic fields. *Astronomy & Astrophysics*, *460*, 583–595.
Wenzler, T., Solanki, S. K., & Krivova, N. A. (2009). Reconstructed and measured total solar irradiance: is there a secular trend between 1978 and 2003? *Geophysical Research Letters*, *36*, L11102. doi:10.1029/2009GL037519.
Willson, R. C., & Mordvinov, A. V. (2003). Secular total solar irradiance trend during solar cycles 21–23. *Geophysical Research Letters*, *30*, 1199. doi:10.1029/2002GL016038.

Willson, R. C., Gulkis, S., Janssen, M., Hudson, H. S., & Chapman, G. A. (1981). Observations of solar irradiance variability. *Science*, *211*, 700–702.

Woods, T. N., Tobiska, W. K., Rottman, G. J., & Worden, J. R. (2000). Improved solar Lyman-α irradiance modeling from 1947 through 1999 based on UARS observations. *Journal of Geophysical Research*, *105*(A12), 27195–27215.

Wright, J. T. (2004). Do we know of any Maunder minimum stars? *The Astronomical Journal*, *128*, 1273–1278.

Zhang, Q., Soon, W. H., Baliunas, S. L., Lockwood, G. W., Skiff, B. A., & Radick, R. R. (1994). A method of determining possible brightness variations of the Sun in past centuries from observations of solar-type stars. *The Astrophysical Journal Letters*, *427*, L111–L114.

Chapter 3
Investigation of Solar Irradiance Variations and Their Impact on Middle Atmospheric Ozone

Mark Weber, Joseph Pagaran, Sebastian Dikty, Christian von Savigny, John P. Burrows, Matt DeLand, Linton E. Floyd, Jerry W. Harder, Martin G. Mlynczak, and Hauke Schmidt

Abstract The satellite spectrometer SCIAMACHY aboard ENVISAT is a unique instrument that covers at a moderately high spectral resolution the entire optical range from the near UV (230 nm) to the near IR (2.4 μm) with some gaps above 1.7 μ. This broad spectral range allows not only the retrieval of several atmospheric trace gases (among them ozone), cloud and aerosol parameters, but also regular daily measurements of the spectral solar irradiance (SSI) with an unprecedented spectral coverage. The following studies were carried out with irradiance and ozone data from SCIAMACHY: a) SCIAMACHY SSI was compared to other solar data from space and ground as well as with SIM/SORCE (Solar Irradiance Monitor, the only other satellite instrument daily measuring the visible and near IR), in order to verify

M. Weber (✉) · J. Pagaran · S. Dikty · C. von Savigny · J.P. Burrows
Institute of Environmental Physics, University of Bremen FB1, PO Box 330 440, 28334 Bremen, Germany
e-mail: weber@uni-bremen.de

M. DeLand
Science System and Applications, Inc (SSAI), 10210 Greenbelt Road, Suite 600, Lanham, MD 20706, USA
e-mail: matthew.deland@ssaihq.com

L.E. Floyd
Interferometrics Inc., 13454 Sunrise Valley Drive, Herndon, Virginia, VA 20171, USA
e-mail: linton4@gmail.com

J.W. Harder
Laboratory for Atmospheric and Space Physics (LASP), University of Colorado, 1234 Innovation Drive, Boulder, CO 80303, USA
e-mail: jerry.harder@lasp.colorado.edu

M.G. Mlynczak
NASA Langley Research Center, Hampton, VA 23681-2199, USA
e-mail: m.g.mlynczak@nasa.gov

H. Schmidt
Max-Planck-Institute for Meteorology, Bundesstr. 53, 20146 Hamburg, Germany
e-mail: hauke.schmidt@zmaw.de

F.-J. Lübken (ed.), *Climate and Weather of the Sun-Earth System (CAWSES)*, Springer Atmospheric Sciences,
DOI 10.1007/978-94-007-4348-9_3, © Springer Science+Business Media Dordrecht 2013

the quality of the SCIAMACHY measurements, b) an empirical solar proxy model, in short the SCIA proxy model, was developed that permits expressing the SCIAMACHY SSI variations by fitting solar proxies for faculae brightening and sunspot darkening, which then allows investigation of solar variability on time scales beyond the instrument life time, e.g. 11-year solar cycle, c) solar cycle SSI variations derived from empirical models (Lean2000, SATIRE, SCIA proxy) and different observations (SBUV composite, SUSIM) were compared for the three most recent solar cycles 21–23, and d) SCIAMACHY ozone limb profiles were analyzed to derive signatures of the 27-day solar rotation on stratospheric ozone. Our studies were complemented by investigations of daytime variations in mesospheric ozone (here data from SABER/TIMED), which were compared to results from the HAMMONIA chemistry climate model.

3.1 Introduction

Regular daily space-borne satellite SSI monitoring started in 1978. The wavelength coverage of early SSI measurements from different satellite instruments was generally limited to below 400 nm (UV), where the largest variations occur over an 11-year solar cycle [*Rottman et al.*, 2004]. A limiting factor for many space spectrometers measuring in the UV is the optical degradation due to hard radiation that makes it challenging to maintain the accuracy over the instrument lifetime which rarely extends to more than a decade [*DeLand et al.*, 2004]. In order to derive estimates for SSI variations over an entire 11-year solar cycle or more one needs to rely on a SSI time series composed of different instruments (UV composite) as done for the UV spectral range [*DeLand and Cebula*, 2008] or use solar proxies, like the Mg II index, that are well correlated with irradiance changes over a large spectral range to extrapolate beyond the instrumental lifetime [*DeLand and Cebula*, 1993; *Viereck et al.*, 2001].

Daily observations of the visible and near-IR started with the three channel SPM (Sun Photometer) of VIRGO/SOHO (1996–2010) at selected wavelength bands [*Fröhlich et al.*, 1997] and were continued with GOME/ERS-2 (Global Ozone Monitoring Experiment) since 1995, covering 240–800 nm [*Weber et al.*, 1998; *Burrows et al.*, 1999], SCIAMACHY/ENVISAT (Scanning Imaging Absorption Spectrometer for Atmospheric Chartography) since 2002, covering 220 nm–2.4 μm [*Bovensmann et al.*, 1999], and SIM/SORCE (Solar Irradiance Monitor) since 2003, 240 nm–3 μm [*Harder et al.*, 2005a, 2005b]. In Fig. 3.1 a sample SCIAMACHY solar irradiance spectrum is shown. Compared to the UV region, daily irradiance measurements simultaneously covering the UV, visible, and the near IR do not cover yet a complete solar cycle. One of the important scientific question is what are the irradiance changes in the visible and near IR during 27-day solar rotations and can we use this information to extrapolate to changes during the 11-year solar cycle.

SCIAMACHY is primarily an atmospheric sounder measuring several trace gases in nadir (column amounts) and limb viewing geometry (vertical profiles)

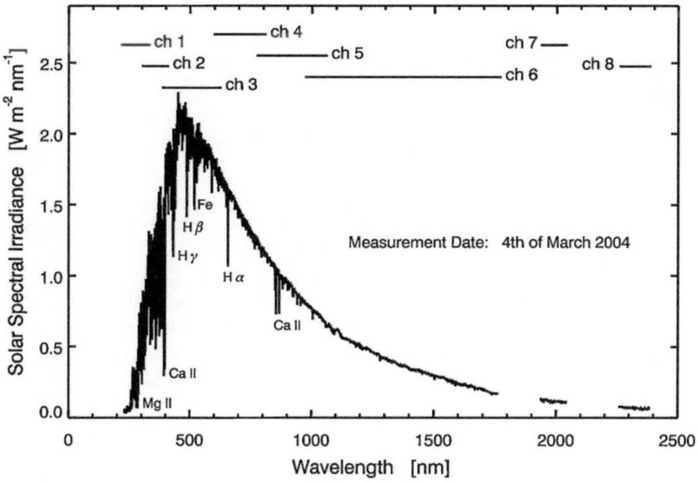

Fig. 3.1 SCIAMACHY full disc solar spectrum measured on March 4, 2004. The eight spectral channels varying in spectral resolution from 0.2 nm to 1.5 nm are indicated. The gaps near 1850 nm as well as 2200 nm are not measured by SCIAMACHY since terrestrial water vapor absorption saturates in the atmospheric observation mode. From *Pagaran et al.* [2009]. Reproduced by permission of the AAS

[*Bovensmann et al.*, 2011]. Global vertical profiles of ozone are measured by SCIAMACHY and cover the altitude range from the tropopause to about 70 km altitude [*von Savigny et al.*, 2005; *Sonkaew et al.*, 2009]. The influence of irradiance variations related to the 27-day mean solar rotation period on upper stratosphere ozone can be investigated using SCIAMACHY ozone data. The upper stratosphere above 30 km is chemically controlled and an immediate radiative influence on the photochemistry is expected [e.g. *Gruzdev et al.*, 2009; *Fioletov*, 2009]. In this study for the first time a wavelet analysis was applied to study the 27-day signature in ozone. This permits the investigation of the time-varying frequency content of the ozone signal.

The non-polar orbit of the TIMED satellite (Thermosphere, Ionosphere, Mesosphere, Energetics and Dynamics) carrying the SABER instrument (Sounding of the Atmosphere using Broadband Emission Radiometry) [*Russell III et al.*, 1999] permits the study of daytime variations in mesospheric ozone that are significantly larger than the 27-day and solar cycle related changes observed in the upper stratosphere [*Huang and Mayr*, 2008]. In this study the daytime variation of mesospheric ozone were compared for the first time with the output of a chemistry climate model [*Dikty et al.*, 2010a].

3.2 SCIAMACHY Spectral Solar Irradiance

SCIAMACHY is a passive remote sensing double spectrometer combining a predispersing prism and eight gratings in separate channels. Silicon and InGaAs detectors

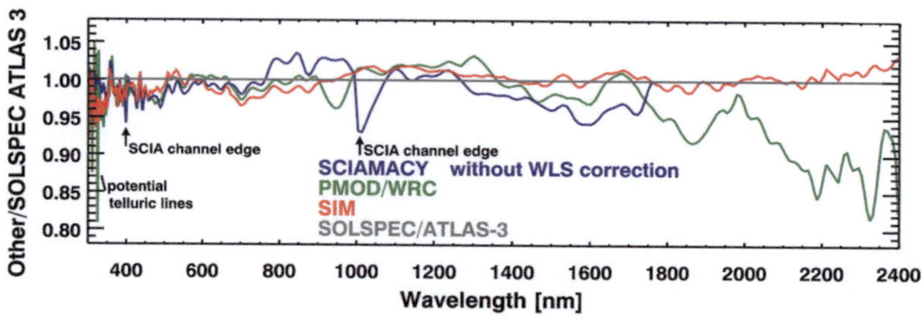

Fig. 3.2 SSI ratios of SCIAMACHY, SIM [*Harder et al.*, 2010], and PMOD/WRC (WRC85) composite [*Wehrli*, 1985] with respect to the SOLSPEC/ATLAS-3 shuttle experiment data [*Thuillier et al.*, 2004]. Close to the channel boundaries of the SCIAMACHY instrument larger deviations are observed due to instrumental artifacts. From *Pagaran et al.* [2011a]. Reproduced with permission ©ESO

are used as linear arrays with 1024 pixels each in Channels 1–5 (UV/visible) and Channels 6–8 (near IR), respectively. A detailed description of SCIAMACHY can be found in *Bovensmann et al.* [1999] and *Pagaran et al.* [2009].

Radiometrically calibrated SSI has been measured by SCIAMACHY since July 2002 once a day. A sample spectrum from March 2004 is shown in Fig. 3.1. The SCIAMACHY SSI has been compared with solar data from other satellites and measurements from the ground [*Skupin et al.*, 2005a, 2005b; *Pagaran et al.*, 2011a]. Figure 3.2 shows the comparison of SCIAMACHY with SIM [*Harder et al.*, 2010], the SOLSPEC/ATLAS-3 shuttle experiment [*Thuillier et al.*, 2004], and the PMOD/WRC (WRC85) composite [*Wehrli*, 1985]. The PMOD/WRC composite (200 nm–10 μm) was derived from various spectra obtained from aircraft, rocket, and balloon experiments as well as ground data from *Neckel and Labs* [1984]. SCIAMACHY agrees to within 5 % (SIM within 4 %) with the SOLSPEC data from 300 to 1600 nm [*Pagaran et al.*, 2011a]. The theoretical precision is usually in the range of 2–3 % based upon radiometric standards [*Bovensmann et al.*, 1999]. A more comprehensive comparison also to other solar data can be found in *Pagaran et al.* [2011a].

In later years of the SCIAMACHY mission the optical degradation in the UV due to the hard radiation environment in space is evident. The agreement of SCIAMACHY with other solar data can be improved when using the white light lamp (WLS) source as a degradation correction, however, the corrections are too strong since WLS itself is optically degrading and therefore this type of correction cannot be applied to the more recent SCIAMACHY data [*Pagaran et al.*, 2011a]. Further investigations are underway to improve upon the in-flight radiometric calibration. For atmospheric studies this is generally not a problem since the degradation cancels out in the sun-normalized earth radiances used in most atmospheric retrievals.

The Mg II core to wing ratio derived from the Mg II Fraunhofer lines at 280 nm (Fig. 3.3) is an index that has been proven to correlate well with UV irradiance changes down to 30 nm [*DeLand and Cebula*, 1993; *Viereck et al.*, 2001]. It is a measure for the chromospheric activity of the sun and describes the plage and faculae

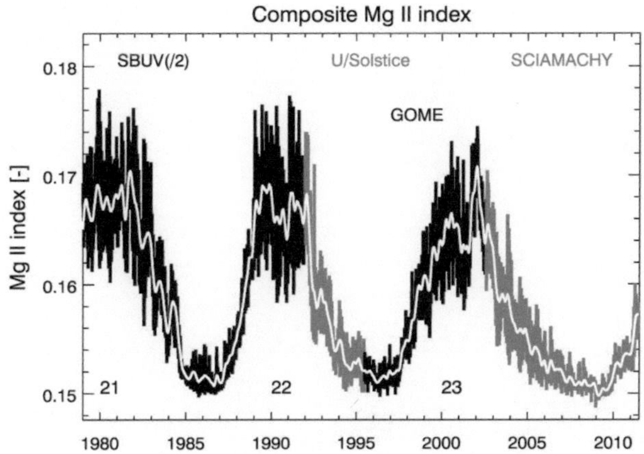

Fig. 3.3 Composite Mg II index measured near 280 nm derived from multiple satellite data and extending from solar cycle 21 to 24. In addition to SBUV(/2), UARS/Solstice, GOME, and SCIAMACHY as indicated here, other data such as SUSIM, the more recent SBUV/2 instruments from NOAA-16 to NOAA-18 as well as SORCE/Solstice have been used to fill daily gaps. The smooth *white line* shows the low-pass filtered time series by applying a 55-day triangular filter to remove the 27-day solar rotation signature

brightening responsible for the UV increase. The Mg II index, defined as a ratio, is insensitive to instrumental degradation and has been derived from many different instruments to provide a long-term time series going back to the late 1970s [*DeLand and Cebula*, 1993; *Viereck and Puga*, 1999; *Viereck et al.*, 2004]. An updated composite Mg II index by adding GOME [*Weber et al.*, 1998; *Weber*, 1999] and recent SCIAMACHY data is shown in Fig. 3.3. It seems that the Mg II index was lower during the recent solar minimum in 2008 than the two solar minima before, but this is not statistically significant. A potential lower solar minimum value could be expected from the very low thermospheric density observed in 2008 [*Emmert et al.*, 2010]. Solar irradiance at extreme ultraviolet (EUV) wavelengths heats the thermosphere, causing it to expand. Low EUV irradiance contracts the thermosphere and decreases the density at a given altitude. The cooling of the upper atmosphere due to increases of greenhouse gases can only explain part of the recent contraction observed [*Emmert et al.*, 2010; *Solomon et al.*, 2011].

3.3 Irradiance Variations from Solar Rotations to Several Solar Cycles

In order to estimate SSI irradiance variations beyond the instrument lifetime and covering several decades a model was developed that uses solar proxies scaled to SCIAMACHY SSI observations. The underlying assumption is that irradiance variations are mainly caused by solar surface magnetic activity [*Fligge et al.*, 2000] and can be expressed in terms of faculae brightening as represented by the Mg II index

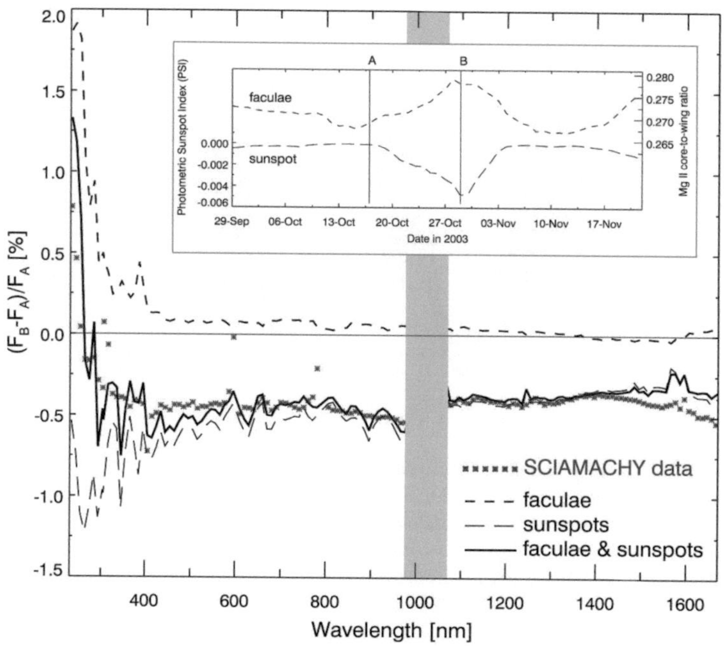

Fig. 3.4 Modeled and observed SCIAMACHY solar irradiance change during the Halloween solar storm in 2003 decomposed into faculae and sunspot contributions. The *inset* shows the Mg II and PSI index with labels A and B indicating dates from which irradiance differences were derived. From *Pagaran et al.* [2009]. Reproduced by permission of the AAS

and sunspot darkening as expressed by the photometric sunspot index (PSI), here taken from *Balmaceda et al.* [2009]. The SSI can then be written as a time series as follows

$$I_\lambda(t) = I_\lambda(t_o) + a_\lambda\left[P_a(t) - P_a(t_o)\right] + b_\lambda\left[P_b(t) - P_b(t_o)\right] + p_\lambda(t), \quad (3.1)$$

where $P_a(t)$ and $P_b(t)$ are the Mg II index and PSI time series, respectively. A similar approach was used to model UV irradiance variations derived from UARS/Solstice [*Lean et al.*, 1997].

A multivariate linear regression is performed to determine the regression coefficients of the solar proxies. In addition to the two solar proxy terms piecewise polynomials, $p_\lambda(t)$, are used to correct for instrument degradation and small jumps following instrument and satellite platform anomalies [*Pagaran et al.*, 2009]. The regression was applied to SCIAMACHY SSI time series over several solar rotations during 2003 and 2004. Regression coefficients, a_λ and b_λ, were determined from 240 nm to 1750 nm (SCIAMACHY channels 1 to 6) in steps of 10 nm [*Pagaran et al.*, 2009]. As a solar reference spectrum, $I_\lambda(t_0)$, the SCIAMACHY SSI from March 4, 2004, (Fig. 3.1) was selected.

The modeled and observed SCIAMACHY solar irradiance change is shown as an example in Fig. 3.4 during the Halloween 2003 solar storm, where the PSI index

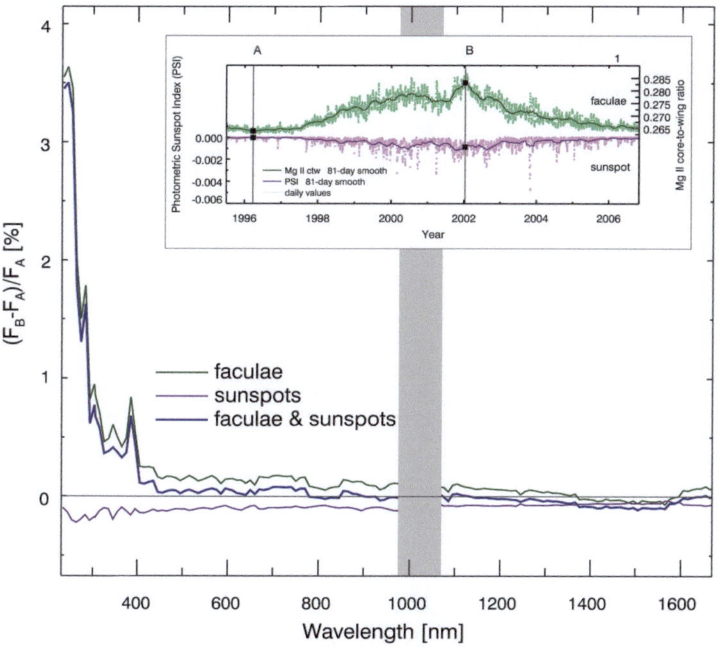

Fig. 3.5 Solar irradiance variations during solar cycle 23 as derived from SCIAMACHY observations and proxy data. Solar maximum and minimum dates were defined by the 81 day boxcar smooth of Mg II index time series (*inset*). Contributions from faculae and sunspots are indicated. From *Pagaran et al.* [2009]. Reproduced by permission of the AAS

value reached the lowest value since 1980 and substantial mesospheric ozone loss (mainly due to solar protons) was observed by SCIAMACHY [*Rohen et al.*, 2005]. The combined faculae and sunspot contributions and SCIAMACHY observations are in qualitative agreement with Fig. 6 in *Lean et al.* [2005]. Across the near-UV, vis, and near-IR spectral range solar irradiance dropped by 0.3 % (near-IR) to 0.5 % (near-UV). This is consistent with a drop of about 0.4 % in the total solar irradiance (TSI) or solar constant. Below 300 nm an irradiance enhancement due to faculae activity was observed reaching +1.3 % near 250 nm.

The SCIAMACHY irradiance time series as well as the SCIA proxy model show the dark faculae effect in the spectral region 1400–1600 nm (near opacity H^- minimum), where both sunspot and faculae contributions are negative in agreement with observations from ground indicating a darkening under enhanced solar activity conditions [*Moran et al.*, 1992]. The SCIA proxy model, nevertheless, underestimates the observed irradiance depletion in this spectral region.

The SCIA proxy model can be used to reconstruct spectral irradiance changes since the late 1970s, where the Mg II index record started, covering nearly three solar cycles. From the SCIA proxy model the UV contribution below 400 nm to TSI changes in solar cycle 23 (~0.1 %) is 55 % [*Pagaran et al.*, 2009] which is higher than the 30 % estimate from solar cycle 22 derived from SOLSTICE observations

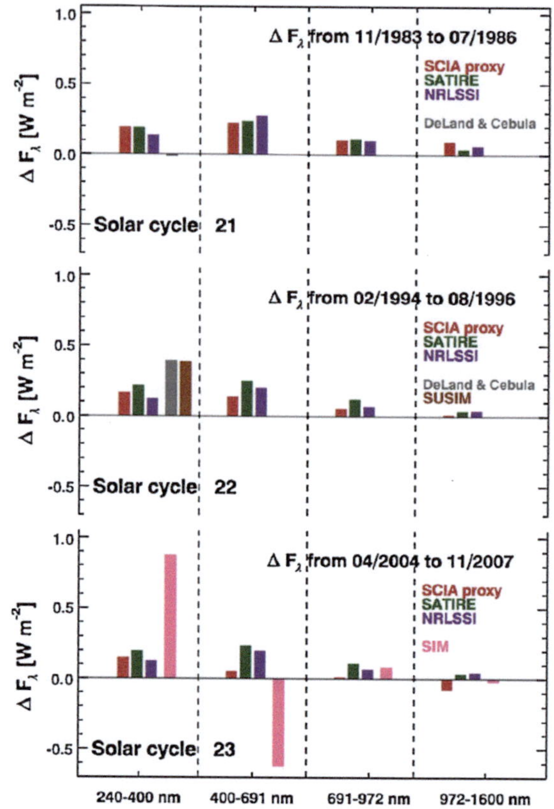

Fig. 3.6 SSI changes during part of the descending phase of solar cycles 21 to 23 (*top* to *bottom*), respectively. Dates near solar maximum and minimum are chosen in such a way that the differences between the Mg II indices are about the same in each solar cycle and correspond to that of the SIM observation period used here. NRLSSI [*Lean*, 2000], SATIRE [*Krivova et al.*, 2009] and SCIA proxy are models. SUSIM, SIM, and UV composite from *DeLand and Cebula* [2008] are direct satellite observations. From *Pagaran et al.* [2011b]

[*Lean et al.*, 1997] and lower than the 63 % derived from the semi-empirical model SATIRE (Spectral and Total Irradiance Reconstructions) [*Krivova et al.*, 2006].

The largest TSI change contribution comes from the near UV (300–400 nm), where the irradiance solar cycle change per wavelength is well below 1 % [*Pagaran et al.*, 2009]. During solar cycles 21 to 23, the dominant contribution to irradiance changes in the UV from solar minimum to maximum comes from the faculae brightening. The sunspot contribution is non-negligible in the near UV and in the visible cancels within the error bars the faculae brightening (see Fig. 3.5 and *Pagaran et al.* [2009, 2011b]). The dark faculae near 1400–1600 nm are again evident at solar maximum in agreement with observations by SIM and results from the SATIRE model [*Unruh et al.*, 2008].

Harder et al. [2009] reported on SIM irradiance changes during the descending phase of solar cycle 23 (April 2004 to November 2007) and found UV changes that are much larger than models like the NRLSSI irradiance model [*Lean*, 2000] indicate. This is also true when comparing to other data sets as shown in Fig. 3.6 where the comparison is extended to the SCIA proxy model, the SATIRE model [*Krivova et al.*, 2009], and the UV composite from *DeLand and Cebula* [2008] as well. Also shown in this figure are the comparison of irradiance changes during the descending phase of prior solar cycles with similar Mg II index change as in solar cycle 23 [*Pagaran et al.*, 2011b].

It appears that current models including the SCIA proxy model that assume that irradiance changes are mostly related to surface magnetic activity are underestimating solar cycle changes in the UV as compared to the SIM observations. Direct observations from SUSIM and the UV composite also see larger UV changes during solar cycle 22 than the models, but are still only about half of SIM's result for solar cycle 23. Such a large UV change as observed by SIM has strong implications on radiative forcing in the upper atmosphere [*Haigh et al.*, 2010; *Oberländer et al.*, 2012] and will remain a matter of debate.

3.4 Solar Rotation (27-Day) Signature in Stratospheric Ozone

The solar variation on the 11-year time scale has been shown to cause 2–3 % variability in tropical ozone at altitudes of approximately 40 km. This has been concluded from different satellite observations [e.g. *Remsberg*, 2008; *Fioletov*, 2009, and references therein] and was confirmed by model studies [e.g. *Langematz et al.*, 2005; *Sekiyama et al.*, 2006; *Marsh et al.*, 2007]. The influence of the 27-day solar rotation on ozone was first investigated by *Hood* [1986] in the 1980s using SBUV ozone measurements. He found the ozone sensitivity at 45 km to be slightly more than 0.4 % per 1 % change in the 205 nm flux. Further investigations with different satellite data sets and model outputs covering other time periods followed [*Gruzdev et al.*, 2009; *Fioletov*, 2009, and references therein]. *Austin et al.* [2007] and *Gruzdev et al.* [2009] compared the 27-day ozone variability determined by chemistry climate models (CCM) with satellite measurements and were able to verify the observations in magnitude (0.4 to 0.5 %/%) but found the maximum ozone sensitivity slightly lower in altitude (approx. 40 km) in the model simulations.

The motivation for this study is to use the new dataset that is available from SCIAMACHY, e.g. global ozone profiles during the descending phase of solar cycle 23 [*von Savigny et al.*, 2005; *Sonkaew et al.*, 2009]. Continuous wavelet transform (CWT), fast Fourier transform (FFT), and cross correlations (CC) have been applied to SCIAMACHY ozone in the tropics ($<20°$ latitude) between 20 and 60 km altitude [*Dikty et al.*, 2010b]. The maximum correlation between the Mg II index and ozone is weaker during the maximum of solar cycle 23 ($r = 0.38$) than in the previous two solar cycles that have been investigated in earlier studies using different data sets. This is in agreement with results from *Fioletov* [2009].

The magnitude of the ozone signals is highly time dependent as revealed by the CWT analysis and may vanish for several solar rotations even close to solar maximum conditions (see Fig. 3.7). The ozone sensitivity (ozone change in percent per percent change in 205 nm solar flux) is on average about 0.2 %/% above 30 km altitude and smaller by about a factor of two compared to earlier studies. For selected three month periods the sensitivity may rise beyond 0.6 %/% in better agreement with earlier studies. The analysis of the 27-day solar forcing was also carried

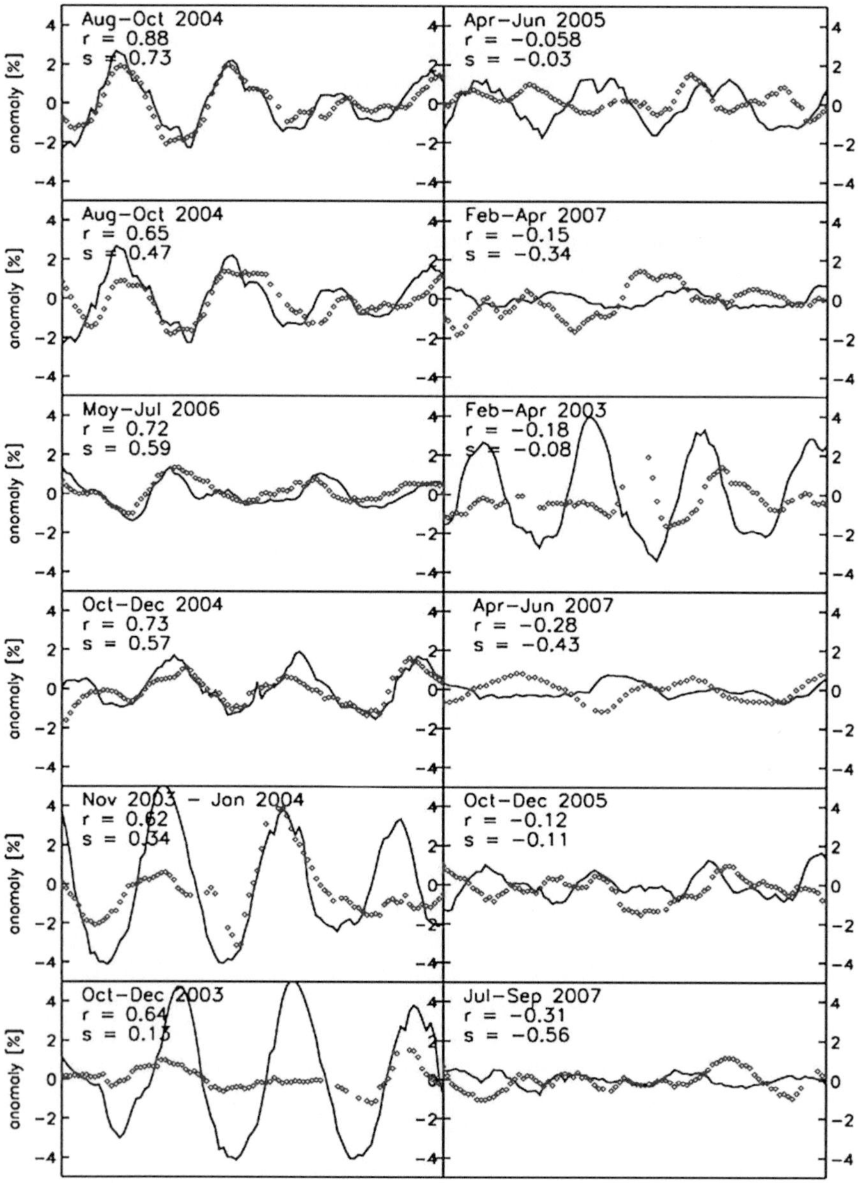

Fig. 3.7 Selected three month periods with high (*left panels*) and low correlation (*right panels*) between ozone (*solid line*) and Mg II index (*circles*). In each panel, the period, correlation (r), and ozone sensitivity (s) is indicated, the latter is defined as the ozone change per Mg II index change in units of %/%. The ozone sensitivity per unit 205 nm solar irradiance change is obtained by multiplying s with 0.61. From *Dikty et al.* [2010b]. Reproduced by permission of American Geophysical Union. ©2010 American Geophysical Union

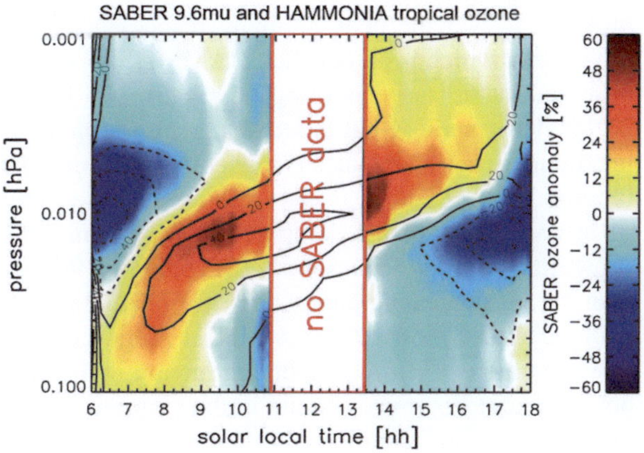

Fig. 3.8 SABER observations at 9.6 μm (*color*) and HAMMONIA model (*contour*) daytime ozone variations between 0.1 and 0.001 hPa expressed as percent deviation from the daytime mean. From *Dikty et al.* [2010a]

out with stratospheric temperatures from the European Centre for Medium-Range Weather Forecasts operational analysis. Although direct radiation effects on temperature are weak in the upper stratosphere, temperature signals with statistically significant periods in the 25–35 day range similar to ozone were found [*Dikty et al.*, 2010b].

3.5 Daytime Variations in Mesospheric Ozone

In comparison with the 27-day solar rotation signal and the 11-year solar cycle response in the stratosphere, the diurnal and daytime variation of UV radiation inflicts a by far greater response in upper atmosphere ozone. The response of ozone above 60 km to variations in UV radiation is less well established. Ozone and temperature data from SABER (Sounding of the Atmosphere using Broadband Emission Radiometry) in its version 1.07 [*Russell III et al.*, 1999] are used to study the daytime pattern of mesospheric ozone. In contrast to SCIAMACHY, SABER aboard TIMED flies in a more inclined orbit allowing measurements at different local times. In our study [*Dikty et al.*, 2010a] a specific sampling of SABER data was preformed to derive daytime pattern in tropical ozone using both the results from the 1.27 μ air glow [*Mlynczak et al.*, 2007] and 9.6 μm thermal emission retrieval [*Rong et al.*, 2008]. Compared to the earlier study on daytime variations by *Huang et al.* [2008] more years of SABER data were used and our results were compared to HAMMONIA (Hamburg Model of the Neutral and Ionized Atmosphere) [*Schmidt et al.*, 2006].

The amplitude of daytime ozone variations is approximately 60 % of the daytime mean for SABER and lower for the model (see Fig. 3.8). The agreement with HAMMONIA is generally better for the 9.6 μm retrieved ozone data than for the 1.27 μm air glow retrieval [*Dikty et al.*, 2010a]. The maximum daytime peak anomaly observed at 0.05 hPa (∼70 km) in the morning shifts its altitude to about 0.007 hPa (∼80 km) in the afternoon. This daytime shift is in very good agreement with the model, however the peak anomaly reaches a maximum of 40–50 % of the daytime mean, which is higher than HAMMONIA (30–40 %). Negative anomalies are observed in the early morning hours at 0.007 hPa and in the late afternoon near 0.015 hPa in quite good agreement with the model. During equinox the daytime maximum ozone abundance is higher than during solstice, especially above 0.01 hPa (approx. 80 km). The seasonal variation is somewhat weaker in HAMMONIA.

In contrast to ozone, temperature data from SABER [*Remsberg et al.*, 2008] show little daytime variations between 65 and 90 km and their amplitudes are less than 1.5 %, suggesting photochemistry playing a dominant role in the mesospheric ozone chemistry. *Marsh et al.* [2002] proposed that the solar diurnal tide brings down atomic oxygen for ozone production in the afternoon (>85 km). The model, however, underestimates ozone in the afternoon above approximately 0.01 hPa, so the remaining difference could be attributed to solar tides. The minimum early in the morning is caused by the direct photolysis of ozone before enough atomic oxygen is produced to counteract the ozone destruction. The ozone rise in the morning hours may also be due to tides transporting ozone rich air from below [*Marsh et al.*, 2002]. The new SABER version 1.08 data will also include water vapor, which will be helpful to constrain the HO_x budget and its influence on daytime ozone.

3.6 Conclusion

SCIAMACHY was the first satellite instrument providing daily spectral solar irradiances (SSI) from the UV, visible and near infrared. The comparisons with other solar data from space and ground showed good agreement to within a few percent up to 1700 nm [*Skupin et al.*, 2005a, 2005b; *Pagaran et al.*, 2011a]. Expressing SCIAMACHY irradiance variations over several solar rotations in terms of solar proxies for sunspot darkening and faculae brightening permits the extrapolation of SCIAMACHY SSI variations to the 11-year solar cycle scales [*Pagaran et al.*, 2009, 2011b]. It was shown that about half of the 0.1 % change in the solar constant over solar cycle 23 has originates from the visible and IR spectral region [*Pagaran et al.*, 2009]. A particular challenge is the solar cycle variation estimate for the near UV (300–400 nm), where recent SIM observations [*Harder et al.*, 2009] indicate changes during solar cycle 23 that are much higher than expected from indirect SCIAMACHY observations and other empirical models (assuming solar surface magnetic activity as primary driver for SSI variations) as well as observations from other satellite data in earlier solar cycles [*Pagaran et al.*, 2011b].

SCIAMACHY limb ozone vertical profiles from 2003 to 2008 were analyzed for signatures of the 27-day solar rotation. It was found that this signature is highly variable in time and that even under solar maximum condition this signal can vanish for several months [*Dikty et al.*, 2010b]. On average the sensitivity above 30 km is a 0.2 % ozone change per percent change in the 205 nm solar flux (important for ozone production) near solar maximum, which is smaller than found in earlier studies and prior solar cycles.

Daytime variations in tropical mesospheric ozone yield changes of up to 60 % from the daytime mean based upon SABER ozone data and peak anomalies are generally higher during equinox. SABER results were compared for the first time with an output of a chemistry climate model, here the HAMMONIA model [*Dikty et al.*, 2010a]. SABER ozone from the 9.6 μm retrieval agrees qualitatively very well with HAMMONIA, however, little agreement was found between modeled and SABER temperatures above 0.01 hPa. The low temperature variations of a few degree during daytime may suggest that photochemical processes are the main driver for daytime ozone variations and to a lesser degree transport related to tides.

Acknowledgements The support by the DFG research project SOLOZON (DFG WE 3647/1-1) within the German CAWSES priority program and the support by the State of Bremen is gratefully acknowledged. The authors JP, MW, LEF, and JWH were part of the International Space Studies Institute (ISSI) team on spectral solar irradiance (http://www.issibern.ch/teams/solarspect/), whose meetings and discussions have greatly benefited this work.

References

Austin, J., Hood, L. L., & Soukharev, B. E. (2007). Solar cycle variations of stratospheric ozone and temperature in simulations of a coupled chemistry-climate model. *Atmospheric Chemistry and Physics*, *7*, 1693–1706.

Balmaceda, L. A., Solanki, S. K., Krivova, N. A., & Foster, S. (2009). A homogeneous sunspot areas database covering more than 130 years. *Journal of Geophysical Research*, *114*, A7104. doi:10.1029/2009JA014299.

Bovensmann, H., Burrows, J. P., Buchwitz, M., Frerick, J., Noël, S., Rozanov, V. V., Chance, K. V., & Goede, A. P. H. (1999). SCIAMACHY: mission objectives and measurement modes. *Journal of the Atmospheric Sciences*, *56*, 127–150.

Bovensmann, H., Aben, I., van Roozendael, M., Kühl, S., Gottwald, M., von Savigny, C., Buchwitz, M., Richter, A., Frankenberg, C., Stammes, P., de Graaf, M., Wittrock, F., Sinnhuber, M., Sinnhuber, B.-M., Schönhardt, A., Beirle, S., Gloudemans, A., Schrijver, H., Bracher, A., Rozanov, A. V., Weber, M., & Burrows, J. P. (2011). SCHIAMACHY's view of the changing earth's environment. In M. Gottwald, & H. Bovensmann (Eds.), *SCIAMACHY—exploring the changing earth's atmosphere*. Dordrecht: Springer. Chap. 10. doi:10.1007/978-90-481-9896-2.

Burrows, J. P., Weber, M., Buchwitz, M., Rozanov, V., Ladstätter-Weißenmayer, A., Richter, A., Debeek, R., Hoogen, R., Bramstedt, K., Eichmann, K.-U., Eisinger, M., & Perner, D. (1999). The global ozone monitoring experiment (GOME): mission concept and first scientific results. *Journal of the Atmospheric Sciences*, *56*, 151–175.

DeLand, M. T., & Cebula, R. P. (1993). Composite Mg II solar activity index for solar cycles 21 and 22. *Journal of Geophysical Research*, *98*, 12809–12823.

DeLand, M. T., & Cebula, R. P. (2008). Creation of a composite solar ultraviolet irradiance data set. *Journal of Geophysical Research*, *113*, A11103. doi:10.1029/2008JA013401.

DeLand, M. T., Floyd, L. E., Rottman, G. J., & Pap, J. (2004). Status of UARS solar UV irradiance data. *Advances in Space Research*, *34*, 243–250.

Dikty, S., Schmidt, H., Weber, M., von Savigny, C., & Mlynczak, M. G. (2010a). Daytime ozone and temperature variations in the mesosphere: a comparison between SABER observations and HAMMONIA model. *Atmospheric Chemistry and Physics*, *10*, 8331–8339. doi:10.5194/acp-10-8331-2010.

Dikty, S., Weber, M., von Savigny, C., Sonkaew, T., Rozanov, A., & Burrows, J. P. (2010b). Modulations of the 27-day solar cycle signal in stratospheric ozone from SCIAMACHY. *Journal of Geophysical Research*, *115*, D00115. doi:10.1029/2009JD012379.

Emmert, J. T., Lean, J. L., & Picone, J. M. (2010). Record-low thermospheric density during the 2008 solar minimum. *Geophysical Research Letters*, *37*, L12102. doi:10.1029/2010GL043671.

Fioletov, V. E. (2009). Estimating the 27-day and 11-year solar cycle variations in tropical upper stratospheric ozone. *Journal of Geophysical Research*, *114*, D02302. doi:10.1029/2008JD010499.

Fligge, M., Solanki, S. K., & Unruh, Y. C. (2000). Modelling irradiance variations from the surface distribution of the solar magnetic field. *Astronomy & Astrophysics*, *353*, 380–388.

Fröhlich, C., Crommelynck, C. A., Wehrli, C., Anklin, M., Dewitte, S., Fichot, A., Finsterle, W., Jiménez, A., Chevalier, A., & Roth, H. (1997). In-flight performance of the Virgo solar irradiance instruments on SOHO. *Solar Physics*, *175*, 267–286.

Gruzdev, A. N., Schmidt, H., & Brasseur, G. P. (2009). The effect of the solar rotational irradiance variation on the middle and upper atmosphere calculated by a three-dimensional chemistry-climate model. *Atmospheric Chemistry and Physics*, *8*, 1113–1158.

Haigh, J. D., Winning, A. R., Toumi, R., & Harder, J. W. (2010). An influence of solar spectral variations on radiative forcing of climate. *Nature*, *467*, 696–699. doi:10.1038/nature09426.

Harder, J. W., Lawrence, G., Fontenla, J., Rottman, G., & Woods, T. (2005a). The spectral irradiance monitor: scientific requirements, instrument design, and operation modes. *Solar Physics*, *230*, 141–167. doi:10.1007/s11207-005-5007-5.

Harder, J. W., Fontenla, J., Lawrence, G., Woods, T., & Rottman, G. (2005b). The spectral irradiance monitor: measurement equations and calibration. *Solar Physics*, *230*, 169–204. doi:10.1007/s11207-005-1528-1.

Harder, J. W., Fontenla, J. M., Pilewskie, P., Richard, E. C., & Woods, T. N. (2009). Trends in solar spectral irradiance variability in the visible and infrared. *Geophysical Research Letters*, *36*, 7801 doi:10.1029/2008GL036797.

Harder, J. W., Thuillier, G., Richard, E. C., Brown, S. W., Lykke, K. R., Snow, M., McClintock, W. E., Fontenla, J. M., Woods, T. N., & Pilewskie, P. (2010). The SORCE SIM solar spectrum: comparison with recent observations. *Solar Physics*, *263*, 3–24. doi:10.1007/s11207-010-9555-y.

Hood, L. L. (1986). Coupled stratospheric ozone and temperature response to short-term changes in solar ultraviolet flux: An analysis of Nimbus 7 SBUV and SAMS data. *Journal of Geophysical Research*, *91*, 5264s–5276.

Huang, F. T., Mayr, H. Russell III, J. M., Mlynczak, M. G., & Reber, C. A. (2008). Ozone diurnal variations and mean profiles in the mesosphere, lower thermosphere, and stratosphere, based on measurements from SABER on TIMED. *Journal of Geophysical Research*, *113*, A04307. doi:10.1029/2007JA012739.

Krivova, N. A., Solanki, S. K., & Floyd, L. (2006). Reconstruction of solar UV irradiance in cycle 23. *Astronomy & Astrophysics*, *452*, 631–639. doi:10.1051/0004-6361:20064809.

Krivova, N. A., Solanki, S. K., Wenzler, T., & Podlipnik, B. (2009). Reconstruction of solar UV irradiance since 1974. *Journal of Geophysical Research*, *114*, D00I04. doi:10.1029/2009JD012375.

Langematz, U., Grenfell, J. L., Matthes, K., Mieth, P., Kunze, M., Steil, B., & Brühl, C. (2005). Chemical effects in 11-year solar cycle simulations with the Freie Universität Berlin Middle Atmospheric Model with online chemistry (FUB-CMAM-CHEM). *Geophysical Research Letters*, *32*, L13803. doi:10.1029/2005GL022686.

Lean, J., Rottman, G., Harder, J., & Kopp, G. (2005). SORCE contributions to new understanding of global change and solar variability. *Solar Physics*, *230*, 27–53. doi:10.1007/s11207-005-1527-2.

Lean, J. L. (2000). Evolution of the sun's spectral irradiance since the Maunder minimum. *Geophysical Research Letters*, *16*, 2425–2428.

Lean, J. L., Rottman, G. J., Kyle, H. L., Woods, T. N., Hickey, J. R., & Puga, L. C. (1997). Detection and parameterization of variations in solar mid- and near-ultraviolet radiation (200–400 nm). *Journal of Geophysical Research*, *102*, 29939–29956.

Marsh, D. R., Skinner, W. R., Marshall, A. R., Hays, P. B., Ortland, D. A., & Yee, J.-H. (2002). High Resolution Doppler Imager observations of ozone in the mesosphere and lower thermosphere. *Journal of Geophysical Research*, *107*, 4390. doi:10.1029/2001JD001505.

Marsh, D. R., Garcia, R. R., Kinnison, D. E., Boville, B. A., Sassi, F., Solomon, S. C., & Matthes, K. (2007). Modelling the whole atmosphere response to solar cycle changes in radiative and geomagnetic forcing. *Journal of Geophysical Research*, *112*, D23306. doi:10.1029/2006JD008306.

Mlynczak, M. G., Marshall, B. T., Martin-Torres, F. J., Russel III, J. M., Thompson, R. E., Remsberg, E. E., & Gordley, L. L. (2007). Sounding of the atmosphere using broadband emission radiometry observations of daytime mesospheric $O_2(^1\Delta)$ 1.27 μm emission and derivation of ozone, atomic oxygen, and solar and chemical energy deposition rates. *Journal of Geophysical Research*, *112*, D15306. doi:10.1029/2006JD008355.

Moran, T., Foukal, P., & Rabin, D. (1992). A photometric study of faculae and sunspots between 1.2 and 1.6 micron. *Solar Physics*, *142*, 35–46.

Neckel, H., & Labs, D. (1984). The solar radiation between 3300 and 12500 Å. *Solar Physics*, *90*, 205–258.

Oberländer, S., Langematz, U., Matthes, K., Kunze, M., Kubin, A., Harder, J., Krivova, N. A., Solanki, S. K., Pagaran, J., & Weber, M. (2012). The influence of spectral solar irradiance data on stratospheric heating rates during the 11 year solar cycle. *Geophysical Research Letters*, *39*, L01801. doi:10.1029/2011GL049539.

Pagaran, J., Weber, M., & Burrows, J. P. (2009). Solar variability from 240 to 1750 nm in terms of faculae brightening and sunspot darkening from SCIAMACHY. *The Astronomical Journal*, *700*, 1884–1895.

Pagaran, J., Harder, G., Weber, M., Floyd, L., & Burrows, J. P. (2011a). Intercomparison of SCIAMACHY and SIM vis-IR irradiance over several solar rotational timescales. *Astronomy & Astrophysics*, *528*(A67). doi:10.1051/0004-6361/201015632.

Pagaran, J., Weber, M., DeLand, M., Floyd, L., & Burrows, J. P. (2011b). Solar spectral irradiance variations in 240–1600 nm during the recent solar cycles 21–23. *Solar Physics*, *XX*, 1–30. doi:10.1007/s11207-011-9808-4.

Remsberg, E. E. (2008). On the response of halogen occultation experiment (HALOE) stratospheric ozone and temperature to the 11-year solar cycle forcing. *Journal of Geophysical Research*, *113*, D22304. doi:10.1029/2008JD010189.

Remsberg, E. E., Marshall, B. T., Garcia-Comas, M., Krueger, D., Lingenfelser, G. S., Martin-Torres, J., Mlynczak, M. G., Russel III, J. M., Smith, A. K., Zhao, Y., Brown, C., Gordley, L. L., Lopez-Gonzales, M. J., Lopez-Puertas, M., She, C.-Y., Taylor, M. J., & Thompson, R. E. (2008). Assessment of the quality of the version 1.07 temperature-versus-pressure profiles of the middle atmosphere from TIMED/SABER. *Journal of Geophysical Research*, *113*, D17101. doi:10.1029/2008JD010013.

Rohen, G., von Savigny, C., Sinnhuber, M., Llewellyn, E. J., Kaiser, J. W., Jackman, C. H., Kallenrode, M.-B., Schröter, J., Eichmann, K.-U., Bovensmann, H., & Burrows, J. P. (2005). Ozone depletion during the solar proton events of October/November 2003 as seen by SCIAMACHY. *Journal of Geophysical Research*, *110*, A09S39. doi:10.1029/2004JA010984.

Rong, P. P., Russell III, J. M., Mlynczak, M. G., Remsberg, E. E., Marshall, B. T., Gordley, L. L., & Lopez-Puertas, M. (2008). Validation of TIMED/SABER v1.07 ozone at 9.6 μm in the altitude range 15–70 km. *Journal of Geophysical Research*, *114*, D04306. doi:10.1029/2008JD010073.

Rottman, G., Floyd, L., & Viereck, R. (2004). Measurement of solar ultraviolet irradiance. *Geophysical Monograph*, *41*, 111–125. doi:10.10219/141GM10.

Russell III, J. M., Mlynczak, M. G., Gordley, L. L., Tansock, J. J. Jr., & Esplin, R. W. (1999). Overview of the SABER experiment and preliminary calibration results. *Proceedings—SPIE*, *3756*(277). doi:10.1117/12.366382.

Schmidt, H., Brasseur, G. P., Charron, M., Manzini, E., Giorgetta, M. A., Diehl, T., Fomichev, V. I., Kinnison, D., Marsh, D., & Walters, S. (2006). The HAMMONIA chemistry climate model: sensitivity of the mesopause region to the 11-year solar cycle and CO_2 doubling. *Journal of Climate*, *19*, 3903–3931.

Sekiyama, T. T., Shibata, K., Deushi, M., Kodera, K., & Lean, J. L. (2006). Stratospheric ozone variation induced by the 11-year solar cycle: recent 22-year simulation using 3-D chemical transport model with reanalysis data. *Geophysical Research Letters*, *33*, L17812. doi:10.1029/2006GL026711.

Skupin, J., Noël, S., Wuttke, M. W., Gottwald, M., Bovensmann, H., Weber, M., & Burrows, J. P. (2005a). SCIAMACHY solar irradiance observation in the spectral range from 240 to 2380 nm. *Advances in Space Research*, *35*, 370–375. doi:10.1016/j.asr.2005.03.036.

Skupin, J., Weber, M., Noël, S., Bovensmann, H., & Burrows, J. P. (2005b). GOME and SCIAMACHY solar measurements: Solar spectral irradiance and Mg II solar activity proxy indicator. *Memorie Della Società Astronomica Italiana*, *76*, 1038–1041.

Solomon, S. C., Qian, L., Didkovsky, L. V., Viereck, R. A., & Woods, T. N. (2011). Causes of low thermospheric density during the 2007–2009 solar minimum. *Journal of Geophysical Research*, *116*, A00H07. doi:10.1029/2011JA016508.

Sonkaew, T., Rozanov, V. V., von Savigny, C., Rozanov, A., Bovensmann, H., & Burrows, J. P. (2009). Cloud sensitivity studies for stratospheric and lower mesospheric ozone profile retrievals from measurements of limb scattered solar radiation. *Atmospheric Measurement Techniques*, *2*, 653–678. doi:10.5194/amt-2-653-2009.

Thuillier, G., Floyd, L., Woods, T. N., Cebula, R., Hilsenrath, E., Hersé, M., & Labs, D. (2004). Solar irradiance reference spectra. In J. M. Pap, P. Fox, C. Fröhlich, J. K. H. S. Hudson, J. McCormack, G. North, W. Sprigg & S. T. Wu (Eds.), *Geophysical monograph: Vol. 141. Solar variability and its effects on climate* (pp. 171–194).

Unruh, Y. C., Krivova, N. A., Solanki, S. K., Harder, J. W., & Kopp, G. (2008). Spectral irradiance variations: comparison between observations and the SATIRE model on solar rotation time scales. *Astronomy & Astrophysics*, *486*, 311–323. doi:10.1051/0004-6361:20078421.

Viereck, R. A., & Puga, L. C. (1999). The NOAA Mg II core-to-wing solar index: construction of a 20-year times series of chromospheric variability from multiple satellites. *Journal of Geophysical Research*, *104*, 9995–10005.

Viereck, R. A., Puga, L., McMullin, D., Judge, D., Weber, M., & Tobiska, W. K. (2001). The Mg II index: a proxy for solar EUV. *Geophysical Research Letters*, *28*, 1343–1346.

Viereck, R. A., Floyd, L. E., Crane, P. C., Woods, T., Knapp, B. G., Rottman, G., Weber, M., Puga, L. C., & DeLand, M. T. (2004). A composite Mg II index spanning from 1978 to 2003. *Space Weather*, *2*, S10005.

von Savigny, C., Rozanov, A., Bovensmann, H., Eichmann, K.-U., Noël, S., Rozanov, V. V., Weber, M., Burrows, J. P., & Kaiser, J. W. (2005). The ozone hole breakup in September 2002 as seen by SCIAMACHY on ENVISAT. *Journal of the Atmospheric Sciences*, *62*, 721–734.

Weber, M. (1999). Solar Activity during solar cycle 23 monitored by GOME. In *ESA special publication: Vol. 161. ESA-WPP 1999: European symposium on atmospheric measurements from space, proc. ESAMS '99* (pp. 611–616). Nordwijk: ESA Publications Division.

Weber, M., Burrows, J. P., & Cebula, R. P. (1998). GOME solar uv/vis irradiance measurements between 1995 and 1997—First results on proxy solar activity studies. *Solar Physics*, *177*, 63–77.

Wehrli, C. (1985). *Extraterrestrial solar spectrum* (Tech. rep.). Physikalisch-Meteorologisches Observatorium, World Radiation Center (PMO/WRC), Publication No. 615.

Chapter 4
Solar Activity, the Heliosphere, Cosmic Rays and Their Impact on the Earth's Atmosphere

Horst Fichtner, Bernd Heber, Klaudia Herbst, Andreas Kopp, and Klaus Scherer

Abstract During recent years it became evident that the climate of the Earth is not only determined by terrestrial, in particular anthropogenic influences, but also by external parameters. An open question is still whether the solar radiation or the cosmic rays are the main agents regarding the external climate driving. An answer to this question requires quantitative modelling of all related processes. The present chapter concentrates on the modelling of the cosmic ray transport from the interstellar medium into the Earth's atmosphere and, thus, on interstellar-terrestrial relations. After a discussion of the significance of the local interstellar spectrum of cosmic rays, first their transport in a dynamical heliosphere is considered. Second, the cosmic ray propagation within the terrestrial magnetosphere is studied in order to determine the so-called cut-off rigidities. And, third, the cosmic ray interaction with the Earth's atmosphere is described, with an emphasis on the ionisation and the production of cosmogenic nuclides. On the basis of the suite of models being discussed in this overview further studies will be possible that should help to quantify the overall effect of cosmic rays on the Earth environment and, particularly, climate.

4.1 Introduction

The Earth, as a small body located within our Galaxy, is exposed to a rush of energetic particles, so-called Galactic Cosmic Rays (GCRs), accelerated e.g. at supernova remnants [*Büsching et al.*, 2005] to energies of several TeV or even higher. On the way towards the Earth, however, these particles encounter three effective protection shields: the outer one is the heliosphere, defined as the bubble around the solar

H. Fichtner (✉) · K. Scherer
Theoretische Physik IV, Ruhr-Universität Bochum, Bochum, Germany
e-mail: hf@tp4.rub.de

B. Heber · K. Herbst · A. Kopp
Institut für Experimentelle und Angewandte Physik, Christian-Albrechts-Universität zu Kiel, Kiel, Germany

B. Heber
e-mail: heber@physik.uni-kiel.de

F.-J. Lübken (ed.), *Climate and Weather of the Sun-Earth System (CAWSES)*, Springer Atmospheric Sciences,
DOI 10.1007/978-94-007-4348-9_4,

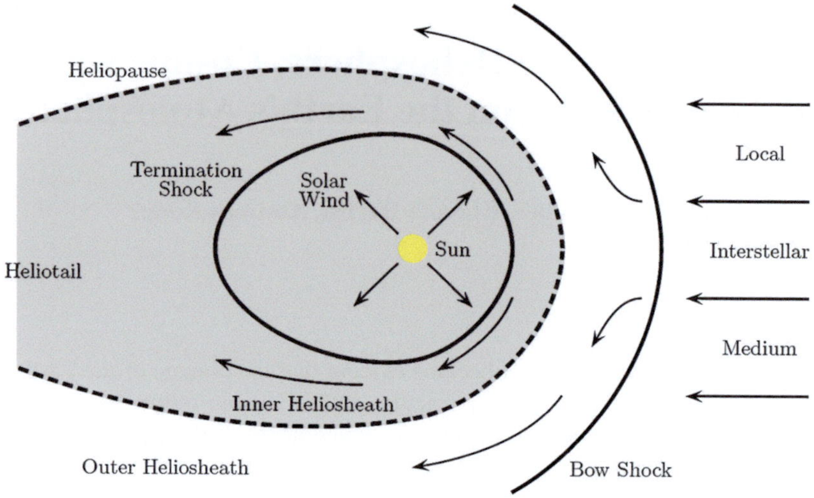

Fig. 4.1 A sketch of the heliosphere indicating the basic plasma streamings as seen in the rest frame of the Sun. The main discontinuity surfaces of the heliosphere (*dark grey region*), i.e., the heliospheric shock, the heliopause, and the bow shock. The *light grey region* depicts the inner heliosheath (taken from *Sternal* [2010])

system being dominated by the solar wind, a supersonic stream of energetic particle away from the Sun. The other two shields are located closer to the Earth: the magnetosphere, which is built up by the interaction of the Earth's magnetic field with the solar wind, and the terrestrial atmosphere. All these shields cause a modulation of the CR spectra. The interaction of CR particles with atmospheric molecules and atoms leads to an ionisation of the upper and middle atmosphere as well as to the formation of cosmogenic nuclides, the most known of which are ^{10}Be and ^{14}C.

In addition to GCRs, the Earth is also exposed to energetic particles from the Sun (Solar Energetic Particles (SEPs)), one indication for solar activity. Besides such isolated events, the Sun shows a periodic behaviour between quiet and active phases, which are correlated with the number of sunspots leading to the well known Schwabe cycle of 11 years. At each solar maximum, the Sun's magnetic field reverses its polarity. Thus the magnetic cycle of the Sun is twice as long as the Schwabe cycle and is called the Hale cycle (22 years).

The heliosphere as a whole is embedded into the Local Interstellar Medium (LISM), the boundary between which is called the heliopause. From inside the heliosphere a further boundary is defined as the region where the solar wind has slowed down to subsonic values, the termination or heliospheric shock, whereas the LISM from outside could form a bow shock, which existence is still to be confirmed by observations. The region between the termination shock and the heliopause is called the inner heliosheath, while the one between heliopause and the bow shock is known as the outer heliosheath. The structure of the heliosphere is shown schematically in Fig. 4.1.

As could be shown very recently by *Scherer et al.* [2011], the above mentioned modulation of GCR spectra does not only take place in the heliosphere and inner heliosheath, but already in the outer heliosheath, i.e. beyond the heliopause. All these long term galactic and solar influences the Earth is exposed to are expected not only to be recorded in terrestrial archives, but should have direct consequences for the atmosphere and may even affect the climate, leading *Scherer et al.* [2006] to the introduction of the term "Interstellar-Terrestrial Relations". The variation of GCR spectra can be measured by satellites in the heliosphere. In contrast, traces found in terrestrial archives, e.g. ice cores, are subject also to the variation of the Earth's magneto- and atmospheres. In order to interpret such ice core data back in time the latter variations need to be taken into account.

Correlations of cosmogenic nuclide abundances with climate tracers give valuable information regarding the question to what extent the terrestrial climate is influenced by external drivers. Candidates for the latter are the variable Sun (solar forcing), perturbations of the Earth's orbit (Milankovitch forcing), the variable CR flux (CR forcing), and the varying atomic hydrogen inflow into the atmosphere of Earth (hydrogen forcing).

A current debate is on solar vs. CR forcing, which are both modulated by solar activity. In recent years various indicators of an influence of solar activity on the terrestrial climate have been identified, see, e.g., *Haigh* [2007]. The exact chain of physical processes, however, that is responsible for an external climate driving is still unclear, and so the question whether the driving is mainly direct (solar forcing) or indirect (CR forcing) remains unanswered. After the revival of the latter idea [*Svensmark and Früs-Christensen*, 1997; *Svensmark*, 1998] the evidence for an influence of CRs on the atmosphere of Earth has increased [e.g. *Usoskin and Kovaltsov*, 2008; *Wissing et al.*, 2010; *Rohrs et al.*, 2010]. This is recognised in the Fourth Intergovernmental Panel on Climate Change (IPCC) Assessment Report 'Climate Change 2007' [*IPCC*, 2007], wherein the potential significance of CRs for the physics of the atmosphere and, thus, climate has been acknowledged. The present Chapter concentrates on the modelling of the CR transport from the local interstellar medium through the helio-, magneto- and atmosphere in order to get a quantitative understanding of the parameter which determines the ionisation and production of cosmogenic nuclides in the Earth's atmosphere. This comprises a study of the local interstellar spectrum (LIS) of galactic CRs (Sect. 4.2), of their transport through the dynamical heliosphere (Sects. 4.3 and 4.4), through the magnetosphere (Sect. 4.5), and their interaction with the atmosphere (Sect. 4.6). The latter interaction manifests in three (partially debated) effects, namely (a) atmospheric ionisation and (b) the production of cosmogenic nuclides (as described in Sect. 4.7 for ^{10}Be).

4.2 Local Interstellar Spectra of Galactic Cosmic Ray Protons

The heliopause is the boundary layer between the interstellar and solar winds. In the paradigm of cosmic ray modulation GCRs are entering our heliosphere and

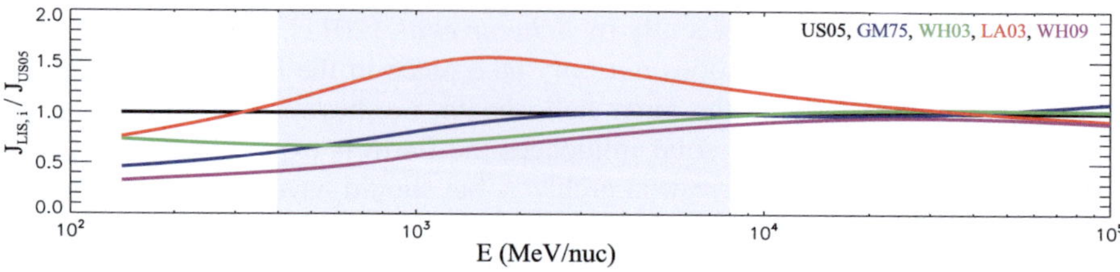

Fig. 4.2 Ratios of different unmodulated LIS proton spectra used in the study by *Herbst et al.* [2010]. The *blue*, *green*, *red*, *magenta* and *black curves* correspond to the ratio of the local interstellar spectra given by *Garcia-Munoz et al.* [1975]; *Webber and Higbie* [2003]; *Langner et al.* [2003]; *Webber and Higbie* [2009], and *Usoskin et al.* [2005] to the one by *Usoskin et al.* [2005], respectively. Accentuated in *light-blue* is the energy range relevant for the ^{10}Be production, 0.4 to 8 GeV/n

encounter the outward-flowing solar wind. The intensity of GCRs is modulated as they traverse the turbulent heliospheric magnetic field embedded into the solar wind. In order to solve the transport equation for these particles numerically the local interstellar spectrum (LIS) has to be specified among other parameters (e.g. the diffusion coefficient in the heliosphere and the modulation volume). From recent Ulysses measurements it became obvious that the LIS of galactic cosmic ray nuclei can not be derived at energies below a few hundred MeV/nucleon [see e.g. *Heber and Potgieter*, 2006; *Scherer et al.*, 2011]. Thus, these spectra need to be specified in an independent way. As an example, *Webber and Higbie* [2009] used a cosmic ray transport model for the galaxy. This model includes, e.g., hadronic interactions and a transport by diffusion in the turbulent galactic magnetic field. The corresponding diffusion coefficients were taken from electron measurements [*Webber and Higbie*, 2008]. By this method they derived a proton LIS. Other authors like *Burger et al.* [2000], *Usoskin et al.* [2005] or *Langner et al.* [2003] used results of more complex models, like the GALactic PROPagation code described in e.g. *Strong and Moskalenko* [1998]. This program calculates the propagation of relativistic charged particles and the diffuse emissions (γ-ray and synchrotron emission) produced during their propagation and incorporates as much realistic astrophysical input as possible together with latest theoretical developments. Several other parameterisations of the proton LIS exist as well [e.g. *Garcia-Munoz et al.*, 1975; *Webber and Higbie*, 2003]. Figure 4.2, taken from *Herbst et al.* [2010], shows the variation of these spectra in an energy subset with respect to the one derived by *Usoskin et al.* [2005]. In their analysis *Usoskin et al.* [2010] used the more recent ones by *Webber and Higbie* [2003, 2009], *Langner et al.* [2003], and the spectrum by *Garcia-Munoz et al.* [1975] derived already in the 1970's. The shaded area shows the energy range which is most important for the atmospheric production of ^{10}Be nuclides. As! can be seen these spectra agree well with each other for proton energies above 10 GeV/n. However, at lower energies differences up to a factor of two exist. Nevertheless, the LIS has not been measured by now and, therefore, each of these models may approximate it correctly at energies below a few hundred MeV.

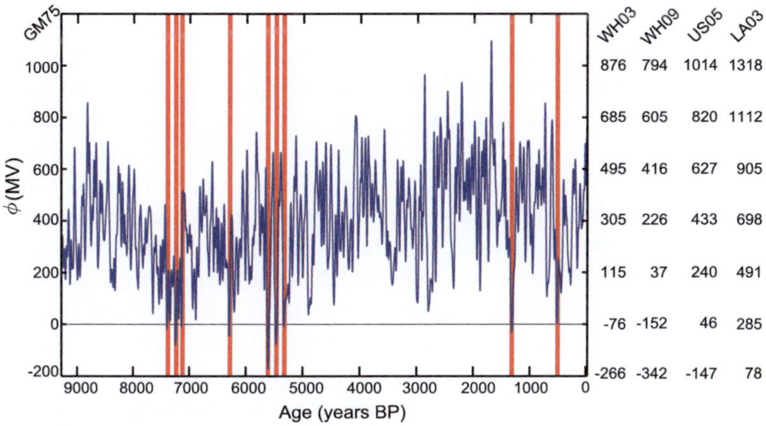

Fig. 4.3 The modulation parameter reconstruction for the past 9300 years based on ^{10}Be mainly measured in the GRIP ice core (see *Steinhilber et al.* [2008] and *Vonmoos et al.* [2006]).The left y-axis show the reconstructed ϕ using the *Garcia-Munoz et al.* [1975] LIS (*blue curve*). For most time intervals the parameter is in the range of presently observed values, but there are times (marked with the *red vertical lines*) where the parameter becomes zero or even negative (e.g. around the years 500, 1300, and 5600 BP). The modulation values are converted to the other LIS used in *Herbst et al.* [2010]. The resulting modulation parameters are shown on the right y-axes

Following *Gleeson and Axford* [1968] and *Caballero-Lopez and Moraal* [2004] the full transport equation, discussed in Sect. 4.4, can be reduced to a simple convection-diffusion equation, if (a) there are no additional sources of CRs within the heliosphere, (b) there is a quasi-steady state, (c) the adiabatic energy loss rate is negligible, and (d) there is no drift. *Caballero-Lopez and Moraal* [2004] showed that the measured proton spectra at 1 AU can reasonably well be approximated by this approach, the so-called force-field solution, during recent solar cycles. The calculated spectrum at 1 AU depends, thus, only on the assumed LIS and the so-called modulation parameter ϕ, describing the solar activity.

Herbst et al. [2010] showed that despite the differences of all these modelled LIS the modulated spectra in the vicinity of the Earth can be adjusted to fit recent PAMELA observations [*Casolino et al.*, 2009] by varying the modulation parameter, ϕ. Although a linear relationship between the LIS dependent modulation parameters was already found by *Usoskin et al.* [2005], *Herbst et al.* [2010] could show that this relationship depends in addition on the energy range of interest. Since we are interested in the production of cosmogenic nuclides, in particular ^{10}Be, we focus our study on the energy range from 0.4 to 8 GeV/n [*Masarik and Beer*, 1999; *Webber and Higbie*, 2003; *McCracken*, 2004; *McCracken and Beer*, 2007; *Masarik and Beer*, 2009].

Figure 4.3 shows the long-term reconstruction of the solar modulation parameter ϕ reconstructed by *Steinhilber et al.* [2008]. The left y-axis displays the corresponding ϕ record using the LIS by *Garcia-Munoz et al.* [1975]. As can be seen, for most of the time, the calculated modulation parameter is in the range of presently observed values. However, in some time periods (marked as red vertical lines) ϕ becomes very small or even negative, for example around the years 500 BP and 1300 BP. Further periods with negative values are found between 7400 and 5000 BP with the lowest values around 5600 BP. *Herbst et al.* [2010] applied different conversion equations in the energy range of 0.4 to 8 GeV/n to the long-term ϕ record by *Steinhilber et al.* [2008]. This record (see Fig. 4.3) shows negative ϕ values for all local interstellar spectra beside the two given by *Usoskin et al.* [2005] and *Langner et al.* [2003]. Negative values are non-physical, and therefore must be explained. Important sources of uncertainty are the atmospheric mixing and the paleo-geomagnetic field reconstruction, which have been considered in the ^{10}Be production calculations of *Masarik and Beer* [1999], and therewith in ϕ. It is not likely, however, that all negative ϕ values could be explained by these uncertainties alone and, thus, other sources of uncertainty, i.e. contribution of heavier elements of cosmic rays to the ^{10}Be production, inner-heliospheric effects and the LISM turbulence, have to be investigated in addition. Nevertheless, the long-term reconstruction of the modulation parameter ϕ provides the potential to derive the low intensity limits of the LIS [see *Herbst et al.*, 2010, for further details] in the energy range of a few hundred MeV.

4.3 The Dynamic Heliosphere

More detailed or even quantitative studies of interstellar-terrestrial relations require the modelling of the global, dynamic heliosphere being located inside the turbulent LISM. The present state-of-the-art of the modelling of a dynamic heliosphere with a self-consistent treatment of the transport of cosmic rays is reviewed in *Florinski et al.* [2009] and *Potgieter* [2010].

4.3.1 Basic Equations

A self-consistent description of the heliosphere and its dynamics requires taking into account the full set of magnetohydrodynamic equations that describe the heliosphere within the framework of a fluid picture, where several different fluids interacting with each other are considered. The basic equations read in normalised form

$$\frac{\partial}{\partial t}\begin{pmatrix}\rho\\ \rho\mathbf{v}\\ e\\ \mathbf{B}\end{pmatrix} + \nabla\cdot\begin{pmatrix}\rho\mathbf{v}\\ \rho\mathbf{v}\mathbf{v}+(p_{th}+\frac{1}{2}B^2)\hat{I}-\mathbf{B}\mathbf{B}\\ (e+p_{th}+\frac{1}{2}B^2)\mathbf{v}-\mathbf{B}(\mathbf{v}\cdot\mathbf{B})\\ \mathbf{v}\mathbf{B}-\mathbf{B}\mathbf{v}\end{pmatrix} = \begin{pmatrix}Q_\rho\\ \mathbf{Q}_{\rho\mathbf{v}}\\ Q_e\\ 0\end{pmatrix} \tag{4.1}$$

for each thermal (charged and neutral) component taken into account. Here, ρ is the mass density, $\mathbf{v}$ the velocity, e the total energy density and p_{th} the thermal pressure of a given component. $\mathbf{B}$ is the magnetic field and $\hat{I}$ the unity tensor. The terms Q_ρ, $\mathbf{Q}_{\rho\mathbf{v}}$ and Q_e describe the exchange of mass, momentum and energy between the

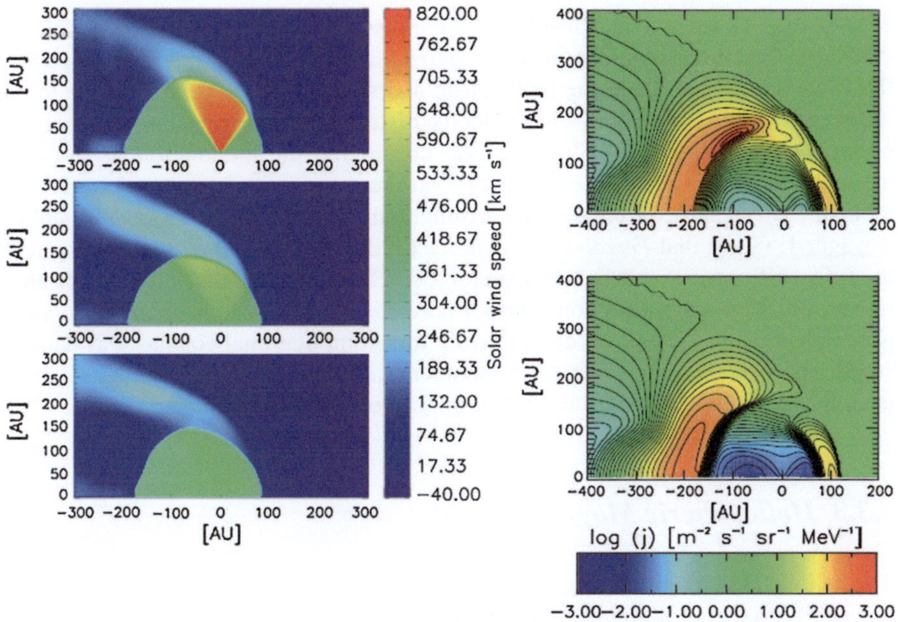

Fig. 4.4 *Left column*: Time evolution of the dynamic heliosphere represented by the solar wind speed. The *red line* indicates the inclination at which Voyager 1 has crossed the termination shock. *Right column*: 30 MeV CR intensities in the meridional plane of the heliosphere. Shown is a snapshot to solar activity minimum (*top*) and one at maximum (*bottom*)

thermal components and with the CRs if present. For details we refer to *Ferreira and Scherer* [2006].

The transport of CRs is described by the Parker transport equation (see Sect. 4.4.1 for more details). The simultaneous solution of the above equations is referred to as hybrid modelling and is discussed briefly in the following.

4.3.2 Hybrid Modelling

The self-consistent solution of the CR transport in a dynamical, i.e. time-varying heliosphere (due to solar activity) requires, so far, a limitation to two spatial dimensions. *Scherer and Ferreira* [2005a] developed the first hybrid model called 'BoPo', combining plasma and CR propagation models, that was validated successfully by fitting spectra during two consecutive 11-year Schwabe cycles [*Scherer and Ferreira*, 2005a] and time series [*Scherer and Ferreira*, 2005b; *Ferreira and Scherer*, 2006] observed by the Pioneer and Voyager spacecraft. A typical result from such modelling is shown in Fig. 4.4.

Hybrid modelling in 3-D is possible but requires, so far, the assumption of a steady-state heliosphere. *Langner et al.* [2006a] and *Florinski and Pogorelov* [2009, e.g.] used 3-D steady-state results as an input to a kinetic transport model and studied in detail the modulation in the (inner) heliosheath, i.e. the region between the termination shock and the heliopause, and found it to be very sensitive to the shocked, i.e. subsonic solar wind velocity distribution. In extension of this modelling, *Langner et al.* [2006b] have also considered the modulation of so-called anomalous protons accelerated at or beyond the termination shock and found that their intensity can peak beyond the termination shock location, as is observed by Voyager 1. Given that *Ferreira and Scherer* [2006] arrived at the same finding with the fully self-consistent 2-D BoPo model it can be concluded that 2-D models are a reasonably well approximation. Furthermore, 2-D models can accommodate additional physical processes at comparatively low computational cost, as was demonstrated in *Ferreira et al.* [2007], where stochastic acceleration as well as adiabatic heating were included to successfully explain Voyager observations.

4.3.3 Heliospheric Magnetic Field

The heliospheric magnetic field, which has its origin in the solar magnetic field that expands with the solar wind, is relevant on a large scale for the transport of CRs and for injecting possibly pre-accelerated particles into the Fermi-I acceleration process at the solar wind termination shock. [*Scherer et al.*, 2010, among others] compared four analytically described fields presently discussed in the literature and studied the associated Fermi-I injection efficiency at the termination shock and found it to be largely insensitive to the different field configurations.

Further important aspects are gradient and curvature drifts (see Sect. 4.4.1) of CRs within the heliospheric magnetic field that depend on the location of the heliospheric current sheet (HCS). The latter is a plasma layer dividing the heliosphere into the regions of different magnetic field polarity. Because of the cyclic solar magnetic field reversals and the diverging and slowing plasma flow in the outer heliosphere, the HCS is expected to have a complex structure during periods around maximum solar activity, and in general in the heliosheath. *Czechowski et al.* [2010] have derived the shape of the heliospheric current sheet at a given time by following the plasma flow lines originating at the neutral line close to the Sun.

4.3.4 Connection with the Local Interstellar Medium

The above studies are triggered by recent observations [*McComas et al.*, 2009; *Richardson and Stone*, 2009] and indicate the growing interest in models of the outer heliosphere that put emphasis on advanced treatments of the LISM including a more realistic interstellar magnetic field [*Pogorelov et al.*, 2009; *Alouani-Bibi et al.*, 2011], more detailed neutral flows [*Lallement et al.*, 2010], and a complex structured outer heliosheath, i.e. a layer of disturbed interstellar plasma beyond the heliopause [*Scherer et al.*, 2011].

Interestingly, there is not only an influence of the LISM on the heliosphere and the CR fluxes within, but also vice versa: the low-energy particles being accelerated in the outer heliosphere contribute to the interstellar CR spectrum [*Scherer et al.*, 2008] and, moreover, can heat and ionise the LISM.

4.4 Propagation of Galactic Cosmic Rays in the Heliosphere

Due to the extremely low densities, galactic cosmic rays (GCRs) do not collide with solar wind particles, but are affected by magnetic field irregularities frozen in and carried along in the solar wind plasma.

4.4.1 Parker's Transport Equation

The transport of GCRs in the heliosphere can be described by Parker's [1965] transport equation. Let $f(\mathbf{r}, R, t)$ be the differential CR distribution function with respect to the particle rigidity R (momentum per charge), then its variation with time t and position r is given by:

$$\frac{\partial f}{\partial t} = -(\underbrace{\mathbf{v}}_{a} + \underbrace{\langle v_D \rangle}_{b}) \cdot \nabla f + \underbrace{\nabla \cdot (\mathbf{K}_{(s)} \cdot \nabla f)}_{c} + \underbrace{\frac{1}{3}(\nabla \cdot \mathbf{v}) \frac{\partial f}{\partial \ln R}}_{d}, \tag{4.2}$$

where terms on the right-hand side represent:

a. Outward convection by the solar wind speed.
b. Gradient and curvature drifts ($\langle v_D \rangle$) in the global heliospheric magnetic field.
c. Inward diffusion ($\mathbf{K}_{(s)}$) through the interplanetary magnetic field irregularities in response to the gradient set up by convection and deceleration. The symmetric part of the tensor $\mathbf{K}_s$ consists of a parallel diffusion coefficient ($K_{\|}$) and two perpendicular diffusion coefficients ($K_{\perp}$). In a spherical coordinate system and averaged over all longitudes the effective radial diffusion coefficient is given by $K_{rr} = K_{||} \cos^2 \Psi + K_{\perp r} \sin^2 \Psi$, with Ψ the angle between the radial direction and the averaged HMF direction. Note that $K_{||}$ dominates K_{rr} in the inner and polar regions and $K_{\perp r}$ dominates in the outer equatorial regions of the heliosphere.
d. Adiabatic energy changes from the divergence of the expanding solar wind.

Although the modulation of GCRs in the heliosphere still depends on all of the processes described by the Parker equation, the simpleforce-field approximation (cf. Sect. 4.2), is often used in the literature [*Usoskin et al.*, 2005; *Herbst et al.*, 2010] and has been used to interpret ^{10}Be-data from ice-cores [e.g. *Steinhilber et al.*, 2008].

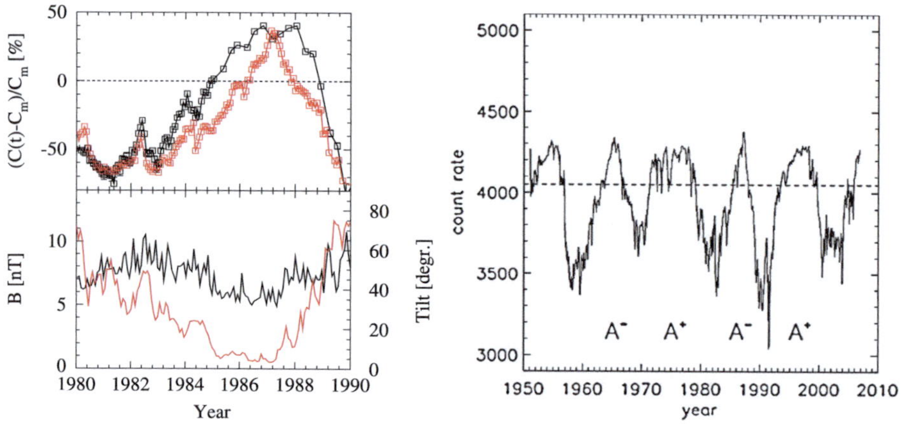

Fig. 4.5 Normalised count rates of 1.2 GV galactic cosmic ray electrons (*black curve*) and helium (*red curve*) from 1980 to 1990. The count rates have been normalised so that a value of −70 % are determined in 1991 [for details see *Heber et al.*, 2009] (*left panel*). Cosmic ray flux as measured via the Climax neutron monitor [*Lopate*, 2006], the *dashed line* indicates a (here arbitrarily chosen) threshold discussed in the text (*right panel*)

4.4.2 The Imprint of the Hale-Cycle

The solar activity affects the CR modulation and is, thus, visible in the signatures of energetic particles. Therewith the 78 day averaged count rate of electrons and α particles from *Heber et al.* [2009], as displayed in the left panel of Fig. 4.5, shows the "flat" and "peaked" maxima in the time profiles manifesting a Hale periodicity of about 22 years in addition to the Schwabe period due to solar activity. While the interpretation of alternating "flat" and "peaked" maxima may be ambiguous the time profiles of negatively and positively charged particles are unambiguously an imprint of particle drifts in the global heliospheric magnetic field.

From the presence of the 11-year Schwabe and 22-year Hale cycles in numerous time series of climate indicators like tree rings, varves, precipitation, droughts or temperatures [*Fichtner et al.*, 2006] it has been concluded that solar activity has an influence on the terrestrial climate. While at present it is unclear whether this influence is direct or indirect, cf. Sect. 4.1, and which processes establish such relation, it is likely that the observed periods do contain valuable information.

Fichtner et al. [2010] have performed a comparative period analysis of solar irradiance and cosmic ray flux. After demonstrating this way that the Hale period is strongly associated to the latter, these authors offer a hypothesis why the Hale period can occur in certain climate-indicative time series despite the insignificance or absence of the Schwabe period.

From the right panel of Fig. 4.5 it is clear that above a certain threshold value (indicated by the horizontal line), there is a significantly higher integral CR flux for the duration of the broader flat maxima as compared to that of the peaked ones. This is true for a whole range of threshold values (and, thus, to a great extent independent of them). The authors demonstrated that the integrated flux is not only systematically higher but that also the duration of high CR flux is, on average, by more than three years longer for an A^+ cycle compared to an A^- cycle, where $A^\pm$ indicates the polarity of the solar magnetic field. This extended duration might facilitate the triggering of climate forcing: during the shorter A^- cycles the integrated CR forcing is too weak.

There are three important aspects to be noted: first, such threshold argument cannot be made for the solar irradiance that is characterised by rather similar intensity maxima. Consequently, the 11-year period could, in principle, be accompanied by higher harmonics or sub-harmonics, but it cannot be expected to be absent when these harmonics exist. Second, while at present the nature of a threshold effect is still unclear, it could well be related to atmospheric ionisation [*Usoskin and Kovaltsov*, 2006] or to the global electric circuit [*Tinsley et al.*, 2007; *Harrison and Usoskin*, 2010]. And, third, in view of the complexity of the terrestrial climate system it should not surprise that this threshold argument does not result in a strict rule, it rather suggests an empirical explanation for the frequent strength of the Hale period in time series of climate indicators. Once the actual processes relating solar activity to the atmosphere and climate are identified, the above hypothesis can be tested quantitatively.

4.5 Propagation of Galactic Cosmic Rays in the Earth's Magnetosphere

The above findings may depend also on the second one of the initially mentioned shields, the Earth's magnetic field. The main reason is the Lorentz force that deflects charged particles perpendicular to the magnetic field direction.

Viewed from larger distances, a planetary magnetic field can be approximated as a (usually) tilted dipole field, which becomes deformed under the influence of the solar wind to form a magnetosphere with a long tail in downwind direction. An inspection of magnetic field measurements within the Earth's atmosphere, however, reveals that the dipole approximation becomes less and less valid when approaching the surface. Already Carl-Friedrich Gauß found in 1838 a representation of the Earth's magnetic field by means of an expansion of Schmidt-Legendre polynomials in spherical coordinates r, ϑ and φ with the coefficients being calculated from observations. A very common representation of the magnetic field approach based on this approach is the International Geomagnetic Reference Field (IGRF, e.g. *Pilchowski et al.* [2010] for details).

A measure for the effectiveness of the magnetic shield is the so-called cut-off rigidity, i.e. the minimal momentum per charge a particle must have in order to penetrate the shield and reach the surface (cf. Sect. 4.6). The higher this value, the more effective is the magnetic protection. Figure 4.6 (left panel) shows a map of the cut-off rigidity for vertically incidenting particles (vertical cut-off rigidity). The figure shows high values in equatorial (magnetic field lines essential parallel to the surface) and low values in polar (perpendicular) regions demonstrating the influence of

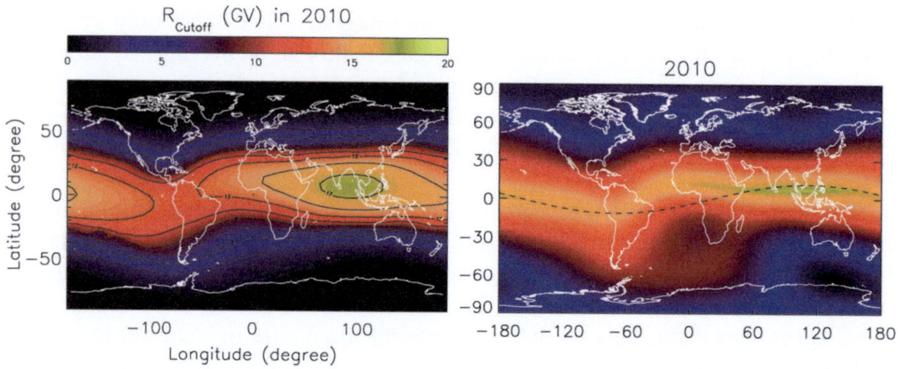

Fig. 4.6 *Left panel*: vertical cutoff rigidities mapped onto the Earth's surface, computed with the PLANETOCOSMICS simulation package (cf. Sect. 4.6) for the IGRF 2010 [*Herbst et al.*, 2011]. *Right panel*: difference between the tangential and the radial components of the magnetic field, $\delta B = \sqrt{B_\vartheta{}^2 + B_\varphi{}^2} - |B_r|$, calculated with the IGRF model. r stands for the radial component, ϑ and φ for the geographic co-latitude and longitude, respectively

the geometry of the magnetic field on the propagation of charged particles. A closer inspection shows in addition also longitudinal variations, in particular a maximum over India, which cannot be seen if the magnetic field is represented by a tilted (and shifted) dipole field. As could be shown by *Pilchowski et al.* [2010] the geometry of the magnetic field may have even larger influence on the cut-off rigidities than the field strength has. To illustrate the influence of the geometry, the right panel of Fig. 4.6 shows the difference, δB, between the tangential and radial components of the IGRF 2010. Even when keeping in mind that this quantity cannot be much more than a rough measure, a comparison shows remarkable similarities between this quantity and the cut-off rigidity.

To go a step further, we investigate in addition temporal changes of the vertical cut-off rigidity and compare these values with those of the magnetic field, as a measure of which we take again the quantity δB. The left panel of Fig. 4.7 shows the temporal evolution of the cut-off rigidity from 1955 up to 2010 at five different places, while that of δB at the same places is shown in the right panel. Both quantities are normalised to the values in 1955. For δB, we used the absolute value and shifted curves afterwards to $+1$, if the value 1955 is negative. These places are marked with (*) in the figure.

We observe a close correlation between the behaviour of the cut-off rigidities and the geometry of the magnetic field with respect to the temporal evolution as well. A second important quantity for the effectiveness of the magnetic shielding is the magnitude $|\mathbf{B}|$ of the magnetic field. The prominent features are again only visible in the IGRF, the most remarkable of which is the South Atlantic Anomaly, a region with reduced magnetic field ranging from the eastern coast of South America to the South African Cape region, where significantly enhanced counting rates of energetic charged particles are detected, so that the magnitude appears to influence the counting rates rather than the cut-off rigidities.

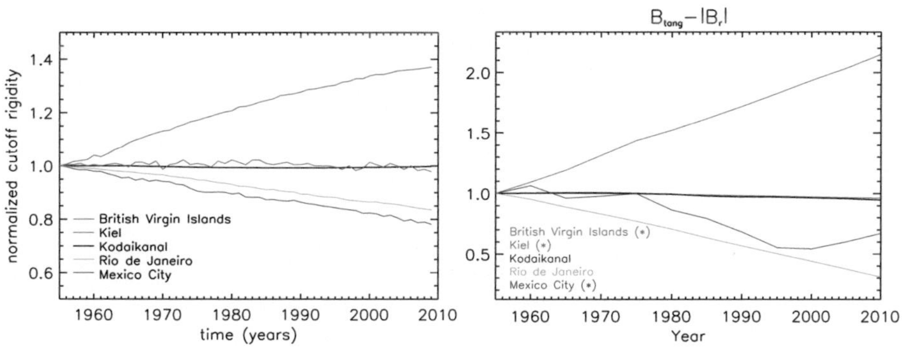

Fig. 4.7 Change of the cut-off rigidity (*left panel*) and of the quantity δB (*right panel*) between 1955 and 2010 at five different places. Both values were normalised to the absolute value in 1955. If the value of δB for 1955 is negative (stations marked with (*) in the *right panel*), the whole curve was shifted from -1 to $+1$

4.6 Energetic Particles in the Atmosphere

Energetic charged particles that penetrate the magnetic field are able to penetrate the third shield, the terrestrial atmosphere where they will lose a significant amount of their energy due to ionisation occurring in the upper atmospheric layers above approximately 25 km (corresponding to a column density of 100 g/cm^2, see also 4.6.1). Deeper inside the atmosphere primary particles are stopped efficiently due to hadronic interactions with the surrounding atmospheric gases by forming a fully developed secondary particle cascade which can be divided into three main components: 1) the "soft" or electromagnetic component consisting mainly of electrons, positrons and photons, 2) the "hard" or muon component and 3) the "hadronic" nucleonic component consisting mostly of supra-thermal protons and neutrons (see 4.6.2). The PLANETOCOSMICS code [*Desorgher*, 2006], a Monte-Carlo simulation code based on the GEANT-4 simulation package [*Agostinelli et al.*, 2003], has been used to calculate the development of atmospheric cascades [*Dorman*, 2004] as well as to model energetic charged particle measurements in the atmosphere successfully [see e.g. *Mironoval et al.*, 2008; *Bazilevskaya et al.*, 2008; *Matthiä et al.*, 2009].

4.6.1 Ion Pair Production in the Atmosphere

Atmospheric measurable quantities like the ionisation as well as the production of secondary neutrons, protons and muons can be computed for any given solar modulation and location within the Earth's atmosphere. Particles entering the atmosphere lose an amount of energy due to the interaction with atmospheric atoms and molecules. An altitudinal dependence is represented by the specific energy loss

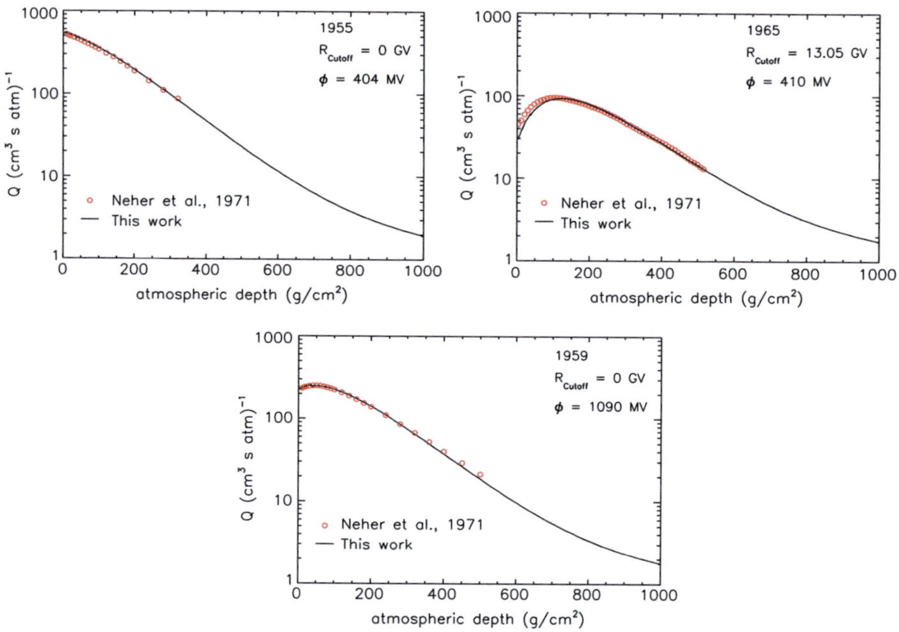

Fig. 4.8 Ion pair production rates as a function of atmospheric depth at Thule (*top left* and *bottom panel*) and at Peru for solar minimum conditions (1955 and 1965) and the solar maximum in 1959. The *open circles* represent the measurements performed by *Neher* [1971], while the *solid lines* show our computations. The simulations by *Herbst* [2011] are in good agreement with the balloon measurements

$\frac{dE}{dx}(x)$ at a specific atmospheric depth x, which is associated with the ion pair production rate $Q_i(x, \phi)$ in the following way:

$$Q_i(x, \phi) = \int_{E_{\text{Cutoff},i}}^{\infty} Y_i(x, E) \cdot J_i(E, \phi)\, dE, \tag{4.3}$$

where $Y_i(x, E)$ is the ionisation yield representing the number of ions produced per mass unit of the surrounding atmospheric environment at a certain location, represented by its cutoff energy E_{Cutoff} and atmospheric depth x due to a single primary particle type i of a certain energy [see *Usoskin et al.*, 2006]. *Herbst* [2011] calculated the ion pair production rate $Q(x, \phi) = \sum_i Q_i(x, \phi)$ for Greenland and Peru during solar minimum conditions in 1955 ($\phi = 404$ MV, left panel of Fig. 4.8) and 1959 ($\phi = 1090$ MV, middle panel of Fig. 4.8), respectively. The right panel of Fig. 4.8 displays the ion pair production rate at Thule during solar maximum. Here the solid lines represent our computations while the open circles are measurements by *Neher* [1967, 1971].

4.6.2 Secondary Particle Production

Since the soft component of the cosmic-ray-induced cascade is only of minor importance in the troposphere we here compare computations of the muonic and hadronic components with atmospheric measurements at different locations as well as modulation levels. The most important source for atmospheric secondary muons is the decay of charged pions produced in inelastic nucleus-nucleus collisions of primary cosmic ray particles and secondary hadrons with particles in the atmosphere. As *Matthiä* [2009] described in detail the calculations are in good agreement with atmospheric measurements by *Kremer et al.* [1999].

4.6.2.1 Computation of the Hadronic Component

Secondary protons and neutrons produced by collisions of CR ions with atmospheric nuclei are the main contributors to the count rates of Neutron Monitors [see *Clem and Dorman*, 2000]. *Goldhagen et al.* [2004] performed balloon flights at different geographic locations characterised by cutoff rigidities of $R_{\text{Cutoff}} = 2.7$ GV and $R_{\text{Cutoff}} = 11.6$ GV, respectively. The weighted energy spectra $\frac{dI}{dE} \cdot E$ for these two measurements, see Fig. 4.9 left and middle panel, are obtained at an altitude of 20.3 km (53.5 g/cm^2) and at the Earth's surface (1030 g/cm^2). The red curves represent the measurements while the black curves display our computations. The measurements and our calculations are in good agreement, with a small overestimation of the experimental data by the model below 1 MeV at 53.5 g/cm^2.

For the second important hadronic component in the atmospheric CR induced radiation field, the protons, we compare measured differential proton intensities at three different atmospheric depths with our calculations (see the right panel of Fig. 4.9). Here the measurements performed by the experiments BESS-2001 (*Abe et al.* [2003], open triangles), CAPRICE98 (*Boezio et al.* [2003], diamonds) as well as the AMS98 (*AMS Collaboration et al.* [2002], squares) at a geomagnetic cutoff rigidity of $R_{\text{Cutoff}} = 4.3$ GV and atmospheric depths of 11.9 g/cm^2, 5.5 g/cm^2 and 0.0023 g/cm^2 are used. As can be seen from Fig. 4.9 our calculations are in good agreement with the measurements.

4.7 Cosmogenic Nuclides

Cosmogenic Nuclides (CNs) are the product of an interaction of GCRs as well as SEPs with the atmospheric gases Oxygen, Nitrogen and Argon. Even if the intensity of the LIS outside the heliosphere is assumed to be constant in time and GCRs are modulated only by the solar magnetic activity and the geomagnetic field strength [see e.g. *Beer*, 2000] the production of CNs is neither temporally nor spatially constant. CNs are produced by mainly three production mechanisms: 1) spallation

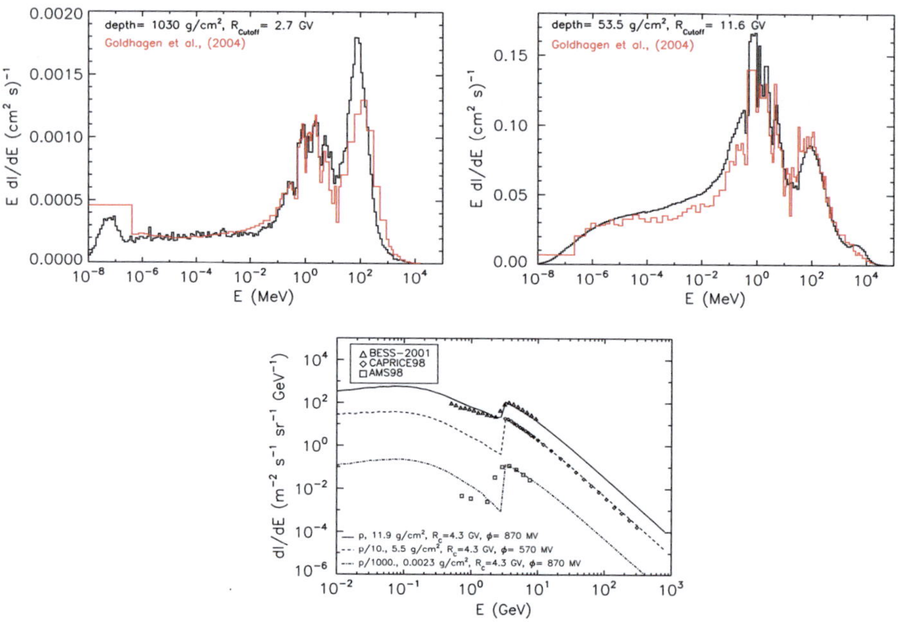

Fig. 4.9 Omnidirectional differential secondary neutron intensities (*top left* and *top right panel*) at different atmospheric depths and geomagnetic locations in comparison to measurements by *Goldhagen et al.* [2004] (*red curves*) as well as differential proton intensities for three atmospheric depths at a cutoff rigidity of 4.3 GV (*bottom panel*) performed by BESS-2001 (*Abe et al.* [2003], *open triangles*), CAPRICE98 (*Boezio et al.* [2003], *diamonds*) as well as the AMS98 experiment (*AMS Collaboration et al.* [2002], *squares*)

reactions, where protons as well as neutrons are sputtered off the atmospheric target nuclei (O, N, Ar) during the interaction with a high energetic secondary neutron, 2) thermal neutron capture, which occurs because most of the neutrons produced in the cascades and sub-cascades are slowed down to thermal energies in the energy range of $1 \cdot 10^{-8} - 7 \cdot 10^{-7}$ MeV [see *Phillips et al.*, 2001], and 3) negative-muon capture, where thermal energetic muons are captured by the atomic ! electron shell of the target atom, cascading towards the lowest electron shell, where they decay or are captured by the nucleus [see *Eidelman et al.*, 2004]. Once cosmogenic nuclides are produced they are subject of atmospheric mixing and transport mechanisms and become attached to atmospheric aerosols. After being transported and distributed by these circulation mechanisms they either are eroded by condensation (^{10}Be and ^{36}Cl), or they become attached to carbon-cycle based organic molecules [^{14}C, see e.g. *Dunai*, 2010] and therewith archived in natural archives like ice, rocks or trees containing and preserving the paleomagnetic informations. Here we will investigate the computation of the production of the cosmogenic nuclide ^{10}Be in more detail.

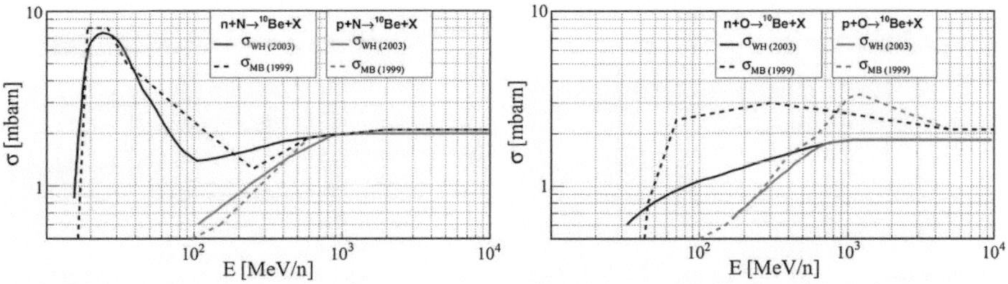

Fig. 4.10 ^{10}Be production cross-sections for neutron and proton projectiles on atmospheric nitrogen (*top*) and oxygen (*bottom*) derived from *Masarik and Beer* [1999] labelled as MB (1999) and *Webber and Higbie* [2003] labelled as WH (2003)

4.7.1 The Production of Cosmogenic Nuclides

In order to describe the probability of a certain nuclear reaction to occur, the (nuclear) cross section σ is used, which strongly depends on the type as well as the relative velocity of the projectile. Most of the cross sections used in order to calculate the cosmogenic nuclide productions are not known very well, and only a few theoretical or measured data exist. The production of ^{10}Be can be described by two sets of cross-sections based on *Masarik and Beer* [1999] and *Webber and Higbie* [2003], which vary significantly from each other, as shown in Fig. 4.10, so that the influence of the used cross-sections on the production of cosmogenic ^{10}Be is investigated in the following.

4.7.2 Modelling the Production of Cosmogenic Nuclides

According to *Masarik and Beer* [2009] the production of cosmogenic nuclides is given by:

$$P_j(\phi, R_{\text{Cutoff}}, \text{x}) = \sum_i N_i \sum_k \int_{E_{\text{Cutoff}}}^{\infty} \sigma_{ijk}(E_k) \cdot J_k(\phi, R_{\text{Cutoff}}, \text{x}, E_k)\, dE_k, \quad (4.4)$$

with N_i as the number of atoms per mass unit of the target nucleus type i, $\sigma_{ijk}(E_k)$ representing the cross section for the production of the cosmogenic nuclide type j from the target element i by the secondary particles of type k with a specific energy E_k. Furthermore $J_k(\phi, R_{\text{Cutoff}}, \text{x}, E_k)$ is the total secondary particle flux of type k with an energy E_k as function of the modulation parameter ϕ, geographic location (R_{Cutoff}) as well as atmospheric depth x.

Figure 4.11 shows the atmospheric-depth-integrated ^{10}Be production rates as function of latitude and longitude for different solar modulation periods using the galactic cosmic ray proton and α-particle model by *Usoskin et al.* [2005]. The upper panels of Fig. 4.11 display the computed depth-integrated ^{10}Be production rates of the two cross section sets, those from *Masarik and Beer* [1999] on the left, the

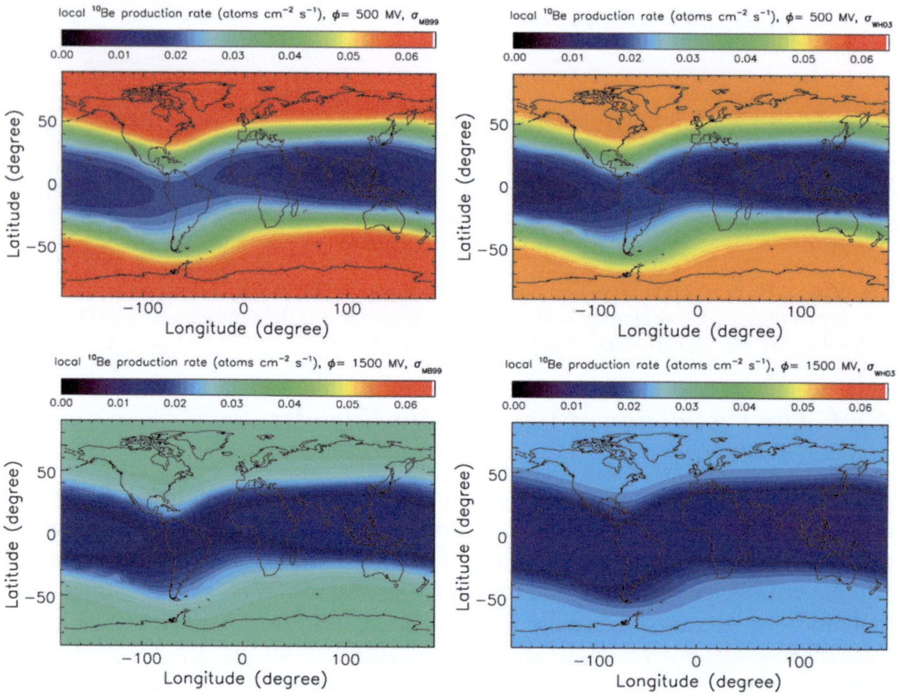

Fig. 4.11 Atmospheric depth integrated ^{10}Be production rates as function of latitude and longitude [*Herbst*, 2011]. The *upper panels* show the distribution according to solar minimum conditions (the *left panel* displays the production rates using the cross sections by *Masarik and Beer* [1999], the *right panels* those by *Webber and Higbie* [2003]), while the *lower panels* display the same, but for solar maximum ($\phi = 1500$ MV)

ones by *Webber and Higbie* [2003] on the right, during a typical solar minimum ($\phi = 500$ MV) while the lower panels display the results for solar maximum conditions ($\phi = 1500$ MV), showing that the ^{10}Be production rates are anti-correlated with the cutoff rigidities (see Fig. 4.6), i.e. locations with high cutoff rigidities show small ^{10}Be production rate values and vice versa. The comparison between the two cross-section sets reveals a variation of up to 25 %, and therewith is not negligible.

4.8 Summary and Conclusion

While the solar-terrestrial relationship has been established to be important for our modern society, that relies on sensitive electronic devices, the importance of the interstellar terrestrial relationship for the Earth environment is discussed seriously only since the 1990's. While most studies are based on correlations with proxies of the Earth climate, the goal of our project was to estimate and even quantify the variation of energetic particle intensities in the Earth's atmosphere caused by Galactic Cosmic Rays. These particles interact with the Earth's atmosphere and produce cosmogenic isotopes and ionise the atmosphere. While the projects by Kallenrode (see Chap. 13) concentrate on the consequences for the atmospheric chemistry due to ionisation, this chapter is focused around the understanding of the long-term variation of the ^{10}Be production rate as the last link in a long chain of processes ranging from galactic to terrestrial scales. This rate is determined by the physics of energetic particles in the atmosphere discussed in Sect. 4.7. We found that only due to the uncertainty of the different cross-sections the resulting uncertainty of the production rate is on the order of 25 %. The radiation (energetic particle) field in the atmosphere consisting mainly of secondaries has been determined in Sect. 4.6 using a Monte-Carlo Simulation. A comparison with measurements results in a similar uncertainty for the flux of secondary particles. The magnetosphere acts as a magnetic filter, which means that the lowest energy for a "primary" cosmic ray hitting the Earth's atmosphere depend on the geomagnetic position as discussed in Sect. 4.5. As a consequence we show in the same section that a detailed knowledge of the Earth magnetic field is essential to understand the variation of ground based measurements. The intensity of cosmic rays outside of the Earth magnetosphere depend on the local interstellar spectrum and its modulation in the heliosphere as discussed in Sects. 4.2 and 4.4, respectively. These topics are tied close together and cannot be discussed separately, since the local interstellar spectra have not been measured in-situ so far. Using the force-field solution we could show that all proposed models are able to reproduce observations close to Earth with different modulation parameters. These parameters, however, depend on numerous physical processes and boundary conditions. One of them is the dynamical heliosphere as discussed in Sect. 4.3. In contrast to last millennium propagation models hybrid models (combining plasma and propagation models) reveal the importance of the outer heliosphere beyond the termination shock for the local interstellar spectrum and long-term modulation. In addition using the full description of the particle transport in the heliosphere we propose a criterion to distinguish between cosmic ray and photon induced variations of parameters describing the Earth's climate: determining the importance of the Hale cycle (22 years, CRs) relative to the Schwabe cycle (11 years, photons) in such parameters.

From our research project the following conclusion can be given: in order to model the variation of the cosmic ray intensities as imprinted in terrestrial archives it is mandatory to achieve a detailed knowledge of:

1. the local interstellar spectra of all elements of interest,
2. the galactic cosmic ray modulation volume and the modulation processes in the different regions of the heliosphere,
3. the Earth's magnetic field and its variation over the last million years in structure as well as magnitude,
4. the secondary particle energy spectra and their modelling in the Earth atmosphere, and
5. the cross-sections for the production of cosmogenic isotopes.

Although already a major step forward has been achieved in our understanding the radiation field causing the different cosmic ray archives, further investigations are needed, i.e. on the atmospheric transport of these tracers and its ability of reflecting the global or local radiation environment.

Acknowledgements The authors like to thank Friedhelm Steinhilber, Daniel Matthiä, Marius Potgieter, Stefan Ferreira and Hans Fahr for valuable discussions. We are grateful to the Deutsche Forschungsgemeinschaft (DFG) for the project "Heliocauses" (FI 706/6-1/2/3 and HE 3279/8-2/3/4) and the Bundesministerium für Forschung und Bildung (BMBF) for supporting the German-South African collaboration (projects SUA 07/013 and 08/011) and the Ulysses/KET project (grant 50 OC 0902) through the Deutsches Zentrum für Luft- und Raumfahrt (DLR). This work profited from the discussions with the participants of the ISSI team meeting "Transport of Energetic Particles in the Inner Heliosphere".

References

Abe, K., Sanuki, T., et al. (2003). Measurements of proton, helium and muon spectra at small atmospheric depths with the BESS spectrometer. *Physics Letters B*, *564*, 8–20. doi:10.1016/S0370-2693(03)00676-2.

Alouani-Bibi, F., Opher, M., Alexashov, D., Izmodenov, V., & Toth, G. (2011). Kinetic versus multi-fluid approach for interstellar neutrals in the heliosphere: exploration of the interstellar magnetic field effects. *The Astrophysical Journal*, *734*, 45.

AMS Collaboration, Aguilar, M., Alcaraz, J., Allaby, J., Alpat, B., Ambrosi, G., Anderhub, H., Ao, L., Arefiev, A., Azzarello, P., et al. (2002). The alpha magnetic spectrometer (AMS) on the International Space Station: Part I – results from the test flight on the space shuttle. *Physics Reports*, *366*, 331–405. doi:10.1016/S0370-1573(02)00013-3.

Bazilevskaya, G. A., Usoskin, I. G., Flückiger, E. O., Harrison, R. G., Desorgher, L., Bütikofer, R., Krainev, M. B., Makhmutov, V. S., Stozhkov, Y. I., Svirzhevskaya, A. K., Svirzhevsky, N. S., & Kovaltsov, G. A. (2008). Cosmic ray induced ion production in the atmosphere. *Space Science Reviews*, 44. doi:10.1007/s11214-008-9339-y.

Beer, J. (2000). Long-term indirect indices of solar variability. *Space Science Reviews*, *94*, 53–66.

Boezio, M., Bonvicini, V., et al. (2003). The cosmic-ray proton and helium spectra measured with the CAPRICE98 balloon experiment. *Astroparticle Physics*, *19*, 583–604. doi:10.1016/S0927-6505(02)00267-0.

Burger, R. A., Potgieter, M. S., & Heber, B. (2000). Rigidity dependence of cosmic ray proton latitudinal gradients measured by the Ulysses spacecraft: implications for the diffusion tensor. *Journal of Geophysical Research*, *105*, 27447–27456. doi:10.1029/2000JA000153.

Büsching, I., Kopp, A., Pohl, M., Schlickeiser, R., Perrot, C., & Grenier, I. (2005). Cosmic-ray propagation properties for an origin in supernova remnants. *The Astrophysical Journal*, *619*, 314–326. doi:10.1086/426537.

Caballero-Lopez, R. A., & Moraal, H. (2004). Limitations of the force field equation to describe cosmic ray modulation. *Journal of Geophysical Research (Space Physics)*, *109*, 1101. doi:10.1029/2003JA010098.

Casolino, M., de Simone, N., & de Pascale, M. P. e. a. (2009). Cosmic ray measurements with Pamela experiment. *Nuclear Physics. B, Proceedings Supplement*, *190*, 293–299. doi:10.1016/j.nuclphysbps.2009.03.102.

Clem, J. M., & Dorman, L. I. (2000). Neutron monitor response functions. *Space Science Reviews*, *93*, 335–359 doi:10.1023/A:1026508915269.

Czechowski, A., Strumik, M., Grygorczuk, J., Grzedzielski, S., Ratkiewicz, R., & Scherer, K. (2010). Structure of the heliospheric current sheet from plasma convection in time-dependent heliospheric models. *Astronomy & Astrophysics*, *516*, A17.

Desorgher, L. (2006). *The planetocosmics code* (Tech. rep.). http://cosray.unibe.ch/~laurent/planetocosmics.

Dorman, L. I. (2004). *Cosmic rays in the Earth's atmosphere and underground, astrophysics and space science library: Vol. 303*. Dordrecht: Kluwer Academic Publishers.
Dunai, T. J. (2010). *Cosmogenic nuclides*. Cambridge: Cambridge University Press.
Eidelman, S., Hayes, K. G., & Olive, K. A. e. a. (2004). Review of particle physics. *Physics Letters B*, *592*, 1–5. doi:10.1016/j.physletb.2004.06.001.
Ferreira, S. E. S., & Scherer, K. (2006). Time evolution of galactic and anomalous cosmic-ray spectra in a dynamic heliosphere. *The Astrophysical Journal*, *642*, 1256–1266. doi:10.1086/501113.
Ferreira, S. E. S., Potgieter, M. S., & Scherer, K. (2007). Transport and acceleration of anomalous cosmic rays in the inner heliosheath. *Journal of Geophysical Research*, *112*, 11101.
Fichtner, H., Scherer, K., & Heber, B. (2006). A criterion to discriminate between solar and cosmic ray forcing of the terrestrial climate. *Atmospheric Chemistry and Physics Discussion*, *6*, 10811–10836.
Fichtner, H., Scherer, K., & Heber, B. (2010). The Hale period and climate forcing. In *38th COSPAR scientific assembly* (Vol. 38, p. 1772). http://adsabs.harvard.edu/abs/2010cosp...38.1722F
Florinski, V., & Pogorelov, N. V. (2009). Four-dimensional transport of galactic cosmic rays in the outer heliosphere and heliosheath. *The Astrophysical Journal*, *701*, 642–651.
Florinski, V., Balogh, A., Jokipii, J. R., McComas, D. J., Opher, M., Pogorelov, N. V., Richardson, J. D., Stone, E. C., & Wood, B. E. (2009). The dynamic heliosphere: outstanding issues. Report of working groups 4 and 6. *Space Science Reviews*, *143*, 57–83.
Garcia-Munoz, M., Mason, G. M., & Simpson, J. A. (1975). The anomalous He-4 component in the cosmic-ray spectrum at below approximately 50 MeV per nucleon during 1972–1974. *The Astrophysical Journal*, *202*, 265–275.
Geant4 Collaboration, Agostinelli, S., Allison, J., & Amako, K. e. a. (2003). Geant4-a simulation toolkit. *Nuclear Instruments and Methods in Physics Research A*, *506*, 250–303.
Gleeson, L. J., & Axford, W. I. (1968). Solar modulation of galactic cosmic rays. *The Astrophysical Journal*, *154*, 1011.
Goldhagen, P., Clem, J. M., & Wilson, J. W. (2004). The energy spectrum of cosmic ray-induced neutrons measured on an airplane over a wide range of altitude and latitude. *Radiation Protection Dosimetry*, *110*, 387–392.
Haigh, J. D. (2007). The Sun and the Earth's climate. *Living Reviews in Solar Physics*, *4*, 2.
Harrison, R. G., & Usoskin, I. (2010). Solar modulation in surface atmospheric electricity. *Journal of Atmospheric and Solar-Terrestrial Physics*, *72*, 176–182. doi:10.1016/j.jastp.2009.11.006.
Heber, B., & Potgieter, M. S. (2006). Cosmic rays at high heliolatitudes. *Space Science Reviews*, *127*, 117–194. doi:10.1007/s11214-006-9085-y.
Heber, B., Kopp, A., Gieseler, J., Müller-Mellin, R., Fichtner, H., Scherer, K., Potgieter, M. S., & Ferreira, S. E. S. (2009). Modulation of galactic cosmic ray protons and electrons during an unusual solar minimum. *The Astrophysical Journal*, *699*, 1956–1963. doi:10.1088/0004-637X/699/2/1956.
Herbst, K. (2011). *The interaction of cosmic rays with the Earth's magnetosphere and atmosphere – modeling the cosmic ray induced ionization and the production of cosmogenic nuclides Earth's magnetic field and atmosphere*. Dissertation, Universität Kiel, in progress.
Herbst, K., Kopp, A., Heber, B., Steinhilber, F., Fichtner, H., Scherer, K., & Matthiä, D. (2010). On the importance of the local interstellar spectrum for the solar modulation parameter. *Journal of Geophysical Research (Space Physics)*, *115*(A14), D00I20. doi:10.1029/2009JD012557.
Herbst, K., Kopp, A., & Heber, B. (2011). On the influence of the geomagnetic field geometry on the propagation of charged energetic particles. *AGU Fall Meeting Abstracts*, *Dec*, A6. http://adsabs.harvard.edu/abs/2011AGUFMSM12A..06H
IPCC (Ed.) (2007). *Intergovernmental panel on climate change: 2007, climate change 2007: Working group I report "The physical science basis"*. Cambridge: Cambridge University Press.
Kremer, J., Boezio, M., & Ambriola, M. L. e. a. (1999). Measurements of ground-level muons at two geomagnetic locations. *Physical Review Letters*, *83*, 4241–4244.
Lallement, R., Quémerais, E., Koutroumpa, D., Bertaux, J.-L., Ferron, S., Schmidt, W., & Lamy, P. (2010). The interstellar H flow: updated analysis of SOHO/SWAN data. *Twelfth International*

Solar Wind Conference, *1216*, 555–558.

Langner, U. W., Potgieter, M. S., & Webber, W. R. (2003). Modulation of cosmic ray protons in the heliosheath. *Journal of Geophysical Research (Space Physics)*, *108*, 14. doi:10.1029/2003JA009934.

Langner, U. W., Potgieter, M. S., Fichtner, H., & Borrmann, T. (2006a). Effects of different solar wind speed profiles in the heliosheath on the modulation of cosmic-ray protons. *The Astrophysical Journal*, *640*, 1119–1134. doi:10.1086/500162.

Langner, U. W., Potgieter, M. S., Fichtner, H., & Borrmann, T. (2006b). Modulation of anomalous protons: effects of different solar wind speed profiles in the heliosheath. *Journal of Geophysical Research (Space Physics)*, *111*, A01106, 1–14. doi:10.1029/2005JA011066.

Lopate, C. (2006). http://ulysses.sr.unh.edu/NeutronMonitor/. Space Physics Data System, University of New Hampshire.

Masarik, J., & Beer, J. (1999). Simulation of particle fluxes and cosmogenic nuclide production in the Earth's atmosphere. *Journal of Geophysical Research*, *104*, 12099–12112. doi:10.1029/1998JD200091.

Masarik, J., & Beer, J. (2009). An updated simulation of particle fluxes and cosmogenic nuclide production in the Earth's atmosphere. *Journal of Geophysical Research (Atmospheres)*, *114*(D13), D11103. doi:10.1029/2008JD010557.

Matthiä, D. (2009). *The radiation environment in the lower atmosphere—a numerical approach*. Dissertation, Universität Kiel.

Matthiä, D., Heber, B., Reitz, G., Meier, M., Sihver, L., Berger, T., & Herbst, K. (2009). Temporal and spatial evolution of the solar energetic particle event on 20 January 2005 and resulting radiation doses in aviation. *Journal of Geophysical Research (Space Physics)*, *114*(A13), A08104. doi:10.1029/2009JA014125.

McComas, D. J., Allegrini, F., Bochsler, P., Bzowski, M., Christian, E. R., Crew, G. B., DeMajistre, R., Fahr, H., Fichtner, H., Frisch, P. C., Funsten, H. O., Fuselier, S. A., Gloeckler, G., Gruntman, M., Heerikhuisen, J., Izmodenov, V., Janzen, P., Knappenberger, P., Krimigis, S., Kucharek, H., Lee, M., Livadiotis, G., Livi, S., MacDowall, R. J., Mitchell, D., Möbius, E., Moore, T., Pogorelov, N. V., Reisenfeld, D., Roelof, E., Saul, L., Schwadron, N. A., Valek, P. W., Vanderspek, R., Wurz, P., & Zank, G. P. (2009). Global observations of the interstellar interaction from the interstellar boundary explorer (IBEX). *Science*, *326*, 959.

McCracken, K. G. (2004). Geomagnetic and atmospheric effects upon the cosmogenic ^{10}Be observed in polar ice. *Journal of Geophysical Research (Space Physics)*, *109*, 4101. doi:10.1029/2003JA010060.

McCracken, K. G., & Beer, J. (2007). Long-term changes in the cosmic ray intensity at Earth, 1428–2005. *Journal of Geophysical Research (Space Physics)*, *112*, 10101. doi:10.1029/2006JA012117.

Mironova, I. A., Desorgher, L., Usoskin, I. G., Flückiger, E. O., & Bütikofer, R. (2008). Variations of aerosol optical properties during the extreme solar event in January 2005. *Geophysical Research Letters*, *35*, 18610. doi:10.1029/2008GL035120.

Neher, H. V. (1967). Cosmic-ray particles that changed from 1954 to 1958 to 1965. *Journal of Geophysical Research*, *72*, 1527.

Neher, H. V. (1971). Cosmic rays at high latitudes and altitudes covering four solar maxima. *Journal of Geophysical Research*, *76*, 1637–1651. doi:10.1029/JA076i007p01637.

Parker, E. N. (1965). The passage of energetic charged particles through interplanetary space. *Planetary and Space Science*, *13*, 9–49.

Phillips, F. M., Stone, W. D., & Fabryka-Martin, J. T. (2001). An improved approach to calculating low-energy cosmic-ray neutron fluxes near the land/atmosphere interface. *Chemical Geology*, *175*(3–4), 689–701. doi:10.1016/S0009-2541(00)00329-6.

Pilchowski, J., Kopp, A., Herbst, K., & Heber, B. (2010). On the definition and calculation of a generalised McIlwain parameter. *Astrophysics and Space Sciences Transactions*, *6*, 9–17. doi:10.5194/astra-6-9-2010.

Pogorelov, N. V., Borovikov, S. N., Zank, G. P., & Ogino, T. (2009). Three-dimensional features of the outer heliosphere due to coupling between the interstellar and interplanetary magnetic fields.

III. The effects of solar rotation and activity cycle. *The Astrophysical Journal*, *696*, 1478–1490.

Potgieter, M. S. (2010). The dynamic heliosphere, solar activity, and cosmic rays. *Advances in Space Research*, *46*, 402–412.

Richardson, J. D., & Stone, E. C. (2009). The solar wind in the outer heliosphere. *Space Science Reviews*, *143*, 7–20.

Rohs, S., Spang, R., Rohrer, F., Schiller, C., & Vos, H. (2010). A correlation study of high-altitude and midaltitude clouds and galactic cosmic rays by MIPAS-Envisat. *Journal of Geophysical Research (Atmospheres)*, *115*(D14), D14212. doi:10.1029/2009JD012608.

Scherer, K., & Ferreira, S. E. S. (2005a). A heliospheric hybrid model: hydrodynamic plasma flow and kinetic cosmic ray transport. *Astrophysics and Space Sciences Transactions*, *1*, 17–27.

Scherer, K., & Ferreira, S. E. S. (2005b). The heliomagnetic and solar-cycle related variations of the cosmic ray flux modeled with the BoPo-hybrid code. *Astronomy & Astrophysics*, *442*, L11–L14.

Scherer, K., Fichtner, H., Borrmann, T., Beer, J., Desorgher, L., Flückiger, E., Fahr, H. J., Ferreira, S. E. S., Langner, U., Potgieter, M. S., Heber, B., Masarik, J., Shaviv, N., & Veizer, J. (2006). Interstellar-terrestrial relations, the variable cosmic environments, the dynamic heliosphere, and their imprints on terrestrial archives. *Space Science Reviews*. doi:10.1007/s11214-006-9126-6.

Scherer, K., Fichtner, H., Ferreira, S. E. S., Büsching, I., & Potgieter, M. S. (2008). Are anomalous cosmic rays the main contribution to the low-energy galactic cosmic ray spectrum? *The Astrophysical Journal Letters*, *680*, L105–L108. doi:10.1086/589969.

Scherer, K., Fichtner, H., Effenberger, F., Burger, R. A., & Wiengarten, T. (2010). Comparison of different analytic heliospheric magnetic field configurations and their significance for the particle injection at the termination shock. *Astronomy & Astrophysics*, *521*, A1.

Scherer, K., Fichtner, H., Strauss, D. T., Ferreira, S. E. S., Potgieter, M., & Fahr, H. J. (2011). On cosmic ray modulation beyond the heliopause: where is the modulation boundary? *The Astrophysical Journal*. doi:10.1088/0004-637X/735/2/128.

Steinhilber, F., Abreu, J. A., & Beer, J. (2008). Solar modulation during the Holocene. *Astrophysics and Space Sciences Transactions*, *4*, 1–6.

Sternal, O. (2010). *Transport of galactic cosmic rays in different heliospheric magnetic field configurations*. Dissertation, Universität Kiel.

Strong, A. W., & Moskalenko, I. V. (1998). Propagation of cosmic-ray nucleons in the galaxy. *Astrophysical Journal*, *509*, 212–228. doi:10.1086/306470. http://adsabs.harvard.edu/abs/1998ApJ...509..212S

Svensmark, H. (1998). Influence of cosmic rays on Earth's climate. *Physical Review Letters*, *81*, 5027–5030.

Svensmark, H., & Friis-Christensen, E. (1997). Variation of cosmic ray flux and global cloud coverage-a missing link in solar-climate relationships. *Journal of Atmospheric and Terrestrial Physics*, *59*, 1225–1232.

Tinsley, B. A., Burns, G. B., & Zhou, L. (2007). The role of the global electric circuit in solar and internal forcing of clouds and climate. *Advances in Space Research*, *40*, 1126–1139.

Usoskin, I. G., & Kovaltsov, G. A. (2006). Cosmic ray induced ionization in the atmosphere: full modeling and practical applications. *Journal of Geophysical Research (Atmospheres)*, *111*, 21206. doi:10.1029/2006JD007150.

Usoskin, I. G., & Kovaltsov, G. A. (2008). Cosmic rays and climate of the Earth: possible connection. *Comptes Rendus. Géoscience*, *340*, 441–450.

Usoskin, I. G., Alanko-Huotari, K., Kovaltsov, G. A., & Mursula, K. (2005). Heliospheric modulation of cosmic rays: monthly reconstruction for 1951–2004. *Journal of Geophysical Research (Space Physics)*, *110*(A9), A12108. doi:10.1029/2005JA011250.

Usoskin, I. G., Solanki, S. K., Kovaltsov, G. A., Beer, J., & Kromer, B. (2006). Solar proton events in cosmogenic isotope data. *Geophysical Research Letters*, *33*, 8107. doi:10.1029/2006GL026059.

Vonmoos, M., Beer, J., & Muscheler, R. (2006). Large variations in Holocene solar activity: constraints from ^{10}Be in the Greenland Ice Core Project ice core. *Journal of Geophysical Research (Space Physics)*, *111*, 10105. doi:10.1029/2005JA011500.

Webber, W. R., & Higbie, P. R. (2003). Production of cosmogenic Be nuclei in the Earth's atmosphere by cosmic rays: Its dependence on solar modulation and the interstellar cosmic ray spectrum. *Journal of Geophysical Research (Space Physics)*, *108*, 1355. doi:10.1029/2003JA009863.

Webber, W. R., & Higbie, P. R. (2008). Limits on the interstellar cosmic ray electron spectrum below 1–2 GeV derived from the galactic polar radio spectrum and constrained by new Voyager 1 measurements. *Journal of Geophysical Research (Space Physics)*, *113*(A12), A11106. doi:10.1029/2008JA013386.

Webber, W. R., & Higbie, P. R. (2009). Galactic propagation of cosmic ray nuclei in a model with an increasing diffusion coefficient at low rigidities: a comparison of the new interstellar spectra with Voyager data in the outer heliosphere. *Journal of Geophysical Research (Space Physics)*, *114*(A13), A02103. doi:10.1029/2008JA013689.

Wissing, J. M., Kallenrode, M.-B., Wieters, N., Winkler, H., & Sinnhuber, M. (2010). Atmospheric ionization module osnabrck (aimos): 2. total particle inventory in the October–November 2003 event and ozone. *Journal of Geophysical Research*, *115*, A02308.

Chapter 5
Do Galactic Cosmic Rays Impact the Cirrus Cloud Cover?

Susanne Rohs, Reinhold Spang, Lars Hoffmann, Franz Rohrer, and Cornelius Schiller

Abstract Atmospheric ions produced through solar-modulated galactic cosmic rays can promote both the nucleation and the growth of aerosols. The potential impact on the cloud cover is subject of current debates. The CAWSES project SAGACITY (SAtellite and model studies of GAlactic cosmic rays and Clouds modulated by solar activITY) focuses on the statistical analysis of this link, using MIPAS-E satellite data. The extinction data, the cloud occurrence frequency, and the cloud index data from MIPAS-E are correlated with the data from the Climax neutron monitor. A superposed epoch analysis of 6 selected Forbush decrease events yields several weak but statistically significant correlations with an excess of positive cloud-GCR correlations. The impact of a 15 % increase in the Climax neutron monitor data is estimated to result in a small decrease in cloud index (corresponding to an increase in cloud opacity) which is most pronounced at 9 km altitude (−9 % to +0.5 %).

5.1 Introduction

The role of the solar activity in the formation of clouds is important to understand the natural climate variability. In the Fourth IPCC Assessment Report [*Solomon et al.*, 2007] the level of scientific understanding of the link between galactic cosmic rays (GCR) and clouds is classified as very low. A direct influence of the change in solar irradiance on climate is too weak to contribute significantly to the observed global warming. However, some researchers claim an indirect influence by GCRs [*Marsh and Svensmark*, 2000]. This high-energy particle radiation from space is the main source for ionisation in the stratosphere and troposphere. At strong solar activity the flux of GCRs, which reach the Earth, is reduced. The idea behind the

---

Cornelius Schiller is deceased.

S. Rohs (✉) · R. Spang · L. Hoffmann · C. Schiller
IEK-7, Forschungszentrum Jülich GmbH, 52425 Jülich, Germany
e-mail: s.rohs@fz-juelich.de

F. Rohrer
IEK-8, Forschungszentrum Jülich GmbH, 52425 Jülich, Germany
e-mail: f.rohrer@fz-juelich.de

F.-J. Lübken (ed.), *Climate and Weather of the Sun-Earth System (CAWSES)*, Springer Atmospheric Sciences,
DOI 10.1007/978-94-007-4348-9_5,

most common GCR-cloud hypothesis—the ion-aerosol "clear sky" hypothesis—is that a decrease in GCR would lead to a decrease in ionisation, and thereby a decrease in aerosol concentration. According to *Yu et al.* [2008] with fewer aerosols, fewer cloud condensation nucleis (CCN) develop causing an increase in droplet size and hence a decrease of cloud reflectivity and cloud lifetime. The effect of this ion induced nucleation (IIN) should be largest in the lower atmosphere where the nucleation rate of aerosols is limited by the ion concentration. Thus, an increase in solar activity would result in fewer low clouds, which have a net cooling effect, and, hence, in global warming. For the upper troposphere (UT) the prediction is less clear [*Enghoff and Svensmark*, 2008; *Arnold*, 2008] since the ionisation rate is high and particle formation is rather limited to the occurrence of H_2SO_4.

A correlation between solar modulated GCRs and low clouds has been derived from decadal changes in the data of the International Satellite Cloud Climatology Project (ISCCP) [*Marsh and Svensmark*, 2000]. Since then, there is an ongoing debate whether this correlation is real or incidentally caused by other climate variability factors. A comprehensive review is given by *Usoskin and Kovaltsov* [2008].

An abrupt decrease in cosmic ray intensity followed by a slow recovery typically lasting for several days is called a Forbush decrease (FD) event. It is caused by coronal mass ejections on the Sun. FD events occur several times per year, depending on solar cycle. They provide an opportunity to study the influence of the solar variability without superposition from other climate variability. Using data from the Moderate-resolution Imaging Spectroradiometer (MODIS) and ISCCP, *Svensmark et al.* [2009] report that the fraction of low clouds decreases after strong FD events. In contrast, no correlation was found between ISCCP data and GCR flux [*Calogovic et al.*, 2010] as well as MODIS data and GCR flux in Southern Hemisphere ocean regions [*Kristjánsson et al.*, 2008].

5.2 Data and Methods

In the CAWSES project SAGACITY we look for a possible connection between GCRs and high altitude clouds (cirrus + polar stratospheric clouds (PSC)) by using measurements of the satellite instrument MIPAS-E (Michelson Interferometer for Passive Atmospheric Sounding). In the analysed period from July 2002 to March 2004 approx. 425000 profiles were obtained. The long limb path through a cloud structure allows the detection of optically thin clouds [*Spang et al.*, 2004]. This and the good vertical resolution of 3 km are advantages compared to the nadir looking instruments used in the ISCCP. However, these advantages are partly compensated by the sparse horizontal sampling of 600 km.

In this study we use the extinction data (Ext), the cloud occurrence frequency (Occ) as well as the cloud index data (CI). CI is defined as the ratio of mean radiances from two different wavelength regions where one microwindow is dominated by trace gases and the other by aerosols and clouds. It is a measure for cloud transparency and decreases rapidly from background values of around 6–10 to 1–4 if a

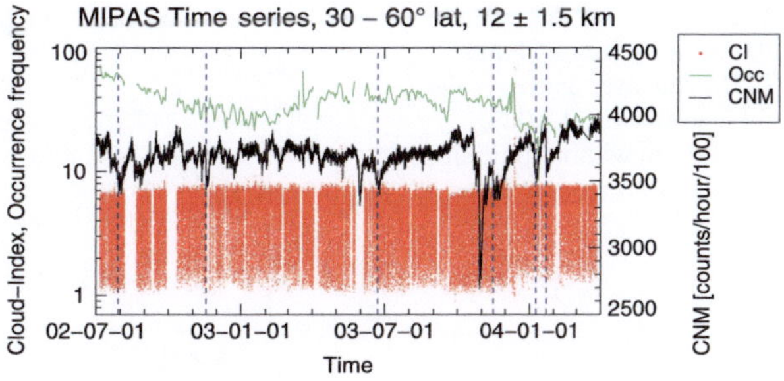

Fig. 5.1 CI, Occ [%] and CNM time series at 30–60°N and 11–13 km altitude from 2002-07 until 2004-03; *vertical lines* indicate the included FD events. This figure is from *Rohs et al.* [2010]

Table 5.1 Date and magnitude of the 6 investigated FD events

Date	Magnitude [%]
2002-07-29	−7.7
2002-11-17	−7.1
2003-06-22	−6.8
2003-11-15	−7.8
2004-01-09	−6.4
2004-01-22	−6.6

cloud is in the field-of-view. In this study CI < 1.5 is chosen as a threshold for optically thick conditions which obscure the view to lower altitudes. CI > 4.5 indicates cloud-free conditions. This method has been used already for analyses of PSC and cirrus clouds in the MIPAS data [e.g. *Spang et al.*, 2004]. We define Occ as the number of cloud events with CI < 4.5 divided by the total number of CI observations within a 3 km altitude and 30° latitude grid box in a time-window of one day.

The Ext data are caused by the background aerosol and by emissions from optically thin or even subvisible clouds. An enhanced extinction corresponds to enhanced aerosol and/or cloud load [*Hoffmann et al.*, 2008]. The retrieval is stopped at the highest tangent altitude where CI is below 2. In our analysis we use the logarithm of the extinction data log(Ext) at a wavenumber region from 832.3–834.4 cm^{-1}.

For a latitudinal resolution of 30° and a vertical resolution of 3 km we correlate these data with the neutron count data (CNM) from the Climax station (39°N, 254°E).

Figure 5.1 shows as an example the CI, Occ and CNM time series between 30°N and 60°N at an altitude interval from 12 ± 1.5 km. The CI data scatter between 1 (thick cirrus) and 10 (cloud free), the Occ data vary between 7 and 80 %.

The focus of this study lies on a superposed epoch (SPE) analysis of the 21 day periods around Forbush decrease (FD) events. A FD event was defined as a >5 % suppression of the hourly count rate of the CNM data with respect to the 100-day running mean. FD events accompanied by solar energetic particle events are omitted from the analysis. During the analysed time period 6 pure FD events occurred (Table 5.1 and blue lines in Fig. 5.1).

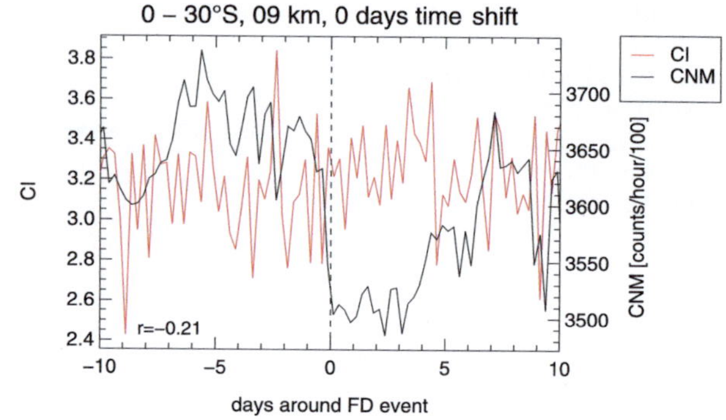

Fig. 5.2 20 day periods of the 6 hourly mean CI and CNM data averaged over the 6 FD events for 0–30°S, 9 ± 1.5 km without time shift. This figure is from *Rohs et al.* [2010]

Based on a variance analysis, we chose 6-hour averages of the CI and log(Ext) data. For the SPE analysis, we calculate the correlation coefficients r between the binned MIPAS cloud parameter and the binned CNM data averaged over all 6 FD events. In order to allow for nucleation and growth of ice particles but also to identify statistical artifacts, these studies were performed with time lags ranging from −5 to 5 days. A time shift of +5 days means that we look for a cloud signal 5 days after the occurrence of a FD event. We regard a significance level > 95 % as significant and a significance level between 90–95 % as marginally significant. For a detailed description see *Rohs et al.* [2010].

5.3 Results

With the cloud index (CI), the logarithm of the extinction (log(Ext)) and the cloud occurrence frequency (Occ) data we have a very sensitive tool for the investigation of high clouds. While CI and Occ are directly related to cloud effects, log(Ext) in the UT/LS region is also sensitive to aerosol effects as precursors for clouds. The CI and Occ data are calculated for an altitude range from 9–18 km altitude except for the midlatitudes and polar summer regions at 15 and 18 km altitude where no clouds exist. The log(Ext) data are calculated for an altitude range from 12–24 km altitude.

In Fig. 5.2 the CI and CNM data—averaged over the 6 FD events—from 10 days prior to the FD event to 10 days past the event, are exemplified for 0–30°S, 9 ± 1.5 km without time shift. Albeit a relation between the CI and CNM data is hardly visible, the SPE analysis delivers a nominally significant correlation of $r = -0.21$. However, due to the low value of r, this relation can only explain 4.8–7.8 % of the observed variance.

CI measures the cloud transparency. Since a low CI belongs to an optically thick cloud cover, a negative CI-CNM correlation coefficient corresponds to a positive cloud-CNM correlation coefficient and vice versa. In Fig. 5.3a the inverse of the correlation is plotted for a better comparison with the log(Ext)-CNM and Occ-CNM correlations. Blue points indicate negative correlations between cloud cover and

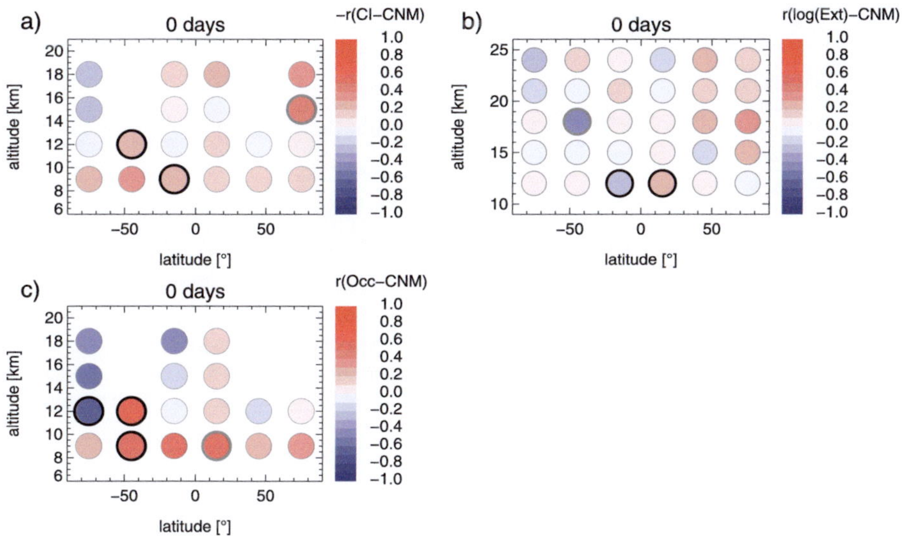

Fig. 5.3 Latitude-altitude distribution of r between (**a**) CI, (**b**) log(Ext), and (**c**) Occ and CNM for the SPE analyses without time shift—*grey circles* indicate marginally significant correlations, *black circles* indicate significant correlations; for better comparison with log(Ext) and Occ the negative of the CI-CNM correlation is plotted. This figure is from *Rohs et al.* [2010]

CNM and red points positive correlations. From the CI-CNM analysis, we derive an excess of positive cloud-GCR correlations with only two significant and one marginally significant positive correlations.

The log(Ext) data are sensitive to enhancement in the aerosol amount and clouds. The SPE analysis of log(Ext) and CNM (Fig. 5.3b) yields positive and negative correlations in nearly the same amount with one positive and one negative significant correlation. Occ is a measure for the degree of cloudiness. For Occ we find higher correlations than for CI and log(Ext) (Fig. 5.3c). However, the Occ-CNM correlations are less reliable than the CI—and log(Ext)-CNM—correlations (Sect. 5.4). The distribution of the Occ correlation coefficients is similar to that of the CI correlation coefficients with an excess of positive correlations. Note that CI and Occ are not independent, because an increasing CI corresponds to a decreasing Occ.

5.4 Discussion

While from previous studies (Sect. 5.1) even the sign of a possible GCR-cloud correlation in the UT remains uncertain, our findings from the CI-CNM correlation and the Occ-CNM correlation indicate that positive correlations prevail. In contrast, the log(Ext)-CNM correlation shows a more balanced picture with nearly equal amounts of positive and negative correlations—also at high altitudes (21–24 km), where log(Ext) is a pure parameter for aerosols. Since a cloud or aerosol occurrence

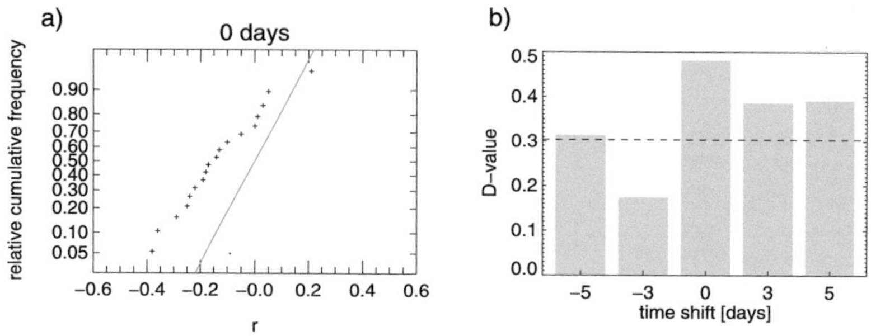

Fig. 5.4 (**a**) Relative cumulative frequency of the CI-CNM correlation coefficients; the *solid line* indicates a random distribution, (**b**) Kolmogorov-Smirnov test statistics D for a confidence level of 95 % for time shifts from –5 to 5 days. The critical D-value for CI is $D_{crit} = 0.304$. This figure is from *Rohs et al.* [2010]

might need some days to respond to a change in GCRs, we investigated time shifts of 0–5 days (not shown). Overall, time shifts do not change the results significantly.

Due to the large amount of MIPAS-E data points scattered over the whole latitude band, meteorological variability should be evened out to a large extent. Only the analysis of the polar latitudes might still be influenced by meteorological variability. So, we compensate the low number of only 6 FD events.

The derived correlations are weak and close to the chosen significance value. Therefore, it is crucial to assess, whether the obtained correlations are robust enough to support the GCR-cloud hypothesis. For this purpose, we perform a Kolmogorov-Smirnov test [*Rohs et al.*, 2010, and references therein].

Figure 5.4a shows the cumulative fraction of the obtained correlation coefficients for CI without time shift in comparison with the random distribution. For negative time shifts all obtained D-values (see Fig. 5.4b) are located below D_{crit} (or only slightly above in case of -5 days time shift). For time shifts of 0 to $+5$ days, however, D-values well above D_{crit} are obtained. Thus, for the CI-CNM and also for the log(Ext)-CNM correlations (not shown), we derive a probability of more than 95 % that for positive time shifts between 0 and 5 days the entity of the correlation coefficients is not randomly distributed. For the Occ-CNM correlation, the Kolmogorov-Smirnov test is less explicit. This might be due to the fact that we have used daily means of the Occ data and, hence, have only 21 instead of 84 data points for each correlation.

Additionally, we performed a time scale analysis of r analogous to the variance analysis. Figure 5.5 shows as an example the evolution of r(CI, CNM) between 30 and 60°N at an altitude of 12 ± 1.5 km. We find that about half of the correlation can be attributed to the low frequency variability during the 20-day windows and half to the high frequency. During the step from 10 days to the full 20 days, which encloses the Forbush decrease, in 18 of 20 cases (including the cases, where r is not significant or even positive), r drops strongly. Hence, the shift of r to negative values which is visible in the Kolmogorov-Smirnov test (see Fig. 5.4a) results from

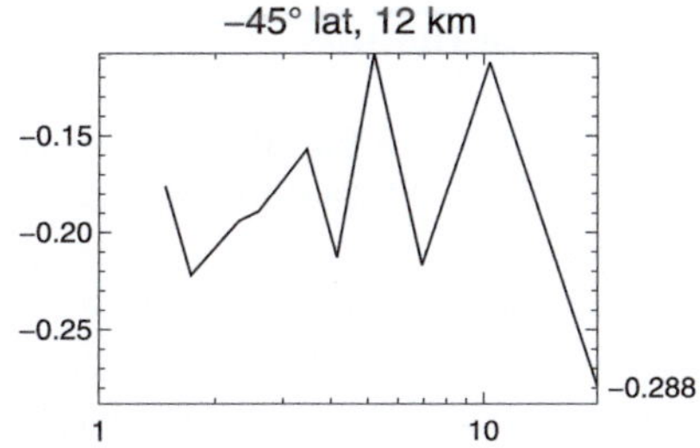

Fig. 5.5 Time scale analysis of the correlation between the 6 h mean CI and CNM data of the SPE analysis between 30 and 60°S at an altitude range from 12 ± 1.5 km

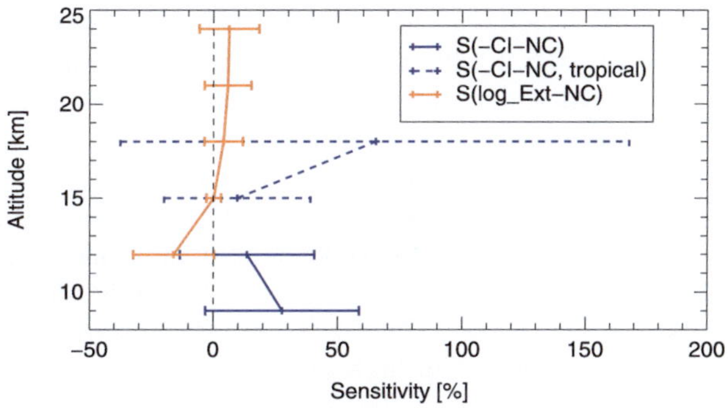

Fig. 5.6 Sensitivity S of the impact of GCRs on the cloud parameters CI and log(Ext); latitudinal range for CI at 9, 12 km altitude: 90°S–90°N, at 15, 18 km altitude: 30°S–30°N; latitudinal range for log(Ext) at all altitudes: 90°S–90°N. This figure is from *Rohs et al.* [2010]

the low frequency variability. This might be an indication that the correlation is not coincidal but stems from a GCR—high cloud interaction.

Whereas the quality of the Occ-CNM correlations is not sufficient, we can estimate the confidence intervals of a possible impact from GCRs on CI and log(Ext). For this purpose we calculate the correlation coefficients of the SPE analysis between CI and CNM at 9 and 12 km altitude and between log(Ext) and CNM at all altitude intervals without separating into latitude bins. For the correlation of CI and CNM at 15 and 18 km altitude, we restrict the latitude range to 30°S–30°N, where high clouds exist. For details see *Rohs et al.* [2010].

The highest sensitivities to GCR flux are found at low altitudes (Fig. 5.6). At 9 km altitude the calculated best estimate of the sensitivity yields $S_{best} = -28\ \%$ with $S_{min}(1\sigma) = -58\ \%$ and $S_{max}(1\sigma) = +3\ \%$. That means, an increase in GCRs which produces a 15 % increase in CNM (typical amplitude of CNM for a solar cycle) would lead to a change in CI between −9 % and +0.5 % with a best estimate of −4.1 %. At 12 km altitude the sensitivity is $S_{best} = -13\ \%$ (S_{min}: −40 %, S_{max} + 13 %). At 15 km altitude we obtain only a weak sensitivity ($S_{best} = -10\ \%$). At 18 km the sensitivity is $S_{best} = -65\ \%$ but with a very broad range. For log(Ext) the best estimate of the sensitivity at 12 km altitude is $S_{best} = -16\ \%$ (S_{min}: −32 %, S_{max} 0 %). So, a 15 % increase in CNM would lead to a −2.4 % (−5–0 %) decrease in log(Ext). This is a discrepancy to the findings from the CI-CNM correlation. At higher altitudes the sensitivity of the log(Ext)-CNM effect shifts towards weak positive values, which is in accordance to the findings from the CI-CNM analysis.

5.5 Conclusions

In this study, we find that a positive correlation between thin high clouds and cosmic rays prevails and that this correlation is especially pronounced at 9 km. In the log(Ext) data at 21 and 24 km altitude, positive and negative correlations occur in nearly equal amounts. With a Kolmogorov-Smirnov test we prove that the weak GCR—high cloud (or high cloud plus aerosol) effect in the MIPAS-E data is real and not coincidental. By calculating the impact from a 15 % increase in CNM (typical amplitude during one solar cycle) on clouds from the MIPAS-E measurements we derive small but not significant decreases in CI (which has low values for optically thick clouds). The most pronounced impact on CI is found at 9 km altitude (-9 % to $+0.5$ %). For the log(Ext) data at 12 km altitude a decrease of -5 to 0 % is calculated which shifts towards weak positive values at higher altitudes. Further investigations with a prolonged MIPAS-E data set should narrow down the range of a possible GCR—high cloud effect. However, albeit a zero-effect can not be excluded, the best estimate of the CI-CNM analyses indicates that—according to our study—an increase in GCRs leads to a small increase in high cloud cover. While an increase of lower clouds results in a cooling, the effect of changing high cloud cover on the radiative forcing is not well known: Most theories derive a warming potential by increasing high clouds, but the magnitude of this effect depends on the microphysical properties of the cirrus and can in some cases even change its sign. Therefore, we cannot provide a quantitative estimate of the radiation impact by the enhanced high cloud cover by GCR, in particular as our statistical analysis of satellite data does not allow to derive more detailed parameters of the nature of GCR-induced high clouds.

Acknowledgements This work was funded by the Deutsche Forschungsgemeinschaft under the SPP 1176 (CAWSES) grant SCHI-872/1-1 (SAGACITY). The Climax-Neutron count data are generated within the NSF Grant ATM-0339527 and the MIPAS level 1b and level 2 data are provided by ESA. With great sadness we mourn the death of our dear colleague Cornelius Schiller, who passed away on 3 March 2012.

References

Arnold, F. (2008). Atmospheric ions and aerosol formation. *Space Science Reviews*, *137*(1–4), 225–239.

Calogovic, J., Albert, C., Arnold, F., Beer, J., Desorgher, L., & Flueckiger, E. O. (2010). Sudden cosmic ray decreases: No change of global cloud cover. *Geophysical Research Letters*, *37*, L03802. doi:10.1029/2009GL041327.

Enghoff, M. B., & Svensmark, H. (2008). The role of atmospheric ions in aerosol nucleation—a review. *Atmospheric Chemistry and Physics*, *8*(16), 4911–4923.

Hoffmann, L., Kaufmann, M., Spang, R., Müller, R., Remedios, J. J., Moore, D. P., Volk, C. M., von Clarmann, T., & Riese, M. (2008). ENVISAT MIPAS measurements of cfc-11: retrieval, validation, and climatology. *Atmospheric Chemistry and Physics*, *8*(13), 3671–3688.

Kristjánsson, J. E., Stjern, C. W., Stordal, F., Fjaeraa, A. M., Myhre, G., & Jonasson, K. (2008). Cosmic rays, cloud condensation nuclei and clouds—a reassessment using MODIS data. *Atmospheric Chemistry and Physics*, *8*(24), 7373–7387.

Marsh, N. D., & Svensmark, H. (2000). Low cloud properties influenced by cosmic rays. *Physical Review Letters*, *85*, 5004–5007. doi:10.1103/PhysRevLett.85.5004.

Rohs, S., Spang, R., Rohrer, F., Schiller, C., & Vos, H. (2010). A correlation study of high-altitude and midaltitude clouds and galactic cosmic rays by MIPAS-ENVISAT. *Journal of Geophysical Research*, *115*, D14212. doi:10.1029/2009JD012608.

Solomon, S., Qin, D., Manning, M., Chen, Z., Marquis, M., & Averyt, M. T. K. B. (Eds.) (2007). *IPCC, 2007: Climate change 2007: the physical science basis. Contribution of working group I to the fourth assessment report of the intergovernmental panel on climate change*. Cambridge: Cambridge University Press.

Spang, R., Remedios, J. J., & Barkley, M. P. (2004). Colour indices for the detection and differentiation of cloud types in infra-red limb emission spectra. *Advances in Space Research*, *33*(7), 1041–1047.

Svensmark, H., Bondo, T., & Svensmark, J. (2009). Cosmic ray decreases affect atmospheric aerosols and clouds. *Geophysical Research Letters*, *36*, L15101. doi:10.1029/2009GL038429.

Usoskin, I. G., & Kovaltsov, G. A. (2008). Cosmic rays and climate of the Earth: Possible connection. *Comptes Rendus. Géoscience*, *340*(7), 441–450.

Yu, F., Wang, Z., Luo, G., & Turco, R. (2008). Ion-mediated nucleation as an important global source of tropospheric aerosols. *Atmospheric Chemistry and Physics*, *8*(9), 2537–2554.

Chapter 6
Laboratory Experiments on the Microphysics of Electrified Cloud Droplets

Daniel Rzesanke, Denis Duft, and Thomas Leisner

Abstract The global electric circuit (GEC) is one of the candidates for a coupling of terrestrial climate with solar activity [Friis-Christensen, in Space Sci. Rev. 94(1–2): 411–421, 2000]. It has been suggested that vertical electric currents in the atmosphere can modify cloud microphysics and thereby alter the properties of the earth's cloud system [Tinsley, in Space Sci. Rev. 98:16889–16891, 2000]. In the framework of the Deutsche Forschungsgemeinschaft (DFG) priority program *CAWSES* we conducted laboratory experiments which quantify the influence of electric charges on the microphysics of cloud droplets in order to assess the atmospheric relevance of this link between the GEC and the cloud system. More specific, we quantify the influence of charges, electric fields and ionizing radiation on the heterogeneous and homogeneous nucleation in cloud droplets. These experiments were carried out on individual electrified cloud droplets using an electrodynamic balance enclosed by a miniaturized climate chamber to provide realistic atmospheric conditions. Several effects that could link the electrical state of the atmosphere to cloud microphysics have been investigated and quantified. While no direct effect of cloud droplet charge on homogeneous freezing was found, we were able to confirm and quantify the enhanced scavenging of aerosol particles by charged cloud droplets. Together with the first direct measurement of size dependent contact freezing probabilities it is now possible to quantify the role of charges for cloud glaciation in cloud models. Additionally, a substantial effect of cloud droplet and ice charge on the vapor pressure of these cloud elements has been found that has so far not been taken into account in cloud modeling.

Glossary

C	*Cunningham* factor
D^*	Diffusivity of water in gas phase (corrected for kinetic effects)
L	Latent heat of evaporation (water)
M_W	Mole mass of water
N_c	Number concentration of aerosol particles

D. Rzesanke · D. Duft · T. Leisner (✉)
Institute for Meteorology and Climate Research – Atmospheric Aerosol Research (IMK-AAF), Karlsruhe Institute of Technology (KIT), P.O. Box 3640, 76021 Karlsruhe, Germany
e-mail: Thomas.Leisner@kit.edu

F.-J. Lübken (ed.), *Climate and Weather of the Sun-Earth System (CAWSES)*, Springer Atmospheric Sciences,
DOI 10.1007/978-94-007-4348-9_6,

P_f	Freeze rate
P_{scav}	Scavenging rate (actual collision rate of aerosol particles with the droplet)
Q_d	droplet charge
Q_p, q	particle charge
R_d, r	Droplet radius
R_p	Particle radius
R	Gas constant
RH_W	Relative Humidity
T_∞	Ambient temperature
$e_{s,i}$	Saturation pressure over plane clean ice surface
$e_{s,l}$	Saturation pressure over plane clean liquid water surface
k_a^*	Heat conductivity of air (corrected for kinetic effects below mean free path)
k_B	Boltzmann's constant
r_{eq}	Terminal radius of a charged droplet in a subsaturated environment
v_0	Relative velocity between air flow and droplet
$\mathbf{p}$	Water dipole moment
ε_0	Vacuum permittivity
η	Scavenging efficiency
ρ_W	Liquid water density
σ_d	Geometrical droplet cross section (πR_d^2)
σ_{scav}	Scavenging cross section (effective droplet cross section)
μ	Viscosity of air

6.1 Introduction

More than two hundred years ago, William Herschel [1796] was the first to propose a link between solar activity and climate. He realized that a high count of visible sunspots coincided with warmer weather and higher yields of wheat crops. Since then, numerous investigations on this topic have been performed. See for example *Dickinson* [1975]; *Friis-Christensen and Lassen* [1991]; *Hoyt and Schatten* [1997] and the references therein. With the advent of highly sensitive satellite measurements [*Sonett et al.*, 1991], it became obvious, that over the solar cycle, the solar irradiance is variable only within 0.1 % and cannot account for the observed changes of the earth's climate. Modern theories of solar driven climate variability usually invoke either effects of the much more variable ultraviolet radiation, the solar wind or the solar magnetic field on the energy budget of the earth [for a review see *Cubasch et al.*, 2000; *Gray et al.*, 2010]. Variations of the solar magnetic field also modulate the flux of galactic cosmic rays (GCR) which reach the earth [*Fisk et al.*, 1998]. However, the energy carried by the solar wind or the GCRs is much too low to have a direct effect on the climate [*Bazilevskaya*, 2000] thus an amplification mechanism has to be invoked for this link. Cloud microphysical processes might serve as such an amplification mechanism and therefore could constitute one of the links between solar variability and climate. Various microphysical mechanisms, which depend on the electrical state of the atmosphere, have been proposed to link cloud processes to solar variability. As only the very high energy GCR particles can penetrate down into the troposphere, these mechanisms can be classified into two groups: *A:* Mechanisms which rely on the direct interaction between GCR and the aerosol and cloud system and *B:* mechanisms which invoke an indirect link between the ionizing radiation and the cloud system. In the former case, the formation of large ions, aerosol particles, cloud droplets and ice crystals is considered as a direct consequence of the effect of the ionizing radiation on the cloud systems [*Marsh and Svensmark*, 2000; *Carslaw et al.*, 2002; *Kirkby et al.*, 2011]. Among the latter, a detailed mechanism has been proposed [*Tinsley and Deen*, 1991; *Tinsley*, 2000] which suggests that clouds are influenced by the vertical electric current in the atmosphere, which itself is modulated by the flux of the ionizing radiation. Some key features of this model are illustrated in Fig. 6.1 which shows a schematic representation of the global electric circuit. The flux of highly energetic particles of galactic origin (Fig. 6.1, blue arrows) is modulated by the variations in the solar magnetic field. This in turn influences the concentration of ion pairs in the upper atmosphere which are created by photoionization and subsequent scavenging of the photoelectrons by molecular oxygen. The ion pairs are separated in the fair weather electric field of the earth, which is created and maintained predominantly by the electrical activity of large storms within the Intertropical Convergence Zone.

Accelerated by the electric field of the earth the positively charged molecular ions or ionic clusters move towards the earth's surface. When the ions encounter a neutral cloud droplet, the interaction with the image charge induced on the droplet results in mutual attraction and if close enough in a collision. This process ultimately leads to an electrification of the droplets on the cloud tops and continues until the attractive image force is counterbalanced by the repulsive Coulomb force between the charged droplets and the ions. Typical charges on stratiform cloud droplets have been measured to be up to several hundred elementary charges [*Pruppacher and Klett*, 2004], while the typical electric fields range between 500 V/m and 10000 V/m which compares to the fair weather electric field of 120 V/m [*MacGorman and Rust*, 1998]. Through this scavenging process, the mobility of the charge carriers is strongly reduced, so that in the presence of clouds the global electric circuit of the earth is locally interrupted.

It is clear from the above, that cloud electrification depends on the strength of the vertical atmospheric current, which itself depends on the amount of ionizing radiation in the upper atmosphere and thereby on solar activity. While this part of the proposed link between solar activity and climate is largely undisputed, it is much less clear which consequences a change in cloud electrification has for cloud microphysics, cloud lifecycle and optical properties and ultimately on climate.

It is assumed, that upon evaporation of cloud droplets on the lee side of clouds, charged cloud condensation nuclei (CCN) are released, and that their ability to induce contact or immersion freezing when reentering another cloud depends on their charge state. Under such circumstances, the flux of ionizing radiation alters the cloud microphysics and indirectly modulates the development, lifetime and optical properties of clouds. Many elementary processes in cloud microphysics like freezing, scavenging and evaporation might depend on the charge state of the aerosol.

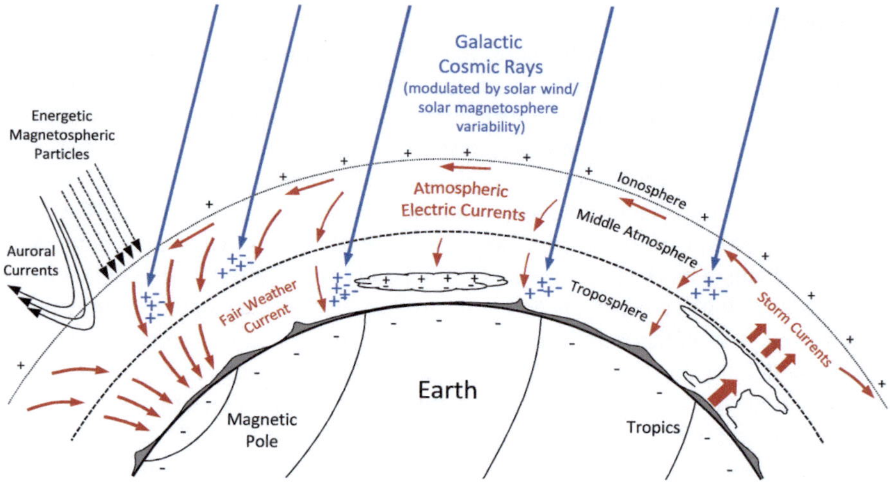

Fig. 6.1 Schematic representation of the global electric circuit (GEC) and its interaction with energetic cosmic particles and the tropospheric cloud system. Galactic Cosmic Rays (GCR) (*blue arrows*) and other highly-energetic particles of solar and cosmic origin (*black arrows*) create ion pairs when penetrating the atmosphere and by that change the ionospheric electrical potential and the atmospheric vertical column resistance. The electrical charges transported by the vertical currents (indicated by *red arrows*) accumulate on airborne aerosols as well as on droplets and ice crystals at cloud tops and bottoms and potentially change the cloud physical properties, e.g. cloud lifetime, cloud optical density or precipitation. Variations in the solar wind and in the extend of the solar magnetosphere modulate the atmospheric flux of energetic cosmic particles and in this way may exert an indirect influence on the earth's cloud system and climate [figure adapted from *Tinsley and Yu*, 2004, see here for more details]

Some possible charge dependent cloud processes are discussed in *Pruppacher and Klett* [2004, Chap. 18.6, p. 827ff.].

In order to assess the importance of charges for the respective efficiencies and process rates, we conducted laboratory experiments that model some of these interactions between aerosol particles and charged cloud droplets. The experiments were performed using individual water droplets levitated in an electrodynamic balance under a range of conditions relevant for tropospheric clouds. As ice formation is one of the most important but also most elusive and sensitive processes that affect cloud development, lifetime and radiative properties, we mainly focused on ice formation and ice properties. After a short review of the experimental techniques, we report the results of four different types of experiments. In Sect. 6.3.1 we consider the charge dependence of homogeneous nucleation rates of supercooled cloud droplets, followed by a quantification of the attraction of aerosol particles by a charged droplet in Sect. 6.3.2. The effect of droplet charge on the vapor pressure of water droplets and ice crystals is detailed in Sect. 6.3.3, while the ice nucleation efficiency of heterogeneous ice nuclei is investigated in Sect. 6.3.4.

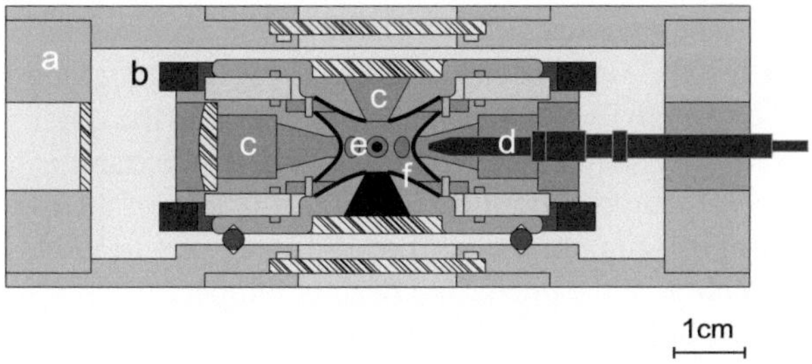

Fig. 6.2 Cross section through the levitator placed in a vacuum chamber (**a**) for electric and thermal insulation. Further noticeable parts are a cooling circuit (**b**), ports for optical and mechanical access (**c**), droplet dispenser (**d**) and hyperbolical shaped electrodes (**f**). The levitated particle (**e**) is centered within the electrodes

6.2 Experimental Setup

All experiments described in this work were performed using individual charged cloud droplets (diameter 20 μm–100 μm) levitated in an electrodynamic balance (EDB, see early works by *Paul and Steinwedel* [1953]; *Fischer* [1959]; *Wuerker et al.* [1959] and reviews of the technique by *March and Hughes* [1989] and *Davis* [1997]). The electrodynamic levitation technique allows a contact free storage of charged particles for prolonged periods of time without contact to any wall or substrate under conditions relevant for tropospheric clouds. This technique is therefore well suited for the investigation of airborne aerosol or cloud particles.

As details of the setup have been given in previous publications [*Duft and Leisner*, 2004a; *Achtzehn et al.*, 2005], only the main features are recalled here. A cross section through the EDB setup is given in Fig. 6.2. The electrodynamic levitator is of the classical type consisting of three rotationally symmetric hyperboloids, which form a toroidal body electrode and two endcap electrodes. The body electrode carries the AC voltage that is essential for the levitation, while a bipolar DC potential is applied to the bottom and top endcap electrodes, which is used to balance the gravitational force on the droplet. In order to ensure a homogeneous temperature field across the device the electrodes are machined from massive copper blocks and electrical insulators with high thermal conductivity are used. The exterior of the levitator has the shape of an octagon and carries eight circular ports which are used as optical and mechanical access points to the interior of the levitator (cf. Fig. 6.2c/d). Bottom and top endcap electrodes carry openings through which a laminar gas flow to and from the trap can be maintained.

Individual droplets in the diameter range between 20 μm–100 μm are generated on demand by a piezo-driven droplet dispenser [*Gast and Fiehn*, 2003] (cf. Fig. 6.2d) at room temperature. The droplets are charged inductively by an electric field at the tip of the dispenser during their creation and are introduced ballistically into the EDB which is held at the constant temperature of interest. By choosing a fixed delay time of droplet injection relative to the phase of the AC voltage each individual droplet can be decelerated and trapped at the center of the levitator.

The body electrode is cooled either by a liquid-nitrogen cooled cold finger (Cryovac Konti and PK2001) or by a closed cycle refrigerator (Julabo FP50-HE). Due to the good heat conductivity between the electrodes and the efficient thermal insulation within a vacuum chamber, the temperature of the levitator is homogeneous to within ± 1 K. Precise temperature measurements on each of the electrodes allow to reconstruct the droplet temperature with even higher precision by thermodynamic modeling. In a stagnant atmosphere, the droplet temperature can be known within an error of ± 0.3 K, while in a gas flow, the temperature uncertainty is estimated to be ± 0.5 K.

Inside the EDB, the droplet is illuminated with a He-Ne laser beam. In order to monitor the vertical position of the droplet, the elastically scattered light is imaged by an optical lens onto a linear CCD array which is mounted vertically outside the vacuum chamber. The droplet position can be locked to the center of the EDB by feeding an error signal from a proportional-integral-derivative controller (PID controller) back to the DC voltage supply connected to the endcap electrodes. From that balance voltage, the charge to mass ratio of the droplet and the optimal frequency and amplitude of the AC trapping voltage are calculated online and fed to the AC voltage generator. Thereby the droplet can be held stable even if its mass or charge changes during the measurement. Light scattered in an angular cone centered around $\pi/4$ is detected by a two-dimensional CCD array. By comparison of the scattering pattern with *Mie* theory [*Bohren and Huffman*, 2004] using tabulated values for the refractive index of supercooled water [*Duft and Leisner*, 2004a], the size of the droplet can be determined to within 1 %.

For the experiments presented in Sects. 6.3.2 and 6.3.4, a well defined aerosol flow through the EDB was realized. As this is a new extension to the EDB, it is described here in some more detail. A schematic diagram of the aerosol system is shown in Fig. 6.3. Aerosol particles can be dry or wet dispersed in an electrospray generator (TSI EAG 3480) an atomizer or a fluidized bed generator (both home build). After diffusive drying and charging into a well-defined charge distribution [*Fuchs*, 1963; *Wiedensohler*, 1988] by an ionizing source (Kr-85 or Po-210) the aerosol particles are size selected in an electrostatic classifier (TSI Electrostatic classifier 3080L). The resulting flow of quasi-monodisperse aerosol is precooled to the temperature of the trap and a controlled fraction of the flow is passed vertically through the trap. Computational fluid dynamics calculations were performed to check for laminar flow conditions and to model the temperature of the droplet suspended within the flow. After passing through the trap, the aerosol number density is monitored by an Condensation Particle Counter (CPC, TSI 3776) or for higher concentrations (more than 10^4 cm^{-3}) by a *Faraday Cup* Electrometer (FCE, TSI 3068B).

Depending on the aerosol particle size range used in a particular experiment, the total flow rate of the aerosol generation and the aerosol preparation section of the flow system had to be varied. For a particular aerosol particle size the total flow rate was held constant through the experiment. The aerosol system was typically operated at a constant total flow rate of 5–10 l/min in the aerosol generation section and 0.1–1.5 l/min in the aerosol preparation section. The flow through the EDB

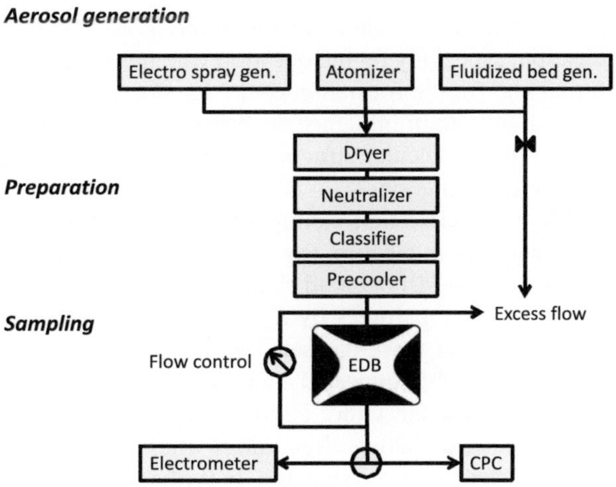

Fig. 6.3 Schematic diagram of the used aerosol system. Different generator types are necessary for wet and dry dispersion of diverse types of airborne particles. For description of the apparatus see text

was limited by the stability of the droplet levitation and could be varied between 0 and 0.3 l/min resulting in a maximum flow speed of 65 cm/s which corresponds to the terminal fall velocity of a 140 μm water droplet. Typical aerosol number concentrations used in our experiments varied between 10 cm^{-3} and 300 cm^{-3} for size selected mineral dust and between 20000 cm^{-3} and 25000 cm^{-3} for silica-sphere particles while aerosol particle diameters ranged from 29 nm to 2 μm.

6.3 Experimental Results

6.3.1 Homogeneous Freezing of Supercooled Cloud Droplets as a Function of Charge

One of the most important and direct cloud microphysical processes that are relevant for the climatic impact of clouds is the freezing of cloud droplets. It controls cloud dynamics by the release of latent heat and by the reduction of the vapor pressure of ice as compared to water which leads to growth of the ice phase on expense of the liquid droplets (*Bergeron-Findeisen*-Process).

Homogeneous freezing is an activated process which proceeds via the formation of a germ, i.e. a nucleus of critical size that represents a barrier in the free energy of the process. Therefore, pure water can be readily supercooled and freezes homogeneously only at temperatures below −36 °C [*Krämer et al.*, 1996]. In the atmosphere, homogeneous freezing can occur mainly at the top of deeply convective clouds, where the temperature can drop below −36 °C or in solution droplets that form cirrus clouds at even lower temperatures. Otherwise, atmospheric ice formation is usually induced by solid particles immersed in supercooled cloud droplets (immersion freezing), or by contact of these droplets with interstitial ice nuclei (contact freezing).

There is considerable debate on the molecular mechanism of homogeneous nucleation, especially on the question whether the surface of the droplet may be a preferred site for the formation of the ice nucleus. In that case, electric charges might have an important influence as they are located at the surface of liquid droplets and modify the molecular structure in their proximity and thus may either promote or inhibit ice formation.

Many experimental examinations were analyzed assuming volume dominated homogeneous nucleation (by *Taborek* [1985] using emulsion samples, by *DeMott and Rogers* [1990] in cloud chambers, by *Wood et al.* [2002] in droplet trains and in levitated single particles by *Krämer et al.* [1996, 1999] and *Stöckel et al.* [2002]). A reanalysis of these datasets by *Tabazadeh et al.* [2002a, 2002b] and *Djikaev et al.* [2002, 2003] suggested that freezing in a small layer of the droplets surface might be dominant under atmospheric conditions. This conclusion was supported by molecular dynamics simulations where *Vrbka and Jungwirth* [2006] found hints for subsurface initiated freezing. In further experiments by *Duft and Leisner* [2004b] and *Benz et al.* [2005] volume dominated homogeneous nucleation was reconfirmed for atmospherically relevant droplets larger than 3 μm. However, the volume dominance of homogeneous nucleation for droplets of smaller sizes is still subject to debate.

Under the presence of very strong external electric fields (several thousand volts per centimeter) some models and experiments have been reported to show electrofreezing, i.e. freezing initiated by strong electric fields or by effects associated with it [*Maybank and Barthakur*, 1967; *Abbas and Latham*, 1969; *Pruppacher*, 1973]. The underlying mechanisms are not well understood [for a review see *Doolittle and Vali*, 1975]. In model simulations a strong enough electric field was able to produce an alignment of water dipoles in a microscopic layer below the water surface [*Svishchev and Kusalik*, 1996; *Zangi and Mark*, 2004; *James et al.*, 2007]. However, in most experiments, the observed enhanced freezing rate might be caused by the attraction of small ice-active particles due to electric forces [*Dufour*, 1862; *Rau*, 1951; *Salt*, 1961; *Dawson and Cardell*, 1973].

In order to assess the importance of charge-induced freezing, we have investigated the rate of homogeneous freezing of levitated pure water droplets as a function of droplet charge. In contrast to most freezing experiments, where ensembles of droplets are subjected to a decreasing temperature ramp, we perform temperature jump experiments where a sequence of droplets is rapidly (within about 100 ms) quenched to the temperature of interest by injecting them into the cold EDB. Due to the stochastic nature of freezing, they then freeze within a variable time interval after the temperature jump. The onset of freezing was determined through analysis of the elastically scattered light as described by *Krämer et al.* [1999]. The freezing time is recorded for each droplet and the fraction of unfrozen to the total number of droplets is determined as a function of residence time at the low temperature. This allows for a direct determination of a freezing probability which is then analyzed with respect to its dependence on droplet charge.

To differentiate small changes in the freezing rate of droplets as a function of their charge, it is necessary to analyze the statistics of the freezing of a large ensemble of droplets under otherwise identical conditions. For that purpose, a fully

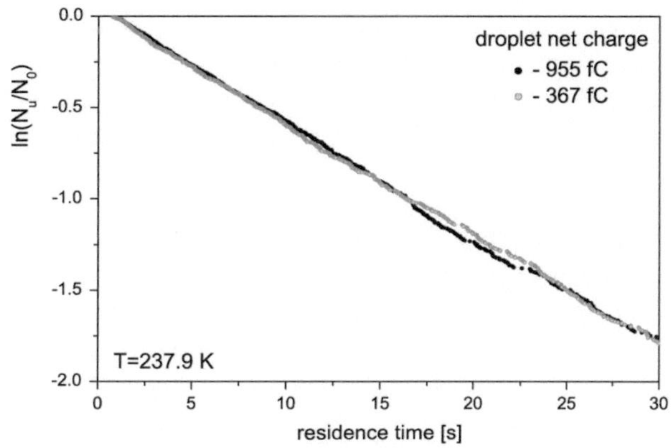

Fig. 6.4 Fraction of unfrozen droplets of two ensembles of charged droplets ($Q_d = -955$ fC, *black circles* and $Q_d = -367$ fC, *gray circles*) as a function of observation time. Within the accuracy of measurement and the statistical quality of the freezing behavior of a large droplet ensemble, both curves result in identical slopes—i.e. no influence of different droplet charges is detectable

automated experimental setup was developed, which allows to observe the freezing of a large number of droplets. While the charge of the droplets has been varied from droplet to droplet in a predefined scheme, all other experimental parameters were kept constant.

With this setup we sequentially observed and analyzed the freezing of several thousands of droplets carrying a charge that was varied from droplet to droplet by randomly selecting the injection and trap parameters from a precompiled list. The resulting database of freezing times and unfrozen fractions was then sorted according to droplet charge and analyzed with respect to the freezing rate. By this strategy, unavoidable drifts in temperature ($\Delta T \sim 0.1$ K) and droplet radius ($\Delta R_d \sim 1$ μm) are distributed equally among all charge states and do not introduce systematic errors.

In Fig. 6.4 a typical result of such an experiment is shown where only two charge states were selected. The logarithm of the fraction of unfrozen droplets to the total number of observed droplets is given as a function of their residence time in the trap which was held at $T = 237.9$ K (normal pressure and 95.6 % RH_W) for the two classes of droplets. No evidence for charge dependent homogeneous freezing can be seen. This finding was confirmed for more charge states and polarities spanning the relevant range of atmospheric charges [*Bourdeau and Chauzy*, 1989; *MacGorman and Rust*, 1998]. We therefore conclude that droplet charge has a negligible influence on the homogeneous freezing rates of cloud droplets under atmospheric conditions.

6.3.2 Aerosol Scavenging Efficiencies of Charged Cloud Droplets

Another important effect of cloud charge may be the modification of scavenging efficiencies for charged or neutral aerosol particles due to electrostatic interaction. Once an aerosol particle is collected by a cloud droplet, it could act as a heterogeneous ice nucleus in contact or immersion mode and therefore modify the glaciation of the cloud with the aforementioned consequences. The influence of solar activity on aerosol charging and as such on the scavenging of aerosol particles with cloud droplets is the prime process that is evoked by some authors [*Tinsley and Deen*, 1991; *Tinsley*, 1993] to establish the link between solar activity and cloud cover. The scavenging of aerosol particles by cloud droplets can occur as a result of *Brownian* motion and turbulent transport of the aerosol particle, or due to the difference in the terminal fall velocity of aerosol particle and droplet. However, the scavenging efficiency is greatly enhanced by coulombic or induced dipole interaction between droplet and aerosol particle [*Kraemer and Johnstone*, 1955; *Fuchs*, 1964; *Pruppacher and Klett*, 2004]. For the situation realized in our experiment diffusive and turbulent processes are negligible and only coulombic and image force interactions are considered along with the drag force associated with a motion in a viscous fluid. The scavenging efficiency η is usually expressed as a dimensionless ratio between the scavenging cross section and the geometric droplet cross section $\eta = \sigma_{scav}/\sigma_d$. The basic analytical formulation for the scavenging efficiency of charged aerosol particles of radius R_p and charge Q_p, in a laminar gas flow of velocity v_0 around an oppositely charged sphere of radius R_d and charge Q_d was given by *Kraemer and Johnstone* [1955]:

$$\eta = -\frac{C Q_d Q_p}{6\pi^2 \varepsilon_0 \mu R_p R_d^2 v_0} \tag{6.1}$$

where C denotes the *Cunningham* correction factor and μ the viscosity of the gas. This formulation includes drag force and static Coulomb interaction only. So far no direct derivation of the scavenging efficiency including the important image force interaction has been obtained; an overview of the semi empirical parameterizations of electro scavenging can be found in *Dhariwal et al.* [1993].

To estimate scavenging efficiencies the differential equation for the aerosol particle motion in the laminar flow around a charged spherical collector has been solved numerically by *Runge-Kutta* method of 4th order, including drag, Coulomb- and image forces following the method described by *Kraemer and Johnstone* [1955] and most recently applied by *Tinsley et al.* [2000]. In brief, the particle trajectory with the largest impact distance that still hit the collector was determined with an iterative method. The squared ratio of this distance and the collector radius is equal to the scavenging efficiency.

To verify the numerical algorithm we compared the scavenging efficiencies computed by our numerical algorithm (excluding image force interaction) with the scavenging efficiencies calculated from Eq. (6.1) over the range of input parameters used in our experiments. The resulting scavenging efficiencies agreed to within 5 % over a range of four orders of magnitude of the scavenging efficiency. This result confirmed the quality of the computational algorithm which was then applied to calculate the scavenging efficiency including image force interactions.

In order to assess in a first experiment the collision rates between aerosol particles and droplets, 29 nm diameter silica spheres (LUDOX®AS40-Grace) were dispersed by the electrospray aerosol generator. The airborne particles were dryed, charged and size selected as described above and then passed through the EDB holding a

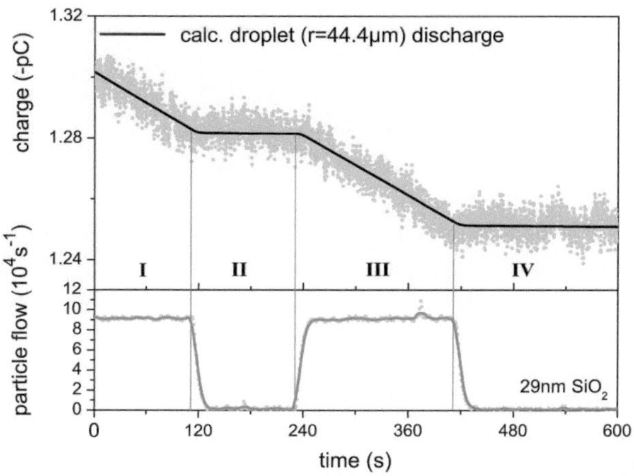

Fig. 6.5 Discharging (*upper panel*) of a charged cloud droplet by collection of singly charged aerosol particles by scavenging from a continuous flow (*lower panel*). A discharging curve is modeled (*solid line*) using observed aerosol flow rates, with the scavenging efficiency as a free parameter. The resulting fitted scavenging efficiency of 8.8 is close to the numerically computed value of 11.9 considering the uncertainty in determination of particle flow conditions (spatial velocity profile of the flow and possible influence of turbulence)

droplet of well characterized size and charge. The electrostatic classifier could also be used to provide for a particle free control flow.

The uptake of aerosols by the levitated droplet changes both its mass and charge. With known aerosol size, charge and flux, the scavenging rate can be calculated from the time dependent charge to mass ratio of the droplet. In Fig. 6.5 such raw data are shown: the droplet charge to mass ratio decreases while exposed to the particle flux but stays constant in absence of the aerosol flux. The continuous line in Fig. 6.5 (upper panel) shows the modeled discharge curve for our situation, using values for the scavenging efficiency calculated with the numerical method described above and observed aerosol flow rate (lower panel), which obviously fits our measurements very well. We therefore conclude, that electrically enhanced scavenging in our range of particle sizes, charges and flow rates is very well described by the procedure.

In one special case, we were additionally able to observe the trajectories of large scavenged ice aerosol particles using an optical microscope equipped with a high speed CCD camera (Vision Research, Phantom v710). From the microscope image the diameter of the particles was estimated to be between 1–3 μm. Trajectories computed by the numerical model described above could be directly fitted to the measured trajectories using the particle size and the impact parameter as free parameters. The fitted particle diameters of 1.7 μm and 2.4 μm are within the range obtained from the microscope image. As demonstrated in Fig. 6.6, the particle trajectory was reproduced very well by including drag, Coulomb and image forces in the numerical model.

From these experiments, we conclude that the computation of the scavenging rates by the numerical method describes the experimental conditions very well. The

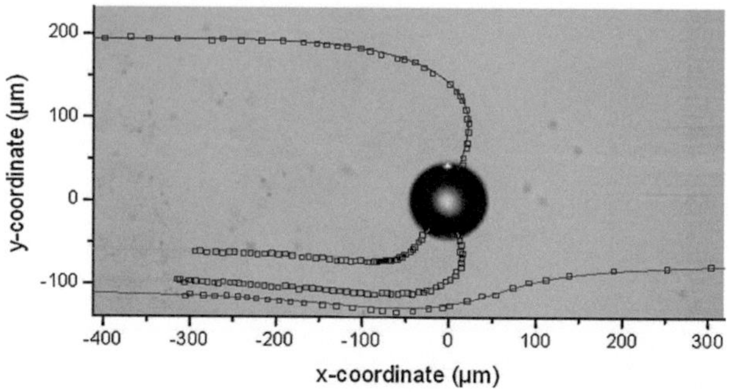

Fig. 6.6 Still image of a charged water droplet ($D = 90$ µm, $Q_d = 1.7$ pC) levitated in the EDB ($T = -9.8$ °C) overlaid with trajectories of neutral and weakly charged ice particles moving from left to right within a laminar stream of air (flow speed 0.015 m/s). The *open rectangles* represent visually tracked particle positions and the *colored lines* follow model trajectory calculations. The *green* and *red lines* correspond to neutral particles attracted to the droplet by induced dipole interaction ($d = 1.7$ µm, all three particles), while the *blue line* follows the track of a charged particle ($d = 2.4$ µm, $Q_p = 33e^-$)

results illustrate the importance to consider not only the droplet charge but also the polarizability of the aerosol substance to predict charge effects in cloud microphysics.

6.3.3 Vapor Pressure of Charged Hydrometeors

In recent field studies in tropical regions [*Nielsen et al.*, 2007; *Chaboureau et al.*, 2007; *Corti et al.*, 2008; *de Reus et al.*, 2009] ice particles have been found in the lower stratosphere where they might have been lifted by the deep convection from tropical thunderstorms. However, measurements [*Khaykin et al.*, 2009] showed that the air mass was subsaturated with respect to ice on its way to the stratosphere which would have rendered the lifetime of the particles too short to be detected. It was then speculated, that charges on the ice particles might have stabilized them sufficiently to survive the subsaturated conditions. Though it is obvious that charges on a particle attract polar water vapor molecules and therefore reduce its water vapor pressure, this effect seems to have escaped the attention of many atmospheric scientists. In principle, a charge effect on hydrometeor vapor pressure can effect cloud evaporation and therefore form a link between global electric circuit and cloud optical thickness. In order to quantify this effect experimentally, we measured the vapor pressure of individual charged water droplets levitated within the EDB at well defined temperatures. The vapor pressure was derived from the speed of evaporation of charged droplets exposed to an ice saturated environment which was established by coating the electrodes of the EDB with ice before the beginning of each experiment. At various temperatures, water droplets of well defined charge were then injected into the levitator and their speed of evaporation was measured as a function

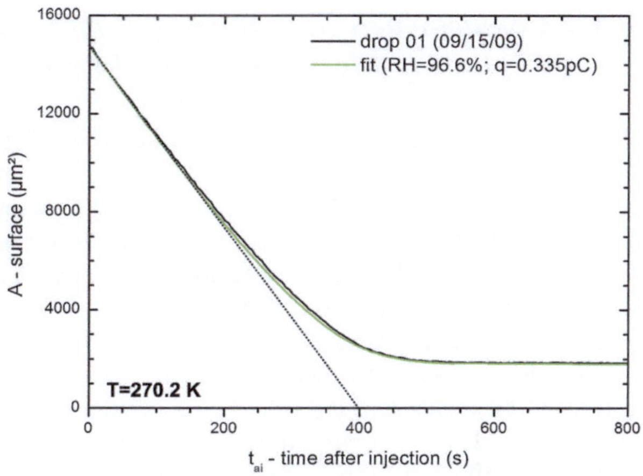

Fig. 6.7 Surface area of an evaporating droplet as a function of time. After some time the droplet reaches a terminal diameter of 50 μm. The *green curve* is generated by numerical solution of the evaporation (Eq. (6.2)). The *dashed line* is an asymptotic fit for $t = 0$, from its slope, the relative humidity within the EDB is calculated. Ambient temperature was 270.2 K

of their size. A typical evaporation curve of a droplet is given in Fig. 6.7. Initially, the droplet surface area decreases linearly with time, as expected for a diffusion limited evaporation process. From the slope of this curve, the water vapor pressure within the levitator can be obtained as the only fitting parameter, as all other quantities are known from independent measurements. As obvious from Fig. 6.7, after some time, the evaporation of the liquid droplet deviates from the expected curve and eventually comes to a complete standstill. At this point in time, the vapor pressure of the charged liquid droplet equals the vapor pressure in the levitator, which is usually determined by that of ice [taken from *Murphy and Koop*, 2005] at the levitator temperature.

The observed evaporation curve can be reproduced perfectly by considering the charge—dipole and charge—polarization interactions between the droplet and the evaporating water molecules:

$$\frac{1}{2}\frac{dr^2}{dt} = \frac{(1+U_d)RH_W - 1}{\frac{\rho_W RT_\infty}{e_{s,l}(T_\infty)D^* M_W} + \frac{L\rho_W}{k_a^* T_\infty}(\frac{LM_W}{RT} - 1)} \tag{6.2}$$

where

$$U_d = \frac{|q\mathbf{p}|}{4\pi\varepsilon_0 k_B T_\infty r^2} \tag{6.3}$$

leading to the liquid drop terminal radius:

$$r_{eq}^2 = \frac{|q\mathbf{p}|}{4\pi\varepsilon_0 k_B T \ln\frac{e_{s,l}}{e_{s,i}}} \tag{6.4}$$

As the thermodynamic properties of water and the molecular polarizability of water are well tabulated, no further fitting parameters are required. The detailed derivation is presented in *Nielsen et al.* [2011].

We found that within the limits of experimental error, the classical theory for the evaporation of charged droplets is confirmed in our experiments. The resulting saturation vapor pressure depressions can be considerable and reached up to 6.5 % under our experimental conditions.

6.3.4 Contact Nucleation Rates of Supercooled Cloud Droplets Exposed to Electrified Aerosol Particles

As discussed in Sect. 6.3.2, charges on cloud droplets enhance their efficiency to scavenge interstitial aerosols which can serve as ice nuclei in contact or immersion mode. In order to assess the importance of this effect on cloud glaciation, the contact freezing probability e_c has to be known, which is defined as the probability of freezing of a supercooled water droplet upon contact with a heterogeneous ice nucleus. This probability is a function of the droplet temperature and the substance, size and possibly shape and structure of the heterogeneous ice nucleus.

In order to arrive at a first assessment, the contact freezing probability can be estimated by measuring the freeze rate of supercooled liquid droplets exposed to aerosol particles in a laminar flow. The contact freezing probability e_c can be calculated as the quotient of the actual measured freeze rate P_f and the scavenging rate P_{scav}:

$$e_c = \frac{P_f}{P_{scav}} = \frac{P_f}{N_c v_0 \sigma_d \eta} \tag{6.5}$$

The scavenging rate depends on the particle concentration N_c, the flow speed at the droplet position v_0, the droplet geometrical cross section σ_d and the scavenging efficiency η.

We measured the contact freezing probability for several of the most common atmospheric mineral dust particles [see *Zimmermann et al.*, 2008, for a compilation of mineral dust components frequently found in the atmosphere] as a function of their size and ambient temperature. The results for two selected (illite and kaolinite) mineral dust components are presented in this section. Size selected aerosol particles were generated by dry dispersion from a dust powder sample in a fluidized bed generator. Large particles were removed aerodynamically and a flow of singly charged size selected particles was generated using a differential mobility analyzer (cf. Fig. 6.3). The rate of collisions between aerosol particles and droplets P_{scav} was derived from the measured aerosol concentration and velocity profile assuming a collection efficiency as derived in Sect. 6.3.2, whereas the freezing rate was determined as reported in Sect. 6.3.1.

Some preliminary results are summarized in Fig. 6.8. In the left chart the contact freezing probability is given as a function of temperature for illite particles of different sizes. It can be seen that the contact freezing probability is a very steep function of temperature and increases with particle size. The latter effect can be explained by assuming the freezing probability to be proportional to the effective area of contact between aerosol particle and supercooled liquid. Similar results have been found for kaolinite particles (Fig. 6.8, right chart). Compared to illite, they have been identified as less efficient contact freezing nuclei which presumably can be attributed to the different chemical composition and especially the different morphology of the two mineral dust components.

Contact freezing probabilities larger than unity, that appear in Fig. 6.8 for large particles and low temperatures are unphysical and represent the range of experimental uncertainty within our experiment. This uncertainty results predominantly from

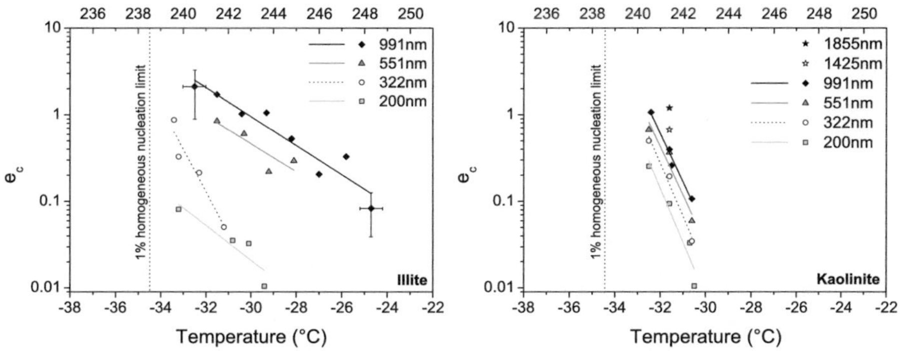

Fig. 6.8 Contact freezing probability e_c of size selected Illite and Kaolinite particles impacting on a supercooled cloud droplet as a function of temperature and particle diameter. Weighted exponential fits to the data sets (*straight lines*) have been added to improve perceivability. The *vertical dashed lines* represent the lowest temperature at which contact nucleation is not substantially masked by homogeneous nucleation events (less than 1 % within 30 s measurement interval and 70 μm typical droplet diameter). The given error bars in the left diagram are representative for the uncertainty of all data points shown

the determination of the aerosol flow conditions, i.e. vertical and horizontal velocity profiles and influence of turbulence within the EDB, which sensitively impacts the scavenging efficiency η.

On the other hand impacted particles that do not trigger freezing immediately can accumulate in the supercooled droplet and act as immersion freezing nuclei at later time. We are able to discriminate between these two modes of ice formation by their different effect on the fraction of frozen droplets as a function of time (cf. Sect. 6.3.1). Details on this will be given in a forthcoming publication.

6.4 Summary

Several effects that could link the electrical state of the atmosphere to cloud microphysics have been investigated and quantified. While no direct effect of cloud droplet charge on homogeneous freezing was found, we were able to confirm and quantify the enhanced scavenging of aerosol particles by charged cloud droplets. Together with a direct measurement of size dependent contact freezing probabilities it will now be possible to incorporate charges in cloud models and to quantify their role in cloud microphysics. Currently we assume, that the effects described here will not have an important impact on climate, as cloud droplets are rarely charged to the high values used in the experiments reported herein.

Acknowledgements The authors would like to thank Johannes Nielsen, Christopher Maus and Jens Nadolny. The work was funded by the German Research Foundation (Deutsche Forschungsgemeinschaft—DFG) within the priority program SPP1176-CAWSES (Grant LE834/2-3).

References

Abbas, M. A., & Latham, J. (1969). The electrofreezing of supercooled water drops. *Journal of the Meteorological Society of Japan*, *47*, 65–74.

Achtzehn, T., Müller, R., Duft, D., & Leisner, T. (2005). The coulomb instability of charged microdroplets: dynamics and scaling. *The European Physical Journal. D, Atomic, Molecular and Optical Physics*, *34*, 311–313.

Bazilevskaya, G. (2000). Observations of variability in cosmic rays. *Space Science Reviews*, *94*, 25–38. doi:10.1023/A:1026721912992.

Benz, S., Megahed, K., Möhler, O., Saathoff, H., Wagner, R., & Schurath, U. (2005). T-dependent rate measurements of homogeneous ice nucleation in cloud droplets using a large atmospheric simulation chamber. *Journal of Photochemistry and Photobiology. A, Chemistry*, *176*, 208–217.

Bohren, C. F., & Huffman, D. R. (2004). *Absorption and scattering of light by small particles*. Weinheim: Wiley-VCH.

Bourdeau, C., & Chauzy, S. (1989). Maximum electric charge of a hydrometeor in the electric field of a thundercloud. *Journal of Geophysical Research*, *94*, 13121–13126. doi:10.1029/JD094iD11p13121.

Carslaw, K. S., Harrison, R. G., & Kirkby, J. (2002). Cosmic rays, clouds, and climate. *Science*, *298*(5599), 1732–1737. doi:10.1126/science.1076964.

Chaboureau, J.-P., Cammas, J.-P., Duron, J., Mascart, P. J., Sitnikov, N. M., & Voessing, H.-J. (2007). A numerical study of tropical cross-tropopause transport by convective overshoots. *Atmospheric Chemistry and Physics*, *7*(7), 1740.

Corti, T., Luo, B. P., de Reus, M., Brunner, D., Cairo, F., Mahoney, M. J., Martucci, G., Matthey, R., Mitev, V., dos Santos, F. H., Schiller, C., Shur, G., Sitnikov, N. M., Spelten, N., Voessing, H. J., Borrmann, S., & Peter, T. (2008). Unprecedented evidence for deep convection hydrating the tropical stratosphere. *Geophysical Research Letters*, *35*, 10. doi:10.1029/2008GL033641.

Cubasch, U., Voss, R., Larkin, A., Haigh, J. D., Djavidnia, S., Marsh, N., Svensmark, H., Tinsley, B. A., van Loon, H., Labitzke, K., Arnold, N. F., Robinson, T. R., Haarsma, R. J., Drijfhout, S. S., Opsteegh, J. D., Selten, F. M., Mende, W., & Stellmacher, R. (2000). Solar variability and the Earth's climate—chapter: Influences. *Space Science Reviews*, *94*, 185–306.

Davis, E. J. (1997). A history of single particle levitation. *Aerosol Science and Technology*, *26*, 212–254.

Dawson, G. A., & Cardell, G. R. (1973). Electrofreezing of supercooled waterdrops. *Journal of Geophysical Research*, *78*, 8864–8866. doi:10.1029/JC078i036p08864.

de Reus, M., Borrmann, S., Bansemer, A., Heymsfield, A. J., Weigel, R., Schiller, C., Mitev, V., Frey, W., Kunkel, D., Kürten, A., Curtius, J., Sitnikov, N. M., Ulanovsky, A., & Ravegnani, F. (2009). Evidence for ice particles in the tropical stratosphere from in-situ measurements. *Atmospheric Chemistry and Physics*, *9*(18), 6775–6792. doi:10.5194/acp-9-6775-2009.

DeMott, P. J., & Rogers, D. C. (1990). Freezing nucleation rates of dilute solution droplets measured between −30 °C and −40 °C in laboratory simulations of natural clouds. *Journal of the Atmospheric Sciences*, *47*, 1056–1064.

Dhariwal, V., Hall, P. G., & Ray, A. K. (1993). Measurements of collection efficiency of single, charged droplets suspended in a stream of submicron particles with an electrodynamic balance. *Journal of Aerosol Science*, *24*(2), 197–209. doi:10.1016/0021-8502(93)90058-H.

Dickinson, R. (1975). Solar variability and the lower atmosphere. *Bulletin of the American Meteorological Society*, *56*, 1240.

Djikaev, Y. S., Tabazadeh, A., Hamill, P., & Reiss, H. (2002). Thermodynamic conditions for the surface-stimulated crystallization of atmospheric droplets. *The Journal of Physical Chemistry. A, Molecules, Spectroscopy, Kinetics, Environment, & General Theory*, *106*, 10247–10253.

Djikaev, Y. S., Tabazadeh, A., & Reiss, H. (2003). Thermodynamics of crystal nucleation in multicomponent droplets: adsorption, dissociation, and surface-stimulated nucleation. *Journal of Chemical Physics*, *118*, 6572–6581.

Doolittle, J. B., & Vali, G. (1975). Heterogeneous freezing nucleation in electric fields. *Journal of the Atmospheric Sciences*, *32*, 375–379.

Dufour, L. (1862). Ueber das Gefrieren des Wassers und über die Bildung des Hagels. *Annalen der Physik*, *190*(12), 530–554. doi:10.1002/andp.18621901203.
Duft, D., & Leisner, T. (2004a). The index of refraction of supercooled solutions determined by the analysis of optical rainbow scattering from levitated droplets. *International Journal of Mass Spectrometry*, *233*, 61–65.
Duft, D., & Leisner, T. (2004b). Laboratory evidence for volume-dominated nucleation of ice in supercooled water microdroplets. *Atmospheric Chemistry and Physics*, *4*, 1997–2000.
Fischer, E. (1959). Die dreidimensionale Stabilisierung von Ladungsträgern in einem Vierpolfeld. *Zeitschrift für Physik*, *156*(1), 1–26.
Fisk, L. A., Wenzel, K.-P., Balogh, A., Burger, R. A., Cummings, A. C., Evenson, P., Heber, B., Jokipii, J. R., Krainev, M. B., Kóta, J., Kunow, H., Le Roux, J. A., McDonald, F. B., McKibben, R. B., Potgieter, M. S., Simpson, J. A., Steenberg, C. D., Suess, S., Webber, W. R., Wibberenz, G., Zhang, M., Ferrando, P., Fujii, Z., Lockwood, J. A., Moraal, H., & Stone, E. C. (1998). Global processes that determine cosmic ray modulation. *Space Science Reviews*, *83*, 179–214.
Friis-Christensen, E. (2000). Solar variability and climate—A summary. *Space Science Reviews*, *94*(1–2), 411–421.
Friis-Christensen, E., & Lassen, K. (1991). Length of the solar cycle: an indicator of solar activity closely associated with climate. *Science*, *254*, 698.
Fuchs, N. A. (1963). On the stationary charge distribution on aerosol particles in a bipolar ionic atmosphere. *Pure and Applied Geophysics*, *56*, 185–193. doi:10.1007/BF01993343.
Fuchs, N. A. (1964). *Mechanics of aerosols*. New York: Pergamon.
Gast, F. U., & Fiehn, H. (2003). Profile. The development of integrated microfluidic systems at GeSiM. *Lab on a Chip*, *3*, 6N–10N.
Gray, L. J., Beer, J., Geller, M., Haigh, J. D., Lockwood, M., Matthes, K., Cubasch, U., Fleitmann, D., Harrison, G., Hood, L., Luterbacher, J., Meehl, G. A., Shindell, D., van Geel, B., & White, W. (2010). Solar influences on climate. *Reviews in Geophysics*, *48*. doi:10.1029/2009RG000282.
Herschel, W. (1796). On the method of observing the changes that happen to the fixed stars; with some remarks on the stability of the light of our sun, to which is added, a catalogue of comparative brightness, for ascertaining the permanency of the lustre of stars. *Philosophical Transactions of the Royal Society of London*, *86*, 166–226.
Hoyt, D. V., & Schatten, K. H. (1997). *The role of the sun in climate change*. Oxford: Oxford University Press.
James, T., Wales, D. J., & Rojas, J. H. (2007). Energy landscapes for water clusters in a uniform electric field. *Journal of Chemical Physics*, *126*, 054506–054519. doi:10.1063/1.2429659.
Khaykin, S., Pommereau, J. P., Korshunov, L., Yushkov, V., Nielsen, J., Larsen, N., Christensen, T., Garnier, A., Lukyanov, A., & Williams, E. (2009). Hydration of the lower stratosphere by ice crystal geysers over land convective systems. *Atmospheric Chemistry and Physics*, *9*(6), 2275–2287.
Kirkby, J., Curtius, J., Almeida, J., Dunne, E., Duplissy, J., Ehrhart, S., Franchin, A., Gagne, S., Ickes, L., Kuerten, A., Kupc, A., Metzger, A., Riccobono, F., Rondo, L., Schobesberger, S., Tsagkogeorgas, G., Wimmer, D., Amorim, A., Bianchi, F., Breitenlechner, M., David, A., Dommen, J., Downard, A., Ehn, M., Flagan, R. C., Haider, S., Hansel, A., Hauser, D., Jud, W., Junninen, H., Kreissl, F., Kvashin, A., Laaksonen, A., Lehtipalo, K., Lima, J., Lovejoy, E. R., Makhmutov, V., Mathot, S., Mikkila, J., Minginette, P., Mogo, S., Nieminen, T., Onnela, A., Pereira, P., Petaja, T., Schnitzhofer, R., Seinfeld, J. H., Sipila, M., Stozhkov, Y., Stratmann, F., Tome, A., Vanhanen, J., Viisanen, Y., Vrtala, A., Wagner, P. E., Walther, H., Weingartner, E., Wex, H., Winkler, P. M., Carslaw, K. S., Worsnop, D. R., Baltensperger, U., & Kulmala, M. (2011). Role of sulphuric acid, ammonia and galactic cosmic rays in atmospheric aerosol nucleation. *Nature*, *476*(7361), 429–433. doi:10.1038/nature10343.
Kraemer, H. F., & Johnstone, H. F. (1955). Collection of aerosol particles in presence of electrostatic fields. *Industrial and Engineering Chemistry*, *47*(12), 2426–2434. doi:10.1021/ie50552a020.

Krämer, B., Schwell, M., Hübner, O., Vortisch, H., Leisner, T., Rühl, E., Baumgärtel, H., & Wösle, L. (1996). Homogeneous ice nucleation observed in single levitated micro droplets. *Berichte der Bunsen-Gesellschaft für Physikalische Chemie*, *100*, 1911–1914.

Krämer, B., Hübner, O., Vortisch, H., Wöste, L., Leisner, T., Schwell, M., Rühl, E., & Baumgärtel, H. (1999). Homogeneous nucleation rates of supercooled water measured in single levitated microdroplets. *Journal of Chemical Physics*, *111*, 6521–6527.

MacGorman, D. R., & Rust, W. D. (1998). *The electrical nature of storms*. Oxford: Oxford University Press.

March, R. E., & Hughes, R. J. (1989). *Quadrupole storage mass spectrometry*. New York: Wiley.

Marsh, N., & Svensmark, H. (2000). Cosmic rays, clouds and climate. *Space Science Reviews*, *94*, 215–230.

Maybank, J., & Barthakur, N. N. (1967). Growth and destruction of ice filaments in an electric field. *Nature*, *216*, 50–52. doi:10.1038/216050a0.

Murphy, D. M., & Koop, T. (2005). Review of the vapour pressures of ice and supercooled water for atmospheric applications. *Quarterly Journal of the Royal Meteorological Society*, *131*(608, Part b), 1539–1565. doi:10.1256/qj.04.94.

Nielsen, J. K., Larsen, N., Cairo, F., Di Donfrancesco, G., Rosen, J. M., Durry, G., Held, G., & Pommereau, J. P. (2007). Solid particles in the tropical lowest stratosphere. *Atmospheric Chemistry and Physics*, *7*(3), 695.

Nielsen, J. K., Maus, C., Rzesanke, D., & Leisner, T. (2011). Charge induced stability of water droplets in subsaturated environment. *Atmospheric Chemistry and Physics*, *11*, 2031–2037. doi:10.5194/acp-11-2031-2011.

Paul, W., & Steinwedel, H. (1953). Ein neues Massenspektrometer ohne Magnetfeld. *Zeitschrift Für Naturforschung. A, a Journal of Physical Sciences*, *8*(7), 448–450.

Pruppacher, H. R. (1973). Electrofreezing of supercooled water. *Pure and Applied Geophysics*, *104*, 623–634. doi:10.1007/BF00875907.

Pruppacher, H. R., & Klett, J. D. (2004). *Microphysics of clouds and precipitation* (2nd ed.). Dordrecht: Kluwer Academic Publishers. doi:10.1007/0-306-48100-6, ebook-edition.

Rau, W. (1951). Eiskeimbildung durch dielektrische Polarisation. *Zeitschrift für Naturforschung Teil A*, *6*, 649.

Salt, R. W. (1961). Effect of electrostatic field on freezing of supercooled water and insects. *Science*, *133*(3451), 458–459. doi:10.1126/science.133.3451.458.

Sonett, C. P., Giampapa, M. S., & Matthews, M. S. (Eds.) (1991). *The sun in time*. Tuscon: University of Arizona Press.

Stöckel, P., Vortisch, H., Leisner, T., & Baumgärtel, H. (2002). Homogeneous nucleation of supercooled liquid water in levitated microdroplets. *Journal of Molecular Liquids*, *96–97*, 153–175.

Svishchev, I. M., & Kusalik, P. G. (1996). Electrofreezing of liquid water: A microscopic perspective. *Journal of the American Chemical Society*, *118*(3), 649–654. doi:10.1021/ja951624l.

Tabazadeh, A., Djikaev, Y. S., Hamill, P., & Reiss, H. (2002a). Laboratory evidence for surface nucleation of solid polar stratospheric cloud particles. *The Journal of Physical Chemistry. A, Molecules, Spectroscopy, Kinetics, Environment, & General Theory*, *106*, 10238–10246.

Tabazadeh, A., Djikaev, Y. S., & Reiss, H. (2002b). Surface crystallization of supercooled water in clouds. *Proceedings of the National Academy of Sciences of the United States of America*, *99*, 15873–15878.

Taborek, P. (1985). Nucleation in emulsified supercooled water. *Physical Review. B, Condensed Matter*, *32*, 5902–5906.

Tinsley, B. A. (1993). Apparent tropospheric response to MeV-GeV particle flux variations: a connection via electrofreezing of supercooled water in high-level clouds: reply to detwiler-comment. *Journal of Geophysical Research*, *98*, 16889–16891.

Tinsley, B. A. (2000). Influence of solar wind on the global electric circuit, and inferred effects on cloud microphysics. *Space Science Reviews*, *94*, 231–258.

Tinsley, B. A., & Deen, G. W. (1991). Apparent tropospheric response to MeV-GeV particle flux variations: a connection via electrofreezing of supercooled water in high-level clouds. *Journal of Geophysical Research*, *96*(D12), 22283–22296. doi:10.1029/91JD02473.

Tinsley, B. A., & Yu, F. (2004). Atmospheric ionization and clouds as links between solar activity and climate. In *Geophysical monograph series: Vol. 141. Solar variability and its effects on climate* (pp. 321–339), American Geophysical Union.
Tinsley, B. A., Rohrbaugh, R. P., Hei, M., & Beard, K. V. (2000). Effects of image charges on the scavenging of aerosol particles by cloud droplets and on droplet charging and possible ice nucleation processes. *Journal of the Atmospheric Sciences*, *57*(13), 2118–2134.
Vrbka, L., & Jungwirth, P. (2006). Homogeneous freezing of water starts in the surface. *Journal of Physical Chemistry. B*, *110*, 18126–18129.
Wiedensohler, A. (1988). An approximation of the bipolar charge distribution for particles in the submicron size range. *Journal of the Atmospheric Sciences*, *19*, 387–389.
Wood, S. E., Baker, M. B., & Swanson, B. D. (2002). Instrument for studies of homogeneous and heterogeneous ice nucleation in freefalling supercooled water droplets. *Review of Scientific Instruments*, *73*, 3988–3996.
Wuerker, R. F., Shelton, H., & Langmuir, R. V. (1959). Electrodynamic containment of charged particles. *Japanese Journal of Applied Physics*, *30*(3), 342–349.
Zangi, R., & Mark, A. E. (2004). Electrofreezing of confined water. *Journal of Chemical Physics*, *120*, 7123–7130. doi:10.1063/1.1687315.
Zimmermann, F., Weinbruch, S., Schuetz, L., Hofmann, H., Ebert, M., Kandler, K., & Worringen, A. (2008). Ice nucleation properties of the most abundant mineral dust phases. *Journal of Geophysical Research*, *113*. doi:10.1029/2008JD010655.

Chapter 7
Investigations of the Solar Influence on Middle Atmospheric Water Vapour and Ozone During the Last Solar Cycle—Analysis of the MPS Data Set

Paul Hartogh, Christopher Jarchow, and Kristoffer Hallgren

Abstract The MPS water vapour and ozone data set was gained with ground-based microwave heterodyne spectrometers operating since mid of the nineties of the last century at two locations: at MPS at mid latitude in Katlenburg-Lindau (51.66°N, 10.13°E) in Germany and at ALOMAR in polar latitude at (69.29°N, 16.03°E) in Northern Norway. The water vapour observations show a pronounced year-to-year variability with annual maxima in summer and minima in winter related to the Lyman-α radiation. In winter we found an anti-correlation of upper mesospheric water vapour with the solar activity. After winter solstice the mesospheric water vapour concentration is strongly influenced by sudden stratospheric warmings (SSWs) appearing more frequently during high solar activity. In the stratopause and lower mesosphere region we find a positive correlation with solar activity during the whole year. Ozone also shows a strong (but different) annual pattern: we find a late summer maximum in the middle and lower mesosphere which is shifted into autumn and winter in the stratopause region, and a distinct nighttime maximum around 72 km during the winter season, whereas no annual maximum occurs there during daytime. A clear annual asymmetry of the nighttime ozone distribution exists in this domain, marked by a decline of the mean ozone values in January/February and an increase to the subsidiary annual maximum a few kilometres higher in March/April. These asymmetric variations result from the asymmetric occurrence rate of SSWs with maxima after winter solstice and the asymmetric annual variation of water vapour with minima around spring equinox.

7.1 Introduction

In the mesosphere, water vapour is the most important minor constituent. The reason is chiefly that water vapour is the main source of the chemically active hydrogen radicals which affect the chemistry of all other chemically active minor constituents.

P. Hartogh (✉) · C. Jarchow · K. Hallgren
Max-Planck-Institut für Sonnensystemforschung, Max-Planck-Str. 2, 37191 Katlenburg-Lindau, Germany
e-mail: hartogh@mps.mpg.de

F.-J. Lübken (ed.), *Climate and Weather of the Sun-Earth System (CAWSES)*, Springer Atmospheric Sciences,
DOI 10.1007/978-94-007-4348-9_7,

Water vapour itself plays a role in various atmospheric phenomena, such as the creation of ice particles responsible for the so-called polar mesospheric summer echoes and the formation of noctilucent clouds (NLCs). Moreover, it determines the production of water cluster ions in the mesopause region and it influences the thermal regime of the atmosphere. Considering both loss and production, its effective lifetime in the middle atmosphere is very long, amounting from several months up to infinity. Below about 65 km the lifetime is even negative, meaning that water vapour is not decomposed but is formed from the source species methane and molecular hydrogen [*Sonnemann et al.*, 2005]. Only in and above the upper mesosphere, water vapour is destroyed and converted into molecular or atomic hydrogen. The lifetime in the lower thermosphere amounts to more than a week. The most important photolyzer is the solar Lyman-α radiation which varies by nearly a factor of 2 from solar minimum to maximum [*Woods et al.*, 2000]. The lifetime of water vapour is on the order of transport time scales even in the lower thermosphere [*Stevens et al.*, 2003] and can be used as tracer for dynamical processes such as planetary waves [*Sonnemann et al.*, 2008].

Besides water vapour, ozone is the most important key constituent for understanding all aeronomic processes of the middle atmosphere. It governs both the chemistry and the dynamics by absorption of solar shortwave radiation and emission of infrared radiation. While the stratospheric ozone was a subject of intense investigation in the context of the anthropogenic decline of the ozone layer, the mesospheric ozone did not attract this attention. The global models describing the behaviour of ozone were often confined only to the stratosphere. However, it becomes more and more clear that the atmosphere can only be understood if it is considered as a unity. There are important exchange processes between the layers, and consequently the mesosphere has recently moved more to the centre of scientific interest.

A powerful tool for continuous monitoring of the mesospheric water vapour and ozone abundance is the microwave heterodyne spectroscopy. At the Max-Planck-Institut für Solar System Research (MPS) ground-based millimetre wave measurements were carried out from April 1993 to October 1995 and from December 1998 until now, detecting the rotational transition of ozone at 142 GHz. Between October 1995 and June 1996, the instrument operated at the ALOMAR facility in Norway. The water vapour measurements at 22 GHz were performed continuously since 1995 at the Arctic Lidar Observatory for Middle Atmospheric Research (ALOMAR) with some interrupts caused by logistical problems at the beginning and maintenance intervals.

7.2 Description of Instruments and Data Analysis Method

The rotational transition of water vapour is detected at 22.235 GHz using the microwave heterodyne technique. A rotating mirror reflects in turns the atmospheric signal as well as signals from the hot (ambient temperature) and cold (60 K) calibration loads into a horn antenna operated at room temperature. The water vapour

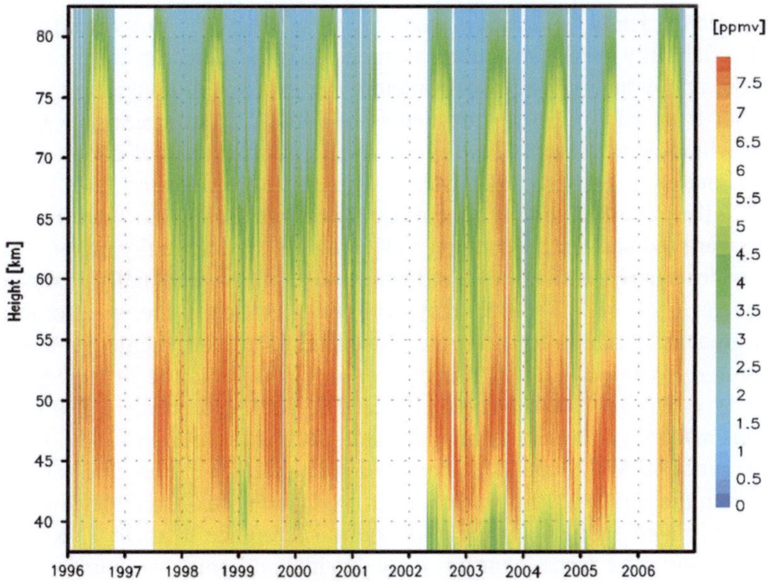

Fig. 7.1 Outline of the water vapour measurements at ALOMAR from 1995 to 2006. Between 2002 and 2005 the lower altitude limit of the instrument was extended by using an additional broadband CTS with 200 MHz bandwidth. As a result the measurement gets more sensitive below about 40 km and clearly resolves the annual pattern there. This figure is from *Hartogh et al.* [2010]

signal is observed with an elevation angle of 18°. A GaAs HEMT (High Electron Mobility Transistor) amplifier cooled to 20 K detects the input signal received by a horn antenna. At the output of the amplifier a steep single sideband waveguide filter selects the water line in the lower sideband. This filtered signal is multiplied with a 22.535 GHz local oscillator signal using a Schottky mixer. The 22.235 GHz signal is down-converted to an IF (intermediate frequency) of 300 MHz and fed into a Chirp Transform Spectrometer (CTS) [*Hartogh and Hartmann*, 1990] with a 40 MHz bandwidth and 20 KHz spectral resolution. Between 2002 and 2005 a second CTS with about 200 MHz bandwidth and 50 kHz spectral resolution was operated [*Villanueva and Hartogh*, 2004; *Villanueva et al.*, 2006], extending the vertical coverage of the water vapour profile to altitudes below 40 km (see Fig. 7.1). The single sideband (SSB) receiver noise temperature of the instrument was 110 K.

In 2009 this instrument was replaced by a new dual-polarisation receiver. An improved sensitivity was achieved by cooling of the horn antenna (not cooled in the system described above), the use of InP HEMT amplifiers and reduction of the cryogenic temperature from 20 K to 10 K. The receiver temperature for each of the polarisations is 30 K. Adding the two polarisations reduces the noise by a factor of the square root of 2, since the two signals are orthogonal and therefore the signals statistically independent. More detailed descriptions of the instruments are given in *Hartogh and Jarchow* [1995], *Seele and Hartogh* [1999], *Hallgren et al.* [2010] and *Straub et al.* [2011]. During the time of the ozone instrument development in the early 1990s, HEMT amplifiers were not yet available at mm-wave frequencies. Instead a Schottky mixer was used as receiver input detector. As in the case of the water vapour heterodyne spectrometers the instrument consists of a so-called radiometer front end and a spectrometer back end. The front end is a heterodyne receiver which detects the 142.175 GHz rotational transition of ozone. The atmospheric signal is first filtered by a Martin-Puplett single sideband filter [*Hartogh et al.*, 1991], then combined with a local oscillator signal using a folded Fabry-Perot, and afterward fed back into a cooled (20 K) single-ended Schottky mixer. The mixer then provides a down-converted signal which is amplified and finally analysed in the spectrometer back end. The ozone system and specifically the spectrometer back end, of great importance for accurate mesospheric measurements, is described in *Hartogh and Hartmann* [1990], *Hartogh et al.* [1991] and a refurbished version is described by *Hartogh and Jarchow* [1995]. The back end consists of a broadband filter bank covering 1.2 GHz bandwidth and a high-resolution (44 kHz) CTS covering the inner 40 MHz with 1024 equidistant channels; i.e., the channel spacing is smaller than the frequency resolution. Data are taken with a fixed elevation angle of 30° and calibrated every 6 s using two external reference loads at 77 K and ambient temperature. The instrument has a single sideband noise temperature of 500 K which allows retrieval of mesospheric ozone profiles every few minutes. However, a longer integration time is usually applied in order to obtain a better altitude resolution. Here we show single day and night averages. The resulting high-resolution spectra allow the retrieval of middle atmospheric ozone profiles from 15 km to approximately 80 km.

The essential component for analysing the data is a radiation transfer model calculating the atmospheric brightness at ground level as a function of the ozone respectively water vapour profiles. Input parameters to the radiative transfer model are the temperature profile, pressure profile, line intensity and partition function according to *Poynter and Pickett* [1985], and pressure broadening coefficients according to *Rosenkranz* [1993]. The inversion of the spectra has been carried out by means of the optimal estimation method given by *Rodgers* [1976]. This algorithm needs the radiation transfer model $y_n = F(x_n)$ which relates to an ozone or water vapour profile x_n, to the corresponding spectrum y_n, to the measured spectrum y, to its covariance matrix S_y, and to a so-called a priori profile x_0 providing just that part of profile information that cannot be derived from measurement. The use of additional a priori information makes this technique especially suited for ill-posed inversion problems. Annual averages of the vertical water vapour and ozone profiles for MPS and ALOMAR have been employed as an a priori profile for all retrievals. The a priori covariance matrix S_0 has been set to reflect a volume mixing ratio uncertainty of ± 1 ppmv at each altitude. For more details, see the reports by *Jarchow and Hartogh* [1995, 1998]. The typical signal-to-noise (SNR) ratios of any retrieved spectrum presented in this study are 500–1000, providing a vertical resolution between approximately 7 km (middle stratosphere) and 10 km (middle and upper mesosphere). This SNR is sufficient to resolve the transition region of collisional and Doppler broadening which appears to be at 83–85 km for water vapour at 22 GHz and at around 75–80 km of the 142 GHz ozone line. Uncertainties of a priori data and temperatures and their temporal evolution are discussed in *Hartogh et al.* [2004, 2010].

7.3 Description and Analysis of the Water Vapour Data Set

Figure 7.1 shows an outline of the water vapour measurements at ALOMAR from the end of 1995 until the end 2006. Some smaller gaps and three interruptions of monitoring in the winters 1996/1997 and 2005/2006 and from spring 2001 to spring 2002 occurred during this period. The annual variations exhibit the well-known patterns. The main water vapour maximum occurs at the stratopause region. The height of this peak varies by about 5 km over the year. The peak altitude is highest in August and occurs at about 50 km. Largest mixing ratios occur during autumn when the peak altitude has declined. A secondary maximum of the water vapour mixing ratio occurs at 65–70 km between the summer months and autumn, but it is absent during the rest of the year. The annual maximum in constant heights propagates downward from the upper mesosphere to the upper stratosphere beginning in late June through November. Particularly after winter solstice in the wake of SSWs the water vapour mixing ratios increase so that the annual minima appear as early as late November/early December and a long-stretched second winter minimum occurs in the lower mesosphere in March/April. The stratospheric warming of February 1998 present in our data set has been analysed in more detail. *Seele and Hartogh* [2000] illustrate the relationship of stratospheric warmings and the increase of water in the middle to upper stratosphere and mesosphere.

Figure 7.2 displays the 7 day sliding average of the mean annual variation of the water vapour mixing ratio for the observations. A secondary water vapour maximum appears between 65 and 70 km. As mentioned before, the annual minimum values occur already in late November and early December before the SSW season starts. In the stratopause region a second slight long-stretched minimum occurs in late winter/early spring. Figure 7.3 displays the monthly mean values at 50, 60, 70, and 80 km altitude. This figure demonstrates that the summer values at 70 km are equal to or often slightly higher than the values at 60 km.

A cursory inspection of Figs. 7.1 and 7.3 reveals an already decreasing tendency of water vapour mixing ratios. Particularly after 2001 the water vapour mixing ratios decreased suddenly with the largest change in the upper mesosphere, whereas during the years prior it seemed to increase slightly or to stagnate. This sudden decrease has been observed by other groups in low and mid latitudes [e.g. *Rosenlof and Reid*, 2008]. We present it for the first time in high latitudes. However, the behaviour is different for the summer and winter months. Figure 7.4 exhibits trend analyses of the ALOMAR measurements for the three summer months June, July and August and the three winter months December, January and February at the height levels 50, 60, 70, and 80 km. The results generally exhibit a decreasing behaviour. At 50 km in summer the correlation coefficient of the trend for the observations is very small with $r = -0.287$ and at 80 km a sudden decline occurred after the interruption

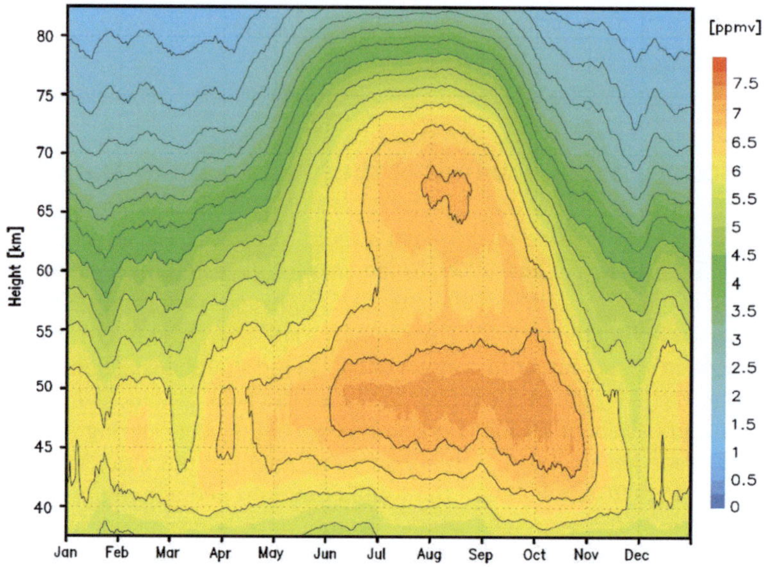

Fig. 7.2 Seven-day sliding mean of the annual variation of the water vapour mixing ratio at ALOMAR averaged over 11 years. This figure is from *Hartogh et al.* [2010]

Fig. 7.3 Monthly mean values of the water vapour mixing ratio measured at ALOMAR at 50, 60, 70 and 80 km altitude. This figure is from *Hartogh et al.* [2010]

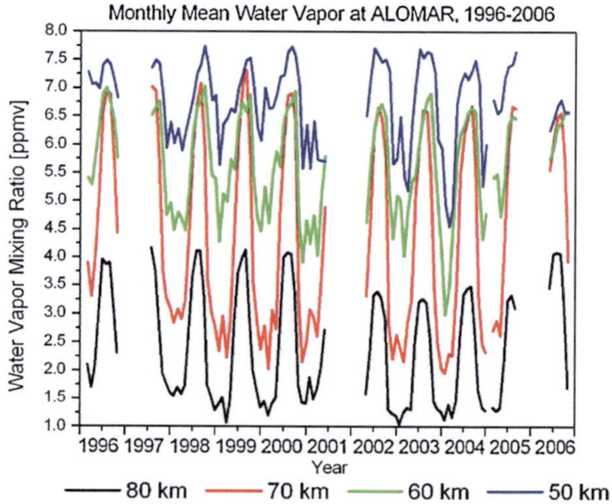

of the monitoring in 2002. In winter the behaviour is essentially clearer, when a general tendency of water vapour decrease exists in the mesosphere. Because the monitoring interval which covers slightly more than a solar cycle begins shortly before the solar activity minimum phase and ends also shortly before the next solar activity minimum phase (the maximum years were around 2001/2002), the declining tendency cannot directly result from the influence of the solar activity. We note that

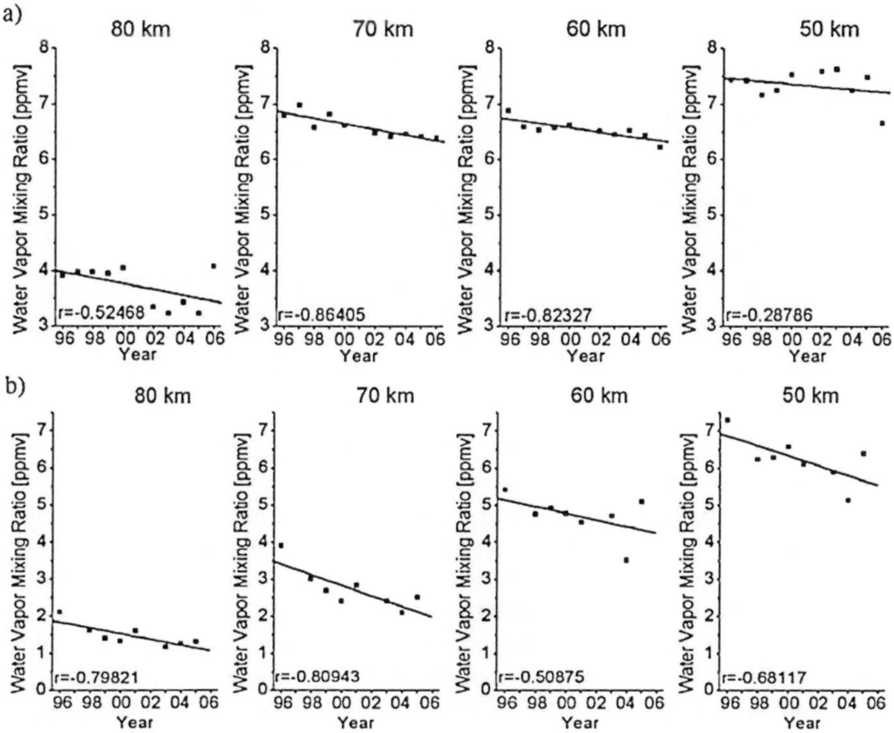

Fig. 7.4 Correlation analyses of the ALOMAR measurements for (**a**) the three summer months June, July, and August and (**b**) the three winter months December, January, and February at the height levels 50, 60, 70 and 80 km. This figure is from *Hartogh et al.* [2010]

the solar activity during the minimum phases does not considerably differ from one solar cycle to the next.

Figure 7.5 depicts the relative change of water vapour mixing ratio (in %) for an increase of the Lyman-α radiation by 1×10^{11} photons $cm^{-2} s^{-1}$ for summer and winter months. The lowest mean Lyman-α flux values did not fall below 3.5×10^{11} photons $cm^{-2} s^{-1}$. The uppermost value did not exceed 6.25×10^{11} photons $cm^{-2} s^{-1}$. As expected the winter values show a clear anti-correlation to the Lyman-α radiation in the middle and upper mesosphere with growing response with increasing height. In the lower mesosphere, however, the response is only weakly negative. One reason of this behaviour in the lower mesosphere could be that SSWs occurred more frequently and were stronger during times of high solar activity [*Sonnemann and Grygalashvyly*, 2007]. SSWs enhance the water vapour mixing ratio particular in the lower mesosphere [*Seele and Hartogh*, 2000]. This behaviour is clearly demonstrated in Fig. 7.1 (e.g. the red to orange narrow spikes in February and December 1998) The spikes in the water vapour mixing ratios in the lower mesosphere in winter result from the impact of SSWs.

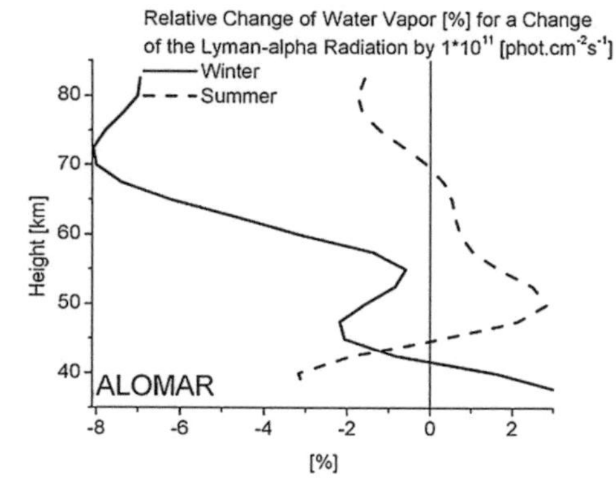

Fig. 7.5 Relative change of water vapour mixing ratio (in percent) for an increase of the Lyman-α radiation by 1×10^{11} photons $\text{cm}^{-2}\,\text{s}^{-1}$ with summer and winter months considered separately (from ALOMAR data). This figure is from *Hartogh et al.* [2010]

The response to the Lyman-α radiation is completely different in summer than in winter. The response is positive below 70 km with only a small negative value between 70 and 80 km. The reason seems to be understandable, as the optical depth of penetration of unity for the Lyman-α radiation ranges around 75 km depending on the solar zenith angle. The Lyman-α radiation is the most important radiation dissociating water vapour in the mesosphere which varies strongly with the solar activity by approximately a factor of 2 from minimum to maximum [*Woods et al.*, 2000]. During winter the downward directed vertical wind conveys air which is poor in water vapour and is impacted by the varying Lyman-α radiation from the mesopause region into the lower domain, whereas the situation is different in summer, when relatively humid air not strongly influenced by the Lyman-α radiation is lifted upward. Below about 70–75 km the effective lifetime of water vapour is extremely large, on the order of several months, and even changes its sign below about 65 km [*Sonnemann et al.*, 2005]. The effective lifetime includes both loss and production of the considered constituent. The largest part of dissociated water vapour returns to water vapour in some so-called zero cycles. But below about 65 km, water vapour is autocatalytically produced from the reservoir of molecular hydrogen and the rest of methane, thus increasing dissociating radiation amplifies this effect. This is the reason why, under conditions of high solar activity, water vapour increases in the domain below 65 km.

The negative trend of the water vapour mixing ratio is amazing because the anthropogenic growth of methane would be expected to increase the middle atmospheric humidity. Methane is not subjected to the freeze-drying at the hygropause and can enter into the stratosphere in the tropics. However, on the one hand the methane increase has seemed to be stopped or slowed down in the recent past [*Khalil et al.*, 1993; *Dlugokencky et al.*, 2003], and on the other hand only one part of the stratospheric variation of water vapour can be contributed to the methane oxidation. The other part results from a natural variability [*Forster and Shine*, 1999] such as the Brewer-Dobson circulation connected with exchange processes between troposphere and stratosphere. The global circulation also influences the water vapour transport in the mesosphere of high latitudes.

Although Fig. 7.5 indicates an influence of the solar activity on the mesospheric water vapour distribution, it is difficult to distinguish between the impact resulting from the variation of the Lyman-α radiation and that part caused by the internal dynamics which is not influenced by the solar activity. As reported by *Randel et al.* [2006] and *Scherer et al.* [2008] the Brewer–Dobson circulation in the tropics changed abruptly after 2001, impacting the water vapour distribution in the lower stratosphere. *Bittner et al.* [2000] and *Höppner and Bittner* [2007] found a slowdown of the planetary wave activity also in middle latitudes. Precisely such a sudden decrease in the water vapour mixing ratio was also observed at ALOMAR determining the trend in the whole period. This sudden change in the middle atmospheric dynamics during and after the solar maximum influences the correlation analysis between water vapour and solar Lyman-α radiation. Above 80 km the dependence is negative for all seasons including the summer months. Deducing from HALOE, *Chandra et al.* [1997] found a variation of the water vapour concentration over a solar cycle by 30–40 % at 80 km and only 1–2 % in the lower mesosphere (60–65 km). Also from HALOE, *Hervig and Siskind* [2006] derived a clear anti-correlation between Lyman-α radiation and water vapour mixing ratio at high latitude at 80 km in summer, as was likewise derived from the ALOMAR observations.

7.4 Description and Analysis of the Ozone Data Set

Figure 7.6 shows the single night averages of the ozone mixing ratio (called nighttime ozone values) at MPS between 50 and 80 km for different heights for the period December 1998 to December 2004 (black points) and the running average (red lines) using a Gaussian function with an 8 day full width at half maximum (FWHM). Figure 7.6 clearly exhibits the recurrence of typical annual patterns, but it also displays pronounced short-term variations, particularly in the middle to upper mesosphere and at the stratopause. There are no strong annual and short-term variations in the mesosphere above 60 km during day (not shown here). Typically, the strongest variations occur around the stratopause instead.

In December 2003 and January 2004 (a period of moderate solar activity) an SSW event occurred which may be responsible for the ozone increase at 75 km and above. Several solar proton events occurred in October and November 2003, forming nitric oxide [*Seppälä et al.*, 2004], but these events had obviously no distinct influence on ozone at MPS. The chemistry in the middle latitude mesosphere is almost pure: an odd oxygen odd hydrogen chemistry [*Crutzen et al.*, 1995]. An ozone decrease at 50 km in December 2003 could be interpreted by positive feedback between ozone and the ozone dissociation rate introduced by *Sonnemann and Hartogh* [2009] and *Hartogh et al.* [2011a]. A maximum of ozone occurs between 65 and 80 km around winter solstice. It is most pronounced at 70 and 75 km. Above 65 km after summer solstice the annual minimum appears. Below 65 km the annual period is subjected to a phase jump of approximately 180°. For lower altitudes the

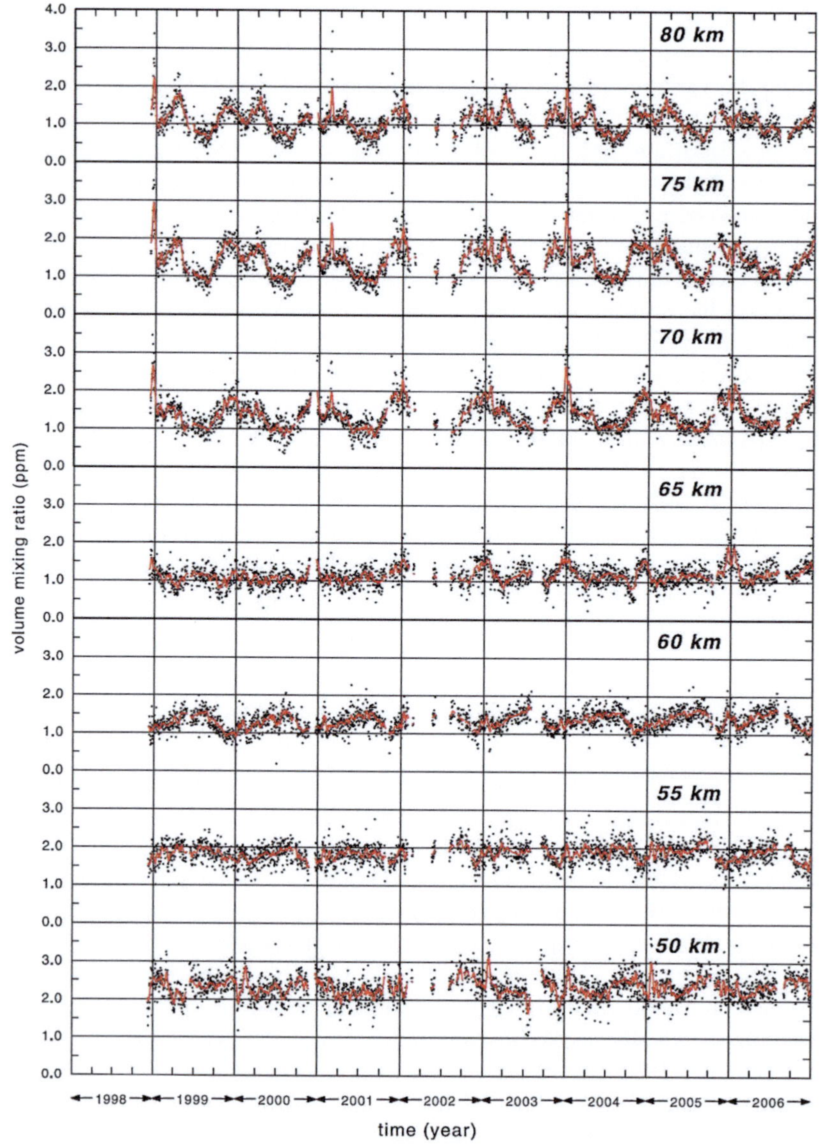

Fig. 7.6 Nighttime ozone mixing ratio at MPS between 50 and 80 km from December 1998 to December 2006. Heights are resolved in 5 km steps. The *black dots* represent the observations, the *red line* a running mean using a Gaussian function with an 8 day FWHM. This figure is from *Hartogh et al.* [2011a]

annual ozone maximum is shifted from summer into fall with decreasing height. The deviations of the current ozone values from the mean annual variation are most marked at the stratopause at 50 km. The annual period is smallest around 65 km.

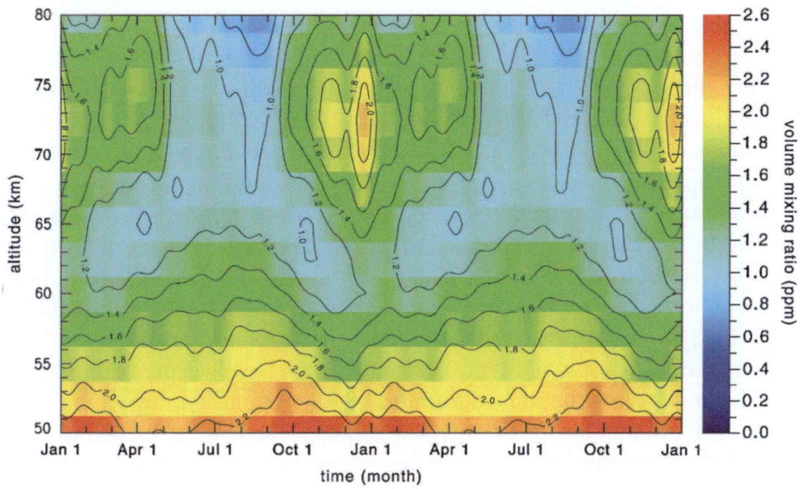

Fig. 7.7 Contour plot of the mean nighttime ozone mixing ratio at MPS using the first 12 Fourier coefficients. The figure clearly shows two annual maxima. This figure is from *Hartogh et al.* [2011a]

Figure 7.7 displays the mean ozone mixing ratio in a contour plot based on a Fourier analysis using the first 12 Fourier coefficients. Compared with Fig. 7.6, it exhibits more details of the intra-annual variations. Figure 7.7 shows one after another two identical years so that the annual variation of the mean ozone-mixing ratio during summer and particularly during winter is good to recognise without a step on the first January. As Fig. 7.7 makes clear, a subsidiary maximum appears around spring equinox. Its altitude lies somewhat higher than that of the main maximum around winter solstice. There is a remarkable asymmetry of the ozone distribution in this region, although the solar insolation is symmetric with respect to the time from the solstices. The ozone values are clearly larger in the domain of the middle mesospheric maximum (MMM) around 72 km (also known as tertiary maximum of ozone) during the first half of the winter season than in the second half. A small premaximum occurs as early as November. The phase jump below 65 km can be clearly recognised. During the summer months the strong ozone depletion at 80 km can be seen. As a simple visual inspection seems to reveal, the subsidiary maximum could result from a decrease of ozone in January/February. However, this does not explain the seasonal asymmetry, marked by a slower decrease of ozone toward the summer minimum than the respective increase toward the winter maximum. Below 65 km the annual variations with the largest values in summer are distinctly recognisable. A faint secondary maximum also occurs in winter. With decreasing height all maxima are shifted from the solstices to later days.

In the middle and upper mesosphere the enhancement of the nighttime ozone mixing ratio marked by oscillatory patterns, as shown in Fig. 7.6, is a regular feature at MPS in the winter season. We define nighttime value as the average between sunrise and sunset. An explanation for the nighttime enhancement of ozone in high latitudes in winter (the MMM) was first given by *Marsh et al.* [2001] and later given in more detail by *Hartogh et al.* [2004]. According to the suggestion of these groups, the general cause of the formation of the MMM is given by the fact that the solar radiation dissociating water vapour, producing hydrogen radicals in this process which destroy ozone, decreases faster with increasing solar zenith angle than that part of radiation dissociating molecular oxygen and thus forming ozone. There is an increasing imbalance with the rising solar zenith angle between ozone production and hydrogen radical formation. The increase of ozone for this increasing solar zenith angle is confined by the fact that the radiation which dissociates ozone is nearly not absorbed, even for grazing incidence of radiation penetrating the domain of the MMM [*Hartogh et al.*, 2004].

Another phenomenon which must be understood and interpreted consists of the seasonal asymmetry of the annual variation of the ozone mixing ratios which are modulated by wave-like oscillations. The strong year-to-year variability indicates that the effect also depends on variable conditions in the mesosphere changing considerably from year to year. The solar insolation is, of course, symmetric with regard to the solstices. The water vapour distribution shows an asymmetric annual variation, and the molecular hydrogen varies inversely to water vapour *Sonnemann and Grygalashvyly* [2005]. Water vapour as source gas for the hydrogen radicals is at a maximum in late summer and a minimum at the time of the subsidiary nighttime ozone maximum [e.g. *Seele and Hartogh*, 1999, 2000; *Körner and Sonnemann*, 2001; *Hartogh et al.*, 2010]. This subsidiary maximum is explained by the decrease of ozone after winter solstice due to the more frequent occurrence of sudden stratospheric warming (SSW) before this period and by the occurrence of the annual minimum of water vapour during this period [*Hartogh et al.*, 2011b]. The SSW events are connected with a cooling of the mesosphere above about 65 km which should increase ozone. However, they are also connected with a drastic change of the dynamic patterns. The zonal wind considerably influences the night-to-day ratio (NDR) of ozone as a result of the photochemical Doppler-effect [*Sonnemann*, 2001]. The magnitude of the MMM is considerably determined by this effect. The change of the zonal wind from a west wind into an east wind system reduces the NDR and reduces the nighttime ozone concentration. The NDR displays a wintertime enhancement modulated by a planetary wave-like variation resembling the winter anomaly of the plasma parameter of the D layer [*Schwentek*, 1971]. A possible reason for these patterns could be linked to the gravity wave activity. The meridional wind is able to disperse the MMM, but for a southward blowing wind it can also transport air rich in ozone from high latitudes into middle latitudes. The annual variation of the water vapour mixing ratio also considerably influences the annual variation of ozone, particularly at the upper levels as water vapour is the source gas for the hydrogen radicals destroying ozone. In the middle to lower mesosphere an ozone maximum occurs in late summer. With decreasing height this maximum is shifted into autumn. A spring minimum and a broad autumn/winter maximum have been observed at the stratopause. The concrete annual ozone variations in different heights depend on the annual variations of the solar insolation, the temperature, and the water vapour concentration. There are pronounced ozone variations, particularly around the stratopause, which are connected with SSW events and water vapour variations. On the whole, the measured values agree fairly well with the Ozone Reference Model [*Keating et al.*, 1990] presenting day values. However, in detail larger differences also occurred.

7.5 Summary and Conclusions

We investigated two long term data sets of mesospheric ozone and water vapour at mid and polar latitudes and tried to disentangle the complex influence and interaction of dynamics and chemistry on the variability of the two species. Although the water vapour concentration was superimposed by strong dynamical signals caused by changes in the Brewer-Dobson circulation after 2001, we found a negative correlation with solar activity in the winter mesosphere and a positive correlation in the summer mesosphere. In winter the Lyman-α signal in water is transported downward by the downward directed vertical wind, while in summer up-welling humid air reflects (upper) stratospheric water concentrations and their changes being not directly related to solar irradiance. The positive summer correlation in the mesosphere is most likely related to autocatalytic water production during high solar activity.

Our measurements brought evidence that the middle mesospheric maximum of ozone extends into mid latitudes. A subsidiary maximum appearing in late winter/spring equinox was explained by the increase of ozone due to a decline of sudden stratospheric warmings which sporadically enhance the mesospheric water concentration and by the lowest values of the annual water cycle during this period. In the middle and lower mesosphere an ozone maximum occurs in late summer. With decreasing height this maximum is shifted into autumn. A spring minimum and a broad winter maximum have been observed at the stratopause. The concrete annual ozone variations in different altitudes depend on the annual variations of the solar insolation, the temperature and the water vapour concentration. There are pronounced ozone variations, particularly around the stratopause, which are connected with SSWs appearing more frequently during phases of high solar activity. The enhanced temperatures and water vapour concentrations appearing in the vicinity of SSWs result in a strong decrease of ozone.

Acknowledgements We appreciate the financial support from the DFG for this project during the last 2 CAWSES funding periods.

References

Bittner, M., Offermann, D., & Graef, H. H. (2000). Mesopause temperature variability above a midlatitude station in Europe. *Journal of Geophysical Research*, *105*, 2045–2058. doi:10.1029/1999JD900307.

Chandra, S., Jackman, C. H., Fleming, E. L., & Russell III, J. M. (1997). The seasonal and long term changes in mesospheric water vapor. *Geophysical Research Letters*, *24*, 639–642. doi:10.1029/97GL00546.

Crutzen, P. J., Groob, J.-U., Bruhl, C., Muller, R., & Russell, J. M. III (1995). A reevaluation of the ozone budget with HALOE UARS data: no evidence for the ozone deficit. *Science*, *268*, 705–708. doi:10.1126/science.268.5211.705.

de F. Forster, P. M., & Shine, K. P. (1999). Stratospheric water vapor changes as a possible contributor to observed stratospheric cooling. *Geophysical Research Letters*, *26*, 3309–3312. doi:10.1029/1999GL010487.

Dlugokencky, E. J., Houweling, S., Bruhwiler, L., Masarie, K. A., Lang, P. M., Miller, J. B., & Tans, P. P. (2003). Atmospheric methane levels off: temporary pause or a new steady-state? *Geophysical Research Letters*, *30*(19), 1992. doi:10.1029/2003GL018126.

Hallgren, K., Hartogh, P., & Jarchow, C. (2010). A new, high-performance, heterodyne spectrometer for ground-based remote sensing of mesospheric water vapour. In A. Bhardwaj, S. A. Haider, P. Hartogh, W.-H. Ip, T. Ito, Y. Kasaba, G. M. Muños Cara & C. Y. R. Wu (Eds.), *Advances in geosciences* (Vol. 19, pp. 569–578). Singapore: World Scientific.

Hartogh, P., & Hartmann, G. K. (1990). A high-resolution chirp transform spectrometer for microwave measurements. *Measurement Science & Technology*, *1*(7), 592–595.

Hartogh, P., & Jarchow, C. (1995). Groundbased detection of middle atmospheric water vapor. In *EUROPTO-series: Vol. 2586. Global process monitoring and remote sensing of ocean and sea ice* (pp. 188–195). Bellingham: SPIE.

Hartogh, P., Hartmann, G. K., & Zimmermann, P. (1991). Simultaneous water vapor and ozone measurements with millimeter-waves in the stratosphere and mesosphere. In *IEEE catalog number 91CH2971-0* (Vol. I, pp. 227–230). Helsinki.

Hartogh, P., Jarchow, C., Sonnemann, G. R., & Grygalashvyly, M. (2004). On the spatiotemporal behaviour of ozone within the upper mesosphere/mesopause region under nearly polar night conditions. *Journal of Geophysical Research*, *109*, D18303. doi:10.1029/2004JD004576.

Hartogh, P., Sonnemann, G. R., Grygalashvyly, M., Berger, U., & Lübken, F.-J. (2010). Water vapor measurements at ALOMAR over a solar cycle compared with model calculations by LIMA. *Journal of Geophysical Research*, *115*, D00I17. doi:10.1029/2009JD012364.

Hartogh, P., Jarchow, C., Sonnemann, G. R., & Grygalashvyly, M. (2011a). Ozone distribution in the middle latitude mesosphere as derived from microwave measurements at Lindau (51.66N, 10.13E). *Journal of Geophysical Research*, *116*, D04305. doi:10.1029/2010JD014393.

Hartogh, P., Sonnemann, G. R., Grygalashvyly, M. & Jarchow, Ch. (2011b). Ozone trends in the mid-latitude stratopause region based on microwave measurements at Lindau (51.66N°, 10.13E°), the ozone reference model, and model calculations. *Advances in Space Research*, *47*, 11. doi:10.1016/j.asr.2011.01.010.

Hervig, M., & Siskind, D. (2006). Decadal and inter-hemispheric variability in polar mesospheric clouds, water vapor, and temperature. *Journal of Atmospheric and Solar-Terrestrial Physics*, *68*, 30–41. doi:10.1016/j.jastp.2005.08.010.

Höppner, K., & Bittner, M. (2007). Evidence for solar signals in the mesopause temperature variability? *Journal of Atmospheric and Solar-Terrestrial Physics*, *69*, 431–448. doi:10.1016/j.jastp.2006.10.007.

Jarchow, C., & Hartogh, P. (1995). Retrieval of data from ground-based microwave sensing of the middle atmosphere: comparison of two inversion techniques. In *EUROPTO-series: Vol. 2586. Global process monitoring and remote sensing of ocean and sea ice* (pp. 196–205). Bellingham: SPIE.

Jarchow, C., & Hartogh, P. (1998). Analysis of forward models using the singular value decomposition algorithm. In J. D. Haigh (Ed.), *Proceedings of SPIE: Vol. 3220. Satellite remote sensing of clouds and the atmosphere II* (pp. 163–173). London.

Keating, G. M., Pitts, M. C., & Young, D. F. (1990). Ozone reference models for the middle atmosphere. *Advances in Space Research*, *10*, 317–355. doi:10.1016/0273-1177(90)90404-N.

Khalil, M. A. K., Rasmussen, R. A., & Moraes, F. (1993). Atmospheric methane at Cape Meares—analysis of a high-resolution data base and its environmental implications. *Journal of Geophysical Research. Atmospheres*, *981*, 14753. doi:10.1029/93JD01197.

Körner, U., & Sonnemann, G. R. (2001). Global three-dimensional modeling of the water vapor concentration of the mesosphere-mesopause region and implications with respect to the noctilucent cloud region. *Journal of Geophysical Research*, *106*, 9639–9652. doi:10.1029/2000JD900744.

Marsh, D., Smith, A., Brasseur, G., Kaufmann, M., & Grossmann, K. (2001). The existence of a tertiary ozone maximum in the high-latitude middle mesosphere. *Geophysical Research Letters*, *28*, 4531–4534. doi:10.1029/2001GL013791.

Poynter, R. L., & Pickett, H. M. (1985). Submillimeter, millimeter, and microwave spectral line catalog. *Applied Optics*, *24*, 2235–2240. doi:10.1364/AO.24.002235.

Randel, W. J., Wu, F., Vömel, H., Nedoluha, G. E., & Forster, P. (2006). Decreases in stratospheric water vapor after 2001: links to changes in the tropical tropopause and the Brewer-Dobson circulation. *Journal of Geophysical Research*, *111*(D10), D12312. doi:10.1029/2005JD006744.

Rodgers, C. D. (1976). Retrieval of atmospheric temperature and composition from remote measurements of thermal radiation. *Reviews of Geophysics and Space Physics*, *14*, 609.

Rosenkranz, P. W. (1993). Absorption of microwaves by atmospheric gases. In M. A. Janssen (Ed.), *Atmospheric remote sensing by microwave radiometry* (p. 37).

Rosenlof, K. H., & Reid, G. C. (2008). Trends in the temperature and water vapor content of the tropical lower stratosphere: sea surface connection. *Journal of Geophysical Research*, *113*(D12), D06107. doi:10.1029/2007JD009109.

Scherer, M., Vömel, H., Fueglistaler, S., Oltmans, S. J., & Staehelin, J. (2008). Trends and variability of midlatitude stratospheric water vapour deduced from the re-evaluated Boulder balloon series and HALOE. *Atmospheric Chemistry and Physics*, *8*, 1391–1402.

Schwentek, H. (1971). Regular and irregular behaviour of the winter anomaly in ionospheric absorption. *Journal of Atmospheric and Terrestrial Physics*, *33*, 1647.

Seele, C., & Hartogh, P. (1999). Water vapor of the polar middle atmosphere: annual variation and summer mesosphere conditions as observed by ground-based microwave spectroscopy. *Geophysical Research Letters*, *26*(11), 1517–1520.

Seele, C., & Hartogh, P. (2000). A case study on middle atmospheric water vapor transport during the February 1998 stratospheric warming. *Geophysical Research Letters*, *27*(20), 3309–3312.

Seppälä, A., Verronen, P. T., Kyrölä, E., Hassinen, S., Backman, L., Hauchecorne, A., Bertaux, J. L., & Fussen, D. (2004). Solar proton events of October-November 2003: ozone depletion in the Northern Hemisphere polar winter as seen by GOMOS/Envisat. *Geophysical Research Letters*, *311*, L19107. doi:10.1029/2004GL021042.

Sonnemann, G. R. (2001). The photochemical effects of dynamically induced variations in solar insolation. *Journal of Atmospheric and Solar-Terrestrial Physics*, *63*, 781–797. doi:10.1016/S1364-6826(01)00010-4.

Sonnemann, G. R., & Grygalashvyly, M. (2005). Solar influence on mesospheric water vapor with impact on nlcs. *Journal of Atmospheric and Solar-Terrestrial Physics*, *67*, 177–190. doi:10.1016/j.jastp.2004.07.026.

Sonnemann, G. R., & Grygalashvyly, M. (2007). The relationship between the occurrence rate of major stratospheric warmings and solar Lyman-alpha flux. *Journal of Geophysical Research*, *112*(D11), D20101. doi:10.1029/2007JD008718.

Sonnemann, G. R., & Hartogh, P. (2009). Upper stratospheric ozone decrease events due to a positive feedback between ozone and the ozone dissociation rate. *Nonlinear Processes in Geophysics*, *16*, 409–418.

Sonnemann, G. R., Grygalashvyly, M., & Berger, U. (2005). Autocatalytic water vapor production as a source of large mixing ratios within the middle to upper mesosphere. *Journal of Geophysical Research*, *110*(D9), D15303. doi:10.1029/2004JD005593.

Sonnemann, G. R., Hartogh, P., Grygalashvyly, M., Li, S., & Berger, U. (2008). The quasi 5-day signal in the mesospheric water vapor concentration at high latitudes in 2003—a comparison between observations at ALOMAR and calculations. *Journal of Geophysical Research*, *113*(D12), D04101. doi:10.1029/2007JD008875.

Stevens, M. H., Gumbel, J., Englert, C. R., Grossmann, K. U., Rapp, M., & Hartogh, P. (2003). Polar mesospheric clouds formed from space shuttle exhaust. *Geophysical Research Letters*, *30*(10), 1546. doi:10.1029/2003GL017249.

Straub, C., Murk, A., Kämpfer, N., Golchert, S. H. W., Hochschild, G., Hallgren, K., & Hartogh, P. (2011). ARIS-Campaign: intercomparison of three ground based 22 GHz radiometers for middle atmospheric water vapor at the Zugspitze in winter 2009. *Atmospheric Measurement Techniques*, *4*, 1979–1994. doi:10.5194/amt-4-1979-2011.
Villanueva, G., & Hartogh, P. (2004). The high resolution chirp transform spectrometer for the Sofia-Great instrument. *Experimental Astronomy*, *18*, 77–91. doi:10.1007/s10686-005-9004-3.
Villanueva, G. L., Hartogh, P., & Reindl, L. M. (2006). A digital dispersive matching network for SAW devices in chirp transform spectrometers. *IEEE Transactions on Microwave Theory and Techniques*, *54*, 1415–1424. doi:10.1109/TMTT.2006.871244.
Woods, T. N., Tobiska, W. K., Rottman, G. J., & Worden, J. R. (2000). Improved solar Lyman α irradiance modeling from 1947 through 1999 based on UARS observations. *Journal of Geophysical Research*, *105*, 27195–27216. doi:10.1029/2000JA000051.

Chapter 8
Influence of Solar Radiation on the Diurnal and Seasonal Variability of O_3 and H_2O in the Stratosphere and Lower Mesosphere, Based on Continuous Observations in the Tropics and the High Arctic

Mathias Palm, Sven H.W. Golchert, Miriam Sinnhuber, Gerd Hochschild, and Justus Notholt

Abstract During the CAWSES DFG (German Research Association) priority program measurements of stratospheric and mesospheric O_3 using ground based millimeterwave radiometry have been established and analyzed. Instruments have been operated at two different locations, at Mérida, Venezuela, a high altitude tropic station and at Ny Ålesund, Spitsbergen, an Arctic station. Additionally, data obtained from the millimeterwave radiometer based at Kiruna, Sweden, have been used for an analysis of the 5-day planetary wave.

Measurements of O_3 have yielded short term variations in the stratosphere and mesosphere, i.e. diurnal variations. Discrepancies between measured and modeled diurnal amplitude have been found and partially explained.

H_2O measurements are more difficult than O_3 measurements, as a result of its weak emission and strong tropospheric absorption. Nevertheless considerable effort has been put into the enhancement of H_2O measurements using ground based millimeterwave radiometry and the suitability of such measurements could be demonstrated in a campaign at the Schneefernerhaus, Germany (Zugspitze).

8.1 General Information

The Institut für Meteorologie und Klimaforschung[1] (IMK) of the Karlsruhe Institute of Technology (KIT) and the Institut für Umweltphysik[2] (IUP) of the Universität

[1] Institute for Meteorology and Climate Research.

[2] Institute of Environmental Research.

M. Palm (✉) · J. Notholt
Institut für Umweltphysik, Universität Bremen, Otto-Hahn-Allee 1, 28359 Bremen, Germany
e-mail: mathias@iup-physik.uni-bremen.de

S.H.W. Golchert · M. Sinnhuber · G. Hochschild
Institut für Klimaforschung, KIT Karlsruhe, Hermann-von-Helmholtz-Platz 1, 76344, Eggenstein-Leopoldshafen, Germany

F.-J. Lübken (ed.), *Climate and Weather of the Sun-Earth System (CAWSES)*, Springer Atmospheric Sciences,
DOI 10.1007/978-94-007-4348-9_8,

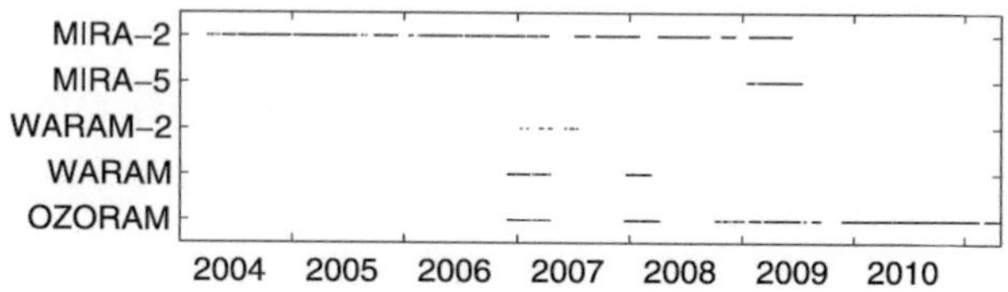

Fig. 8.1 A time series of all measurements obtained during this project

Bremen have been cooperating for several years in the development and deployment of millimeterwave radiometers in order to measure thermally excited radiation emitted from the atmosphere.

The IMK operated a tropical station on the Pico Espejo (8N, 71W, 4765 m altitude) near Mérida, Venezuela, in cooperation with the Universidad de los Andes (ULA). Two instruments, MIRA2 and WARAM2, measured emissions from O_3 and H_2O, respectively. Unfortunately, the station on top of the Pico Espejo became inaccessible due to a severe malfunction of the cable car, which is the only practicable working access to the mountain top. The instruments could be salvaged in 2010 but another suitable high mountain station in the tropics is yet to be found.

Additionally, the IMK developed KIMRA, a millimeterwave radiometer located in Kiruna, Sweden (67.8N, 20.2E) [*Raffalski et al.*, 2005], now operated by the Institutet för rymdfysik (IRF) Kiruna.[3] While this instrument was not part of the CAWSES project, data obtained from its measurements have been used towards studies of CAWSES related topics in Sect. 8.5.5.

The IUP operates and maintains the millimeterwave radiometer OZORAM on the AWIPEV research base at Ny Ålesund , Spitsbergen, Norway (78.9N, 11.9E). The AWIPEV research base is jointly run by the **A**lfred **W**egener **I**nstitute (AWI)[4] and the **I**nstitut polaire français **P**aule **E**mile **V**ictor (IPEV),[5] the polar research institutions of Germany and France, respectively.

This chapter presents the outcome of the project SACOSAT, which was collaboratively worked on by the MWR group of the IMK-ASF of the KIT Karlsruhe and the RAM group within the IUP of the Universität Bremen.

Figure 8.1 shows the days during which measurements of the respective instruments are available. The OZORAM, the WARAM, both at Spitsbergen and WARAM-2 at the Pico Espejo have been developed and operated by the Universität Bremen. MIRA2 and MIRA5 have been developed and operated by the KIT Karlsruhe.

[3]Institute of Space Physics, www.irf.se.

[4]www.awi.de.

[5]www.institut-polaire.fr.

8.2 Measurements of O_3 from the Ground Using Millimeterwave Radiometry

Since the 1970s measurements of millimeterwaves emitted from the atmosphere have been used to gain information about atmospheric quantities.

The most important applications have been the measurements of temperature [*Waters*, 1973] and trace gas species, e.g. O_3 [*Penfield et al.*, 1976], H_2O [*Nedoluha et al.*, 1995], CO [*Waters et al.*, 1976; *Kunzi and Carlson*, 1982], ClO [*Parrish et al.*, 1981; *Klein et al.*, 2000, 2002], N_2O [*Connor et al.*, 1987], HCN [*Jaramillo et al.*, 1988] and HNO_3 [*Kuntz et al.*, 1999]. At their outset, the measurements were technically demanding, no long term data sets have been derived. In the last years the technology has evolved in order to make very stable and long term measurements feasible, e.g. H_2O measurements [*Nedoluha et al.*, 1995] or O_3-measurements [*Boyd et al.*, 2007].

Millimeterwave measurements are performed at specific stations around the world which are organized in the Network for Detection of Atmospheric Composition Change (NDACC).[6]

8.2.1 *The OZORAM Instrument on Ny Ålesund, Spitsbergen*

The following section summarizes the description of the OZORAM instrument and the data set derived from its measurements published by *Palm et al.* [2010]. For a detailed account please refer this publication.

OZORAM is a heterodyne single side-band receiver tuned to the 142.176 GHz emission line ($10_{1,9} \rightarrow 10_{0,10}$) of O_3. The bandwidth is 800 MHz and the frequency resolution is approx. 60 kHz. The measurement time for one spectrum is about 12 min. 5 consecutive spectra are averaged in order to improve the signal-to-noise ratio (SNR). Hence, spectra are measured with a time resolution of 1 h. See Fig. 8.2 for an example of a calibrated spectrum recorded in winter 2009 and Fig. 8.3 for the profile (in black) retrieved from it. The night time profile exhibits an enhancement in mesospheric O_3 as is expected in the absence of solar illumination. A day time profile is plotted (in green) for comparison, retrieved from a spectrum taken 12 h later. The a priori profile is plotted in red.

8.2.1.1 Measurement Geometry

Figure 8.4 depicts the line of sight of the instrument OZORAM. The azimuth of the line of sight is 113° and the elevation angle 20°. This information has to be taken into account for comparison with models and/or other instruments, especially during winter, when strong gradients in all atmospheric parameters may exist due to the presence of the polar vortex and for studies related to the day-night terminator.

---

[6] www.ndacc.org.

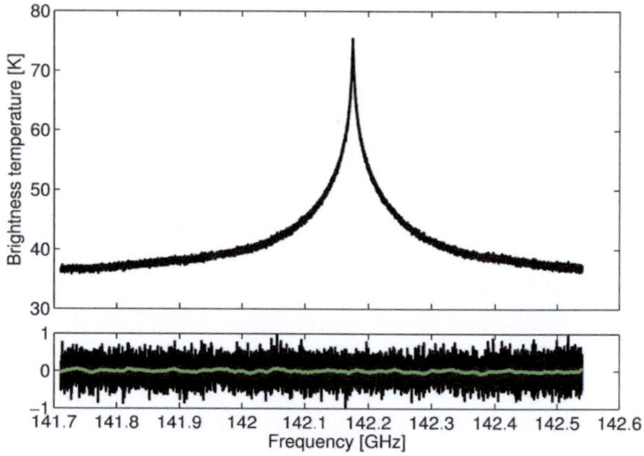

Fig. 8.2 A spectrum measured on the 25 February 2009, mid night (UTC). The *lower figure* shows the residuum after profile retrieval (see Fig. 8.3). A running mean (*green line*) has been applied to the residuum. Figure reproduced from *Palm et al.* [2010]

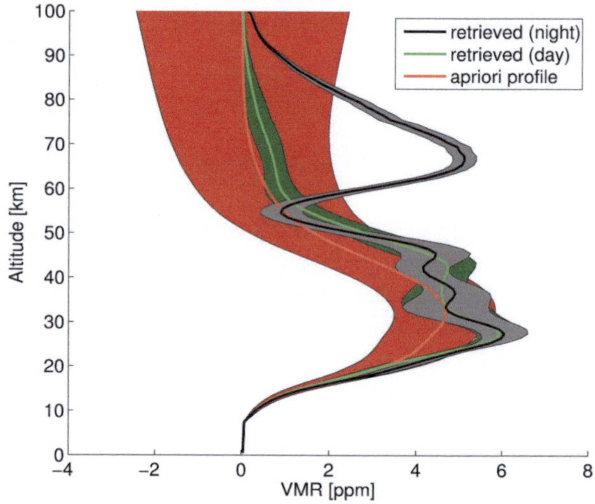

Fig. 8.3 O_3 profiles retrieved during night (*black*) and day (*green*) and the a priori profile (*red*). The shaded regions (*green*, *black*) are the uncertainties caused by the total error as derived in *Palm et al.* [2010]. The *red shaded region* is the a priori standard deviation used in the retrieval. Figure reproduced from *Palm et al.* [2010]

8.2.2 The MIRA2 Instrument in Mérida, Venezuela

The millimeterwave observations at Mérida used the ground-based millimeterwave Radiometer MIRA2. It was constructed at the Forschungszentrum Karlsruhe and measures radiation emitted from O_3, ClO, HNO_3 and N_2O in the frequency range from 268 to 280 GHz. At Mérida, MIRA2 measures in northward direction because of the building situation. A cooled Schottky diode mixer converts the signal to an intermediate frequency range centered at 2.1 GHz with a width of 1.3 GHz. The single side-band receiver noise temperature is about 700 K. The spectral analysis is performed by an acousto-optical spectrometer with a resolution of about 1.2 MHz. For balanced calibration an internal adjustable reference load is used. A detailed description of the system is given in *Berg et al.* [1998]. The measured spectra are integrated until either the noise in the resulting spectrum is at least ten times smaller

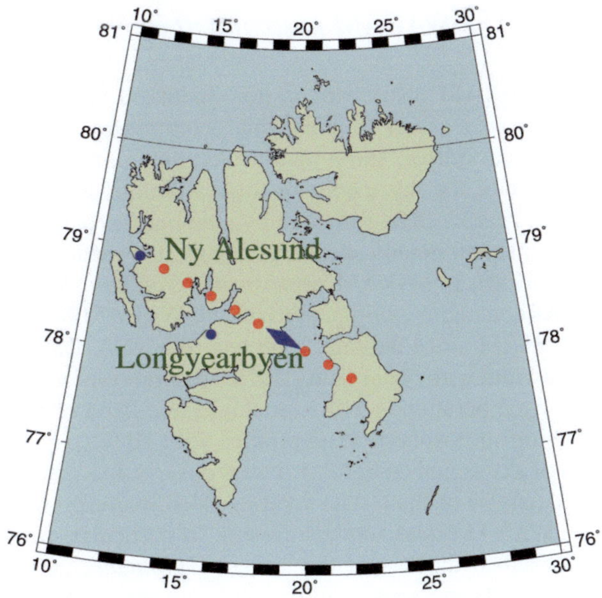

Fig. 8.4 Location and measurement geometry of the millimeterwave radiometer OZORAM at Ny Ålesund, Spitsbergen. Every dot denotes the intersection of the line of sight with the atmosphere at 10, 20, 30,... km. The *blue diamond* is the center of the profile measurement at 60 km altitude. Figure reproduced from *Palm et al.* [2010]

than the intensity of the signature of the desired trace gas or until the noise is significantly smaller than baseline artefacts. Integration times depend on the species under observation and the tropospherical conditions and range typically 1 h for O_3.

8.2.3 The KIMRA Instrument in Kiruna, Sweden

The basic design of the KIMRA follows the same conception as MIRA2, but differs in the fact, that the frequency range has been focused on the 204 GHz ClO line and that the tuning range is enlarged [*Raffalski et al.*, 2005]. In the meantime, this instrument has been successfully modified for CO measurements at 230 GHz [*Hoffmann et al.*, 2011], in particular by adding a complementary digital spectrometer with a high frequency resolution.

8.3 Measurements of H_2O Using Millimeterwave Radiometry

The authors observe middle atmospheric water vapor with ground-based millimeterwave radiometers. The instruments record thermal emission from the rotational transition at 22.235 GHz ($6_{1,6} \rightarrow 5_{2,3}$). The general measurement and retrieval setup is discussed by *Golchert* [2010].

8.3.1 The WARAM Instrument in Ny-Ålesund, Spitsbergen

The WARAM instrument started operation at Ny-Ålesund, Spitsbergen, in 1999. It employs a rather simple design, comprising a corrugated horn antenna, a single ellipsoidal mirror, and a revolvable plane mirror that selects the calibration target either to be the sky, a hot or a cold calibration load. The receiver operates an amplifier at 22.235 GHz, followed by side-band separation and heterodyne mixing. For more general details about this design, the reader is referred to a discussion of the near-identical WARAM2 instrument [*Golchert*, 2010].

After a major equipment overhaul, WARAM has been measuring in parallel with OZORAM since autumn 2006 (see Fig. 8.1). Data have been recorded at 1 GHz bandwidth with better than 20 kHz spectral resolution at the line center. The receiver noise temperature is 210 K in single side-band operation. This typically allows retrieving stratospheric H_2O profiles with 10 % (0.5 ppmv) observational error from 7 h of sky signal (roughly 1 d total observation time).

Analysis of these data soon revealed major problems that could be attributed to the WARAM quasi-optical system. In particular, the antenna appeared to be badly matched to the rest of the system [*Hoffmann*, 2008], probably because of it not meeting design specifications. This introduces highly variable spurious signals, in terms of so-called standing waves, that prohibit retrieval of geophysical parameters from the measurements in an automated, uniform fashion. These problems have encouraged a new water vapor radiometer design undertaken by the KIT team, which yielded a stable working instrument (see Sect. 8.3.3).

8.3.2 The WARAM2 Instrument in Mérida, Venezuela

The WARAM2 instrument has been derived from the WARAM blueprints and only differs from these in the signal conditioning (the so-called intermediate frequency chain) and spectrometer. It has been deployed to Pico Espejo (near Mérida, Venezuela, 8°N 71°W, 4765 m above sea level) in spring 2004. Albeit sharing its quasi-optical design with the WARAM instrument, WARAM2 measurements suffer considerably less interference from spurious signals. *Golchert* [2010] presents an operational retrieval for these data.

Continuous operation of WARAM2 has been severely hindered by limited access to the high-alpine measurement site, which caused long instrument down-times even after minor hardware break-downs. A continuous period of measurements has been performed from January to July 2007. These data feature 183 K single side-band receiver noise temperature, 1.2 GHz bandwidth and 1.1 MHz spectral resolution. Stratospheric H_2O profiles with 10 % observational error may be retrieved from 4 h of sky signal, or 14 h total observation time.

A gradual degeneration in the local oscillator rendered measurements after July 2007 invalid. In Aug 2008, the research site became completely inaccessible after the cable car (Teleférico) to Pico Espejo was diagnosed to have reached the end of its lifetime.

8.3.3 The MIRA5 Instrument at Zugspitze, Germany

The MIRA5 instrument forms a completely new and versatile set-up. Since an extensive discussion is yet to be published, some detail will be given here. MIRA5 is designed to easily allow different modes of calibration and operation, and study the effects of these on parameter retrieval. This may be used in clarifying characterization issues of the WARAM and WARAM2 instruments and with respect to newer and smaller designs planned for the future.

To add the required flexibility, MIRA5 quasi-optics comprise multiple focusing elements. This is different to all other 22 GHz receivers in regular operation [for an overview, see *Golchert*, 2010]. The MIRA5 corrugated horn antenna and first ellipsoidal mirror are mounted on a bench that can be moved back and forth towards the second ellipsoidal mirror and thus forms a path length modulator. The second mirror then switches the beam between the sky and different calibration branches. The sky branch includes one additional ellipsoidal mirror during Zugspitze operation, or several more in experimental set-ups. A revolvable plane mirror eventually determines the observation angle into the sky.

MIRA5 carries three calibration loads (microwave absorbers at well-defined physical temperatures). One of these is mounted in a vacuum Dewar and cooled by a closed-cycle helium refrigerator. This optical branch holds two more ellipsoidal mirrors. The one inside the Dewar is specially coated with infrared filter to protect the absorber from radiative heating. The refrigerator maintains a physical temperature of 15 K. The other two absorbers are held at liquid nitrogen temperature (75 K for the Zugspitze measurements) and slightly above laboratory temperature (310 K). The brightness of the refrigerated load is determined by calibration against the other two loads, and checked in monthly intervals. This results in a Rayleigh Jeans equivalent brightness temperature of 32 K, and hardly varies over time.

The described set-up allows calibrating measurements in a variety of modes. For total-power calibration, one can freely select from all three calibration loads. For balanced calibration [*Krupa et al.*, 1998], the system deploys an additional revolvable grid that combines radiation from the refrigerated cold load with radiation from another microwave absorber at ambient temperature. Reference beam calibration, which is common among the H_2O millimeterware community, has not yet been implemented, but requires little effort. It uses the sky at zenith plus an additional absorber as the reference.

Throughout Zugspitze operation, the measured signal has been recorded by an acousto-optical spectrometer (made at Universität zu Köln) and an Acqiris AC240 digital fast Fourier transform spectrometer (FFTS). These result in a bandwidth of 1.3 GHz with 62 kHz spectral resolution at the line center. Later during the Zugspitze campaign an RPG FFTS with 212 kHz spectral resolution has been added to the system (cf. *Straub et al.* [2011]). MIRA5 operates at a single side-band receiver noise temperature of 140 K. This allows retrieving a stratospheric H_2O profile with less than 10 % observational error from 2.3 h of sky signal (5–8 h total observation time, depending on calibration mode). Figure 8.5 presents a 2.3 h spectrum, obtained by averaging measurements from 26 February 2009, 16:16 UTC, until 0:04 UTC the next day.

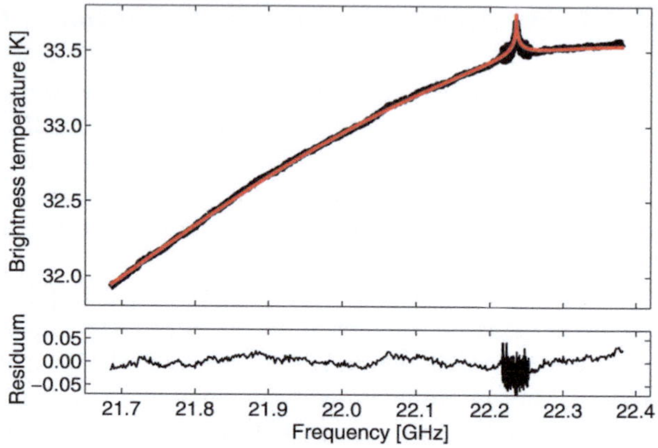

Fig. 8.5 Average of MIRA5 spectra obtained from 26 February 2009, 16:16 UTC, until 0:04 UTC the next day. The *bottom panel* shows the spectral residuum after profile retrieval (see Fig. 8.6). Data around 22.235 GHz are taken at higher spectral resolution and thus more noisy

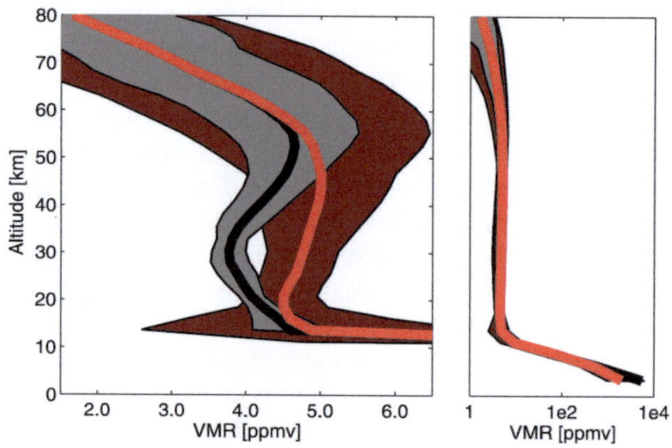

Fig. 8.6 (*black*) H_2O profile retrieved from the spectrum shown in Fig. 8.5. The a priori profile is given in *red*. The shaded regions (*black*, *red*) denote the uncertainties connected to the a priori standard deviation (see text) and total error as discussed in *Straub et al.* [2011]

Retrieval of volume mixing ratio (VMR) profiles is performed using the optimal estimation method [*Rodgers*, 2000] with a Marquardt-Levenberg iteration scheme, implemented by the ARTS/Qpack environment [*Bühler et al.*, 2005; *Eriksson et al.*, 2005]. For the results presented here, spectroscopic parameters are adopted from an earlier ground-based millimeterwave validation study [*Haefele et al.*, 2009], with an added H_2O continuum contribution according to *Rosenkranz* [1998]. A scaled FASCOD midlatitude summer scenario H_2O profile [*Anderson et al.*, 1986] forms the a priori distribution, covariance is set to 50 % in the troposphere, 10 % in the lower stratosphere, and increasing to 100 % in the upper mesosphere. Temperature and geopotential height have been taken from EOS MLS satellite instrument. A typical retrieval example is given by Figs. 8.6 and 8.7.

8.4 Chemistry of Stratospheric and Mesospheric O_3

O_3 is crucial to life on earth. It absorbs most of the UV, which is harmful to life outside water, in the stratosphere. The O_3-layer and its development is therefore a

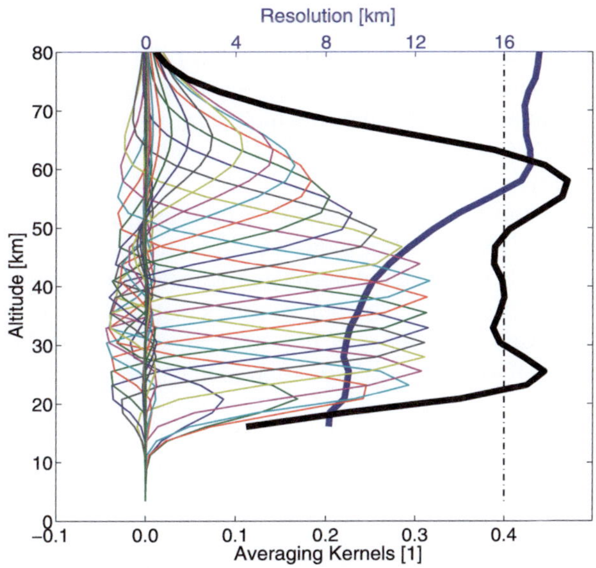

Fig. 8.7 Averaging kernels (AVK) for the result given in Fig. 8.6. *Thin line colors* denote nominal layer altitudes of the retrieval grid, starting at 16 km. The *thick black line* indicates measurement response times 0.4 (100 % measurement response is marked by *thin dash-dotted black line*). Height resolution, in terms of the AVK full width at half maximum, is given by the *thick blue line*

particular important research topic. Following the detection of the so-called O_3-hole [*Farman et al.*, 1985] science was quickly pointing out the harmfulness of a widely used class of chemicals, the chlorofluorocarbons (CFC's). The strong link which could be drawn from the release of those chemicals to the destruction of the O_3 layer led to a worldwide ban on those chemicals. The effectiveness of this ban is still a research topic and will continue to be throughout the next decade.

The atmospheric O_3 interacts strongly with solar radiation. The O_3 content of the atmosphere is largely governed by the Chapman cycle of odd oxygen [*Chapman*, 1930]:

$$O_2 + h\nu(\lambda \leq 242.4\ \mathrm{nm}) \rightarrow O + O \tag{8.1}$$

$$O + O_2 \rightarrow O_3 + M \tag{8.2}$$

O_3 is photo-dissociated by

$$O_3 + h\nu(\lambda \geq 320\ \mathrm{nm}) \rightarrow O_2(^3\Sigma_g^-) + O(^3P) \tag{8.3}$$

$$O_3 + h\nu(\lambda \leq 320\ \mathrm{nm}) \rightarrow O_2(^1\Delta_g) + O(^1D) \tag{8.4}$$

where $O(^3P)$ is the ground state and $O(^1D)$ the first excited state of atomic oxygen, respectively. Most of $O(^1D)$ is quickly quenched to $O(^3P)$ by collision with O_2 and N_2. $O(^3P)$ reacts swiftly with O_2 to reform O_3. Thus, O_3, $O(^3P)$ and $O(^1D)$ form the odd oxygen family which has a chemical lifetime several orders of magnitude larger than its constituents in the stratosphere and lower mesosphere [*Brasseur and Solomon*, 2005].

The reactions (8.2) to (8.4) are swift in the upper stratosphere and lower mesosphere (order of minutes). During sunlit conditions the balance in the O_x family is shifted to atomic oxygen. As soon as the sun sets, all odd oxygen is completely converted to O_3, compare Fig. 8.12 for the resulting O_3 time series.

There exist several sinks for the constituents of the odd oxygen family, namely reactions with Cl radicals which stem from anthropogenically released chemicals but also N and H radicals, which are partly of natural origin [*Brasseur and Solomon*, 2005].

In the presence of radicals additional paths of O_x-destruction open [*Solomon et al.*, 1981]. Two of them are presented here, because they become relevant for the study of the sun-atmosphere interaction. Charged particle radiation entering the atmosphere above the polar cusps create HO_x and NO_x radicals via positive ion chemistry [*Sinnhuber et al.*, 2003; *Brasseur and Solomon*, 2005]. While HO_x is most effective in O_3-destruction in the lower mesosphere via

$$OH + O_3 \rightarrow HO_2 + 2O_2 \tag{8.5}$$

$$HO_2 + O(^3P) \rightarrow OH + O_2 \tag{8.6}$$

$$\text{Net: } O(^3P) + O_3 \rightarrow 2O_2 \tag{8.7}$$

[*Lary*, 1997] it is also very short lived (compare Sect. 8.5.6). NO_x radicals, on the other hand, do not cause such severe destruction of O_x in the mesosphere, but they are stable in the absence of solar light [*Solomon et al.*, 1984; *Russell III et al.*, 1984]. Due to the down transport above the winter pole they may reach the stratosphere where they form the most effective O_x destruction reaction [*Lary*, 1997] via

$$NO + O_3 \rightarrow NO_2 + O_2 \tag{8.8}$$

$$NO_2 + O(^3P) \rightarrow NO + O_2 \tag{8.9}$$

$$\text{NET: } O(^3P) + O_3 \rightarrow 2O_2 \tag{8.10}$$

This cycle is most effective in the middle and upper stratosphere and may contribute significantly to spring O_x-destruction above the poles [*Randall et al.*, 2005].

The O_3 chemistry is encoded in most of the models to date. For studies in connection with the microwave instruments the three dimensional chemistry transport model B3DCTM (refer to Sect. 16.2.2), the two dimensional prognostic B2DM model (refer to Sect. 16.2.1) and the three dimensional KASIMA (refer to Sect. 15.2.2) have been used.

8.5 Results

8.5.1 General Observation Obtained from OZORAM Measurements

Figure 8.8 shows the measured O_3 throughout the year 2010. Some typical features are readily observed: The year starts with the large scale descent of stratospheric and mesospheric air-masses above the winter pole. The mesospheric O_3 maximum

Fig. 8.8 Measurements of strato- and mesospheric O_3 measurements obtained with microwave radiometry above Spitsbergen

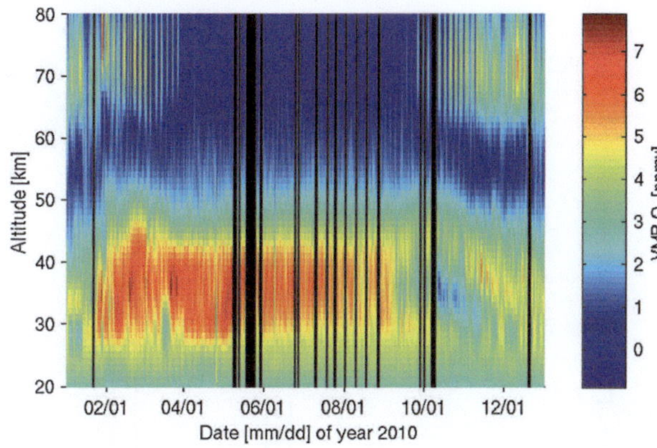

is clearly discernible. The descent has started with the advent of the polar night in October 2009. With the raising of the sun above horizon in the end of January, the mesospheric diurnal cycle appears. This cycle stops with the end of the transition period from polar night to polar day in the mid of April. Now only a faint diurnal cycle in the upper stratosphere can be observed. In the end of September the transition to the polar night sets in and with it the descent of the air masses, the development of the mesospheric diurnal cycle and the night maximum of O_3.

In the end of January a stratospheric warming took place which can be recognized in the uplift of stratopause minimum. The polar vortex is not above Ny Ålesund throughout February and March except in the middle of March where a distinct O_3 minimum reaches up to 35 km. Figure 8.9 shows the comparison of OZORAM measurements to two time series of O_3 obtained from satellite measurements, EOS-MLS on AURA and SABER on TIMED. The comparison shows an agreement of better than 20 % in the stratosphere and better than 30 % in the mesosphere [*Palm et al.*, 2010]. Taking into account the specific location this is comparable to similar instruments [see *Hocke et al.*, 2007; *Boyd et al.*, 2007, for recent studies].

8.5.2 *The Diurnal Variation of Stratospheric O_3 During Polar Day Above Spitsbergen*

Figure 8.10 shows the diurnal variation of middle stratospheric O_3 at the end of April 2010 during the beginning of the polar day. It is evident, that the O_3-VMR is closely linked to the SZA.

This may be explained by the different wave length sensitivities of the photolysis of O_2 (react. (8.1)) and the photolysis of O_3 (reacts. (8.3) and (8.4)). Because the light travels through the atmosphere it can be expected that it is attenuated depending on the atmospheric layers which are traversed, which is different for different wave lengths. This leads to the observed diurnal dependency of the stratospheric O_3 even in polar summer when the sun is above horizon all day.

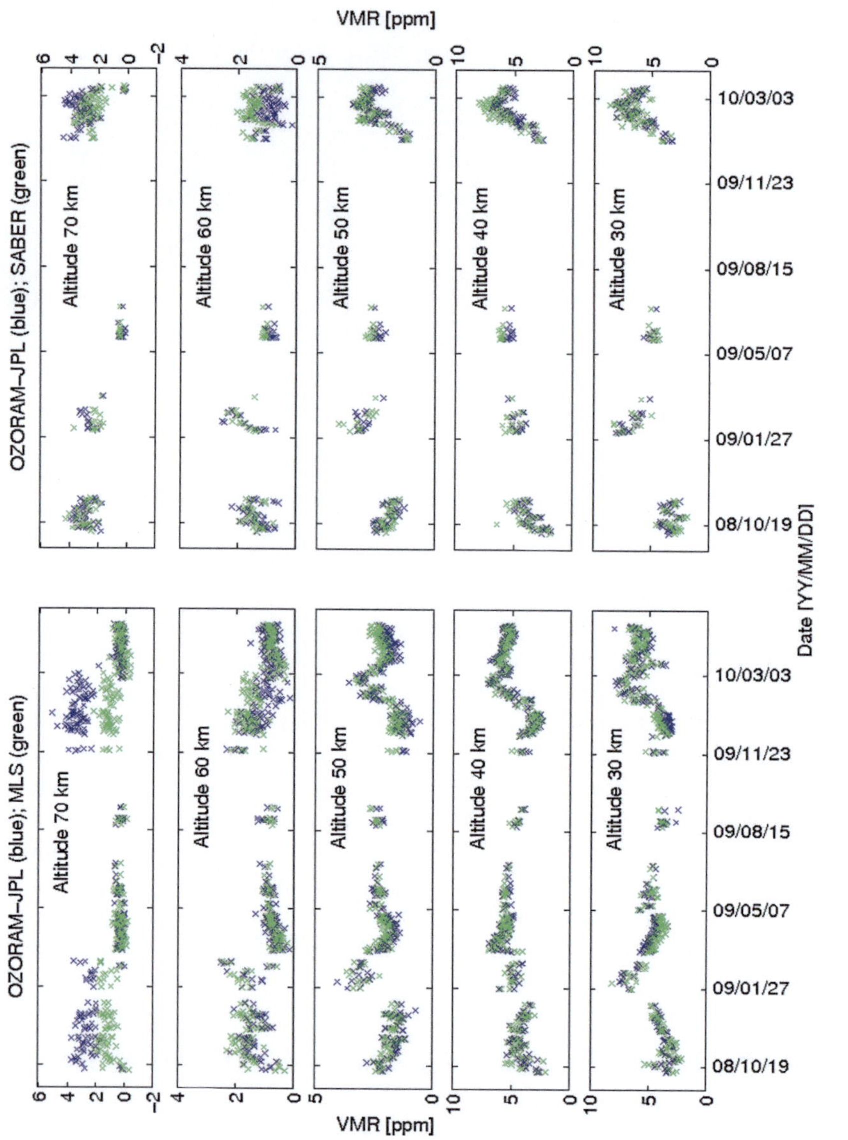

Fig. 8.9 Comparison of time series of OZORAM measurements and two satellite instruments, SABER and EOS-MLS. Figure reproduced from *Palm et al.* [2010]

The anti-correlation (shown in Fig. 8.11) of the (detrended) measurements to the SZA, shifted by 5 h is significant. The time lag is explained because the solar illumination influences the change of O_3 through the production and destruction of various terms within the O_x. The change of O_3 is the derivative of the measured O_3 which has its minima and maxima at the point of highest and lowest solar illumination.

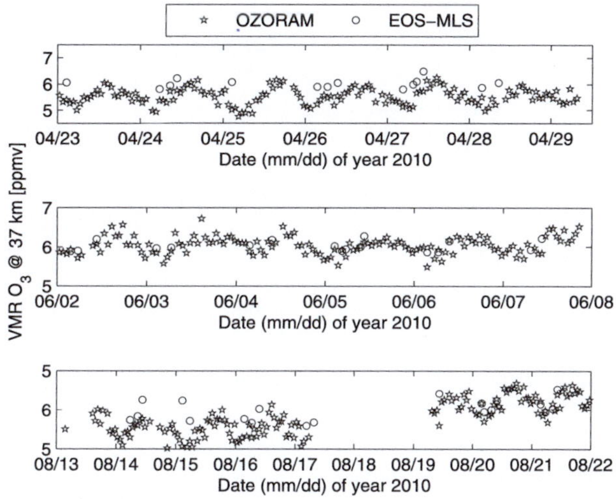

Fig. 8.10 Diurnal variation of O_3 above Spitsbergen measured by OZORAM and EOS-MLS during three periods of the polar day 2010

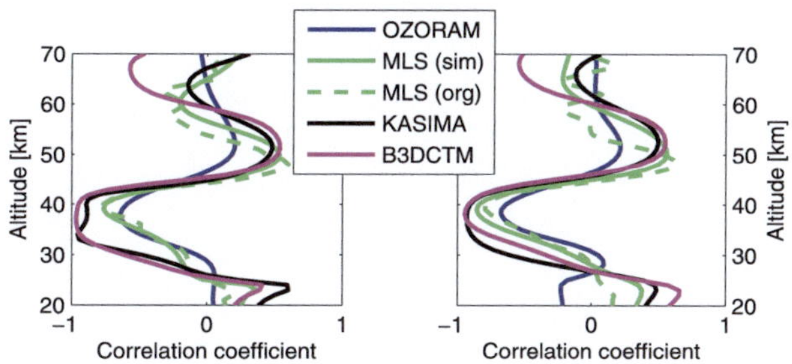

Fig. 8.11 Correlation of measurements and modeled data sets in the year 2009 (*left figure*) and 2010 (*right figure*) to the SZA (shifted by 5 h). The data-set have been high-pass filtered with a cut off frequency of 1 d^{-1}

The differences in higher altitudes, where the models as well as the measurements show a larger anti-correlation, is due to the noise on the measurements. Because the VMR value is rather small between 55 and 80 km altitude, the correlation is dominated by noise.

8.5.3 The Diurnal Variation of Mesospheric O_3 Above Spitsbergen

Above Spitsbergen a mesospheric diurnal variation is observed during January till April and September till November when the sun crosses the 95°SZA threshold, i.e. there is day and night in the mesosphere. The balance $O \leftrightarrow O_3$ is on the left hand side during sunlit conditions and on the right side during dark conditions (compare Sect. 8.4). Figure 8.12 shows the measured and modeled (B3DCTM, see Sect. 16.2.2) mesospheric O_3-VMR above Spitsbergen. During day, the O_3 is photolysed. Apart from a small residual, no O_3 is present. During night, the O_x is entirely in the form of O_3 forming the night maximum in the mesosphere. Note, this is the weighted mean over some 20 km which is defined by the averaging kernel

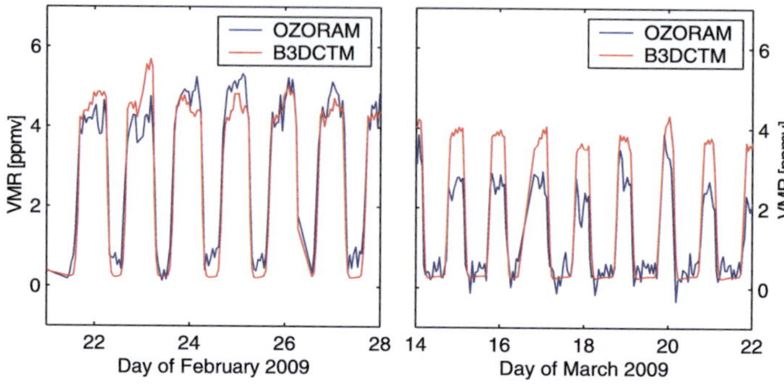

Fig. 8.12 Diurnal variation of mesospheric O_3 above Spitsbergen as measured by OZORAM and modeled by B3DCTM Chap. 8

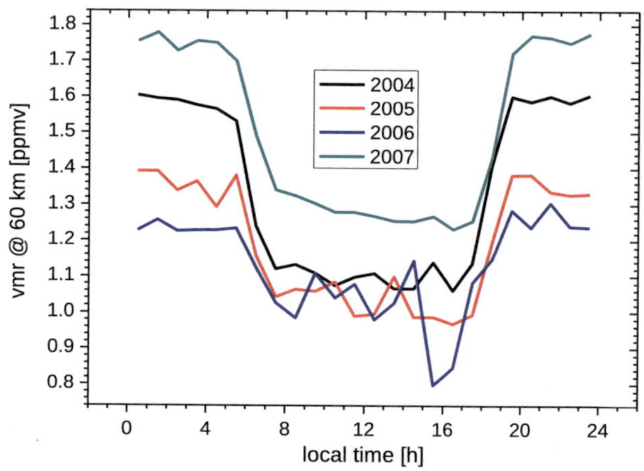

Fig. 8.13 The mean diurnal cycle during November 2004–2007 as seen by the MIRA2 instrument [*Kopp et al.*, 2009]

of the measurement [*Palm et al.*, 2010]. The variability of the night time O_3 is due to dynamical change in the mesosphere and change in weather conditions, the first of which alters the total O_x content and the second alters the sensitivity of the instrument. While during the first period (February 2009, Fig. 8.12 left) the modeled and the measured O_3 VMR are almost perfectly matching, the diurnal variation is overestimated by the model during March 2009 (Fig. 8.12 right). The reason for this failure of the model to reproduce the diurnal variation of mesospheric O_3 is unknown at the time of the production of this book.

8.5.4 The Diurnal Variation of Mesospheric O_3 Above Mérida

The following section summarizes the publication of *Kopp et al.* [2009].

Figure 8.13 shows the mean diurnal cycle over the years 2004 to 2007 above Mérida. While in all years a clear diurnal cycle can be seen in some years it is

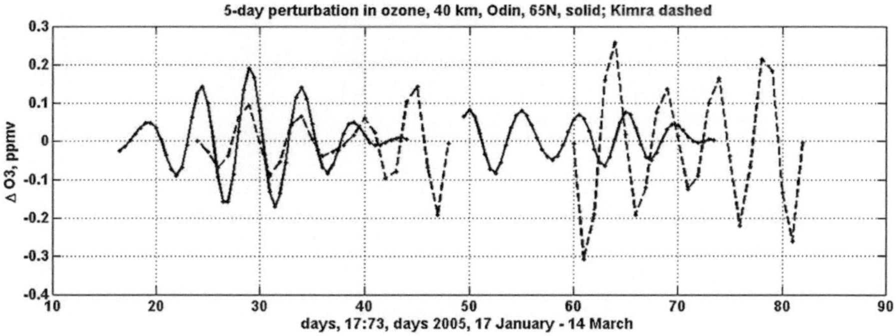

Fig. 8.14 The 5-day migrating wave as seen by the ODIN satellite and the ground-base millimeterwave radiometer KIMRA. Figure reproduced from *Belova et al.* [2008]

modified by a second strong signal, which cannot be explained by the chemistry (compare Sect. 8.4) alone.

In the years 2004 and 2007 the diurnal variation is very distinct, showing higher night values and almost constant daytime values. The years 2005 and 2007 show a lower nighttime O_3 value and a more variable day time value. A closer analysis shows that the variable daytime value is caused by a steady decrease in the O_3 content of the lower mesosphere. The cause of this steady decline is not clear, auxiliary studies of the atmospheric dynamics would be needed.

8.5.5 Measurements of a 5-Day Planetary Wave Above Kiruna

The following section summarizes the publication of *Belova et al.* [2008].

Waves are a common phenomena in the atmosphere and exert strong influences on the atmospheric state. They may change the local temperature considerably and induce or reduce winds via wave breaking.

Waves in the atmosphere are excited by several causes, first of all by the diurnal cycle of the solar energy input. But also mountain ridges, thunderstorms, gradients in potential temperature cause waves, which travel horizontally and vertically over great distances.

A frequently studied wave is the 5-day wave [*Belova et al.*, 2008, and references therein], which is a westward traveling free planetary wave. The study of *Belova et al.* [2008] uses the satellite instrument SMR on ODIN and the ground-based millimeterwave instrument KIMRA in Kiruna (67.83°N, 20.4°E) to analyze O_3 data for the signature of this wave.

In Fig. 8.14 the 5-day component of O_3 measurements of both the ODIN satellite and the KIMRA instrument are shown. While in the first period, both instruments show the 5-day wave component in phase this is different in the second period. *Belova et al.* [2008] attribute this effect to other effects in the KIMRA measurements

Fig. 8.15 Response of O_3 above Spitsbergen to the Halloween solar storms on the 28th of October and the 3rd of November 2003. See Sect. 8.5.7 for an interpretation of the data

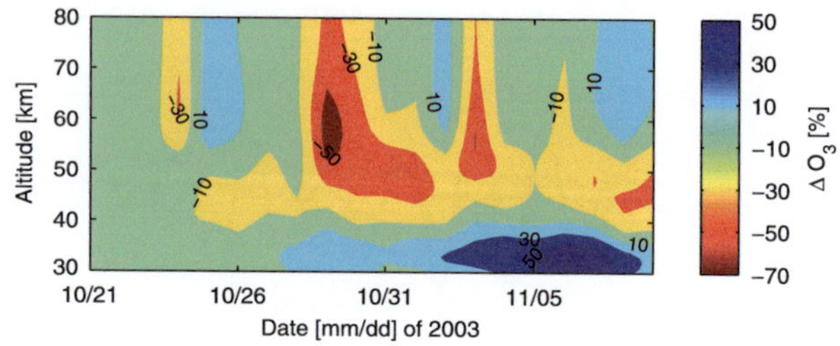

like a eastward traveling 5-day wave. An instrument like the KIMRA, measuring stationary in space, is principally not able to distinguish such signals.

8.5.6 *Response of Arctic O_3 to SPE*

The following section has been published as part of *Palm* [2006].

In October and November 2003 a large solar proton event (SPE) took place. This so called Halloween event has been subject of numerous studies. It is special because it took place just before the Polar night set in and thus preserved many chemical compounds which are usually quickly photolysed. Together with the strong descend of air-masses large amounts of NO_x radicals could enter the stratosphere where they become effective in depleting O_3 [*Randall et al.*, 2005; *Lary*, 1997]. In Fig. 8.15 the response of stratospheric and mesospheric O_3 to the Halloween solar storm in October 2003 can be seen. The data are relative to the mean O_3 content during the 21th till 25th of October 2003. Because of the measurement properties, the O_3 content above 60 km is a mixture of columnar information and profile information.

On the 28th of October 2003 the first of a series of particle precipitation events took place. A few days later, on the 3rd of November, a second, somewhat smaller, shower of protons hit the polar mesosphere. Both events can clearly be seen in Fig. 8.15. In the instant of the event, the HO_x catalytic cycle (8.5) to (8.7) of O_3-destruction leads to large losses of O_3 in the mesosphere. The HO_x is itself quickly removed as soon as the event stops and the O_x is replenished by the residual circulation. At the altitude of the stratopause and below the effect persists longer because the O_3 is not replaced as quickly as in the mesosphere.

The measurements present a mixture of dynamical and chemical effects which cannot be untangled using the ground based measurements alone. This can be done using a model run with, x_{WI}, and without, x_{WO} the ion precipitation causing the O_3-destruction. Figure 8.16 shows the difference, Δx, of both model runs as calculated by

$$\Delta x = \frac{x_{WI} - x_{WO}}{x_{WO}} \tag{8.11}$$

Figure 8.16 is interpreted as showing the evolution of the disturbed atmosphere following a large SPE. The two SPE events can clearly be recognized as well as

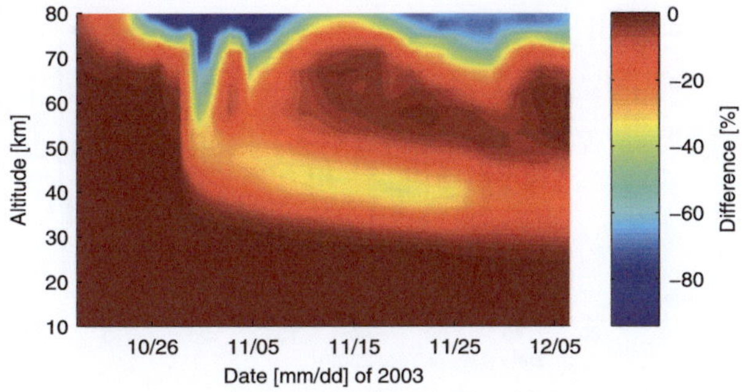

Fig. 8.16 Difference the O_3 VMR modeled with and without the ion precipitation due the Halloween SPE in October/November 2003 using the model B2DM (Sect. 16.2.1). Refer to text for more details

the quick (dynamic) replenishment of O_3 as soon as the production of HO_x due to the particle precipitation stops. During both events, however, NO_x radicals are also produced. Although they are not destructing O_3 in the mesosphere, they are also long lived and are transported down into the stratosphere [*Randall et al.*, 2005], where they become indeed very effective catalysts, causing O_3 destruction [*Lary*, 1997].

Figure 8.16 demonstrates the importance of considering ion chemistry in modeling the atmosphere disturbed by precipitation of high energy particles.

8.5.7 General Observations Obtained from MIRA5 Measurements

This section refers to results that have been obtained in winter 2009 during the Alpine Radiometer Intercomparison at the Schneefernerhaus (ARIS). The Schneefernerhaus is located near the Zugspitze summit (47°N 11°E, 2650 m above sea level). *Straub et al.* [2011] give an overview of the campaign results.

Figure 8.17 compares H_2O profiles retrieved from daily averaged MIRA5 measurements with correlative data from the EOS MLS instrument. Collocations are required to fall within ±2° (±220 km) latitude and ±5° (±390 km) longitude of the measurement site. For comparison with the ground-based data, *Straub et al.* [2011] calculate average profiles from all MLS data that meet this criterion in a given overpass. Here, the comparison is instead presented for the individual MLS profiles matching closest in space.

The four panels of Fig. 8.17 refer to 4 layers evenly spaced in log(p) between 3 and 0.03 hPa. The profile information for both MIRA5 and MLS data is averaged across these layers, weighted with particle density. With regard to H_2O evolution over time, the data exhibit good general agreement. However, the data series do indicate some bias between the two instruments, most notably in the 0.3–1.0 hPa pressure range. For a more detailed discussion, the reader is referred to *Straub et al.* [2011].

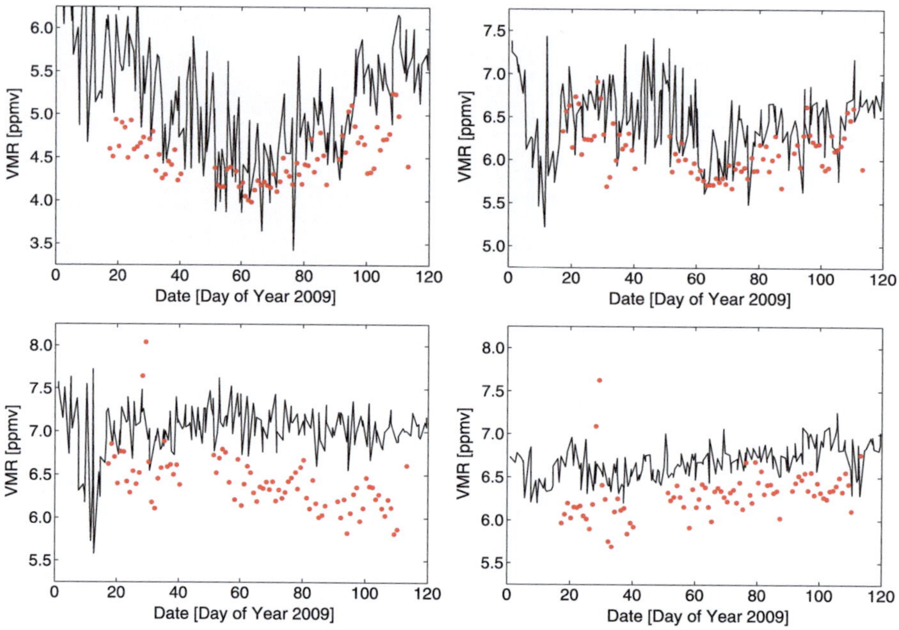

Fig. 8.17 Time series of middle atmospheric H_2O as obtained from MIRA5 (*red dots*) and EOS MLS (*solid black line*). The plots display volume mixing ratios averaged over certain pressure ranges, (*top left*) 0.03–0.1 hPa, (*top right*) 0.1–0.3 hPa, (*bottom left*) 0.3–1.0 hPa, (*bottom right*) 1.0–3.0 hPa

8.5.8 Response of Mid-latitude H_2O to the 2009 SSW

The MIRA5 data presented in Sect. 8.5.7 above exhibit a period of increased water vapor due to the stratospheric warming (SSW) event in January 2009 [*Manney et al.*, 2009]. Figure 8.18 takes a closer look at the results for weeks 3–5 of 2009. The 20th January marks the onset of the polar vortex breakdown in the upper stratosphere during the SSW [*Manney et al.*, 2009]. The middle panel presents stratospheric temperatures south of Zugspitze (MIRA5 field of view) from ECWMF ERA-40 data. They exhibit considerable warming ($\sim$40 K) in the upper stratosphere and cooling ($>$20 K) in the lower stratosphere at the end of January. These features are connected to descending air-masses from lower latitudes in the upper stratosphere and uplifted air-masses from polar latitudes below [*Manney et al.*, 2009; *Flury et al.*, 2009].

MIRA5 data for this time period (Fig. 8.18, upper panel) show a distinct increase of 20 % (1.5 ppmv) in upper stratospheric H_2O in presence of the warmer air. *Flury et al.* [2009] have found a similar signal for the 2008 SSW with the ground-based 22 GHz spectroradiometer MIAWARA above Zimmerwald, Switzerland (300 km west of the Zugspitze). Contrary to this, EOS MLS data south of Zugspitze (Fig. 8.18, lower panel) show a decrease of 10 % (0.6 ppmv) in H_2O for the 2009 SSW. A pronounced increase of 20 % (1.5 ppmv) is instead found in the

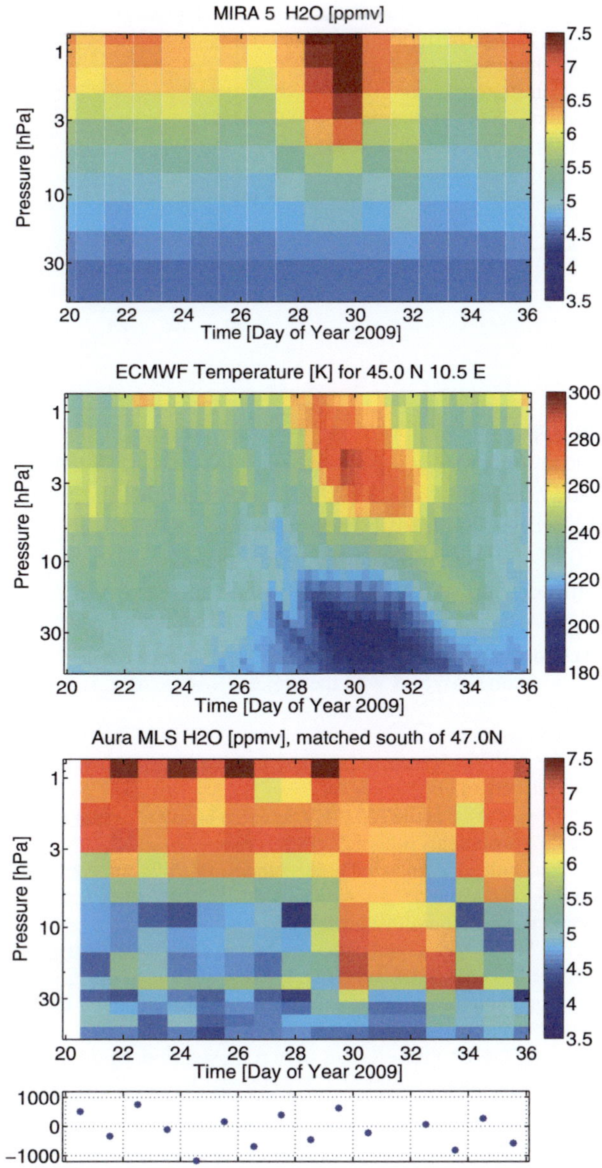

Fig. 8.18 (*top*) Stratospheric H_2O south of Zugspitze as seen by MIRA5 during 17 d after the onset of the 2009 sudden stratospheric warming. (*middle*) Corresponding temperature data from ECMWF ERA-40. (*bottom*) EOS MLS H_2O data matched to the ground-based results. The *small panel* below gives the distances of the individual satellite footprints (negative values: west of the MIRA5 field of view, positive values: east). Only data from ascending (poleward) MLS overpasses are considered here

lower stratosphere. This is more consistent with the notion of (younger) sub-tropical air above and (older) polar air below.

The limited sensitivity of the ground-based measurement to lower stratospheric H_2O and the coarser altitude resolution may cause profile information to be shifted to higher layers. To identify such effects, Aura MLS data have been convolved with MIRA 5 averaging kernels as seen in Fig. 8.19. In the convolved data, the increase in H_2O appears more than half a decade in pressure higher up. However, the decrease

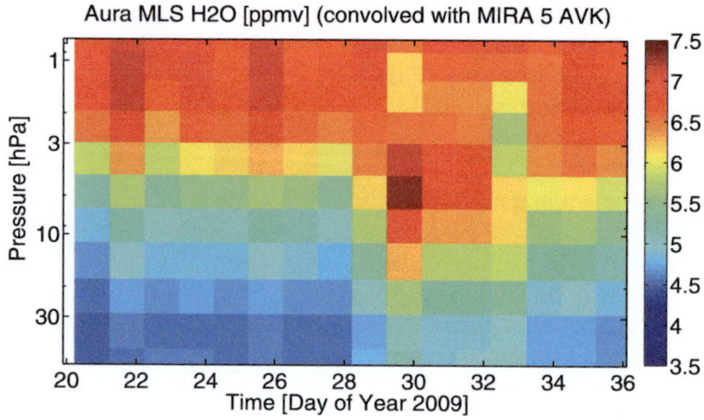

Fig. 8.19 EOS MLS H_2O above Zugspitze during 17 d after the onset of the 2009 sudden stratospheric warming. Same data as in Fig. 8.18, but convolved with MIRA5 averaging kernels

in upper stratospheric H_2O is still present in the convolved data. It turns out that the quality of the spectroscopic data is significant point of concern. In a separate study, *Golchert et al.* [2010] have used the H_2O full model by *Rosenkranz* [1998] for MIRA 5 retrieval. In this case an SSW related increase in H_2O is found down to 10 hPa, in better agreement with the satellite data. Yet the strong increase at higher altitudes still prevails in the ground-based results and remains an open question.

Acknowledgements The work presented in this chapter was supported by the DFG in the priority program CAWSES with the projects SACOSAT I, II and III and by the EU within the TASTE Project.

The German polar research institute, AWI, and the French polar research institute, IPEV, operate the AWIPEV research base and provide logistic support for the millimeterwave measurements on Spitsbergen.

The personal at the AWIPEV research base performed numerous tasks in maintaining and keeping the instruments running.

The authors acknowledge the support of Environmental Research Station Schneefernerhaus—UFS, Universidad de Los Andes (Mérida, Venezuela), Sistema Teleférico de Mérida and Instituto Nacional de Parques (Venezuela).

ECMWF operational data used in this work have been provided by ECMWF.

The EOS MLS data used in this study were acquired as part of the activities of NASA's Science Mission Directorate, and are archived and distributed by the Goddard Earth Sciences (GES) Data and Information Services Center (DISC).

The TIMED-SABER data were retrieved from ftp://saber.gats-inc.com/. We thank the science and data processing team of TIMED-SABER.

References

Anderson, G. P., Clough, S. A., Kneizys, F. X., Chetwynd, J. H., & Shettle, E. P. (1986).*AFGL atmospheric constituent profiles (0–120 km)* (Environmental Research Papers, No. 954, Technical Report AFGL-TR-86-0110). Air Force Geophysics Laboratory.

Belova, A., Kirkwood, S., Raffalski, U., Kopp, G., Hochschild, G., & Urban, J. (2008). Five-day planetary waves as seen by the Odin satellite and the ground-based Kiruna millimeter wave radiometer in January–March 2005. *Canadian Journal of Physics*, *86*, 459–466. doi:10.1139/P07-172.

Berg, H., Krupa, R., Hochschild, G., Kopp, G., & Kuntz, M. (1998). Millimeterwave radiometer with adjustable internal calibration load for high resolution measurements of stratospheric constituents. In *Proceedings of 2nd ESA workshop on millimetre wave technology and applications: antennas, circuits and systems* (pp. 372–377).

Boyd, I. S., Parrish, A. D., Froidevaux, L., von Clarmann, T., Kyrölä, E., Russell III, J. M., & Zawodny, J. M. (2007). Ground-based microwave ozone radiometer measurements compared with Aura-MLS v2.2 and other instruments at two Network for Detection of Atmospheric Composition Change sites. *Journal of Geophysical Research, 112*, D24S33. doi:10.1029/2007JD008720.

Brasseur, G., & Solomon, S. (2005). *Aeronomy of the middle atmosphere* (3rd ed.). Berlin: Springer.

Bühler, S., Eriksson, P., Kuhn, T., von Engeln, A., & Verdes, C. (2005). ARTS, the atmospheric radiative transfer simulator. *Journal of Quantitative Spectroscopy & Radiative Transfer, 91*, 65–93. doi:10.1016/j.jqsrt.2004.05.051.

Chapman, S. (1930). On ozone and atomic oxygen in the upper atmosphere. *Philosophical Magazine, 10*(64), 369–383.

Connor, B. J., de Zafra, R. L., Solomon, P. M., Parrish, A., Barrett, J. W., & Jaramillo, M. (1987). Nitrous oxid in the tropical middle atmosphere, observed by ground-based mm-wave spectrometry. *Geophysical Research Letters, 14*, 1254–1257.

Eriksson, P., Jiménez, C., & Bühler, S. A. (2005). Qpack, a general tool for instrument simulation and retrieval work. *Journal of Quantitative Spectroscopy & Radiative Transfer, 91*, 47–64. doi:10.1016/j.jqsrt.2004.05.050.

Farman, J. C., Gardiner, B. G., & Shanklin, J. D. (1985). Large losses of total ozone in Antarctica reveal seasonal ClOx/NOx interaction. *Nature, 315*, 207–210. doi:10.1038/315207a0.

Flury, T., Hocke, K., Haefele, A., Kämpfer, N., & Lehmann, R. (2009). Ozone depletion, water vapor increase, and psc generation at midlatitudes by the 2008 major stratospheric warming. *Journal of Geophysical Research, 114*, D18302. doi:10.1029/2009JD011940.

Golchert, S., Hochschild, G., & Gross, J. (2010). Middle-atmospheric water vapour profiles above the Zugspitze obtained with a new ground-based 22 GHz spectroradiometer. In *11th specialist meeting on microwave radiometry and remote sensing of the environment (MicroRad)* (pp. 152–154). doi:10.1109/MICRORAD.2010.5559570.

Golchert, S. H. W. (2010). *Stratospheric water vapour in the tropics*. Ph.D. thesis, Universität Bremen.

Haefele, A., De Wachter, E., Hocke, K., Kämpfer, N., Nedoluha, G. E., Gomez, R. M., Eriksson, P., Forkman, P., Lambert, A., & Schwartz, M. J. (2009). Validation of ground-based microwave radiometers at 22 GHz for stratospheric and mesospheric water vapor. *Journal of Geophysical Research, 114*, D23305. doi:10.1029/2009JD011997.

Hocke, K., Kämpfer, N., Ruffieux, D., Froidevaux, L., Parrish, A., Boyd, I., von Clarmann, T., Steck, T., Timofeyev, Y. M., Polyakov, A. V., & Kyrölä, E. (2007). Comparison and synergy of stratospheric ozone measurements by satellite limb sounders and the ground-based microwave radiometer SOMORA. *Atmospheric Chemistry and Physics, 7*, 4117–4131.

Hoffmann, C. (2008). *Optimierung der Mikrowellenradiometer OzoRAM und WaRAM*. Master's thesis, Universität Bremen.

Hoffmann, C. G., Raffalski, U., Palm, M., Funke, B., Golchert, S. H. W., Hochschild, G., & Notholt, J. (2011). Observation of strato-mesospheric CO above Kiruna with ground-based microwave radiometry—-retrieval and satellite comparison. *Atmospheric Measurement Techniques, 4*, 2389–2408. doi:10.5194/amt-4-2389-2011.

Jaramillo, M., de Zafra, R. L., Parrish, A., & Solomon, P. M. (1988). MM-wave observations of stratospheric HCN at tropical latitudes. *Geophysical Research Letters, 15*, 265–268.

Klein, U., Barry, B., Lindner, K., Wohltmann, I., & Künzi, K. F. (2000). Winter and spring observations of stratospheric chlorine monoxide from Ny-Ålesund, Spitsbergen in 1997/98 and 1998/99. *Geophysical Research Letters, 27*(24), 4093–4097.

Klein, U., Wohltmann, I., Lindner, K., & Künzi, K. F. (2002). Ozone depletion and chlorine activation in the Arctic winter 1999/2000 observed in Ny-Ålesund. *Journal of Geophysical Research,*

107(D20), 8288. doi:10.1029/2001JD000543.

Kopp, G., Calderón, S. M., Gross, M. J., Hochschild, G. G., Hoffmann, C. G. P., Notholt, J. J., & Sinnhuber, M. (2009). Inner-tropical ozone measurements at the Mérida Atmospheric Research Station (MARS) using ground-based microwave radiometry. *International Journal of Remote Sensing*, *30*(15–16), 4019–4032.

Krupa, R., Hochschild, G., Fischer, H., & Wiesbeck, W. (1998). Balanced calibration technique with an internal reference load for ground based millimeter wave radiometry. In *IEEE international geoscience and remote sensing symposium proceedings* (pp. 387–389).

Kuntz, M., Kopp, G., Berg, H., Hochschild, G., & Krupa, R. (1999). Joint retrieval of atmospheric constituent profiles from ground-based millimeterwave measurements: ClO, HNO_3, N_2O, and O_3. *Journal of Geophysical Research*, *104*(D11), 13981–13992.

Künzi, K. F., & Carlson, E. R. (1982). Atmospheric CO mixing ratio profiles determined from ground based measurements of the $J = 1 > 0$ and $J = 2 > 1$ emission lines. *Journal of Geophysical Research*, *87*, 7235–7241.

Lary, D. J. (1997). Catalytic destruction of stratospheric ozone. *Journal of Geophysical Research*, *102*(D17), 21515–21526.

Manney, G. L., Schwartz, M. J., Krüger, K., Santee, M. L., Pawson, S., Lee, J. N., Daffer, W. H., Fuller, R. A., & Livesey, N. J. (2009). Aura microwave limb sounder observations of dynamics and transport during the record-breaking 2009 arctic stratospheric major warming. *Geophysical Research Letters*, *36*, L12815. doi:10.1029/2009GL038586.

Nedoluha, G. E., Bevilacqua, R. M., Gomez, R. M., Thacker, D. L., Waltman, W. B., & Pauls, T. A. (1995). Ground-based measurements of water vapor in the middle atmosphere. *Journal of Geophysical Research*, *100*(D2), 2927–2939.

Palm, M. (2006). *Extension and combination of existing remote-sensing instruments*. Ph.D. thesis, Universität Bremen.

Palm, M., Hoffmann, C. G., Golchert, S. H. W., & Notholt, J. (2010). The ground-based MW radiometer OZORAM on Spitsbergen—description and status of stratospheric and mesospheric O_3-measurements. *Atmospheric Measurement Techniques*, *3*, 1533–1545. doi:10.5194/amt-3-1533-2010.

Parrish, A., de Zafra, R. L., Solomon, P. M., Barrett, J. W., & Carlson, E. R. (1981). Chlorine oxide in the stratospheric ozone layer: ground-based detection and measurement. *Science*, *211*, 1158–1161.

Penfield, H., Litvak, M. M., Gottlieb, C. A., & Lillley, A. E. (1976). Mesospheric ozone measured from ground-based millimeter wave observations. *Journal of Geophysical Research*, *81*(34), 6115–6120.

Raffalski, U., Hochschild, G., Kopp, G., & Urban, J. (2005). Evolution of stratospheric ozone during winter 2002/2003 as observed by a ground-based millimetre wave radiometer at Kiruna, Sweden. *Atmospheric Chemistry and Physics*, *5*, 1399–1407.

Randall, C. E., Harvey, V. L., Manney, G. L., Orsolini, Y., Codrescu, M., Sioris, C., Brohede, S., Haley, C. S., Gordley, L. L., Zawodny, J. M., & Russell III, J. M. (2005). Stratospheric effects of energetic particle precipitation in 2003–2004. *Geophysical Research Letters*, *32*, L05802. doi:10.1029/2004GL022003.

Rodgers, C. D. (2000). *Inverse methods for atmospheric sounding*. Singapore: World Scientific.

Rosenkranz, P. W. (1998). Water vapor microwave continuum absorption: a comparison of measurements and models. *Radio Science*, *33*, 919–928. doi:10.1029/98RS01182, correction in 34, 1025, 1999.

Russell III, J. M., Solomon, S., Gordley, L. L., Remsberg, E. E., & Callis, L. B. (1984). The variability of stratospheric and mesospheric NO_2 observed by LIMS in the polar winter night. *Journal of Geophysical Research*, *89*(D5), 7267–7275.

Sinnhuber, M., Burrows, J. P., Chipperfield, M. P., Jackman, C. H., Kallenrode, M.-B., Künzi, K. F., & Quack, M. (2003). A model study of the impact of magnetic field structure on atmospheric composition during solar proton events. *Geophysical Research Letters*, *30*(15), 1818. doi:10.1029/2003GL017265.

Solomon, S., Rusch, D., Gerard, J., Reid, G., & Crutzen, P. (1981). The effect of particle-precipitation events on the neutral and ion chemistry of the middle atmosphere. 2. Odd hydrogen. *Planetary and Space Science*, *29*(8), 885–892.

Solomon, S., Mount, G. H., & Zawodny, J. M. (1984). Measurements of stratospheric NO_2 from the solar mesosphere explorer satellite. 2. General morphology of observed NO_2 and derived N_2O_5. *Journal of Geophysical Research*, *89*(D5), 7317–7321.

Straub, C., Murk, A., Kämpfer, N., Golchert, S. H. W., Hochschild, G., Hallgren, K., & Hartogh, P. (2011). ARIS-Campaign: intercomparison of three ground based 22 GHz radiometers for middle atmospheric water vapor at the Zugspitze in winter 2009. *Atmospheric Measurement Techniques*, *4*, 1979–1994. doi:10.5194/amt-4-1979-2011.

Waters, J. W. (1973). Ground-based measurement of millimetre-wavelength emission by upper stratospheric O_2. *Nature*, *242*.

Waters, J. W., Wilson, W. J., & Shimabukuro, F. I. (1976). Microwave measurement of mesospheric carbon-monoxide. *Science*, *191*(4232), 1174–1175.

Chapter 9
Data Assimilation and Model Calculations to Study Chemistry Climate Interactions in the Stratosphere

Björn-Martin Sinnhuber, Gregor Kiesewetter, John P. Burrows, and Ulrike Langematz

Abstract The variability of polar stratospheric ozone is investigated through analysis of a new long-term global stratospheric ozone data set, and model calculations. A 29-year stratospheric ozone data set was generated through assimilation of satellite ozone observations into a chemical transport model. An analysis of this data set shows patterns of positive and negative ozone anomalies at high latitudes during winter; the anomalies often develop in the mid- to upper stratosphere in early winter and descend into the lower stratosphere during the following winter months, persisting for up to five months. These ozone anomalies are often related to anomalies in the Annular Modes. Another less frequent class of ozone anomalies is related to solar proton events (SPEs). Although anomalies in Arctic ozone during winter often persist for several months, descending from the mid- to the lower stratosphere, we find no further evidence for a close correlation between mid-stratospheric ozone in autumn and total ozone in spring, in agreement with an analysis of long-term calculations from the chemistry climate model EMAC. In order to test how ozone anomalies in early winter may influence stratospheric dynamics, we performed a set of sensitivity runs with the EMAC model with imposed ozone anomalies. These studies show that the initial perturbation and conditions in autumn do have a long term effect on the behaviour of the polar vortex and the troposphere as a whole but it is not strong enough to prevent tropospheric events propagating and impacting on the stratosphere.

B.-M. Sinnhuber (✉) · G. Kiesewetter · J.P. Burrows
Institute of Environmental Physics, University of Bremen, Otto-Hahn-Allee, 28359 Bremen, Germany
e-mail: Bjoern-Martin.Sinnhuber@kit.edu

B.-M. Sinnhuber
Now at Karlsruhe Institute of Technology, Karlsruhe, Germany

U. Langematz
Freie Universität Berlin, Berlin, Germany
e-mail: Ulrike.Langematz@met.fu-berlin.de

F.-J. Lübken (ed.), *Climate and Weather of the Sun-Earth System (CAWSES)*, Springer Atmospheric Sciences,
DOI 10.1007/978-94-007-4348-9_9,

9.1 Introduction

Stratospheric ozone has shown large decreases over past decades at high latitudes over both hemispheres. The large ozone losses over the southern hemisphere high latitudes in spring leading to the Antarctic Ozone Hole have been clearly attributed to chemical ozone depletion resulting from the anthropogenic emissions of chlorine and bromine containing substances, most importantly the chlorofluorocarbons (CFCs) and halons [*World Meteorological Organization*, 2010]. Similarly, large losses of stratospheric ozone have been observed in the Arctic during individual winters. However, there exists a large inter-annual variability in springtime total ozone over the Arctic as a result of inter-annual changes in the meteorological conditions during the northern hemisphere winter. Understanding the processes behind the inter-annual variability of polar ozone is important in order to predict if and when polar ozone will recover as ozone depleting substances reduce whilst simultaneously being influenced by a changing climate.

It is now well accepted that the inter-annual variability of total ozone during spring in the Arctic is closely related to the activity of planetary waves during winter, propagating from the troposphere into the stratosphere. Years with higher than average wave activity during winter are associated with higher stratospheric temperatures and higher than average total ozone during spring, and vice versa [*Newman et al.*, 2001; *Weber et al.*, 2003]. Planetary waves influence stratospheric ozone in three ways: (a) by modulating the poleward and downward transport of ozone into high latitudes through the wave driven Brewer-Dobson circulation, (b) by affecting mixing between polar and mid-latitude air and (c) indirectly through changes in stratospheric temperature which affects polar chemical ozone depletion. Years with low wave activity exhibit a slowing down of the Brewer-Dobson circulation and consequently colder polar stratospheric temperatures, which facilitates chemical ozone depletion in the lower polar stratosphere. For example, *Weber et al.* [2003] have shown that for both hemispheres the activity of planetary waves in a given winter (expressed by the upward component of the Eliassen-Palm flux through the mid-latitude lower stratosphere) and total ozone in spring are closely correlated.

The focus of this chapter is to investigate the variability of Arctic stratospheric ozone during winter and identify possible mechanisms explaining the interaction of ozone and atmospheric dynamics in the polar winter stratosphere. More specifically, *Kawa et al.* [2005] have shown that apparently an unexpectedly close correlation between ozone in the Arctic mid-stratosphere during autumn and total column ozone in the following spring exists. A similar relationship was reported by us [*Sinnhuber et al.*, 2006] based on ozone sonde observations at Ny-Ålesund, Spitsbergen (79°N, 12°E) (Fig. 9.1). However, while a very close correlation between mid-stratospheric ozone during October to December and total column ozone during March existed for the first ten years of the observations at Ny-Ålesund (with correlation coefficient $R = 0.94$), this correlation between ozone in autumn and total ozone in spring breaks down during the most recent decade of observations ($R = 0.08$). Nevertheless, these observations show a robust and close correlation

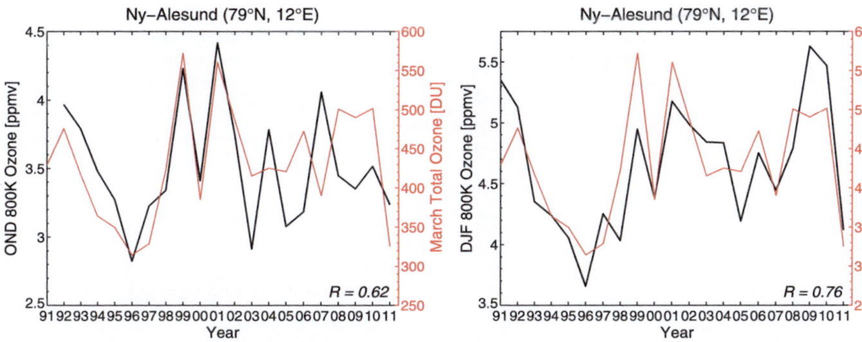

Fig. 9.1 The relation between early and mid-winter ozone, respectively, to total ozone in spring from ozone sonde observations at Ny-Ålesund (79°N, 12°E). *Left*: Previous years October to December ozone at 800 K potential temperature (about 30 km altitude) and March total column ozone. *Right*: December to February ozone at 800 K and March total column ozone. Updated from *Sinnhuber et al.* [2006]

between mid-winter ozone (December to February) and total ozone in spring, indicating that spring time ozone anomalies are usually accompanied by similar anomalies during the previous months. In the following sections of this chapter we analyse in more detail the development and persistence of ozone anomalies during polar winter. We show that ozone anomalies in the Arctic persist for several months, descending from the mid- to the lower stratosphere (Sect. 9.3, Figs. 9.9 and 9.10). The variability of the Annular Modes (Northern Annular Mode, NAM, and Southern Annular Mode, SAM) explains much of the ozone variability during winter. A different class of ozone anomalies is associated with Solar Proton Events (SPEs) and discussed in Sect. 9.3.2. These results are derived from a new long-term stratospheric ozone data set, that was generated from assimilation of satellite observations into a chemical transport model as part of the project DACCS. This data set, its construction, characteristics and validation with external observations is discussed in Sect. 9.2.

In Sect. 9.4 we analyse the relationship between ozone in autumn and total ozone in spring by using a transient simulation of the chemistry climate model ECHAM-MESSy (EMAC). Sensitivity calculations with the EMAC model with imposed ozone anomalies are presented in Sect. 9.5.

9.2 A Long-Term Stratospheric Ozone Data set from Assimilation of Satellite Observations

The investigation of the variability and development of anomalies of stratospheric ozone during polar winter has thus far been limited by a lack of appropriate data sets. The longest currently available global ozone profile data set comes from the Solar Backscatter Ultra-Violet (SBUV and SBUV/2) satellite instruments, covering more than three decades now. However, the SBUV observations require sunlight and are thus not available during polar night. Passive infra-red or microwave

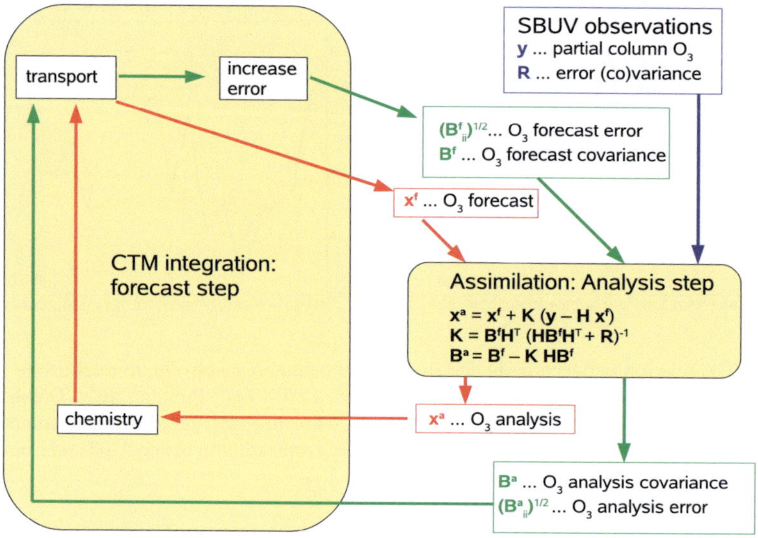

Fig. 9.2 The sequential assimilation scheme (Kalman filter) used in this study. The Kalman filter is divided into two steps which are repeated periodically: the forecast step (CTM integration) and the analysis step, in which all satellite observations made during the forecast interval are assimilated. *Red* and *green arrows* refer to the evolution of the ozone tracer and its covariance, respectively. For technical reasons, the square root of the variance, or 'error', is transported in the forecast step

sounding instruments on the other hand cover only a few winters (although the situation is improving with the now several years long record of observations from MIPAS/ENVISAT and MLS). Here we have constructed a long-term (29-year) global stratospheric ozone data set by assimilating SBUV(/2) ozone observations into a chemical transport model (CTM). This results in a daily global stratospheric ozone data set that is closely constrained by SBUV(/2) observations in sunlit regions, where measurements are available and fills the gaps at high latitudes by transporting information into polar night, where ozone is rather long-lived. A similar approach was already carried out as part of the European Centre for Medium-Range Weather Forecasts' ERA-40 reanalysis project [*Dethof and Hólm*, 2004]. However, the ERA-40 assimilated ozone data set suffered from a number of problems that limit its usefulness for the investigation of polar ozone variability. In order to avoid similar problems here, we restricted the observations used for the assimilation process to only SBUV(/2) ozone profile observations. This avoids problems related to potential biases between different instruments and possible ambiguities, associated with the assimilation of total column ozone observations. This is in particularly important here, as we mostly focus on the development of ozone anomalies. A somewhat different approach was followed by *Hassler et al.* [2009], who constructed a long-term satellite and sonde based vertically resolved ozone data set by applying a regression technique.

The assimilation method used here is a sequential assimilation scheme (simplified Kalman filter), based on *Khattatov et al.* [2000] and *Chipperfield et al.*

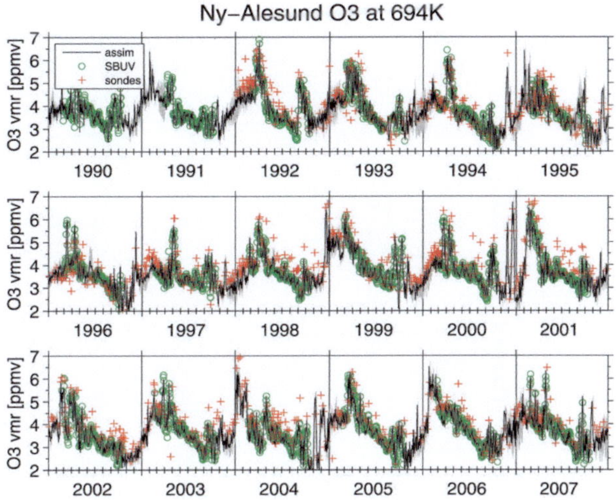

Fig. 9.3 Ozone time series at Ny-Ålesund, Spitsbergen. Assimilated data set (*black line*), sondes (*red* +), and SBUV satellite observations located within 250 km (*green* ∘) are shown. 1σ uncertainty limits as provided by the ozone error tracer for the assimilated data set are indicated as *grey shadow*

[2002] (Fig. 9.2), implemented into our isentropic CTM [e.g. *Sinnhuber et al.*, 2003]. We use SBUV(/2) version 8 ozone profiles from the Nimbus-7, NOAA-9, NOAA-11 and NOAA-16 satellites, covering the period from October 1978 to December 2007. The CTM is driven by temperatures and wind fields from the ECMWF ERA-40 reanalysis from 1978 to 2000 and from the ERA-Interim reanalysis from 1990 to 2007 with a ten year overlap between both data sets. A detailed description of the method and the data set is given by *Kiesewetter et al.* [2010a].

As an example, Fig. 9.3 shows a time series of ozone at Ny-Ålesund at one altitude from our assimilated data set, in comparison with the independent sonde observations. The assimilated data set is in very good agreement with independent observations. Figure 9.4 shows the mean bias between the assimilated data set and independent sonde observations at two Arctic and two Antarctic sites, averaged over the whole period where sonde observations are available (1991–2007). There is a consistent bias, with differences mostly well within ±10 %, except for the lowermost stratosphere. The ozone time series and anomalies (i.e., ozone time series with mean annual cycle removed) from our assimilated data set correlate well with independent ozone sonde observations (Fig. 9.5), even in altitude regions where there is a significant bias. As can be seen from Fig. 9.5 our assimilated data set reproduces the high latitude ozone variability much better than ozone from the ECMWF ERA-40 reanalysis [*Dethof and Hólm*, 2004], which significantly underestimates high latitude ozone variability.

Even though no total column ozone observations were included in the assimilation process, integrated ozone columns from the assimilated data set agree well with independent total columns from the merged GOME/SCIAMACHY/GOME2 (GSG) data set (Fig. 9.6). In order to enable comparison of modelled column ozone to observations, the ozone column below the lower boundary of the CTM has to be taken into account. This is taken here from the climatology of *Fortuin and Kelder*

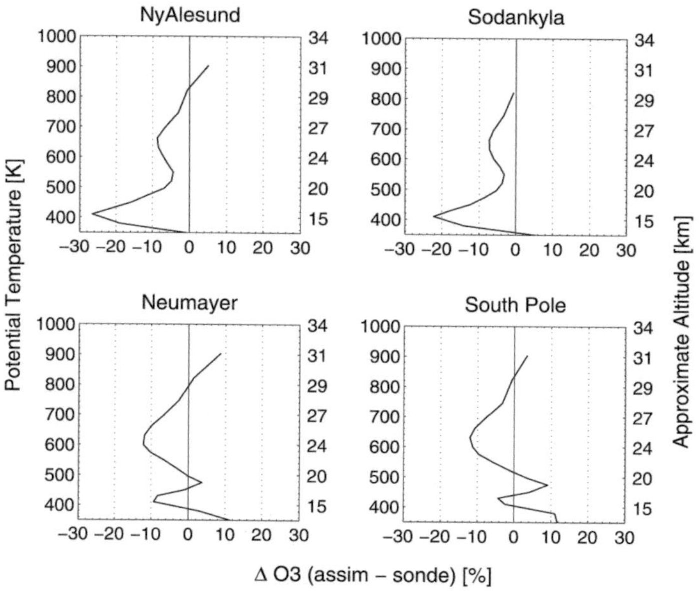

Fig. 9.4 Differences between assimilated ozone vmr and sonde observations from high-latitude stations (NH: Ny-Ålesund, Sodankylä; SH: Neumayer, South Pole), averaged over the whole available sonde time series (1991–2007) and expressed as fraction of sonde ozone vmr. From *Kiesewetter et al.* [2010a]. © 2010 American Geophysical Union. Reproduced by permission of American Geophysical Union

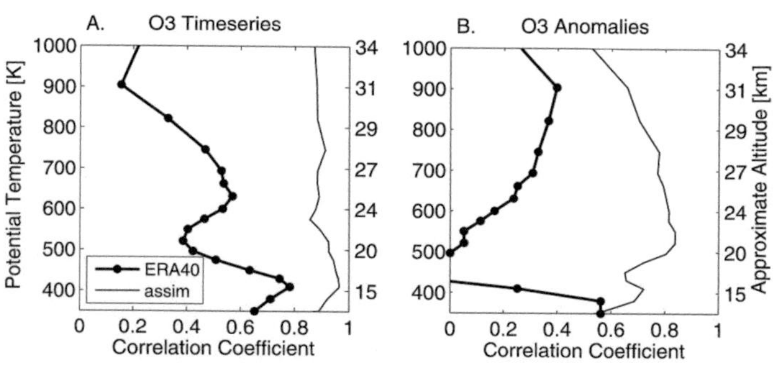

Fig. 9.5 Correlation of ERA-40 ozone to Ny-Ålesund sondes (*solid black line with symbols*), as compared to assimilated ozone (*thin black line*) for the period of 1991–2001. (**A**) Correlation of monthly mean ozone vmr (including the annual cycle), (**B**) Correlation of monthly mean ozone vmr anomalies (annual cycle subtracted). High-latitude profile ozone variability seems underrepresented in the ERA-40 data set. From *Kiesewetter et al.* [2010a]. © 2010 American Geophysical Union. Reproduced by permission of American Geophysical Union

[1998] and added to total ozone, calculated from the assimilated data set (Fig. 9.6d). Inter-annual variations correlate very well with the GSG data set (Fig. 9.6b). Both data sets show excellent agreement within less than 5 % at most latitudes and

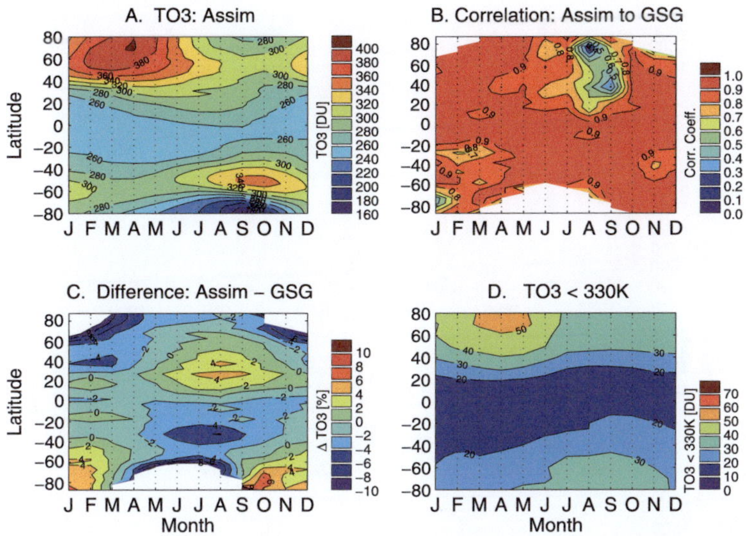

Fig. 9.6 Comparison of column ozone in the assimilated data set to GOME/SCIAMACHY/GOME2 (GSG) observations [*Weber et al.*, 2007], using monthly zonal means: (**A**) TO3 in the assimilated data set (including climatological column below the CTM boundary), (**B**) Correlation of the assimilated TO3 to GSG TO3, (**C**) Difference between TO3 in the assimilated data set and GSG, expressed as fractions of GSG TO3, (**D**) Climatological Ozone Column below 330 K (lower model boundary) that has been included in (**A**)–(**C**). Updated from *Kiesewetter et al.* [2010a]. © 2010 American Geophysical Union. Modified by permission of American Geophysical Union

months, with maximal offsets of +6 % in Antarctic spring and −8 % in Arctic spring (Fig. 9.6c). Taking into account that column ozone itself is not assimilated, the agreement to independent observations is remarkable.

9.3 Analysis of Polar Winter Ozone Anomalies

A central aim of this project is the analysis of Arctic ozone anomalies and the persistence of anomalies during winter. We calculate anomalies as differences from the mean annual cycle, calculated over the entire 29-year time period. Figure 9.7 shows Arctic ozone anomalies averaged over 70–90°N equivalent latitude as a function of time and potential temperature or altitude. (Equivalent latitude is a vortex following coordinate system, defined as an isoline of potential vorticity enclosing the same area as the corresponding latitude coordinate would.) Alternating patterns of positive and negative patterns are observed (with up to 40 % anomalies) that originate in the mid- and upper stratosphere and descend to the lower stratosphere during winter. Many of these anomalies persist for several months. Downward propagation of large positive anomalies is, for example, visible during the 1980s (1979/80, 1980/81, 1981/82). These structures develop from October to December at potential temperatures of about 1000 K (about 34 km) and more, and then descend to about 500 K

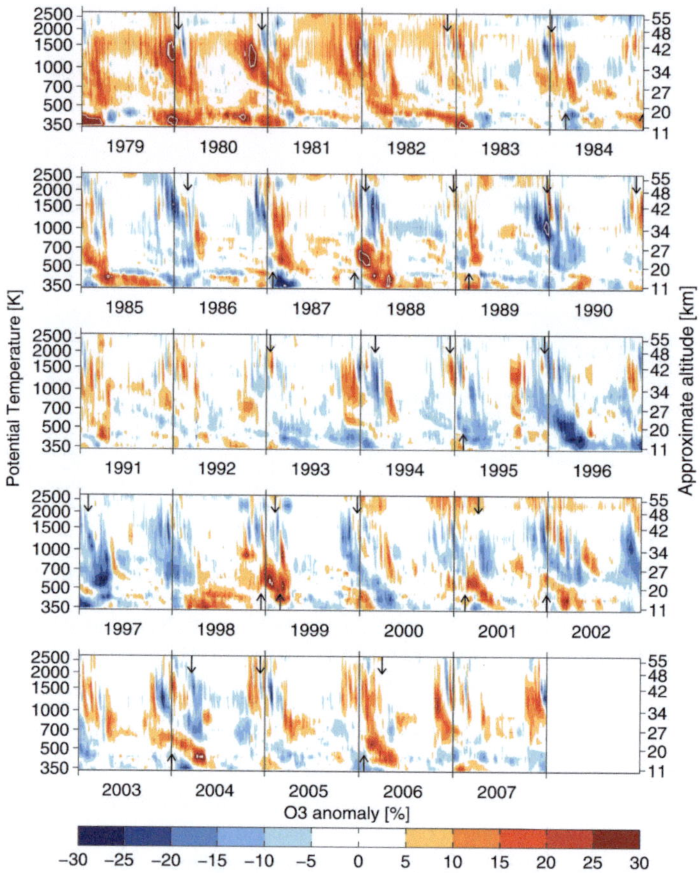

Fig. 9.7 Ozone anomalies relative to the annual cycle, area weighted average over 70–90°N equivalent latitude. In addition to potential temperature (*left axis*), also the corresponding approximate geopotential height values are given (*right axis*). Values exceeding the colour scale are indicated by the *white contour lines*. A 3 day running mean has been applied for smoother viewing. Dates of weak and strong vortex events (see Sect. 9.3.1) are indicated as *arrows* near the lower (weak vortex events) and the upper (strong vortex events) boundaries of the vertical range. Plot updated from *Kiesewetter et al.* [2010a]. © 2010 American Geophysical Union. Modified by permission of American Geophysical Union

(about 20 km) during the winter months, from where they slowly descend further and remain visible for up to one year. Similar, albeit somewhat weaker, anomalies are observed in 1987/88, 2005/06. During the 1990s, negative anomalies dominate. Large negative anomalies in the winters 1995/96 and 1996/97 are pronounced. The negative anomaly developing at the end of 1995 remains in the lower stratosphere almost unchanged for more than a year. Other examples for descending negative anomalies may be found in 1989/90, 1994/95, 2002/03, 2004/05, some of which also show long residence times in the lower stratosphere.

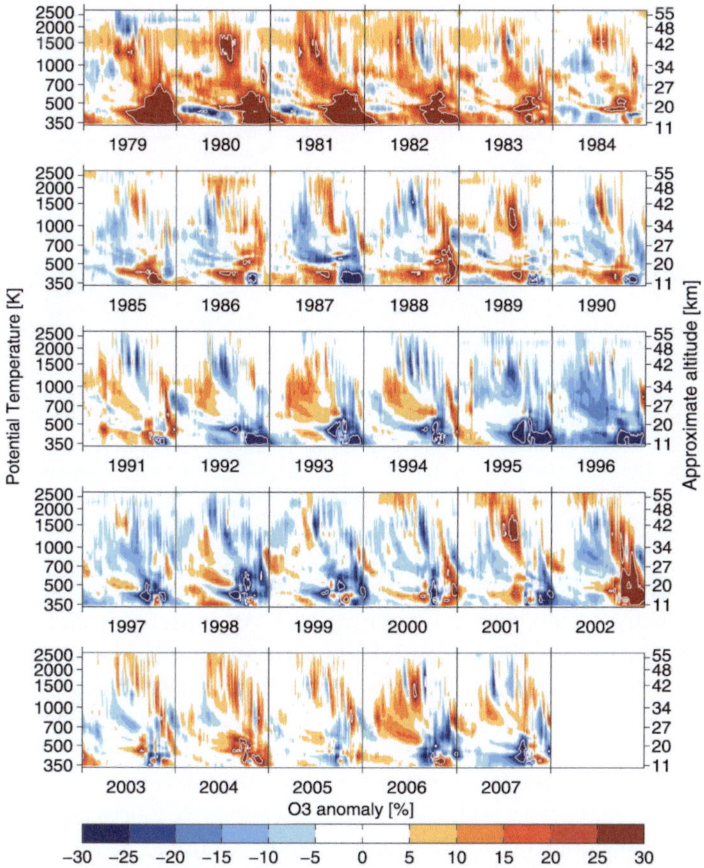

Fig. 9.8 Ozone anomalies relative to the annual cycle, area weighted average over 70–90°S equivalent latitude. The whole available time series is shown, split up into six year stretches. In addition to the model-inherent potential temperature (*left axis*), also the corresponding approximate geopotential height values are given (*right axis*). Values exceeding the colour scale are indicated by the *white contour lines*. A 3 day running mean has been applied for smoother viewing. Plot updated from *Kiesewetter et al.* [2010a]. © 2010 American Geophysical Union. Modified by permission of American Geophysical Union

A very similar behaviour is observed in the southern hemisphere high latitudes: Fig. 9.8 showing the corresponding picture for 70–90°S. (Throughout the first years of the assimilation, large positive anomalies are visible in the lower stratosphere in spring. These appear as positive anomalies here due to the persistent formation of the Antarctic ozone hole in later years.)

To provide a more quantitative picture of the persistence and development of ozone anomalies, the autocorrelation of anomalies with respect to a reference level at 631 K (~25 km) is shown in Fig. 9.9 in the zonal as well as the equivalent latitude frame. The level of 631 K is chosen as a representative of the connection range between the higher stratosphere, where many ozone anomalies originate, and

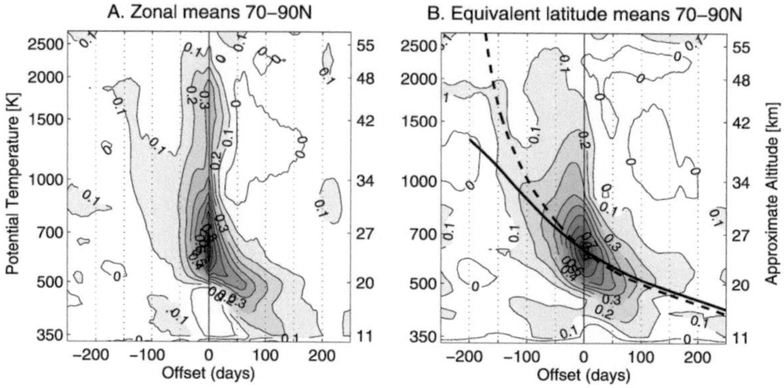

Fig. 9.9 Autocorrelation of relative ozone anomalies, with respect to relative ozone anomalies at 631 K (∼25 km). Area weighted averages over 75–90°N latitude (**A**) as well as equivalent latitude (**B**) are shown. The *solid black line* in (**B**) displays the diabatic descent path expected from heating rates averaged year-round; the *dashed line* represents the same but following winter (DJF) heating rates

the lower stratosphere, where they then subside and remain. The result does not depend critically on the chosen reference level. As expected from the qualitative picture described so far, the autocorrelation structure displays a consistent pattern descending from the upper (>1000 K) to the lower stratosphere. The correlation coefficient itself declines relatively fast both in the past and future, attaining values of 0.5 or less 2 months before and after day zero. Nonetheless, the whole picture seems consistent, and due to the long time series, even low correlation coefficients are significant. For comparison, an average descent curve calculated from forward and backward integrated diabatic heating rates is indicated in Fig. 9.9. One sees that the autocorrelation in general follows the average descent, in particular for the equivalent latitude average. In addition, a faster descending mode is present as well, most clearly seen in the ordinary zonal mean picture. This fast descending mode is associated with anomalies in the annular modes, to be discussed in greater detail in the next section.

We now analyse composites of high and low ozone anomalies (Fig. 9.10), which are identified here by the date when the ozone anomaly at 631 K exceeds 1.5 standard deviations (of the whole record). In order to assure that anomalies are not counted more than once, we require that the ozone anomaly stays below the threshold value for 60 days before the onset date. With these parameters, 17 positive and 16 negative ozone anomaly events are detected.

Both positive and negative anomalies generally appear at around 1000 K (about 30 km altitude) about one month before the onset at 631 K, intensify to an anomaly of about 10 % and descend into the lower stratosphere over the following three to five months. The long persistence of the ozone anomalies is not in contradiction to known photochemical life times, but is remarkable. *Tegtmeier et al.* [2008] have reported an unexpectedly long persistence of ozone anomalies in the middle stratosphere at mid-latitudes, and hypothesised that this persistence is connected to

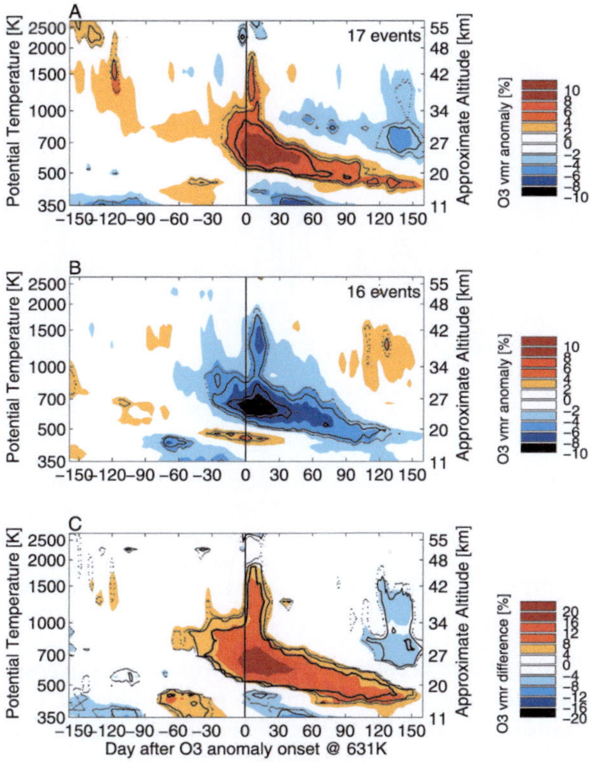

Fig. 9.10 Composite plots of positive (**A**) and negative (**B**) ozone anomalies, and their difference (**C**). Day 0 indicates the day when the ozone anomaly at 631 K exceeds 1.5 standard deviations; the evolution of ozone anomalies is shown for 150 days before and after day 0. All values are area-weighted averages north of 65°N equivalent latitude. (**A**), (**B**) Composites of ozone anomalies. *Colours* indicate anomalies in terms of fractions of annual cycle values, *black contour lines* indicate the significance of the anomalies: *dotted* 90 %, *thin black* 95 %, *solid black* 99 %. (**C**) Difference of (**A**) and (**B**), divided by the mean ozone at the respective day. *Black contour lines* indicate the significance of the difference: *dotted* 90 %, *thin black* 95 %, *solid black* 99 %. Plot modified from *Kiesewetter et al.* [2010a]. © 2010 American Geophysical Union. Modified by permission of American Geophysical Union

transport-induced anomalies in odd nitrogen (NO_y). We hypothesise that transport-induced NO_y anomalies may at least partly play a role for the long persistence of polar ozone anomalies here as well. This possibility requires further testing.

9.3.1 Relation to Annular Mode Variability

Much of the variability in polar stratospheric ozone during winter can be related to the variability of the Annular Modes (NAM and SAM, respectively) [*Baldwin and Dunkerton*, 2001; *Baldwin and Thompson*, 2009]. As an example Fig. 9.11 shows anomalies of ozone and the NAM index at around 24 km altitude for the two-year

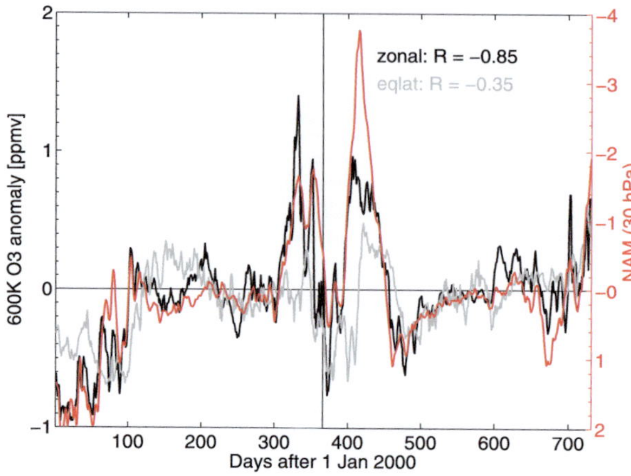

Fig. 9.11 Polar ozone anomalies in the mid-stratosphere (600 K, ∼24 km) versus NAM phase (at 30 hPa, ∼24 km), for the years 2000 and 2001. *Black solid line, left axis*: Zonal ozone anomalies averaged over 75–90°N. *Gray line, left axis*: Equivalent latitude ozone anomalies averaged over 75–90°N. *Red line, right axis*: NAM index. Note the reversed NAM axis. Plot updated from *Kiesewetter et al.* [2010a]. © 2010 American Geophysical Union. Modified by permission of American Geophysical Union

period 2000 to 2001. Ozone is enhanced during weak vortex events, negative NAM index, that disturb the polar vortex and mix in mid-latitude air. Conversely, strong vortex events, positive NAM index, are associated with a more stable, colder and more isolated vortex leading to photochemically reduced ozone. Composite plots of ozone anomalies according to weak and strong vortex events are shown in Fig. 9.12. Positive and negative anomalies occur almost instantaneously over much of the mid-stratosphere and then slowly descend into the lower stratosphere, where they are clearly visible for four to five months.

A detailed investigation shows that different interaction processes of chemistry and dynamics are responsible for the distinct shape of NAM-related ozone anomalies. The initial shape of the ozone anomaly is induced by an anomaly in meridional advection associated with a disturbed or strengthened vortex. As the meridional ozone gradient reverses at around 500 K (20 km), a weak vortex event is associated with advection of ozone rich air and a positive vortex ozone anomaly above 500 K, and mixing in of ozone poor air masses resulting in a negative ozone anomaly below 500 K. The inverse is the case for strong vortex events. However, while the anomaly in transport explains the initial shape of the ozone anomaly for the weeks following the onset of the vortex events, where the dynamical perturbation quickly descends from the upper to the lower stratosphere, it is not able to explain the long residence time of the ozone anomaly in the lower stratosphere, and the large negative ozone anomaly that arises in the upper stratosphere (1000–1500 K) one to two months after a strong vortex event. Here, the temperature anomalies associated with the vortex events play a role: reduced temperatures in the lower stratosphere in the course of a strong vortex event lead to more widespread formation of polar stratospheric clouds (PSCs) and stronger heterogeneous destruction of ozone, while the

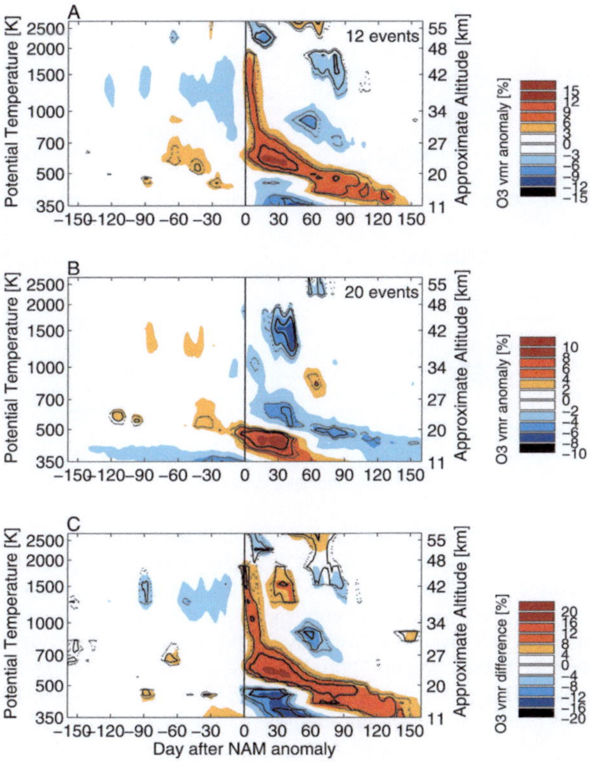

Fig. 9.12 Ozone anomalies after weak (**A**) and strong (**B**) vortex events, and their difference (**C**). Day 0 is the onset of the vortex event, i.e. the day when the NAM index at 10 hPa first crosses the threshold value ($+1.5/-3.0$) in 60 days. All values are equivalent latitude averages north of 65°N. (**A**), (**B**) Composites of ozone anomalies. *Colours* indicate anomalies in terms of fractions of annual cycle values, *black contour lines* indicate the significance of the anomalies: *dotted* 90 %, *thin black* 95 %, *solid black* 99 %. (**C**) Difference of (**A**) and (**B**), divided by the mean ozone at the respective day. *Black contour lines* indicate the significance of the difference: *dotted* 90 %, *thin black* 95 %, *solid black* 99 %. Plot updated from *Kiesewetter et al.* [2010a]. © 2010 American Geophysical Union. Modified by permission of American Geophysical Union

opposite is the case during weak vortex events. Hence, lower stratospheric temperature anomalies act to enhance the ozone anomalies induced by advection. In the upper stratosphere, a secondary positive temperature anomaly that is delayed by a month from the strong vortex event is responsible for shifting the photochemical gas phase reaction equilibria towards a state of lower ozone, leading to the strong negative and apparently delayed signal in upper stratospheric ozone.

9.3.2 The Impact of Solar Proton Events

A very different class of ozone anomalies is related to solar proton events (SPEs). Ionisation by energetic particles during SPEs lead to enhanced levels of NO_y in the

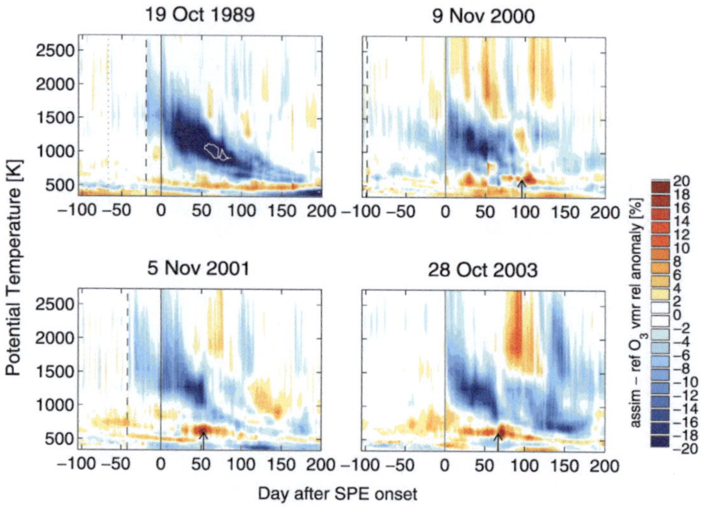

Fig. 9.13 Anomalies of ozone differences between assimilated ozone and reference runs, averaged poleward of 70°N in the vicinity of four different SPEs and expressed as percentages of assimilated ozone vmr. An additional contour line at the −25 % level is shown as *white line* (only relevant for winter 1989/90). Day zero (*black line*) corresponds to the first day of enhanced proton fluxes measured by GOES. For October 1989, November 2000, and November 2001, also the preceding events of 30 September 1989, 2 August 2000, and 24 September 2001 are shown as *dashed lines*. In 1989, the SPE in August is shown as *dotted line*. Weak vortex events occurring in three of the winters are shown as *black arrows* near the abscissa (13 Feb 2001, 28 Dec 2001, 3 Jan 2004)

upper stratosphere that can result in enhanced catalytic destruction of stratospheric ozone (see also Chaps. 15, 16 and 17). Figure 9.13 shows Arctic ozone anomalies for four of the largest SPEs that occurred during the time period of our assimilated ozone data set. Large negative anomalies of up to −20 % are clearly visible that slowly descend from the upper to the lower stratosphere. In three out of the four cases multiple SPEs occurred within weeks, leading already to reduced ozone before the onset of the main SPE. The negative anomalies are accompanied by positive anomalies in the lower stratosphere, possibly as a result of the so-called self-healing effect: A reduction of ozone at higher levels leads to increased ozone at lower levels resulting from the enhanced UV radiation. The negative ozone anomalies can be identified for almost half a year following the SPE [*Jackman et al.*, 2008].

9.3.3 Long-Term Variability

Although the focus of this project is on Arctic ozone variability on short to seasonal time scales (days to months), the assimilated ozone data set also compares well with independent data sets over long time scales (years to decades), making it valuable for the attribution of long-term trends and decadal scale variability. A separate

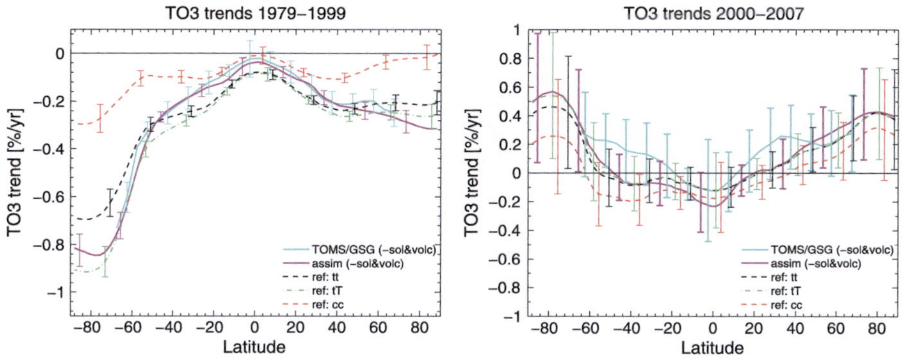

Fig. 9.14 Trends in total ozone (TO3) for 1979–1999 (*left*) and 2000–2007 (*right*). TO3 trends in the assimilated data set ('assim'), the TOMS/GOME/SCIAMACHY/GOME2 merged data sets, and CTM sensitivity runs ('tt' and 'tT': time-dependent concentrations of ozone depleting substances with different assumptions on the scaling of polar ozone loss with chlorine loading; 'cc' constant year-2000 loading of ozone depleting substances), as obtained from the piecewise linear trend regression model for the whole time series. Adapted from *Kiesewetter et al.* [2010b]

trend analysis over the periods 1979–1999 and 2000–2007 shows excellent agreement of trends in total column ozone, calculated from the assimilated data set and independent TOMS and GOME/SCIAMACHY/GOME2 observations (Fig. 9.14). Additional sensitivity calculations with the CTM without assimilation of satellite ozone observations can be used for an attribution of observed ozone trends due to changes in the concentration of ozone depleting substances and changes in transport and dynamics [*Kiesewetter et al.*, 2010b]. In the period 1979–1999, modelled TO3 trends at mid-latitudes are almost equally caused by changes in gas-phase chemistry, changes in polar heterogeneous chemistry (including dilution of ozone depleted air masses), and meteorological changes. In most seasons, in situ gas-phase chemistry is the single largest contributor to mid-latitude trends. As a result of the large seasonality of polar ozone destruction through heterogeneous processes on PSCs, strong effects of export of ozone depleted air masses to mid-latitudes are seen only in spring (NH, SH) and partly in summer (SH). Changes in meteorology contribute around 35 % to mid-latitude TO3 trends, with strong differences between different seasons. Over the time period 2000–2007 positive linear trends in total column ozone are dominated by changes in meteorology, as to be expected for the yet small decrease in stratospheric halogen loading over this period [*Kiesewetter et al.*, 2010b].

The good agreement between the long-term trends derived from our assimilated ozone data set and independent observations suggests that this data set may also be useful for the detection of decadal scale signals related to solar variability (Fig. 9.15). Solar cycle effects are not included in the CTM chemistry scheme used here, and may only be induced by the assimilation. We can thus use a comparison of the assimilated data set to the unconstrained CTM to diagnose the solar signal in the assimilated data set. A linear least squares regression model containing a mean offset, a volcanic proxy and the solar F10.7 radio flux is then applied to deseasonalized differences between assimilated data set and the unconstrained CTM. Since

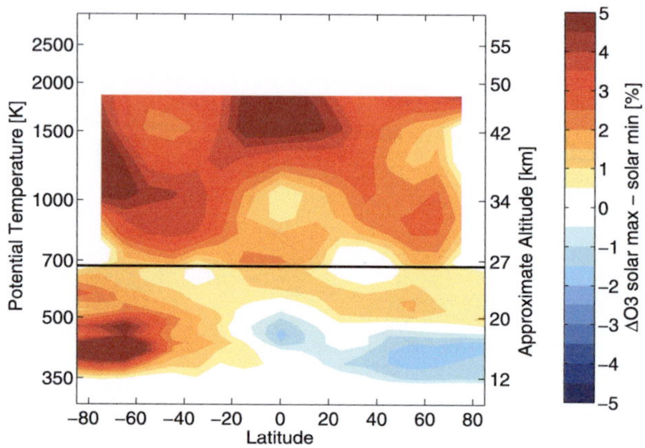

Fig. 9.15 Vertical distribution of the solar signal (in %) in the assimilated data set (below 700 K) and SBUV observations (above 700 K). Contour levels correspond to differences in regressed solar-related ozone vmr offset (assim-tt), taken between solar maximum and solar minimum

the content of satellite information in the assimilated data set decreases in the upper stratosphere as a result of the short photochemical lifetime of ozone, much of the solar signal in the upper stratosphere is lost in the assimilation. Figure 9.15 therefore uses a combination of satellite data (the SBUV merged ozone data set, above 700 K) and the assimilation (below 700 K) to cover the whole stratosphere. Our analysis shows a solar signal of about 5 % in ozone between solar maximum and minimum, somewhat larger than that derived by *Soukharev and Hood* [2006] from a direct regression of SBUV(/2) and SAGE II satellite observations (see also Chaps. 3, 29 and 31).

9.4 Relation Between Early Winter Ozone and Spring Total Ozone in Chemistry Climate Model Runs

To understand the mechanisms leading to the possible relationship between upper stratospheric ozone in autumn and total ozone in the following spring, as reported by *Kawa et al.* [2005] and *Sinnhuber et al.* [2006], multi-year simulations with the ECHAM5-MESSy Atmospheric Chemistry (EMAC) model have been analysed. EMAC is a chemistry-climate model (CCM) that includes modules to calculate chemical ozone production and loss processes as well as the transport of ozone [e.g. *Roeckner et al.*, 2006; *Joeckel et al.*, 2006]. To account for the effects of solar variability on stratospheric ozone EMAC includes a solar radiation scheme with improved spectral resolution [*Nissen et al.*, 2007]. Figure 9.16 shows the time series of upper stratospheric ozone in autumn together with total ozone in the following spring from a transient simulation of the period 1960 to 2000 performed at MPI for Chemistry (see also Chap. 25). The simulation considers, according to the recommendations for a REF-B1 simulation by the Chemistry-Climate Model Validation (CCMVal) initiative [*Eyring et al.*, 2008], measured changes of external forcings of the atmosphere, such as the concentrations of greenhouse gases, ozone depleting substances and aerosols, sea-surface temperatures and sea-ice concentration, solar

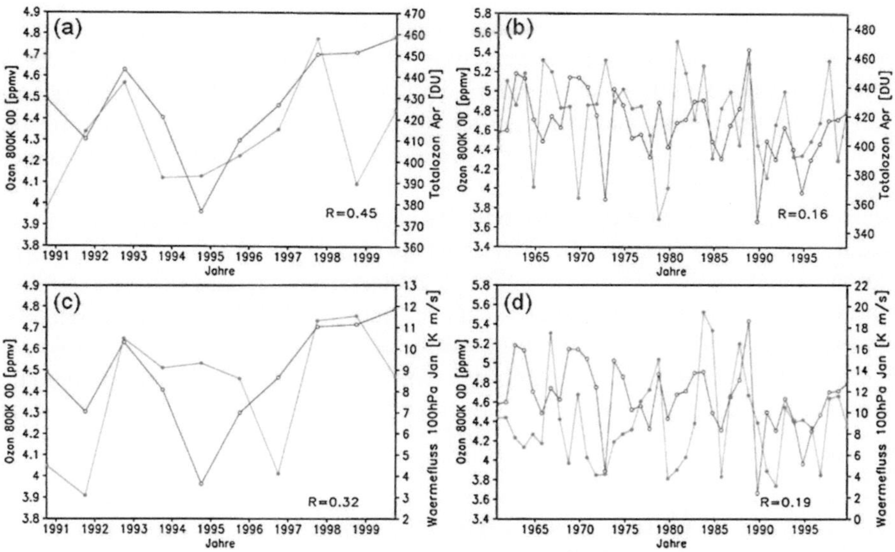

Fig. 9.16 The relation between early-winter ozone to total ozone in spring and to planetary wave activity in winter from the EMAC REF-B1 simulation at a grid point close to Ny-Ålesund (79°N, 12°E). (**a**) October and December ozone at 800 K potential temperature (about 30 km altitude) and April total column ozone for 1991–2000. (**b**) same like (**a**) but for 1960–2000, (**c**) October and December ozone at 800 K and zonal mean eddy heat flux at 100 hPa in January for 1991–2000, (**d**) same like (**c**) but for 1960–2000. All quantities are plotted for October, i.e. for spring total ozone the corresponding year is obtained by adding 1 year. From *Wersing* [2010]

spectral irradiance and an internally generated Quasi-Biennial Oscillation (QBO) of the zonal wind in the tropical stratosphere. This simulation represents the most realistic reproduction of the atmosphere with our model and thus allows us to compare directly with Fig. 5 of *Sinnhuber et al.* [2006] and Fig. 9.1 of this chapter. During the period in which observations and model years overlap (i.e. 1991–2000), the model tends to confirm the observed correlation over Ny-Ålesund, with high total ozone in spring following high upper stratospheric ozone in the preceding autumn, and vice versa, although the correlation in the model ($R = 0.45$) is somewhat lower than observed ($R = 0.78$) (Fig. 9.16a) [*Wersing*, 2010]. In individual years, such as in the early 1990s, discrepancies between the model and observations are obvious, which is however expected, as the model is a free running climate model developing its own internal dynamics even with prescribed external drivers. In single model years, planetary wave activity in the mid-latitude lower stratosphere in winter varies consistently with polar total ozone in spring (Fig. 9.16c) as shown e.g. by *Weber et al.* [2003]. Moreover, in these years increases in winter planetary wave activity and spring total ozone may follow autumns with high upper stratosphere ozone. However, in general the correlations between the zonal mean eddy heat flux and total ozone do not reach the observed level.

For the complete simulation period 1960–2000 the correlations between upper stratosphere ozone in autumn and total ozone in spring (Fig. 9.16b) as well as the correlation between winter lower stratosphere planetary wave activity and spring total ozone (Fig. 9.16d) become very weak ($R < 0.20$). Thus, our simulations do not confirm the hypothesis of an enduring influence throughout the winter of stratospheric autumn ozone anomalies on spring ozone. Similar conclusions were obtained from additional analyses for the whole Arctic region (instead of Ny-Ålesund), the Antarctic and from supplementary time-slice simulations, performed with constant external forcings and prescribed solar irradiance for either minimum or maximum solar activity (not shown). Given that also for the observations the high correlations found for the 1990s and beginning 2000s have weakened in the most recent past (see Fig. 9.1, left), the good coincidence between upper stratospheric ozone in autumn and spring total ozone for the earlier years appears to be an effect that was confined to a limited period as part of the well-known inter-annual variability in the northern stratosphere.

9.5 Impact of Ozone Anomalies on Mid-winter Stratospheric Dynamics from GCM Calculations

A potential pathway for the impact of stratospheric ozone anomalies in autumn on spring total ozone could be a modification of the propagation and dissipation of planetary waves in the stratosphere during the following winter. *Kodera and Kuroda* [2002] for example showed that zonal wind anomalies in the autumn and winter stratosphere, induced by solar cycle induced irradiance variations, are able to influence the planetary wave driving of the stratosphere. This kind of mechanism was tested here in a series of EMAC simulations that were performed with prescribed ozone climatologies. The chemistry and transport modules were switched off, i.e. EMAC was applied like a general circulation model (GCM). Polar stratospheric ozone anomalies (north of 60°N centred at 10 hPa, about 30 km altitude) were derived from the assimilated ozone data set introduced in Sect. 1.2 and added both as positive and negative bias to a zonally monthly mean climatological ozone field each October (Fig. 9.17a), November and December of two 20-year simulations. In comparison with the reference simulation using the ozone climatology, the polar lower and middle stratosphere cools by up to 0.08 K/d in October in the experiment with negative ozone bias as a result of weaker absorption of solar radiation (Fig. 9.17b). This effect decreases in amplitude and latitude with southward propagation of the terminator towards the winter. Consistently, the temperature decreases by about 1–3 K in the polar lower stratosphere during October, November and December, when ozone is reduced, and the polar night jet increases, as exemplified for December in Fig. 9.17c. However these changes are not statistically significant, nor is the change in the Eliassen-Palm (EP) flux vector and its divergence (Fig. 9.17d) indicating that a significant modification of the planetary wave driving does not take place. The strengthening of the polar night in December implies favourable conditions for a potential reduction in upward planetary wave propagation from the stratosphere for the remaining winter period. However, in January strongly increased

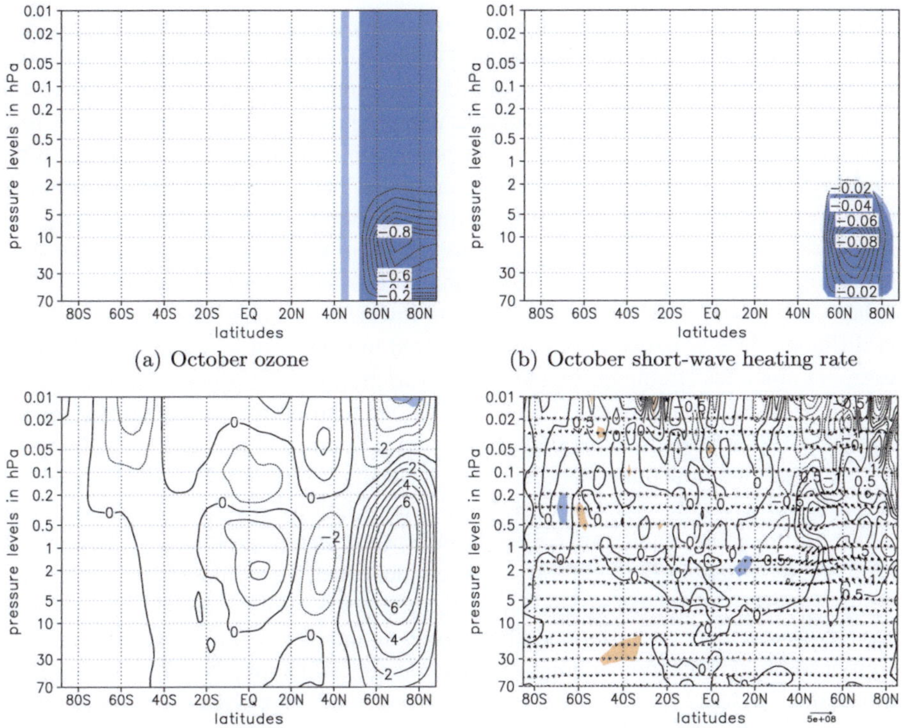

(a) October ozone (b) October short-wave heating rate

(c) December zonal mean zonal wind (d) December EP-flux vector/divergence

Fig. 9.17 The response of the stratosphere to prescribed early winter polar stratospheric ozone anomalies in EMAC GCM-type sensitivity simulations. (**a**) Prescribed ozone anomalies (in ppmv) in October, (**b**) short-wave heating rate anomalies (in K/d) in October, (**c**) anomalies in zonal mean zonal wind (in m/s) in December, and (**d**) anomalies in EP-flux vector (arrows) and the EP Flux divergence (in m/s/d) in December. *Light* (*dark*) shading denotes significant results at the 95 (99) % levels

EP-fluxes and convergence indicate enhanced planetary wave activity leading to a significant warming and concurrent deceleration of the polar night jet in the upper stratosphere. In summary, the results of these sensitivity simulations suggest that the initial radiative perturbation of stratospheric ozone anomalies in autumn has a nearly simultaneous, direct effect on stratospheric temperature and wind, however this effect is not strong enough to prevent sufficiently strong tropospheric dynamical disturbances from penetrating into the stratosphere during mid-winter.

9.6 Conclusions and Outlook

A long-term (29-year) data set of global stratospheric ozone was generated in this project from assimilation of satellite observations into a chemical transport model.

This data set agrees well with independent observations; in particular ozone variability during high latitude winter is well captured, making this data set well suited for the analysis of high latitude ozone changes during winter. An analysis of the assimilated data set during high latitude winter shows alternating patterns of positive and negative ozone anomalies that propagate downward from the mid- to the lower stratosphere during winter. These anomalies have a remarkably long lifetime, e.g. several months. Much of the variability is related to dynamical variability in the Northern Annular Mode (NAM). Extreme phases of the NAM lead to the formation of strong and distinctly shaped ozone anomalies that traverse most of the stratosphere within days and then remain in the lower stratosphere for several months. A different class of ozone anomalies is related to solar proton events, which, however, are much less frequent. In addition to variability on intra-seasonal time scales, we show that the assimilated ozone data set can also be used successfully for the analysis of long-term changes such as long-term trends and decadal scale changes.

Although we find that ozone anomalies during Arctic winter often persist for several months, no further evidence for a close relation between mid-stratospheric ozone during autumn and total ozone during spring, as first proposed by *Kawa et al.* [2005] and *Sinnhuber et al.* [2006], has been identified. This is consistent with a correlation analysis using calculations from the chemistry climate model EMAC. GCM sensitivity calculations with prescribed ozone anomalies in autumn (OND) that are similar to observed ozone anomalies show a direct effect on stratospheric temperature and wind, however this effect is not strong enough to prevent sufficiently strong tropospheric dynamical disturbances from penetrating into the stratosphere during mid-winter.

One PhD thesis [*Kiesewetter*, 2011], one diploma thesis [*Wersing*, 2010] and one bachelor thesis [*Höweling*, 2010] were successfully completed as part of the project.

Further studies are planned to investigate the possible role of NO_y anomalies for the development and persistence of ozone anomalies, using observational data as well as model calculations. We will also continue to analyse the assimilated data set for decadal scale ozone changes and further investigate the factors influencing ozone variability during autumn and winter.

Acknowledgements We thank Martyn Chipperfield for making the assimilation code available to us and Michael Prather and Juno Hsu for providing the linearised ozone chemistry scheme used in our CTM. We thank Peter von der Gathen for sharing the Ny-Ålesund ozone sonde data. Claudia Wersing analysed the EMAC-REFB1 simulations and Markus Kunze performed the EMAC-GCM simulations.

References

Baldwin, M. P., & Dunkerton, T. J. (2001). Stratospheric harbingers of anomalous weather regimes. *Science*, *294*, 581–584.

Baldwin, M. P., & Thompson, D. W. J. (2009). A critical comparison of stratosphere–troposphere coupling indices. *Quarterly Journal of the Royal Meteorological Society*, *135*, 1661–1672.

Chipperfield, M. P., Khattatov, B. V., & Lary, D. J. (2002). Sequential assimilation of stratospheric chemical observations in a three-dimensional model. *Journal of Geophysical Research, 107*(D21), 4585. doi:10.1029/2002JD002110.

Dethof, A., & Hólm, E. V. (2004). Ozone assimilation in the ERA-40 reanalysis project. *Quarterly Journal of the Royal Meteorological Society, 130*, 2851–2872. doi:10.1256/qj.03.196.

Eyring, V., Chipperfield, M. P., Giorgetta, M. A., Kinnison, D. E., Manzini, E., Matthes, K., Newman, P. A., Pawson, S., Shepherd, T. G., & Waugh, D. W. (2008). Overview of the new CCMVal reference and sensitivity simulations in support of upcoming ozone and climate assessments and the planned SPARC CCMVal. *SPARC Newsletter, 30*, 20–26.

Fortuin, J. F. P., & Kelder, H. (1998). An ozone climatology based on ozonesondes and satellite measurements. *Journal of Geophysical Research, 103*, 31709–31734.

Hassler, B., Bodeker, G. E., Cionni, I., & Dameris, M. (2009). A vertically resolved, monthly mean, ozone database from 1979 to 2100 for constraining global climate model simulations. *International Journal of Remote Sensing, 30*, 4009–4018.

Höweling, D. (2010). *Assimilierung von Ozonmessungen aus dem SCIAMACHY-Instrument in ein Chemie-Transportmodell*. Bachelorarbeit, Institute of Environmental Physics, University of Bremen.

Jackman, C. H., Marsh, D. R., Vitt, F. M., Garcia, R. R., Fleming, E. L., Labow, G. J., Randall, C. E., Lopez-Puertas, M., Funke, B., von Clarmann, T., & Stiller, G. P. (2008). Short- and medium-term middle atmospheric constituent effects of very large solar proton events. *Atmospheric Chemistry and Physics, 8*, 765–785.

Jöckel, P., Tost, H., Pozzer, A., Brühl, C., Buchholz, J., Ganzeveld, L., Hoor, P., Kerkweg, A., Lawrence, M. G., Sander, R., Steil, B., Stiller, G., Tanarhte, M., Taraborrelli, D., van Aardenne, J., & Lelieveld, J. (2006). The atmospheric chemistry general circulation model ECHAM5/MESSy1: consistent simulation of ozone from the surface to the mesosphere. *Atmospheric Chemistry and Physics, 6*, 5067–5104.

Kawa, S. R., Newman, P. A., Stolarski, R. S., & Bevacquila, R. M. (2005). Fall vortex ozone as a predictor of springtime total ozone at high northern latitudes. *Atmospheric Chemistry and Physics, 5*, 1655–1663.

Khattatov, B. V., Lamarque, J.-F., Lyjak, L. V., Menard, R., Levelt, P., Tie, X., Brasseur, G. P., & Gille, J. C. (2000). Assimilation of satellite observations of long-lived chemical species in global chemistry transport models. *Journal of Geophysical Research, 105*(D23), 29135–29144.

Kiesewetter, G. (2011). *The variability of stratospheric ozone in a 29 year assimilated data set sensitivity calculations*. Ph.D. thesis, University of Bremen, Bremen, Germany.

Kiesewetter, G., Sinnhuber, B.-M., Vountas, M., Weber, M., & Burrows, J. P. (2010a). A long-term stratospheric ozone dataset from assimilation of satellite observations: high-latitude ozone anomalies. *Journal of Geophysical Research, 115*, D10307. doi:10.1029/2009JD013362.

Kiesewetter, G., Sinnhuber, B.-M., Weber, M., & Burrows, J. P. (2010b). Attribution of stratospheric ozone trends to chemistry and dynamics: a modelling study. *Atmospheric Chemistry and Physics, 10*, 12073–12089. doi:10.5194/acp-10-12073-2010.

Kodera, K., & Kuroda, Y. (2002). Dynamical response to the solar cycle. *Journal of Geophysical Research, 107*, 4749. doi:10.1029/2002JD002224.

Newman, P. A., Nash, E. R., & Rosenfield, J. E. (2001). What controls the temperature of the Arctic stratosphere during the spring? *Journal of Geophysical Research, 106*, 19999–20010. doi:10.1029/2000JD000061.

Nissen, K., Matthes, K., Langematz, U., & Mayer, B. (2007). Towards a better representation of the solar cycle in general circulation models. *Atmospheric Chemistry and Physics, 7*, 5391–5400.

Roeckner, E., Brokopf, R., Esch, M., Giorgetta, M., Hagemann, S., Kornblueh, L., Manzini, E., Schlese, U., & Schulzweida, U. (2006). Sensitivity of simulated climate to horizontal and vertical resolution in the ECHAM5 atmosphere model. *Journal of Climate, 19*, 3771–3791.

Sinnhuber, B.-M., Weber, M., Amankwah, A., & Burrows, J. P. (2003). Total ozone during the unusual Antarctic winter of 2002. *Geophysical Research Letters, 30*, 1580. doi:10.1029/2002GL016798.

Sinnhuber, B.-M., von der Gathen, P., Sinnhuber, M., Rex, M., König-Langlo, G., & Oltmans, S. J. (2006). Large decadal scale changes of polar ozone suggest solar influence. *Atmospheric Chemistry and Physics*, *6*, 1835–1841.

Soukharev, B. E., & Hood, L. L. (2006). Solar cycle variation of stratospheric ozone: multiple regression analysis of long-term satellite data sets and comparisons with models. *Journal of Geophysical Research*, *111*, D20314. doi:10.1029/2006JD007107.

Tegtmeier, S., Fioletov, V. E., & Shepherd, T. G. (2008). Seasonal persistence of northern low- and middle latitude anomalies of ozone and other trace gases in the upper stratosphere. *Journal of Geophysical Research*, *113*, D21308.

Weber, M., Dohmse, S., Wittrock, F., Richter, A., Sinnhuber, B.-M., & Burrows, J. P. (2003). Dynamical control of NH and SH winter/spring total ozone from GOME observations 1995–2002. *Geophysical Research Letters*, *30*, 1583. doi:10.1029/2002GL016799.

Weber, M., Lamsal, L. N., & Burrows, J. P. (2007). Improved SCIAMACHY WFDOAS total ozone retrieval: steps towards homogenising long-term total ozone datasets from GOME, SCIAMACHY, and GOME2. In *Proc. 'Envisat symposium 2007'*, Montreux, Switzerland, 23–27 April 2007, ESA SP-636.

Wersing, C. (2010). *Einfluss des Ozons der frühwinterlichen Stratosphäre auf das Totalozon im Frühjahr in hohen Breiten*. Diplomarbeit, Institut für Meteorologie. Freie Universität Berlin.

World Meteorological Organization (2011). *Scientific assessment of ozone depletion*: *2010* (Global Ozone Research and Monitoring Project—Report No. 52).

Chapter 10
The Response of Atomic Hydrogen to Solar Radiation Changes

Martin Kaufmann, Manfred Ern, Catrin Lehmann, and Martin Riese

Abstract The combination of satellite born SCIAMACHY hydroxyl and GOMOS ozone limb measurements allows for the derivation of the global distribution of atomic hydrogen abundance and instantaneous chemical heating rates in the mesopause region. Chemical heating rates show maximum values of 5–10 K/day at 85–90 km; atomic hydrogen densities are 1–5 $\cdot$ 10^8 cm^{-3}. Signatures of equatorial Kelvin waves, Rossby-gravity waves and Rossby waves are clearly visible in the data. A pronounced latitudinal structure with maxima at the equator and at mid latitudes is observed. Between 2002 and 2008 chemical heating rates decreased and atomic hydrogen density increased, in accordance with model simulations of the 11-year solar cycle.

10.1 Introduction

The upper mesosphere/lower thermosphere energy balance is still a frontier in atmospheric sciences. The input of solar radiation, the output of energy in the form of infrared and airglow emission, storage of energy in latent chemical form, conversion of energy, and energy transport are current topics of research. A primary source of energy in this region is the absorption of solar UV radiation by molecular oxygen and ozone. However, about 70 % of the absorbed solar UV energy is stored as chemical energy, i.e. the energy required to break the bonds of the absorbing species is stored as chemical potential energy in the product species. These products can be in excited photochemical states or produce excited states in subsequent exothermic reactions. These excited products may radiate, so that part of their energy is not available for heating the atmosphere. They may also be involved in non-local thermodynamic equilibrium processes and act as cooling agents, transferring heat from the ambient atmosphere to internal energy of molecules, which, in turn, lose part of this energy by radiation such that the atmosphere is cooled. Prominent examples of the former are the OH and O_2 afterglows, while the excitation and subsequent

M. Kaufmann (✉) · M. Ern · C. Lehmann · M. Riese
Institute for Energy and Climate Research – Stratosphere (IEK-7), 52425 Jülich, Germany
e-mail: m.kaufmann@fz-juelich.de

F.-J. Lübken (ed.), *Climate and Weather of the Sun-Earth System (CAWSES)*, Springer Atmospheric Sciences,
DOI 10.1007/978-94-007-4348-9_10,

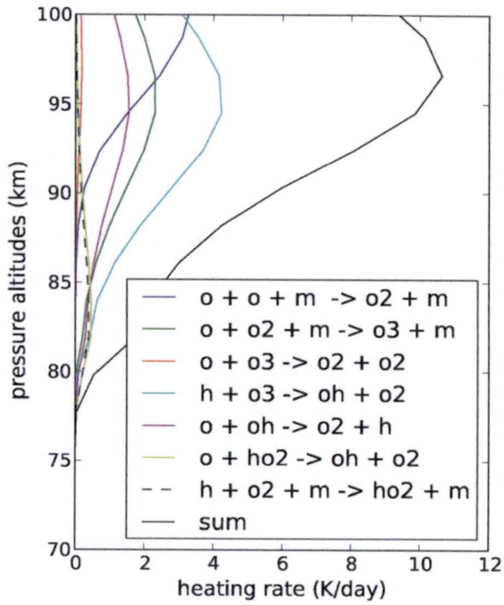

Fig. 10.1 Chemical heating rates for equinox conditions at the equator for midnight, as calculated by the NCAR ROSE model

radiation of CO_2 and NO by atomic oxygen illustrate the latter process. Products of photo-dissociation such as atomic oxygen or the active hydrogen chemical family ($HO_x = H + OH + HO_2$) have chemical lifetimes of several weeks or months in the mesopause region, thus chemical energy can be transported over planetary scales and may be released at different latitudes and altitudes far away from the place of production of these species. Recent examples of this process are a lifted stratopause or mesospheric ozone anomalies as a consequence of enhanced down-welling of thermospheric atomic oxygen [*Smith et al.*, 2009] and associated heat release.

The reaction of atomic hydrogen with ozone forming highly excited hydroxyl molecules is the most relevant heating process in the nighttime upper mesosphere. The importance of the hydrogen family for the chemical heating in the upper mesosphere has been identified by *Brasseur and Offermann* [1986] on the basis of rocket sounding observations. Figure 10.1 illustrates the seven most important heating terms in the upper mesosphere as calculated by means of the NCAR ROSE model [*Marsh et al.*, 2001; *Smith and Marsh*, 2005] for the equator at midnight. The reaction of atomic hydrogen and ozone contributes nearly 50 % (4 K/day) to the total chemical heating in this region. The knowledge of atomic hydrogen and ozone is essential to model this important heating term, in particular during nighttime [cf. *Riese et al.*, 1994; *Mlynczak and Solomon*, 1993].

General circulation models [*Schmidt et al.*, 2006; *Marsh et al.*, 2007] show that the upper mesosphere and lower thermosphere region is significantly affected by solar variability. The extent of this influence depends on the location, the season of the year and on the quantity observed. The solar UV-forcing in this region is highly variable, changing by more than 50 % for Lyman-α and 10 %–20 % for the Schumann-Runge-continuum during the 11-year solar cycle. Atomic oxygen will increase due to enhanced O_2 photolysis by about 20 % in the mesopause region during solar maximum compared to solar minimum, resulting in larger chemical heating and CO_2 cooling at the same time. Enhanced O_3 photolysis results in an expansion of the atmosphere, shifting isobars by 300–500 m in the mesopause region [*Schmidt et al.*, 2006]. During solar maximum, ozone mixing ratios increase by about 20 % in the vicinity of the secondary ozone maximum, while ground-state hydroxyl and water vapor decrease by about 10 % and 40 %, respectively. The zonal mean circulation changes due to solar cycle depending wind filtering of gravity waves, which affects the latitudinal distribution of several constituents.

Long term measurements of upper mesospheric constituents such as atomic oxygen or hydrogen are virtually not existent. Temperature observations covering at least a full solar cycle exist only for ground based observations. Most observations show a positive signal in the annual mean solar response [*Beig et al.*, 2008], but with significant dependencies on the latitude and season of the year. This points to the importance of dynamics for the interpretation and intercomparison of these measurements.

The recent ESA Environmental satellite (Envisat) has given a significant advancement of our knowledge about the response of the middle and upper atmosphere to solar forcing. This article focuses on chemical heating rates and atomic hydrogen abundance obtained from measurements of the Scanning Imaging Absorption spectroMeter for Atmospheric CHartographY (SCIAMACHY) [*Bovensmann et al.*, 1999] and from the Global Ozone Monitoring by Occultation of Stars (GOMOS) instrument [*Bertaux et al.*, 2010; *Kyrölä et al.*, 2010], which are both part of the Envisat satellite. Uniquely, SCIAMACHY is able to measure nearly all vibrationally excited states of mesospheric hydroxyl (OH^*). Hydroxyl emissions are a valuable proxy for chemical heating associated to the reaction of atomic hydrogen and ozone. In combination with mesospheric ozone measurements from GOMOS, these measurements can be used to obtain the atomic hydrogen abundance.

10.2 SCIAMACHY Hydroxyl Measurements

SCIAMACHY is capable of performing spectroscopic observations of the Earth's atmosphere in nadir viewing mode, limb viewing mode, as well as in solar and lunar occultation modes. For this study, SCIAMACHY limb observations are employed. In the limb observation mode the Earth's limb is scanned from the surface up to the lower thermosphere in steps of about 3 km. The instantaneous field of view spans 2.6 km vertically and 110 km horizontally at the tangent altitude. After each tangent height step an azimuthal scan is performed covering a distance of about 960 km at the tangent point. To increase the signal to noise ratio, measurements from one horizontal sweep are averaged in this analysis. SCIAMACHY consists of 8 channels, each with its own grating, covering the spectral range between 220 nm and 2.4 μm with a wavelength-dependent spectral resolution of 0.1 nm to 1.5 nm. The instrument covers nearly all vibrational transitions of $OH(\upsilon \leq 9)$—from the UV/visible into the near infrared region. In this work, we utilize the near-infrared OH^* emissions recorded in SCIAMACHY channel 6, covering 1000–1750 nm with 1.5 nm

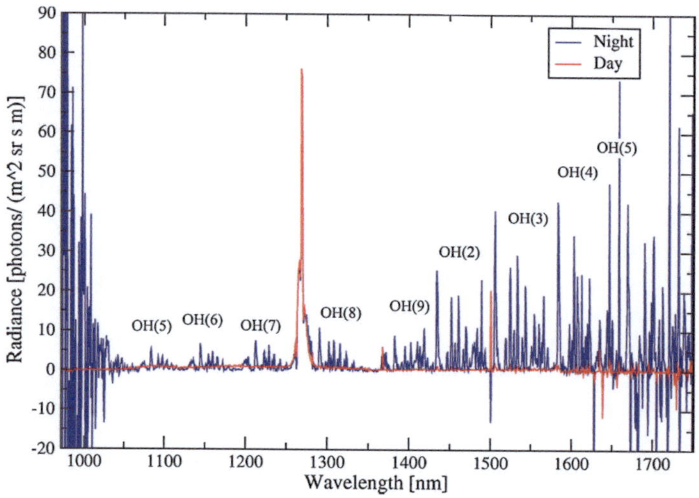

Fig. 10.2 Individual SCIAMACHY channel 6 daytime (*red*) and nighttime (*blue*) limb emission spectra at 90 km altitude for mid latitudes. OH $\Delta v = 3$ (<1400 nm) and $\Delta v = 2$ (>1400 nm) vibrational bands are marked by the upper vibrational state. At 1270 nm the $O_2(^1\Delta)$ is visible. Increased instrument noise appears at the channel boundaries

resolution. The processing of this data is described in *Kaufmann et al.* [2008]. In channel 6, SCIAMACHY is able to observe $\Delta v = 2$ limb emissions from vibrational states $v = 5$ to $v = 2$ and $\Delta v = 3$ limb emissions from $v = 9$ to $v = 5$ (Fig. 10.2). During nighttime, these emissions are about one order of magnitude larger than during daytime, which is a consequence of the nighttime increase of ozone.

The latitudinal region covered by SCIAMACHY limb observations depends on the season of the year and extends from 30°S–80°N (Fig. 10.3). It depends on the seasonality of solar illumination and various calibration measurements performed at the night side of the satellite orbit.

Radiance altitude profiles from SCIAMACHY are illustrated in Fig. 10.4 for two spectral regions covering the OH(3–1) and OH(9–6) emissions. A clear emission layer centered at about 85 km (for the 3–1 band) and 88 km (for the 9–6 band) is observed. The altitude of the radiance maximum is slightly lower for the OH(3–1) band, resulting from the balance between the chemical energy release and the collision frequency.

The latitudinal distribution of the nocturnal OH^* emission layer (not shown) indicates highest emission rates at the equator, and a local minimum around 30° latitude. This latitudinal structure was also seen in the ISAMS observations [*Zaragoza et al.*, 2001] and in SABER data [*Marsh et al.*, 2006] and is caused by the modulation of atomic oxygen abundance and temperature by atmospheric tides.

The radiometric uncertainties of SCIAMACHY channel 6 radiances are less than 4 % [*Lichtenberg et al.*, 2006; *Noel et al.*, 2006]. The long term stability is assured by regular measurements of the solar spectrum and applying monitoring factors to account for instrument degradation [*Bramstedt et al.*, 2010], resulting in a channel-6

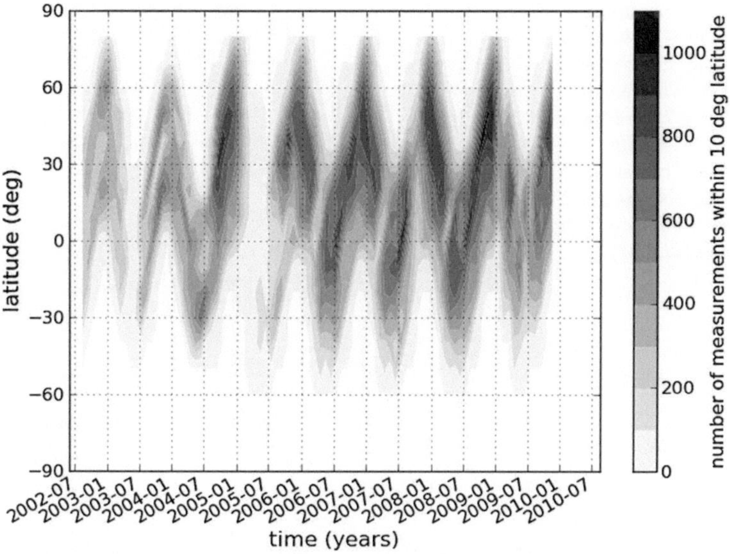

Fig. 10.3 Spatio-temporal distribution of coincident SCIAMACHY and GOMOS data; the *shading* marks the number of coincident profiles per 10° latitude and month. Coincident criteria for the miss-distance of the two datasets are <5° latitude, <45° longitude, and the miss-times are <4 h local time and <1 day universal time. Figure adopted from *Kaufmann et al.* [2010]

radiometric stability better than 1 %. The pointing accuracy is 250 m [*von Savigny et al.*, 2005].

10.3 Modeling

Excited hydroxyl molecules are produced in the reaction of atomic hydrogen and ozone:

$$\mathbf{H} + \mathbf{O_3} \stackrel{\mathbf{k}}{\rightarrow} \mathbf{OH(v)} + \mathbf{O_2} \tag{10.1}$$

The rate constant for this reaction is $k = 1.4 \cdot \exp(-470\ \mathrm{K}/T)$ [*Sander et al.*, 2010], where T is the kinetic temperature. There is sufficient energy released to populate OH up to vibrational mode $\upsilon = 9$. Subsequently, the OH molecules radiate strongly through vibration-rotation transitions.

The modeling of OH airglow is performed with a scheme considering the various production and loss mechanisms of vibrational states of OH: (1) the quasi-nascent production of OH vibrational states just after formation of the OH molecule, (2) the radiative transitions between all states, (3) state-to-state collisional transitions with different collision partners, and (4) the chemical loss of vibrationally excited OH by reaction with atomic oxygen.

Assuming statistical equilibrium, the density of OH in vibrational state υ can be calculated from the ratio of production and loss of each individual state:

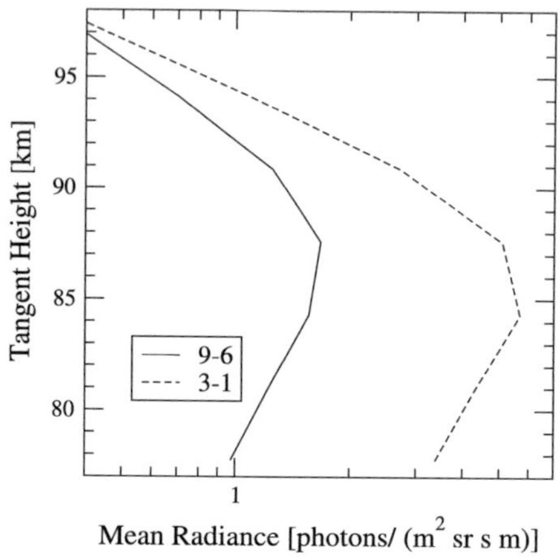

Fig. 10.4 Mean altitude profiles of the OH(9–6) and OH(3–1) vibrational bands, covering the wavelength range from 1375 to 1400 nm and 1515 to 1545 nm, respectively. The data is for mid latitudes, November 25, 2005. Figure adopted from *Kaufmann et al.* [2008] (page 1915)

$$[\mathrm{OH}(v)] = \frac{f_v P_C + \sum_{v'} A_{v',v}[\mathrm{OH}(v')] + \sum_{v'} (k^{N2}_{v',v}[N_2] + k^{\mathrm{O}_2}_{v',v}[\mathrm{O}_2] + k^{\mathrm{O}}_{v',v}[\mathrm{O}])[\mathrm{OH}(v')]}{\sum_{v'} A_{v,v'} + \sum_{v'} (k^{N2}_{v,v'}[N_2] + k^{\mathrm{O}_2}_{v,v'}[\mathrm{O}_2] + k^{\mathrm{O}}_{v,v'}[\mathrm{O}]) + k_c'[\mathrm{O}]} \tag{10.2}$$

$P_c = k[O_3][H]$ describes the formation of vibrationally excited OH by reaction (10.1). f_v is the quasi-nascent fraction of molecules produced in vibrational state v ($f_v = 0.40, 0.29, 0.21, 0.10$ for $v = 9, 8, 7, 6$; lower vibrational states are not populated, cf. *Klenerman and Smith* [1987], *Mlynczak and Solomon* [1993]). $[O_3]$ and [H] are the ozone and atomic hydrogen number density, respectively. The other two terms in the numerator describe the production of OH(v) by relaxation of higher excited vibrational states v' by spontaneous emissions ($A_{v',v}$ is the Einstein-coefficient), and collisional processes ($k^X_{v',v}$ is the rate coefficient for collisions with $X = N_2$, O_2, and O; the concentration of these species is indicated by brackets). The denominator consists of radiative, collisional and chemical loss rates. k_c' is the chemical loss rate coefficient for the reaction of OH(v) with atomic oxygen [*Adler-Golden*, 1997] and [O] is the atomic oxygen density. The collisional deactivation of OH(v) by O_2 is one or two orders of magnitude more efficient than by N_2. There is an ongoing debate on the interaction of OH(v) and O_2 [e.g. *Adler-Golden*, 1997; *McDade and Llewellyn*, 1987]. Possible pathways are described by a sudden-death model, assuming that OH(v) is multi-quantum quenched to OH($v = 0$), a step ladder single-quantum relaxation model, and a cascade model as a mixture of both. The choice of the model is most important for mid- and low-v vibrational states, because these states are not populated directly in the H + O_3 reaction, such that the collisional model determines the amount of chemical energy reaching these levels significantly.

A comparison between our model calculations and SCIAMACHY measurements for two spectral windows at 87 km altitude is illustrated in Fig. 10.5 for the OH(8–5) and OH(9–6) vibrational bands. For a more comprehensive comparison refer to

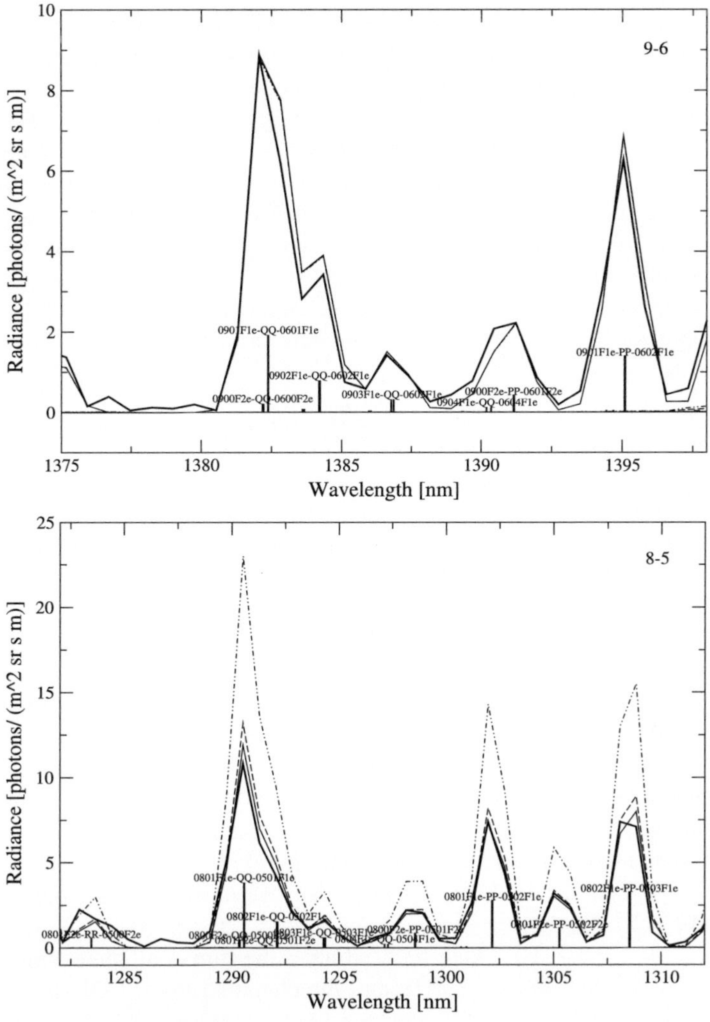

Fig. 10.5 Nighttime limb OH emission spectra at 87 km altitude for northern mid latitudes in November in different wavelength regimes. *Thick black lines* represent SCIAMACHY data, other lines model calculations. The dominant vibrational bands are labeled in the upper right corner of each plot. Line styles mark different relaxation schemes. *Solid lines*: sudden death model; *dashed lines*: cascade model; *dashed-dotted lines*: single quantum relaxation model. The *vertical bars* mark the position of individual lines and the text gives their quantum numbers. Format: AABBCCE-FF-GGHHIIJ: AA(GG): upper(lower) vibrational mode, BB(HH): rotational quantum number minus 0.5, CC(DD): spin state, E(J): Λ-type doubling, FF: branch symbol excluding/including nuclear spin. Figures adopted from *Kaufmann et al.* [2008] (page 1917)

Kaufmann et al. [2008]. The atomic hydrogen abundance was adjusted to fit measurement and model calculations for the OH(9–6) vibrational band. The difference between the various model calculations for the OH(8–5) band is caused by the fact, that $OH(v = 8)$ is not populated by chemical excitation, only, but also by radiative and collisional de-excitation of $OH(v = 9)$. Consequently, the choice of the relaxation scheme is relevant here. Since the sudden death model de-excites vibrational levels into the ground state, their vibrational energy is lost completely. Hence, spectra calculated with this model show lowest intensities. The cascade model constitutes the other extreme: the energy loss per collision is as small as possible and the corresponding spectra show highest intensities. The cascade model seems to overestimate measured OH(8–5) radiances significantly, whereas the other two models both fit the measurements well.

The energy deposition or heating rate $\frac{d\Theta}{dt}$ associated with reaction (10.1) is directly proportional to the chemical production term P_c in Eq. (10.2) and can be calculated as follows:

$$\frac{d\Theta}{dt} = \frac{2}{7} \frac{1}{k_B[M]} \Delta E \, P_c \varepsilon \tag{10.3}$$

ε is the heating efficiency taking into account the portion of exothermicity appearing as the vibrational energy of the product OH [about 90 %; *Charters et al.*, 1971; *Ohoyama et al.*, 1985] under consideration of radiative losses [less than 10 %; *Mlynczak and Solomon*, 1993]. $\varepsilon = 1$ is assumed here. ΔE is the enthalpy of the reaction (77 kcal/mol), k_B the Boltzmann constant, $[M]$ the total density, and $2/7$ the ratio of the specific heat capacity of air at constant pressure to the specific gas constant of air.

For the retrieval of heating rates and atomic hydrogen densities, we select the 1375–1400 nm spectral range showing ro-vibrational lines from the OH(9–6) vibrational transition. Since the population of the $v = 9$ vibrational mode is virtually independent of the population of all other vibrational states (except for the calculation of the vibrational partition sum), the retrieval of heating rates depends on four model parameters, only: The number of OH molecules produced in $OH(v = 9)$ after the $H + O_3$ reaction (for details see above), the total radiative loss $A_9 = \sum_{v'<9} A_{9,v'} = 215\ s^{-1}$ [*Smith et al.*, 2010], the total collisional loss $k_9 = \sum_{v'<9} k_{v',9} = 3.1 \cdot 10^{-11}\ cm^3\, s^{-1}$, which is the sum of all quenching processes of $OH(v = 9)$ [*Adler-Golden*, 1997], and the chemical loss of $OH(v)$ by reaction with atomic oxygen $k_c = 2 \cdot 10^{-10}\ cm^3\, s^{-1}$ [*Adler-Golden*, 1997; *Copeland*, 2002]. At 90 km (95 km), the ratio of quenching of $OH(v = 9)$ with O_2 to reaction with atomic oxygen to spontaneous radiative loss is 77:5:18 (58:14:28).

The accuracy of kinetic and spectroscopic parameters used in the modeling of vibrationally excited OH are discussed in *Mlynczak et al.* [1998]. An important source of error for the derivation of atomic hydrogen is the 20 % uncertainty of reaction rate (10.1) [*Sander et al.*, 2010], which maps linearly into the atomic hydrogen error. However, this uncertainty is not relevant for the derivation of heating rates, because $\frac{d\Theta}{dt}$ is directly proportional to the OH(9–6) radiance/volume emission rate. This dependency is given, because all production terms in Eq. (10.2) except for P_C can be neglected for OH(9). The accuracy of other model parameters affect atomic hydrogen and heating rates by about 20 %. The uncertainty of GOMOS ozone profiles mapping linearly into the uncertainty of atomic hydrogen is estimated to be less than 20 %.

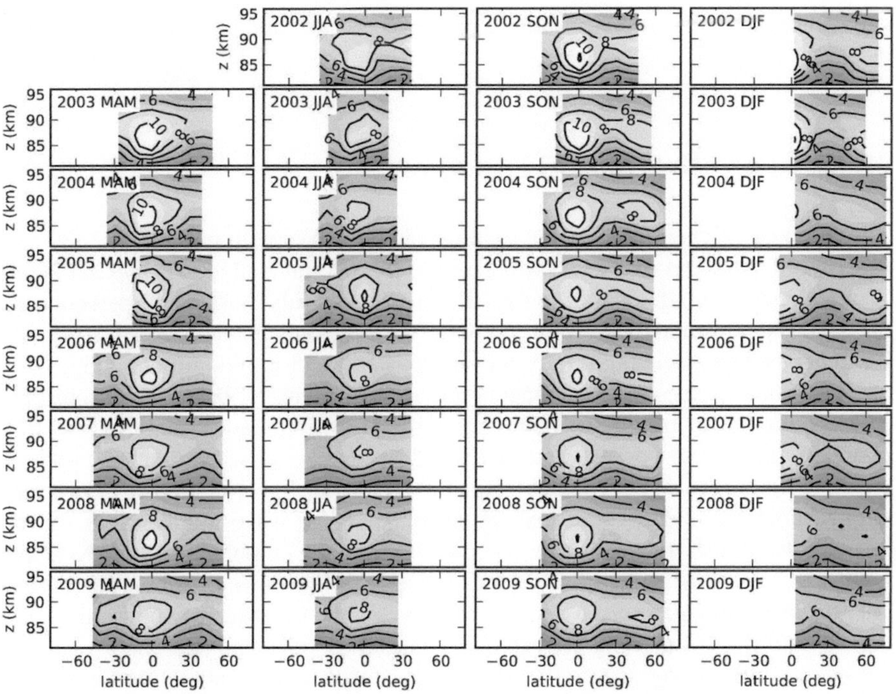

Fig. 10.6 Heating rate profiles (in K/day) for 2002–2009 (rows) for March-April-May (*1st column*), June-July-August (*2nd column*), September-October-November (*3rd column*) and December-January-February (*4th column*). The contour line interval is 2 K/day

The inversion of radiances to heating rates is performed by means of an optimal estimation approach [*Rodgers*, 2000]. A priori profiles of heating rates as well as other atmospheric parameters (temperature, N_2, O_2, and O abundances) are taken from the NCAR ROSE model.

10.4 Heating Rates

Seasonal mean heating rates as derived from SCIAMACHY OH* measurements are shown in Fig. 10.6 for the period 2002–2009. Values are larger than 2 K/day in a vertical layer extending by about 15 km. Values at the maximum range from 10 K/day at the equator to 6 K/day at high latitudes. Absolute values are comparable to 2-D model calculations of *Mlynczak* [2000], but larger than the ROSE 3-D model results presented in Fig. 10.1. A comparison of SABER OH* radiances and ROSE model calculations [*Marsh et al.*, 2006] showed an underestimation of equatorial measurements as well and was explained by an insufficient tidal forcing in the model. The importance of the diurnal migrating tide is visible in both the semi-annual cycle of equatorial heating rates, as well as in the latitudinal structure of heating rates, which shows largest values at the equator, a minimum at 30°, and a secondary maximum at 60°.

10.4.1 Space Time Spectral Analysis

The response of chemical heating to atmospheric waves with periods of 2–30 days is analyzed by means of a windowed space-time spectral analysis [*Ern et al.*, 2009]. A time window length of 31 days for 10° latitude intervals is used. This accounts for the seasonal variations visible in the data (see previous chapter). Results are analyzed for the equatorial region, for the sub-tropics, and for mid latitudes (Fig. 10.7).

The strongest spectral components are at frequencies of 0 and ±1 cycles/day. These components are due to stationary planetary waves and tides as well as their aliases, which show up periodically shifted by one zonal wavenumber and 1 cycle/day [e.g. *Salby*, 1982a, 1982b]. Since these strong spectral features also cause stronger aliasing and spectral leakage at higher zonal wave numbers, we cut the zonal wavenumber range of the spectra at zonal wavenumber 4. SCIAMACHY nighttime data allows to resolve periods larger than two days unambiguously, since only the ascending part of the satellite orbit is covered by observations. At the equator, we find not only strong contributions caused by tides and stationary planetary waves, but also strong spectral signatures for zonal wavenumber 1 at frequencies between 0.2 and 0.4 cycles/day, corresponding to periods between 2.5 and 5 days (Fig. 10.7a). These spectral signatures are due to equatorial Kelvin waves, which are confined to the equatorial region between about 20°S and 20°N, in good agreement with previous findings [e.g. *Salby et al.*, 1984; *Garcia et al.*, 2005]. The latitude-time distribution of wave variances due to Kelvin waves is shown in Fig. 10.7b. Wave variances are obtained from the single 31-day space-time analyses by applying a band pass filter for zonal wavenumber 1 and periods between 2.5 and 10 days, i.e., the location of the Kelvin wave peak in the average spectra shown in Fig. 10.7a. Kelvin wave variances are found only in the latitude band between about 20°S and 20°N, as expected. Rossby gravity wave activity (periods of 1.4–2.5 days) is expected to peak at subtropical latitudes. And indeed, Fig. 10.7c exhibits a clear spectral signature at these periods, peaking at 20°S and 20°N (Fig. 10.7d). At 40°N (Fig. 10.7e), tidal signatures and signatures due to stationary planetary waves are weaker than at the equator. The spectral peak due to equatorial Rossby-gravity waves is no longer visible, and there are also no more indications of equatorial Kelvin waves. Instead, a strong spectral peak has appeared at zonal wave numbers 1–2 and frequencies in the range between about −0.2 and 0 cycles/day. Obviously, this spectral feature is caused by westward traveling long period planetary (Rossby) waves with periods in the range of about 5–30 days. This period range is also in good agreement with the range expected from previous observations. The wave variances due to Rossby waves are obtained by applying a band pass for zonal wavenumber 1 and frequencies between −0.25 and −0.05 cycles/day (periods between 4 and 20 days). Rossby waves at 87 km altitude mainly appear at mid and high latitudes, as expected (Fig. 10.7f).

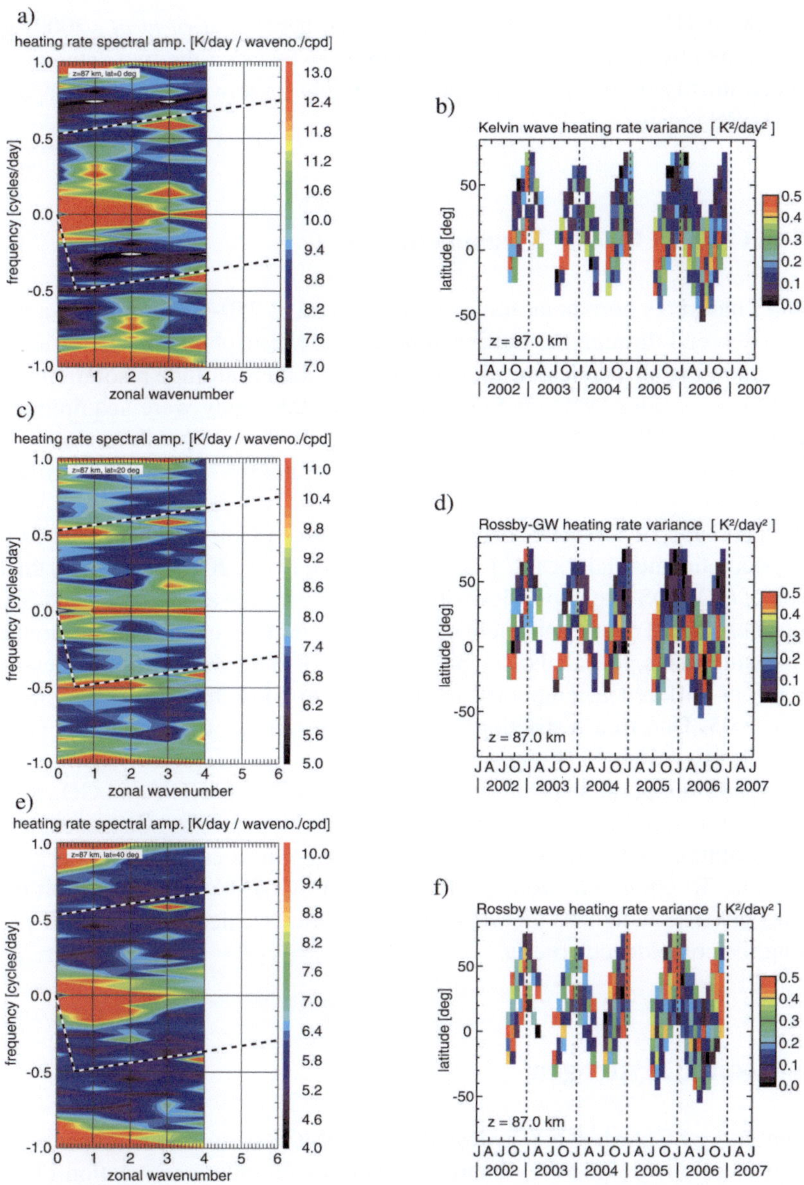

Fig. 10.7 *First column*: space-time spectra averaged over the period 2002–2006 of residual SCIAMACHY heating rates at 87 km altitude at the equator (**a**), 20°N (**c**) and 40°N (**d**). Positive frequencies denote eastward and negative frequencies westward traveling waves, respectively. The spectral range that can be resolved unambiguously by the SCIAMACHY sampling are rectangles rotated with respect to the zonal wavenumber/frequency coordinates (*black* and *white dashed lines*). *Second column*: variances due to Kelvin waves (**b**), Rossby-gravity waves (**d**), and Rossby waves (**f**). Only latitude-time intervals with more than 200 data points are considered. The median spectral background (determined from zonal wave numbers >2) is subtracted. Figures adopted from *Ern et al.* [2009] (pages 5 and 6)

SCIAMACHY and SABER data (not shown here, cf. *Ern et al.*, 2009) give a consistent picture of the wave modes. Differences between the two datasets can be explained mostly by the spatio-temporal sampling pattern and by data gaps in the SCIAMACHY data.

10.5 GOMOS Ozone Measurements

GOMOS measures horizontal transmission spectra at 250–956 nm while a star appears to descend through the Earth's atmosphere. Mesospheric ozone retrievals are performed in the ultraviolet wavelength range, and their altitude resolution is about 3 km. Ozone profiles (Version 5.00) selected for this study were acquired at solar zenith angles $>100°$ by the measurement of stars with a temperature >6000 K and a magnitude <1.9 as well as of stars with a temperature >7000 K [cf. *Kyrölä et al.*, 2010].

The precision of GOMOS ozone in the vicinity of the OH^* layer is 5–10 %, with systematic uncertainties of 1–2 % [*Tamminen et al.*, 2010]. However, the comparison of GOMOS ozone profiles with MIPAS [*Verronen et al.*, 2005] and SABER [*Smith et al.*, 2008] ozone data, both of which are based on ozone mid-infrared emission measurements, revealed discrepancies up to 20 % in vicinity of the mesopause.

GOMOS suffered a temporary malfunction in May–June 2003 and January–August 2005. Due to a restriction of the pointing range, the amount of GOMOS data has decreased by 35 % since 2005.

The spatio-temporal distribution of co-located SCIAMACHY and GOMOS profiles is illustrated in Fig. 10.3. Coincidence criteria (typical miss-distances) are $<5°$ (1°–4°) latitude, $<45°$ (0°–5°) longitude, <4 h (0–1 h) local time and <1 day universal time. To obtain an ozone profile for each SCIAMACHY dark limb profile, a mean, distance-weighted GOMOS ozone profile is calculated from all profiles meeting the coincidence criteria.

10.6 Atomic Hydrogen

By combining SCIAMACHY hydroxyl and GOMOS ozone measurements, global maps of atomic hydrogen are obtained. A latitude altitude cross section of atomic hydrogen is shown in Fig. 10.8 for summer 2008. Atomic hydrogen decreased monotonically from $4\text{–}6 \cdot 10^8$ cm^{-3} at 86 km to about $1 \cdot 10^8$ cm^{-3} at 95 km. Largest values appear at the equator, similar to the heating rate profiles. During equinox (not shown), atomic hydrogen is reduced by 30 %–50 %. Absolute values are similar to the atomic hydrogen profiles as derived from the Solar Mesosphere Explorer [SME, *Thomas*, 1990], although a detailed comparison is difficult since *Thomas* [1990] gives atomic hydrogen volume mixing on pressure levels rather than atomic hydrogen abundance on geopotential altitudes, as in this study.

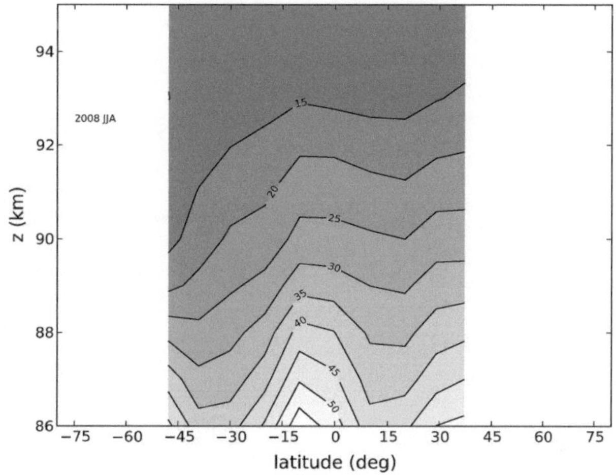

Fig. 10.8 Atomic hydrogen densities for June, July and August 2008. The contour line interval is $5 \cdot 10^7$ cm^{-3}

The temporal evolution of atomic hydrogen abundance at 90 km altitude is shown in Fig. 10.9. The data shows a pronounced semiannual cycle with amplitudes of about $7 \cdot 10^7$ cm^{-3}. In addition, a linear increase of about 14 % between 2002 and 2008 is visible.

To determine this increase quantitatively, the temporal (t) dependence is fitted using a constant (y_0), a linear (y_1) as well as annual (amplitude y_2 and phase y_3) and semiannual harmonics (y_4 and y_5):

$$\frac{d[\mathrm{H}]}{dt} = y_0 + y_1 \cdot t + y_2 \cdot \sin\left(\frac{t}{2\pi \cdot 1yr} + y_3\right) + y_4 \cdot \sin\left(\frac{t}{2\pi \cdot 1/2yr} + y_5\right). \tag{10.4}$$

The linear term characterizes long term changes as well as the dependence on solar flux, which cannot be discriminated so far due to the temporal extent of the data set.

The fit parameters are obtained applying a modified Levenberg-Marquardt algorithm [*Moré et al.*, 1980]. The cost function is calculated by weighting measurement-model differences by the standard deviation of the measurement data in each altitude/latitude bin.

The uncertainty of the fitting parameters is dominated by the occurrence of several data gaps in the time series. These data gaps are related to the latitudinal sampling pattern and instrument failures or specific instrument operations. We estimate the magnitude of this uncertainty by creating several synthetic data sets exhibiting the same spatio-temporal behavior and noise distribution as in the measurement data. Data gaps as present in the measurements are randomly redistributed within the given time period and mask the synthetic data. The resulting data sets are analyzed in the same way as the original data and the scatter of the resultant fit parameters is taken as an estimate for the uncertainty of these parameters with respect to data gaps. For the linear component (y_1) the uncertainty estimated by this method is about 4 %.

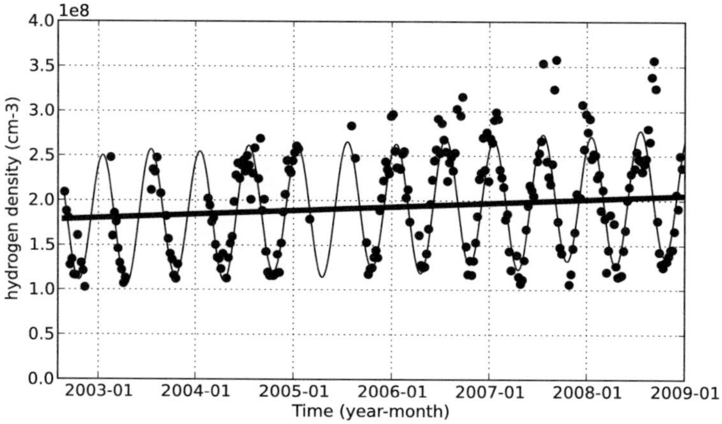

Fig. 10.9 Atomic hydrogen abundance for 0°N (±5°), 90 km altitude. *Dots* represent weekly mean data of atomic hydrogen abundance, the sinusoidal line is a fit according to Eq. (10.4), and the *thick black line* is calculated using the first two components (y_0 and y_1) of the fit

Fig. 10.10 Atomic hydrogen increase (in %) between September 2002 and December 2008; for details see text

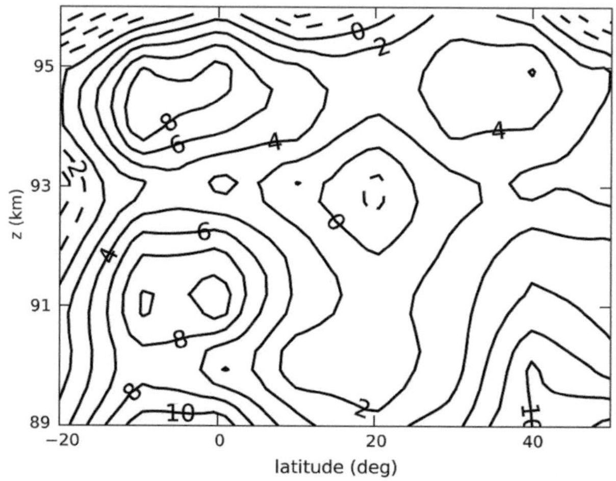

Other sources of uncertainty, such as noise of the measurement data, correlation of the fit parameters or inadequacy of the fit function are negligible in comparison to uncertainties related to the data gaps.

Atomic hydrogen zonal mean differences between 2008 and 2002 according to the linear term in Eq. (10.4) are shown in Fig. 10.10. Atomic hydrogen increases by 0–5 % at mid latitudes, and up to 10 % around the equator.

This increase has been corrected by 5 % to take into account the 11-year solar cycle dependence of atomic and molecular oxygen, which is not considered in the quenching of OH* and in the retrieval of atomic hydrogen, respectively. Since quenching by molecular oxygen is the most important relaxation mechanism of OH* below 95 km, uncertainties in the quenching partner map largely into the atomic hydrogen abundance. Although molecular oxygen mixing ratios do not change below 95 km during the 11-year solar cycle, molecular oxygen abundance observed on geopotential altitudes is indeed affected by solar insolation due to the expansion or shrinking of the atmosphere. General circulation models [e.g. *Schmidt et al.*, 2006] predict a shrinking of the upper mesosphere by several hundred meters during solar minimum. This leads to a decrease of O_2 and total density by 5–7 % at 90–95 km geopotential altitude (Schmidt, personal communication). Atomic oxygen abundance decreases by about 10 % accordingly. This leads to decreased quenching of OH^* in 2008 (solar minimum) in comparison to 2002 (solar maximum), which is not considered in the model and therefore the quenching of OH^* is overestimated in 2008 in comparison to 2002. The retrieval compensates overestimated OH^* quenching in the model by an increase of atomic hydrogen abundance, leading to an artificial overestimation of atomic hydrogen by about 5 % in 2008 in comparison to 2002. The data shown in Fig. 10.10 has been compensated for this effect.

Figure 10.10 indicates a small increase (0–6 % $\pm$ 4 %) of atomic hydrogen abundance at mid latitudes and a slightly larger increase of 4–8 % ($\pm$4 %) at equatorial latitudes. This anti-correlation with solar flux is predicted by model calculations (Schmidt, personal communication). It is caused by an increase of H_2O due to a reduced photolytic destruction of this species in the entire middle and upper mesosphere, also affecting the upward flux of H_2O by eddy diffusion. In total, this increase of water vapor is over-compensating the reduction of the photolytic solar flux producing atomic hydrogen in the upper mesosphere during solar minimum.

10.7 Summary

To quantify and understand the response of the upper atmosphere to long term changes and solar activity requires the measurement of temperature and constituent data. Atomic hydrogen as the most prominent member of the odd hydrogen family is of particular interest in this context. The combination of two instruments (SCIAMACHY and GOMOS) on Envisat gives an excellent opportunity to obtain a time series of this data ranging from 2002 until today in the 85–95 km altitude regime. Instantaneous heating rates due to the $H + O_3$ reaction reach 6–10 K/day, depending on latitude. A pronounced semi-annual cycle is evident in the data, pointing to the importance of atmospheric tides for the vertical distribution of constituent and temperature data. Signatures of equatorial Kelvin waves, Rossby-gravity waves and Rossby waves are clearly visible in the data. The long term evolution of atomic hydrogen indicates an increase by 0–8 % since 2002, which is in accordance with 11-year solar cycle model calculations.

Acknowledgements We gratefully acknowledge H. Schmidt for useful discussions related to HAMMONIA climate simulations. We also like to thank B. Funke for fruitful discussions related to the non-LTE model. We thank ESA for providing the SCIAMACHY and GOMOS data within the Cat-1 project 2515 and S. Niemejer (S&T Corporation) for assistance with the BEAT toolbox. This work has been supported by DFG under grants RI1546/1-1 and RI1546/1-2 within the SPP 1176 'CAWSES'.

References

Adler-Golden, S. (1997). Kinetic parameters for OH nightglow modeling consistent with recent laboratory measurements. *Journal of Geophysical Research*, *102*(A9), 19969–19976.

Beig, G., Scheer, J., Mlynczak, M. G., & Keckhut, P. (2008). Overview of the temperature response in the mesosphere and lower thermosphere to solar activity. *Reviews of Geophysics*, *46*(3). doi:10.1029/2007RG000236.

Bertaux, J. L., Kyrölä, E., Fussen, D., Hauchecorne, A., Dalaudier, F., Sofieva, V., Tamminen, J., Vanhellemont, F., Fanton d'Andon, O., Barrot, G., Mangin, A., Blanot, L., Lebrun, J. C., Pérot, K., Fehr, T., Saavedra, L., Leppelmeier, G. W., & Fraisse, R. (2010). Global ozone monitoring by occultation of stars: an overview of GOMOS measurements on ENVISAT. *Atmospheric Chemistry and Physics*, *10*(24), 12091–12148. doi:10.5194/acp-10-12091-2010.

Bovensmann, H., Burrows, J., Buchwitz, M., Frerick, J., Noel, S., Rozanov, V., Chance, K., & Goede, A. (1999). Global measurement systems for atmospheric composition. *Journal of the Atmospheric Sciences*, *56*, 127.

Bramstedt, K., Noel, S., Bovensmann, H., Burrows, J., Lerot, C., Tilstra, G. L., Lichtenberg, G., Dehn, A., & Fehr, T. (2010). SCIAMACHY monitoring factors: observation and end-to-end correction of instrument performance degradation. In *Proc. ESA living planet symposium (ESA SP686)*.

Brasseur, G., & Offermann, D. (1986). Recombination of atomic oxygen near the mesopause: interpretation of rocket data. *Journal of Geophysical Research*, *91*, 10818–10824.

Charters, P. E., Macdonald, R. G., & Polanyi, J. C. (1971). Formation of vibrationally excited OH by the reaction $H + O_3$. *Applied Optics*, *10*, 1747–1754.

Copeland, R. A. (2002). Recommended OH vibrational energy transfer rate constants based on laboratory data. *Eos Translations AGU*, *83*(47). Abstract SA72A-0510, Fall Meet. Suppl.

Ern, M., Lehmann, C., Kaufmann, M., & Riese, M. (2009). Spectral wave analysis at the mesopause from SCIAMACHY airglow data compared to SABER temperature spectra. *Annales Geophysicae*, *27*(1), 407–416. doi:10.5194/angeo-27-407-2009.

Garcia, R. R., Lieberman, R., Russell III, J. M., & Mlynczak, M. G. (2005). Large-scale waves in the mesosphere and lower thermosphere observed by SABER. *Journal of the Atmospheric Sciences*, *62*, 4384–4399.

Kaufmann, M., Lehmann, C., & Riese, M. (2010). Solar cycle 23 in SCIAMACHY mesospheric chemical heating rates. In *Proc. ESA living planet symposium (ESA SP686)*.

Kaufmann, M., Lehmann, C., Hoffmann, L., Funke, B., Lopez-Puertas, M., Von Savigny, C., & Riese, M. (2008). Chemical heating rates derived from SCIAMACHY vibrationally excited OH limb emission spectra. *Advances in Space Research*, *41*(11) 1914–1920. doi:10.1016/j.asr.2007.07.045.

Klenerman, D., & Smith, I. W. M. (1987). Infrared chemiluminescence studies using a sisam spectrometer. Reactions producing vibrationally excited OH. *Journal of the Chemical Society, Faraday Transactions*, *2*(83), 229–241. doi:10.1039/F29878300229.

Kyrölä, E., Tamminen, J., Sofieva, V., Bertaux, J. L., Hauchecorne, A., Dalaudier, F., Fussen, D., Vanhellemont, F., Fanton d'Andon, O., Barrot, G., Guirlet, M., Fehr, T., & Saavedra de Miguel, L. (2010). GOMOS O_3, NO_2, and NO_3 observations in 2002–2008. *Atmospheric Chemistry and Physics*, *10*(16), 7723–7738. doi:10.5194/acp-10-7723-2010.

Lichtenberg, G., Kleipool, Q., Krijger, J. M., van Soest, G., van Hees, R., Tilstra, L. G., Acarreta, J. R., Aben, I., Ahlers, B., Bovensmann, H., Chance, K., Gloudemans, A. M. S., Hoogeveen, R. W. M., Jongma, R. T. N., Noël, S., Piters, A., Schrijver, H., Schrijvers, C., Sioris, C. E., Skupin, J., Slijkhuis, S., Stammes, P., & Wuttke, M. (2006). SCIAMACHY level 1 data: calibration concept and in-flight calibration. *Atmospheric Chemistry and Physics*, *6*(12), 5347–5367. doi:10.5194/acp-6-5347-2006.

Marsh, D., Smith, A., Brasseur, G., Kaufmann, M., & Grossmann, K. (2001). The existence of a tertiary ozone maximum in the high-latitude middle mesosphere. *Geophysical Research Letters*, *28*, 4531–4534.

Marsh, D. R., Smith, A. K., Mlynczak, M. G., & Russell, J. M. III (2006). SABER observations of the OH Meinel airglow variability near the mesopause. *Journal of Geophysical Research*, *111*, A10S05.

Marsh, D. R., Garcia, R. R., Kinnison, D. E., Boville, B. A., Sassi, F., Solomon, S. C., & Matthes, K. (2007). Modeling the whole atmosphere response to solar cycle changes in radiative and geomagnetic forcing. *Journal of Geophysical Research*, *112*(D23). doi:10.1029/2006JD008306.

McDade, I., & Llewellyn, E. (1987). Kinetic parameters related to sources and sinks of vibrationally excited OH in the nightglow. *Journal of Geophysical Research*, *92*, 7643–7650.

Mlynczak, M. G. (2000). *A contemporary assessment of the mesospheric energy budget. Geophysical monograph series: Vol. 123*. Washington: AGU.

Mlynczak, M. G., & Solomon, S. (1993). A detailed evaluation of the heating efficiency in the middle atmosphere. *Journal of Geophysical Research*, *98*(D6), 10517–10541.

Mlynczak, M. G., Zhou, D. K., & Adler-Golden, S. M. (1998). Kinetic and spectroscopic requirements for the inference of chemical heating rates and atomic hydrogen densities from OH Meinel band measurements. *Geophysical Research Letters*, *25*(5), 647–650.

Moré, J. J., Garbow, B. S., & Hillstrom, K. E. (1980). *User guide for MINPACK-1*. ANL-80-74, Argonne National Laboratory.

Noel, S., Kokhanovsky, A. A., Jourdan, O., Gerilowski, K., Pfeilsticker, K., Weber, M., Bovensmann, H., & Burrows, J. P. (2006). SCIAMACHY reflectance and solar irradiance validation. In *Proc. third workshop on the atmospheric chemistry validation of ENVISAT (ACVE-3)*.

Ohoyama, H., Kasai, T., Yoshimura, Y., Kimura, H., & Kuwata, K. (1985). Initial distribution of vibration of the OH radicals produced in the $\mathrm{H} + \mathrm{O}[3] \rightarrow \mathrm{OH}(\mathrm{X}[2]\Pi[1/2, 3/2]) + \mathrm{O}[2]$ reaction. Chemiluminescence by a crossed beam technique. *Chemical Physics Letters*, *118*, 263.

Riese, M., Offermann, D., & Brasseur, G. (1994). Energy released by recombination of atomic oxygen and related species at mesopause heights. *Journal of Geophysical Research*, *99*, 14585–14593.

Rodgers, C. D. (2000). *Inverse methods for atmospheric sounding: theory and practice. Series on atmospheric, oceanic and planetary physics: Vol. 2*. Singapore: World Scientific.

Salby, M. L. (1982a). Sampling theory for asynoptic satellite observations, Part I: space-time spectra, resolution, and aliasing. *Journal of the Atmospheric Sciences*, *39*, 2577–2600.

Salby, M. L. (1982b). Sampling theory for asynoptic satellite observations, Part II: fast Fourier synoptic mapping. *Journal of the Atmospheric Sciences*, *39*, 2601–2614.

Salby, M. L., Hartmann, D. L., Bailey, P. L., & Gille, J. C. (1984). Evidence for equatorial Kelvin modes in Nimbus-7 LIMS. *Journal of the Atmospheric Sciences*, *41*, 220–235.

Sander, S. P., Friedl, R. R., Golden, D. M., Barker, J. R., Wine, P. H., Kurylo, M. J., Huie, R. E., Orkin, V. L., Moortgat, G. K., Kolb, C. E., Abbatt, J. P. D., & Burgholder, J. B. (2010). *Chemical kinetics and photochemical data for use in stratospheric studies, evaluation number 17*. JPL Publication 10-6.

Schmidt, H., Brasseur, G. P., Charron, M., Manzini, E., Giorgetta, M. A., Diehl, T., Fomichev, V. I., Kinnison, D., Marsh, D., & Walters, S. (2006). The HAMMONIA chemistry climate model: sensitivity of the mesopause region to the 11-year solar cycle and CO_2 doubling. *Journal of Climate*, *19*(16), 3903–3931.

Smith, A. K., & Marsh, D. R. (2005). Processes that account for the ozone maximum at the mesopause. *Journal of Geophysical Research*, *110*, D23305.

Smith, A. K., Marsh, D. R., Russell, I. J. M., Mlynczak, M. G., Martin-Torres, F. J., & Kyrola, E. (2008). Satellite observations of high nighttime ozone at the equatorial mesopause. *Journal of Geophysical Research*, *113*, D17312. doi:10.1029/2008JD010066.

Smith, A. K., Lopez-Puertas, M., Garcia-Comas, M., & Tukiainen, S. (2009). SABER observations of mesospheric ozone during NH late winter 2002–2009. *Geophysical Research Letters*, *36*(23), L23804. doi:10.1029/2009GL040942.

Smith, A. K., Marsh, D. R., Mlynczak, M. G., Mast, J. C. (2010). Temporal variations of atomic oxygen in the upper mesosphere from SABER. *Journal of Geophysical Research*, *115*, D18309. doi:10.1029/2009JD013434.

Tamminen, J., Kyrölä, E., Sofieva, V. F., Laine, M., Bertaux, J.-L., Hauchecorne, A., Dalaudier, F., Fussen, D., Vanhellemont, F., Fanton-d'Andon, O., Barrot, G., Mangin, A., Guirlet, M., Blanot, L., Fehr, T., Saavedra de Miguel, L., & Fraisse, R. (2010). GOMOS data characterisation and error estimation. *Atmospheric Chemistry and Physics, 10*(19), 9505–9519. doi:10.5194/acp-10-9505-2010.

Thomas, R. J. (1990). Atomic-hydrogen and atomic oxygen density in the mesopause region—global and seasonal-variations deduced from Solar Mesosphere Explorer near-infrared emissions. *Journal of Geophysical Research, 95*(D10), 16457–16476.

Verronen, P. T., Kyrölä, E., Tamminen, J., Funke, B., Gil-López, S., Kaufmann, M., López-Puertas, M., von Clarmann, T., Stiller, G. P., Grabowski, U., & Höpfner, M. (2005). A comparison of night-time GOMOS and MIPAS ozone profiles in the stratosphere and mesosphere. *Advances in Space Research, 36*(5), 958–966. doi:10.1016/j.asr.2005.04.073.

von Savigny, C., Kaiser, J. W., Bovensmann, H., Burrows, J. P., McDermid, I. S., & Leblanc, T. (2005). Spatial and temporal characterization of SCIAMACHY limb pointing errors during the first three years of the mission. *Atmospheric Chemistry and Physics, 5*(10), 2593–2602. doi:10.5194/acp-5-2593-2005.

Zaragoza, G., Taylor, F., & Lopez-Puertas, M. (2001). Latitudinal and longitudinal behavior of the mesospheric OH nightglow layer as observed by the improved stratospheric and mesospheric sounder on UARS. *Journal of Geophysical Research, 106*, 8027–8033.

Chapter 11
High-Latitude Thermospheric Density and Wind Dependence on Solar and Magnetic Activity

Hermann Lühr and Stefanie Marker

Abstract Processes in the high-latitude thermosphere are strongly controlled by the activity of the sun and by the geomagnetic field geometry. The CHAMP satellite, with its sensitive tri-axial accelerometer, provided detailed information about thermospheric dynamics over its mission life-time (2000–2010), thus contributing significantly to the CAWSES (Climate And Weather of the Sun-Earth System) programme. In this chapter, studies on thermospheric winds and density anomalies at high magnetic latitudes are presented. Thermospheric winds above the poles are directed predominantly from day to night side. Observations, however, reveal a distinct difference between winds on the dawn and dusk sides at auroral latitudes. While on the dawn side fast zonal winds towards night are prevailing, an anti-cyclonic vortex is formed on the dusk side. For the explanation of these local time dependent features various thermodynamic and electrodynamic influences have to be considered. As an example for mass density variation the cusp-related density anomaly is studied. The amplitude of this prominent local peak in mass density is influenced by the level of solar flux (F10.7) and by the solar wind input into the magnetosphere as quantified by the electric field caused by reconnection. A prerequisite for the appearance of density anomalies is the presence of soft-energy particle precipitation. By combining CHAMP and EISCAT measurements, it has been shown that Joule heating, fuelled predominantly by small-scale field-aligned currents (FACs), causes a strong increase in temperature at altitudes below 200 km. As a consequence molecular-rich air is up-welling. A density anomaly is recorded at 400 km altitude. Combining different observations and numerical model results provides a plausible chain of processes leading to the observed cusp-related density anomaly.

H. Lühr (✉)
GFZ German Research Centre for Geosciences, Telegrafenberg, 14473 Potsdam, Germany
e-mail: hluehr@gfz-potsdam.de

S. Marker
ILS Kraftfahrzeuge, Technische Universität Berlin, Gustav-Meyer-Allee 25, 13355 Berlin, Germany
e-mail: stefanie.marker@tu-berlin.de

F.-J. Lübken (ed.), *Climate and Weather of the Sun-Earth System (CAWSES)*, Springer Atmospheric Sciences,
DOI 10.1007/978-94-007-4348-9_11,

11.1 Introduction

The term thermosphere is generally used for the upper atmospheric layer in the range of about 100 to 1000 km altitude. Embedded in it is the exobase at around 600 km above which altitude lightweight particles do not behave as a fluid anymore and start to follow ballistic trajectories. There is a steep temperature increase over the lower part of the thermosphere, 100 to 300 km, due to absorption of solar extreme ultraviolet (EUV) and x-ray radiation. Above, the temperature approaches asymptotically the exospheric temperature of about 1000 K. For further general characteristics of the thermosphere the reader is referred to *Prölss* [2004]. Solar EUV is the most important energy input to the thermosphere. The expansion and contraction of the upper atmosphere in response to solar activity over a solar cycle is quite prominent. The other important energy source is the solar wind. When the charged particles of the solar wind interact with the geomagnetic field, a part of the wind energy is converted to electric currents which are routed down along the magnetic field lines to the ionosphere at high latitudes. These currents partly dissipate and partly drive plasma convection vortices in polar regions. Another part of the solar wind input is converted to mechanical energy accelerating electrons and ions earthward. When these particles hit the atmospheric atoms and molecules they are excited and subsequently emit light, the well-known aurora. This process also deposits an appreciable amount of energy in the thermosphere. In a comprehensive review, *Prölss* [2011] has described the perturbations of the upper atmosphere caused by the dissipation of solar wind energy.

This chapter deals with thermospheric phenomena at high magnetic latitudes. In particular, observation and interpretation of the wind system and local density anomalies are presented. In auroral regions the neutral particles in the thermosphere are strongly coupled with the plasma of the ionosphere. Therefore, both thermodynamic and electrodynamic processes have to be considered simultaneously for a proper interpretation.

In the past, detailed studies of the thermosphere were suffering from a lack of sufficient and accurate measurements. This situation has improved significantly with the availability of air drag measurements on board of satellites like CHAMP (CHAllenging Minisatellite Payload) and GRACE (Gravity Recovery And Climate Experiment) by means of sensitive tri-axial accelerometers. In particular, the accelerometer data from CHAMP collected over a period of 10 years have been used to identify a number of new phenomena in the upper atmosphere. In this chapter we provide an overview and assessment of the main results published during recent years. For most of them the long and continuous CHAMP data set is used for revealing insight into the wind patterns at polar latitudes and to investigate the prominent mass density anomaly related to the ionospheric cusp.

11.2 Measuring the Thermosphere

Opposed to the ionospheric properties, neutral particle dynamics is difficult to be sensed remotely from ground. There have been rather few satellites in orbit carrying

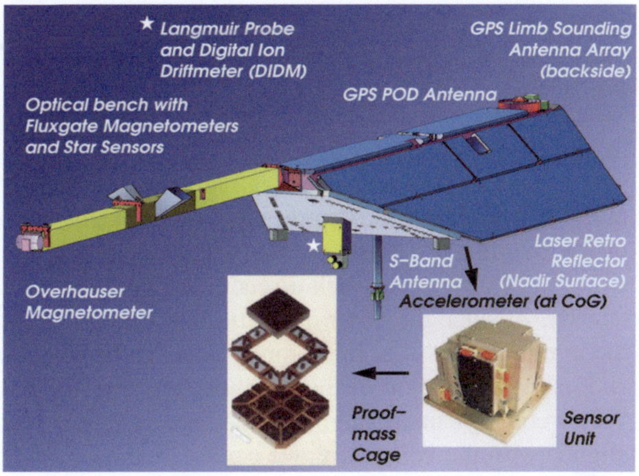

Fig. 11.1 Schematic illustration of the CHAMP satellite with its scientific instruments

mass spectrometers (e.g., Atmospherics Explorer-C, ESRO-4). Important features of the thermosphere have been deduced from precise orbital tracking of space objects like satellites, rockets, and debris. Since the effects on the ephemeris are small, integration has to be performed over fairly large arcs. For that reason only gross features of the thermosphere had been deduced. For example, the dependence of the mass density on external forcing, like solar EUV flux or magnetic activity, was determined from orbit-average effects. That means regional or latitudinal variations cannot simply be resolved. In comprehensive studies, *Emmert* [2009] and *Emmert and Picone* [2010] have deduced a globally-averaged climatology of the thermospheric mass density based on orbital elements of a large number of space objects, covering the years 1967–2007. From them thermospheric features like seasonal and solar cycle variations, magnetic activity dependence and long-term trends have been derived.

Detailed observations of thermospheric phenomena have become possible since sensitive accelerometers are operated on spacecraft like CHAMP and GRACE. They have sampled the upper atmosphere on a global scale since 2000. The *in situ* measurements enable the resolution of local features. A first overview on the mass density distribution derived from CHAMP data was given by *Liu et al.* [2005]. The authors presented mass density anomalies both at low and high latitudes. In this chapter we focus on thermospheric characteristics at high latitudes deduced from CHAMP measurements.

CHAMP, a German satellite, was launched on 15 July 2000 into a circular, near-polar (inclination: 87.2°) orbit at 456 km altitude [*Reigber et al.*, 2002]. Over its 10-year life time the orbit slowly decayed, and the spacecraft re-entered the atmosphere on 19 September 2010 at an altitude of 150 km. A schematic picture of CHAMP with all the science instruments is shown in Fig. 11.1. Of particular interest for the topics addressed here is the accelerometer, which is accommodated at the satellite's centre of gravity. The measurement principle is based on a proof-mass of about 100 g that is kept floating in the centre of a vacuum cage by electrostatic forces. Non-gravitational forces acting on the satellite are deduced from the accelerometer readings. From the air drag experienced by the satellite in orbit the thermospheric mass density and wind can be retrieved. The basis equation is

$$\mathbf{a} = -\frac{1}{2}\rho\frac{\mathbf{C}_d}{m}A_{\mathrm{eff}}v^2 \tag{11.1}$$

where **a** is the measured acceleration vector, ρ is the mass density, m is the satellite mass, A_{eff} denotes the effective cross-section area in ram direction and v the velocity of the satellite. $\mathbf{C}_d$ represents the drag coefficient vector with different values for along-track and cross-track directions. Equation (11.1) can be solved for the mass density, ρ, since the other quantities are known or measured. Further details about the interpretation of accelerometer data can be found in *Doornbos et al.* [2010].

For the estimation of the thermospheric winds we make use of the acceleration vector components. Under the assumption that the experienced acceleration, **a**, is aligned with the velocity, **v**, we can relate the components of the two vectors:

$$-\frac{a_y}{a_x} = \frac{v_y}{v_x} \tag{11.2}$$

where the x component is aligned with the spacecraft-fixed along-track axis and y with the cross-track axis. Vertical wind contributions are not considered because (1) they are generally weak (except in the auroral zone during active periods) and (2) the vertical component of CHAMP's accelerometer did not work properly. Since CHAMP had a polar orbit, the zonal wind velocity can be derived from the effect on the cross-track axis.

$$v_y = -v_x\frac{a_y}{a_x} \tag{11.3}$$

when v_x is assumed to be the orbital velocity (7.6 km/s), the cross-track wind velocity, v_y, can be calculated. A more sophisticated approach for deriving wind was later developed by *Doornbos et al.* [2010]. The new data reveal wind speeds that are in general lower by a few percent, but the relative variations stay practically the same.

CHAMP data of the 4 years 2002–2005 have been used to study thermospheric features at high latitudes.

11.3 Wind Distribution at Polar Regions

A first statistical study of zonal winds at low latitudes, based on CHAMP data, was presented by *Liu et al.* [2006]. Typical features emerged from it, such as the strong eastward wind during evening and night hours and the westward wind during morning and noon time. Later the zonal wind study was extended by *Liu et al.* [2009] up to subauroral latitudes. These authors found that the thermosphere at the equator rotates on average faster than the planet. The topic of super-rotation has been discussed controversely in the literature [see *Rishbeth*, 2002, for an overview]. *Liu et al.* [2009] report a diurnally averaged eastward wind of about 35 m/s, which is in good agreement with the prediction (30–40 m/s) of *Rishbeth* [2002] based on the electrodynamics at the magnetic equator. Conversely, at higher latitudes the thermosphere rotates slower than the planet. For an outside observer, the upper atmosphere exhibits a differential rotation like, for example, the sun.

The wind distribution in the polar region, as observed by CHAMP, was first studied by *Lühr et al.* [2007]. Twenty years earlier *Thayer et al.* [1987] presented a similar study based on DE-2 satellite data. Due to the short DE-2 mission duration and sparse sampling, detailed features could not be resolved at that time. At high latitudes all the different cross-track measurements are combined in a statistically-significant way such that the average horizontal wind vector can be derived (for more details see *Lühr et al.* [2007]). A prominent result of such an approach is the strong day-to-night wind over the polar cap with speeds in excess of 600 m/s. In this region hydrodynamic forces and plasma drifts point approximately in the same direction. The primary cause is the cross-polar cap potential difference which drives the plasma into the anti-sunward direction at speeds of more than 1 km/s. The plasma drift velocity, v, in the ionospheric F region can be expressed as $v = \mathbf{E} \times \mathbf{B}/B^2$. At CHAMP altitude we obtain for a typical polar cap electric field of $E = 50$ mV/m and an ambient magnetic field, $B = 50000$ nT, a plasma velocity of $v = 1$ km/s. The relevant terms of the equation describing the momentum transfer from ions to neutral particles by collisions [e.g., *Rishbeth*, 2002] can be written

$$\rho \frac{d\mathbf{u}}{dt} = -\nabla P + \rho v_{i,n}(\mathbf{v} - \mathbf{u}) \tag{11.4}$$

where ρ is the mass density, $\mathbf{u}$ is the wind velocity, P is the thermal pressure and $v_{i,n}$ is the ion/neutral collision frequency. From this equation we note that the acceleration of neutral particles is proportional to the collision frequency and to the velocity difference between ions and neutrals.

The average wind distribution in the polar region for the four months centred on June solstice 2003 is shown in Fig. 11.2. For the presentation a magnetic latitude vs. magnetic local time frame has been chosen. We find similar distributions at both hemispheres, but in the northern (summer) hemisphere winds are stronger and signatures are clearer. The higher plasma and neutral densities during summer cause a higher collision frequency thus make the coupling between ions and neutrals more efficient. At auroral latitudes there is a clear asymmetry between the dawn and dusk sides. On the dawn side we observe fast winds from day to night. This effect is attributed to a combined action of Coriolis and centrifugal forces. *Fuller-Rowell and Rees* [1984] argue that particles starting at noon and moving westward experience an equatorward centrifugal force, but at the same time a poleward Coriolis force. The velocity at which the two forces according to *Fuller-Rowell and Rees* [1984] cancel depends on the polar distance, θ

$$\sin\theta = v/985 \text{ (m/s)} \tag{11.5}$$

where 985 m/s is twice the Earth rotation speed at the equator. At a geographic latitude of 65° ($\theta = 25°$) we obtain $v = 416$ m/s for the critical velocity. This value is close to the wind speed observed on the dawn side (cf. Fig. 11.2). Obviously, many air particles move on the stable paths in this latitude range. Note that magnetic

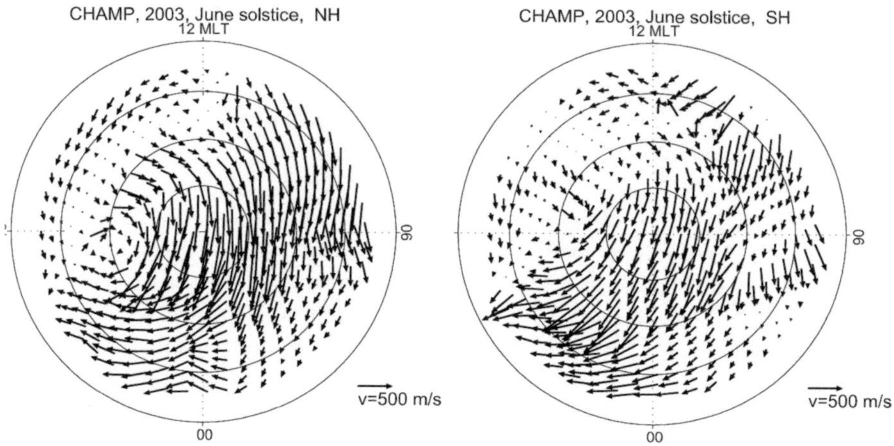

Fig. 11.2 Distribution of mean thermospheric wind vectors in the northern (summer) (*left*) and southern (winter) (*right*) hemispheres for June solstice 2003. Concentric rings mark magnetic latitude at 10° spacing (80°, 70°, 60°, 50°) (from *Lühr et al.* [2007], Fig. 4)

latitudes are shown in Fig. 11.2, while the force balance discussed above is valid in geographic coordinates. But over many Earth rotations the latitude difference averages out.

Conversely, on the dusk side the Coriolis force and centrifugal forces act in the same direction deflecting particles into an anti-cyclonic spiral motion. Such a feature appears in Fig. 11.2 near 70° of magnetic latitude and 18:00 local time. In the early evening time sector the sunward plasma flow along the auroral oval causes a stagnation of the wind. Particularly clear is the effect of plasma drift on thermospheric winds during times of Subauroral Polarization Streams (SAPS) events. *Wang et al.* [2011] report that SAPS, occurring preferentially in the evening sector, cause the neutral wind to switch from anti-sunward to sunward at magnetic latitudes around 60° to 65°. A typical occurrence rate of SAPS is 30 %. SAPS-type phenomena are not observed on the dawn side. This may also contribute to the dawn/dusk difference. On the night side the wind is deflected westward. This is consistent with the Coriolis force acting on air parcels moving away from the pole.

The high-latitude plasma drift pattern is closely controlled by the interplanetary magnetic field (IMF) components B_y and B_z. *Förster et al.* [2008] clearly showed that the wind direction over the polar regions depends a lot on the plasma dynamics. Over the northern hemisphere polar cap highest wind speeds are observed for a combination of negative IMF B_y and B_z components. In the southern hemisphere wind speeds maximise for positive IMF B_y and negative B_z. Under these conditions the plasma flow is best aligned with the cross-polar cap wind at high latitudes on the dawn side. Furthermore, the clockwise wind vortex on the dusk side increases in the northern hemisphere when IMF B_y is positive, while the opposite B_y polarity causes the vortex to grow in the south. This behaviour also proves the close control of the plasma drift on thermospheric winds.

11.4 Cusp-Related Mass Density Anomaly

In their initial paper on the global mass density distribution *Liu et al.* [2005] reported also about thermospheric density anomalies at high magnetic latitudes both on the day and night sides. The night side features were associated with magnetospheric substorms since their intensity increased with magnetic activity. This suggestion was later confirmed by a dedicated study about thermospheric responses on substorms by *Ritter et al.* [2010]. The day-side feature was identified as the cusp-related density anomaly that had been introduced earlier by *Lühr et al.* [2004]. These authors reported that CHAMP detected quite frequently a localised enhancement in air drag whenever it passed the polar cusp region. When interpreting this signal in terms of mass density they obtained local enhancements up to a factor of 2. In a quest for the drivers of these air density peaks at 400 km altitude they suggested as one cause Joule heating by intense small-scale FAC. Figure 11.3 shows CHAMP observations of cusp-related density anomalies collocated with intense bursts of kilometre-scale FACs for three examples. Nowhere along the orbit have FACs been observed with comparable amplitudes. In order to test the Joule heating hypothesis *Demars and Schunk* [2007] tried to reproduce the density anomaly with their high-resolution thermosphere model. They had to increase the heating in the E region by an unrealistic factor of greater than 100, in order to achieve a local doubling of the mass density near the cusp. Otherwise the model simulated quite realistically the up-welling of air in the heated region and the subsequent poleward and equatorward divergence. Motivated by the open questions concerning the root causes of the cusp-related density anomalies *Rentz and Lühr* [2008] performed a dedicated study on the climatological properties of the anomaly. Important findings of that study will be summarised below.

11.4.1 Climatology of the Density Anomaly

For a systematic survey of the mass density distribution at polar regions, data have been sorted into 918 equal-area bins of 222 × 232 km size. For the binning of the CHAMP measurements we considered the range 60°–90° of magnetic latitude (MLat) in both hemispheres. A detailed description of the binning approach is given in *Lühr et al.* [2007]. From Fig. 11.4 it is evident that the density distribution at auroral latitudes is well-ordered when sorting the data by geomagnetic latitude and magnetic local time (MLT). Conversely, a data representation in geographic latitude and solar local time does not show a clear maximum around noon [see *Rentz*, 2009]. As a consequence, the geomagnetic frame was used throughout the analysis.

The density enhancement at cusp latitudes is superposed on top of the large-scale (diurnal, latitudinal) density variations. The considered effect has been isolated by removing the background according to the NRLMSISE-00 atmospheric model density. In addition a systematic difference between model and observation was removed by allowing for a bias term and a latitude-dependent trend. As a result

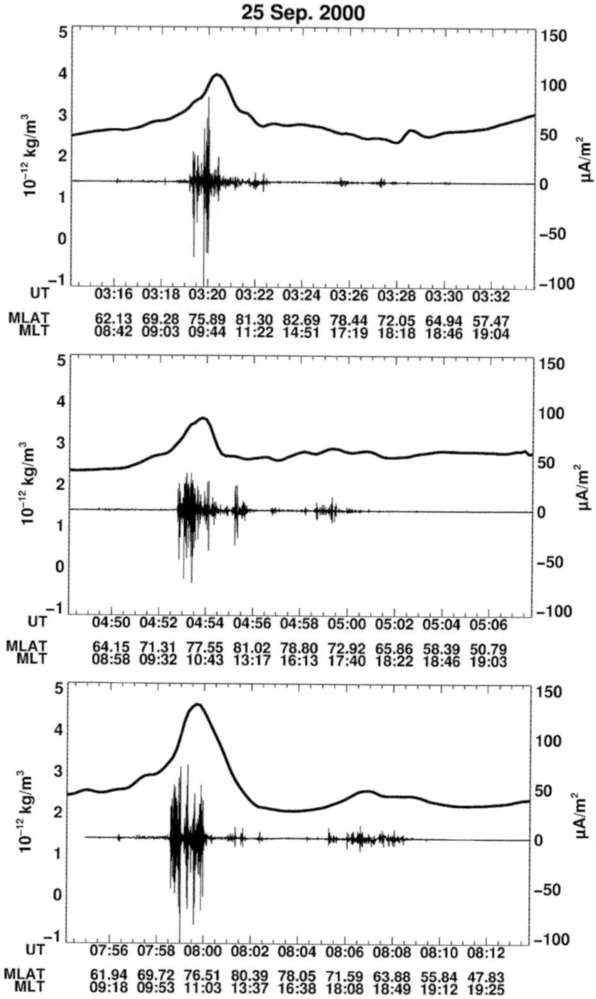

Fig. 11.3 Three examples of cusp density anomalies observed by CHAMP. Collocated with the air density enhancements intense bursts of kilometre-scale field-aligned currents appear (from *Lühr et al.* [2004], Fig. 3)

the density anomaly, $\Delta\rho$, is obtained [for details of the approach see *Rentz and Lühr*, 2008]. The climatological properties of the density anomaly were derived from a systematic survey over four years (2002–2005) of CHAMP air drag measurements. Figure 11.4 shows, for example, the average anomaly location and amplitude, as deduced from 2003 data. In both hemispheres density peaks cluster in the latitude range 70°–80°MLat and the 12 ± 2 h MLT time sector. It is interesting to note that the amplitude of the anomalies is smaller by a factor of 1.4 in the south compared to the northern hemisphere. *Rentz and Lühr* [2008] suggest that the obtained hemispheric difference is primarily caused by the larger offset between the magnetic and geographic pole in the south. Over the course of a day the cusp moves from the South Pole down to $-60°$ geographic latitude and back again. Largest density anomalies appear typically at lowest latitudes. But CHAMP was sampling

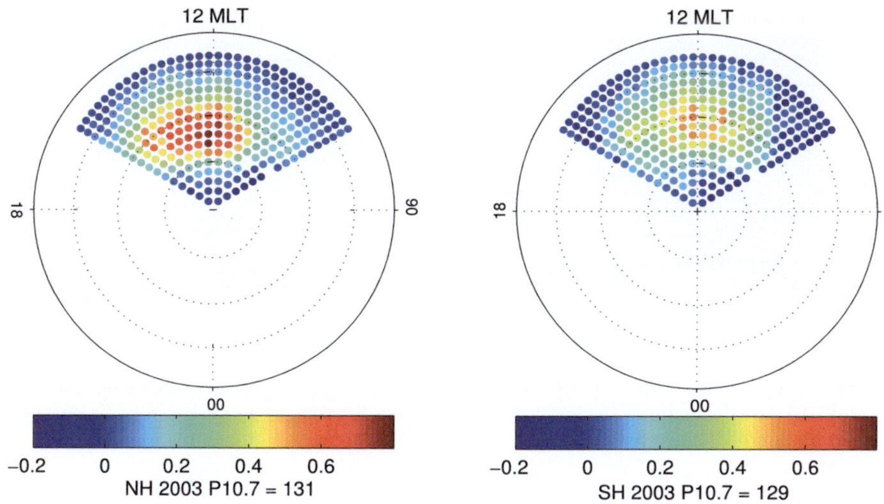

Fig. 11.4 Statistical survey of the density anomaly amplitude (in 10^{-12} kg/m^3) within the day side polar region of the northern (*left panel*) and the southern (*right panel*) hemisphere in 2003. Yearly averaged solar flux values (P10.7) are listed below the plots

the region around the pole more than 10 times denser than the $-60°$ latitude zone, where the larger effects occur. This selective sampling accounts for a good part of the north/south amplitude difference.

Since the anomaly location coincides rather well with the mean position of the ionospheric cusp, an energy source near the day side magnetopause is suggested to be an important driver of this phenomenon. We investigated the solar wind input, quantified by the electric field caused by reconnection at the magnetopause, commonly termed merging electric field, E_m. In our study we have defined $E_m = v_{sw}(B_y^2 + B_z^2)^{0.5} \sin^2(\theta/2)$, where B_y and B_z are the IMF components in GSM coordinates, v_{sw} is the solar wind speed, and θ is the IMF clock angle ($\tan\theta = B_y/B_z$).

The amplitude of the anomaly depends also on the solar flux index, as can be seen in Fig. 11.5 (left frame). The index is defined as P10.7 = 0.5(F10.7 + F10.7a), where F10.7 is the daily solar flux index and $F10.7a$ is its 81-average. It seems that below a certain flux level, e.g. P10.7 < 100, no clear anomalies form anymore. A similar effect can be found for $E_m < 0.5$ mV/m. A superposed epoch analysis yields that E_m at the magnetopause is enhanced on average for about an hour before the detection of the anomaly [*Rentz and Lühr*, 2008]. This time delay might be explained by two effects, partly by the upward propagation of the density front from the lower thermosphere to orbital altitudes (ca. 400 km) and partly by the infrequent sampling of the cusp once per orbit (93 min). For the upward propagation one may assume the thermal speed which ranges around 600 m/s, resulting in a travel time of order 10 min. Also the background air density seems to play a role for the anomaly amplitude. In order to test that Fig. 11.5 (right frame) shows the relative size of the anomaly with respect to the ambient density. In this case the dependence on P10.7 is

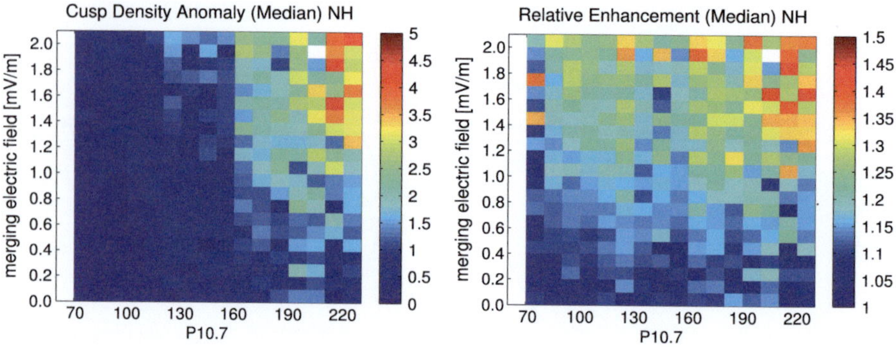

Fig. 11.5 Dependence of the northern hemispheric cusp-related density anomaly on the solar flux index, P10.7, and the merging electric field. (*left*) Amplitude of anomaly (in 10^{-12} kg/m^3), (*right*) relative amplitude of anomaly with respect to background density

much reduced because the solar flux controls directly the background density. From this perspective the important role of E_m for the cusp-related anomaly becomes even clearer.

11.4.2 CHAMP-EISCAT Campaign on Joule Heating

A dedicated CHAMP-EISCAT campaign was scheduled and executed in order to observe simultaneously thermospheric characteristics with CHAMP and ionospheric parameters with the EISCAT radar facility. The aim was to monitor directly the Joule heating in the ionosphere that may be responsible for driving the neutral fountain effect. The campaign lasted from 1 to 13 October 2006 and included the ESR (EISCAT Svalbard Radar, 78.1°N, 16.0°E, 75.6°MLat) at Longyearbyen and the VHF (Very High Frequency, 69.5°N, 19.2°E, 66.9°MLat) radar near Tromsø. When CHAMP passed close to the radars the ESR 42 m antenna measured the ionospheric parameters along the magnetic field line while the ESR 32 m and VHF radars were used to track the cusp position. The EISCAT measurement programme was nearly the same as the one described by *Schlegel et al.* [2005].

Most favourable results were obtained on 13 October 2006. Particularly outstanding on that day is a band of strongly enhanced ion temperature, up to 4000 K, observed by the VHF radar. As can be seen in Fig. 11.6, the apparent height of the high temperature region varies with time. At the start of the measurements, 07 UT, it is detected near 600 km, 1.5 hours later, around 08:30 UT, it has reached its lowest altitude at 260 km. Subsequently it rises again until reaching 550 km at 10 UT, after which the high temperature feature disappears. A similar structure is observed for the electron temperature, but in that case the hot electrons are observed up to great heights everywhere. For electrons the field lines are very good temperature conductors. Due to the low elevation of the VHF radar beam (30° above horizon),

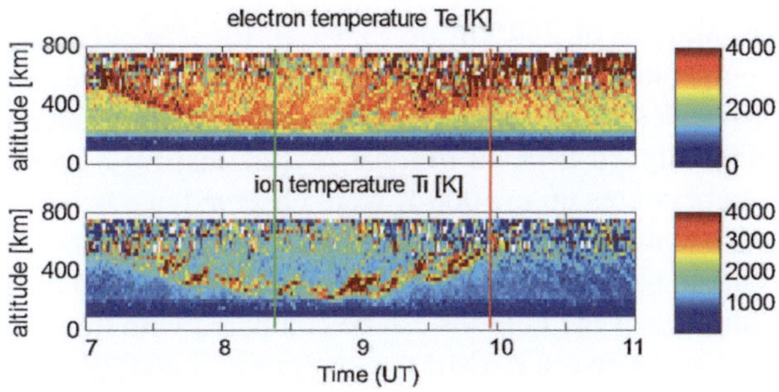

Fig. 11.6 Electron and ion temperature derived by EISCAT VHF on 13 October 2006. The *vertical lines* mark the times of CHAMP overpasses (*green* orbit 6, *red* orbit 7). VHF was looking northward at an elevation of 30°

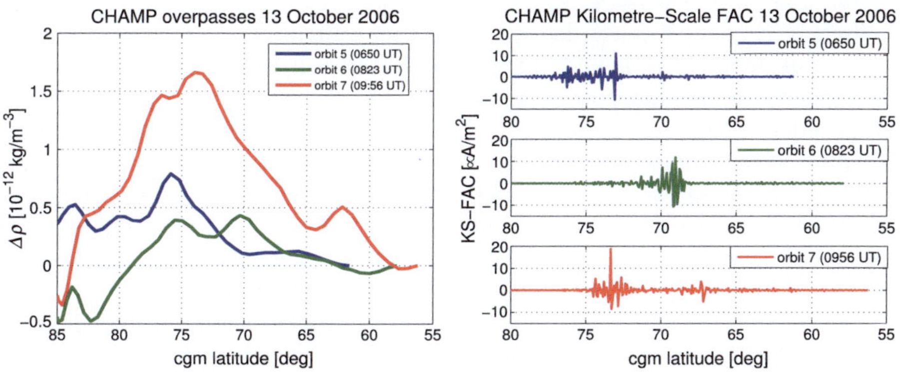

Fig. 11.7 CHAMP-observed density anomaly, $\Delta\rho$ (*left*) and kilometre-scale field-aligned currents (*right*) along three passes near the EISCAT facilities on 13 October 2006. Density peaks are accompanied by FAC bursts. The separation of CHAMP passes from ESR are 12°, 9°, 30° in longitude for the orbits 5, 6, 7, respectively

echoes plotted at different heights come from different latitudes. For the following interpretation apparent heights are converted to latitudes.

CHAMP crossed on that day the magnetic latitude of ESR three times at longitudes 28.0°E, 6.7°E, −14.5°E on orbits 5, 6, 7, respectively. Times of closest approach are marked by vertical lines in Fig. 11.6 (green: orbit 6, red: orbit 7). Figure 11.7 shows CHAMP measurements. In the left frame recorded mass density anomalies are shown. Most prominent is the density peak of orbit 7 appearing near 74°MLat. But also during the earlier passes density peaks, although smaller, were recorded at 75° and 70°MLat on the orbits 5 (blue) and 6 (green), respectively. In the right frame of Fig. 11.7 CHAMP derived field-aligned currents are shown. It is interesting to note that intense small-scale FAC bursts with current densities up to 10 $\mu A/m^2$ appear right at the latitudes of the mass density peaks in all three cases. This observation lends support to the suggestion of *Lühr et al.* [2004] that small-scale FACs are commonly accompanying cusp density anomalies.

By comparing the EISCAT and CHAMP observations new insight into the driving processes may be expected. CHAMP passed by the EISCAT facility on orbit 6 shortly before 08:30 UT. At that time the high temperature echoes came from an altitude of 260 km or 73.5°Lat. (cf. Fig. 11.6). During the next orbit, 7, high temperature was detected at 550 km altitude corresponding to 78°Lat. This latitude coincides with the position of the ESR radar. Inspecting the measurements from the field-aligned ESR 42 m radar reveals that for a short time around 10 UT high ion temperatures up to 4000 K were observed in a height range from 120 up to 200 km. This confirms the reliability of interpreting VHF height ranges in terms of latitudes. One may speculate that the ion heating is caused by intense FACs. To test that the latitudes of the hot ions can be compared with those of the small-scale FAC bursts.

On the two CHAMP passes (orbits 6 and 7) marked in Fig. 11.6 we observed hot ions at geographic latitudes of 73.5°N and 78°N and intense FACs at magnetic latitudes of 70°N and 74°N, respectively. In the Scandinavian sector the magnetic latitude is by about 3° lower than the geographic latitude. The remaining small difference of about 1° in latitude between CHAMP and EISCAT observations can be explained by the mapping of the signal along the field lines. Since field lines are inclined equatorward, CHAMP detects the FACs somewhat southward of their footprint in the E region. A similar inference on collocation between small-scale FACs (~75°MLat) and high ion temperature can be made for orbit 5. Although CHAMP passed 10 minutes before EISCAT was switched on, the latitude of hot ions (~79°Lat) can be estimated. These three examples demonstrate the close relation between intense FAC bursts and local ionospheric heating.

A viable process for a local heating of the thermosphere is Joule heating. In order to see where it takes place we examine the Pedersen conductivity. Figure 11.8 shows vertical profiles of the Pedersen conductivity as derived from ESR measurements on 13 October 2006 in the left frame. It can be noted that the peak conductivity continuously increased until 10 UT. Furthermore, it is obvious that the high-conductivity layer appears at greater altitude, reaching up to 150 km. Typically the Pedersen conductivity peaks in the E-layer at about 115 km. Joule heating can be expressed as $W_J = \sigma_P E^2$. Since the electric field, E, varies little with altitude, the vertical distribution of heating is determined by the height profile of the conductivity, σ_P. The vertical profile of the local ion temperature enhancement observed by ESR at 10 UT fits well the expected heating profile, in particular, when considering the thermospheric density decay with altitude. *Rentz* [2009] has deduced numbers for the heating rate by estimating the electric field from CHAMP current measurements. She obtained 14×10^{-7} W/m^3 for the heating at 140 km altitude around 10 UT and 16 mW/m^2 for the height-integrated heating rate. Both these values are by a factor of about 5 larger than earlier reports [e.g., *Schlegel et al.*, 2005] although her measurements took place close to solar minimum. Previous authors did not take into account the important contribution from small-scale E-field (FAC) variations.

11.4.3 Comparison with Numerical Modelling

A remaining question concerns the elevated altitude of maximal Pedersen conductivities. Here, particle precipitation may play a role. The ionospheric response to

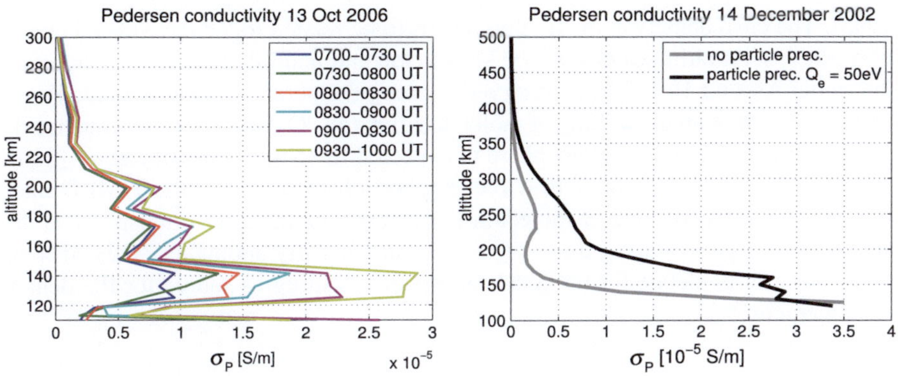

Fig. 11.8 Vertical profiles of Pedersen conductivity from EISCAT measurements (*left*) and SHL model (*right*). Minor peaks in the observed profiles are not considered as significant

precipitating particles varies markedly with altitude: the higher the electron energy the lower the altitude of penetration and peak ionisation. Ion precipitation in the cusp can lead to an elevated conductivity at lower altitudes, especially at the equatorward edge of the cusp. The influence of particle precipitation has earlier been addressed by numerical modelling. With the CTIP (Coupled Thermosphere-Ionosphere-Plasmasphere)-style Sheffield High Latitude (SHL) model, *Millward et al.* [1999] have simulated the effect of particle precipitation on the electron density distribution in the cusp region using typical particle populations, i.e. a Maxwellian energy distribution of 50 eV electrons and 500 eV ions. From their obtained electron density profiles Pedersen conductivity and Joule heating rate have been calculated. The modelled example matches very well the geophysical conditions of 14 December 2002 (Kp = 2.7, F10.7 = 167, winter solstice). Figure 11.8 (right frame) shows results for the cases with particle precipitation (black) and without precipitation (grey). It is quite evident that the Pedersen conductivity is significantly enhanced in case of particle precipitation, especially in the height range above 150 km. When comparing the modelled conductivity curves with the profiles derived from EISCAT observations (left frame), in particular the profile at 10 UT, reasonable agreement is found with the modelled case of precipitation. There is a systematic difference in altitude between the conducting layers that can be explained by the difference in solar activity. While the model run considers solar maximum conditions (F10.7 = 167), the solar flux on 13 October 2006 was quite low, F10.7 = 72.9. Furthermore, there is a seasonal difference between the observed and modelled event. Both the solar insolation and the intensity of small-scale FACs vary with season. *Rother et al.* [2007] showed that kilometre-scale FACs in the cusp region are three times more intense during equinoxes than during winter season.

The region of enhanced ion temperature does not appear as a continuous band in Fig. 11.6, but as a series of patches. High temperatures within a patch have the tendency to move to more northward range gates with time. This is a typical ionospheric signature of pulsed reconnection at the Earth's magnetopause [e.g., *Lockwood et al.*,

1993]. The cusp density anomaly thus seems to be related to the day side reconnection process. At altitudes below 300 km the ion and neutral temperatures are much the same. Around 10 UT ESR observed a strong temperature enhancement. This is assumed to cause an up-welling of molecular-rich (N_2, O_2) air. As a consequence the composition of the upper atmosphere is expected to change. *Crowley et al.* [2008] reported that GUVI (Global Ultraviolet Imager) observations on the TIMED satellite indicate an up-welling of molecularly rich air into the upper atmosphere at day side auroral regions during times of cusp-related density anomalies. This adds another piece of evidence for the local ionospheric heating as source for the density enhancement.

11.4.4 Resumé on Cusp-Related Density Anomaly

From the observations and modelling results, a rather complete picture of the mechanisms responsible for the described high-latitude density peaks emerges. The chain of processes driving the thermospheric anomaly in the cusp region starts with the reconnection of solar and terrestrial magnetic field lines at the day side magnetopause. This process accelerates solar wind plasma earthward. As a consequence high fluxes of soft particle precipitation occur in the cusp region. In particular, electrons with energies below 100 eV cause significant ionisation at altitudes above the E region. During high solar activity when the thermosphere is expanded, ionisation occurs at higher altitude than during low activity in a shrunk atmosphere. The reconnection process, in addition, drives strong field-aligned currents which tend to break up in filaments with small transverse scales ($<$10 km). FACs of this size are very efficient in dissipating their energy in the ionosphere [e.g., *Vogt*, 2002]. Also the precipitating particles deposit energy in the thermosphere. *Clemmons et al.* [2008] claim that these might even contribute the major part to energy input. As a consequence of the various kinds of heating the ion temperature is observed to rise strongly and we suggest that this rises also the neutral temperature in the height range 150–200 km. This causes an up-welling of molecular-rich air from the lower thermosphere and reduces the O/N_2 ratio column density, as diagnosed by GUVI. Satellites at 400 km and above (CHAMP and GRACE) record significant mass density anomalies when passing the cusp region.

The amplitude of the density anomaly is found to increase with the solar flux level (F10.7). An increase of EUV radiation causes larger thermospheric scale heights and correspondingly, a reduction in penetration depth of the precipitating particles. The heating per particle is higher at greater altitudes, thus causing a larger temperature increase. Consequently, the density anomaly in the upper thermosphere becomes larger. The other important controlling parameter is the electric field. This quantity is a measure for the reconnection rate at the day side magnetopause, thus a proxy for the flux of precipitating particles. Furthermore, E_m reflects the transfer of momentum to the ionospheric plasma. In regions of high ion/neutral collisions (here caused by precipitation) kinetic energy of the convecting plasma is converted into heat. The confinement of the density anomaly occurrence to the vicinity of the cusp can be explained by its magnetic connection to the major reconnection sites at the magnetopause. *Watermann et al.* [2009] showed that the cusp at ionospheric altitudes can be characterised by soft particle precipitation and intense small-scale field-aligned currents. These two phenomena are regarded as a key for generating cusp-related density anomalies.

11.5 Summary

In this chapter CHAMP and EISCAT measurements of ionospheric and thermospheric parameters have been considered, together with numerical model results, for explaining the mechanisms that cause the prominent cusp-related mass density anomalies. Main findings are:

(1) The amplitude of the cusp density anomalies depends on both the solar wind input (reconnection rate) and the solar flux level (F10.7). When considering the relative density enhancement with respect to ambient mass density the dependence on solar flux largely disappears. This shows that primarily the reconnection rate and the ambient neutral density are responsible for the anomaly amplitude.
(2) The cusp-related density anomalies are commonly accompanied by bursts of small-scale field-aligned currents. These contribute considerably to Joule heating in the ionosphere. Another feature of the cusp is the intense precipitation of soft particles. These facilitate an enhanced ionospheric conductivity.
(3) Numerical simulations confirm that soft particle precipitation can create a prominent conducting layer at about 150 km altitude. Here the Joule heating rate per particle (or the temperature increase) is larger by about an order of magnitude compared to the E-layer. This causes an up-welling of molecular rich air.
(4) Composition estimates derived from GUVI measurements on TIMED confirm decreased O/N_2 ratios during times of enhanced magnetic activity (density enhancement) in the cusp/cleft region.

With the presented chain of processes we offer an explanation why the cusp region is predestined for local thermospheric mass density anomalies. Some of the indirect conclusions such as the intense small-scale Poynting flux and the co-located ion (neutral) temperature enhancement should be confirmed by direct measurements, for example, with the help of the upcoming Swarm mission.

Acknowledgements We thank W. Köhler for pre-processing the accelerometer data. The operational support of the CHAMP mission by the German Aerospace Centre (DLR) and the financial support for the data processing by the Federal Ministry of Education and Research (BMBF), as part of the Geotechnology Programme, are gratefully acknowledged. EISCAT is an international association supported by research organisations in China (CRIRP), Finland (SA), France (CNRS, till end 2006), Germany (DFG), Japan (NIPR and STEL), Norway (NFR), Sweden (VR), and the United Kingdom (PPARC). One the authors, S. Marker (previously Rentz), was supported by the

Deutsche Forschungsgemeinschaft DFG through the DFG Priority Programme "CAWSES", SPP 1176 (Lu 446/8).

References

Clemmons, J. H., Hecht, J. H., Salem, D. R., & Strickland, D. J. (2008). Thermospheric density in the Earth's magnetic cusp as observed by the Streak mission. *Geophysical Research Letters, 35*, L24103. doi:10.1029/2008GL035972.

Crowley, G., Reynolds, A., Thayer, J. P., Lei, J., Paxton, L. J., Christensen, A. B., Zhang, Y., Meier, R. R., & Strickland, D. J. (2008). Periodic modulations in thermospheric composition by solar wind high speed streams. *Geophysical Research Letters, 35*, L21106. doi:10.1029/2008GL035745.

Demars, H. G., & Schunk, R. W. (2007). Thermospheric response to ion heating in the dayside cusp. *Journal of Atmospheric and Solar-Terrestrial Physics, 69*, 649–660. doi:10.1016/j.jastp.2006.11.002.

Doornbos, E., van den IJssel, J., Lühr, H., Förster, M., & Koppenwallner, G. (2010). Neutral density and crosswind determination from arbitrarily oriented multiaxis accelerometers on satellites. *Journal of Spacecraft and Rockets, 47*(4), 580–589. doi:10.2514/1.48114.

Emmert, J. T. (2009). A long-term data set of globally averaged thermospheric total mass density. *Journal of Geophysical Research, 114*, A09326. doi:10.1029/2009JA014102.

Emmert, J. T., & Picone, J. M. (2010). Climatology of globally averaged thermospheric mass density. *Journal of Geophysical Research, 115*, A09326. doi:10.1029/2010JA015298.

Förster, M., Rentz, S., Köhler, W., Liu, H., & Haaland, S. E. (2008). IMF dependence of high-latitude thermospheric wind pattern derived from CHAMP cross-track measurements. *Annals of Geophysics, 26*, 1581–1595. doi:10.5194/angeo-26-1581-2008.

Fuller-Rowell, T. J., & Rees, D. (1984). Interpretation of an anticipated long-lived vortex in the lower thermosphere following simulation of an isolated substorm. *Planetary and Space Science, 32*, 69–86.

Liu, H., Lühr, H., Henize, V., & Köhler, W. (2005). Global distribution of the thermospheric total mass density derived from CHAMP. *Journal of Geophysical Research, 110*, A04301. doi:10.1029/2004JA010741.

Liu, H., Lühr, H., Watanabe, S., Köhler, W., Henize, V., & Visser, P. (2006). Zonal winds in the equatorial upper thermosphere: Decomposing the solar flux, geomagnetic activity, and seasonal dependencies. *Journal of Geophysical Research, 111*, A07307. doi:10.1029/2005JA011415.

Liu, H., Watanabe, S., & Kondo, T. (2009). Fast thermospheric wind jet at the Earth's dip equator. *Geophysical Research Letters, 36*, L08103. doi:10.1029/2009GL037377.

Lockwood, M., Denig, W. F., Farmer, A. D., Davda, V. N., Cowley, S. W. H., & Lühr, H. (1993). Ionospheric signatures of pulsed reconnection at the Earth's magnetopause. *Nature, 361*, 424. doi:10.1038/361424a0.

Lühr, H., Rother, M., Köhler, W., Ritter, P., & Grunwaldt, L. (2004). Thermospheric up-welling in the cusp region: Evidence from CHAMP observations. *Geophysical Research Letters, 31*, L06805. doi:10.1029/2003GL019314.

Lühr, H., Rentz, S., Ritter, P., Liu, H., & Häusler, K. (2007). Average thermospheric wind patterns over the polar regions, as observed by CHAMP. *Annals of Geophysics, 25*, 1093–1101. doi:10.5194/angeo-25-1093-2007.

Millward, G. H., Moffett, R. J., Balmforth, H. F., & Rodger, A. S. (1999). Modeling the ionospheric effects of ion and electron precipitation in the cusp. *Journal of Geophysical Research, 104*(A11), 24603–24612. doi:10.1029/1999JA900249.

Prölss, G. W. (2004). *Physics of the Earth's space environment: an introduction*. Berlin: Springer.

Prölss, G. W. (2011). Density perturbations in the upper atmosphere caused by the dissipation of solar wind energy. *Surveys in Geophysics, 32*, 101–195. doi:10.1007/s10712-010-9104-0.

Reigber, C., Lühr, H., & Schwintzer, P. (2002). CHAMP mission status. *Advances in Space Research*, *30*, 129–134. doi:10.1016/S0273-1177(02)00276-4.

Rentz, S. (2009). *The upper atmospheric fountain effect in the polar cusp region*. Ph.D. thesis, TU Braunschweig. doi:10.2312/GFZ.b103-09050.

Rentz, S., & Lühr, H. (2008). Climatology of the cusp-related thermospheric mass density anomaly, as derived from CHAMP observations. *Annals of Geophysics*, *26*, 2807–2823. doi:10.5194/angeo-26-2807-2008.

Rishbeth, H. (2002). Whatever happened to superrotation? *Journal of Atmospheric and Solar-Terrestrial Physics*, *64*, 1351–1360. doi:10.1016/S1364-6826(02)00097-4.

Ritter, P., Lühr, H., & Doornbos, E. (2010). Substorm-related thermospheric density and wind disturbances derived from CHAMP observations. *Annals of Geophysics*, *2008*, 1207–1220. doi:10.5194/angeo-28-1207-2010.

Rother, M., Schlegel, K., & Lühr, H. (2007). CHAMP observation of intense kilometer-scale field-aligned currents, evidence for an ionospheric Alfvén resonator. *Annals of Geophysics*, *25*, 1603–1615. doi:10.5194/angeo-25-1603-2007.

Schlegel, K., Lühr, H., St.-Maurice, J.-P., Crowley, C., & Hackert, G. (2005). Thermospheric density structure over the polar regions observed with CHAMP. *Annals of Geophysics*, *23*, 1659–1672. doi:10.5194/angeo-23-1659-2005.

Thayer, J. P., Killeen, T. L., McCormac, F. G., Tschan, C. R., Ponthieu, J.-J., & Spencer, N. W. (1987). Thermospheric neutral wind signatures dependent on the east-west component of the interplanetary magnetic field for Northern and Southern Hemispheres, as measured from Dynamics Explorer-2. *Annals of Geophysics*, *5*, 363–368.

Vogt, J. (2002). Alfvén wave coupling in the auroral current circuit. *Surveys in Geophysics*, *23*, 335–377. doi:10.1023/A:1015597724324.

Wang, H., Lühr, H., Häusler, K., & Ritter, P. (2011). Effect of subauroral polarization streams on the thermosphere: A statistical study. *Journal of Geophysical Research*, *116*, A03312. doi:10.1029/2010JA016236.

Watermann, J., Stauning, P., Lühr, H., Newell, P. T., Christiansen, F., & Schlegel, K. (2009). Are small-scale field-aligned currents and magnetosheath-like particle precipitation signatures of the same low-altitude cusp? *Advances in Space Research*, *43*, 41–46. doi:10.1016/j.asr.2008.03.031.

Chapter 12
Global Sporadic E Layer Characteristics Obtained from GPS Radio Occultation Measurements

Christina Arras, Jens Wickert, Christoph Jacobi, Georg Beyerle, Stefan Heise, and Torsten Schmidt

Abstract Sporadic E layers have been observed on a global scale using GPS radio occultation measurements performed by the CHAMP, GRACE and FORMOSAT-3/COSMIC satellites. This has been done by analysing signal amplitude variations of the GPS L_1 signal. The global distribution of sporadic E layer occurrence shows strong variations on different time scales with highest sporadic E occurrence rates during the afternoon at the midlatitudes of the summer hemisphere. Furthermore, sporadic E occurrence depends on several geophysical parameters such as dynamic conditions in the lower thermosphere, intensity of Earth's magnetic field and meteor flux. The connection between the global sporadic E occurrence in dependence on these parameters is discussed in this chapter.

12.1 Introduction

In the last decade radio occultation (RO) measurements have been established as a powerful technique to observe the Earth's atmosphere and ionosphere on a global scale [e.g., *Wickert et al.*, 2009]. Since the launch of the CHAMP (CHAllenging Minisatellite Payload) satellite in July 2000, a continuous time series of, on average, $\sim$200 RO measurements per day is available to date although the RO experiment on the CHAMP satellite ended in October 2008. The available RO time series was continued by the measurements of the GRACE (GRAvity recovery and Climate Experiment) [*Wickert et al.*, 2009] and the six FORMOSAT-3/COSMIC (FORMOsa SATellite mission-3/Constellation Observing System for Meteorology, Ionosphere and Climate) satellites. Thus, continuous global analysis of geophysical parameters such as, e.g., global tropopause heights [*Schmidt et al.*, 2004, 2008], gravity wave activity [*Ratnam et al.*, 2004; *de la Torre et al.*, 2006; *Namboothiri et al.*, 2008] and

C. Arras (✉) · J. Wickert · G. Beyerle · S. Heise · T. Schmidt
GFZ German Research Centre for Geosciences, Telegrafenberg, 14473 Potsdam, Germany
e-mail: arras@gfz-potsdam.de

C. Jacobi
Institute for Meteorology, University of Leipzig, Stephanstr. 2, 04103 Leipzig, Germany
e-mail: jacobi@rz.uni-leipzig.de

F.-J. Lübken (ed.), *Climate and Weather of the Sun-Earth System (CAWSES)*, Springer Atmospheric Sciences,
DOI 10.1007/978-94-007-4348-9_12,

water vapour content in the lower and middle troposphere [*Heise et al.*, 2006] is possible. Especially the COSMIC mission increases the available daily soundings to an amount of ∼2500 per day. This fact allows increased spatial and temporal resolution of obtained data fields.

Neutral atmospheric soundings are used to calculate profiles of temperature, density, pressure and water vapour content in the troposphere and stratosphere. By now the RO profiles are assimilated by several weather prediction centres to improve weather forecasts [*Anthes et al.*, 2008; *Poli et al.*, 2009]. Beside these familiar applications for the neutral atmosphere, also ionospheric soundings are recorded that are used to calculate vertical electron density profiles [*Hajj and Romans*, 1998; *Schreiner et al.*, 1999; *Jakowski et al.*, 2002] for ionospheric monitoring and to identify electron density anomalies.

Since RO profiles have a high vertical resolution of only several metres, they allow to detect thin structures as, e.g., sporadic E (E_s) layers. Sporadic E layers appear in the E region of Earth's ionosphere and consist of relatively high electron densities compared to the background ionisation. They consist of metallic ions [*Ernest*, 1997] that originate from meteroids entering Earth's atmosphere.They occur between 90 and 120 km altitude with a thickness of usually 0.5–5.0 km and a horizontal extent of 10–1000 km [*Wu et al.*, 2005].

In the early 1960s, Whitehead [1961] described the mechanisms responsible for sporadic E formation at midlatitudes. He suggested that sporadic E formation is a complex process of interactions between ion-neutral collisions, the Earth's magnetic field and vertical shears in the lower thermospheric wind field. The mechanism may be described by the following Eq. (12.1).

$$w = \frac{\cos I \sin I}{1+r^2} U + \frac{r \cos I}{1+r^2} V \qquad (12.1)$$

Here, w represents the vertical ion drift, U, V the neutral winds in geomagnetic southward and eastward direction, I is the magnetic inclination and the term r denotes the ratio of ion-neutral collision frequency and ion gyrofrequency. The formation of a sporadic E layer is most likely in regions where $w = 0$ and the vertical gradient $\frac{dw}{dz}$ is large and negative, i.e. with upward drift below and downward drift above the level of consideration. At altitudes below ∼125 km, the plasma drift is collision dominated. Then r becomes large and the first term on the right-hand side of Eq. (12.1) becomes small compared to the second one. Consequently, the behaviour of the zonal wind shear, which is mainly in parallel to the geomagnetic eastward wind shear, is the most important factor for sporadic E layer formation at midlatitudes in the altitude range between 95 km and 125 km. Thus, sporadic E layers are most likely to be expected when the zonal wind shear is negative. Above, E_s formation depends mainly on the meridional wind shear [e.g., *Haldoupis*, 2010].

Recent investigations from Incoherent Scatter Radar observations indicate that sporadic E layers are quasi-continuous features, and thus rather non-sporadic phenomena [*Haldoupis et al.*, 2007]. Especially during winter and at night, however, they are very weak. Therefore, it is often not possible to observe them with standard techniques like ionosondes and GPS RO.

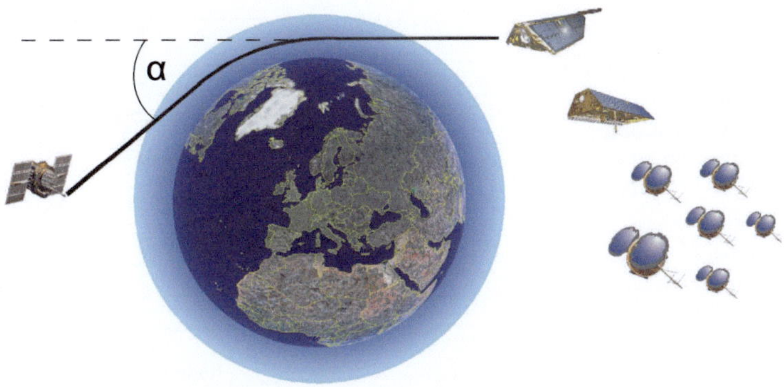

Fig. 12.1 Scheme of the radio occultation principle. The low-Earth orbiting satellites CHAMP (*top*), GRACE-A (*middle*) and the COSMIC constellation (*bottom*) receive signals emitted from rising or setting GPS satellites (*left hand side*) after passing Earth's atmosphere. Atmospheric refraction induces a bending of the GPS signals which is represented by the bending angle, α. The bending angle is a key observable for the GPS radio occultation technique

Relatively strong E_s events cause intense disturbances in radio wave propagation due to their strong vertical electron density gradients. The resulting disturbance in radio waves depends on the waves transmission frequency. In the past, observations mostly came from ground based remote sensing or from rare in-situ rocket measurements [*Whitehead*, 1989; *Mathews*, 1998, and references therein]. Ionosonde measurements allow to observe an E_s layer for a comparatively long time over a single location and to detect temporal variations of the layer. The main advantage of recent E_s observations by the satellite based GPS RO technique is the global data coverage with a relatively high spatial, but low temporal resolution.

12.2 GPS Radio Occultation Principle, Data Base and Sporadic E Detection

The GPS RO technique relies on accurate measurements of the two GPS signals ($f_{L1} = 1.575$ GHz and $f_{L2} = 1.227$ GHz) by a Low-Earth orbiting (LEO) satellite. Due to their low altitude of only a few hundred kilometres, the LEO spacecraft orbits the Earth faster than the GPS satellites which fly at significantly higher altitudes of 20 200 km. During its revolution, the LEO spacecraft tracks the signals of GPS satellites rising or setting behind the Earth's limb. The signals are received after their propagation through the Earth's ionosphere and atmosphere, which modify the signals due to their refractive index which finally leads to a bending of the signal paths (cf. Fig. 12.1). The bending angle α is the key observable of this technique. After an inversion process, it can be converted into pressure, density, temperature and water vapour profiles in the troposphere and stratosphere and into electron density profiles at ionospheric altitudes.

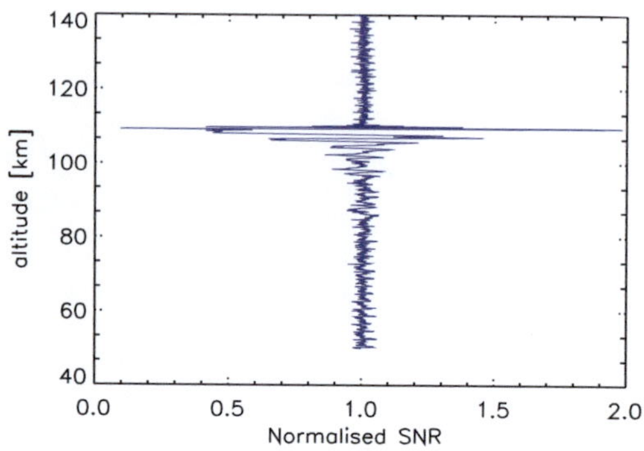

Fig. 12.2 Example of a normalised GPS L_1 SNR profile measured by the CHAMP satellite on 1 January 2003. The profile includes the signature of a sporadic E layer at ~110 km altitude

Table 12.1 Total number of RO measurements from CHAMP, GRACE and COSMIC collected between 2001 and 2009

Satellite	Total number of RO	Mean daily number of RO
CHAMP	541 527	~250
GRACE	203 455	~200
COSMIC	2 575 701	~2200
Total	3 320 683	

These converted profiles rely on several assumptions such as spherical symmetry, which is not valid for sporadic E layers, it was decided to use Signal-to-Noise Ratio (SNR) profiles of the GPS L_1 occultation links to identify sporadic E layer signatures from RO measurements [*Wickert et al.*, 2004; *Wu*, 2006; *Zeng and Sokolovskiy*, 2010]. These data are recorded directly onboard the satellite and do not require additional data treatment or further assumption before being analysed in order to extract information on the presence of an E_s layer. The disadvantage of this method is that only quantitative information on E_s are received, without any knowledge on the electron density of the single events.

But, in presence of strong vertical gradients in the electron density and the corresponding bending angles, as it is expected for sporadic E layers, the SNR shows strong fluctuations (cf. Fig. 12.2). These fluctuations are caused by signal divergence and convergence which leads to an increase or decrease of the signal intensity at the receiving antenna, respectively. The signal disturbances can easily be identified by applying a bandpass filter which only accepts fluctuations that stretch vertically between 1–12.5 km. The altitude of the signal perturbation maximum is taken to be the altitude at which the E_s layer is located. Note that this altitude may slightly differ from the height of maximum electron density. Since also the latitude, longitude and time of every single event are known, it is possible to perform statistical analysis of global sporadic E occurrence in dependence on these variables.

The CHAMP, GRACE and COSMIC satellites provided more than 2500 RO profiles per day during the years 2007–2009. Table 12.1 reveals the amount of measurements contributed by each satellite mission.

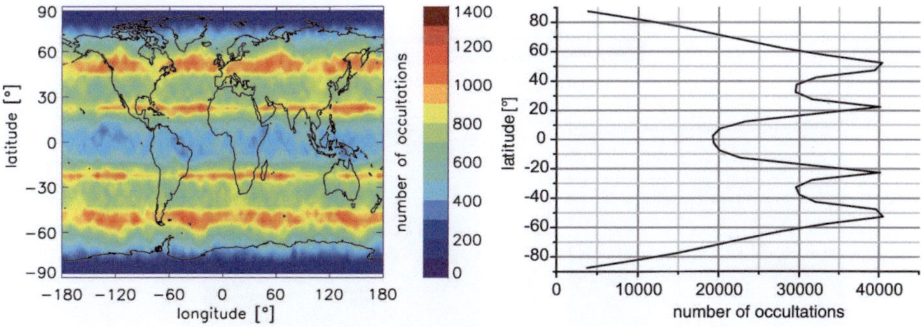

Fig. 12.3 *Left*: global distribution of radio occultation measurements performed by the CHAMP, GRACE and COSMIC satellites during the year 2008 summed up in a 5° × 5° grid. *Right*: latitude dependent number of RO profiles in a 5° resolution

Figure 12.3 includes all measured RO profiles during the year 2008 in a 5° × 5° latitude/longitude resolution. The recorded soundings are not uniformly distributed over the globe. Four latitude-parallel maxima are spanning around the globe at approximately 25°N/S and 45°N/S and a broad minimum is observed in the equatorial region. This effect originates from the LEO spacecraft positions in relation to the GPS satellite constellation. Nevertheless, all grid cells include a sufficient amount of measurements to perform statistical analysis. But, the inhomogeneous data distribution may distort the results concerning the global E_s distribution. Consequently, the ratio of confirmed sporadic E events and the total number of measurements in each grid cell, referred to as E_s occurrence rate, is considered in the following sections.

12.3 Global Sporadic E Occurrence

The occurrence of sporadic E layers shows variations on several time scales. Figure 12.4 displays the global E_s occurrence rates for the northern hemisphere summer (including the months of June, July, August) and winter (including the months of December, January and February) conditions. The plots are based on RO measurements from GRACE and COSMIC during the year 2009. The measurements are binned in a 5° × 5° latitude/longitude grid with 1 km height resolution. Since the interannual E_s variability is relatively low compared to diurnal and seasonal variations, the plots in Fig. 12.4 appear very similar in other years (not shown here).

As global and annual mean, a sporadic E occurrence rate of approximately 14 % is observed. Generally, E_s occurs mainly during daytime at midlatitudes of the summer hemisphere. E_s rates up to 50 % are observed in single grid cells. In the winter hemisphere E_s rates are very low except for few E_s events observed at low latitudes. During equinox a moderate E_s activity is observed at the low latitudes of both hemispheres. During all seasons a slim line where no sporadic E events are detected is found in the equatorial region (right panels). This line follows exactly the course of

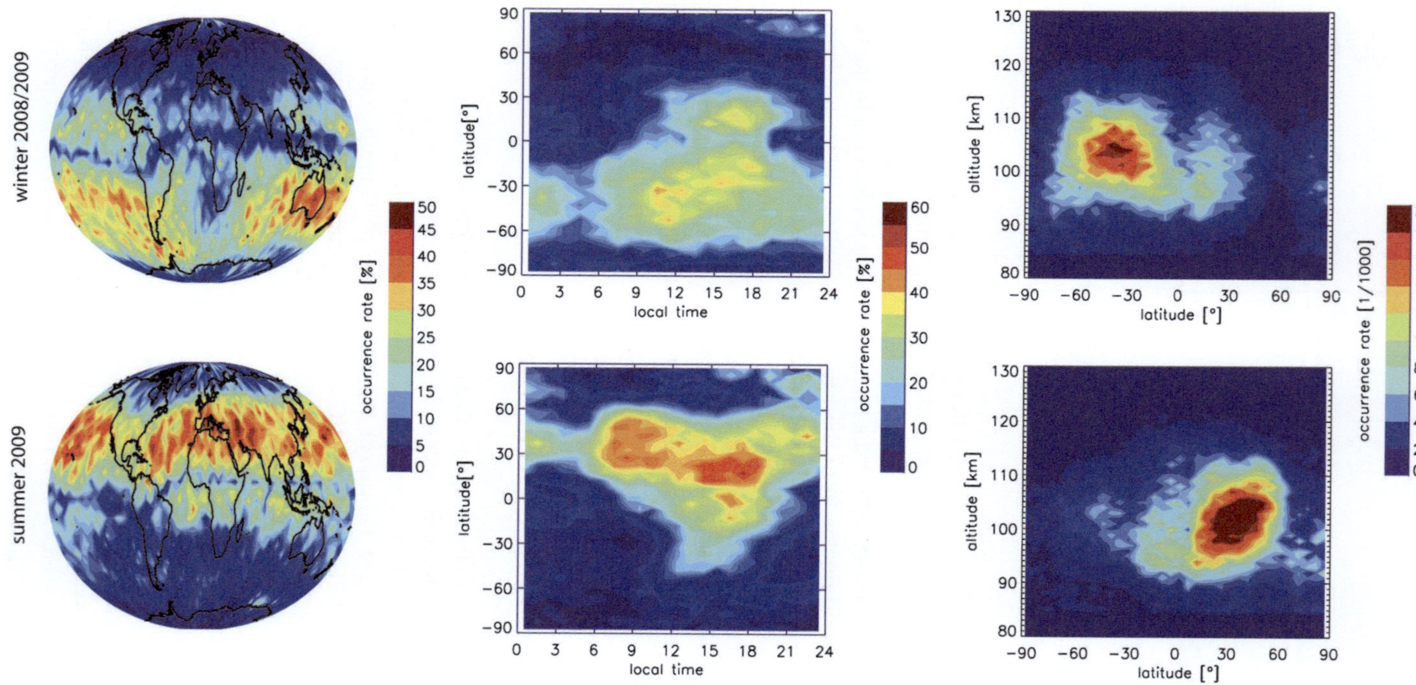

Fig. 12.4 Global sporadic E occurrence for boreal winter 2008/2009 (*upper panels*) and summer 2009 (*lower panels*). *Left*: vertical and time-integrated geographic distribution of E_s occurrence rates. *Middle*: zonal mean integrated occurrence rates. *Right*: zonal and time means

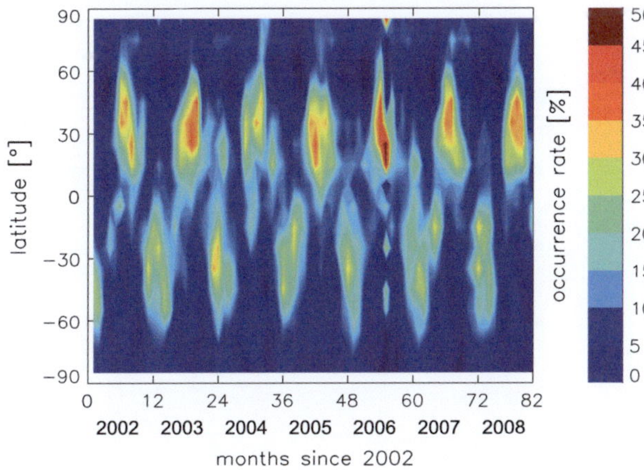

Fig. 12.5 Time series of latitude dependent monthly sporadic E rates observed by the CHAMP satellite

the magnetic equator [*Arras et al.*, 2008]. The band of high E_s rates is interrupted in the Atlantic and African sector of the southern hemisphere. This gap stretches across approximately 90° in longitude. It is caused by the South Atlantic Anomaly of Earth's magnetic field. This point will be discussed further in the following section.

As visible in the middle panel of Fig. 12.4, E_s layers form predominantly during daytime. In the summer hemisphere E_s rates show two maxima. The first one appears before noon at higher latitudes (~45°N/S), while the second one appears in the afternoon at ~30°N/S. During winter, a weak E_s activity is observed in the early afternoon at low latitudes. The layers are preferably seen in an altitude range between 90 and 120 km with maximum rates around 105 km. They form at slightly higher altitudes during summer compared to winter conditions.

The RO technique is a relatively new method and thus long-term results of sporadic E behaviour including long-term trends are not yet available. Currently, CHAMP provides the longest available consistent time series of GPS RO measurements comprising a period of nearly eight years. The CHAMP data set was used for variability analysis of global sporadic E occurrence rates. Figure 12.5 shows the latitudinal structure of time series of monthly mean E_s rates measured by CHAMP between January 2002 and October 2008. The latitudinal resolution is 10°. The individual summer maxima are interrupted by low E_s rates in spring and autumn. Weak secondary maxima are found at low latitudes in the winter hemisphere. In general, E_s rates in the southern hemisphere are 25 % smaller than the ones in the northern hemisphere. The typical value for E_s occurrence during summer maximum is about 40 % in the northern hemisphere and about 30 % in the southern hemisphere. This is a consequence resulting from the sporadic E rate minimum in the South Atlantic magnetic anomaly region.

The summer maxima vary from year to year in intensity, duration, and latitudinal distribution. The underlying processes responsible for these variations have not been fully identified yet.

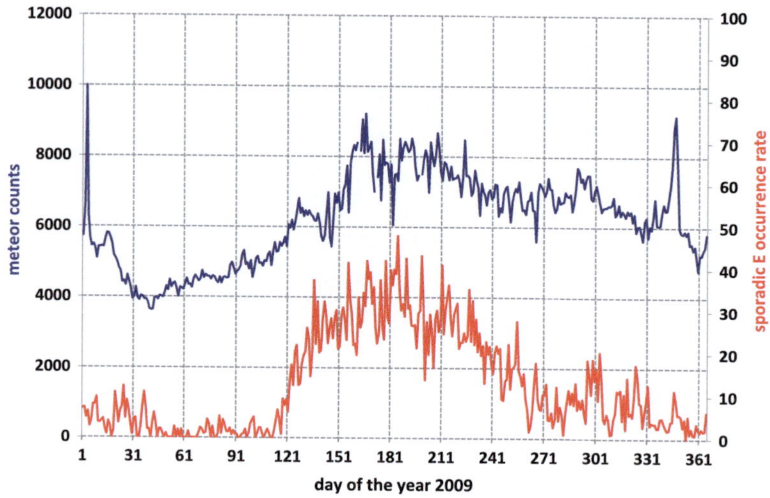

Fig. 12.6 Annual cycle of daily meteor rates (*blue*) measured with the Collm meteor radar and sporadic E rates (*red*) based on COSMIC RO measurements at northern midlatitudes (50°–55°N) during the year 2007

12.4 Coupling Mechanisms

As indicated in the previous sections, the sporadic E layer formation process depends on several geophysical parameters inducing the variations of E_s occurrence in time and space. For sporadic E formation, several components have to work together. First of all, a certain amount of metallic ions is needed. These ions originate from meteors evaporating in Earth's atmosphere. Their daily number shows a characteristic annual cycle. The mechanism that seems to be most effective to compress the metallic ions into thin layers is based on wind shears produced by tidal winds at low and midlatitudes. Thus, upper atmospheric dynamics are an important factor for E_s formation.

The global E_s distribution and the wind shear theory imply that Earth's magnetic field and especially its horizontal intensity plays an important role in E_s formation and its global distribution which can already be recognised in Fig. 12.4.

12.4.1 Connection Between Annual E_s Occurrence and Meteor Flux

Metallic ions at lower ionospheric altitudes are deposited by meteoroids evaporating in Earth's atmosphere. The metallic ion concentration is thus dependent on the intensity of the meteor influx. The meteor influx shows a clear seasonal cycle displayed in Fig. 12.6 (blue). The plot is based on daily meteor count rates detected with the meteor radar (MR) at Collm (51.3°N, 13.0°E) during the year 2007 covering the altitude range between 80 and 100 km. Strong annual variations in the daily meteor counts are observed with largest values detected in June, July and August and lowest rates in February and March. After passing the winter minimum, the meteor rates increase steeply to the summer maximum. The summer rates are about twice as large compared to the winter months. But the summer rates decrease less abruptly during the autumn months. This behaviour leads to an asymmetric course during the year. In January and December, two sharp peaks occur. They can be attributed to the strong meteor showers called Quadrantids (January) and Geminids (December).

Because sporadic E layers consist mainly of metallic ions, it is not surprising that the course of the annual E_s rates is, in general, similar to the one of the meteor rates *Haldoupis et al.* [2007]. A qualitative comparison between daily meteor rates and daily E_s rates (red line) is displayed in Fig. 12.6. The E_s rates are based on COSMIC RO measurements performed during the year 2007 in a latitude band between 50°N and 55°N. The minimum in E_s occurrence is observed in February and March, which is in good agreement with the observed meteor rates. In spring, they suddenly increase to the maximum rates in June, July and August. Afterwards, they decrease quickly in September but showing a secondary maximum at the end of October.

When comparing the course of both curves in Fig. 12.6 one can see that the shape of the annual cycle of E_s occurrence is more symmetric than the one of the meteor rates. Both curves show the steep increase in spring and the maximum rates during the whole summer. However, starting from the end of August the shapes of the curves differ. The meteor rates are still high during September and October while the E_s rates behave differently. In order to explain this phenomenon, the vertical zonal wind shears have to be taken into account. Since the amplitudes of the semidiurnal tide, which is the main source of vertical wind shears in northern midlatitudes, are comparably weak during autumn months, it is quite reasonable that also the E_s rates are low during that time.

The presented curves of daily meteor counts and sporadic E rates in Fig. 12.6 are only valid for the higher midlatitudes in the northern hemisphere. Since meteors are of astronomic origin moving predominantly in the ecliptic plane, their annual cycle behaves proportional to the rising of the apex sporadic meteor source above the horizon. It leads to the presented cycle in the northern midlatitudes. Along the equator the annual cycle amplitude diminishes completely and in the southern hemisphere the annual cycle appears again but with a minimum in July and August and the maximum around February and March.

12.4.2 Connection with Atmospheric Tides

According to wind shear theory [*Whitehead*, 1961; *Axford*, 1963], sporadic E layer formation depends mainly on the vertical shears of the horizontal winds at midlatitudes. Atmospheric tides produce considerably stronger shears than the background circulation. They, especially the semidiurnal and the diurnal components, are the strongest oscillations within the lower thermosphere. The largest amplitudes of the diurnal tide are found at lower latitudes, while the semidiurnal tide shows stronger activity at higher midlatitudes [e.g. *Pancheva et al.*, 2002]. Tidal amplitudes may reach values of more than 40 m/s [*Manson et al.*, 2002a; *Jacobi et al.*, 2009; *Jacobi*, 2011]. Therefore, it is expected that the daily variations in E_s occurrence and its height dependence may result from changes in the lower thermospheric wind field caused by atmospheric tides [*Christakis et al.*, 2009]. In the following the correlation between tidal winds and E_s occurrence are qualitatively investigated. Hereby, the focus is set to measurements performed by the Collm meteor radar located at the Collm Observatory (51.3°N, 13°E).

The MR technique is based on measurements of the radial winds through the Doppler shift of radar signals reflected from ionised meteor trails. The MR at Collm operates at 36.2 MHz in all-sky configuration. Hourly horizontal winds are calculated through projecting the mean wind on the radial winds from individual meteors and minimising the squared differences. The method is described by *Hocking et al.* [2001] in detail. The data are binned in 6 height gates centred at 82, 85, 88, 91, 94, and 98 km, and the hourly mean shear (vertical wind shear, S) is calculated via $S = \Delta U/\Delta h$, where ΔU represents the difference in zonal wind velocity between two adjacent height gates and its vertical distance is given by Δh. From the hourly shear values, monthly means are calculated forming a monthly mean diurnal cycle.

Maximum amplitudes of the semidiurnal tide (SDT) appear in winter. In summer, large amplitudes are only found at higher altitudes. This behaviour of SDT is well known from available climatologies [e.g. *Manson et al.*, 2002b; *Kürschner and Jacobi*, 2005; *Jacobi*, 2011].

Following the wind shear theory, it is expected that sporadic E layers appear in zones of negative zonal wind shear, i.e. in regions of decreasing/increasing eastward/westward wind shear. Figure 12.7 compares the wind shears with the altitude-local time dependent sporadic E occurrence. The sporadic E rates are based on seasonal mean CHAMP, GRACE and COSMIC E_s rates, that are centred around the months of January, April, July 2007 and October 2006, within a latitude range of 50°–55°N. The rates are binned into hourly intervals of local time to get a comparable view on the correlation between sporadic E and zonal wind shear. Figure 12.7 shows height-local time cross sections of E_s rates (displayed in colour code) in comparison to the zonal wind shears (isolines) for four different seasons. The E_s rates are given in $1/1000$. As expected, the highest E_s probability is found in summer with values of up to 5.6 %. Although the rates are very weak (especially during January), a clear semidiurnal structure is recognisable in all seasons. Additionally, a descending structure of E_s rates with local time is visible in all panels. During all seasons this descending structure corresponds with the negative wind shears indicated as solid isolines in the lower part of the panels. At first glance, there is an intense contradiction. Strongest wind shears, required for E_s formation, appear in winter whereas the highest sporadic E rates are measured in summer where the shears are distinctly smaller. However, in addition to the vertical zonal wind shears, the meteor rates have to be considered for the interpretation of this phenomenon [*Haldoupis et al.*, 2007].

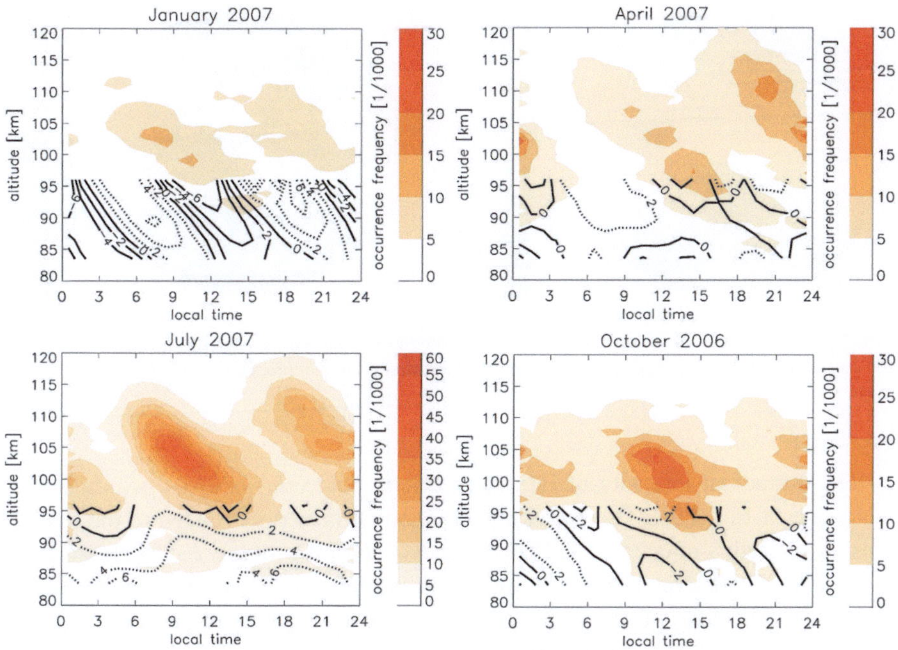

Fig. 12.7 Altitude/local time cross-sections of sporadic E rates (*red contours*) in the latitude range 50°–55°N and zonal wind shear from the MR Collm (*black isolines*). The shear is given in $\mathrm{ms^{-1}\,km^{-1}}$. Positive shear values are indicated by dotted isolines, negative ones by solid isolines. The plots represent the conditions during the period autumn 2006 to summer 2007. Note the different scaling for summer E_s rates compared to the other seasons. Adapted from *Arras et al.* [2009]

Further, there is also a diurnal component in the E_s rates in Fig. 12.7. During all seasons, one of the two descending maxima is stronger than the other one. Possible influences contributing to this signal could be a diurnal tide in the wind field producing additional wind shears [*Haldoupis et al.*, 2006], and an increased background ionisation during daytime.

12.4.3 Connection with Earth's Magnetic Field

The narrow dark (no-E_s) belt centred around the magnetic equator in the global maps of E_s occurrence (Fig. 12.4), as well as the low E_s rates over the South Atlantic, prompted the investigation of the Earth's magnetic field role in the observed global sporadic E distribution. Magnetic field values are calculated using the 10th version of the International Geomagnetic Reference Field (IGRF) model released in 2004 [*Macmillan and Maus*, 2005]. In the following, the applied IGRF data represent year 2007 conditions.

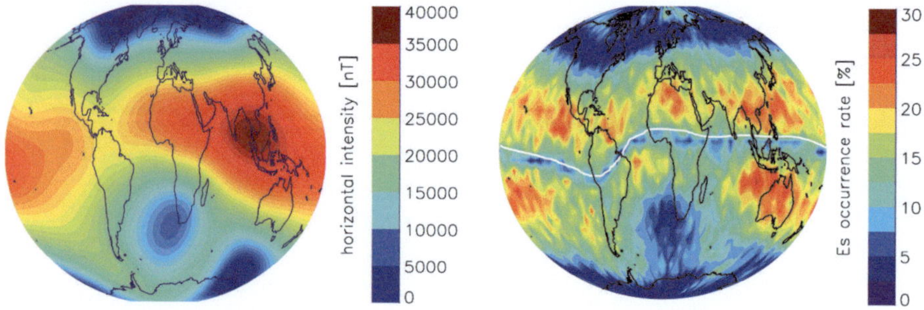

Fig. 12.8 *Left*: Horizontal intensity of Earth's magnetic field. *Right*: Sporadic E occurrence rate for complete year 2009. The magnetic equator is indicated as a *white line*

The horizontal intensity of Earth's magnetic field (B_h) is displayed on the left hand side of Fig. 12.8. The horizontal component is closely related to the total intensity of Earth's magnetic field (B_0) via $B_h = B_0 \cos I$ where I denotes the inclination. As it was already pointed out, at midlatitudes the E_s formation depends on the zonal wind shear mechanism (second term on the right hand side of Eq. (12.1)). This term also includes a dependence on the horizontal component of Earth's magnetic field. Consequently, we expect a close relation between sporadic E occurrence in midlatitudes and the B_h values and will focus on this relation in the following.

On the right hand side of Fig. 12.8 the global E_s distribution for the complete year 2009 is shown. The magnetic equator is indicated as a white line in order to demonstrate that the zone without sporadic E layers at low latitudes exactly follows the magnetic equator. When comparing both panels of Fig. 12.8 it is qualitatively obvious that both parameters are strongly correlated. In regions of strong horizontal magnetic field values sporadic E rates are generally large. Also the footprint of the South Atlantic Anomaly (low B_h values in the South Atlantic and South African region) is well represented in the E_s map. A correlation analysis confirmed the visible connection between both parameters. At midlatitudes the correlation between horizontal intensity of Earth's magnetic field and E_s occurrence rates are generally very high with values around 0.8. The highest correlation of 0.94 is found in a band 25–35°S.

The absence of sporadic E layers along the geomagnetic equator is attributed to the strong magnetisation of electrons [*Kelley*, 1989]. The electrons are confined to magnetic field lines that are exactly horizontal along the equator. Therefore, they are not able to follow the converging ions moving perpendicular to the neutral wind shears. If this would be the case, a strong electric field would build up prohibiting the E_s formation. Also in polar regions electric fields are too strong to enable E_s formation in line with the wind shear process. Nevertheless, sporadic E layers are found in polar regions. Their formation is associated with auroral activity and particle precipitation [*Maehlum*, 1962; *Nygren et al.*, 1984].

12.5 Summary and Conclusions

This study is based on GPS radio occultation measurements performed by the low-Earth orbiting satellites CHAMP, GRACE and FORMOSAT-3/COSMIC between 2001 and 2009. The large number of daily GPS RO measurements, their global distribution and the high vertical resolution of the profiles is an excellent basis for investigating the lower ionosphere on a global scale. Data supplied by an ensemble of satellites are used to receive a global and comprehensive picture of sporadic E layer occurrence and the associated coupling process between neutral atmosphere and ionosphere.

E_s occurrence underlies variations on several time scales. The observed diurnal fluctuations with higher rates during daytime and lower rates at night are interfered by an annual cycle where the largest values are measured in summer. The variations are caused by geophysical parameters controlling the sporadic E formation. The interaction of these parameters, namely the deposition of metallic ions in the lower ionosphere, the horizontal intensity of Earth's magnetic field and vertical wind shear, is summarised in the wind shear theory that describes the E_s formation process at midlatitudes. The results presented here support the theoretically expected relationship between the sporadic E formation process and the discussed parameters.

Acknowledgements The authors acknowledge the CHAMP and GRACE teams at GFZ and JPL for their work. We like to thank UCAR (Boulder, US) and NSPO (Taiwan) for the provision of the FORMOSAT-3/COSMIC data sets and related support. This work was partly funded by grant WI 2634/2-1 within the German DFG priority research program 1176 related to CAWSES.

References

Anthes, R. A., Bernhardt, P. A., Chen, Y., Cucurull, K., Dymond, K. F., Ector, S., Healy, S. B., Ho, S.-P., Hunt, D. C., Kuo, Y.-H., Liu, H., Manning, K., McCormick, C., Meehan, T. K., Randel, W. J., Rocken, C., Schreiner, W. S., Sokolovskiy, S. V., Syndergaard, S., Thompson, D. C., Trenberth, K. E., Wee, T.-K., Yen, N. L., & Zhang, Z. (2008). The COSMIC/FORMOSAT-3 mission: early results. *Bulletin of the American Meteorological Society*, *89*(3), 313–333. doi:10.1175/BAMS-89-3-313.

Arras, C., Wickert, J., Jacobi, C., Heise, S., Beyerle, G., & Schmidt, T. (2008). A global climatology of ionospheric irregularities derived from GPS radio occultation. *Geophysical Research Letters*, *35*, L14809. doi:10.1029/2008GL034158.

Arras, C., Jacobi, C., & Wickert, J. (2009). Semidiurnal tidal signature in sporadic E occurrence rates derived from GPS radio occultation measurements at higher midlatitudes. *Annals of Geophysics*, *27*, 2555–2563.

Axford, W. I. (1963). The formation and vertical movement of dense ionized layers in the ionosphere. *Journal of Geophysical Research*, *68*, 769.

Christakis, N., Haldoupis, C., Zhou, Q., & Meek, C. (2009). Seasonal variability and descent of mid-latitude sporadic E layers at Arecibo. *Annals of Geophysics*, *27*, 923–931.

de la Torre, A., Schmidt, T., & Wickert, J. (2006). A global analysis of wave potential energy in the lower stratosphere derived from 5 years of GPS radio occultation data with CHAMP. *Geophysical Research Letters*, *33*, L24809. doi:10.1029/2006GL027696.

Ernest, K. (1997). On the abundance of metal ion in the lower ionosphere. *Journal of Geophysical Research, 102*, 9667–9674.
Hajj, G. A., & Romans, L. J. (1998). Ionospheric electron density profiles obtained with the global positioning system: results from the GPS/MET experiment. *Radio Science, 33*, 175–190.
Haldoupis, C. (2010). A tutorial review on Sporadic E layers. In *IAGA special sopron book series: Vol. 2. Aeronomy of the Earth's atmosphere and ionosphere* (pp. 381–394). Chap. 29.
Haldoupis, C., Meek, C., Christakis, N., Pancheva, D., & Bourdillon, A. (2006). Ionogram height-time-intensity observations of descending sporadic E layers at mid-latitudes. *Journal of Atmospheric and Solar-Terrestrial Physics, 68*, 539–557.
Haldoupis, C., Pancheva, D., Singer, W., Meek, C., & MacDougall, J. (2007). An explanation for the seasonal dependence of midlatitude sporadic E layers. *Journal of Geophysical Research, 112*, A06315. doi:10.1029/2007JA012322.
Heise, S., Wickert, J., Beyerle, G., Schmidt, T., & Reigber, C. (2006). Global monitoring of tropospheric water vapor with GPS radio occultation aboard CHAMP. *Advances in Space Research, 27*, 2222–2227.
Hocking, W. K., Fuller, B., & Vandepeer, B. (2001). Real-time determination of meteor-related parameters utilizing modern digital technology. *Journal of Atmospheric and Solar-Terrestrial Physics, 63*, 155–169.
Jacobi, C. (2011). 6 year mean prevailing winds and tides measured by VHF meteor radar over Collm (51.3°N, 13.0°E). *Journal of Atmospheric and Solar-Terrestrial Physics*. doi:10.1016/j.jastp.2011.04.010.
Jacobi, C., Arras, C., Kürschner, D., Singer, W., Hoffmann, P., & Keuer, D. (2009). Comparison of mesopause region meteor radar winds, medium frequency radar winds and low frequency drifts over Germany. *Advances in Space Research, 43*, 247–252.
Jakowski, N., Wehrenpfennig, A., Heise, S., Reigber, C., Lühr, H., Grunwaldt, L., & Meehan, T. K. (2002). GPS radio occultation measurements of the ionosphere from CHAMP: early results. *Geophysical Research Letters, 29*(10), 1457. doi:10.1029/2001GL014364.
Kelley, M. C. (1989), *The Earth's ionosphere: plasmaphysics and electrodynamics*. San Diego: Academic Press.
Kürschner, D., & Jacobi, C. (2005). The mesopause region wind field over Central Europe in 2003 and comparison with a long-term climatology. *Advances in Space Research, 35*, 1981–1986.
Macmillan, S., & Maus, S. (2005). International geomagnetic reference field—the tenth generation. *Earth, Planets and Space, 57*, 1135–1140.
Maehlum, B. (1962). The sporadic E auroral zone. In *Geofysiske publikasjoner* (Vol. XXIII, pp. 1–32). Oslo: Oslo University Press.
Manson, A. H., Luo, Y., & Meek, C. (2002a). Global distribution of diurnal and semi-diurnal tides: observations from HRDI-UARS of the MLT region. *Annals of Geophysics, 20*, 1877–1890.
Manson, A. H., Meek, C., Hagan, M. E., Franke, S., Fritts, D., Hall, C., Hocking, W., Igarashi, K., MacDougall, J., Riggin, D., & Vincent, R. (2002b). Seasonal variations of the semi-diurnal and diurnal tides in the MLT: multi-year MF radar observations from 2–70°N, modelled tides (GSWM, CMAM). *Annals of Geophysics, 20*, 661–677.
Mathews, J. D. (1998). Sporadic E: current views and recent progress. *Journal of Atmospheric and Solar-Terrestrial Physics, 60*, 413–435.
Namboothiri, S. P., Jiang, J. H., Kishore, P., Igarashi, K., Ao, C. O., & Romans, L. J. (2008). CHAMP observations of global gravity wave fields in the troposphere and stratosphere. *Journal of Geophysical Research, 113*, D07102. doi:10.1029/2007JD008912.
Nygren, T., Jalonen, L., Oksman, J., & Turunen, T. (1984). The role of electric fields and neutral wind direction in the formation of sporadic E-layers. *Journal of Atmospheric and Terrestrial Physics, 46*, 373–381.
Pancheva, D., Merzlyakov, E., Mitchell, N. J., Portnyagin, Y., Manson, A. H., Jacobi, C., Meek, C. E., Luo, Y., Clark, R. R., Hocking, W. K., MacDougall, J., Muller, H. G., Kürschner, D., Jones, G. O. L., Vincent, R. A., Reid, I. M., Singer, W., Igarashi, K., Fraser, G. I., Fahrutdinova, A. N., Stephanov, A. M., Poole, L. M. G., Malinga, S. B., Kashcheyev, B. L., & Oleynikov, A. N. (2002). Global-scale tidal variability during the PSMOS campaign of June–August 1999:

interaction with planetary waves. *Journal of Atmospheric and Solar-Terrestrial Physics*, *64*, 1865–1896.

Poli, P., Moll, P., Puech, D., Rabier, F., & Healy, S. B. (2009). Quality control, error analysis, and impact assessment of FORMOSAT-3/COSMIC in numerical weather prediction. *Journal Terrestrial Atmospheric and Oceanic Sciences*, *20*, 101–113. doi:10.3319/TAO.2008.01.21.02(F3C).

Ratnam, M. V., Tetzlaff, G., & Jacobi, C. (2004). Global and seasonal variations of stratospheric gravity wave activity deduced from the Challenging Minisatellite Payload (CHAMP)-GPS satellite. *Journal of the Atmospheric Sciences*, *61*, 1610–1620.

Schmidt, T., Wickert, J., Beyerle, G., & Reigber, C. (2004). Tropical tropopause parameters derived from GPS radio occultation measurements with CHAMP. *Journal of Geophysical Research*, *109*, D13105. doi:10.1029/2004JD004566.

Schmidt, T., Wickert, J., Beyerle, G., & Heise, S. (2008). Global tropopause height trends estimated from GPS radio occultation data. *Geophysical Research Letters*, *35*. doi:10.1029/2008GL034012.

Schreiner, W. S., Sokolovskij, S. V., Rocken, C., & Hunt, D. C. (1999). Analysis and validation of GPS/MET radio occultation data in the ionosphere. *Radio Science*, *34*, 949–966.

Whitehead, J. (1961). The formation of the sporadic-E layer in the temperate zones. *Journal of Atmospheric and Terrestrial Physics*, *20*, 49–58.

Whitehead, J. (1989). Recent work on mid-latitude and equatorial sporadic-E. *Journal of Atmospheric and Terrestrial Physics*, *51*, 401–424.

Wickert, J., Pavelyev, A., Liou, Y., Schmidt, T., Reigber, C., Igarashi, K., Pavelyev, A., & Matyugov, S. (2004). Amplitude variations in GPS signals as a possible indicator of ionospheric structures. *Geophysical Research Letters*, *31*, L24801. doi:10.1029/2004GL02607.

Wickert, J., Michalak, G., Schmidt, T., Beyerle, G., Cheng, C. Z., Healy, S. B., Heise, S., Huang, C. Y., Jakowski, N., Köhler, W., Mayer, C., Offiler, D., Ozawa, E., Pavelyev, A. G., Rothacher, M., Tapley, B., & Arras, C. (2009). GPS radio occultation: results from CHAMP, GRACE and FORMOSAT-3/COSMIC. *Journal Terrestrial Atmospheric and Oceanic Sciences*, *1*, 35–50. doi:10.3319/TAO.2007.12.26.01(F3C).

Wu, D. L. (2006). Small-scale fluctuations and scintillations in high-resolution GPS/CHAMP SNR and phase data. *Journal of Geophysical Research*, *68*, 999–1017.

Wu, D. L., Ao, C. O., Hajj, G. A., de la Torre Juarez, M., & Mannucci, A. J. (2005). Sporadic E morphology from GPS-CHAMP radio occultation. *Journal of Geophysical Research*, *110*, A01306. doi:10.1029/2004JA010701.

Zeng, Z., & Sokolovskiy, S. (2010). Effect of sporadic E clouds on GPS radio occultation signals. *Geophysical Research Letters*, *37*, L18817. doi:10.1029/2010GL044561.

Chapter 13
Atmospheric Ionization Due to Precipitating Charged Particles

Jan Maik Wissing, Jan Philipp Bornebusch, and May-Britt Kallenrode

Abstract Precipitating charged particles contribute to the natural variations in the Earth's atmosphere such as ionization, electron density, and composition of e.g. NOx and Ozone. Precipitating solar energetic and magnetospheric particles show a highly dynamic behavior in space and time. We present a 3D ionization model considering the relevant particle species (electrons, protons, and alpha particles) as well as precipitation areas (polar cap and auroral oval): the Atmospheric Ionization Model OSnabrück AIMOS, and discuss some of the atmospheric consequences of precipitating particles. We present the limitations of direct comparisons between EISCAT and precipitating particles and give comparisons between incoherent scatter measurements and a combination of AIMOS and the HAMMONIA GCM to demonstrate the consistency in both methods.

13.1 Introduction

The most obvious effect of energetic charged particles on the atmosphere is the aurora: precipitating magnetospheric particles excite and ionize e.g. oxygen in the upper atmosphere. Speculations about an aurora related production of NOx date back into the second half of the 20th century, however, only the studies following the 1972 solar energetic particle (SEP) event led *Heath et al.* [1977] and *Crutzen et al.* [1975] to develop the relevant chemical models.

Studies of particle precipitation into the atmosphere in general were limited to solar protons precipitating into the polar cap [e.g. *Jackman et al.*, 2001, 2005; *Randall et al.*, 2007; *Rohen et al.*, 2005; *Verronen et al.*, 2002] with the contribution of solar electrons thought to be less than 10 % [*McPeters and Jackman*, 1985]. The energy/height range under study was defined by the particle instrument being in use; combinations of instruments such as in *Mewaldt et al.* [2005] are rarely used. Other

With contributions from Jens Kieser, Hauke Schmidt, Miriam Sinnhuber, Holger Winkler, Nadine Wieters.

J.M. Wissing · J.P. Bornebusch · M.-B. Kallenrode (✉)
University of Osnabrück, Barbarastr. 7, 49076 Osnabrück, Germany
e-mail: mkallenr@uos.de

F.-J. Lübken (ed.), *Climate and Weather of the Sun-Earth System (CAWSES)*, Springer Atmospheric Sciences,
DOI 10.1007/978-94-007-4348-9_13,

studies considered magnetospheric electrons precipitating into the auroral oval either uniformly [*Callis et al.*, 1996a, 1996b, 1998] or in a spatial pattern determined by parametrization of a geomagnetic index [*Fang et al.*, 2008; *Marsh et al.*, 2007]. Magnetospheric protons were mostly neglected and again, the energy range of the particle detectors defined the height range under study.

The limitation to selected particle species or particle sources as in these previous modeling approaches can be considered as incomplete for two reasons: (1) solar energetic as well as magnetospheric particles consist of different particle species: protons and electrons and to a lesser extend alpha particles and heavier atoms. (2) Precipitating particles of solar and magnetospheric origin can coexist in time: a coronal mass ejection (CME) might accelerate SEPs close to the Sun and in interplanetary space. On hitting the magnetosphere, it can trigger a geomagnetic disturbance, which causes precipitation of magnetospheric particles. Thus atmospheric consequences are more complex than indicated in a simple solar protons vs. magnetospheric electron paradigm; even within this simple paradigm, it is not always simple to disentangle the atmospheric effects of both particle sources in the NOx data [*Seppälä et al.*, 2007].

13.2 Atmospheric Ionization Model OSnabrück AIMOS

AIMOS is designed to give a more complete picture of precipitating particles. The basic idea behind the Atmospheric Ionization Module OSnabrück AIMOS [*Wissing and Kallenrode*, 2009] is to create 3D ionization rates over a wide range of heights/particle energies. The model is based on satellite observations of energetic particles and consists of two parts: a sorting algorithm creating a 2D horizontal pattern of the particle spectra on the top of the atmosphere and a Monte-Carlo simulation to determine the vertical pattern of ionization within each horizontal bin. The flow chart of AIMOS is shown in Fig. 13.1.

The model's vertical extent is based on (a) the model atmosphere and (b) the available particle data. Since it uses the atmosphere of the HAMMONIA GCM [*Schmidt et al.*, 2006] it extends from ground up to 1.7×10^{-5} Pa, corresponding to an upper boundary between 250 and 600 km; it is spaced into 67 in pressure logarithmically equidistant vertical layers. The horizontal grid is $3.6° \times 3.6°$. Since data from one Geostationary Operational Environmental Satellite (GOES) and two Polar Orbiting Environmental Satellites (POES) are used, the energy range is determined by the particle detectors on these satellites. The GOES satellite measures protons from 4 to 500 MeV and Alpha particles from 4 to 500 MeV/particle using the SEM instrument [*NASA*, 1996]. The POES satellites are flying at 850 km altitude. And as being equipped with MEPED and TED instruments [*Evans and Greer*, 2004], the energy range of protons enlarges down to 150 eV. In addition, electrons are detected between 150 eV and 2.5 MeV (which is extrapolated to 5 MeV in AIMOS). In total 24 energy channels on 3 satellites are used.

AIMOS data are available at http://aimos.physik.uos.de. Examples for the use of AIMOS data within the HEPPA initiative are given in Chap. 15; energetic particle

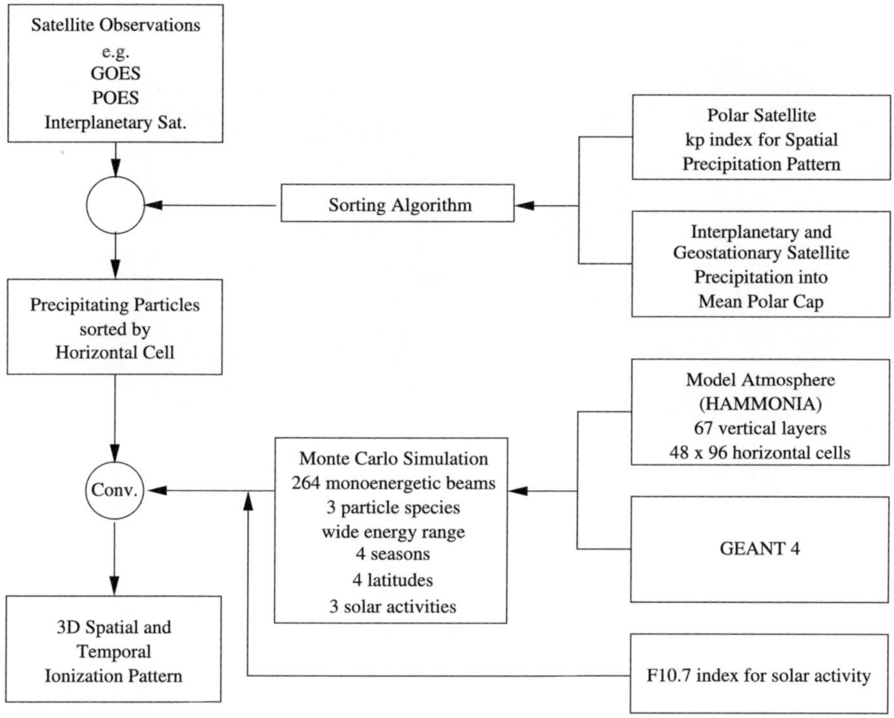

Fig. 13.1 Flow chart of the Atmospheric Ionization Module OSnabrück AIMOS [*Wissing and Kallenrode*, 2009]. Reprinted with kind permission from American Geophysical Union

precipitation impacts on the middle atmosphere are also discussed in Chaps. 8, 16, and 17.

The derivation of the horizontal precipitation pattern is the main challenge because it requires some assumptions on how to distinguish between a rather uniformly filled polar cap [*Bornebusch et al.*, 2010] and the auroral oval with particle fluxes showing a strong dependence on local time [*Wissing et al.*, 2008]. In addition, the transition from cap to oval depends on energy and is highly variable with geomagnetic activity [*Leske et al.*, 2001]. A rather complex retrieval mechanism is required to extract the horizontal patterns from the satellite date: since particle data are in-situ measurements, only particle fluxes along the satellite orbit can be measured as indicated on the left hand side of Fig. 13.2; snapshots of particle spectra at the top of the atmosphere are not available. The combined data from at least two polar orbiting spacecraft can be used to derive a map of the relative flux of magnetospheric particles, leading to a horizontal precipitation pattern: the globe is split into 14 regions with similar precipitation patterns with four subdivisions depending on local time [*Wissing et al.*, 2008].

The maps depend on particle rigidity because the geomagnetic cutoff depends on rigidity. They vary in time because polar cap and auroral oval expand equatorward

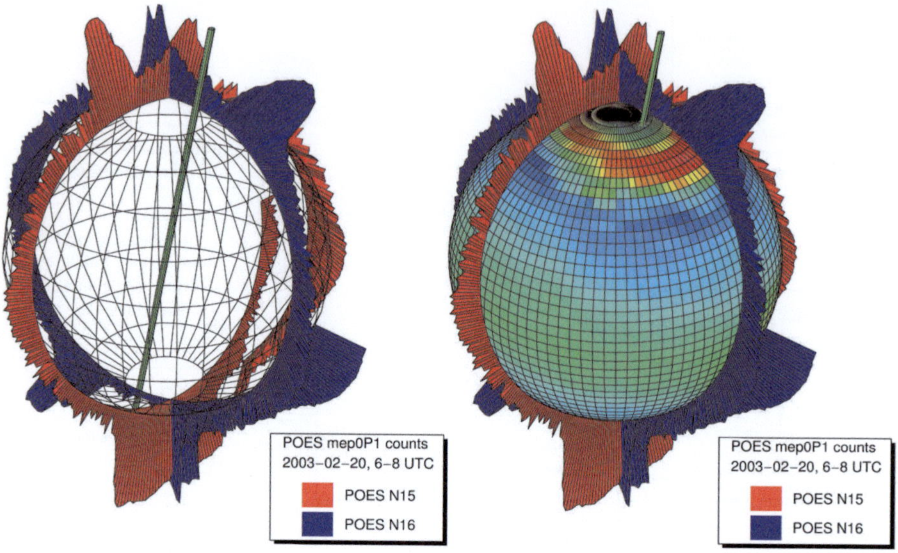

Fig. 13.2 Particle fluxes of 30–80 keV protons measured by two POES satellites in roughly perpendicular orbits (*left*) are combined to give a precipitation matrix (*right*) [*Wissing and Kallenrode*, 2009]. Reprinted with kind permission from American Geophysical Union

with increasing geomagnetic activity. The maps give patterns: the actual flux of magnetospheric particles can be obtained by scaling the flux measured along the orbit with the corresponding matrix. Precipitating solar energetic particles inside the polar cap are taken from geostationary satellites or spacecraft in interplanetary space; the polar cap as the void in the matrices described above is filled uniformly with these particles. An example for the resulting map of both particle species is shown in the right hand side of Fig. 13.2. In the flow chart this corresponds to the rectangle labeled "Precipitating particles sorted by horizontal cell".

The interaction between the precipitating particles and the atmosphere—and thus the third dimension of the model—is described by a Monte-Carlo simulation using the GEANT 4 toolkit (Agostinelli et al., 2003). A Monte-Carlo simulation is used instead of the simpler Bethe–Bloch or Berger–Seltzer algorithms for three reasons: (a) secondaries and their interaction with the atmosphere can be considered, (b) multiple scattering of electrons can be considered, and (c) the generation of Bremsstrahlung by electrons can be taken into account. The first two processes are important in particular in the rarefied upper atmosphere, the latter leads to ionization of the atmosphere well beyond the range of the primary electron. The simulation used within AIMOS is an extended version of the one introduced by *Schroeter et al.* [2006], in particular because it is based on an atmosphere taken from the HAMMONIA GCM [*Schmidt et al.*, 2006]. Runs were performed for four different geographic latitudes, the four seasons, and three different levels of solar activity.

Processes considered in the Monte-Carlo simulation include continuous energy loss, multiple scattering, generation of Bremsstrahlung by primaries as well as secondaries, photoelectric effect, Compton scattering and pair production—although the latter is extremely rare because of the rather low energies under study. Hadronic interactions are not considered for the same reason. The algorithm has been tested against the results of *Berger et al.* [1970], *Callis et al.* [1998], *Jackman et al.* [2005] and *Schröter et al.* [2006].

The Monte-Carlo simulation is performed with 40 logarithmically equidistant monoenergetic particle beams with 9 different equally spaced directions of incidence to take into account the angular distribution of the particles. The ionization profiles than are derived as a convolution with the incident particle spectrum on top of the atmosphere.

Since AIMOS is the first model to combine solar and magnetospheric particles and to consider all relevant particle species, it can be used to study the influence of different particle populations and species at different levels of solar and or geomagnetic activity. *Wissing and Kallenrode* [2009] applied AIMOS at quiet times, during a simple solar energetic particle event and during the complex series of SEPs and shocks in October/November 2003. Although spatial and temporal variations in ionization rates are rather complex, the authors identified some general features.

During quiet times (no SEPs, no geomagnetic disturbance), ionization inside both polar cap and auroral oval is mainly by electrons; the ionization inside the auroral oval exceeds that inside the polar cap. Thus for quiet times, a limitation to magnetospheric electrons is a reasonable approximation.

During a simple solar particle event (no shock, no strong geomagnetic activity) the ionization in the meso- and stratosphere mainly is due to protons while in the (lower) thermosphere the ionization by electrons accelerated in the flare dominates. But even in the mesosphere, solar electrons can contribute up to 30 % to the ionization. While solar electrons correctly have been neglected in the study of stratospheric effects of precipitating particles their influence should not be ignored at higher altitudes [*Wissing and Kallenrode*, 2009].

Now we consider a more complex situation, when after a solar event a shock hits the magnetosphere. Here the relative importance of particle precipitation inside the polar cap and the auroral oval depends on geomagnetic activity and the spectrum of the solar energetic particles. At least in the lower thermosphere, both particle populations have comparable influence on ionization rates. In the mesosphere, ionization by magnetospheric electrons often exceeds that by solar electrons; it also can exceed that by solar protons. Consequently, the consideration of both particle populations with all particle species modifies the estimated ionization rates. These modifications are not large enough to change the essential results of earlier studies but they are large enough to be considered in comparisons between atmospheric models and measurements of atmospheric constituents.

The main vertical advantage of AIMOS compared to a previous study by *Jackman et al.* [2005] is given in Fig. 13.4. Showing a 2 hour interval within the October event 2003 both models agree around 50 km altitude, but below 35 km the smaller upper energy range of the Jackman model results in a decreasing ionization rate. Above 65 km the AIMOS model again shows higher ionization rates due to low energetic particles and which are neglected in the Jackman model.

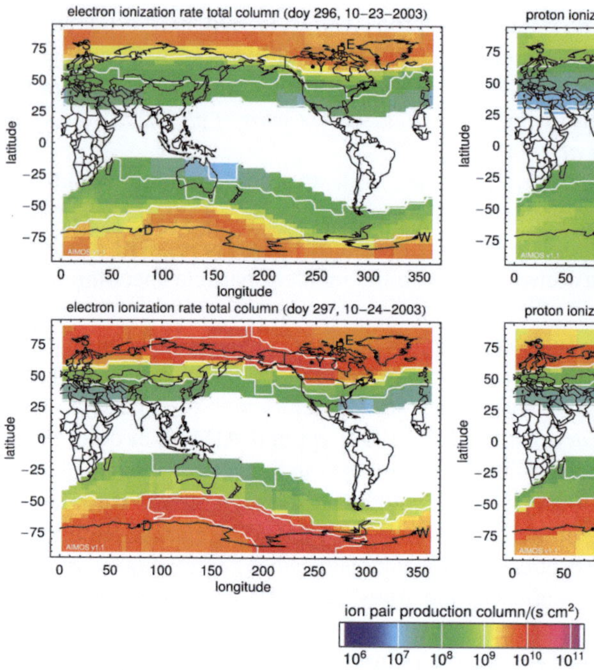

Fig. 13.3 Total electron production (TEP) rate for a quiet time period (*top*) and a geomagnetic active period (*bottom*) for electrons (*left*) and protons (*right*) [*Wissing et al.*, 2010]. Reprinted with kind permission from American Geophysical Union

Fig. 13.4 Comparison of an ionization profile due to proton precipitation from AIMOS and *Jackman et al.* [2005]. A longer time series of both models as well as a more detailed discussion can be found in *Wissing and Kallenrode* [2009] and *Jackmanet al.* [2005]

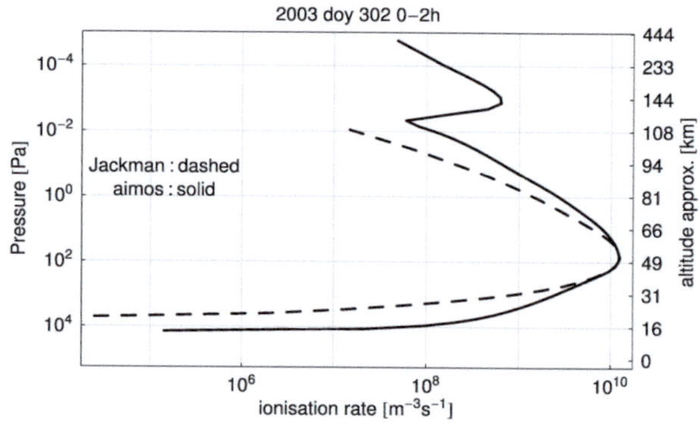

The second main advantage of AIMOS is the horizontal pattern which can be seen e.g. in the relative contributions of electrons and protons to the total electron production TEP, that is the height-integrated ion pair production rate. TEP is shown in Fig. 13.3: the contribution of electrons to TEP rates exceeds that of protons at almost all locations, that is within the auroral oval as well as within the polar cap. This holds for quiet times (top) as well as for times of high geomagnetic activity (bottom). Only during the early phase of a solar energetic particle event and inside the polar cap the contribution of protons to TEP exceeds that of electrons [*Wissing et al.*, 2010].

To illustrate the consequences of the consideration of the total particle inventory atmospheric chemistry modeling, *Wissing et al.* [2010] applied a combination of AIMOS and the Bremen 3D chemistry and transport model [*Sinnhuber et al.*, 2003a, 2003b; *Winkler et al.*, 2008] to the October/November 2003 event. This study shows that the consideration of electrons in addition to protons leads to a significant increase in atmospheric ionization in the mesosphere and less so in the stratosphere. This is reflected in changes in the chemical composition, such as NOx production and ozone depletion. Some of the results are discussed in Chap. 16.

One may argue that the horizontal ionization pattern in the upper atmosphere will be quickly blurred by transport processes. According to *Bhattacharya and Gerrard* [2009] the mean wind speed at 85 km altitude is about 130 km/h; the maximum daily mean during the observation time was less than 290 km/h. As particle precipitation takes place in circles around the geomagnetic pole along the geomagnetic latitudes, zonal (East-West) transport will not impact the atmospheric ionization (electron density) significantly. Meridional (North-South) transport however may have a significant impact. The grid resolution in meridional direction is 400 km. In combination with the 2 h AIMOS model time resolution the "smearing out" effect is limited to one meridional cell at maximum and even less in general.

13.3 Comparison to EISCAT Results

While in the previous DFG-project (ARTOS-AIMOS, described above) the consequences of particle precipitation were modeled with respect to atmospheric chemistry and also can be compared to observations as demonstrated e.g. in Chap. 15, in this DFG-project (MAIONO) we will compare particle precipitation to electron densities as inferred from incoherent scatter radar. Two approaches are chosen, a direct comparison and a comparison using the combined AIMOS/HAMMONIA model chain.

The European Incoherent Scatter Radar (EISCAT) near Tromsø works with frequencies far above the plasma frequency. Therefore the emitted signal is not totally reflected but scattered by single electrons in the ionospheric plasma. The backscattered signal spectrum is analyzed by using plasma theory, which allows to determine plasma parameters *Farley* [1996] e.g. electron density.

For the October/November 2003 event data coverage at EISCAT is excellent. In the comparison between precipitating solar energetic particles and electron densities as measured by EISCAT we used one energy channel of POES (or GOES) at a time, determined the corresponding stopping height and compared to EISCAT measurements at this height. We tried to take into account in the EISCAT data the diurnal effects, solar flare effects due to increased hard electromagnetic radiation, and shocks/geomagnetic disturbances. Detailed analysis shows that electron densities observed by EISCAT show a pronounced diurnal variation below 75 km, are not

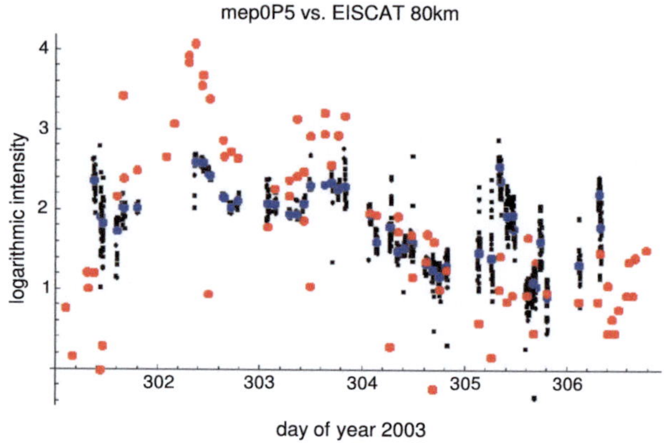

Fig. 13.5 Direct comparison between EISCAT (*red*) and POES (*blue*) data for protons with energies 2.6–6.9 MeV limited to times when POES is magnetically connected to the vicinity of the EISCAT station. Curves are shifted in log intensity to allow for a better comparison. The *black dots* give all POES measurements in a crossing while the *blue dot* indicates a mean

correlated to the fluxes of precipitating solar protons, do not react directly to shocks but show some correlation with substorms, at least at heights above about 90 km. This implies that our originally planned research design had been too optimistic. Nonetheless, the relation to substorms gives a clue on how to proceed.

To gain a better understanding on the relation between precipitating particles and electron densities we dropped the source related approach (limitation to solar energetic particles) and choose a more local one: EISCAT electron densities were compared to the particle fluxes obtained when POES was magnetically connected to the vicinity of the radar station. This means the foot-of-fieldline coordinates should be within a rectangle defined by $\pm 2.5°$ in latitude and $\pm 10°$ in longitude around the radar station. Owing to the fast motion of POES, a smaller rectangle would not lead to any data coverage. This approach has the advantage that precipitating particles are considered independent of their origin. Again, the comparison is done by comparing fluxes in energy ranges to electron densities at the corresponding stopping heights.

Figure 13.5 shows a time series for this comparison. Although there is no perfect correlation between the electron densities and the proton fluxes, similarities in the temporal pattern are apparent in this example of energy-height match. Such correlations strongly depend on energy/height. In addition, Fig. 13.5 suggests that a correlation, if it exists, certainly would not be a simple linear one. To get a better impression of the correlation we decided to perform a rank correlation between all energy channels and height-averaged electron densities in 10 km height bins. We found that correlation coefficients increase with increasing particle energy and decreasing height and tend to be higher during night. However, any correlation at 60 km during nighttime disappears.

In sum, apparently it is more sensible to correlate EISCAT measurements with geographically adjusted fluxes of precipitating energetic particles than with precipitating solar energetic particles.

The horizontal resolution of AIMOS provides the necessary accuracy to allow a comparison with local measurements such as EISCAT. A direct comparison nevertheless is inaccurate as already indicated in Fig. 13.5: AIMOS determines ionization rates while EISCAT measures electron densities. Thus a chemistry model is

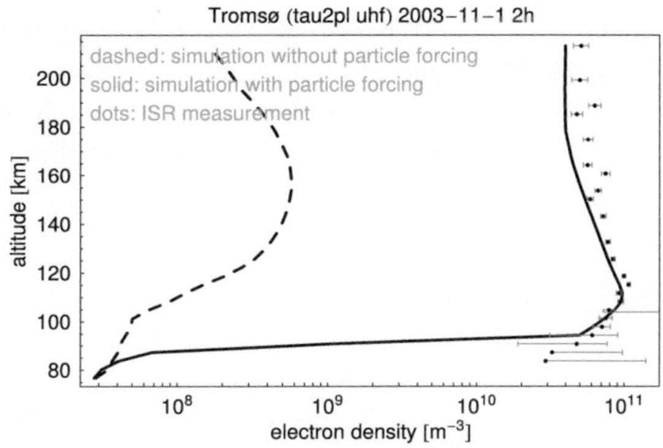

Fig. 13.6 Comparison of a simulation of electron densities with and without particle forcing compared to incoherent scatter radar measurements [*Wissing et al.*, 2011]. Below 90 km the GCM uses parameterization instead of full ion chemistry, therefore this is the lower boundary of the comparison. Reprinted with permission from American Geophysical Union

required to consider recombination and other losses. To model electron densities for the comparison to EISCAT observations we used the combination of AIMOS and the HAMMONIA GCM.

The HAMburg MOdel of the Neutral and Ionized Atmosphere (HAMMONIA) [*Schmidt et al.*, 2006] is a superset of the middle atmosphere version of the ECHAM5 general circulation model (GCM) [e.g., *Manzini et al.*, 2006] interactively coupled to the MOZART3 chemistry scheme [*Kinnison*, 2007]. In the version used here, HAMMONIA includes e.g. photochemical reactions as well as reactions associated due to particle precipitation. The reaction rates are given by the ionization rates from AIMOS combined with branching ratios from *Roble and Ridley* [1987] and *Verronen* [2006]. The new chemistry module is limited to altitudes above 90 km.

Figure 13.6 shows a comparison of the simulated electron densities to the ones measured by EISCAT with and without particle forcing. As day-time electron densities are affected by photoionization—which is comparable in size—we focus on nighttime here. From the figure it is evident that at least during night particle forcing must be considered to simulate reliable electron densities with the HAMMONIA GCM as it gives a benefit in the order of a factor of 100 to 1000. A more detailed study by *Wissing et al.* [2011] shows that this AIMOS/HAMMONIA model chain is able to produce useful electron density fields in terms of absolute values as well as their dependence on geomagnetic latitude, geomagnetic activity and local time. Since interpretation of incoherent scatter radar data relies on assumptions this comparison cannot be used as a validation for AIMOS, however, the consistency in the results at least lends confidence to AIMOS and in particular to the sorting algorithm because in this comparison the spatial resolution of AIMOS has been tested.

13.4 Summary

AIMOS is a combined empirical/Monte-Carlo model to simulate the 3D ionization patterns of the atmosphere for solar and magnetospheric particles over a wide range of energies. Parametric studies show that the inclusion of all relevant particles yields not only different ionization rates but also affects the modeling of atmospheric chemistry. A combination of AIMOS and the HAMMONIA GCM allows to determine electron densities which are in reasonable agreement with the observations from incoherent scatter radar; although this does not validate AIMOS it increases our confidence in the AIMOS sorting algorithm.

References

Agostinelli, S., et al. (2003). Geant4: a simulation toolkit. *Nuclear Instruments & Methods in Physics Research. Section A, Accelerators, Spectrometers, Detectors and Associated Equipment*, *506*, 250–303.

Berger, M. J., Seltzer, S. M., & Maeda, K. (1970). Energy deposition by auroral electrons in the atmosphere. *Journal of Atmospheric and Solar-Terrestrial Physics*, *32*, 1015–1045.

Bhattacharya, Y., & Gerrard, A. J. (2009). Mesospheric winds and polar vortex motion. *Atmospheric Chemistry and Physics Discussion*, *9*, 16549–16562.

Bornebusch, J. P., Wissing, J. M., & Kallenrode, M.-B. (2010). Solar particle precipitation into the polar atmosphere and their dependence on hemisphere and local time. *Advances in Space Research*, *45*, 632–637. doi:10.1016/j.asr.2009.11.008.

Callis, L. B., et al. (1996a). Precipitating electrons: evidence for effects on mesospheric odd nitrogen. *Geophysical Research Letters*, *23*, 1901–1904.

Callis, L. B., Baker, D. N., Natarajan, M., Blake, J. B., Mewaldt, R. A., Selesnick, R. S., & Cummings, J. R. (1996b). A 2-d model simulation of downward transport of noy into the stratosphere: effects on the austral spring o3 and noy. *Geophysical Research Letters*, *23*, 1905–1908.

Callis, L. B., Natarajan, M., Evans, D. S., & Lambeth, J. D. (1998). Solar atmospheric coupling by electrons (solace): 1. Effects of the may 12, 1997 solar event on the middle atmosphere. *Journal of Geophysical Research*, *103*, 28405–28419.

Crutzen, P. J., Isaksen, I. S. A., & Reid, G. C. (1975). Solar proton events: stratospheric sources of nitric oxide. *Science*, *189*, 457–459.

Evans, D. S., & Greer, M. S. (2004). *Polar orbiting environmental satellite, space environment monitor—2: Instrument descriptions and archive data documentation* (Technical memorandum). Space Environ. Cent., NOAA. http://www.ngdc.noaa.gov/stp/NOAA/docs/SEM2v1.4b.pdf.

Fang, X., Randall, C. E., Lummerzheim, D., Solomon, S. C., Mills, M. J., Marsh, D. R., Jackman, C. H., Wang, W., & Lu, G. (2008). Electron impact ionization: a new parameterization for 100 ev to 1 mev electrons. *Journal of Geophysical Research*, *113*, A09311. doi:10.1029/2008JA013384.

Farley, D. (1996). Incoherent scatter radar probing. In H. Kohl, R. Ruster & K. Schlegel (Eds.), *Modern ionospheric science* (pp. 415–439). Katlenburg-Lindau: European Geophysical Society Publications.

Heath, D. F., Krueger, A. J., & Crutzen, P. J. (1977). Solar proton event: influence in stratospheric ozone. *Science*, *197*, 886–889.

Jackman, C. H., McPeters, R. D., Labow, G. J., Fleming, E. L., Praderas, C. J., & Russell, J. M. (2001). Northern hemisphere atmospheric effects due to the July 2000 solar proton event. *Geophysical Research Letters*, *28*, 1886–2883.

Jackman, C. H., DeLand, M. T., Labow, G. J., Flemming, E. L., Weisenstein, D. K., Ko, M. K. W., Sinnhuber, M., & Russell, J. M. (2005). Neutral atmospheric influences of the solar events in October–November 2003. *Journal of Geophysical Research*, *110*, A09S27. doi:10.1029/2004JA010888.

Kinnison, D. E., Brasseur, G. P., Walters, S., Garcia, R. R., Marsh, D. R., Sassi, F., Harvey, V. L., Randall, C. E., Emmons, L., Lamarque, J. F., Hess, P., Orlando, J. J., Tie, X. X., Randel, W., Pan, L. L., Gettelman, A., Granier, C., Diehl, T., Niemeier, U., & Simmons, A. J. (2007). Sensitivity of chemical tracers to meteorological parameters in the mozart-3 chemical transport model. *Journal of Geophysical Research, 112*. doi:10.1029/2006JD007879.

Leske, R. A., Mewaldt, R. A., Stone, E. C., & von Rosenvinge, T. T. (2001). Observations of geomagnetic cutoff variations during solar energetic particle events and implications for the radiation environment at the space station. *Journal of Geophysical Research, 106*, A08305.

Manzini, E., Giorgetta, M. A., Esch, M., Kornblueh, L., & Roeckner, E. (2006). The influence of sea surface temperatures on the northern winter stratosphere: ensemble simulations with the maecham5 model. *Journal of Climate, 19*(16), 3863–3881.

Marsh, D. R., Garcia, R. R., Kinnison, D. E., Boville, B. A., Sassi, F., Solomon, S. C., & Matthes, K. (2007). Modeling the whole atmosphere response to solar cycle changes in radiative and geomagnetic forcing. *Journal of Geophysical Research, 112*, D23306. doi:10.1029/2006JD008306.

McPeters, R. D., & Jackman, C. H. (1985). The response of ozone to solar proton events during solar cycle 21: the observations. *Journal of Geophysical Research, 90*, 7945–7954.

Mewaldt, R. A., et al. (2005). Proton, helium, and electron spectra during the large solar particle events of October–November 2003. *Journal of Geophysical Research, 110*, A09S18. doi:10.1029/2005JA011038.

NASA (1996). *GOES I-M DataBook*. National Aeronautics and Space Administration, Goddard Space Flight Center, Greenbelt, Maryland 20771, USA, revision 1st ed. http://rsd.gsfc.nasa.gov/goes/text/goes.databook.html.

Randall, C. E., Harvey, V. L., Singleton, C. S., Bailey, S. M., Bernath, P. F., Codrescu, M., Nakajima, H., & Russell III, J. M. (2007). Energetic particle precipitation effects on the southern hemisphere stratosphere in 1992–2005. *Journal of Geophysical Research, 112*, D08308. doi:10.1029/2006JD007696.

Roble, R. G., & Ridley, E. C. (1987). An auroral model for the ncar thermospheric general circulation model (tgcm). *Annales Geophysicae, 5A*, 369–382.

Rohen, G., et al. (2005). Ozone depletion during the solar proton events of October/November 2003 as seen by sciamachy. *Journal of Geophysical Research, 110*, A09S39. doi:10.1029/2004JA010984.

Schmidt, H., Brasseur, G. P., Charron, M., Manzini, E., Giorgetta, M. A., & Diehl, T. (2006). The hammonia chemistry climate model: sensitivity of the mesopause region to the 11-year solar cycle and co2 doubling. *Journal of Climate, 19*, 3902–3931.

Schröter, J., Heber, B., Steinhilber, F., & Kallenrode, M.-B. (2006). Energetic particles in the atmosphere: a Monte Carlo approach. *Advances in Space Research, 37*(8), 1597–1601.

Seppälä, A., Clilverd, M. A., & Rodger, C. J. (2007). Nox enhancements in the middle atmosphere during 2003–2004 polar winter: relative significance of solar proton events and the aurora as a source. *Journal of Geophysical Research, 207*. doi:10.1029/2006JD008326.

Sinnhuber, B.-M., Weber, M., Amankwah, A., & Burrows, J. P. (2003a). Total ozone during the unusual antarctic winter of 2002. *Geophysical Research Letters, 30*. doi:10.1029/2002GL016798.

Sinnhuber, M., Burrows, J. P., Chipperfield, M. P., Jackman, C. H., Kallenrode, M.-B., Kunzi, K. F., & Quack, M. (2003b). A model study of the impact of magnetic field structure on atmospheric composition during solar proton events. *Geophysical Research Letters, 30*. doi:10.1029/2003GL017265.

Verronen, P. T. (2006). *Ionosphere-atmosphere interaction during solar proton events*. Doctoral dissertation, Ph.D. thesis, University of Helsinki.

Verronen, P. T., Turunen, E., Ulich, T., & Kyrölä, E. (2002). Modelling the effects of the October 1989 solar proton event on mesospheric odd nitrogen using a detailed ion and neutral chemistry model. *Annales Geophysicae, 20*, 1967–1976.

Winkler, H., Sinnhuber, M., Notholt, J., Kallenrode, M.-B., Steinhilber, F., Vogt, J., Zieger, B., Glassmeier, K.-H., & Stadelmann, A. (2008). Modeling impacts of geomagnetic field variations on middle atmospheric ozone responses to solar proton events on long timescales. *Journal of Geophysical Research, 113*, D02302. doi:10.1029/2007JD008574.

Wissing, J. M., & Kallenrode, M.-B. (2009). Atmospheric ionization module osnabrück (aimos): a 3-d model to determine atmospheric ionization by energetic charged particles from different populations. *Journal of Geophysical Research, 114*, A06104. doi:10.1029/2008JA013884.

Wissing, J. M., Bornebusch, J. P., & Kallenrode, M.-B. (2008). Variation of energetic particle precipitation with local magnetic time. *Advances in Space Research, 41*, 1274–1278. doi:10.1016/j.asr.2007.05.063.
Wissing, J. M., Kallenrode, M.-B., Wieters, N., Winkler, H., & Sinnhuber, M. (2010). Atmospheric ionization module osnabrück (aimos): 2. Total particle inventory in the October–November 2003 event and ozone. *Journal of Geophysical Research, 115*, A02308. doi:10.1029/1009JA014419.
Wissing, J. M., Kallenrode, M.-B., Kieser, J., Schmidt, H., Rietveld, M. T., Strømme, A., & Erickson, P. J. (2011). Atmospheric ionization module osnabrück (aimos): 3. Comparison of electron density simulations by aimos/hammonia and incoherent scatter radar measurements. *Journal of Geophysical Research, 116*. doi:10.1029/2010JA016300.

Chapter 14
EISCAT's Contributions to High Latitude Ionosphere and Atmosphere Science Within CAWSES in Germany

Jürgen Röttger and Norbert Engler

Abstract This article presents a very short overview on EISCAT and the German involvement during the past decade, including overview statistics of operating hours, publications etc. In particular we summarize highlights of scientific research using the EISCAT facilities, which have been performed by German research groups as part of the CAWSES Priority Program of the Deutsche Forschungsgemeinschaft between 2007 and 2011. The purpose of this article is to only extract some highlights, as seen by the authors, since exhaustive descriptions of the complete research results are published elsewhere or in this book. The highlights cover observation of thermosphere density enhancements measured by satellite in the polar cusp region related to ionosphere heating, deduction of electron density profiles by EISCAT in comparison with the AIMOS/HAMMONIA model, Polar Mesosphere Summer Echoes (PMSE), their reflectivity and the consequent deduction of microphysical particle parameters, the radar (PMSE) observation of a Kelvin-Helmholtz-Instability as well as discrimination of turbulence effects on PMSE and the detection of the scatter particle line delineating the presence of meteoric smoke in the mesosphere. Beside PMSE also Polar Mesosphere Winter Echoes were detected and interpreted. Two new projects supported by EISCAT measurements have been launched, the effects of the Heating modified ionosphere on Global Navigation Satellite Systems and coupling processes by winds and waves. Finally we briefly introduce the next generation of high latitude radars, namely a view on the planned EISCAT_3D project and its scientific yield expected by potential German users.

J. Röttger (✉)
Max-Planck-Institute of Solar System Research, Max-Planck-Straße 2, 37191 Katlenburg-Lindau, Germany
e-mail: roettger@mps.mpg.de

N. Engler
Leibniz-Institute of Atmospheric Physics, Schlossstr. 6, 18225 Kühlungsborn, Germany
e-mail: engler@iap-kborn.de

F.-J. Lübken (ed.), *Climate and Weather of the Sun-Earth System (CAWSES)*, Springer Atmospheric Sciences,
DOI 10.1007/978-94-007-4348-9_14,

Fig. 14.1 The 42 m field-aligned dish (*left*) of the EISCAT Svalbard Radar near Longyearbyen and the steerable 32 m dish (*right*). The facility started operation in 1996 and several research projects have been using the results of this radar facility during the period of the CAWSES Priority Program. (Picture: EISCAT Association)

14.1 A Historical Synopsis and Introduction

In the mid 1970s the academies and research councils of six European countries Finland, France, Germany, Norway, Sweden and the United Kingdom joined forces in order to exploit the incoherent scatter radar technique to explore the aurora and the high latitude ionosphere. An organization, the EISCAT (European Incoherent Scatter) Scientific Association, was established with state of the art radar installations in northern Scandinavia, and later on Spitsbergen, Svalbard. In 1996 Japan and at the end of 2006 the P. R. China joined the association.

The incoherent scatter radar (ISR) technique is the most powerful ground-based tool for active remote sensing of the Earth's upper atmosphere and ionosphere as well as coupling processes between near Earth space and the solar wind. It can probe plasma phenomena, which are particularly pronounced at high latitudes of the Earth's atmosphere, where the northern lights, also known as aurora, impressively display the impact of solar processes on the Earth's atmosphere. Applying this technique demands interdisciplinary synergy between plasma physics, physics and dynamics of the atmosphere, informatics and electrical engineering, as well as between experimenters, engineers theoreticians, and modelers.

EISCAT operates three incoherent scatter radars, one on 931 MHz (UHF radar in Tromsø, Kiruna and Sodankylä), one on 500 MHz (UHF EISCAT Svalbard Radar ESR, see Fig. 14.1), one on 224 MHz (VHF radar in Tromsø) and an ionosphere modification "Heating" facility. An overview about the most important technical parameters can be found on the EISCAT website[1] and some parameters are also outlined in Chap. 19 (see Table 19.1). The main operation modes are Common Programs (CP) and Special Programs (SP) covering most of the user requests. Common Programs include coordinated observations of the ionosphere and magnetosphere such as World days of simultaneous measurements of ISR over the northern hemisphere and long term observations during the International Polar Year

[1] http://www.eiscat.se/groups/Documentation/BasicInfo/about/specifications.

(IPY) or the International Heliospheric Year (IHY). Special programs are designed for user defined requests to investigate the dynamics and structure of the ionosphere/magnetosphere and plasma parameters conducted e.g. in coordinated campaigns of several instruments, rocket launches, or satellite overpasses.

On March 14–15, 2002 a workshop "EISCAT in der Zukunft" (EISCAT in Future) was held at the Max-Planck-Institute for Aeronomy in Katlenburg-Lindau, Germany, to discuss EISCAT research and to establish a view of how German institutes and universities could use EISCAT in the future. Twenty-eight scientists from ten institutions participated in this workshop. In total nineteen papers were presented covering the wide range from magnetospheric convection to energetic particle precipitation to middle and lower atmosphere research. The intense discussions yielded valuable material for a follow-up document "New Solar-Terrestrial Physics Research with High-Latitude Facilities—A proposal for future support and evolution of EISCAT-related science in Germany" [*Hagfors et al.*, 2002].

In order to streamline this research initiative a working meeting was held on April 6, 2004 at the GeoForschungszentrum in Potsdam, attended by 18 German scientists, which resulted in the so-called "Potsdamer Resolution" (archived documents). The purpose was to safeguard the German participation in EISCAT after the end of 2006 when the German associate in EISCAT, the Max-Planck-Gesellschaft, was to terminate its membership in the EISCAT association. It was adopted that a suitable new form of organization in Germany would be under the Copernicus Gesellschaft, which supports scientific associations, societies and organizations in the fulfillment of their tasks and in their relation amongst each other.[2]

In 2004 the Scientific Committee on Solar-Terrestrial Physics (SCOSTEP) started the international program "Climate and Weather of the Sun-Earth System" (CAWSES), which resulted in the national priority program CAWSES funded by the German Research Foundation—Deutsche Forschungsgemeinschaft (DFG). Research with EISCAT was regarded as an important part in the CAWSES project in general.

It was soon noticed that a significant number of proposed research projects related to CAWSES would gain from the use of EISCAT, such as the effects of increased ionization by energetic particles on the middle atmosphere, the investigation of the influence of charged aerosol particles on the scattering of radar waves using EISCAT, the solar variability and trend effects in layers and trace gases in the upper atmosphere, the climatology of planetary waves seen in the ionospheric F-region perturbations using total electron content (TEC) measurements obtained from global navigation satellite systems (GNSS), observations and modeling of the upper atmospheric density and winds and their dependence on the geomagnetic activity, and investigations of quasi-periodic fluctuations in auroral phenomena using the EISCAT radar and satellite data.

In 2006 the German Research Foundation (DFG) decided to include EISCAT research in the CAWSES Priority Program and consequently followed the Max-Planck-Gesellschaft as full member in the EISCAT Scientific Association

---

[2]http://www.eiscat.de/images/stories/activities_and_meetings/service_abteilung_eiscat.pdf.

on January 1, 2007. An annual fee of 180 kEUR was negotiated to be paid to EISCAT allowing continuing the operation with some 100 hours of special program time per year and common program data access by German scientists. The DFG also delegated two members into the Council (J. Kowol-Santen, H. Boos, K. Zach, J. Röttger), which is the policy-making body of EISCAT, and one member into the EISCAT Scientific Oversight Committee—SOC (J. Röttger, M. Rapp). A support office was established at the Copernicus Gesellschaft in Katlenburg-Lindau under the name EISCAT-CAWSES-Copernicus (ECCo) to inform about the ongoing and planned activities, to coordinate, and to support scientific activities using the EISCAT facilities. ECCo supported personal contacts, the homepage http://www.eiscat.de as well as several informative flyers, presentations, and the distribution of status reports. After 2009 the office support was partially located at the Leibniz-Institute of Atmospheric Physics in Kühlungsborn.

14.2 German EISCAT Activities During the CAWSES Priority Program

In the years 2007–2010 ECCo published three status reports,[3] participated in several workshops and conferences presenting oral papers and posters to inform about the EISCAT operation and science. At least five German research groups with more than 15 scientists were directly involved in using EISCAT. The results have been published in peer-review journals (24 till the middle of 2011), as conference papers (62 till the middle of 2011) and at least six PhD and Master theses included results from EISCAT measurements. During the years 2007 to 2011 three doctoral theses had already been defended [*Strelnikova*, 2009; *Rentz*, 2009; *Kieser*, 2011] and two more are submitted [*Li*, 2011; *Wissing*, 2011] which used EISCAT data. The statistics of the publications is displayed in the left panel of Fig. 14.2. Complete lists of these publications are found in the ECCo status reports.[4] These publications resulted from a total of about 500 hours EISCAT special program operations during the CAWSES period (see right panel of Fig. 14.2). The allocated user time for German special programs was, thus, successfully used during the period of the CAWSES priority program. Most of the corresponding observations were performed during campaigns and as part of international projects under CAWSES, such as ECOMA [e.g. *Rapp and Strelnikova*, 2009], CHAMP-EISCAT [*Rentz*, 2009, *Rother et al.*, 2010], CharPa,[5] ISSI[6] ESR-PMSE [*Li and Rapp*, 2011; *Li*, 2011], TequilaSunrise (ref.), and others.

[3] http://www.eiscat.de/newsletter.html.

[4] http://www.eiscat.de/newsletter.html.

[5] Charge state of mesospheric smoke Particles—support to the REXUS-5 student rocket launch: Strelnikova et al., 2009.

[6] ISSI Bern: http://www.issibern.ch/ and the corresponding publications cited therein.

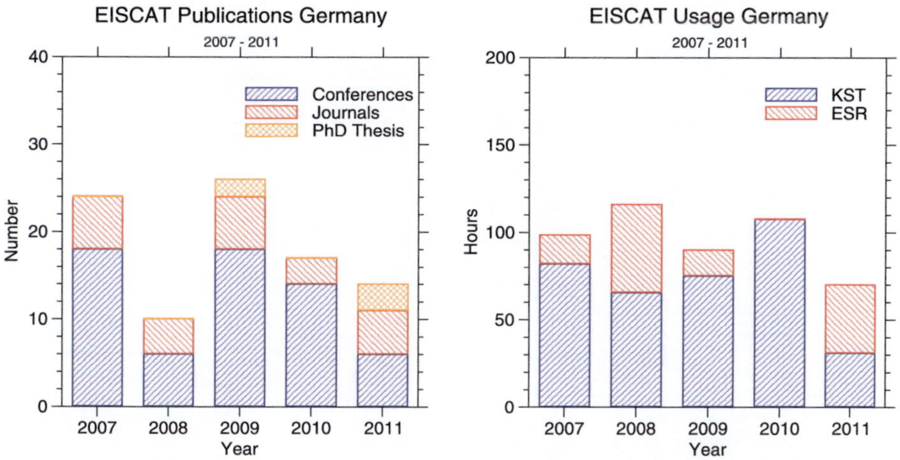

Fig. 14.2 *Left*: Number of publications during funding period of EISCAT in CAWSES. *Right*: Time usage of German EISCAT users per year separated for the EISCAT Svalbard radar as well as the VHF and UHF radars near Tromsø. The budget of about 100 h have been successfully exploited for ionospheric research

14.3 Summary of Selected Projects and Observations

The involvement of EISCAT in the scientific research during the CAWSES Priority Program funded by the DFG is now presented as very brief summary on some particular topics which appeared as highlights. For scientific and technical support ECCo coordinated and mostly accompanied the research groups in running the experiments and the data analysis. Figure 14.3 shows schematically the network of individual EISCAT user groups formed under the CAWSES Priority Program with the support and coordination from ECCo. Detailed descriptions of these are found in the corresponding publications, which are listed in the ECCo status reports[7] and presented in this book.

CHAMP-EISCAT Campaign on Joule Heating During October 1–13, 2006 a dedicated CHAMP-EISCAT campaign was scheduled. The results of simultaneous measurements of thermospheric characteristics with the CHAMP satellite and ionospheric parameters with EISCAT in the polar cusp region are discussed for Oct. 13, 2006 in Chap. 11.4. During this day three orbits were investigated where CHAMP passed close to the EISCAT facilities on Svalbard and near Tromsø. Electron and ion temperatures were derived from the EISCAT measurements (Fig. 11.6) as well as the vertical profiles of the Pedersen conductivity has been determined which were compared to model results (Fig. 11.8). We refer to Chap. 11 which explains the physical details and summarizes the results.

[7] http://www.eiscat.de.

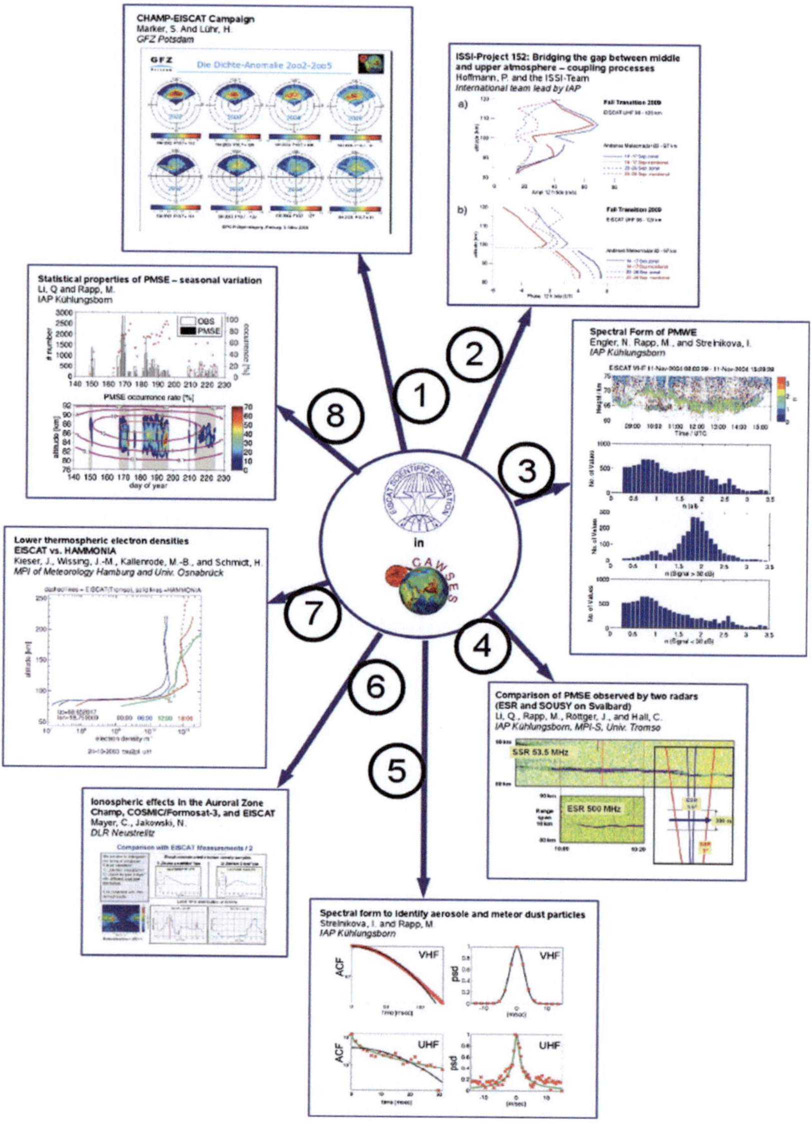

Fig. 14.3 The EISCAT-CAWSES-Copernicus Consortium supported numerous research activities. The network formed within the CAWSES Priority Program is also interacting with each other. The numbers refer to the related projects and are discussed in the relevant chapters in this book. (**1**) CHAMP-EISCAT Campaign (Marker, Lühr: Chap. 11), (**2**) ISSI-Project lead by P. Hoffmann, (**3**) Spectral form of PMWE (Engler, Rapp, Strelnikova: Chap. 19), (**4**) PMSE observed with two frequencies (Li, Rapp, Röttger, et al.: Chap. 19), (**5**) Spectra to identify dust particles (Strelnikova, Rapp: Chap. 19), (**6**) Ionospheric effect in the ionosphere with EISCAT and satellites (Mayer, Jakowski), (**7**) Electron densities with EISCAT and HAMMONIA (Kieser, Schmidt, Wissing, Kallenrode: Chaps. 13 and 31), (**8**) Seasonal variations of PMSE (Li, Rapp: Chap. 19)

There were also supported projects where auroral ionospheric effects have been studied using the CHAMP and COSMIC satellites together with EISCAT (Mayer and Jakowski, 2008 personal communication) to gain more information about the influence of ionospheric disturbances on radio wave propagation and communication as well as GPS signal accuracy. The effect of ionospheric heating on Global Navigation Satellite System (GNSS) have recently been investigated by Jakowski et al., 2010 (personal comm.) using the electron density measurements from EISCAT and the EISCAT heating facility [*Rietveld et al.*, 2003] located close to the radar in Tromsø. These experiments under well-defined and predictable conditions provide conclusions on the accuracy and reliability of GPS signals propagating through a disturbed environment.

AIMOS/Hammonia Comparison to EISCAT Electron Density Measurements Even though, the modeling of the ionosphere-thermosphere at global- and small-scale had produced a variety of results during the past decades the inclusion of recent research results into model experiments is improving the understanding of interactions between the earth-sun system as well as the coupling processes between middle atmospheric and thermospheric layers. The HAMMONIA-HAMburg Model Of the Neutral and Ionized Atmosphere (Chap. 31 and *Schmidt et al.* [2006]) is capable to cover altitudes from the surface to the thermosphere and it contains its coupling with the ionized atmosphere. Recently it had been further upgraded by including the effects of energetic particle precipitation by combining HAMMONIA with the AIMOS code—Atmospheric Ionization Module Osnabrück (Chap. 13 and *Wissing et al.* [2011]). Direct comparison of EISCAT electron densities to particle fluxes observed with POES are shown in Fig. 13.5 for the time period Oct./Nov. 2003. Simulated electron density profiles using the combined AIMOS/HAMMONIA GCM are compared to EISCAT measurements in Fig. 13.6 explaining that particle forcing has to be considered to simulate reliable electron densities. Chapter 13 summarizes the modeling activities and the comparison to EISCAT radar measurements.

Mesospheric Aerosols Studied with the EISCAT Radars The nature of Polar Mesosphere Summer Echoes (PMSE) has been studied further by analyzing the spectral characteristics of the signals obtained from measurements with the EISCAT UHF and VHF radars [*Strelnikova and Rapp*, 2010, 2011]. The shape of the autocorrelation functions as a function of the scattering mechanism has been investigated which is discussed in detail in Chap. 19. Characteristic spectral parameters were determined which prove the existence of charged aerosols and meteor smoke particles at mesospheric altitudes [*Rapp et al.*, 2007; *Strelnikova et al.*, 2007; *Strelnikova and Rapp*, 2010]. EISCAT radar observations have proven the existence of charged mesospheric meteor smoke particles which has been studied in joined rocket and radar campaigns during ECOMA [*Rapp et al.*, 2007; *Strelnikova et al.*, 2009]. The analysis of the data is still ongoing and more promising results are expected in the next years.

The microphysical properties of Polar Mesosphere Summer Echoes (PMSE) were studied using multiple frequency radar soundings by *Li et al.* [2010]. The calibrated backscattered signals from the EISCAT Svalbard Radar (ESR $-\lambda_B = 30$ cm) and the SOUSY Svalbard Radar (SSR – $\lambda_B = 2.8$ m) were used to determine the Schmidt number which is related to the particle sizes inside the PMSE layers. This proved to be in excellent agreement with independent observations relying on well established optical techniques [*Li et al.*, 2010]. Statistical properties of PMSE and the physical mechanism generating strong mesospheric echoes during the polar summer, namely enhanced radar reflectivity in the presence of charged aerosol and ice particles, have been obtained by using the EISCAT UHF and the VHF radar observations by *Rapp et al.* [2008] and *Li and Rapp* [2011]. Also characteristics of Kelvin-Helmholtz instabilities could be detected in PMSE [*Röttger et al.*, 2007].

Observations of Polar Mesospheric Winter Echoes (PMWE) have been analyzed by *Lübken et al.* [2007] and *Engler and Rapp* [2011] where the theory, the spectral properties and the physical understanding of these radar echoes were discussed. A detailed discussion about the physical background, the analysis of spectra, and the extraction of physical relevant parameters and their interpretation is provided in Chap. 19 in this book. A more comprehensive study about the properties of PMWE is prepared by *Engler and Rapp* [2011]. However, most of the observations have been conducted before 2007 because of the minimum of solar fluxes of electrons, protons, and X-rays. In 2011 more observations had been scheduled where the results are expected after the publication of this book.

Chapter 19 provides the description of the results obtained from these analyses of spectral characteristics during the occurrence of mesospheric radar echoes. The physical interpretation includes the existence of charged aerosol particles interacting with the plasma. The radar measurements with EISCAT have essentially improved the understanding of the processes leading to the occurrence of mesospheric radar echoes.

More Topics Using EISCAT Within the CAWSES Priority Program The topics mentioned above are some highlights of the research activities conducted under CAWSES. There have been a few further projects using the EISCAT facilities where German researchers are involved to investigate the ionosphere from the D-region up to the F-region. Out of that one interesting international activity is briefly mentioned in the following.

The ISSI-Project 152 "Bridging the gap between the middle and upper atmosphere: Coupling processes due to winds and waves over an extended altitude range" has been established (Hoffmann, personal communication). The scientific questions concern the coupling processes over a large altitude range using EISCAT for furnishing neutral winds in the ionospheric E-region which is an important upward extension of existing continuous wind measurements by medium frequency (MF) and meteor radars (MR) in the mesosphere/lower thermosphere region. The propagation of atmospheric waves and the coupling of atmospheric layers by these waves is studied e.g. during Sudden Stratospheric Warming (SSW) events [e.g. *Hoffmann et al.*, 2007]. A group of about 10 international scientists have been meeting regularly to discuss new observations by radars, lidars, microwave radiometry, and also the inclusion of models like HAMMONIA (see Chap. 31). Some of the results on gravity wave and tidal coupling as well as on SSW are presented in Chap. 22.

The community of German researchers has gained much scientific success from the coordination of EISCAT measurements within the CAWSES Priority Program funded by the DFG. The following section will discuss the future opportunities of EISCAT_3D and should encourage the German user community to continue the research of the ionosphere using the radar facility provided by the EISCAT Scientific Association.[8]

14.4 The Next Generation of High Latitude Radars: A View on EISCAT_3D

In order to answer the newly arising questions resulting from the research summarized above, it appears worth to continue these research areas in Germany and to ensure the availability of EISCAT and other facilities with the concomitant data access and ability to perform special experiments. One major step was done already by the Leibniz-Institute of Atmospheric Physics by recently building and operating the Middle Atmosphere Alomar Radar System (MAARSY, *Latteck et al.* [2010] and *Rapp et al.* [2011]) which is primarily designed for 3-dimensional imaging observations of the polar summer mesopause environment.

It is also expected that further new science can be achieved by the novel project EISCAT_3D (three-dimensional volumetric imaging of the ionosphere and atmosphere[9]) in a medium to long-term view. During the time of the CAWSES DFG period 2007–2011 we had noticed three major directions in Germany, where the new concept of EISCAT for three-dimensional imaging (E_3D) could be used for new science. The following constitutes a partial coverage only, and is mainly related to present German research areas:

(a) The research on Polar Mesosphere Summer Echoes will remain an important subject in viewing the very fine structure of plasma, neutral turbulence and heavy particles (cluster ions, aerosols and meteoric smoke particles) in the polar mesopause region. It is foreseen that some progress can be made along this direction of mesospheric studies with the present EISCAT systems, which are likely to be mostly and exhaustively exploited in the coming years. However, these are ultimately limited by the present EISCAT system parameters. Much higher temporal and particularly spatial resolutions as well as larger simultaneous spatial coverage of the observations are needed to achieve attractive and new science. This is intended to be solved with the construction of the EISCAT_3D system.

(b) Extend the reliability of observations of the lower D-(C-) region and improve the signal detectability in total (i.e. increase substantially the power-aperture product of radar systems or investigate the possibility to focus the radiation in

[8] After ECCo Status Bericht 2008.

[9] http://www.eiscat3d.se.

given volumes) in order to allow/improve the energetic particle precipitation effects into this still existing "ignorosphere" of radar observations and potentially cover the full mesosphere-stratosphere-troposphere (MST) range.

(c) It is suggested that, taking into account the early proposal of remote site digital beam forming (EISCAT Technical Note, 1990, access through EISCAT Headquarters) and applying the knowledge gained from the ongoing technological development of EISCAT_ 3D, a multi-static receiving system should be added to the EISCAT Svalbard Radar (ESR). This will greatly enhance the capability of medium-scale imaging (about some 10 km resolution) of the ionosphere in the cusp region. This will offer a new dimension of magnetosphere-ionosphere-neutral atmosphere coupling research.

We are aware that these views demand an extremely versatile and powerful radar system. The construction and operating costs will be tremendous. The implementation is expected to be possible only through international collaboration and the continuation of a competent science team.

14.5 Summary

The activities in the research using the EISCAT facility within the CAWSES Priority Program have contributed much to understanding of the physical processes at atmospheric/thermospheric altitudes and have attributed for a continuation of the scientific contributions in the future which can be obtained from observations using this research facility. When collecting material for this article it became most obvious to us that radar observations of the ionosphere and higher atmosphere still yield high-level, novel research results. It is also evident that the symbiosis and synergy of experiments with radars, ionospheric heating, rocket measurements and satellite observations, as well as the inclusion of model calculations, are very important and constitute an efficient break-through in understanding the polar mesosphere and thermosphere-ionosphere structure and dynamics.

Germany has been a member in EISCAT over 35 years from 1976 to 2011. The past five years, under the funding of the CAWSES DFG Priority Program, have been very fruitful. The future of German activities using EISCAT will be determined by the needs and ideas of the researchers who contributed or will contribute to the scientific achievements using high-power radar facilities at polar latitudes.

Acknowledgements The use of EISCAT was made possible under the Schwerpunkt Program CAWSES of the Deutsche Forschungsgemeinschaft (DFG-SPP 1176). We thank Johanna Kowol-Santen, Heike Boos, and Karin Zach of the DFG for their constant care of the EISCAT activities. In particular we acknowledge the stimulating support by the DFG-SPP Chief Coordinator Franz-Josef Lübken. We also appreciate the support by the Copernicus Gesellschaft through Arne Richter and Martin Rasmussen, the Max-Planck-Institute for Solar System Research as well as the assistance of Renate Scherer, and in particular Mike T. Rietveld who was substantial in guiding the preparations, operations, analysis and interpretation of EISCAT experiments. The EISCAT Scientific Association is funded and operated by research councils of Norway, Sweden, Finland, Japan, China, the United Kingdom and Germany (the latter until 2011).

References

Engler, N., & Rapp, M. (2011). Characterization of polar lower mesosphere echoes observed with the EISCAT VHF radar. *Journal of Atmospheric and Solar-Terrestrial Physics*. To be submitted.

Hagfors, T., Rietveld, M. T., & Röttger, J. (2002). New solar-terrestrial physics research with high-latitude facilities—a proposal for future support an evolution of EISCAT-related science in Germany. Personal document.

Hoffmann, P., Singer, W., Keuer, D., Hocking, W., Kunze, M., & Murayama, Y. (2007). Latitudinal and longitudinal variability of mesospheric winds and temperatures during stratospheric warming events. *Journal of Atmospheric and Solar-Terrestrial Physics*, *69*(17–18), 2355–2366.

Kieser, J. (2011). *The influence of precipitating solar and magnetospheric energetic charged particles on the entire atmosphere: simulations with HAMMONIA*. Ph.D. thesis, University of Hamburg, Germany.

Latteck, R., Singer, W., Rapp, M., & Renkwitz, T. (2010). MAARSY—the new MST radar on Andøya/Norway. *Advances in Radio Science*, *8*, 219–224. doi:10.5194/ars-8-219-2010.

Li, Q. (2011). *Multi-frequency radar observations of polar mesosphere summer echoes: statistical properties and microphysical results*. Ph.D. thesis, University of Rostock, Germany.

Li, Q., & Rapp, M. (2011). PMSE-observations with the EISCAT VHF and UHF radars: statistical properties. *Journal of Atmospheric and Solar-Terrestrial Physics*, *73*(9), 944–956. doi:10.1016/j.jastp.2010.05.015.

Li, Q., Rapp, M., Röttger, J., Latteck, R., Zecha, M., Strelnikova, I., Hervig, M., Hall, C., & Tsutsumi, M. (2010). Microphysical parameters of mesospheric ice clouds derived from calibrated observations of polar mesosphere summer echoes at Bragg wavelengths of 2.8 m and 30 cm. *Journal of Geophysical Research*, *115*, D00I13. doi:10.1029/2009JD012271.

Lübken, F.-J., Singer, W., Latteck, R., & Strelnikova, I. (2007). Radar measurements of turbulence, electron densities, and absolute reflectivities during polar mesosphere winter echoes (PMWE). *Advances in Space Research*, *40*(6), 758–764. doi:10.1016/j.asr.2007.01.015.

Rapp, M., & Strelnikova, I. (2009). Measurements of meteor smoke particles during the ECOMA-2006 campaign: 1. Particle detection by active photoionization. *Journal of Atmospheric and Solar-Terrestrial Physics*, *71*(3–4), 477–485. doi:10.1016/j.jastp.2008.07.011.

Rapp, M., Strelnikova, I., & Gumbel, J. (2007). Meteoric smoke particles: evidence from rocket and radar technique. *Advances in Space Research*, *40*(6), 809–817. doi:10.1016/j.asr.2006.11.021.

Rapp, M., Strelnikova, I., Latteck, R., Hoffmann, P., Hoppe, U.-P., Häggström, I., & Rietveld, M. T. (2008). Polar mesosphere summer echoes (pmse) studied at Bragg wavelengths of 2.8 m, 67 cm, and 16 cm. *Journal of Atmospheric and Solar-Terrestrial Physics*, *70*(7), 947–961. doi:10.1016/j.jastp.2007.11.005.

Rapp, M., Latteck, R., Stober, G., Hoffmann, P., Singer, W., & Zecha, M. (2011). First 3-dimensional observations of polar mesosphere winter echoes: resolving space-time ambiguity. *Journal of Geophysical Research*. Submitted.

Rentz, S. (2009). *The upper atmospheric fountain effect in the polar cusp region*. Ph.D. thesis, Technische Universität Carolo-Wilhelmina zu Braunschweig.

Rietveld, M. T., Kosch, M. J., Blagoveshchenskaya, N. F., Kornienko, V. A., Leyser, T. B., & Yeoman, T. K. (2003). Ionospheric electron heating, optical emissions and striations induced by powerful HF radio waves at high latitudes: aspect angle dependence. *Journal of Geophysical Research*, *108*(A4), SIA 2-1–SIA 2-16. doi:10.1029/2002JA009543.

Rother, M., Schlegel, K., Ühr, H., & Cooke, D. (2010). Validation of CHAMP electron temperature measurements by incoherent scatter radar data. *Radio Science*, *45*(6), RS6020. doi:10.1029/2010RS004445.

Röttger, J., Rapp, M., Trautner, J., Serafimovich, A., & Hall, C. (2007). New PMSE observations with the EISCAT 500-MHz Svalbard Radar and the SOUSY 53.5-MHz Svalbard Radar. In *MacMillan adv. res. series. Proc. 11th international workshop on technical and scientific aspects of MST radar*, Gadanki, India (pp. 136–140).

Schmidt, H., Brasseur, G. P., Charron, M., Manzini, E., Giorgetta, M. A., Diehl, T., Formichev, V. I., Kinnison, D., Marsh, D., & Walters, S. (2006). The HAMMONIA chemistry climate model:

sensitivity of the mesopause region to the 11-year solar cycle and CO_2 doubling. *Journal of Climate*, *19*, 3903–3931.
Strelnikova, I. (2009). *Mesospheric aerosol particles: evidence from rocket and radar techniques*. Ph.D. thesis, University of Rostock, Germany.
Strelnikova, I., & Rapp, M. (2010). Studies of polar mesosphere summer echoes with the EISCAT VHF and UHF radars: information contained in the spectral shape. *Advances in Space Research*, *45*(2), 247–259. doi:10.1016/j.asr.2009.09.007.
Strelnikova, I., & Rapp, M. (2011, in press). Majority of PMSE spectral widths at UHF and VHF are compatible with a single scattering mechanism. *Journal of Atmospheric and Solar-Terrestrial Physics*. doi:10.1016/j.jastp.2010.11.025.
Strelnikova, I., Rapp, M., Raizada, S., & Sulzer, M. (2007). Meteor smoke particle properties derived from Arecibo incoherent scatter radar observations. *Geophysical Research Letters*, *34*(15), L15815. doi:10.1029/2007GL030635.
Strelnikova, I., Rapp, M., Strelnikov, B., Baumgarten, G., Brattli, A., Svenes, K., Hoppe, U.-P., Friedrich, M., Gumbel, J., & Williams, B. P. (2009). Measurements of meteor smoke particles during the ECOMA-2006 campaign: 2. Results. *Journal of Atmospheric and Solar-Terrestrial Physics*, *71*(3–4), 486–496. doi:10.1016/j.jastp.2008.07.011.
Wissing, J., Kallenrode, M.-B., Kieser, J., Schmidt, H., Rietveld, M., Strømme, A., & Erickson, P. (2011). Atmospheric ionization module OSnabrück (AIMOS) 3: comparison of electron density simulations by AIMOS/HAMMONIA and incoherent scatter radar measurements. *Journal of Geophysical Research*. doi:10.1029/2010JA016300.
Wissing, J. M. (2011). *Analysis of particle precipitation and development of the atmospheric ionization module OSnabrück—AIMOS*. Ph.D. thesis, University Osnabrück, Germany.

Chapter 15
The Influence of Energetic Particles on the Chemistry of the Middle Atmosphere

Thomas Reddmann, Bernd Funke, Paul Konopka, Gabriele Stiller, Stefan Versick, and Bärbel Vogel

Abstract Energetic particle precipitation (EPP) during solar and geomagnetic active periods causes chemical disturbances in the lower thermosphere and in the middle atmosphere. Additional HO_x (H, OH, HO_2) and NO_x (N, NO, NO_2) are produced in the middle atmosphere, and enhancements of NO_x produced in these events can be transported to the winter stratosphere. These trace species take part in ozone chemistry and, by chemical-radiative coupling, the dynamical state in the middle atmosphere can be altered. There is evidence both from observations and from chemistry-climate models that the EPP induced signal in the middle atmosphere may then propagate into the troposphere. Thus particle precipitation could connect to possible climate effects. The first step in this functional chain is the impact of EPP on the chemical composition in the middle atmosphere and lower thermosphere, and the downward transport in the polar winter middle atmosphere. The general objective of this project was to assess quantitatively the chemical composition change in the middle atmosphere by combining model simulations and observations. The study relays mainly on the observations of the MIPAS instrument on the ENVISAT satellite, whose data set has been expanded in the context of this project by a newly developed retrieval of the gas H_2O_2, a reservoir for the members of the HO_x family. Simulations have been carried out with the two chemical transport models CLaMS and KASIMA, which cover chemistry and transport effects in the stratosphere up to the mesosphere/lower thermosphere region. The impact on the global NOy budget

T. Reddmann (✉) · G. Stiller · S. Versick
Karlsruhe Institute of Technology (KIT), Inst. for Meteorology and Climate Research, Karlsruhe, Germany
e-mail: thomas.reddmann@kit.edu

P. Konopka · B. Vogel
Institute for Energy and Climate Research – Stratosphere (IEK-7) Forschungszentrum Jülich GmbH (FZJ), Jülich, Germany

B. Vogel
e-mail: b.vogel@fz-juelich.de

B. Funke
Instituto de Astrofisica de Andalucia (CSIC), Granada, Spain
e-mail: bernd@iaa.es

F.-J. Lübken (ed.), *Climate and Weather of the Sun-Earth System (CAWSES)*, Springer Atmospheric Sciences,
DOI 10.1007/978-94-007-4348-9_15,

and (the resulting) total ozone change are assessed in these studies. In addition, the ion reaction mechanism for the conversion of N_2O_5 to HNO_3 based on positive ion chemistry was refined. The detailed comparison of model results and observation for the SPE 2003 showed that models can simulate the impact of EPP on ozone chemistry but deficiencies exist for some minor species.

15.1 Introduction

The interaction of solar coronal mass ejections and the solar wind with the interplanetary plasma and processes in the Earth's magnetosphere accelerate particles as protons, electrons, or He nuclei to energies up to the GeV range. Some of these energetic particles penetrate the atmosphere mainly at high geomagnetic latitudes and lose their kinetic energy by cascades of inelastic collisions. Via ionization, dissociation and ion reactions this energy deposition ultimately produces reactive molecules like NO, NO_2, OH and others. This significant enhancement of NO_x ($= NO + NO_2 + N$) and HO_x ($= H, OH, HO_2$) causes additional ozone loss and disturbs other trace gas distributions.

During polar night and the absence of photochemistry the impact of energetic particle precipitation (EPP) on the chemical state of the atmosphere is most pronounced per se. But the interhemispheric diabatic circulation with its downward branch in the polar winter hemisphere in addition allows to propagate the result of this interaction, which mainly takes part in the thermosphere and mesosphere, to the stratosphere. The key processes for the impact of EPP on the middle atmosphere are therefore the combination of disturbed chemistry and transport in the winter polar middle atmosphere.

After its maximum in the year 2000/2002 the solar cycle 23 exhibited prolonged activity which gave rise to several extraordinary manifestations of solar-terrestial connections in the Earth's middle atmosphere. Several strong flare events and several strong geomagnetic storms were responsible for remarkable chemical disturbances in the middle atmosphere in both hemispheres. Evidence for regular long range NO_x descent had already been observed in several satellite experiments [*Callis et al.*, 1996; *Randall et al.*, 1998, 2001; *Rinsland*, 1996], for example from the Halogen Occultation Experiment (HALOE) instrument on UARS, and from the ATMOS and POAM experiments. The late solar cycle 23 fell in a period where new and most capable instruments on satellites gave a wealth of new information for these processes, i.e. the instruments for the determination of atmospheric composition on the ENVISAT satellite, launched in early 2002, SCIAMACHY, GOMOS and MIPAS, the instrument ACE FTS, MLS on Aura, Odin's SMR instrument. The new observational results offered the possibility for detailed and comprehensive model studies to test and improve our understanding of chemical and dynamical processes in the middle atmosphere as a whole. Together with the exceptional active sun, the conditions to study EPP related processes were excellent.

Since July 2002, the Michelson Interferometer for Passive Atmospheric Sounding (MIPAS) onboard the satellite ENVISAT of the European Space Agency (ESA)

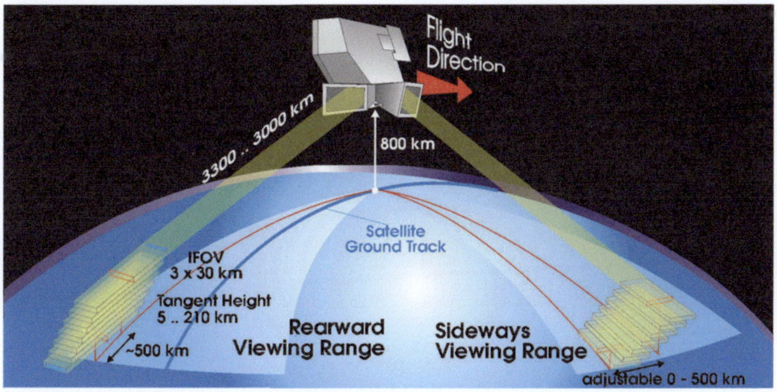

Fig. 15.1 Observation scheme of the MIPAS instrument on the ENVISAT satellite (source ESA)

observes the middle atmosphere from the upper troposphere up to the mesosphere. In Fig. 15.1 a schematic view of MIPAS in space is shown, focusing on observation modes. MIPAS is a Fourier transform spectrometer for the measurement of high-resolution gaseous emission spectra at the Earth's limb [*Fischer and Oelhaf*, 1996; *Fischer et al.*, 2008]. It allows to retrieve global distributions of trace gases as for example O_3, HNO_3, CH_4, N_2O and many other substances. These observations represent one of the most complete data sets for studying the influence of energetic particle induced changes in the middle atmosphere as they cover the middle atmosphere during polar night. This is an important advantage of the MIPAS/ENVISAT dataset compared to other observations, e.g. from UV-VIS instruments or from solar occultation instruments, which both depend on solar light. Reactive and reservoir gases, and tracers for the transport in the polar winter middle atmosphere are included in the dataset. Figure 15.2 shows as an example the time series of NO_2 night-time observations derived from an operational MIPAS data-product provided by ESA (see next paragraph).

Data are available for example from the operational ESA level 2 MIPAS/ ENVISAT data product. An advanced data record with more substances and also dealing with non-LTE effects, has been generated at the Institute for Meteorology and Climate Research—Atmospheric Trace Gases and Remote Sensing at KIT and the Instituto de Astrofisica de Andalucia (CSIC) (see *Lacoste-Francis* [2010]). By now, this advanced data record contains more than 20 substances and spans the period from July 2002 to 2010. From July 2002 to March 2004 MIPAS observed with a spectral resolution of $0.025\ cm^{-1}$, afterwards the spectral resolution was reduced, but the vertical resolution improved [*von Clarmann et al.*, 2009]. MIPAS could observe in detail the solar storms of October/November 2003 known as 'Halloween' storms as well as NO_x intrusions in the years 2004 and 2009 in the NH, observations which gave rise to the detection of new effects related with EPP as documented in a number of publications [*Jackman et al.*, 2005, 2008; *Orsolini et al.*, 2005; *Lopez-Puertas et al.*, 2005a, 2005b; *Funke et al.*, 2005, 2008; *von Clarmann et al.*, 2005].

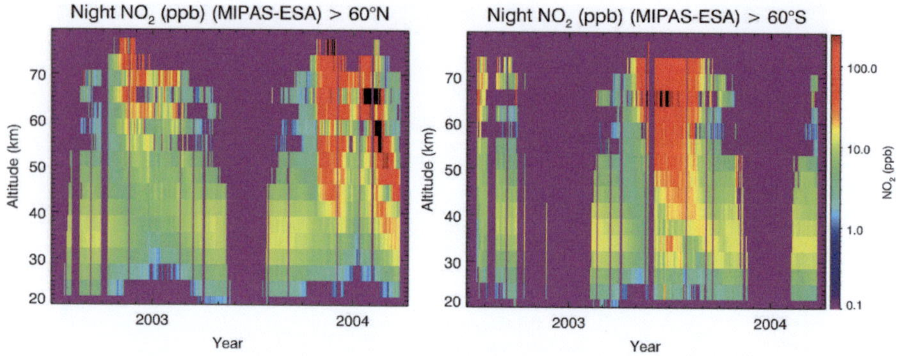

Fig. 15.2 NO_2 night-time observations of the MIPAS instrument on the European ENVISAT satellite, (**a**) Northern hemisphere, (**b**) Southern hemisphere between July 2002 and March 2004. In the NH the weak (2002/3) and strong (2003/4) intrusion from the mesosphere/lower thermosphere are seen together with the NO_x enhancement after the Halloween storms in November 2003, in the SH the pronounced intrusion from the MLT in the course of the Antarctic winter can be seen (from *Reddmann et al.* [2010], reproduced by permission of American Geophysical Union)

In several model studies principal effects of EPP have been studied in the past (e.g. *Siskind et al.* [2000], see *Jackman and McPeters* [2004] for an overview of related work). 3D-model studies of the effects of EPPs have been performed with chemistry climate models (CCMs) applying artificial NO_x enhancements [*Langematz et al.*, 2005; *Rozanov et al.*, 2005] or modules calculating NO_x and HO_x production from prescribed ionization rates (e.g. *Jackman et al.* [2008]). *Baumgaertner et al.* [2009] adapted results of *Randall et al.* [2007] to estimate effects of NO_x produced by low energy electrons in the middle atmosphere. Many model simulations qualitatively reproduce various effects connected with EPPs, but the necessary tight connection of chemical disturbances and transport is still a challenge for current 3D-models of the middle atmosphere. Model simulations using actual meteorological conditions comparing their results with corresponding observations are therefore a valuable tool to test our understanding of the relevant processes in these events. This project focuses on such comparisons.

The first approach within the project to assess the effects caused by EPP was to use detailed observations, primarily of the MIPAS instrument on the ESA satellite ENVISAT, as a boundary condition for the additional NO_x in model simulations. Two chemical transport models took part in this approach, the CLaMS and the KASIMA model. The observations of MIPAS used for the studies cover the period from July 2002 till March 2004 and include the Antarctic winter 2003 with strong NO_x enhancements originating in the upper mesosphere and lower thermosphere, the strong SPE event and the following geomagnetic storms (Halloween storms) in October/November 2003, and the NO_x intrusion following a stratospheric warming in Arctic mid-winter 2003/2004. From these simulations the effect on ozone chemistry could be deduced, e.g. the additional ozone loss and changes in other chemical substances could be quantified. Some results of this work are presented in Sect. 15.2, details can be found in *Vogel et al.* [2008] and *Reddmann et al.* [2010].

The second approach concentrated on the solar proton event in October/November 2003 and had therefore the direct effects of the EPP-atmosphere interaction in the focus, e.g. the amount of produced NO_x and HO_x, and changes of several substances in a short period during and after the ionization event. The very comprehensive observations of the MIPAS instrument of the species NO, NO_2, H_2O_2, O_3, N_2O, HNO_3, N_2O_5, HNO_4, ClO, HOCl, and $ClONO_2$, CO, CH_4, and H_2O allow a profound test of the chemistry implemented in models together with their transport properties in the polar winter middle atmosphere. An international model-data intercomparison project (High Energy Particle Precipitation in the Atmosphere, HEPPA) was established including both chemical transport models, and chemistry-climate models [*Funke et al.*, 2011]; of the nine participating models, five were also involved in the CAWSES SPP, as was the derivation of ionization rates with the AIMOS model (see also Chaps. 9 and 13). The HEPPA intercomparison initiative has lately been invited to become part of the SPARC SOLARIS initiative. For these comparison, the KASIMA model was extended to include a module for NO_x and HO_x production by ionizing particles. Ionization rates have been calculated within the KASIMA model or using the precalculated ionization rates provided by the AIMOS calculations (*Wissing and Kallenrode* [2009], see also Chap. 13. A short overview of results for the KASIMA model is given in Sect. 15.3, for details see *Funke et al.* [2011]).

Rather few observations exist which show the effects of EPP for the HO_x family [*Verronen et al.*, 2006]. Whereas the long-term effects of EPP produced HO_x are probably small, HO_x related fast reactions during the particle interaction seem to be the least understood in the models. A new retrieval setup for MIPAS/ENVISAT observations was therefore developed to assess possible enhancements of H_2O_2 after EPP. H_2O_2 serves as a reservoir of HO_x and it was expected that H_2O_2 concentrations should be enhanced during and after SPEs, and could indeed be detected by the new retrieval scheme. A climatology of H_2O_2 and EPP related enhancements is presented in Sect. 15.4. These data were also used as an additional species in the HEPPA intercomparison.

A strong impact on the partitioning within the NOy family has been observed when NO_x intrusions reach the upper stratosphere resulting in a secondary maximum of HNO_3 distribution in th upper stratosphere [*Kawa et al.*, 1995; *Stiller et al.*, 2005]. The proposed conversion of N_2O_5 to HNO_3 can be explained via a mechanism involving protonized water vapor clusters. A new parameterization was developed within the project for this reaction, and it was found that this conversion may have also an impact in the lower stratosphere. The inclusion of this reaction gives better agreement between HNO_3 observations and model results there (see Sect. 15.5).

Finally, the observations and model results clearly showed that not the strong but rare SP events are the main contributors to EPP related NO_x in the middle atmosphere, but the seemingly more regular but weaker auroras and geomagnetic storms. Most probably, very efficient transport from the lower thermosphere is the key process for these dramatic NO_x enhancements. This efficient transport seems to be related to specific dynamic situations. This finding connects the rather restricted study of chemical EPP effects to the more general question how the MLT region interacts with the lower atmosphere.

15.2 Model Studies with Imposed NO_x Disturbance

Within this project two different chemical transport models were applied to study the chemical effects of NO_x intrusions in the stratosphere, namely the CLaMS model and the KASIMA model. The models use a quite different model architecture and focus on different aspects. The Chemical Lagrangian Model of the Stratosphere (CLaMS) is a Lagrangian chemical transport model, which uses 3-dimensional deformations of the large-scale winds to parameterize mixing and is very well suited to study horizontal transport and mixing processes as shown in previous studies [*McKenna et al.*, 2002a, 2002b; *Konopka et al.*, 2007a]. KASIMA is a combination of a Euler model using analyzed winds in the lower domain, and a mechanistic model on top, and covers the vertical domain up to the lower thermosphere. It focuses on long-term transport processes including the mesospheric branch of the residual circulation and chemical processes in the middle atmosphere.

Both simulations show that the NO_x intrusions can have a significant impact on the ozone budget in the middle stratosphere maximizing in the middle stratosphere where additional ozone loss between 30 %–50 % is deduced. The effect on total ozone was found to be non-negligible but restricted to high latitudes within a range of about 10 DU.

15.2.1 The Arctic Winter 2003/4: The CLaMS Perspective

To study the impact of the downward transport of NO_x-rich air masses from the mesosphere into the lower stratosphere on stratospheric ozone loss, CLaMS simulations were performed for the Arctic winter 2003/04. This winter includes the Halloween storms at beginning of the winter in late October and the NO_x intrusions descending in late winter from the lower thermosphere/mesosphere to the stratosphere. In late December, a stratospheric warming occurred, and low latitude air masses were transported to high latitudes.

CLaMS is based on a Lagrangian formulation of the tracer transport and, unlike Eulerian CTMs, considers an ensemble of air parcels on a time-dependent irregular grid that is transported by use of the 3d-trajectories. The irreversible part of transport, i.e. mixing, is controlled by the local horizontal strain and vertical shear rates with mixing parameters deduced from observations (e.g. *Konopka et al.* [2003], *Grooß et al.* [2005]). CLaMS therefore is especially suited to study horizontal transport processes at the boundary of the polar vortex, and possible loss of the SPE or MLT enriched air masses by the mixing processes occurring during the stratospheric warming.

The simulations cover the altitude range from 350–2000 K potential temperature (approx. 14–50 km). The horizontal and vertical transport is driven by ECMWF winds and heating/cooling rates are derived from a radiation calculation. The mixing procedure uses the mixing parameter as described in *Konopka et al.* [2004] with a horizontal resolution of 200 km and a vertical resolution increasing from 3 km around 350 K potential temperature to approximately 13 km around 200 K potential temperature according to the model set up *Konopka et al.* [2007b]. The halogen, NO_x, and HO_x chemistry mainly based on the current JPL evaluation *Sander et al.* [2006] is included.

Before the first SPE has occurred, the model is initialized once at 4 October 2003 with mainly MIPAS observations (V3O) from 3–5 Oct 2003 (CH_4, CO, N_2O, O_3, NO, NO_2, N_2O_5, HNO_3, H_2O, and $ClONO_2$) processed at IMK and at the Instituto de Astrofisica de Andalucia (IAA), Granada, Spain. In the model, the flux of enhanced NO_x from the mesosphere is implemented in form of the upper boundary conditions at 2000 K potential temperature (50 km altitude) which are updated every 24 hours. The NOy constituents NO_x, N_2O_5, HNO_3 and the tracers CH_4, CO, H_2O, N_2O, and O_3 at the upper boundary are taken from results of a long-term simulation performed with the KASIMA model (Karlsruhe Simulation Model of the middle Atmosphere), see next section. In this KASIMA simulation, enhanced NO_x concentrations in the mesosphere during disturbed periods have been derived above 55 km from MIPAS observations (provided by the European Space Agency (ESA)).

In addition to this ClaMS reference model run (referred to as 'ref' run) a model simulation without an additional NO_x-entry at the upper boundary is performed. In addition, to estimate the possible maximum impact of NO_x on stratospheric ozone loss, we derive a maximum NO_x-entry at the upper boundary condition from several satellite measurements of NO and NO_2 by the MIPAS, HALOE, and ACE-FTS instruments. For each day at the upper boundary the NO_x mixing ratios for equivalent latitudes greater or equal 60°N are replaced by the maximum NO_x value observed by any satellite instruments within 60° and 90°N at 2000 K potential temperature at that day, respectively, by the maximum NO or NO_2 value if no NO_x measurement exists in this 24 hour period. For days where no satellite observations are available we take the maximum NO_x derived by KASIMA at that day. This model simulation is referred to as 'max NO_x' run.

Clams results of NOx enhancements and resulting ozone loss are shown for polar latitudes in Fig. 15.3. For the Arctic polar region (Equivalent Lat. >70°N) we found that enhanced NO_x caused by SPEs in Oct–Nov 2003 is transported downward into the middle stratosphere to about 800 K potential temperature ($\approx$30 km) until end of December 2003. The mesospheric NO_x intrusion due to downward transport of upper atmospheric NO_x produced by auroral and precipitating electrons [*Funke et al.*, 2007] affects NO_x mixing ratios down to about 700 K potential temperature ($\approx$27 km) until March 2004. A comparison of the reference run with a simulation without an additional NO_x source at the upper boundary shows that O_3 mixing ratios are affected by transporting high burdens of NO_x down to about 400 K (17–18 km) during the winter (see Fig. 15.3). Locally, an additional ozone loss of the order 1 ppmv is simulated for January between 850–1200 K potential temperature during the period of the strong polar vortex in the middle stratosphere.

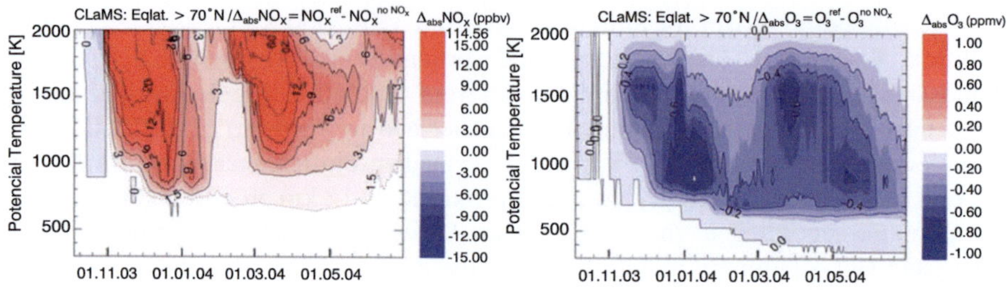

Fig. 15.3 Additional NO_x (ΔNO_x) and ozone reduction ΔO_3 poleward of 70°N equivalent latitude due to mesospheric NO_x intrusions over the course of the winter 2003–2004 for the reference model run (from *Vogel et al.* [2008])

An intercomparison of simulated NO_x and O_3 mixing ratios with satellite observations by ACE-FTS and MIPAS shows that the NO_x mixing ratios at the upper boundary derived from KASIMA simulations are in general too low (see Fig. 15.4). Therefore a model simulation with higher NO_x mixing ratios at the upper boundary derived from satellite measurements was performed. For this model run ('max NO_x') the simulated NO_x mixing ratios and the total area of enhanced NO_x mixing ratios are in general higher compared to satellite observations and so provide so an upper limit for the impact of mesospheric NO_x on stratospheric ozone chemistry. The comparison between simulated and observed ozone mixing ratios confirm the 'max NO_x' run is an upper limit which overestimates the NO_x entry at the upper boundary and underestimates O_3 due to stronger O_3 destruction. Moreover in the 'ref' run the simulated O_3 is in very good agreement with satellite measurements (see Fig. 15.4).

For the different runs ozone loss was calculated from the simulated O_3 values minus the passively transported ozone. Halogen-induced ozone loss plays a minor role in the Arctic winter 2003/2004 because the lower stratospheric temperatures were unusually high. Therefore the ozone loss processes in the Arctic winter stratosphere 2003/2004 are mainly driven by NO_x chemistry. Further, in addition to the transport of NO_x-rich mesospheric air masses in the stratosphere due to SPEs in Oct–Nov 2003 and the downward transport of upper atmospheric NO_x produced by auroral and precipitating electrons in early 2004, the ozone loss processes are also strongly affected by meridional transport of subtropical air masses. Likewise, in this case air masses rich in both ozone and NO_x are transported during the major warming in December 2003 and January 2004 into the lower polar stratosphere.

Our findings show that in the Arctic polar vortex (Equivalent Lat. >70°N) the accumulated column ozone loss between 350–2000 K potential temperature (~14–50 km) caused by the SPEs in Oct–Nov 2003 in the stratosphere is up to 3.3 DU with an upper limit of 5.5 DU until end of November (see Fig. 15.5). Further we found that about 10 DU but lower than 18 DU accumulated ozone loss additionally occurs until end of March 2004 caused by the transport of mesospheric NO_x-rich air in early 2004. In the lower stratosphere (350–700 K, approx. 14–27 km) the SPEs of Oct–Nov 2003 have negligible small impact on ozone loss processes until end of

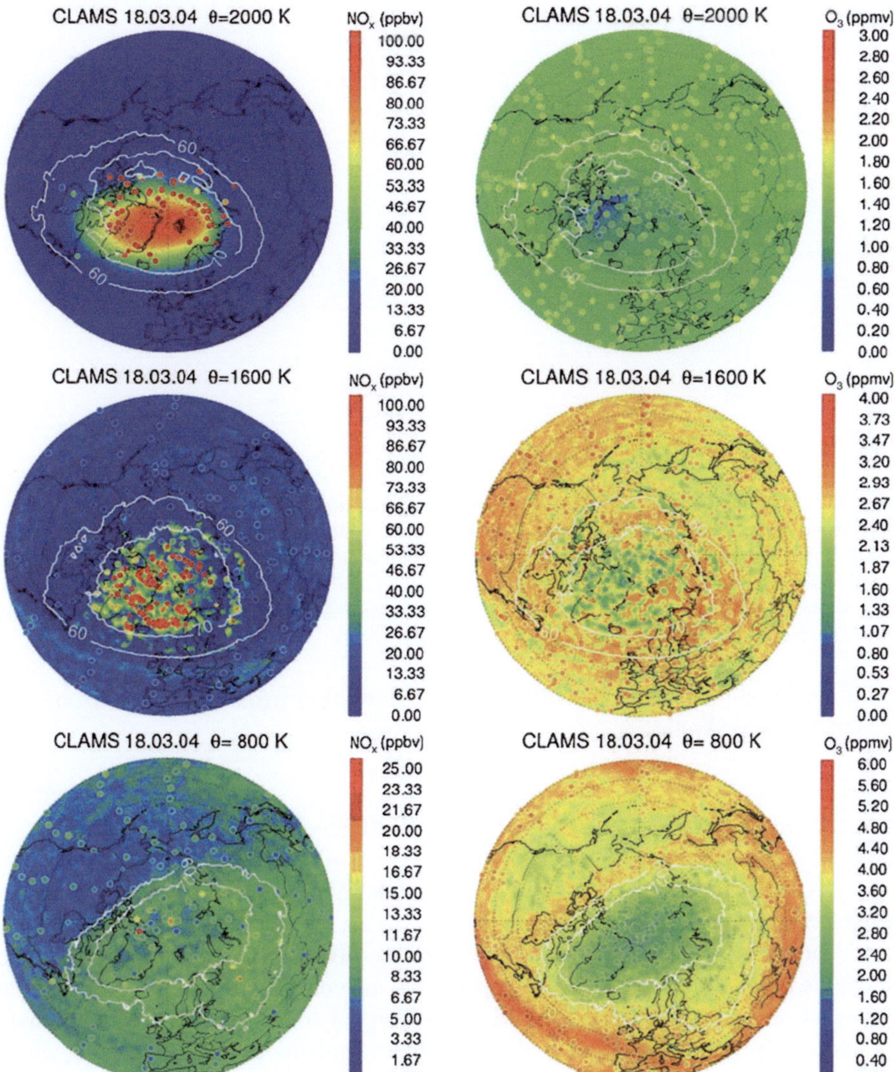

Fig. 15.4 Horizontal view of NO_x and O_3 at 2000 K, 1600 K, and 800 K potential temperature for the reference model run on March 18, 2004. The isolines for the equivalent latitude at 60°N and 70°N are marked by *white lines*. The model results and satellite observations (IMK/IAA-MIPAS: *circles* and ACE-FTS: *diamonds*) are shown for noon time. We note that all NO_x mixing ratios beyond the scale are also plotted in *red* (from *Vogel et al.* [2008])

November and the mesospheric NO_x intrusions in early 2004 yield ozone loss about 3.5 DU, but clearly lower than 6.5 DU until end of March. Overall, the downward transport of NO_x-rich air masses from the mesosphere into the stratosphere is an

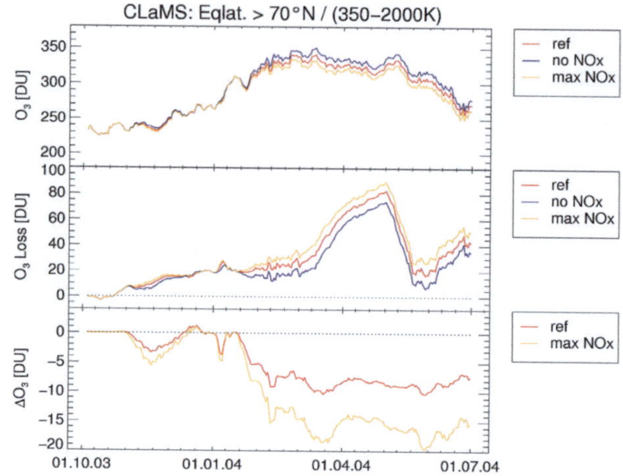

Fig. 15.5 Column ozone (in Dobson units) with (*red*), without (*blue*), and with a maximum (*yellow*) mesospheric NO_x sources integrated over the entire simulated altitude range from 350 K until 2000 K (~14–50 km) for equivalent latitudes poleward of 70°N. *Middle panel*: Ozone loss in DU for these model runs. *Bottom panel*: ΔO_3 caused by the additional NO_x-source at the upper boundary. Shown is the ozone column of the 'no NO_x' run minus the ozone column of the 'ref' (*red*) and the 'max NO_x' run (*green*) (from *Vogel et al.* [2008])

additional and non-negligible variability to the existing variations of the ozone loss observed in the Arctic.

15.2.2 KASIMA Studies: The Period 2002–2004

The KASIMA model is a 3D mechanistic model of the middle atmosphere which can be coupled to specific meteorological situations by using analyzed lower boundary conditions and nudging terms for vorticity, divergence and temperature. The model is based on the solution of the primitive meteorological equations in spectral formulation and uses the pressure altitude as vertical coordinate. Its vertical domain spans the upper troposphere up to the lower thermosphere. In the MLT region, winds are calculated using physical parameterizations for heating rates and gravity wave drag based on a Lindzen scheme. The model has been used for investigations of transport and chemistry in the middle atmosphere [*Kouker*, 1993; *Reddmann et al.*, 1999, 2001; *Ruhnke et al.*, 1999; *Kouker et al.*, 1999; *Khosrawi et al.*, 2009]. The model simulates long-term transport in the stratosphere reasonably well. This has been shown by comparisons with the inert tracer SF_6 and derived mean age of air [*Engel et al.*, 2006; *Stiller et al.*, 2008].

The rationale of this model study was to use two model runs, one standard run, and one with a changed upper boundary condition for NO_x from EPP to derive its impact on the chemical state in the middle atmosphere. No feedback of the chemistry is included to the ozone field used for the calculation of heating rates in the run, so both simulations use the same dynamics for transport and allow to single

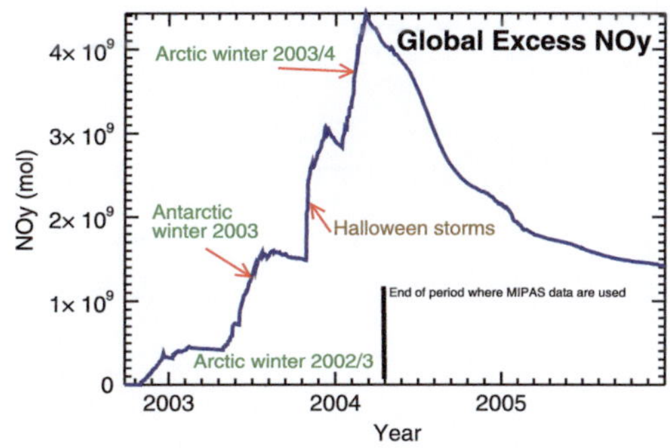

Fig. 15.6 Global excess NO_y from energetic particle production (from *Reddmann et al.* [2010]), modified by permissions of the American Geophysical Union)

out the change esp. of ozone for the disturbed chemical conditions. With a good representation of the mean transport times, the model can therefore be used to study the effects of NO_x intrusions on a longer time-scale. Against the background of the undisturbed stratosphere, the NO_x enhancements in the observations and in the model can be clearly traced as seasonal events restricted to polar latitudes, leaving the rest of the stratosphere mostly unaffected. This allows to estimate the chemical effects of NO_x intrusions by comparing the disturbed and the undisturbed model run, and to analyze the individual events which show different characteristics between intrusions starting in the upper mesosphere and the SPE with enhancements down to the stratosphere. Coupling effects between chemical changes due to EEPs and dynamics take place in the real atmosphere, and are therefore implicitly included in the dynamical fields used in the nudged model. In this sense, the chemical effects are also influenced by the coupling, as the reference dynamical state does not correspond to a real undisturbed one. Nevertheless, for the purpose of quantifying the strength of the NO_x intrusions on chemistry this is only an effect of minor importance.

For our study we used reprocessed operational MIPAS NO_2 data which had a nearly daily coverage [*Ridolfi et al.*, 2000; *Carli et al.*, 2004]. NO_x has a photochemical lifetime of a few days in the mesosphere and during night most of NO_x is in the form of NO_2. Nighttime observations of NO_2 are taken therefore as a representative for the concentration of NO_x.

Figure 15.2 shows time-height cross sections of the mean NO_2 volume mixing ratio of the northern and southern polar cap (geographical latitudes polewards of 60°) in ppb, from the ESA MIPAS/ENVISAT nighttime observations as described above. NO_x values from the MIPAS/ENVISAT observations using nighttime NO_2 as a proxy for NO_x, are imposed on the model values during times of diagnosed increased levels of NO_x in the polar mesosphere and during the SPE in October/November 2003. The ESA dataset has been preferred over other data sets based on MIPAS/ENVISAT observations as it provided a rather complete coverage for the MIPAS/ENVISAT phase I observation period, but it lacks the accuracy of the IMK-IAA data set as it does not include non-LTE effects and does not consider horizontal gradients of mixing ratios in the retrieval (see *Wetzel et al.* [2007]).

The sum of all inorganic nitrogen-containing species except N_2 and the source gases like N_2O is called NO_y, enhanced after EPP [*Brasseur and Solomon*, 2005].

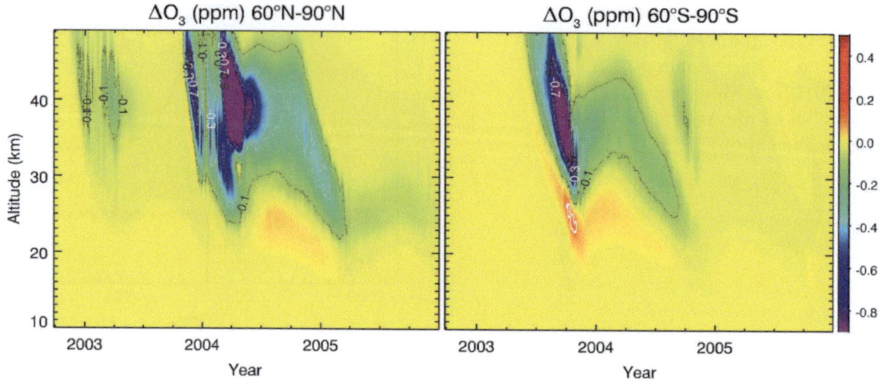

Fig. 15.7 Ozone change through EPP effects (from *Reddmann et al.* [2010], reproduced by permission of American Geophysical Union)

Figure 15.6 shows the global excess NO_y in mol for the period July 2002 to December 2005. The auroral winter intrusions in the Arctic winter 2002/2003 and Antarctic winter 2003, the SP event in fall 2003 and the strong intrusion in the Arctic winter 2003/2004 are clearly discernible as single events adding to the total NO_y background. The characteristic time-scale for the decay of the additional NO_y in the model lies in the range of about 2 years.

We estimate the additional NO_y in Arctic winter 2002/2003 at 0.4 Gigamol (GM), in Antarctic winter 2003 at about 1.4 GM, the NO_y from the SPE at about 1.5 GM and in January 2004 at about 2 GM. As in the Arctic winter both intrusions overlap, they cannot clearly be separated. The value estimated for the Antarctic winter 2003 is about 75 % of the value *Funke et al.* [2005] deduced from their observation based study with MIPAS/ENVISAT data. *Randall et al.* [2007] give a range of 1 to 2 GM estimated from HALOE observations for this winter, depending on the use of average or maximum observed NO_x values. Globally, the additional NO_y in the model which we derive amounts at its maximum to about 5 % of the total 70 GM NO_y in the middle atmosphere from the oxidation of N_2O.

The effects of the additional NO_y in the atmosphere on the formation of particles in the cold polar vortex (for example by changing the saturation vapor pressure) is small, but still significant. The mass of NO_y in NAT-particles increases in the Antarctic winter 2004 and the Arctic winter 2004/2005 by about 5 %.

Figure 15.7 shows the effects on ozone for the whole simulation period again for 75°S and 75°N. Additional ozone loss lasts for the following year in the Southern Hemisphere at the level of about 0.2 ppm. At the end of 2004 a small ozone change can be found above the band related to the Antarctic winter 2003. Closer inspection of time-latitude cross sections of NO_y at different altitudes show that during summer 2004 NO_y is transported from the Northern to the Southern Hemisphere at the stratopause level (not shown) exceeding 1 ppb, causing the additional ozone loss. During the Arctic winter 2002/2003 the minor amounts of additional NO_x cause only small ozone changes. The maximum changes occur during spring 2004, exceed 30 % and follow closely the downward transport of the additional NO_x. The changes caused by the SPE in fall 2003 can be traced until spring 2004 below 30 km but are lost to lower latitudes. The enormous amounts of additional NO_x from January cause ozone changes of several percent even in the following spring at altitudes of about 30 km.

The NO_x induced additional ozone loss causes also changes in total ozone. In the Southern Hemisphere ozone changes are restricted to the Antarctic region. In the Northern Hemisphere a decrease of total ozone of several Dobson units reach the mid-latitudes. At high latitudes changes of about 5 DU can be found even in spring 2005.

15.3 The Solar Proton Events of Oct/Nov 2003 and Direct Effects

This section deals with the second approach where HO_x and NO_x enhancements are calculated in the models directly from ionization rates. This allows chemistry climate models to include EPP effects as a function of solar and geomagnetic activity. The role of CTMs is here again to validate the results of such calculations with the help of detailed observations. Within the CAWSES project the KASIMA model was expanded to include these direct effect using particle flux data from satellites as GOES. Again, the very well observed HALLOWEEN storms have been used to make these comparisons using the MIPAS data set. These comparisons were carried out in the framework of an international model measurement intercomparison initiative of CCMs and CTMs [*Funke et al.*, 2011]. For KASIMA as for most of the other models taking part in that inter-comparison the general agreement between observations and model is reasonable but by far not perfect.

As the HO_x cycle is not well represented in observations a new retrieval strategy was implemented in the MIPAS analysis to derive H_2O_2 concentrations in the middle atmosphere. For the first time, a H_2O_2 climatology could be derived and enhancements during the SPE effect in 2003 could be observed. The comparison with the model shows a qualitative but not a quantitative agreement, hinting at deficits of our understanding of the chemical processes related to H_2O_2 in general as well as and in particular during and after atmospheric ionization events.

15.3.1 Ionization Rates and HO_x and NO_x Production

Different kind of particles contribute to the ionization in the atmosphere. Enhanced fluxes of energetic protons can be included in models with reasonable accuracy by an approximate description, for example applying the Bethe-Bloch formula and assuming that the protons fill the geomagnetic polar cap more or less homogeneously, as it is the case for solar proton events. The spectral flux distribution of the protons has to be derived from measurements of particle detectors on satellites (GOES data series) which only give integrated counts for a few channels. In KASIMA we applied a particle flux of the form $J(E) = A \cdot E^{-\delta}$, where A and δ have been derived

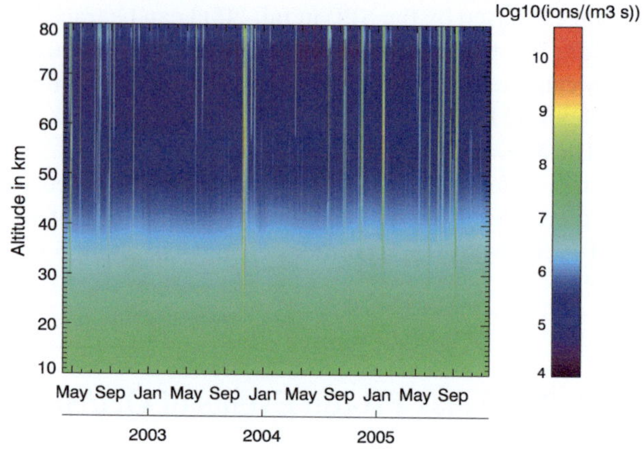

Fig. 15.8 Ionization rate from protons and cosmic rays calculated in the KASIMA SPE module (from *Versick*, 2011)

by a fit procedure (see *Versick* [2011]). We also included a component from cosmic rays using the formalism of *Heaps* [1978]. Figure 15.8 shows ionization rates calculated as described above. It shows distinct and sporadic solar proton events in the upper stratosphere and, approaching solar minimum conditions, a slowly increasing flux of cosmic rays.

To include ionization by electrons a much more complex calculation is necessary, as electron flux is much more dependent on the state of the magnetosphere and the radiation belts. We therefore used the AIMOS data set to include ionization caused by electrons, see the corresponding Chap. 13 of J.M. Wissing and M. Kallenrode. Additional to precipitating electrons, this data-set also considers protons and α-Particles, based on measured particle fluxes. We applied both data sets for the solar proton event in Oct/Nov 2003, and found better agreement of observed and simulated NO_x when applying our module. With the limitations of the observations of the particle flux and other assumptions to derive a energy flux spectrum, this underlines that more detailed observations of incoming particle fluxes are necessary.

15.3.2 Model Intercomparison

The SP event in Oct/Nov 2003 is an ideally suited testbed for simulations of direct effects in the middle atmosphere caused by energetic particles as the detailed observations of the MIPAS instrument allow to validate the response of all the most important chemical species in the models. MIPAS showed changes after the event for species such as NO, NO_2, H_2O_2, O_3, N_2O, HNO_3, N_2O_5, HNO_4, ClO, HOCl, and $ClONO_2$. The HEPPA model data intercomparison initiative brought together scientists involved in atmospheric modeling using state-of-the art CCMs and chemistry-transport models on one hand and scientists involved in the analysis and generation of MIPAS IMK/IAA data on the other hand. The objective of this community effort was (i) to assess the ability of state-of-the-art atmospheric models to reproduce

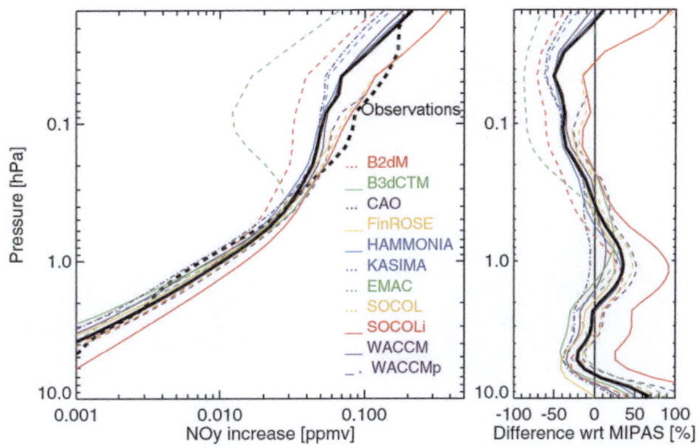

Fig. 15.9 Area-weighted averages (40–90N) of observed and modeled NOy enhancements during 30 October 1 November with respect to 26 October (*left*) and relative deviations of modeled averages from the MIPAS observations (*right*). *Thick solid* and *dashed lines* represent model multi-model mean average and MIPAS observations, respectively. WACCMp denotes the WACCM simulation including proton ionization only (excluded from multi-model-mean) (from *Funke et al.* [2011])

SPE-induced composition changes, (ii) to identify and—if possible—remedy deficiencies in chemical schemes, and (iii) to serve as a platform for discussion between modelers and data producers. This was achieved by a quantitative comparison of observed and modeled species abundances during and after SPEs, as well as by intercomparing the simulations performed by the different models. The initiative focused in its first phase on the inter-comparison of IMK/IAA generated MIPAS/Envisat data obtained in the aftermath of the October/November 2003 SPE with model results.

We give here a few examples of results of the intercomparison for the KASIMA model which extend the studies presented in Sect. 15.2. We refer to the extensive paper of *Funke et al.* [2011] for the complete inter-comparison and the discussion of the results.

An important parameter for subsequent chemical effects caused by the EPP is the amount of additional reactive nitrogen compounds which is produced in the event, expressed as additional NOy. Figure 15.9 shows the observed NOy enhancement and the results of the models directly after the SPE. The agreement seems to be satisfactory, but note the logarithmic scale. As most other models, KASIMA overestimates NOy below about 0.5 hPa and shows some underestimation above. This discrepancy hints to difficulties in describing the ionization rate, see discussion in the section above.

Figure 15.10 top shows the development of CO in observations and the model. Obviously KASIMA overestimates the downward transport compared to observations. A too fast descent in the Northern polar winter stratosphere was also found in the analysis discussed in *Reddmann et al.* [2010]. Figure 15.10 shows an example for substances from the chlorine family. Whereas HOCl changes are well represented in the model, $ClONO_2$ changes are underestimated to a great extent which is probably related to lower ClO values in the models compared to MIPAS even

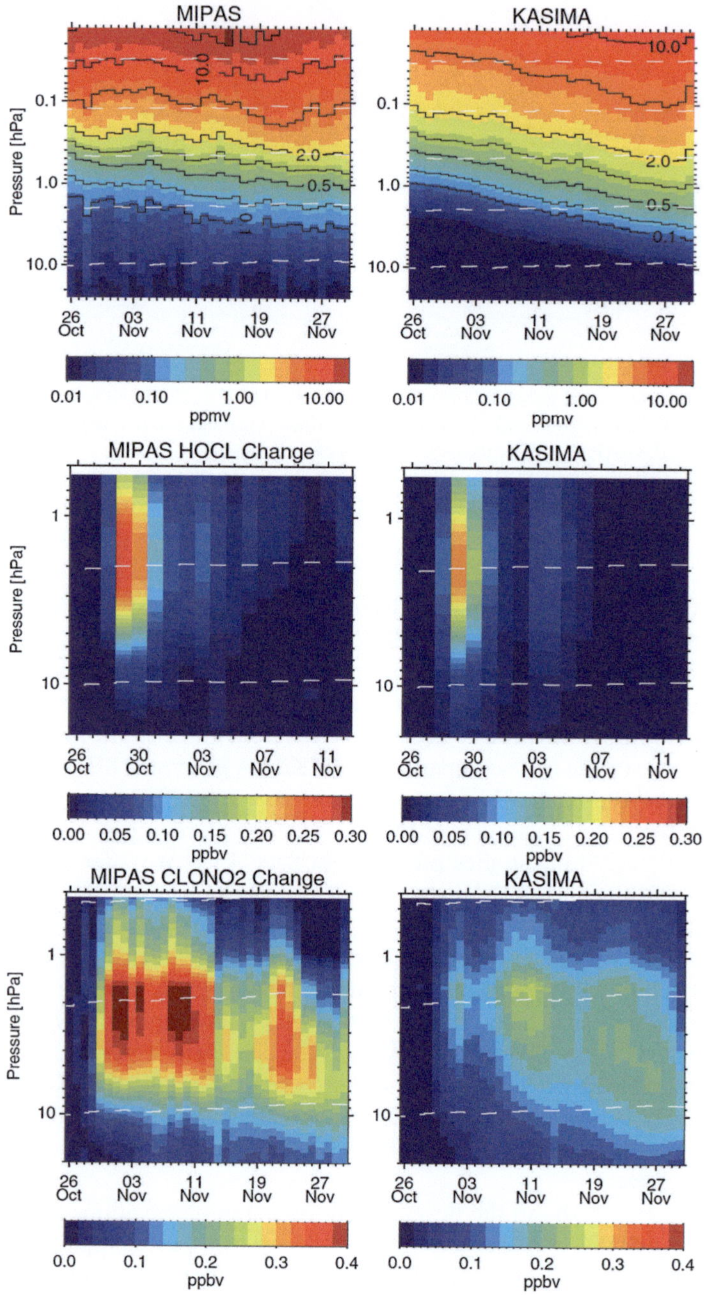

Fig. 15.10 Examples from the HEPPA intercomparison showing KASIMA results. *Top row*: CO time series derived from MIPAS and from KASIMA (ppm). Time series of HOCl and $ClONO_2$ derived from MIPAS observations and difference to KASIMA (ppb)

during background conditions. Other differences between models and observations concern HNO_3 buildup during and shortly after the event, It was shown by *Verronen et al.* [2008] that this is most likely due to missing ion chemistry in most of the models.

15.4 Hydrogen Peroxid as an Indicator for HO_x Production

Besides the production of NO_x, also HO_x is produced in the course of energetic particle precipitation. Changes in trace gases like HOCl, HNO_4 or HNO_3 are dependent on changes in the concentration of the HO_x family. It is therefore worth to try to assess members of the HO_x family directly by observations and to document the changes caused by the EPP. But only the gas H_2O_2, which acts as a reservoir gas for HO_x shows sufficiently strong spectral features in the spectral range of the MIPAS instrument. Therefore, the MIPAS/ENVISAT full and reduced resolution spectra were analyzed to develop a retrieval approach for H_2O_2.

The main source of H_2O_2 is the HO_2 self-reaction:

$$HO_2 + HO_2 \longrightarrow H_2O_2 + O_2 \tag{15.1}$$

Of minor importance is the three-body reaction:

$$OH + OH + M \longrightarrow H_2O_2 + M \tag{15.2}$$

The main sink in the stratosphere is through photolysis:

$$H_2O_2 + h\nu \longrightarrow 2OH \tag{15.3}$$

The reactions with OH and atomic oxygen destroy H_2O_2 to a lesser extent.

15.4.1 Hydrogen Peroxide Spectral Signatures and Retrieval Set-Up

In the mid-infrared spectral region, which is covered by the spectral range of MIPAS, hydrogen peroxide shows weak emission lines between about 1210 cm^{-1} and 1320 cm^{-1} (Fig. 15.11). All these lines belong to the H_2O_2 ν_6 band centered at 1266 cm^{-1}. The spectral signatures are taken from the latest update for H_2O_2 (based on measurements from *Perrin et al.* [1995] and *Klee et al.* [1999]) in the HITRAN 2004 molecular spectroscopic database [*Rothman et al.*, 2005]. The challenge of the hydrogen peroxide retrieval is the very weak signal of the emission lines in comparison to the instrumental noise which is much higher in this spectral region than the H_2O_2 signal (Fig. 15.11).

For the retrieval, 19 narrow spectral regions (microwindows) were selected between 1220 cm^{-1} and 1265 cm^{-1}, which is the lower end of MIPAS channel AB and includes the R-branch of the H_2O_2 ν_6 band. Those microwindows are used up to tangent altitudes of 47 km.

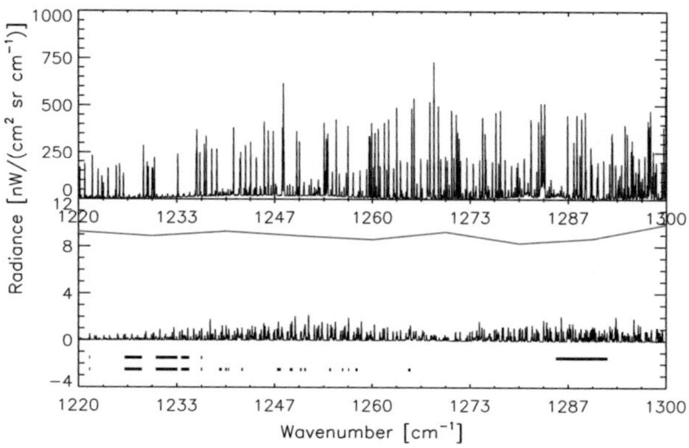

Fig. 15.11 Spectral range of the H_2O_2 ν_6 rotations-vibration band; *Top*: emission of all gases; *Bottom*: emission of H_2O_2 (*black*) and typical noise in MIPAS-spectra (*red*); *black lines* mark used spectral windows, lower row for heights below 44.5 km, top row above (from *Versick* [2011])

The main criterion for the selection was high sensitivity to hydrogen peroxide and low interference by other gases. Unfortunately the P-branch can not be used in the lower stratosphere for our retrieval because the lines of the interfering gases are too dense. Above 47 km an additional microwindow from 1285 cm^{-1} to 1292 cm^{-1} is used. The microwindows between 1237 cm^{-1} and 1265 cm^{-1} are not used in this altitude regions due to potentially bad absorption cross sections for N_2O_5 at low pressures.

Since the hydrogen peroxide contribution is so small, the contribution of other gases still needs to be considered and HOCl, CH_4, N_2O, N_2O_5 and COF_2 have to be retrieved jointly. Other gases had to be retrieved before the hydrogen peroxide retrieval and the results had to be used in our H_2O_2 retrieval. These gases are: H_2O, O_3, $ClONO_2$ and HNO_3. For all the other gases in this spectral region climatological values were used.

The retrieval procedure follows a scheme analog to that described by *Rodgers* [2000]:

$$\mathbf{x}_{i+1} = \mathbf{x}_i + \left(\mathbf{K}_i^{\mathbf{T}}\mathbf{S}_{\mathbf{y}}^{-1}\mathbf{K}_i + \mathbf{R}\right)^{-1} \times \left[\mathbf{K}_i^{\mathbf{T}}\mathbf{S}_{\mathbf{y}}^{-1}\left(\mathbf{y} - \mathbf{F}(\mathbf{x}_i)\right) - \mathbf{R}(\mathbf{x}_i - \mathbf{x}_{\mathbf{a}})\right] \tag{15.4}$$

where $\mathbf{x}$ is the retrieval vector, $\mathbf{K}$ the partial derivatives of the spectral grid points with respect to the retrieval vector (Jacobian), $\mathbf{S_y}$ the covariance matrix due to the measurement noise, $\mathbf{R}$ the regularization or constraint matrix, $\mathbf{y}$ the measurement vector, $\mathbf{F}$ the forward model, $\mathbf{x_a}$ the a priori profile, and i the iteration index.

Due to the very high noise in comparison to the signal we had to choose a rather high regularization strength giving us a low number of degrees of freedom. The regularization is stronger in the upper stratosphere. It is weakest in the middle stratosphere where we expect the maximum volume mixing ratio of H_2O_2. Lowest measurements used were around 25 km because otherwise oscillations in our profiles occurred due to error propagation from below which caused subsequent faults in other altitudes. The retrieval setup was verified by retrieval of H_2O_2 from synthetic spectra. These spectra were calculated with the Karlsruhe optimized and Precise Radiative transfer Algorithm (KOPRA) [*Stiller*, 2000]. The error analysis showed

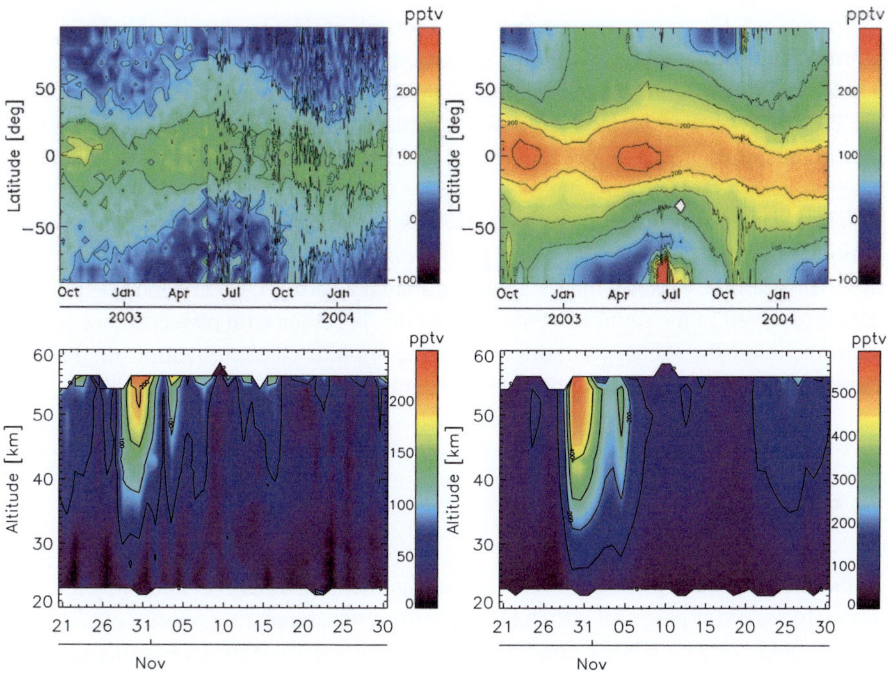

Fig. 15.12 Retrieval (*left*) and model (*right*) results of H_2O_2: top temporal evolution of the daily zonal means at 30 km; KASIMA results are shown on MIPAS geolocations convolved with the MIPAS AK. (*Bottom*): Time-height cross section of H_2O_2 during the Halloween storms; MIPAS averaging kernels has been applied on KASIMA results (from *Versick* [2011])

that concentrations of H_2O_2 can be retrieved between 20–60 km. The corresponding vertical resolution is about 8 km in the lower stratosphere and about 35 km in the upper stratosphere. Comparison of H_2O_2 MIPAS measurements with models must be done by convolving the model results with the MIPAS averaging kernel.

15.4.2 Distribution of H_2O_2 Under Normal and Disturbed Conditions

The temporal evolution of the H_2O_2 stratospheric distribution in KASIMA is very similar to the temporal evolution in the MIPAS measurements (see Fig. 15.12). Both show the highest vmr in the inner tropics shortly after equinox. In the tropics and subtropics the H_2O_2 volume mixing ratio is following the position of the overhead sun. Higher volume mixing ratios are reached in the summer hemisphere. Also the lower volume mixing ratio in the end of 2003 and beginning of 2004 are represented by KASIMA. The absolute value of the mixing ratio vmr in KASIMA is almost twice that of MIPAS. Sensitivity studies show that this could be related to uncertainties of associated reaction rate-constants.

Figure 15.12 shows the temporal evolution of H_2O_2 during the Halloween storms. The retrieval clearly shows enhancements related to the ionization events. The model grossly overestimates the observed concentrations during the SPE. From the HEPPA inter-comparison similar results have been found for other models. Further details of the retrieval procedure and results can be found in *Versick* [2011] and *Versick et al.* [2011].

The reduced resolution mode of MIPAS after March 2004 makes regular observations of H_2O_2 more difficult, and global distributions could not be derived until now. But after the exceptional SP events in January 2005, H_2O_2 enhancements could also be derived in the reduced resolution mode. Together with observations of MLS instrument on the AURA satellite, a more complete characterization of the HO_x family after SPE events was possible (for details see *Jackman et al.*, 2011).

15.5 HNO_3 Enhancements

In the middle stratosphere, the reactive nitrogen compounds NO and NO_2 are converted to reservoir gases, of which HNO_3 is the most abundant. *Orsolini et al.* [2005] found much higher HNO_3 concentrations observed by MIPAS/ENVISAT when compared to their model for the winter 2003/2004, *Stiller et al.* [2005] analyzed the Antarctic winter 2003, when a distinct secondary maximum of HNO_3 was found in MIPAS/ENVISAT data at about 40 km. These observations are in line with findings from earlier satellite missions [*Austin et al.*, 1986; *Kawa et al.*, 1995]. The latter explained this discrepancy as a result of reactions of N_2O_5 with water cluster ions. This reaction, originally proposed by *Böhringer et al.* [1983], had been combined with heterogeneous reactions on sulfate aerosols by *de Zafra and Smyshlyaev* [2001] in order to understand HNO_3 satellite and ground based observations in polar winters.

First comparisons of the HNO_3 observations with the KASIMA model also showed a pronounced underestimation of HNO_3 in the late polar winter when high NO_x concentrations indicate strong NO_x intrusions. We therefore included the parameterization of *de Zafra and Smyshlyaev* [2001] in our chemistry module and got reasonable agreement with the observations (see *Reddmann et al.* [2010]).

The parameterization of *de Zafra and Smyshlyaev* [2001] uses a fixed profile for protonized water clusters. Using a formulation for the cluster concentration according to *Aplin and McPheat* [2005] a variant of this parameterization was developed where the concentration of water clusters is dependent on the actual ionization rate. Figure 15.13 compares versions of the KASIMA model with no additional conversion, the implementation of *de Zafra and Smyshlyaev* [2001] and the new approach. The conversion to HNO_3 is strongest in the new version, and results in too low N_2O_5 concentrations compared to the MIPAS results. Interestingly, through the ionization by cosmic rays we found a pronounced effect of this reaction in the lower stratosphere, bringing the regular maximum of HNO_3 at about 25 km in closer agreement to the observations.

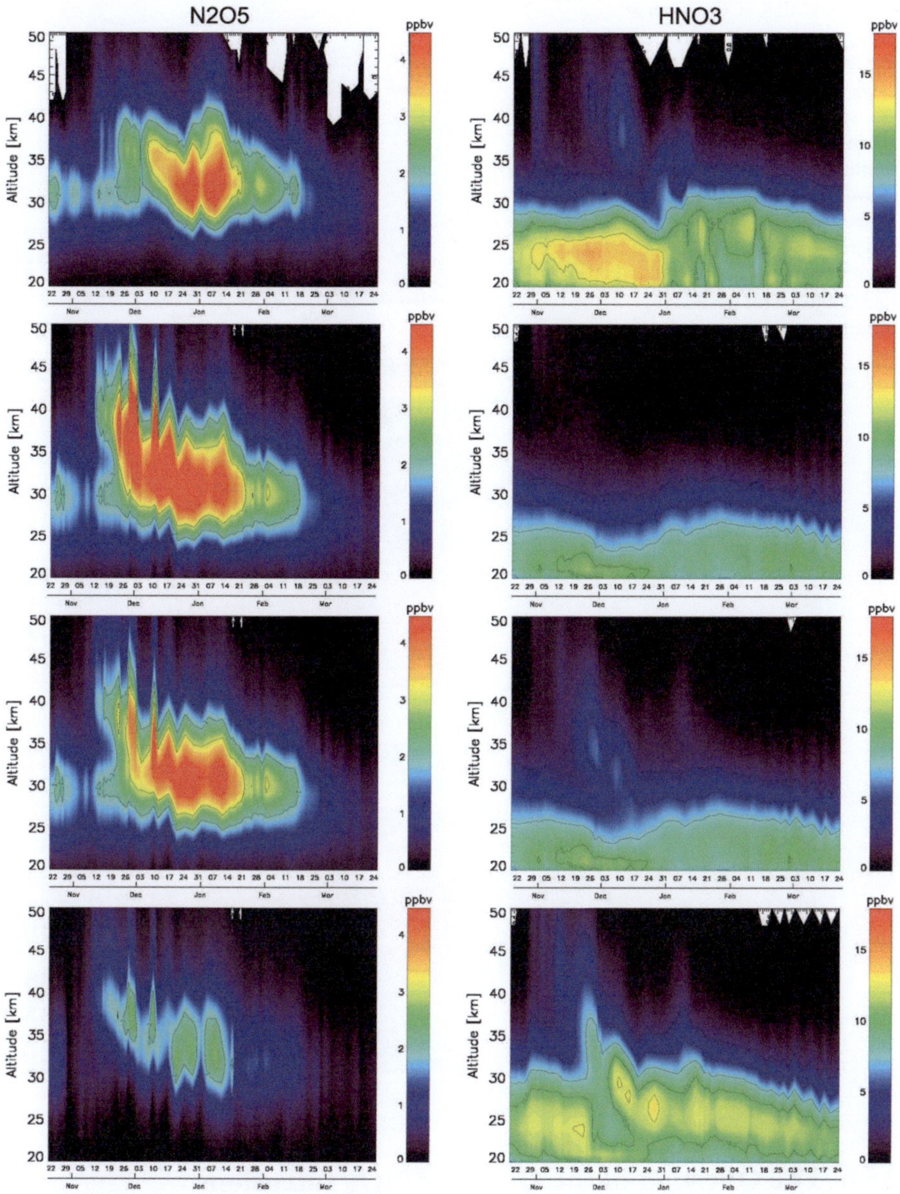

Fig. 15.13 N_2O_5 and HNO_3 time-height cross sections for the Arctic winter 2003/4. From top MIPAS observations, KASIMA simulations without reaction with protonized water clusters, KASIMA including the reaction parameterized according *de Zafra and Smyshlyaev* [2001], and reaction with ionization rate dependent concentration of water clusters

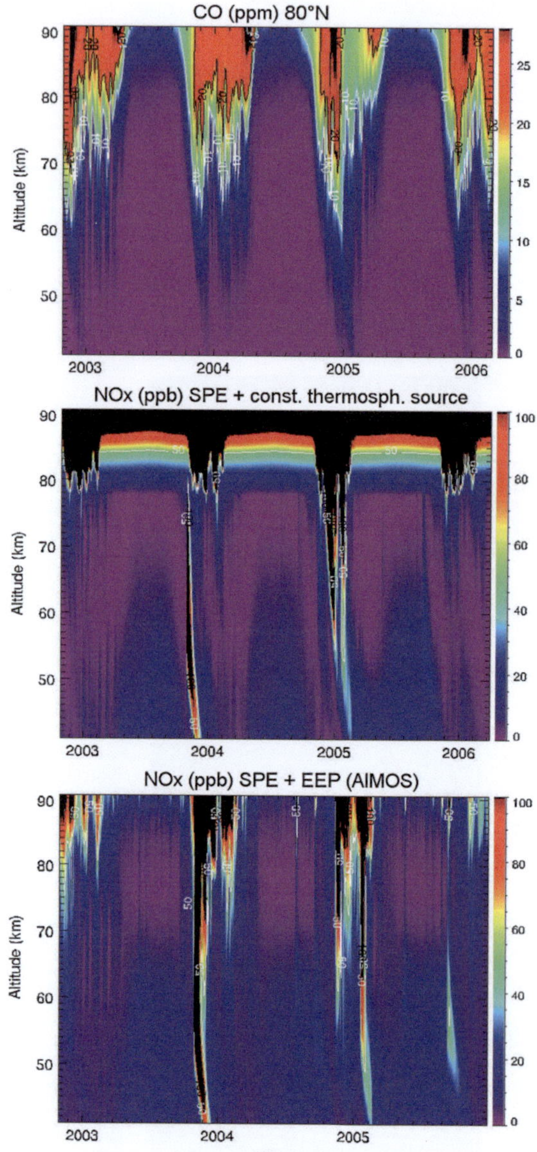

Fig. 15.14 KASIMA simulations of the NH winters 2002/3–2005/6. CO serves as an upper mesospheric tracer, the artificial thermospheric NO_x probes possible transport from the lower thermosphere, and the simulation including the AIMOS electron component probes possible ionization in the mesosphere from auroral sources. The winter 2003/4 and 2006/7 with observed strong intrusion are marked

15.6 The Question of the Origin of NO_x Intrusions

In winter 2003/4 the analysis showed that it is not the SPE which caused most of long-term impacts on ozone but the NO_x enhancements transported from above about 80 km. The question where the massive NO_x intrusions observed in the Arctic winter 2003/4 (see Sect. 15.2), 2005/6 and 2008/9 have its origin is currently under debate. Figure 15.14 shows output of a KASIMA simulation where the downward transport and ionization via electrons is probed. The model shows especially for the 2003/4 and 2005/6 no pronounced downward transport for CO, here used as a tracer for mesospheric air masses. This agrees with the fact that also an artificial thermospheric NO_x source in the model does not contribute to massive NO_x enhancements in contrast to the observations. But also the inclusion of auroral electron ionization according precalculated ionization rates through the AIMOS model does not improve this deviation from observations. Many models seem to fail to simulate the massive NO_x enhancements but it is not clear for what reasons. It was suggested that due to the limited energy range and resolution of the particle counters, precipitation of highly relativistic electrons might not be detected. The fact that the massive enhancements observed in the past Arctic winters occurred after strong stratospheric warmings and accompanied by an elevated stratopause *Smith et al.* [2009] strongly suggests however, that dynamical processes as the propagation of gravity waves play an important role, as the propagation of gravity waves. As it is well known that gravity wave drag is represented in models only in a parameterized form, a plausible reason for the failure of the model is some deficit in the representation of processes connected to gravity waves and their deposition of energy and momentum. Further model studies and observations, specifically in the MLT region are necessary to solve this question.

References

Aplin, K., & McPheat, R. (2005). Absorption of infra-red radiation by atmospheric molecular cluster-ions. *Journal of Atmospheric and Solar-Terrestrial Physics*, *67*(8–9), 775–783. doi:10.1016/j.jastp.2005.01.007. 1st General Meeting of the European-Geosciences-Union, Nice, France, Apr 25, 2004.

Austin, J., Garcia, R. R., Russell III, J. M., Solomon, S., & Tuck, A. F. (1986). On the atmospheric photochemistry of nitric acid. *Journal of Geophysical Research*, *91*, 5477–5485.

Baumgaertner, A. J. G., Joeckel, P., & Bruehl, C. (2009). Energetic particle precipitation in ECHAM5/MESSy1-Part 1: downward transport of upper atmospheric NOx produced by low energy electrons. *Atmospheric Chemistry and Physics*, *9*(8), 2729–2740.

Böhringer, H., Fahey, D. W., Fehsenfeld, F. C., & Ferguson, E. E. (1983). The role of ion-molecule reactions in the conversion of N_2O_5 to HNO_3 in the stratosphere. *Planetary and Space Science*, *31*, 185–191. doi:10.1016/0032-0633(83)90053-3.

Brasseur, G. P., & Solomon, S. (2005). *Aeronomy of the middle atmosphere*. Berlin: Springer.

Callis, L., Baker, D., Natarajan, M., Blake, J., Mewaldt, R., Selesnick, R., & Cummings, J. (1996). A 2-D model simulation of downward transport of NOy into the stratosphere: effects on the 1994 austral spring O_3 and NOy. *Geophysical Research Letters*, *23*(15), 1905–1908.

Carli, B., Alpaslan, D., Carlotti, M., Castelli, E., Ceccherini, S., Dinelli, B. M., Dudhia, A., Flaud, J. M., Hoepfner, M., Jay, V., Magnani, L., Oelhaf, H., Payne, V., Piccolo, C., Prosperi, M., Raspollini, P., Remedios, J., Ridolfi, M., & Spang, R. (2004). First results of MIPAS/ENVISAT with operational level 2 code. *Advances in Space Research*, *33*(7), 1012–1019.

de Zafra, R., & Smyshlyaev, S. (2001). On the formation of HNO3 in the Antarctic mid to upper stratosphere in winter. *Journal of Geophysical Research*, *106*(D19), 23115–23125.

Engel, A., Mobius, T., Haase, H. P., Bonisch, H., Wetter, T., Schmidt, U., Levin, I., Reddmann, T., Oelhaf, H., Wetzel, G., Grunow, K., Huret, N., & Pirre, M. (2006). Observation of mesospheric air inside the arctic stratospheric polar vortex in early 2003. *Atmospheric Chemistry and Physics*, *6*, 267–282.

Fischer, H., & Oelhaf, H. (1996). Remote sensing of vertical profiles of atmospheric trace constituents with MIPAS limb-emission spectrometers. *Applied Optics*, *35*(16), 2787–2796.

Fischer, H., Birk, M., Blom, C., Carli, B., Carlotti, M., von Clarmann, T., Delbouille, L., Dudhia, A., Ehhalt, D., Endemann, M., Flaud, J. M., Gessner, R., Kleinert, A., Koopman, R., Langen, J., Lopez-Puertas, M., Mosner, P., Nett, H., Oelhaf, H., Perron, G., Remedios, J., Ridolfi, M., Stiller, G., & Zander, R. (2008). MIPAS: an instrument for atmospheric and climate research. *Atmospheric Chemistry and Physics*, *8*(8), 2151–2188.

Funke, B., Lopez-Puertas, M., von Clarmann, T., Stiller, G. P., Fischer, H., Glatthor, N., Grabowski, U., Hopfner, M., Kellmann, S., Kiefer, M., Linden, A., Tsidu, G. M., Milz, M., Steck, T., & Wang, D. Y. (2005). Retrieval of stratospheric nox from 5.3 and 6.2 μm nonlocal thermodynamic equilibrium emissions measured by Michelson interferometer for passive atmospheric sounding (MIPAS) on envisat. *Journal of Geophysical Research. Atmospheres*, *110*(D9), D09302.

Funke, B., López-Puertas, M., Fischer, H., Stiller, G., von Clarmann, T., Wetzel, G., Carli, B., & Belotti, C. (2007). Comment on "origin of the January-April 2004 increase in stratospheric NO_2 observed in northern polar latitudes" by Jean-Baptist Renard et al. *Geophysical Research Letters*, *34*, 107813. doi:10.1029/2006GL027518.

Funke, B., López-Puertas, M., García-Comas, M., Stiller, G. P., von Clarmann, T., & Glatthor, N. (2008). Mesospheric N_2O enhancements as observed by MIPAS on Envisat during the polar winters in 2002–2004. *Atmospheric Chemistry and Physics*, *8*, 5787–5800.

Funke, B., Baumgaertner, A., Calisto, M., Egorova, T., Jackman, C. H., Kieser, J., Krivolutsky, A., López-Puertas, M., Marsh, D. R., Reddmann, T., Rozanov, E., Salmi, S.-M., Sinnhuber, M., Stiller, G. P., Verronen, P. T., Versick, S., von Clarmann, T., Vyushkova, T. Y., Wieters, N., & Wissing, J. M. (2011). Composition changes after the "halloween" solar proton event: the high-energy particle precipitation in the atmosphere (heppa) model versus MIPAS data intercomparison study. *Atmospheric Chemistry and Physics Discussions*, *11*(3), 9407–9514. doi:10.5194/acpd-11-9407-2011.

Funke, B., Lopez-Puertas, M., Gil-Lopez, S., von Clarmann, T., Stiller, G., Fischer, H., & Kellmann, S. (2005). Downward transport of upper atmospheric NOx into the polar stratosphere and lower mesosphere during the Antarctic 2003 and Arctic 2002/2003 winters. *Journal of Geophysical Research*, *110*(D24). doi:10.1029/2005JD006463.

Grooß, J.-U., Konopka, P., & Müller, R. (2005). Ozone chemistry during the 2002 Antarctic vortex split. *Journal of the Atmospheric Sciences*, *62*(3), 860–870.

Heaps, M. (1978). Parameterization of cosmic-ray ion-pair production-rate above 18 km. *Planetary and Space Science*, *26*(6), 513–517.

Jackman, C., & McPeters, R. (2004). The effects of solar proton events on ozone and other constituents. *Geophysical Monograph*, *141*, 305–319.

Jackman, C. H., DeLand, M. T., Labow, G. J., Fleming, E. L., Weisenstein, D. K., Ko, M. K. W., Sinnhuber, M., & Russell, J. M. (2005). Neutral atmospheric influences of the solar proton events in October-November 2003. *Journal of Geophysical Research*, *110*(A9), A09S27.

Jackman, C. H., Marsh, D. R., Vitt, F. M., Roble, R. G., Randall, C. E., Bernath, P. F., Funke, B., López-Puertas, M., Versick, S., Stiller, G. P., Tylka, A. J., & Fleming, E. L. (2011). Northern hemisphere atmospheric influence of the solar proton events and ground level enhancement in January 2005. *Atmospheric Chemistry and Physics Discussion*, *11*(3), 7715–7755. doi:10.5194/acpd-11-7715-2011.

Jackman, C. H., Marsh, D. R., Vitt, F. M., Garcia, R. R., Fleming, E. L., Labow, G. J., Randall, C. E., Lopez-Puertas, M., Funke, B., von Clarmann, T., & Stiller, G. P. (2008). Short- and medium-term atmospheric constituent effects of very large solar proton events. *Atmospheric Chemistry and Physics*, *8*(3), 765–785.

Kawa, S. R., Kumer, J. B., Douglass, A. R., Roche, A. E., Smith, S. E., Taylor, F. W., & Allen, D. J. (1995). Missing chemistry of reactive nitrogen in the upper stratospheric polar winter. *Geophysical Research Letters*, *22*, 2629–2632. doi:10.1029/95GL02336.

Khosrawi, F., Mueller, R., Proffitt, M. H., Ruhnke, R., Kirner, O., Joeckel, P., Grooss, J. U., Urban, J., Murtagh, D., & Nakajima, H. (2009). Evaluation of CLaMS, KASIMA and

ECHAM5/MESSy1 simulations in the lower stratosphere using observations of Odin/SMR and ILAS/ILAS-II. *Atmospheric Chemistry and Physics*, *9*(15), 5759–5783.

Klee, S., Winnewisser, M., Perrin, A., & Flaud, J.-M. (1999). Absolute line intensities for the ν_6 band of H_2O_2. *Journal of Molecular Spectroscopy*, *195*, 154–161.

Konopka, P., Grooß, J. U., Günther, G., McKenna, D. S., Müller, R., Elkins, J. W., Fahey, D., & Popp, P. (2003). Weak impact of mixing on chlorine deactivation during SOLVE/THESEO2000: Lagrangian modeling (CLaMS) versus ER-2 in situ observations. *Journal of Geophysical Research*, *108*, 8324. doi:10.1029/2001JD000876.

Konopka, P., Steinhorst, H.-M., Grooß, J.-U., Günther, G., Müller, R., Elkins, J. W., Jost, H.-J., Richard, E., Schmidt, U., Toon, G., & McKenna, D. S. (2004). Mixing and ozone loss in the 1999–2000 Arctic vortex: simulations with the 3-dimensional chemical Lagrangian model of the stratosphere (CLaMS). *Journal of Geophysical Research*, *109*, D02315. doi:10.1029/2003JD003792.

Konopka, P., Günther, G., Müller, R., dos Santos, F. H. S., Schiller, C., Ravegnani, F., Ulanovsky, A., Schlager, H., Volk, C. M., Viciani, S., Pan, L. L., McKenna, D.-S., & Riese, M. (2007a). Contribution of mixing to upward transport across the tropical tropopause layer (TTL). *Atmospheric Chemistry and Physics*, *7*(12), 3285–3308.

Konopka, P., Engel, A., Funke, B., Müller, R., Grooß, J.-U., Günther, G., Wetter, T., Stiller, G., von Clarmann, T., Glatthor, N., Oelhaf, H., Wetzel, G., López-Puertas, M., Pirre, M., Huret, N., & Riese, M. (2007b). Ozone loss driven by nitrogen oxides and triggered by stratospheric warmings may outweigh the effect of halogens. *Journal of Geophysical Research*, *112*, D05105. doi:10.1029/2006JD007064.

Kouker, W. (1993). Evaluation of dynamical parameters with a 3-D mechanistic model of the middle atmosphere. *Journal of Geophysical Research*, *98*, 23165–23191.

Kouker, W., Offermann, D., Kull, V., Ruhnke, R., Reddmann, T., & Franzen, A. (1999). Streamers observed by the CRISTA experiment and simulated in the KASIMA model. *Journal of Geophysical Research*, *104*, 16405–16418.

Lacoste-Francis, H. (Ed.) (2010). *MIPAS observations of stratospheric and upper tropospheric trace gases: an overview: Vol. ESA SP-686*. CD-ROM. ESA Publications Division, ESTEC, Postbus 299, 2200 AG Noordwijk, The Netherlands.

Langematz, U., Grenfell, J., Matthes, K., Mieth, P., Kunze, M., Steil, B., & Bruhl, C. (2005). Chemical effects in 11-year solar cycle simulations with the Freie Universität Berlin climate middle atmosphere model with online chemistry (FUB-CMAM-CHEM). *Geophysical Research Letters*, *32*(13). doi:10.1029/2005GL022686.

Lopez-Puertas, M., Funke, B., Gil-Lopez, S., von Clarmann, T., Stiller, G. P., Hopfner, M., Kellmann, S., Fischer, H., & Jackman, C. H. (2005a). Observation of nox enhancement and ozone depletion in the northern and southern hemispheres after the October-November 2003 solar proton events. *Journal of Geophysical Research*, *110*(A9), A09S43.

Lopez-Puertas, M., Funke, B., Gil-Lopez, S., von Clarmann, T., Stiller, G. P., Hopfner, M., Kellmann, S., Tsidu, G. M., Fischer, H., & Jackman, C. H. (2005b). HNO_3, N_2O_5, and $ClONO_2$ enhancements after the October-November 2003 solar proton events. *Journal of Geophysical Research*, *110*(A9), A09S44.

McKenna, D. S., Konopka, P., Grooß, J.-U., Günther, G., Müller, R., Spang, R., Offermann, D., & Orsolini, Y. (2002a). A new chemical Lagrangian model of the stratosphere (CLaMS): 1. Formulation of advection and mixing. *Journal of Geophysical Research*, *107*(D16), 4309. doi:10.1029/2000JD000114.

McKenna, D. S., Grooß, J.-U., Günther, G., Konopka, P., Müller, R., Carver, G., & Sasano, Y. (2002b). A new chemical Lagrangian model of the stratosphere (CLaMS): 2. Formulation of chemistry scheme and initialization. *Journal of Geophysical Research*, *107*(D15), 4256. doi:10.1029/2000JD000113.

Orsolini, Y. J., Manney, G. L., Santee, M. L., & Randall, C. E. (2005). An upper stratospheric layer of enhanced hno3 following exceptional solar storms. *Geophysical Research Letters*, *32*(12), L12S01.

Perrin, A., Valentin, A., Flaud, J.-M., Camy-Peyret, C., Schriver, L., Schriver, A., & Arcas, P. (1995). The 7.9-μm band of hydrogen peroxide: line positions and intensities. *Journal of Molecular Spectroscopy*, *171*, 358–373.

Randall, C., Siskind, D., & Bevilacqua, R. (2001). Stratospheric NO_x enhancements in the southern hemisphere vortex in winter/spring 2000. *Geophysical Research Letters*, *28*, 2385–2388.

Randall, C., Rusch, D., Bevilacqua, R., Hoppel, K., & Lumpe, J. (1998). Polar ozone and aerosol measurement (POAM) II stratospheric NO_2, 1993–1996. *Journal of Geophysical Research*, *103*(D21), 28361–28371.

Randall, C. E., Harvey, V. L., Singleton, C. S., Bailey, S. M., Bernath, P. F., Codrescu, M., Nakajima, H., & Russell, J. M. (2007). Energetic particle precipitation effects on the southern hemisphere stratosphere in 1992–2005. *Journal of Geophysical Research*, *112*(D11), 8308. doi:10.1029/2006JD007696.

Reddmann, T., Ruhnke, R., & Kouker, W. (1999). Use of coupled ozone fields in a 3-D circulation model of the middle atmosphere. *Annales Geophysicae*, *17*, 415–429.

Reddmann, T., Ruhnke, R., & Kouker, W. (2001). Three-dimensional model simulations of SF_6 with mesospheric chemistry. *Journal of Geophysical Research*, *106*, 14525–14537.

Reddmann, T., Ruhnke, R., Versick, S., & Kouker, W. (2010). Modeling disturbed stratospheric chemistry during solar-induced NOx enhancements observed with MIPAS/ENVISAT, *Journal of Geophysical Research*, *115*. doi:10.1029/2009JD012569.

Ridolfi, M., Carli, B., Carlotti, M., von Clarmann, T., Dinelli, B. M., Dudhia, A., Flaud, J. M., Hopfner, M., Morris, P. E., Raspollini, P., Stiller, G., & Wells, R. J. (2000). Optimized forward model and retrieval scheme for mipas near-real-time data processing. *Applied Optics*, *39*(8), 1323–1340.

Rinsland, C. E. A. (1996). ATMOS measurements of $H_2O + 2CH_4$ and total reactive nitrogen in the November 1994 Antarctic stratosphere: dehydration and denitrification in the vortex. *Geophysical Research Letters*, *23*, 2397–2400.

Rodgers, C. D. (2000). Inverse methods for atmospheric sounding: theory and practice. In F. W. Taylor (Ed.), *Series on atmospheric: Vol. 2. Oceanic and planetary physics* (p. 238), Singapore: World Scientific.

Rothman, L. S., Jacquemart, D., Barbe, A., Benner, D. C., Birk, M., Brown, L. R., Carleer, M. R., Chackerian Jr., C., Chance, K., Coudert, L. H., Dana, V., Devi, V. M., Flaud, J.-M., Gamache, R. R., Goldman, A., Hartmann, J.-M., Jucks, K. W., Maki, A. G., Mandin, J.-Y., Massie, S. T., Orphal, J., Perrin, A., Rinsland, C. P., Smith, M. A. H., Tennyson, J., Tolchenov, R. N., Toth, R. A., Vander Auwera, J., Varanasi, P., & Wagner, G. (2005). The HITRAN 2004 molecular spectroscopic database. *Journal of Quantitative Spectroscopy & Radiative Transfer*, *96*, 139–204.

Rozanov, E., Callis, L., Schlesinger, M., Yang, F., Andronova, N., & Zubov, V. (2005). Atmospheric response to NOy source due to energetic electron precipitation. *Geophysical Research Letters*, *32*(14). doi:10.1029/2005GL023041.

Ruhnke, R., Kouker, W., & Reddmann, T. (1999). The influence of the $OH + NO_2 + M$ reaction on the NO_y partitioning in the late Arctic winter 1992/1993 as studied with KASIMA. *Journal of Geophysical Research*, *104*, 3755–3772.

Sander, S. P., Friedl, R. R., Golden, D. M., Kurylo, M. J., Moortgat, G. K., Keller-Rudek, H., Wine, P. H., Ravishankara, A. R., Kolb, C. E., Molina, M. J., Finlayson-Pitts, B. J., Huie, R. E., & Orkin, V. L. (2006). *Chemical kinetics and photochemical data for use in atmospheric studies* (JPL Publication 06-2).

Siskind, D., Nedoluha, G., Randall, C., Fromm, M., & Russell III, J. (2000). An assessment of southern hemispheric stratospheric NO_x enhancements due to transport from the upper atmosphere. *Geophysical Research Letters*, *27*, 329–332.

Smith, A. K., Lopez-Puertas, M., Garcia-Comas, M., & Tukiainen, S. (2009). SABER observations of mesospheric ozone during NH late winter 2002–2009. *Geophysical Research Letters*, *36*. doi:10.1029/2009GL040942.

Stiller, G. P. (Ed.) (2000). *The Karlsruhe optimized and precise radiative transfer algorithm (KOPRA). Wissenschaftliche Berichte: Vol. FZKA 6487*. Forschungszentrum Karlsruhe.

Stiller, G. P., Tsidu, G. M., von Clarmann, T., Glatthor, N., Hopfner, M., Kellmann, S., Linden, A., Ruhnke, R., Fischer, H., Lopez-Puertas, M., Funke, B., & Gil-Lopez, S. (2005). An enhanced HNO_3 second maximum in the Antarctic midwinter upper stratosphere 2003. *Journal of Geophysical Research, 110*(D20), D20303.

Stiller, G. P., von Clarmann, T., Hoepfner, M., Glatthor, N., Grabowski, U., Kellmann, S., Kleinert, A., Linden, A., Milz, M., Reddmann, T., Steck, T., Fischer, H., Funke, B., Lopez-Puertas, M., & Engel, A. (2008). Global distribution of mean age of stratospheric air from MIPAS SF_6 measurements. *Atmospheric Chemistry and Physics, 8*(3), 677–695.

Verronen, P. T., Seppala, A., Kyrola, E., Tamminen, J., Pickett, H. M., & Turunen, E. (2006). Production of odd hydrogen in the mesosphere during the January 2005 solar proton event. *Geophysical Research Letters, 33*(24). doi:10.1029/2006GL028115.

Verronen, P. T., Funke, B., Lopez-Puertas, M., Stiller, G. P., von Clarmann, T., Glatthor, N., Enell, C. F., Turunen, E., & Tamminen, J. (2008). About the increase of HNO_3 in the stratopause region during the Halloween 2003 solar proton event. *Geophysical Research Letters, 35*(20). doi:10.1029/2008GL035312.

Versick, S. (2011). *Ableitung von H_2O_2 aus MIPAS/ENVISAT-Beobachtungen und Untersuchung der Wirkung von energetischen Teilchen auf den chemischen Zustand der mittleren Atmosphäre*. Ph.D. thesis, Karlsruher Institut für Technologie.

Versick, S., Stiller, G., von Clarmann, T., Reddmann, T., Glatthor, N., Grabowski, U., Hopfner, M., Kellmann, S., Kiefer, M., Linden, A., Ruhnke, R., & Fischer, H. (2011). Global stratospheric hydrogen peroxide distribution from mipas-envisat full resolution spectra compared to kasima model results. doi:10.5194/acp-12-4923-2012.

Vogel, B., Konopka, P., Grooß, J.-U., Müller, R., Funke, B., López-Puertas, M., Reddmann, T., Stiller, G., von Clarmann, T., & Riese, M. (2008). Model simulations of stratospheric ozone loss caused by enhanced mesospheric NO_x during Arctic Winter 2003/2004. *Atmospheric Chemistry and Physics, 8*(17), 5279–5293.

von Clarmann, T., Glatthor, N., Hopfner, M., Kellmann, S., Ruhnke, R., Stiller, G. P., Fischer, H., Funke, B., Gil-Lopez, S., & Lopez-Puertas, M. (2005). Experimental evidence of perturbed odd hydrogen and chlorine chemistry after the October 2003 solar proton events. *Journal of Geophysical Research, 110*(A9), A09S45.

von Clarmann, T., Hoepfner, M., Kellmann, S., Linden, A., Chauhan, S., Funke, B., Grabowski, U., Glatthor, N., Kiefer, M., Schieferdecker, T., Stiller, G. P., & Versick, S. (2009). Retrieval of temperature, H_2O, O_3, HNO_3, CH_4, N_2O, $ClONO_2$ and ClO from MIPAS reduced resolution nominal mode limb emission measurements. *Atmospheric Measurement Techniques, 2*(1), 159–175.

Wetzel, G., Bracher, A., Funke, B., Goutail, F., Hendrick, F., Lambert, J.-C., Mikuteit, S., Piccolo, C., Pirre, M., Bazureau, A., Belotti, C., Blumenstock, T., de Mazière, M., Fischer, H., Huret, N., Ionov, D., López-Puertas, M., Maucher, G., Oelhaf, H., Pommereau, J.-P., Ruhnke, R., Sinnhuber, M., Stiller, G., van Roozendael, M., & Zhang, G. (2007). Validation of MIPAS-ENVISAT NO_2 operational data. *Atmospheric Chemistry & Physics, 7*, 3261–3284.

Wissing, J. M., & Kallenrode, M. B. (2009). Atmospheric Ionization Module Osnabruck (AIMOS): a 3-D model to determine atmospheric ionization by energetic charged particles from different populations. *Journal of Geophysical Research, 114*. doi:10.1029/2008JA013884.

Chapter 16
The Impact of Energetic Particle Precipitation on the Chemical Composition of the Middle Atmosphere: Measurements and Model Predictions

Miriam Sinnhuber, Nadine Wieters, and Holger Winkler

Abstract We investigate the impact of energetic particle precipitation on the chemical composition of the middle atmosphere by developing models, and combining model results with observations of the chemical response to particle precipitation events. We show that in the upper stratosphere and lower mesosphere, negative ion chemistry plays a role in addition to the well-known NOx and HOx production due to positive ion chemistry, releasing chlorine from its reservoir, and re-partitioning NOy. Model results also show a large direct impact of energetic electron precipitation on the chemical composition of the upper stratosphere and mesosphere, both during large solar events and during and after geomagnetic storms. Observations show that the indirect impact of energetic electron precipitation events on the middle atmosphere composition can be much larger than the impact of even large solar particle events. However, observations have not shown clear evidence for a direct impact of energetic electron precipitation at altitudes below 80 km so far; if there is a direct impact of energetic electron precipitation on the lower mesosphere and upper stratosphere as suggested by the model results, then it is small compared to the direct contribution of large solar events, or to the indirect impact of energetic electron precipitation due to downward propagation of mesospheric or thermospheric air during polar winter.

M. Sinnhuber · N. Wieters · H. Winkler
Institute of Environmental Physics, University of Bremen, Otto-Hahn Allee. 1, 28359 Bremen, Germany

N. Wieters
e-mail: nwieters@iup.physik.uni-bremen.de

H. Winkler
e-mail: hwinkler@iup.physik.uni-bremen.de

M. Sinnhuber (✉)
Institute for Meteorology and Climate Research, Karlsruhe Institute of Technology, Hermann-von Helmholtz Platz. 1, 76344 Leopoldshafen, Germany
e-mail: miriam.sinnhuber@kit.edu

F.-J. Lübken (ed.), *Climate and Weather of the Sun-Earth System (CAWSES)*, Springer Atmospheric Sciences,
DOI 10.1007/978-94-007-4348-9_16,

16.1 Introduction: Energetic Particle Precipitation

Energetic particles that precipitate into the Earth's atmosphere come from different sources which display different relations to the 11-year solar cycle depending on how particle flux and spectrum are modulated by solar activity.

High-energy particles, mainly protons of 1 MeV to several 100 MeV that can precipitate into the upper stratosphere, are associated with solar coronal mass ejections or solar flares which occur mainly around the solar maximum. These are called *Solar Proton Events* or *Solar Particle Events* (SPEs) as they are associated with an increase of proton fluxes of several orders of magnitude as measured by particle counters onboard geostationary satellites. As the terrestrial atmosphere is shielded against charged particles by its magnetic field, solar particles can precipitate into the Earth's atmosphere only in the region of the polar caps, the area typically $>60°$ geomagnetic latitude.

Galactic cosmic rays (GCRs) are particles of even higher energies which precipitate into the troposphere everywhere. They originate from outside the solar system and provide a continuous particle flux that is moderated by the varying strength of the solar magnetic field throughout the solar cycle.

Solar wind particles can be coupled into the terrestrial geomagnetic field in the magnetotail. There they can either be accelerated into the interior field, forming the source of auroral particles, or be trapped in the magnetosphere, forming the radiation belts. Auroral particles, mainly electrons with energies up to 10 keV, precipitate into the lower thermosphere (≥ 90 km) in the auroral oval, at the inner boundary of the polar caps ($\approx 65°$ geomagnetic latitude). Particles trapped in the radiation belts can be accelerated during *geomagnetic storms* to energies ranging from tens of keV to several MeV, and precipitate into the atmosphere in so-called *Energetic Electron Precipitation* (EEP) events. These particles precipitate into the atmosphere in geomagnetic latitudes connecting to the radiation belts ($\approx 59°$–$68°$ geomagnetic latitude [*Horne et al.*, 2009]). Geomagnetic storms are initiated by disturbances in the interplanetary plasma, which can be due to, e.g., fast solar wind streams or solar coronal mass ejections. They can occur in all phases of the solar cycle but are more frequent during solar maximum and in the transition from solar maximum to solar minimum, and more rare during the deep solar minimum.

Energetic Particle Precipitation (EPP) into the atmosphere leads to decomposition and ionization of the most abundant species (N_2, O_2, H_2O, O, or NO, depending on altitude). Ionization of the atmosphere leads to fast ion chemistry in which large cluster ions are formed from the primary O^+, O_2^+, N_2^+ and NO^+ ions, and chemically relatively inert H_2O and N_2 are transformed into the chemically active radicals H, OH [*Swider and Keneshea*, 1973; *Solomon et al.*, 1981], N, NO [*Crutzen et al.*, 1975; *Porter et al.*, 1976; *Rusch et al.*, 1981] and O [*Porter et al.*, 1976]. Both HOx (H, OH, HO_2) and NOx (N, NO, NO_2) can destroy ozone in catalytic cycles, HOx mainly at altitudes above 45 km, NOx more effectively at altitudes below 45 km [*Lary*, 1997].

Both excess NOx and ozone loss have been observed during and after large solar proton events, and are reproduced by chemical models reasonably well, e.g.,

[*Solomon et al.*, 1983; *Jackman et al.*, 2001, 2005b; *Rohen et al.*, 2005]. While HOx is short-lived, and HOx recovers quickly after the event, NOx can be very long-lived in the polar middle atmosphere especially during polar winter, when it can also be transported down into the stratosphere and destroy ozone there [*Sinnhuber et al.*, 2003b; *Jackman et al.*, 2005a; *Winkler et al.*, 2008]. This is called the *indirect effect* of particle precipitation. Enhanced NOx values have indeed been observed in the mid-stratosphere after the July 2000 solar proton event [*Randall et al.*, 2001]. These observations can be explained by downward propagation of particle-induced NOx, and are quite well reproduced by chemistry-transport models of the middle atmosphere, e.g., [*Sinnhuber et al.*, 2003c]. Thus, the impact of large SPEs on NOx and ozone loss seems to be qualitatively well understood, but not much is known about the impact of atmospheric ionization on other trace gases besides NOx and ozone.

The impact of energetic electron precipitation directly into the middle atmosphere is not as well investigated as that of the large solar particle events. It has been emphasized by some authors that EEPs can have a similar large impact on the chemical composition of the mesosphere and stratosphere as SPEs [*Callis et al.*, 1998; *Siskind and Russel III*, 1996; *Siskind et al.*, 2000]. Measurements of ozone in high polar latitudes suggest a large influence of magnetospheric electrons of relativistic energies on ozone concentrations in the mid-stratosphere winter [*Sinnhuber et al.*, 2006]. However, observations of the *direct effect* of energetic electron precipitation during an EEP—i.e., the local production of NOx and HOx, and subsequent ozone loss during the particle event—have been shown to be more complicated than for SPEs, possibly because the magnetospheric electrons precipitate into a much smaller area. In recent years, direct observations of large NOx productions due to EEPs have been reported two times [*Renard et al.*, 2006; *Clilverd et al.*, 2009]. These, however, have been interpreted by other authors as downward propagation of NOx from the upper mesosphere or lower thermosphere, probably produced by auroral precipitation [*López-Puertas et al.*, 2006; *Funke et al.*, 2007]. Enhanced OH values have been reported to be correlated with geomagnetic storms in a recent paper by *Verronen et al.* [2011]. However, in this case, significant enhancements were observed only above 70 km, and it is, to date, not clear whether energetic electrons can directly impact the stratosphere and lower mesosphere.

Thus, it seems that the NOx production and subsequent ozone loss during large solar events are reasonably well understood. However, two questions are still open:

- Are other constituents besides NOx and ozone affected by atmospheric ionization, and if yes, by how much?
- How does the impact of EEPs compare to that of SPEs, i.e. is the impact on the chemical composition of the middle atmosphere and particularly on stratospheric ozone comparable to that of large solar events?

These two questions will be addressed in the following using a combination of models of different complexity with observations of middle atmosphere constituents during and after large particle precipitation events.

In Sect. 16.2, the models used for the investigation are described. Section 16.3 describes the Heppa model versus MIPAS data intercomparison initiative, a multi-model intercomparison of chemical changes during and after the October/November

2003 solar particle event, and discusses the most important results of this initiative. In Sect. 16.4, the impact of negative ions on the atmospheric composition is discussed using results from the UBIC ion chemistry model; in Sect. 16.5, the impact of energetic electron precipitation is investigated.

Energetic particle precipitation impacts on the middle atmosphere are also discussed in Chaps. 8, 9, 15, and 17.

16.2 Models

We use models of different complexity to address different aspects of the chemical changes and dynamical couplings related to energetic particle precipitation, ranging from the one-dimensional box-model of middle atmosphere ion chemistry (UBIC, Sect. 16.2.3) to global models of chemistry and transport in the middle atmosphere either driven by prescribed temperatures and wind-fields (B3dCTM, Sect. 16.2.2) or free-running (B2dM, Sect. 16.2.1).

The neutral models share the same description of chemistry, which is based on the SLIMCAT chemistry code [*Chipperfield*, 1999]. This considers 58 neutral trace gases and about 180 gas phase, photochemical, and heterogeneous reactions between those trace species. Reaction rates and absorption cross sections are prescribed by the JPL recommendation of *Sander et al.* [2006].

Atmospheric ionization due to energetic particle precipitation is provided by the AIMOS model, which calculates global altitude-dependent three-dimensional distributions of atmospheric ionization based on observed proton and electron fluxes [*Wissing and Kallenrode*, 2009], see also Chap. 13. The impact of positive ion chemistry and particle induced decomposition of N_2 and O_2 is parameterized based on *Rusch et al.* [1981]; *Porter et al.* [1976] and *Solomon et al.* [1981]. Hence 1.25 NOx are produced (of which 45 % are N, and 55 % NO) as well as up to 2 HOx constituents depending on altitude and ionization rate, and 1.15 O per ion pair.

16.2.1 The Bremen 2-Dimensional Model B2dM

The Bremen 2-dimensional model (B2dM) has been developed originally as a combination of the THIN AIR 2-dimensional general circulation model [*Kinnersley*, 1996] and the chemistry code of the SLIMCAT model [*Chipperfield*, 1999]. The model calculates temperature, pressure and wind fields on isentropic surfaces driven by prescribed sea-surface temperatures and the Montgomery potential. The vertical extent in its present setting is from the surface to about 100 km with a vertical spacing of about 3 km; the horizontal resolution is rather poor, with 19 grid boxes equally distributed in latitude from pole to pole (about 9°). The model uses a family approach considering Ox ($O_3 + O(^3P) + O(^1D)$), NOx ($N + NO + NO_2$), ClOx ($Cl + ClO + 2Cl_2O_2$), BrOx ($Br + BrO$), HOx ($H + OH + HO_2$), and CHOx

(CH_3, CH_3O_2, CH_3OOH, CH_3O, CH_2O and HCO) in the stratosphere, and a non-family version using exactly the same reactions and rates in the mesosphere above $\approx$55 km. The Bremen 2-dimensional model has been used in a number of studies to investigate the impact of energetic particle precipitation on the middle atmosphere in the past [*Sinnhuber et al.*, 2003b, 2003c; *Rohen et al.*, 2005; *Winkler et al.*, 2008].

16.2.2 The Bremen 3-Dimensional Chemistry and Transport Model B3dCTM

The Bremen 3-dimensional Chemistry and Transport Model (B3dCTM) is a combination of the chemistry-transport model CTM-B [*Sinnhuber et al.*, 2003a] with the chemistry code of the Bremen 2-dimensional model of the stratosphere and mesosphere [*Sinnhuber et al.*, 2003b; *Winkler et al.*, 2008]. The B3dCTM is a global 3-dimensional model with a horizontal resolution of 3.75° in longitude and 2.5° in latitude, which is forced by prescribed temperatures and wind-fields, thus, the vertical range of this model is restricted by the availability of these data. Advection is calculated by using the second order moments scheme of *Prather* [1986]. Two versions of this model have been used in this investigation:

The first version of the model uses isentropic surfaces. The lower model boundary is limited by the use of potential temperature as vertical coordinate. Vertical transport perpendicular to the isentropes is derived from diabatic heating and cooling rates calculated using the MIDRAD radiation scheme [*Shine*, 1987]. This will be called the *stratospheric* model version in the following. Here results from model runs driven by European Centre for Medium-Range Weather Forecasts (ECMWF) (ERA Interim [*Simmons et al.*, 2006; *Dee et al.*, 2011]) reanalysis are used, which restricts the model upper boundary to 0.1 hPa (55–60 km), with a vertical resolution of about 1 km in the lower stratosphere, increasing to about 4 km at 60 km altitude.

Additionally, a model version has been developed which runs on isobaric surfaces; this enables us to extend the vertical range of the model based on the availability of meteorological data. At the moment, data from the LIMA general circulation model [*Berger*, 2008] are used, which is nudged to ECMWF ERA 40 in the lower stratosphere, and covers the vertical range from the surface to $\approx$130–140 km. This will be called the *LIMA* model version in the following; it currently runs on 30 isobaric surfaces from 247.8 to 0.00016 hPa (approximately 10 to 100 km) with a vertical resolution of approximately 3 km. The transport is calculated by vertical and horizontal wind fields as provided by the LIMA model.

Both model versions use the same family chemistry scheme as the 2-dimensional model in the stratosphere (Sect. 16.2.1); however, a non-family version for the mesospheric chemistry has been developed for the LIMA model version. Both model versions use a parameterization for NOx and HOx production due to positive ion chemistry as discussed above; additionally, parameterizations for negative ion chemistry developed based on results from the UBIC model (Sect. 16.2.3) can be used (see Sect. 16.4, Fig. 16.5). Results from the same model family based on CTM-B are

Table 16.1 Positive ions, negative ions, and neutral species implemented in the UBIC model

Cations	N^+, N_2^+, NO^+, NO_2^+, $O^+(^4S)$, $O^+(^2D)$, $O^+(^2P)$, O_2^+, $O_2^+(a^4)$, O_4^+, O_5^+, H^+, CO^+, CO_2^+, HCO^+, H_2O^+, $O_2^+(H_2O)$, $H^+(H_2O)_{n=1...7}$, $H^+(H_2O)(OH)$, $H^+(H_2O)(CO_2)$, $H^+(H_2O)_2(CO_2)$, $H^+(H_2O)(N_2)$, $H^+(H_2O)_2(N_2)$, $H^+(CH_3CN)$, $H^+(CH_3CN)(H_2O)_{n=1...6}$, $H^+(CH_3CN)_2$, $H^+(CH_3CN)_2(H_2O)_{n=1...4}$, $H^+(CH_3CN)_3$, $H^+(CH_3CN)_3(H_2O)_{n=1,2}$, $NO^+(H_2O)$, $NO^+(H_2O)_2$, $NO^+(H_2O)_3$, $NO^+(CO_2)$, $NO^+(N_2)$, $NO^+(H_2O)(CO_2)$, $NO^+(H_2O)_2(CO_2)$, $NO^+(H_2O)(N_2)$, $NO^+(H_2O)_2(N_2)$, $NO_2^+(H_2O)_{n=1,2}$
Anions	e, O^-, O_2^-, O_3^-, O_4^-, OH^-, NO_2^-, NO_3^-, CO_3^-, CO_4^-, CH_3^-, HCO_3^-, $O^-(H_2O)$, $O_2^-(H_2O)_{n=1,2}$, $O_3^-(H_2O)_{n=1,2}$, $OH^-(H_2O)_{n=1,2}$, $NO_2^-(H_2O)_{n=1,2}$, $NO_3^-(H_2O)_{n=1,2}$, $CO_3^-(H_2O)_{n=1,2}$, $NO_3^-(HNO_3)_{n=1...4}$, $NO_3^-(HNO_3)(H_2O)$, $NO_3^-(HNO_3)_2(H_2O)$, $H_2SO_4^-$, $HSO_4^-(H_2SO_4)_{n=1,2}$, $HSO_4^-(H_2SO_4)(H_2O)$, $HSO_4^-(HNO_3)_{n=1,2}$, $HSO_4^-(HNO_3)(H_2O)$, $HSO_4^-(HNO_3)_2(H_2O)$, $HSO_4^-(H_2SO_4)(HNO_3)$, $HSO_4^-(H_2SO_4)(HNO_3)(H_2O)$, Cl^-, Cl_2^-, Cl_3^-, ClO^-, $ClO^-(HCl)$, $ClO^-(H_2O)$, $ClO^-(CO_2)$, $ClO^-(HO_2)$, $NO_3^-(HCl)$
Neutrals	$N(^4S)$, $N(^2D)$, N_2, $O(^3P)$, $O(^1D)$, O_2, O_3, H, H_2, OH, HO_2, NO, NO_2, NO_3, N_2O, H_2O, CH_4, CH_3, CO_2, CO, HCO_3, HNO_3, HNO_2, N_2O_5, H_2SO_4, CH_3CN, Cl, Cl_2, ClO, $ClNO_2$, $ClONO_2$, HCl, HOCl

also shown in Chap. 9. The B3dCTM is also used to investigate diurnal variations of ozone measured above Ny Ålesund, Spitsbergen, by a ground-based microwave instrument (see Chap. 8). The *B3dCTM LIMA* model version as well as the combination with UBIC parameterizations have been developed in the framework of the priority program CAWSES funded by the German funding agency *Deutsche Forschungsgemeinschaft* DFG.

16.2.3 The University of Bremen Ion Chemistry Model UBIC

The University of Bremen Ion Chemistry (UBIC) model [*Winkler*, 2007; *Winkler et al.*, 2009] has been developed to study the ion chemistry of the middle atmosphere, and the interaction with neutral species in detail. In particular, it can be used to simulate the impact of energetic particle precipitation on middle atmosphere chemistry. UBIC is an ion chemistry box model which simulates the time evolution of 138 charged and uncharged species considering more than 600 reactions using the semi-implicit symmetric integration method [Ramarson, 1989]. Table 16.1 lists all charged and neutral species considered in UBIC. The model accounts for photo-ionization of NO by Lyman-α radiation, as well as photo-dissociation and photo-detachment of electrons. Particle impact ionizations are considered by the means of external ionization rates, e.g. provided by the AIMOS model [*Wissing and Kallenrode*, 2009], see also Chap. 13.2; additionally, a parameterization for Galactic cosmic rays is implemented [*Heaps*, 1978]. The ionization is distributed on the main atmospheric constituents N_2 and O_2 according to their abundance and ionization cross sections [*Rusch et al.*, 1981; *Porter et al.*, 1976; *Zipf et al.*, 1980]. The model's set of reactions is a combination of the reactions assumed to govern the ion chemistry in the stratosphere, mesosphere, and lower thermosphere, taken from *Brasseur and Chatel* [1983]; *Viggiano et al.* [1994]; *Kopp* [1996]; *Rees* [1998]; *Kazil* [2002]; *Verronen* [2006]. UBIC can either be used on-line with a neutral chemistry model or it can operate as an equilibrium model to calculate plasma concentrations and production rates of uncharged species due to ionizations. UBIC is also used to calculate parameterizations of ion chemistry impacts on the neutral atmosphere for the use in global models (see Sect. 16.4). UBIC has been developed within CAWSES based on an older model version considering only a simple positive ion scheme to investigate in detail the impact of ion chemistry on atmospheric composition during large energetic particle events (see, e.g., Sect. 16.4).

16.3 The Heppa Model Versus MIPAS Data Intercomparison Study

On October 29, 2003, one of the largest solar proton events recorded so far occurred. This event was exceptional both for its size, and for the fact that it was covered very well by global observations from several satellite instruments detecting both the expected NOx increase [*Jackman et al.*, 2005a; *López-Puertas et al.*, 2005a] and ozone loss [*Jackman et al.*, 2005a; *Rohen et al.*, 2005]. Additionally, MIPAS/ENVISAT observed a number of trace species that have not been observed during an SPE before, e.g., HNO_3, N_2O_5, $ClONO_2$, HOCl, ClO, and others [*López-Puertas et al.*, 2005b; *von Clarmann et al.*, 2005]. These observations provided a unique natural experiment to test our understanding of the chemical changes during and after large atmospheric ionization events.

A model-measurement intercomparison study was set up involving nine models of different complexity including a 1D/2D model, 3-dimensional global chemistry-transport models, and 3-dimensional global coupled chemistry-climate models, the so-called *Heppa model versus MIPAS data intercomparison study*, or *Heppa intercomparison*, [*Funke et al.*, 2011]. All nine models carried out model experiments for the time-period of the October/November 2003 SPE using the same set of ionization rates considering both protons and electrons provided by the AIMOS model (*Wissing and Kallenrode* [2009], see also Chap. 13). Model results of temperature, CH_4, CO, NO, NO_2, N_2O, HNO_3, N_2O_5, HNO_4, O_3, H_2O_2, ClO, HOCl, and $ClONO_2$ were provided on the geolocation and local time of MIPAS overpass, and compared against the MIPAS observations. Both the B2dM and the B3dCTM participated in the Heppa intercomparison study (see *Funke et al.* [2011]). B3dCTM provided hourly model output. As B2dM considers a zonally averaged state, and thus cannot consider longitudinal inhomogeneity of the polar cap which is a result of the displacement of the geomagnetic pole to the geographic pole, model runs were carried out for different longitudes with B2dM. 1-dimensional model runs were initialized from these B2dM model runs at the positions of individual MIPAS measurements, and model results were thus provided at the local time and geolocation of every individual MIPAS observation.

Results from the Heppa intercomparison initiative considering results from all nine contributing models have been published recently (see *Funke et al.* [2011]). Here we will give an overview of the most important results concerning our understanding of processes during and after particle precipitation events. Results from the Heppa initiative are also given in Chap. 15.

In the multi-model average, there is good agreement with observed changes both for NOy production and ozone loss during the event at most altitudes (see Fig. 15.9 in Chap. 15 for NOy changes). Systematic differences between MIPAS measurements and model results are observed around 1 hPa, where NOy is overestimated by all models, and above 0.2 hPa, where NOy production is underestimated by all models compared to the observations; this may be due to systematic features in the ionization rates. However, there is also a very large variability and spread between models, and between individual models and the observations. This variability is apparently related to the different dynamics of the participating models, especially to the vortex strength and stability, which varies greatly from model to model. Results of the intercomparison of NOy are also discussed in more detail in Chap. 15.

During the October/November 2003 SPE, an increase of HNO_3 was observed for the first time as a response to a large atmospheric ionization event. The increasing NOx and HOx concentrations after an energetic particle precipitation event can lead to formation of nitric acid through the gas phase reaction

$$NO_2 + OH + M \rightarrow HNO_3 + M. \tag{16.1}$$

However, comparison of model results with MIPAS measurements show that the observed HNO_3 increase cannot be explained by neutral reactions due to the NOx and HOx increase alone as shown in Fig. 16.1 exemplarily for B2dM and B3dCTM. Two additional pathways have been proposed to explain the HNO_3 increase due to atmospheric ionization, both including ion chemistry reactions: Conversion of N_2O_5 to HNO_3 in reactions with positive ion clusters [*Boehringer et al.*, 1983], and HNO_3 production through recombination of positive water clusters with negative NO_3^--containing ions [*Verronen et al.*, 2008]. *Verronen et al.* [2008] have investigated these observations using the SIC ion chemistry model, and found that the HNO_3 increase observed during the event can be interpreted qualitatively by recombination reactions between water cluster ions and NO_3^--containing ions; however, the observed increase is overestimated by the model during polar night up to a factor of 2.5 depending on altitude. Similar results as shown in Fig. 16.1 have been obtained by all models not considering HNO_3 formation due to ion chemistry. However, two models did consider additional HNO_3 formation due to ion chemistry: the FinRose model uses a parameterization of HNO_3 production based on the reaction pathway described in *Verronen et al.* [2008] and overestimates the HNO_3 increase during the event by more than a factor of 4; and the KASIMA model uses a parameterization of the water cluster ion chain [*Boehringer et al.*, 1983], see also Chap. 15. KASIMA also underestimates the HNO_3 increase during the event, but has better agreement for the second HNO_3 enhancement in late November than the other models (see also *Funke et al.* [2011]). This comparison shows that to understand the observed changes to HNO_3, ion chemistry has to be taken into account. Qualitatively, the ion

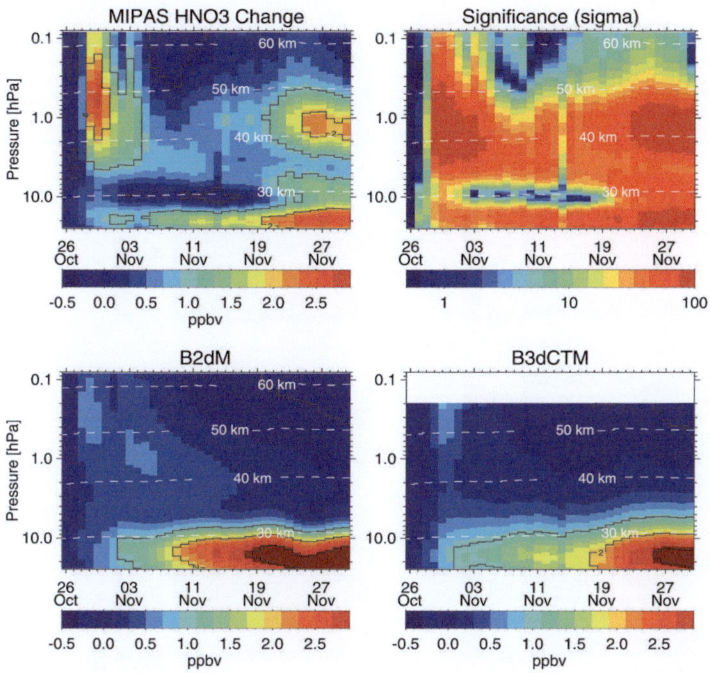

Fig. 16.1 MIPAS HNO_3 change averaged over 70–90°N during the October/November 2003 SPE, compared to modeled HNO_3 changes by the B2dM and B3dCTM models using AIMOS ionization rates considering both protons and electrons, sampled at the geolocations and local times of the MIPAS observations. Also shown is the 1-sigma significance of the observations, the ratio of the average values to the standard deviation. Adapted from *Funke et al.* [2011]

chemistry processes appear to be understood, however, there are still problems to reproduce them quantitatively within the uncertainty of the observations.

Also observed for the first time during a large solar event were a number of chlorine species, namely ClO, HOCl, and $ClONO_2$. Both HOCl and $ClONO_2$ were found to increase during the SPE, while ClO increased at the vortex edge, but decreased in the vortex core. The decrease of ClO via the reaction ClO + NO_2 is reproduced by the models qualitatively, but the absolute values are underestimated by about a factor of 2–4 as shown exemplarily for B2dM and B3dCTM in the left panel of Fig. 16.2; the increase of ClO at the vortex edge is not reproduced by the models, hinting at an additional chemical process not considered in the models (see *Funke et al.* [2011]). The increases of HOCl and $ClONO_2$ were reproduced qualitatively by most models, however, the absolute values especially of $ClONO_2$ were underestimated by most models (see also right panel of Fig. 16.2). This might be either due to an underestimation of ClO by the models already before the particle event (see *Funke et al.* [2011]), or to additional ion chemistry involving chlorine which is not included in the models. This latter possibility is investigated in more detail in the following Sect. 16.4.

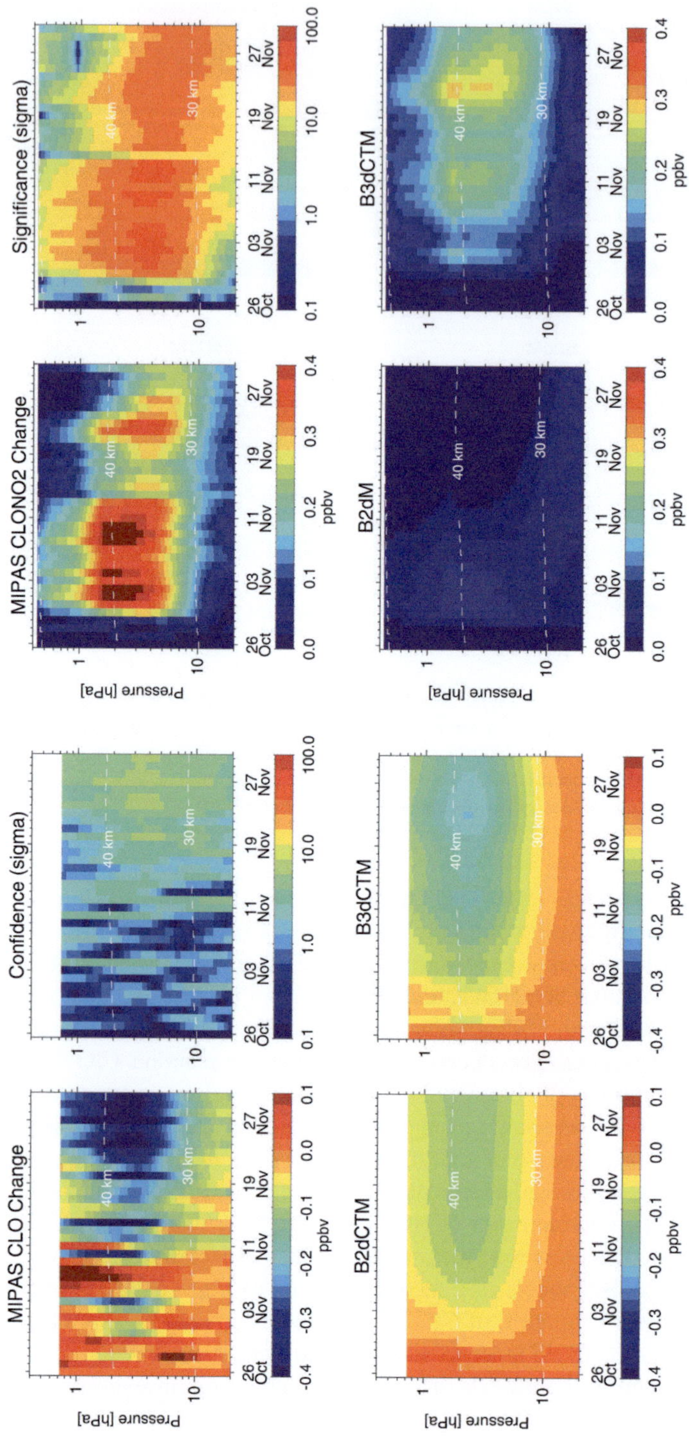

Fig. 16.2 MIPAS ClO (*left*) and $ClONO_2$ (*right*) change averaged over 70–90°N during the October/November 2003 SPE, compared to modeled changes by B2dM and B3dCTM sampled to the MIPAS geolocations and local times. Also shown are the significances of the ClO and $ClONO_2$ observations (see Fig. 16.1). Right panel adapted from *Funke et al.* [2011]

16.4 The Role of Negative Ion Chemistry for Chlorine Activation During SPEs

The effects of solar particle events on NOx and ozone appear to be qualitatively well understood. They can be reproduced by atmospheric models considering the production of NOx and HOx parameterized as a result of positive ion chemistry. However, there have been considerable differences between model predictions and measurements concerning other chemical compounds, in particular nitrogen and chlorine species (see Sect. 16.3). It has been pointed out by *Solomon and Crutzen* [1981] that SPEs might influence chlorine chemistry due to the HOx and NOx increase by transforming hydrogen chloride into reactive chlorine:

$$HCl + OH \rightarrow Cl + H_2O, \tag{16.2}$$

followed by the formation of chlorine nitrate at the expense of reactive radicals, especially at stratospheric altitudes:

$$ClO + NO_2 + M \rightarrow ClONO_2 + M. \tag{16.3}$$

Model simulations which only account for NOx and HOx production due to positive ion chemistry fail to reproduce the observed chlorine perturbations (e.g. *Funke et al.* [2011], see also Sect. 16.3). Apparently, there are impacts on the neutral chemistry in addition to the well-known release of NOx and HOx. Therefore, a detailed consideration of the ion chemical processes is necessary to understand the chemical effects of EPPs. Here, we investigate the impact of negative ion chemistry on chlorine species in the upper stratosphere and lower mesosphere.

Under geomagnetically quiet conditions, negative chlorine species are a significant fraction of the total anion density in the mesosphere, see e.g. *Chakrabarty and Ganguly* [1989]; *Fritzenwallner and Kopp* [1998]. Therefore, it can be assumed that during EPPs reactions of negative ions influence the chlorine chemistry. Hydrogen chloride reacts with several anions to produce Cl^-:

$$HCl + X^- \rightarrow Cl^- + \text{Rest} \tag{16.4}$$

where X^- can be $\{O_2^-, O^-, CO_3^-, OH^-, NO_2^-, NO_3^-\}$. The most abundant chlorine ions in the mesosphere are Cl^- and $Cl^-(H_2O)$, and while reactions of both species with atomic hydrogen re-release HCl, recombination reactions with cations can lead to a production of Cl, ClO, $ClNO_2$, and Cl_2, depending on the detailed reaction pathways. In the latter case, the reservoir compound HCl is partly converted to active chlorine species, similar as in reaction (16.2). Additionally, there is the reaction of chlorine nitrate with the carbon trioxide ion which is one of the most abundant anions in the lower mesosphere:

$$ClONO_2 + CO_3^- \rightarrow ClO + NO_3^- + CO_2. \tag{16.5}$$

This reaction also transforms chlorine from a reservoir to a radical. Note that this process counteracts reaction (16.3). For further details on the atmospheric ion chemistry of chlorine see *Kopp* [1996]; *Kopp and Fritzenwallner* [1997].

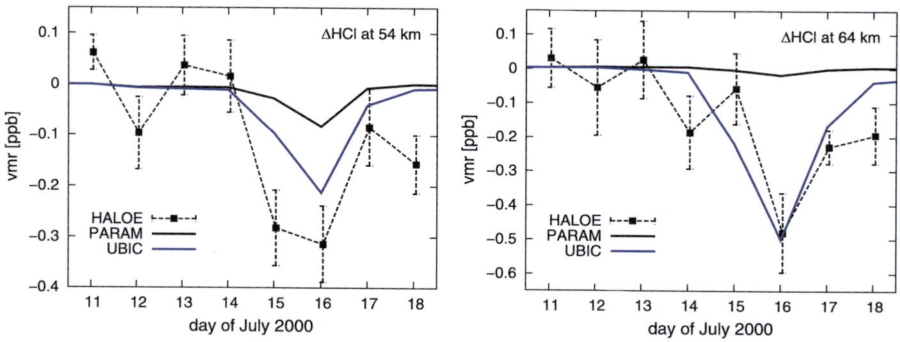

Fig. 16.3 Differences of zonally averaged HCl mixing ratios during the solar proton event in July 2000 at two different altitudes for sunrise conditions (differences with respect to the mean HCl sunrise values of 10–12 July). Shown are HALOE data and simulation results from the atmospheric model at 66.5° North. The HALOE error bars represent one standard deviation. PARAM indicates the model with parametrized production rates for HOx and NOx, and UBIC the model with full ion chemistry. Figure adapted from *Winkler et al.* [2011]

In order to study the effect of the negative ion chemistry on chlorine species during an EPP event we have performed model simulations of the major SPE in July 2000, and compared them to HCl measurement data from the UARS HALOE instrument [*Russel et al.*, 1993]. The HALOE observations of HCl loss during the July 2000 SPE are also discussed in *Kazeminejad* [2009]. The results of the model study have been published in *Winkler et al.* [2009, 2011], and a brief summary is given here. For the purpose of our study, the 2-dimensional model B2dM was used (see Sect. 16.2.1) in combination with the UBIC ion chemistry model described in Sect. 16.2.3. Additional to the UBIC simulations, model runs with parameterized production of NOx and HOx due to the positive ion chemistry during the SPE have been carried out (assuming 1.25 NOx per ionization with 45 % $N(^4S)$ and 55 % $N(^2D)$ according to *Porter et al.* [1976]; *Rusch et al.* [1981], and up to two HOx constituents per ionization in the height region of interest according to *Solomon et al.* [1981]). These simulations are called PARAM model runs in the following. Figure 16.3 shows that the simulations with the UBIC model yield significantly larger HCl losses than the PARAM model runs, and they agree much better with the HALOE measurements. At $\approx$64 km altitude the HCl decrease predicted by the UBIC model for the main event phase (July 16) is in the order of 500 ppt (for sunrise conditions) which agrees with the HALOE observations of HCl decrease within error bars (actually, on this altitude and day the agreement is better than 4 %). The PARAM model is unable to reproduce the effect on HCl at that altitude. If all negative ion chemistry reactions involving chlorine species are switched off (not shown), the simulation results do not differ significantly from the PARAM results. Therefore, the differences arise from the negative ion chemistry of the chlorine species. At lower altitudes, the observed decrease of HCl mixing ratios is smaller and also the difference between the UBIC and PARAM results gets smaller. This is due to the fact that in the lower parts of the middle atmosphere the negative ion chemistry

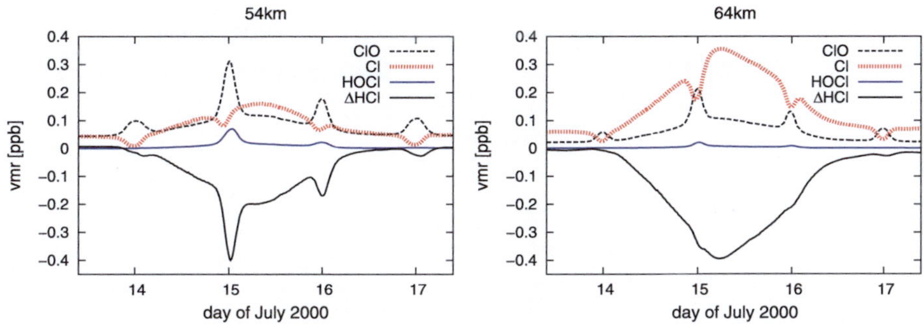

Fig. 16.4 UBIC modeled ΔHCl in comparison with other chlorine species during the solar proton event in July 2000 at 54 km and 64 km at 66.5° North. Figure adapted from *Winkler et al.* [2009]

is no longer dominated by chlorine species but rather by HCO_3^-, CO_3^-, NO_3^-, and their hydrates. The short-lived decrease of HCl during the SPE in July 2000 corresponds to increasing amounts of other chlorine species. Figure 16.4 shows the enhanced mixing ratios of Cl, ClO, and HOCl in comparison with the HCl loss; at these altitudes, all activated chlorine is transferred either to Cl, ClO, or HOCl, and the partitioning between the active chlorine species, and at 54 km also the amount of chlorine activated, depend on the solar zenith angle.

Systematic UBIC model runs have been performed to identify the key parameters on which the impacts on chlorine species depend. These parameters are: Ionization rate, solar zenith angle, pressure altitude, HCl, Cl + ClO, $ClONO_2$, NOx, and H_2O. From UBIC simulations considering these dependencies, a parameterization of the impact of the negative ion chemistry on chlorine activation and chlorine partitioning has been developed. The resulting lookup-table can be used by global three-dimensional models of the middle atmosphere to account for the chemical impact of negative ions on chlorine partitioning without running the time consuming ion chemistry model. This parameterization has been implemented in the *stratospheric* version of B3dCTM (see Sect. 16.2.2) to investigate the impact of negative chlorine chemistry on atmospheric composition during the large solar particle event of October/November 2003 on a global scale. First results are presented in Fig. 16.5. Given are the relative difference between model runs with and without ionization impacts, and the relative difference between model runs considering atmospheric ionization, with and without parameterization for negative chlorine ion chemistry. Generally speaking, the additional change of ClO, HOCl and $ClONO_2$ due to negative chlorine ion chemistry appears to be small compared to the chlorine activation due to the HOx increase at the latitudes, local times, and altitudes considered here. The largest changes—an increase of about 160 ppt—are observed for HOCl at altitudes above 40 km during the event. A small but significant decrease of $ClONO_2$ is observed during the event at altitudes around 40 km, and ClO also increases by about 80–100 ppt above 50 km during the event, leading to a maximal release of active chlorine of about 260 ppt above 50 km, comparable to the observed loss of HCl during the large SPE of July 2000 (see Fig. 16.3). This additional chlorine activation

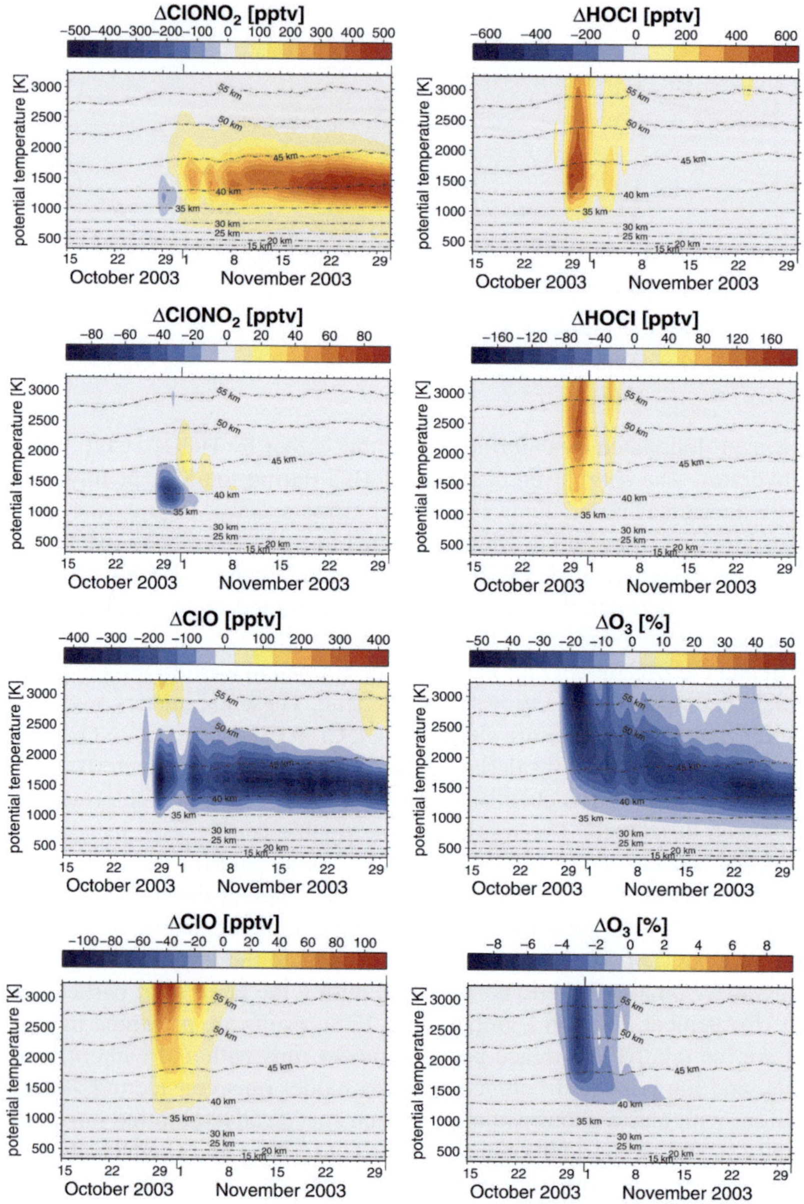

Fig. 16.5 Impact of the SPE of October/November 2003 on chlorine species and ozone, northern polar cap mean values (70°N–90°N). Model results from the *stratospheric* version of B3dCTM using a parameterization of chlorine activation based on UBIC results. The *top* and *third* panel from the top show differences of the SPE model run considering HOx, NOx, and ClOy (UBIC parameterization) production and an undisturbed model run. The *second* panel from the top and the *lowest* panel show the differences due to the impacts of the negative chlorine chemistry (UBIC parameterization compared to the SPE run with HOx and NOx production only)

leads to additional ozone loss of several percent, with maximal values of up to 6 % during the event in an altitude range of 50–55 km.

16.5 The Role of Energetic Electron Precipitation

The potential impact of energetic electron precipitation into the atmosphere has been investigated in two ways: by carrying out model studies with the B3dCTM driven by AIMOS ionization rates with and without electron ionization, and by analyzing global observations of NOx on longer time-scales. Results from the first approach are discussed in Sect. 16.5.1, results from the second approach are discussed in Sect. 16.5.2.

16.5.1 Impact of Energetic Electrons on the Upper Stratosphere Based on Model Experiments

Model runs with the B3dCTM *stratosphere* version were carried out for the period October/November 2003 considering a 'base' situation without atmospheric ionization, a model scenario with atmospheric ionization considering protons only, and a model scenario considering atmospheric ionization by protons as well as electrons. Atmospheric ionization rates considering protons and electrons were provided by the AIMOS model (see Chap. 13). The period October/November 2003 was chosen because it contains a very large solar event on October 29/30, but also very high levels of geomagnetic activity before and after the solar event. Thus, both the additional impact of precipitating electrons during an SPE and the impact of electron precipitation during geomagnetic storms can be studied. Results from these model experiments have been published in *Wissing et al.* [2010], see also Chap. 13.

In Fig. 16.6, results for a day of strong geomagnetic activity before the solar event are shown relative to the 'base' model run. Comparison of the model run with protons only and the model run with protons and electrons show significant ozone losses expected based on the AIMOS ionization rates from precipitating electrons above $\approx$45 km at high latitudes in both hemispheres, with maximum values of more than 15 % above 55 km.

In Fig. 16.7, NOx production during and after the large SPE are compared for model scenarios considering protons only, and considering protons and electrons. Considering electrons leads to larger NOx production both in Southern mid-latitudes, and in high polar regions in both hemispheres.

In Fig. 16.8, modeled ozone loss at the position of Ny Ålesund are shown considering protons and electrons, relative to a model run without atmospheric ionization. Here, results from the *LIMA* version of B3dCTM are shown, extending the altitude range to $\approx$100 km. Fairly significant ozone losses of more than 80 % are observed during the solar event, and areas of low ozone (losses of more than 10 % compared

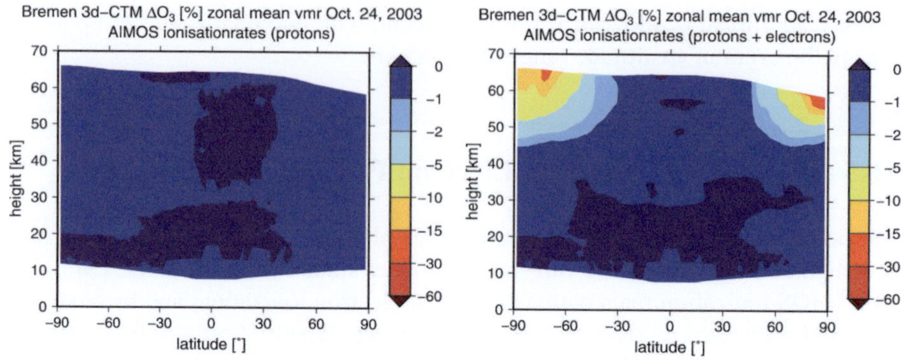

Fig. 16.6 Modeled O_3 changes due to atmospheric ionization using B3dCTM (stratospheric version) on a day of strong geomagnetic activity, considering protons only (*left*) and protons + electrons (*right*). Loss calculated relative to model run without atmospheric ionization. Adapted from *Wissing et al.* [2010]

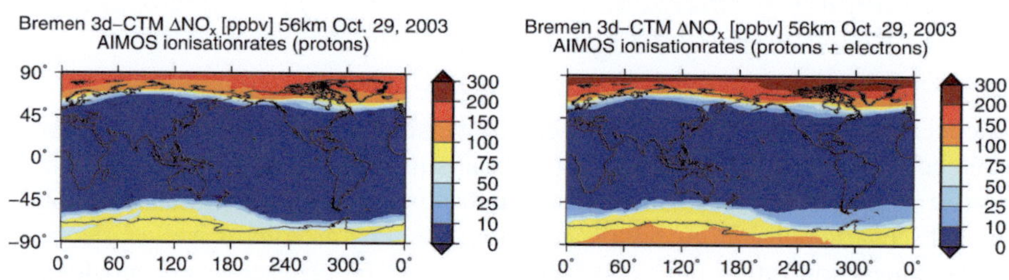

Fig. 16.7 Modeled NOx changes due to the October/November 2003 SPE using B3dCTM (stratospheric version), considering protons only (*left*) and protons + electrons (*right*). Loss and formation calculated relative to model run without SPE. NOx formation globally on October 29 at 56 km. Adapted from *Wissing et al.* [2010]

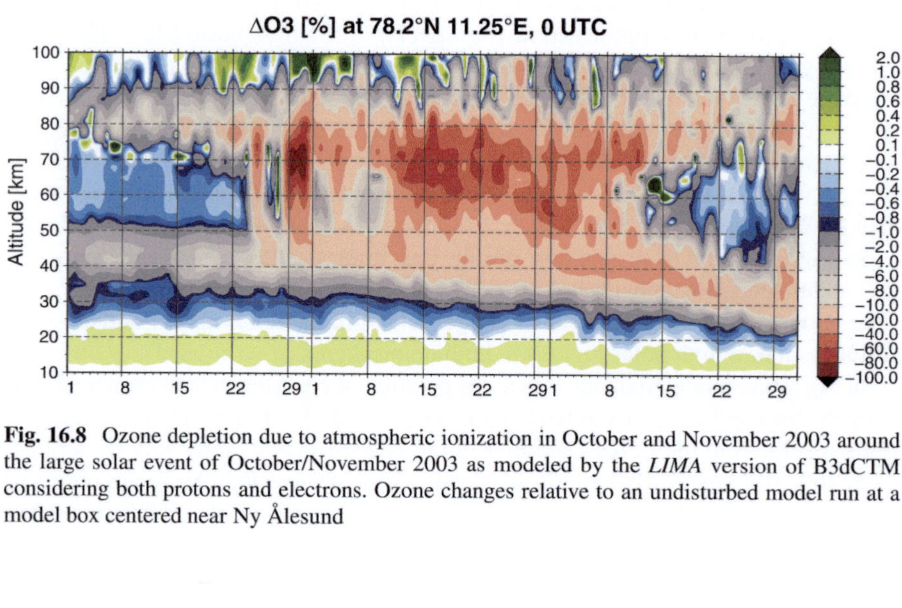

Fig. 16.8 Ozone depletion due to atmospheric ionization in October and November 2003 around the large solar event of October/November 2003 as modeled by the *LIMA* version of B3dCTM considering both protons and electrons. Ozone changes relative to an undisturbed model run at a model box centered near Ny Ålesund

to the model run without ionization) propagate downward in the course of the following weeks. However, the strong electron event on October 24 already shown in Fig. 16.6 is also clearly visible. Significant ozone losses of 10–20 % are observed around October 24 at altitudes between ≈50–80 km, and are transported down to 40 km altitude in the following days.

These model studies show that using ionization rates based on observed electron and proton fluxes, significant impacts of energetic electron precipitation are expected at altitudes below 60 km, both during large geomagnetic storms, and during and after large solar events. However, until now, no unambiguous observations have been published to show how realistic these model predictions are.

16.5.2 Impact of Energetic Electrons from Observations

Two data-sets of NOx (NO and NO_2) observations from satellite-based instruments were used to investigate a possible impact of energetic electron precipitation on the middle atmosphere: HALOE/UARS [*Russel et al.*, 1993] and MIPAS/ENVISAT [*Fischer et al.*, 2008]. HALOE has been observing NO and NO_2 in an altitude range of 10–130 km from 1991 to the end of 2005 in solar occultation mode; MIPAS/ENVISAT was launched in 2002, and is expected to continue measurements until 2013. MIPAS observes NO and NO_2 in nominal limb mode in an altitude range from ≈13–68 km. In this investigation, only measurements from October 2003 to March 2004 are used. As a solar occultation instrument, HALOE has a limited spatial resolution, taking 15 sun-rise and 15 sun-set observations in a very limited latitudinal range every day. However, HALOE data still present the longest continuous data-set of middle-atmosphere NOx. MIPAS has a much better spatial coverage than HALOE, but the time-series is yet not as long as the HALOE data-set, and also has been interrupted due to technical problems for nearly a year in 2004.

HALOE/UARS In Fig. 16.9, time-series of HALOE NOx in two different altitudes—60 and 80 km—are shown both for Northern and Southern high latitudes. Shown are daily averages poleward of 40° for both sun-rise and sun-set mode. At 60 km altitude, several solar particle events (July 2000 and October/November 2003) are clearly visible as large NOx enhancements; however, no clear response of NOx to energetic electron precipitation or geomagnetic storms is observed. However, a strong year-to-year variation of winter-time NOx is observed especially at 80 km, with highest values apparently during the transit from solar maximum to solar minimum (see also *Kazeminejad* [2009]). To investigate whether this interannual variation is due to energetic electron precipitation or geomagnetic activity, winter-time values of NOx (NDJF in the Northern hemisphere, MJJA in the Southern hemisphere) have been compared to the fluxes of energetic electrons both precipitating into the atmosphere from POES, and within the radiation belts from GOES, as well as with the Ap index, an indicator of geomagnetic activity, averaged over the same period of time. Both POES and GOES electrons are in the relativistic energy range

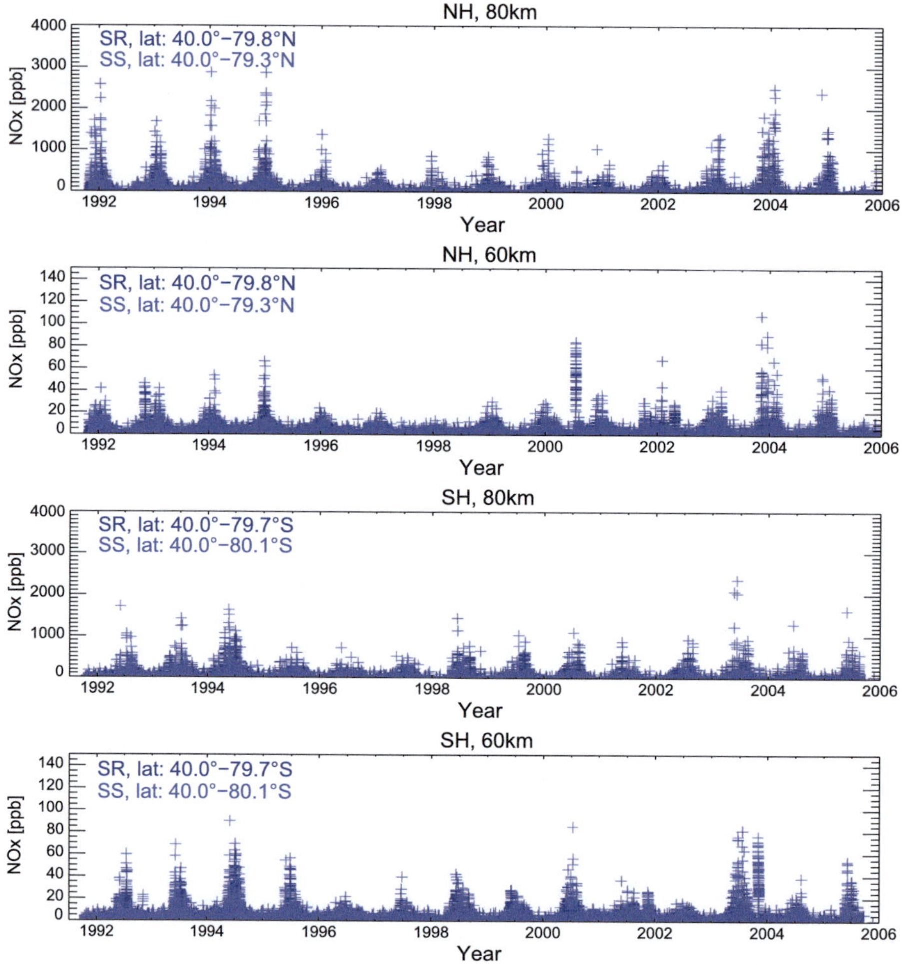

Fig. 16.9 Time-series of daily averaged sun-rise and sun-set NOx (NO + NO_2) as observed by HALOE/UARS. From *top* to *bottom*: 80 km in the Northern hemisphere; 60 km in the Northern hemisphere; 80 km in the Southern hemisphere; and 60 km in the Southern hemisphere. Figure adapted from *Kazeminejad* [2009]

(POES: >300 keV, GOES: >2 MeV) expected to precipitate into the lower mesosphere or even stratosphere.

Resulting correlation coefficients for both hemispheres are shown in Fig. 16.10 (see also *Kazeminejad* [2009]). Years with strong SPEs have been omitted here (2000 and 2003 in the Northern hemisphere, 2000 and 2005 in the Southern hemisphere). Error bars have been calculated as the result of a bootstrap method calculating correlation coefficients of 500 random permutations of the data-sets. A very high correlation is found between both geomagnetic activity and the fluxes of precipitating electrons (POES >300 keV) in both hemispheres, with values with a statistical

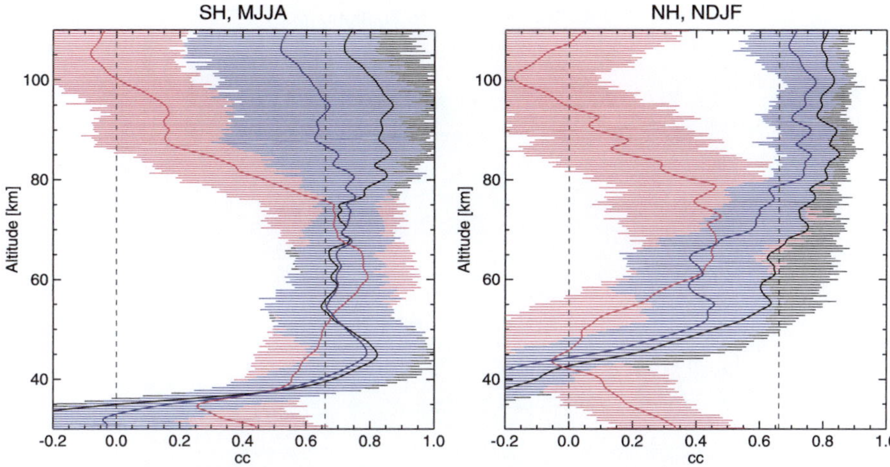

Fig. 16.10 Correlation coefficients of HALOE NOx during polar winter (NDJF in the NH, MJJA in the SH) with Ap index (*black*), precipitating electrons of energies >300 keV as measured by POES (*blue*), and radiation belt electrons of >2 MeV as measured by GOES (*red*) averaged over the same time-period, for the Southern hemisphere (*left*) and the Northern hemisphere (*right*). Error bars are results from a bootstrap method. *Dashed vertical lines* mark a correlation coefficient of zero and 0.63, the latter referring to a significance of 1 % for a statistical ensemble of 14 data-points. Figure adapted from *Kazeminejad* [2009]

significance of better than 1 % at altitudes between ≈40–100 km in the Southern hemisphere, and 70–130 km in the Northern hemisphere. A significant correlation is also found between radiation belt electrons (GOES >2 MeV) and NOx in the Southern hemisphere between ≈60–80 km; in the Northern hemisphere, a positive correlation is found as well, but has a significance considerably lower than 1 %. The correlation of energetic electrons with NOx extends farther up into the atmosphere than expected from the energy range of the electrons especially for the POES electrons, probably because these electron fluxes are highly correlated to fluxes of electrons of lower energies, and also to geomagnetic activity. The correlation between radiation belt electrons and NOx is smaller than shown in *Kazeminejad* [2009] apparently because years with especially strong solar events have not been omitted there.

The observed correlation between NOx and geomagnetic activity respectively electron fluxes implies a significant impact of energetic electron precipitation on the composition of the middle atmosphere. However, it is not clear which altitudes are affected directly, as during winter-time, NOx will be transported downward in the polar vortex very efficiently. To investigate which altitudes are affected directly, summer-time values should be investigated. The annual variability of the correlation has been investigated further by *Sinnhuber et al.* [2011] in a project funded by the University of Bremen, and it was found that the direct impact of energetic electron precipitation and geomagnetic activity appears to be restricted to altitudes

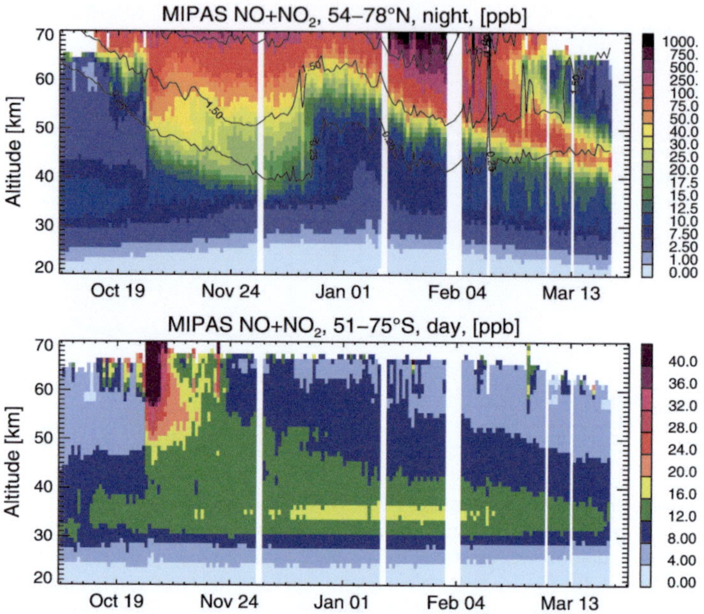

Fig. 16.11 *Top*: MIPAS night-time NOx (NO + NO_2) at high Northern latitudes (54–78°) from October 2003 to end of March 2004. *Solid lines* are the CO 0.25, 1.5, and 5.0 ppm isolines, an indicator for vertical motion within the polar vortex. Results for the Northern hemisphere are similar to those shown by *López-Puertas et al.* [2006]. *Bottom*: MIPAS day-time NOx at high Southern latitudes (51–75°S). MIPAS data courtesy B. Funke, Istituto de Astrofisica de Andalucia, and G. Stiller, KIT

above ≈80 km, while the impact of radiation belt electrons appears to be both less significant and not very robust.

MIPAS/ENVISAT In Fig. 16.11, daily averages of MIPAS NOx in the altitude range of 20–70 km are shown at high latitudes for both hemispheres from October 1, 2003 to March 31, 2004. This time-period was chosen because geomagnetic activity was very high throughout the complete time-series, and additionally, one of the strongest solar events occurred during this time in late October and early November. Also shown are isolines of CO, which is formed by photolysis of CO_2 in the upper mesosphere and thermosphere; as CO has a strong vertical gradient and is chemically inert during polar night, it can be used as a tracer of vertical motion in polar winter in conditions of low solar illumination. The impact of the large solar event is observed clearly in both hemispheres. In the Southern summer-time hemisphere, where the NOx lifetimes are short, this impact decreases continuously; in the Northern hemisphere, NOx values continue to stay high until December, when the signal is diluted quickly after a major stratospheric warming. The vortex reforms strongly after the warming, and a second NOx increase is observed in January and February in the Northern hemisphere which lasts until the end of the time-series in the upper stratosphere. This second NOx increase is likely due to transport from the upper mesosphere or lower thermosphere as shown by the good correlation with CO isolines (see Fig. 16.11); no evidence for direct production of NOx is observed during this time in the Northern hemisphere, and the exact source-region of the NOx values is not clear from these observations. However, it should be pointed out that the NOx mixing ratios transported down into the stratosphere in January/February 2004 in the Northern hemisphere are nearly one order of magnitude larger than the amounts of NOx produced due to the solar event in October/November 2003, which was one of the largest solar events on record! Some short-lived increases of NOx are observed in the Southern hemisphere at altitudes above 60 km, most notably before the large solar event in mid-October 2003 and on November 22. These short-lived increases may be connected to energetic electron precipitation (e.g., due to the geomagnetic storms before and after the solar event, as also discussed in 16.5.1). However, it should be pointed out that the values observed are much smaller than during the solar event, and nearly two orders of magnitude smaller than values transported down from the upper mesosphere or lower thermosphere in the polar winter.

To summarize, there is an impact of energetic electron precipitation on the middle atmosphere down to the upper stratosphere. It clearly exceeds the impact of even very large solar events in some winters. However, it is to date not clear in which altitudes these very large NOx values are produced—either in the upper mesosphere or lower thermosphere; direct production due to energetic electron precipitation below $\approx$80 km altitude is small both compared to the direct impact of large solar events, and compared to the indirect contribution of NOx from the upper mesosphere and lower thermosphere.

Acknowledgements This work was funded within the framework of the priority program *Climate and Weather of the Sun-Earth System* CAWSES by the *Deutsche Forschungsgemeinschaft* as project *SI-1088/1-3*. The authors gratefully acknowledge the work of S. Kazeminejad, who is now at the *German Space Agency* DLR. M. Sinnhuber also gratefully acknowledges funding by the University of Bremen. MIPAS data were kindly provided by B. Funke, Istituto de Astrofisica de Andalucia, and G. Stiller, KIT. The authors would like to thank Bernd Funke for initiating and coordinating the Heppa intercomparison initiative, and U. Berger for providing the LIMA data.

References

Berger, U. (2008). Modeling of middle atmosphere dynamics with LIMA. *Journal of Atmospheric and Solar-Terrestrial Physics*, *70*, 1170–1200.

Boehringer, H., Fahey, D. W., Fehsenfeld, F. C., & Ferguson, E. E. (1983). The role of ion-molecule reactions in the conversion of N_2O_5 to HNO_3 in the stratosphere. *Planetary and Space Science*, *31*, 185–191. doi:10.1016/0032-0633(83)90053-3.

Brasseur, G., & Chatel, A. (1983). Modelling of stratospheric ions: a first attempt. *Annales Geophysicae*, *1*, 173–185.

Callis, L. B., Natarajan, M., Lambeth, J. D., & Baker, D. N. (1998). Solar-atmospheric coupling by electrons (SOLACE) 2. Calculated stratospheric effects of precipitating electrons, 1979–1988. *Journal of Geophysical Research*, *103*, 28421–28438.

Chakrabarty, D. K., & Ganguly, S. (1989). On significant quantities of negative ions observed around the mesopause. *Journal of Atmospheric and Solar-Terrestrial Physics*, *51*, 983–989.

Chipperfield, M. (1999). Multiannual simulations with a three-dimensional chemical transport model. *Journal of Geophysical Research, 104*(D1), 1781–1805.

Clilverd, M. A., Seppälä, A., Rodger, C. J., Mlynczak, M. G., & Kozyra, J. U. (2009). Additional stratospheric NOx production by relativistic electron precipitation during the 2004 spring NOx descent event. *Journal of Geophysical Research, 114*. doi:10.1029/2008JA013472.

Crutzen, P. J., Isaksen, I. S., & Reid, G. C. (1975). Solar proton events: stratospheric sources of nitric oxide. *Science, 189*, 457–458.

Dee, D. P., Uppala, S. M., Simmons, A. J., Berrisford, P., Poli, P., Kobayashi, S., Andrae, U., Balmaseda, M. A., Balsamo, G., Bauer, P., Bechtold, P., Beljaars, A. C. M., van de Berg, L., Bidlot, J., Bormann, N., Delsol, C., Dragani, R., Fuentes, M., Geer, A. J., Haimberger, L., Healy, S. B., Hersbach, H., Holm, E. V., Isaksen, L., Kallberg, P., Köhler, M., Matricardi, M., McNally, A. P., Monge-Sanz, B. M., Morcrette, J.-J., Park, B.-K., Peubey, C., de Rosnay, P., Tavolato, C., Thepaut, J.-N., & Vitart, F. (2011). The era-interim reanalysis: configuration and performance of the data assimilation system. *Quarterly Journal of the Royal Meteorological Society, 137*, 553–597.

Fischer, H., Birk, M., Blom, C., Carli, B., Carlotti, M., von Clarmann, T., Delbouille, L., Dudhia, A., Ehhalt, D., Endemann, M., Flaud, J. M., Gessner, R., Kleinert, A., Koopmann, R., Langen, J., Lopez-Puertas, M., Mosner, P., Nett, H., Oelhaf, H., Perron, G., Remedios, J., Ridolfi, M., Stiller, G., & Zander, R. (2008). Mipas: an instrument for atmospheric and climate research. *Atmospheric Chemistry and Physics, 8*, 2151–2188.

Fritzenwallner, J., & Kopp, E. (1998). Model calculations of the negative ion chemistry in the mesosphere with special emphasis on the chlorine species and the formation of cluster ions. *Advances in Space Research, 21*, 891–894.

Funke, B., López-Puertas, M., Fischer, H., Stiller, G., von Clarmann, T., Wetzel, G., Carli, B., & Belotti, C. (2007). Comment on 'Origin of the January-April 2004 increase in stratospheric NO_2 observed in northern polar latitudes' by Jean-Baptiste Renard et al. *Geophysical Research Letters, 34*. doi:10.1029/2006GL027518.

Funke, B., Baumgaertner, A. J. G., Calisto, M., Egorova, T., Jackman, C. H., Kieser, J., Krivolutsky, A., López-Puertas, M., Marsh, D. R., Reddmann, T., Rozanov, E., Salm, S.-M., Sinnhuber, M., Stiller, G., Verronen, P. T., Versick, S., von Clarmann, T., Vyushkova, T. Y., Wieters, N., & Wissing, J.-M. (2011). Composition changes after the "Halloween" solar proton event: the High-Energy Particle Precipitation in the Atmosphere (HEPPA) model versus MIPAS data intercomparison study. *Atmospheric Chemistry and Physics, 11*, 9089–9139.

Heaps, M. G. (1978). Parameterization of the cosmic ray ion-pair production rate above 18 km. *Planetary and Space Science, 20*, 513–517.

Horne, R. B., Lam, M. M., & Green, J. C. (2009). Energetic electron precipitation from the outer radiation belt during geomagnetic storms. *Geophysical Research Letters, 36*. doi:10.1029/2009GL040236.

Jackman, C., McPeters, R., Labow, G., Praderas, C., & Fleming, E. (2001). Northern hemisphere atmospheric effects due to the July 2000 solar proton events. *Geophysical Research Letters, 28*, 2883–2886.

Jackman, C., DeLand, M., Labow, G., Fleming, E., Weisenstein, D., Ko, M., Sinnhuber, M., Anderson, J., & Russell, J. (2005a). Neutral atmospheric influences of the solar proton events in October-November 2003. *Journal of Geophysical Research, 110*, A09S27. doi:10.1029/2004JA01088.

Jackman, C. H., DeLand, M. T., Labow, G. J., Fleming, E. L., Weisenstein, D. K., Ko, M. K. W., Sinnhuber, M., Anderson, J., & Russell, J. M. (2005b). The influence of the several very large solar proton events in years 2000–2003 on the neutral middle atmosphere. *Advances in Space Research, 35*, 445–450.

Kazeminejad, S. (2009). *Analysis of the middle atmosphere's response to energetic particle events*. Ph.D. thesis, University of Bremen.

Kazil, J. (2002). *The University of Bern atmospheric ion model: time-dependent ion modeling in the stratosphere, mesosphere and lower thermosphere*. Ph.D. thesis, University of Bern.

Kinnersley, J. S. (1996). The climatology of the stratospheric 'THIN AIR' model. *Quarterly Journal of the Royal Meteorological Society*, *122*(529, Part A), 219–252.
Kopp, E. (1996). Electron and ion densities. In W. Dieminger, G. K. Hartman & R. Leitinger (Eds.), *The upper atmosphere, data analysis and interpretation* (pp. 620–630). Berlin: Springer.
Kopp, E., & Fritzenwallner, J. (1997). Chlorine and bromine ions in the D-region. *Advances in Space Research*, *20*, 2111–2155.
Lary, D. J. (1997). Catalytic destruction of stratospheric ozone. *Journal of Chemical Physics*, *102*, 21515–21526.
López-Puertas, M., Funke, B., Gil-López, S., von Clarmann, T., Stiller, G. P., Höpfner, M., Kellmann, S., Fischer, H., & Jackman, C. H. (2005a). Observation of NOx enhancements and ozone depletion in the Northern and Southern Hemispheres after the October-November 2003 solar proton events. *Journal of Geophysical Research*, *110*, A09S43. doi:10.1029/2005JA01105.
López-Puertas, M., Funke, B., Gil-López, S., von Clarmann, T., Stiller, G. P., Höpfner, M., Kellmann, S., Tsidu, G. M., Fischer, H., & Jackman, C. H. (2005b). HNO_3, N_2O_5, and $ClONO_2$ enhancements after the October-November 2003 solar proton events. *Journal of Geophysical Research*, *110*, A09S44. doi:10.1029/2005JA011051.
López-Puertas, M., Funke, B., von Clarmann, T., Fischer, H., & Stiller, G. P. (2006). The stratospheric and mesospheric NOy in the 2002–2004 polar winters as measured by MIPAS/ENVISAT. *Space Science Reviews*, *125*, 403–416. doi:10.1007/s11214-006-9073-2.
Porter, H. S., Jackman, C. H., & Green, A. E. S. (1976). Efficiencies for production of atomic nitrogen and oxygen by relativistic proton impact in air. *Journal of Chemical Physics*, *65*, 154–167.
Prather, M. J. (1986). Numerical advection by conservation of second-order moments. *Journal of Geophysical Research*, *91*(D6), 6671–6681.
Ramarson, R. A. (1989). *Modelisation locale, á une et trois dimensions des processus photochimiques de l'atmosphére moyenne*. Ph.D. thesis, Université Paris VI.
Randall, C. E., Siskind, D. E., & Bevilaqua, R. M. (2001). Stratospheric NOx enhancements in the southern hemisphere vortex in winter/spring of 2000. *Geophysical Research Letters*, *28*, 2385–2388.
Rees, M. H. (1998). *Physics and chemistry of the upper atmosphere*. Cambridge: Cambridge University Press.
Renard, J.-B., Blelly, P.-L., Bourgeois, Q., Chartier, M., Goutail, F., & Orsolini, Y. J. (2006). Origin of the January-April 2004 increase in stratospheric NO_2 observed in the northern polar latitudes. *Geophysical Research Letters*, *33*. doi:10.1029/2005GL025450.
Rohen, G. J., von Savigny, C., Sinnhuber, M., Eichmann, K.-U., Llewellyn, E. J., Kaiser, J. W., Jackman, C. H., Kallenrode, M.-B., Schroeter, J., Bovensmann, H., & Burrows, J. P. (2005). Ozone depletion during the solar proton events of Oct./Nov. 2003 as seen by SCIAMACHY. *Journal of Geophysical Research*, *110*, A09S39.
Rusch, D. W., Gerard, J.-C., Solomon, S., Crutzen, P. J., & Reid, G. C. (1981). The effect of particle precipitation events on the neutral and ion chemistry of the middle atmosphere, 1. Odd nitrogen. *Planetary and Space Science*, *29*, 767–774.
Russell, J. M., Gordley, L. L., Park, J. H., Drayson, S. R., Hesketh, W. D., Cicerone, R. J., Tuck, A. F., Frederick, J. E., Harries, J. E., & Crutzen, P. J. (1993). The HaLogen Occultation Experiment. *Journal of Geophysical Research*, *98*, 10777–10797.
Sander, S. P., Friedl, R. R., Ravishankara, A. R., Golden, D. M., Kolb, C. E., Kurylo, M. J., Molina, M. J., Moortgat, G. K., Keller-Rudek, H., Finlayson-Pitts, B. J., Wine, P., Huie, R. E., & Orkin, V. L. (2006). Chemical kinetics and photochemical data for use in atmospheric studies—evaluation number 15. *JPL Publication*, *06*(2).
Shine, K. P. (1987). The middle atmosphere in the absence of dynamical heat fluxes. *Quarterly Journal of the Royal Meteorological Society*, *113*, 603–633.
Simmons, A., Uppala, S., Dee, D., & Kobayashi, S. (2006). ERA-Interim: new ECMWF reanalysis products from 1989 onwards. *ECMWF Newsletter*, *110*. Winter 2006/2007.
Sinnhuber, B.-M., Weber, M., Amankwah, A., & Burrows, J. P. (2003a). Total ozone during the unusual Antarctic winter of 2002. *Geophysical Research Letters*, *30*, 11. doi:10.1029/

2002GL016798.
Sinnhuber, B.-M., von der Gathen, P., Sinnhuber, M., Rex, M., König-Langlo, G., & Oltmans, S. J. (2006). Large decadal scale changes of polar ozone suggest solar influence. *Atmospheric Chemistry and Physics*, *6*, 1835–1841.
Sinnhuber, M., Burrows, J. P., Künzi, K. F., Chipperfield, M. P., Jackman, C. H., Kallenrode, M.-B., & Quack, M. A. (2003b). A model study of the impact of magnetic field structure on atmospheric composition during solar proton events. *Geophysical Research Letters*, *30*, L01818. doi:10.1029/2003GL017265.
Sinnhuber, M., Jackman, C. H., & Kallenrods, M.-B. (2003c). The impact of large solar proton events on ozone in the polar stratosphere—a model study. In C. Zerefos (Ed.), *Proceedings quadrennial ozone symposium*, 1–8 June 2004, Kos, Greece.
Sinnhuber, M., Kazeminejad, S., & Wissing, J. M. (2011). Interannual variation of NOx from the lower thermosphere to the upper stratosphere in the years 1991–2005. *Journal of Geophysical Research*, *116*. doi:10.1029/2010JA015825.
Siskind, D., Nedoluha, G., Randall, C., Fromm, M., & Russell III, J. M. (2000). An assessment of southern hemisphere stratospheric NOx enhancements due to transport from the upper atmosphere. *Geophysical Research Letters*, *27*, 329–332.
Siskind, D. E., & Russel III, J. M. (1996). Coupling between middle and upper atmospheric NO: constraints from HALOE observations. *Geophysical Research Letters*, *23*, 137–140.
Solomon, S., & Crutzen, P. J. (1981). Analysis of the August 1972 solar proton event including chlorine chemistry. *Journal of Geophysical Research*, *86*, 1140–1146. doi:10.1029/JC086iC02p01140.
Solomon, S., Rusch, D. W., Gerard, J.-C., Reid, G. C., & Crutzen, P. J. (1981). The effect of particle precipitation events on th neutral and ion chemistry of the middle atmosphere II: odd hydrogen. *Planetary and Space Science*, *29*, 885–892.
Solomon, S., Reid, G. C., Rusch, D. W., & Thomas, R. J. (1983). Mesospheric ozone depletion during the solar proton event of July 13, 1983, part II: comparison between theory and measurements. *Geophysical Research Letters*, *10*, 257–260.
Swider, W., & Keneshea, T. J. (1973). Decrease of ozone and atomic oxygen in the lower mesosphere during a PCT event. *Planetary and Space Science*, *21*, 1969–1973.
Verronen, P. T. (2006). *Ionosphere-atmosphere interaction during solar proton events*. Ph.D. thesis, University of Helsinki.
Verronen, P. T., Funke, B., López-Puertas, M., Stiller, G. P., von Clarmann, T., Glatthor, N., Enell, C.-F., Turunen, E., & Tamminen, J. (2008). About the increase of HNO3 in the stratopause region during the Halloween 2003 solar proton event. *Geophysical Research Letters*, *35*, L20809. doi:10.1029/2008GL035312.
Verronen, P. T., Rodger, C. J., Clilverd, M. A., & Wang, S. (2011). First evidence of mesospheric hydroxyl response to electron precipitation from the radiation belts. *Journal of Geophysical Research*, *116*, D07307. doi:10.1029/2010JD014965.
Viggiano, A. A., Morris, R. A., & Doren, J. M. V. (1994). Ion chemistry of $ClONO_2$ involving NO_3^- core ions: a detection scheme for $ClONO_2$ in the atmosphere. *Journal of Geophysical Research*, *99*, 8221–8224.
von Clarmann, T., Glatthor, N., Höpfner, M., Kellmann, S., Ruhnke, R., Stiller, G. P., Fischer, H., Funke, H., Gil-López, S., & López-Puertas, M. (2005). Experimental evidence of perturbed odd hydrogen and chlorine chemistry after the October 2003 solar proton events. *Journal of Geophysical Research*, *110*, A09S45. doi:10.1029/2005JA011053.
Winkler, H. (2007). *Response of middle atmospheric ozone to solar proton events in a changing geomagnetic field*. Ph.D. thesis, University of Bremen.
Winkler, H., Sinnhuber, M., Notholt, J., Kallenrode, M.-B., Steinhilber, F., Vogt, J., Zieger, B., Glassmeier, K.-H., & Stadelmann, A. (2008). Modelling impacts of geomagnetic field variations on middle atmospheric ozone responses to solar proton events on long time scales. *Journal of Geophysical Research*, *113*, D02302. doi:10.1029/2007JD008574.
Winkler, H., Kazeminejad, S., Sinnhuber, M., Kallenrode, M.-B., & Notholt, J. (2009). The conversion of mesospheric HCl into active chlorine during the solar proton event in July 2000

in the northern polar region. *Journal of Geophysical Research*, *114*, D00I03. doi:10.1029/2008JD011587.

Winkler, H., Kazeminejad, S., Sinnhuber, M., Kallenrode, M.-B., & Notholt, J. (2011). Correction to “Conversion of mesospheric HCl into active chlorine during the solar proton event in July 2000 in the northern polar region”. *Journal of Geophysical Research*, *116*, D17303. doi:10.1029/2011JD016274.

Wissing, J.-M., & Kallenrode, M.-B. (2009). Atmospheric Ionisation Module Osnabrück (AIMOS): A 3-D model to determine atmospheric ionization by energetic charged particles from different populations. *Journal of Geophysical Research*, *114*, A06104. doi:10.1029/2008JA013884.

Wissing, J.-M., Kallenrode, M.-B., Wieters, N., Winkler, H., & Sinnhuber, M. (2010). Atmospheric Ionisation Module Osnabrück (AIMOS) 2: Total particle inventory in the October / November 2003 event and ozone. *Journal of Geophysical Research*, *115*, A02308. doi:10.1029/2009JA014419.

Zipf, E. C., Espy, P. J., & Boyle, C. F. (1980). The excitation and collisional deactivation of metastable N(2P) atoms in auroras. *Journal of Geophysical Research*, *85*, 687–694.

Chapter 17
Simulation of Particle Precipitation Effects on the Atmosphere with the MESSy Model System

Andreas J.G. Baumgaertner, Patrick Jöckel, Alan D. Aylward, and Matthew J. Harris

Abstract Focusing on particle precipitation into the atmosphere we present whole-atmosphere model developments as well as scientific results using the Modular Earth Submodel System (MESSy). Parameterizations for Solar Proton Events as well as low-energy-electron precipitation are described and the implementation as MESSy submodels is outlined. Direct and indirect effects found in simulations using MESSy with the basemodel ECHAM5, a general circulation model, are discussed. As an additional development, the middle and upper atmosphere model CMAT2 was implemented as a MESSy basemodel and as a submodel for ECHAM5/MESSy to create a whole atmosphere model. This opens up new perspectives for particle precipitation modeling and some initial results are shown.

17.1 Introduction

Solar variability manifests itself in several different ways. Variations in ultraviolet irradiance, which can exceed 50 % at some wavelengths, have been shown to lead

A.J.G. Baumgaertner (✉) · P. Jöckel
Max Planck Institute for Chemistry, J.-J.-Becher-Weg. 27, 55020 Mainz, Germany
e-mail: work@andreas-baumgaertner.net

P. Jöckel
e-mail: patrick.joeckel@dlr.de

A.J.G. Baumgaertner
Project Management Agency, Deutsches Zentrum für Luft-und Raumfahrt (DLR), 53227 Bonn, Germany

P. Jöckel
Institut für Physik der Atmosphäre, Deutsches Zentrum für Luft-und Raumfahrt (DLR), Oberpfaffenhofen, 82234 Weßling, Germany

A.D. Aylward · M.J. Harris
University College London, Atmospheric Physics Laboratory, London, UK

A.D. Aylward
e-mail: a.aylward@ucl.ac.uk

F.-J. Lübken (ed.), *Climate and Weather of the Sun-Earth System (CAWSES)*, Springer Atmospheric Sciences,
DOI 10.1007/978-94-007-4348-9_17,

to stratospheric ozone changes and induce temperature variations over the 11 year solar cycle [e.g. *Austin et al.*, 2008] as well as other periods. Additionally, solar flares and coronal mass ejections, as well as solar wind variations can lead to particle precipitation into the Earth's atmosphere and can therefore also be important for Sun-Earth connection studies. The former events lead to vastly increased fluxes of high-energy particles. Depending on the position of the Earth relative to the ejection, the particles can reach the Earth's atmosphere, which is then called a Solar Proton Event (SPE). An SPE causes ionizations, dissociations, dissociative ionizations, and excitations in the atmosphere. This results in the production of HO_x and NO_y in the middle and upper atmosphere. Such events are interesting natural experiments that allow us to test and improve our understanding of atmospheric chemistry, our observational capabilities, and numerical models. The effects of SPEs on the atmosphere have been measured and modeled extensively for almost four decades [see *Jackman and McPeters*, 2004, for a review]. A prominent example which has attracted a lot of attention was the October/November 2003 Halloween event [*Jackman et al.*, 2005; *Rohen et al.*, 2005], which is also discussed here.

On longer timescales, NO_y production can also result from solar wind variations: When the solar wind interacts with the Earth's magnetosphere, the magnetosphere can experience a loss of charged particles, mainly electrons, through precipitation into the Earth's upper and middle atmosphere at high geomagnetic latitudes (Clilverd et al., 2006). There, the energetic electron precipitation (EEP) leads to NO_x production by dissociation of nitrogen. It has been found that the amount of NO_x produced corresponds to the level of geomagnetic activity [*Randall et al.*, 2007, and references therein], ultimately linking it to solar variability. During polar winter both SPE and geomagnetic activity related NO_x enhancements can live long enough to be transported into the stratosphere and engage in catalytic ozone destruction. A recent study by *Verronen et al.* [2011] indicates also a production of hydroxyl in response to electron precipitation. Note that energetic particle precipitation impacts on the middle atmosphere are also discussed in Chaps. 8, 15, and 16.

In Sect. 17.2 the employed model system MESSy (Modular Earth Submodel System) and the newly developed components are introduced. Also, the simulated direct impact of the SPE and EEP parameterizations on the atmosphere are shown. We discuss the consequences of the EEP-NO_x enhancements and the associated ozone loss under a climate change scenario in Sect. 17.3.1, and possible effects on polar surface air temperature variability in Sect. 17.3.2. In Sect. 17.4 the vertical extension of MESSy using CMAT2 is outlined.

17.2 The Modular Earth Submodel System (MESSy)

The Modular Earth Submodel System (MESSy) "is an open, multi-institutional project providing a strategy for developing comprehensive Earth System Models with highly flexible complexity" [from *Jöckel et al.*, 2010]. ECHAM5/MESSy, where the general circulation model ECHAM5 [*Roeckner et al.*, 2006] serves as a basemodel, is a numerical chemistry and climate simulation system that includes submodels describing tropospheric and middle atmosphere processes and their interaction with oceans and land and the influence of human activity [*Jöckel et al.*, 2005, 2006, 2010]. The model has been shown to consistently simulate key atmospheric tracers such as ozone [*Jöckel et al.*, 2006], water vapor [*Lelieveld et al.*, 2007], and lower and middle stratospheric NO_y [*Brühl et al.*, 2007]. The discussions here are based on simulations in the T42L90MA-resolution, i.e., with a spherical truncation of T42 (corresponding to a quadratic Gaussian grid of approx. 2.8° by 2.8° in latitude and longitude) with 90 vertical hybrid pressure levels up to 0.01 hPa. The setups used here are based on the setup used in the model evaluation studies by *Jöckel et al.* [2006 and 2010]. However, the chosen chemistry scheme for the configuration of the submodel MECCA [*Sander et al.*, 2005, 2011] is simpler, and several additional submodels are used which are described below (Sects. 17.2.1 and 17.2.2). The middle and upper atmosphere code CMAT2 [*Harris*, 2001; *Harris et al.*, 2002; *Cnossen et al.*, 2009] can serve as a submodel to ECHAM5/MESSy but can also be run as basemodel (CMAT/MESSy) with a prescribed lower atmosphere. This development is outlined in Sect. 17.4, however CMAT is not used for the results presented here but may become an option for such studies in the future.

17.2.1 Solar Proton Events: The SPE Submodel

As introduced above, SPEs lead to an elevated flux of energetic protons into the Earth's polar upper and middle atmosphere. There, they mainly lead to the dissociation of N_2 and H_2 through ionization and dissociation processes, and subsequent production of NO_x and HO_x. The ionization altitude profile is mainly determined by the type and number of precipitating particles and their energy distributions, which then also determines the SPE's significance in terms of associated ozone depletion through catalytic destruction by the HO_x and NO_x products.

The MESSy submodel SPE parameterizes the effects of energetic particles by calculating ionization rates from satellite instrument particle flux measurements. For the calculation of ionization rates a method based on *Vitt and Jakman* [1996] was employed. The proton flux measurements by GOES (Geostationary Operational Environmental Satellite) or IMP (Interplanetary Monitoring Platform) are interpolated to a logarithmic energy spectrum. The energy deposited in any model layer is calculated directly using the thicknesses of the considered layer and an energy-range relationship. Since the magnetic field lines guide the charged particles toward the geomagnetic poles, full ionization was applied poleward of 60° geomagnetic latitude, and linearly increasing ionization between 55° and 60° geomagnetic latitude. The production of H and OH by ionization was parameterized using an approximation to data published in *Solomon et al.* [1981].

The production of N and NO is commonly parameterized using values of 0.55 $N(^4S)$ atoms and 0.7 NO molecules per ion pair [e.g., *Jackman et al.*, 2005] in models that do not incorporate excited state chemistry of atomic nitrogen. In

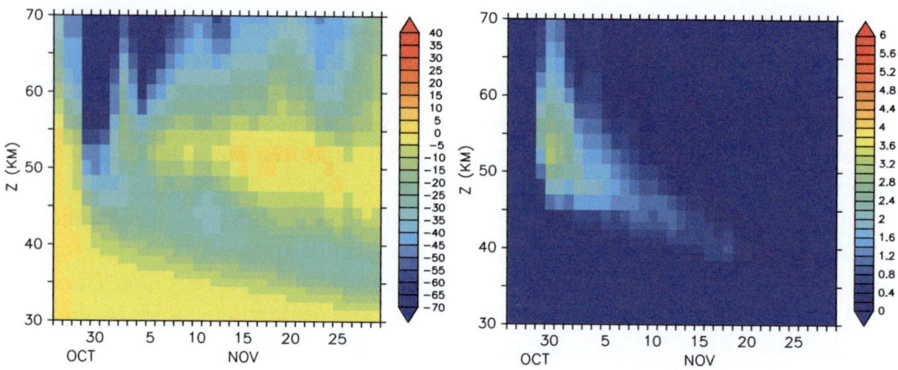

Fig. 17.1 Ozone mixing ratio percentage change (*left*) and N_2O (*right*) mixing ratio change (nmol/mol) with respect to a simulation without an SPE for the 2003 Halloween event

an attempt to improve the agreement between satellite and model results [especially concerning N_2O, see also *Funke et al.*, 2008] we tested a height dependent N/NO production efficiency. This appears to be a manifest approach because the 0.55 N/0.7 NO branching ratio has also been derived by "best fit" between models and observations in the past. The MESSy boxmodel CAABA [*Sander et al.*, 2011] was used to simulate the SPE at MIPAS [Michelson Interferometer for Passive Atmospheric Sounding on board the ENVISAT satellite, see e.g. *Funke et al.*, 2005a] measurement locations for N and NO production per ion pair ranging from 0 to 1.2. Then, the percentage deviations from the MIPAS values of NO_2 and N_2O were calculated. From this, the N and NO productions per ion pair minimizing the total deviation was determined. Generally, there appears to be a strong height dependence, with an almost negligible production of N atoms below 45 km which is in disagreement with the generally assumed production parameterization of 0.55 N atoms. Above 55 km the production of NO is also in disagreement with the previously used value of 0.7. This is mainly due to the fact that MIPAS yields only small increases of NO_2 and N_2O at these altitudes, despite the large ionization rates present there. The production of N and NO per ion pair, as well as the N/NO production ratio appear to maximize at approximately the height of the stratopause ($\sim$0.25 N atoms and $\sim$1.15 NO molecules), possibly indicating a temperature dependent production mechanism.

We used the values obtained from this procedure in a 3-D ECHAM5/MESSy simulation of the Halloween event in 2003, which was the fourth largest period of SPEs in the past 40 years. Changes of ozone (left) and N_2O (right) with respect to a simulation without an SPE for 70–90°N are shown in Fig. 17.1. Short-term ozone loss above the stratopause around the time of the SPE due to OH, as well as longer lasting loss of up to 40 % in the stratosphere due to NO_x are evident. N_2O enhancements of up to 3.5 nmol/mol are found to last for several weeks.

More details on the parameterization, the implementation in MESSy, and its evaluation can be found in *Baumgaertner et al.* [2010a].

17.2.2 Low-Energy Electron Precipitation: The SPACENOX Submodel

While SPEs usually only last for a maximum period of a few days, solar wind fluctuations leading to magnetospheric particle loss (mainly electrons) into the atmosphere can be a source of middle and upper atmospheric NO_x enhancements for several months. *Randall et al.* [2007], hereafter R07, and others have shown that the geomagnetic activity index Ap is in general sufficient to describe measured interannual variations of NO_x in the southern polar vortex. Therefore, the parameterization of energetic electron precipitation (EEP) NO_x flux into the model domain from above discussed here is based on the Ap index. Seasonal variations of NO_x downward transport in the polar lower thermosphere are parameterized using a sinusoidal time dependency. Interannual variations of this transport are not considered, based on the fact that a good agreement between interannual variations of Ap and stratospheric NO_x enhancements was found by other authors. Note, however, that for the Northern Hemisphere, where dynamic variability is much more pronounced than in the Southern Hemisphere (SH), less evidence exists for a simple relationship between the Ap index and stratospheric NO_x enhancements.

R07 derived estimates of EEP NO_x from HALOE measurements for one solar cycle. For the development of the parameterization the following information are needed: (1) NO_x excess densities derived by fitting to the data of R07, (2) an average vertical velocity, (3) a loss factor which accounts for transport out of the polar night region. Instead of making assumptions about the latter two, which would be error-prone, a trial-and-error approach was chosen in order to get results at 45 km that match the observations. Therefore, a value for the combination of the latter factors was established through a series of test simulations. Model excess NO_x densities at 45 km were calculated by subtracting densities from a reference simulation with the SPACENOX submodel turned off. Then, model excess NO_x were compared to the results of R07 (their Fig. 7). The required flux of NO_x into the top two model levels was then determined to be

$$F = {A_p}^{2.5} \cdot c \cdot 2.20 \times 10^5 \ \mathrm{cm}^{-2}\,\mathrm{s}^{-1}, \qquad (17.1)$$

where $c = 0.23$ (0.45) for "average (maximum) excess NO_x" (for a definition see R07). Details about the derivation can be found in *Baumgaertner et al.* [2009].

Figure 17.2 depicts the simulated mixing ratios of NO_x south of 60°S during 2003. Strong enhancements with mixing ratios of more than 300 nmol/mol are evident in the mesosphere. They are related to the NO_x produced by the SPACENOX submodel due to the large Ap index during most of the southern winter 2003. Also clearly distinguishable is a sudden enhancement at the end of October. This is related to the SPE discussed above. It should be noted that measurements of stratospheric NO_x are only available for a limited number of years. The geomagnetic Ap index in comparison has been measured since 1932. Therefore, the fact that the presented parameterization only requires the Ap index as an input function, and is not relying on satellite measurements, is of great advantage for model simulations spanning several decades.

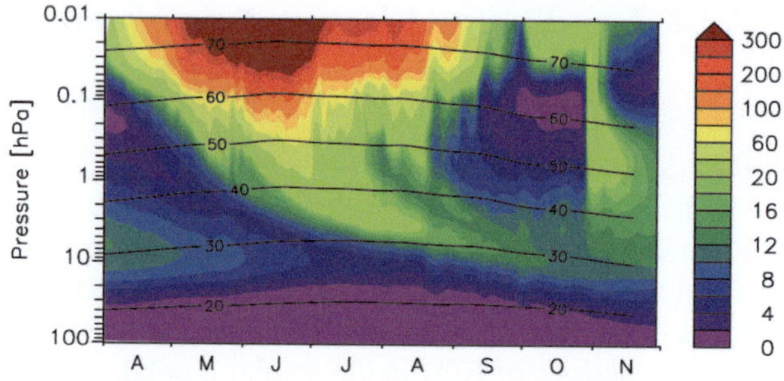

Fig. 17.2 Model NO_x mixing ratios (nmol/mol) averaged over 60°–90°S for 2003. The *contour lines* denote the altitude in km

An in-depth description of the submodel and its evaluation can be found in *Baumgaertner et al.* [2009].

17.3 Effects of Particle Precipitation on the Atmosphere

17.3.1 Low-Energy-Electron Precipitation in a Climate Change Scenario

In the mesosphere, the mean meridional circulation transports air from the summer to the winter hemisphere driven by gravity wave energy and momentum deposition as well as radiative heating and cooling. In the polar winter, this circulation can transport air, including EEP induced NO_x enhancements, from the mesosphere into the stratosphere. In the polar stratosphere, further downward transport is controlled by the Brewer-Dobson circulation (BDC). The BDC is responsible for the meridional transport of air in the stratosphere: It mainly consists of poleward transport in the middle and upper stratosphere, with rising air in the tropics and downwelling air in the polar regions.

As climate change is expected to increase the strength of the Brewer-Dobson circulation [*Butchart et al.*, 2010] including extratropical downwelling, the enhancements of EEP NO_x concentrations are expected to be transported to lower altitudes in extratropical regions, becoming more significant in the ozone budget. There is, however, no clear evidence for such BDC strengthening from measurements so far [*Engel et al.*, 2009]. Similarly, changes in the mesospheric residual circulation might also be subject to changes in a modified climate, but are also still under discussion: *Schmidt et al.* [2006] found a weakening using model simulations, but measurements have not yet identified a trend [*Baumgaertner et al.*, 2005].

The SPACENOX parameterization presented above was used to investigate the effects of a strengthened residual circulation on the transport of EEP NO_x. For full details see *Baumgaertner et al.* [2010b].

Simulations were performed for present-day and for year-2100 conditions with repeating boundary conditions, each for 12 model years. The present-day simulations use the climatological sea ice and sea surface temperature (SST) data set without El Niño or La Niña events. The setup is otherwise equivalent to *Jöckel et al.*

[2010], with the exception that no nudging has been applied. For the simulations with year-2100 conditions in a climate change scenario, the IPCC SRES A2 scenario [*Nakicenovic et al.*, 2000] was chosen. This is the most drastic scenario, with a near doubling of CO_2 resulting in a surface temperature increase of approx. 4 K [*IPCC*, 2007] with respect to 1990. SST and sea ice coverage data were taken from an IPCC AR4 simulation including an interactive ocean model, ECHAM5/MPI-OM [*Jungclaus*, 2006]. Initial concentrations as well as the prescribed surface concentrations of long-lived trace gases were scaled to the expected concentrations of the trace gases in the year 2100 (SRES A2 scenario). Chlorine and bromine containing substances as well as ozone precursors were left unchanged compared to the present-day simulation. While this is unlikely to be realistic, it would be very difficult to distinguish the effects of e.g. changed halogen loading and circulation changes.

The EEP strength for production of NO_x in the mesosphere and lower thermosphere was set to 2003 with repeating monthly Ap values as input. The SH winter 2003 experienced strong enhancements of EEP NO_x [*Funke et al.*, 2005b], the May–July average Ap value of 23.1 exceeds that of all other years since 1958 except for 1991. This "worst case" scenario will make it possible to identify most clearly the effects focused on here.

Because in the simulation for the year 2100 climate change has affected the mean state of the atmosphere such that the induced EEP NO_x changes are difficult to distinguish from other changes in the NO_x and ozone distributions, four simulations were carried out. The basic EEP related changes are calculated separately for both year-2100 conditions and present conditions. Then, the result obtained for present-day is subtracted from the year-2100 result, yielding only the changes in EEP effects due to climate change.

In the SH, the circulation changes and associated changes in horizontal mixing lead to a stronger downward transport of EEP NO_x in the polar winter stratosphere, yielding a surplus of up to 4.7 nmol/mol in the upper stratosphere (not shown).

Figure 17.3 (left) shows the climatological difference of SH winter average (June, July, August, JJA) ozone mixing ratios between the year-2100 and the present-day simulation when no EEP effect is considered. A decrease of up to 0.5 μmol/mol is found in the tropical and subtropical lower stratosphere. These changes are consistent with a strengthening of the BDC. In the upper stratosphere, centered around 40°, ozone increased by up to 1.6 μmol/mol. This is due to a stronger poleward transport of ozone from lower latitudes caused by a strengthened BDC, as well as the increase of greenhouse gas concentrations, which leads to a cooling of the stratosphere [not shown, see e.g. *Jonsson et al.*, 2004], which in turn slows down the temperature dependent photochemical ozone loss reactions [e.g. *Haigh and Pyle*, 1982].

Figure 17.3 (middle) also depicts the winter average ozone change in the year 2100 compared to present-day, but including the EEP effect. In contrast to Fig. 17.3 (left) the enhancement in the upper stratosphere at high southern latitudes is smaller. Figure 17.3 (right) displays a high-latitude decrease of ozone exceeding 0.4 μmol/mol, approximately reflecting the NO_x changes in areas where sunlight and thus atomic oxygen is present, which allows the catalytic destruction of ozone to proceed. Overall it can be concluded that in the SH the EEP NO_x effect, i.e.

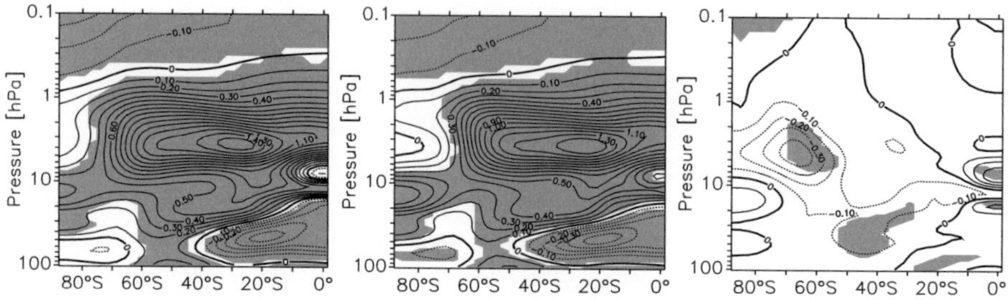

Fig. 17.3 Effect of climate change on ozone without EEP (*left*) and with EEP (*middle*) during SH winter (JJA average). *Right*: Change of EEP effect on ozone due to climate change (difference between left and middle panel). Units are μmol/mol. The *shaded area* indicates statistical significance at the 99 % level

a reduction of ozone, in the presented climate change scenario is approximately compensated by the increase of ozone caused by the climate change induced stratospheric cooling.

17.3.2 Geomagnetic Activity Related NO_x Enhancements and Polar Surface Air Temperature Variability

Seppälä et al. [2009] (hereafter S09) recently suggested that in the ECMWF ERA-40 reanalysis data set wintertime polar surface air temperatures (SAT) have different patterns during years of high and low geomagnetic activity. ECHAM5/MESSy with the SPACENOX submodel is used to investigate this effect.

A comprehensive simulation (hereafter denoted as S-TRANSIENT) covering the period 1960 to 2004 was carried out and an analogous analysis to S09 performed in that the dataset was separated according to the winter averaged Ap index, used as proxy for the overall geomagnetic activity level. The analysis is done separately for the NH and SH. The NH SAT difference $\Delta T = \text{High Ap} - \text{Low Ap}$ for the winter season DFJ (December-January-February) is presented in Fig. 17.4 for the reanalysis (left) and the simulation results (middle) revealing some similarities. Both the reanalysis data and the model show a negative anomaly of about 2 K over the North Atlantic, and a positive anomaly over the Arctic Sea, especially pronounced east of Greenland. The strongest warm anomaly is situated over Siberia in the reanalysis data, the model, however, shows this warm anomaly approximately centered around Svalbard. It should be noted that a perfect pattern match between the model and the reanalysis is not expected because the synoptic situation in the model is different to the situation in the reanalysis at any given time. The model was however driven with observed sea surface temperatures (SST) and sea ice masks [HadISST1, see *Rayner et al.*, 2003], which are expected to influence the SAT. In order to eliminate the possibility of aliasing of SSTs and other influencing factors such as UV variability, sensitivity simulations using boundary conditions from a single year were carried

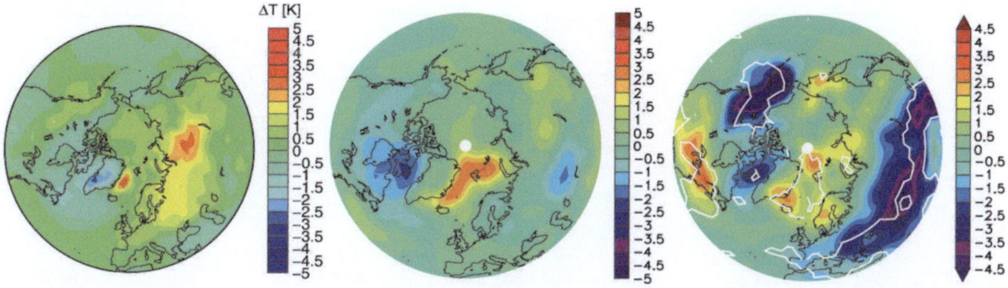

Fig. 17.4 *Left*: Northern Hemisphere DJF surface temperature difference between years of high and low geomagnetic activity. *Left*: from ERA-40 reanalysis data, taken from [*Seppälä et al.*, 2009, their Fig. 2]. *Middle*: from a transient simulation. *Right*: Difference between a simulation with and without EEP, years with sudden stratospheric warmings excluded. *Red/blue* colors indicate positive/negative differences

out. Similar to the previous section, the EEP NO_x source was turned on in one simulation (denoted S-EEP), and turned off in the other (S-noEEP). Each simulation was performed for nine model years after a joint spin up period. The SAT difference (years with sudden stratospheric warmings excluded) is shown in Fig. 17.4 (right), yielding a qualitatively similar result.

Therefore, in the model there appears to be a robust relationship between EEP NO_x and SAT. In order to investigate the underlying mechanism, an analysis of the difference in ozone distribution between the sensitivity simulations is carried out. Stratospheric ozone is reduced by up to 1 μmol/mol in the middle and upper stratosphere (approximately 20 % in the upper stratosphere, 10 % at 20 hPa) in the polar area (not shown). Since ozone is an important radiatively active gas, in general stratospheric ozone concentration changes lead to effects in temperatures. During polar winter, the affected region is mostly dark, so effects caused by the absorption of solar short-wave radiation are expected to be small. However, ozone is also a radiative coolant, and it absorbs longwave radiation from the surface. Ozone changes can therefore potentially lead to temperature changes even during the polar night. Ozone depletion effects on temperature and dynamics have been subject to intensive research in the past [*Langematz et al.*, 2003, and others], mainly focusing on CFC induced ozone depletion. In general, such responses have been shown to be dependent on latitude, season, and on the vertical profile of ozone loss. For the analysis here, a relevant study was conducted by *Langematz et al.* [2003]. Using sensitivity simulations with prescribed ozone loss and a control simulation, they found a heating above the stratopause and cooling below for the NH polar winter (see their Fig. 8a). Using additional radiative transfer calculations, which show only a warming throughout the stratosphere (their Fig. 9), they concluded that the decrease in ozone radiative cooling is responsible for the warming in the simulation with ozone depletion. The cooling below the stratopause was attributed to a decreased mean meridional circulation.

The left panel of Fig. 17.5 presents the temperature differences between the two sensitivity simulations. The figure shows that polar lower stratospheric temperatures between 200–5 hPa decrease by up to 4 K in the S-EEP simulation, and above 4 hPa

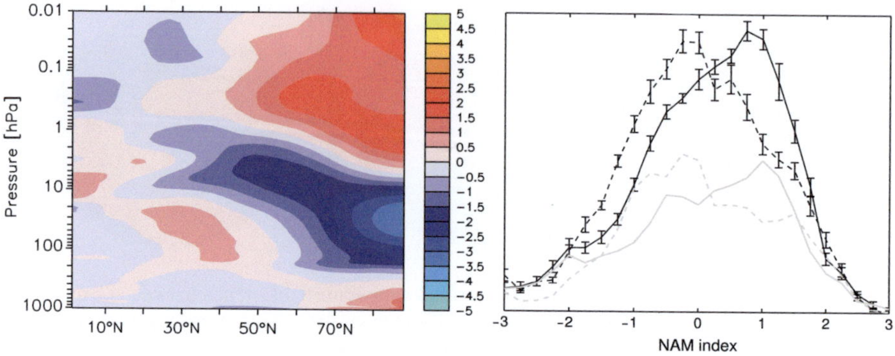

Fig. 17.5 *Left*: Climatological DJF change of temperature (K), $\Delta T = T^{\text{S-EEP}} - T^{\text{S-noEEP}}$. *Right*: NAM index histogram for the altitude region 10 hPa–200 hPa for DJF. *Solid black*: S-EEP, *dashed black*: S-noEEP. The error bars show the standard deviation calculated from the individual years. The *gray lines* show the analogous results for only those years where no SSWs occurred

polar temperatures increase in the S-EEP simulation, indicating that the two-fold response described by *Langematz et al.* [2003] is likely to be also responsible for the effects observed in the ECHAM5/MESSy simulations.

Polar stratospheric temperature changes during winter are likely to have an effect on the polar vortex. A quantity that is often used to describe the strength of the vortex is the Northern Annular Mode (NAM) index [*Baldwin and Dunkerton*, 2001], with positive NAM index values indicating a strong polar vortex and negative index values a weaker vortex. We calculated the NAM index for the model results from the geopotential heights using the zonal mean EOF (empirical orthogonal function) method described by *Baldwin and Thompson* [2009]. Histograms of the NAM indices between 10 hPa and the tropopause for DJF are shown in the right panel of Fig. 17.5. The solid (dashed) black line is the histogram for the simulation S-EEP (S-noEEP). While the tails of the distribution are similar, the probability for NAM indices within the range ± 2 is different between the two simulations: positive NAM indices are more likely in the S-EEP simulation, thus indicating a stronger Arctic polar vortex in the S-EEP simulation. The error bars show the standard deviations resulting from the individual years, indicating that the result is robust. It was investigated if the occurrence of SSWs (Sudden Stratospheric Warmings) in some of the simulated years (four in S-noEEP and three in S-EEP) impacts these results by doing the histogram analysis only for the years where no SSWs occurred. The results are shown as the solid (dashed) gray lines for the S-EEP (S-noEEP) simulation. Qualitatively the result does not change, highlighting the robustness of the result.

In the troposphere NAM anomalies are related to weather anomalies [*Hurrell et al.*, 2003] including specific temperature anomalies. The general characteristic of a positive NAM anomaly is a warmer than average Northern Eurasia, a colder than average Eastern North Atlantic, and a warmer than average United States (see e.g. *Hurrell et al.*, 2003, Chap. 1, Fig. 13). This corresponds to the observed temperature patterns found in the SAT difference between S-EEP and S-noEEP (Fig. 17.4, left). A colder than average Eastern North Atlantic and a warmer than average

United States are also found in the simulations when SSWs years are excluded, see Fig. 17.4, right.

A more comprehensive discussion of this effect including issues concerning statistics and significance can be found in *Baumgaertner et al.* [2011].

17.4 Further Developments Towards a Whole Atmosphere Model

In order to raise the model's upper boundary into the thermosphere, and to include the multitude of processes that govern the upper atmosphere, the CMAT2 model was implemented as a MESSy submodel. CMAT2 [e.g. *Harris*, 2001; *Harris et al.*, 2002; *Cnossen et al.*, 2009] is a modular model system describing the coupled middle atmosphere and thermosphere and is based on the University College London time dependent Three-Dimensional Coupled Thermosphere Ionosphere Plasmasphere (CTIP) model. It includes radiative cooling from NO, O, LTE and non-LTE CO_2 radiative emission, and O_3 radiative emission. Thermospheric heating, photodissociation, and photoionization due to solar X-ray, EUV and UV are considered. Further respected processes are ion drag, molecular viscosity, exothermic chemical heating, particle precipitation, and several ionosphere models.

ECHAM5/MESSy with the CMAT submodel is therefore a whole atmosphere model and is well suited to study solar variability effects that require upper atmospheric coupling to be taken into account from the surface to the thermosphere. This approach enables us to study the effects of solar variability, including particle precipitation, on different atmospheric regions that occur either directly or trough coupling processes. The modularity simplifies cause-and-effect studies by choosing subsets of processes. As part of MESSy, the extension is available to all future development and applications in the framework of Earth System modeling (e.g. the ongoing coupling with an ocean model). Therefore, the combination of CMAT and ECHAM5/MESSy is expected to become one of the most comprehensive atmospheric models available.

The basemodel can either be CMAT itself, ECHAM5, CAABA, or simple environments for testing. For CMAT this means that it can be run in a similar configuration to the original CMAT2 code—but with all the advantages of MESSy such as the flexible input/output control.

The chemistry code of the original CMAT2 is not used in MESSy; instead, the submodel MECCA [*Sander et al.*, 2011] was further enhanced by including electrons, ions, and reactions important in the upper atmosphere, by considering exothermic reactions, and by interfacing with the photolysis and radiation module from CMAT. Also, multiple instances of MECCA can now be operated at the same time in different domains such that ion chemistry reactions do not need to be considered in the lower atmosphere, and higher hydrocarbon chemistry does not need to be computed in the CMAT domain.

When ECHAM5 is used as basemodel, the feedback between the two domains is controlled by the CMAT sub-submodel VGLUE using the nudging (Newtonian

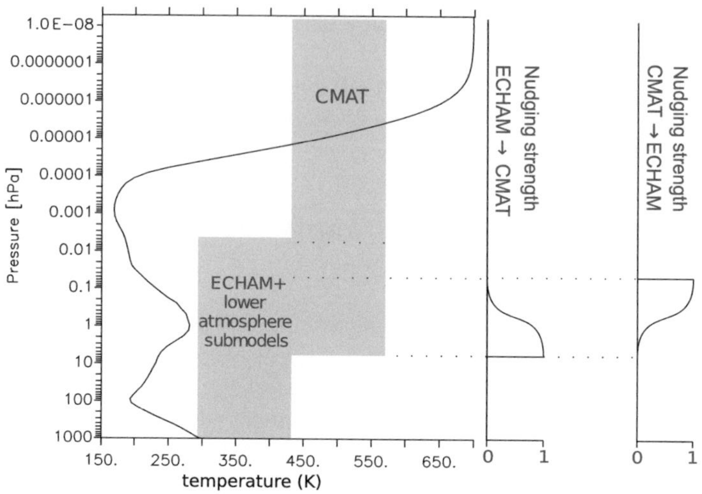

Fig. 17.6 July zonal average temperature averaged over 20°S–20°N and schematic of model domains and nudging

relaxation) technique. Any variable that exists on both grids can be nudged using variable nudging coefficients. Only a specific altitude range (defined either via pressure or special levels such as the stratopause) is nudged, using a given "nudging profile". Generally, a variable in the lower atmosphere domain is nudged towards the upper atmosphere value with a high nudging coefficient at the upper lid, and smaller coefficient towards lower altitudes. The opposite is true when the upper atmosphere variable is nudged towards the lower atmosphere value: A high nudging coefficient is used at the lower boundary, getting smaller with height.

For the gravity wave parameterizations the whole-atmosphere wind profiles are considered in order to yield a consistent wave drag throughout the atmosphere. Several parameterizations, including the original ECHAM5 Hines gravity wave parameterization, are available, but detailed studies are necessary to determine their advantages and disadvantages.

As an example, a temperature profile from a simulation with the ECHAM5 basemodel and CMAT submodel is shown in Fig. 17.6.

A technical documentation of these developments including an evaluation will be presented elsewhere.

17.5 Summary

Substantial extensions of the Modular Earth Submodel System (MESSy) make it possible to study more precisely solar effects on the atmosphere. Solar Proton Events as well as geomagnetic activity related low-energy-electron precipitation have been parameterized and implemented as submodels. With the embedding of the middle and upper atmosphere model CMAT2 as a MESSy submodel,

ECHAM5/MESSy can now be used as a whole-atmosphere model. Scientific studies conducted here show that in a climate change scenario a Brewer-Dobson circulation strengthening will lead to stronger NO_x enhancements in the polar winter middle atmosphere, but any ozone loss may be approximately compensated by a stronger poleward transport of ozone from lower latitudes as well as stratospheric cooling leading to slower photochemical ozone loss reactions. A study of previously proposed surface effects of geomagnetic activity suggests that a possible mechanism for such an effect could be related to shifts in the Northern Annular Mode. The new model tools developed within these projects enable one to perform further studies of this phenomenon.

Acknowledgements The SPACENOX and SPE parameterizations were developed during the CAWSES/ProSECCO project (see Chap. 29), the remaining work was performed in the CAWSES/TIES project. Funding from the DFG for both projects is gratefully acknowledged. The simulations were performed at the Rechenzentrum Garching of the Max Planck Society and the DKRZ (Deutsches Klimarechenzentrum). The development of the ECHAM/MESSy-CMAT coupling was supported in the project WEStSiDe (Workflow environment for Earth System Simulations and Model Development) by the DEISA2 (Distributed European Infrastructure for Supercomputing Applications 2) initiative, funded by the EU FP-7 program (RI-222919). The Ferret program (http://www.ferret.noaa.gov) from NOAA's Pacific Marine Environmental Laboratory was used for creating some of the graphics. Thanks go to all MESSy developers and users for their support.

References

Austin, J., Tourpali, K., Rozanov, E., Akiyoshi, H., Bekki, S., Bodeker, G., Brühl, C., Butchart, N., Chipperfield, M., Deushi, M., Fomichev, V. I., Giorgetta, M. A., Gray, L., Kodera, K., Lott, F., Manzini, E., Marsh, D., Matthes, K., Nagashima, T., Shibata, K., Stolarski, R. S., Struthers, H., & Tian, W. (2008). Coupled chemistry climate model simulations of the solar cycle in ozone and temperature. *Journal of Geophysical Research, 113*, D11306. doi:10.1029/2007JD009391.

Baldwin, M. P., & Dunkerton, T. J. (2001). Stratospheric harbingers of anomalous weather regimes. *Science, 294*, 581–584. doi:10.1126/science.1063315.

Baldwin, M. P., & Thompson, D. W. J. (2009). A critical comparison of stratosphere-troposphere coupling indices. *Quarterly Journal of the Royal Meteorological Society, 135*, 1661–1672. doi:10.1002/qj.479.

Baumgaertner, A. J. G., McDonald, A. J., Fraser, G. J., & Plank, G. E. (2005). Long-term observations of mean winds and tides in the upper mesosphere and lower thermosphere above Scott Base, Antarctica. *Journal of Atmospheric and Solar-Terrestrial Physics, 67*(16), 1480–1496. doi:10.1016/j.jastp.2005.07.018.

Baumgaertner, A. J. G., Jöckel, P., & Brühl, C. (2009). Energetic particle precipitation in ECHAM5/MESSy1—Part 1: downward transport of upper atmospheric NO_x produced by low energy electrons. *Atmospheric Chemistry and Physics, 9*, 2729–2740. doi:10.5194/acp-9-2729-2009.

Baumgaertner, A. J. G., Jöckel, P., Riede, H., Stiller, G., & Funke, B. (2010a). Energetic particle precipitation in ECHAM5/MESSy—Part 2: solar proton events. *Atmospheric Chemistry and Physics, 10*, 7285–7302. doi:10.5194/acp-10-7285-2010.

Baumgaertner, A. J. G., Jöckel, P., Dameris, M., & Crutzen, P. J. (2010b). Will climate change increase ozone depletion from low-energy-electron precipitation? *Atmospheric Chemistry and Physics, 10*(19), 9647–9656. doi:10.5194/acp-10-9647-2010.

Baumgaertner, A. J. G., Seppälä, A., Jöckel, P., & Clilverd, M. A. (2011). Geomagnetic activity related NO_x enhancements and polar surface air temperature variability in a chemistry climate model: modulation of the NAM index. *Atmospheric Chemistry and Physics*, *11*, 4521–4531. doi:10.5194/acp-11-4521-2011.

Brühl, C., Steil, B., Stiller, G., Funke, B., & Jöckel, P. (2007). Nitrogen compounds and ozone in the stratosphere: comparison of MIPAS satellite data with the chemistry climate model ECHAM5/MESSy1. *Atmospheric Chemistry and Physics*, *7*(21), 5585–5598. doi:10.5194/acp-7-5585-2007.

Butchart, N., Cionni, I., Eyring, V., Shepherd, T. G., Waugh, D. W., Akiyoshi, H., Austin, J., Brühl, C., Chipperfield, M. P., Cordero, E., Dameris, M., Deckert, R., Dhomse, S., Frith, S. M., Garcia, R. R., Gettelman, A., Giorgetta, M. A., Kinnison, D. E., Li, F., Mancini, E., McLandress, C., Pawson, S., Pitari, G., Plummer, D. A., Rozanov, E., Sassi, F., Scinocca, J. F., Shibata, K., Steil, B., & Tian, W. (2010). Chemistry-climate model simulations of twenty-first century stratospheric climate and circulation changes. *Journal of Climate*, *23*, 5349–5374. doi:10.1175/2010JCLI3404.1.

Clilverd, M. A., Rodger, C. J., & Ulich, T. (2006). The importance of atmospheric precipitation in storm-time relativistic electron flux drop outs. *Geophysical Research Letters*, *33*, L01102. doi:10.1029/2005GL024661.

Cnossen, I., Harris, M. J., Arnold, N. F., & Yiğit, E. (2009). Modelled effect of changes in the CO_2 concentration on the middle and upper atmosphere: sensitivity to gravity wave parameterization. *Journal of Atmospheric and Solar-Terrestrial Physics*, *71*, 1484–1496. doi:10.1016/j.jastp.2008.09.014.

Engel, A., Möbius, T., Bönisch, H., Schmidt, U., Heinz, R., Levin, I., Atlas, E., Aoki, S., Nakazawa, T., Sugawara, S., Moore, F., Hurst, D., Elkins, J., Schauffler, S., Andrews, A., & Boering, K. (2009). Age of stratospheric air unchanged within uncertainties over the past 30 years. *Nature Geoscience*, *3*(2), 28–31. doi:10.1038/ngeo388.

Funke, B., López-Puertas, M., von Clarmann, T., Stiller, G. P., Fischer, H., Glatthor, N., Grabowski, U., Höpfner, M., Kellmann, S., Kiefer, M., Linden, A., Mengistu Tsidu, G., Milz, M., Steck, T., & Wang, D. Y. (2005a). Retrieval of stratospheric NO_x from 5.3 and 6.2 µm nonlocal thermodynamic equilibrium emissions measured by Michelson Interferometer for Passive Atmospheric Sounding (MIPAS) on Envisat. *Journal of Geophysical Research*, *110*, D09302. doi:10.1029/2004JD005225.

Funke, B., López-Puertas, M., Gil-López, S., von Clarmann, T., Stiller, G. P., Fischer, H., & Kellmann, S. (2005b). Downward transport of upper atmospheric NO_x into the polar stratosphere and lower mesosphere during the Antarctic 2003 and Arctic 2002/2003 winters. *Journal of Geophysical Research*, *110*, D24308. doi:10.1029/2005JD006463.

Funke, B., García-Comas, M., López-Puertas, M., Glatthor, N., Stiller, G. P., von Clarmann, T., Semeniuk, K., & McConnell, J. C. (2008). Enhancement of N_2O during the October/November 2003 solar proton events. *Atmospheric Chemistry and Physics*, *8*, 3805–3815.

Haigh, J. D., & Pyle, J. A. (1982). Ozone perturbation experiments in a two-dimensional circulation model. *Quarterly Journal of the Royal Meteorological Society*, *108*, 551–574. doi:10.1256/smsqj.45704.

Harris, M. J. (2001). *A new coupled middle atmosphere and thermosphere circulation model: studies of dynamic, energetic and photochemical coupling in the middle and upper atmosphere*. Ph.D. thesis, University of London.

Harris, M. J., Arnold, N. F., & Aylward, A. D. (2002). A study into the effect of the diurnal tide on the structure of the background mesosphere and thermosphere using the new coupled middle atmosphere and thermosphere (CMAT) general circulation model. *Annales Geophysicae*, *20*, 225–235. doi:10.5194/angeo-20-225-2002.

Hurrell, J., Kushnir, Y., Ottersen, G., & Visbeck, M. (2003). The North Atlantic oscillation: climate significance and environmental impact. In *An overview of the North Atlantic oscillation* (pp. 1–35). Washington: AGU. doi:10.1029/134GM01.

IPCC (2007). *IPCC fourth assessment report: climate change 2007*. Cambridge: Cambridge University Press.

Jackman, C. H., & McPeters, R. D. (2004). Solar variability and its effects on climate. In *The effect of solar proton events on ozone and other constituents* (pp. 305–319). Washington: Am. Geophys. Union.
Jackman, C. H., DeLand, M. T., Labow, G. J., Fleming, E. L., Weisenstein, D. K., Ko, M. K. W., Sinnhuber, M., & Russell, J. M. (2005). Neutral atmospheric influences of the solar proton events in October-November 2003. *Journal of Geophysical Research, 110*(a9), A09S27. doi:10.1029/2004JA010888.
Jöckel, P., Sander, R., Kerkweg, A., Tost, H., & Lelieveld, J. (2005). Technical note: The Modular Earth Submodel System (MESSy)—a new approach towards Earth System Modeling. *Atmospheric Chemistry and Physics, 5*, 433–444. doi:10.5194/acp-5-433-2005.
Jöckel, P., Tost, H., Pozzer, A., Brühl, C., Buchholz, J., Ganzeveld, L., Hoor, P., Kerkweg, A., Lawrence, M. G., Sander, R., Steil, B., Stiller, G., Tanarhte, M., Taraborrelli, D., van Aardenne, J., & Lelieveld, J. (2006). The atmospheric chemistry general circulation model ECHAM5/MESSy1: consistent simulation of ozone from the surface to the mesosphere. *Atmospheric Chemistry and Physics, 6*, 5067–5104. doi:10.5194/acp-6-5067-2006.
Jöckel, P., Kerkweg, A., Pozzer, A., Sander, R., Tost, H., Riede, H., Baumgaertner, A., Gromov, S., & Kern, B. (2010). Development cycle 2 of the Modular Earth Submodel System (MESSy2). *Geoscientific Model Development, 3*(3), 717–752. doi:10.5194/gmd-3-717-2010.
Jonsson, A. I., de Grandpré, J., Fomichev, V. I., McConnell, J. C., & Beagley, S. R. (2004). Doubled CO_2-induced cooling in the middle atmosphere: photochemical analysis of the ozone radiative feedback. *Journal of Geophysical Research, 109*, D24103. doi:10.1029/2004JD005093.
Jungclaus, J. (2006). IPCC-AR4 MPI-ECHAM5_T63L31 MPI-OM_GR1.5L40 SRESA2 run no.3: ocean monthly mean values MPImet/MaD Germany. World Data Center for Climate. CERA-DB "OM-GR1.5L40_EH5-T63L31_A2_3_MM".
Langematz, U., Kunze, M., Krüger, K., Labitzke, K., & Roff, G. L. (2003). Thermal and dynamical changes of the stratosphere since 1979 and their link to ozone and CO_2 changes. *Journal of Geophysical Research, 108*, 4027. doi:10.1029/2002JD002069.
Lelieveld, J., Brühl, C., Jöckel, P., Steil, B., Crutzen, P. J., Fischer, H., Giorgetta, M. A., Hoor, P., Lawrence, M. G., Sausen, R., & Tost, H. (2007). Stratospheric dryness: model simulations and satellite observations. *Atmospheric Chemistry and Physics, 7*, 1313–1332. doi:10.5194/acp-7-1313-2007.
Nakicenovic, N., Alcamo, J., Davis, G., de Vries, B., Fenhann, B., Gaffin, S., Gregory, K., Grübler, A., Jung, T. Y., Kram, T., La Rovere, E. L., Michaelis, E. L., Mori, S., Morita, T., Pepper, W., Pitcher, H., Price, L., Raihi, K., Roehrl, A., Rogner, H.-H., Sankovski, A., Schlesinger, M., Shukla, P., Smith, S., Swart, R., van Rooijen, S., Victor, N., & Dadi, Z. (2000). *IPCC special report on emissions scenarios*. Cambridge: Cambridge University Press.
Randall, C. E., Harvey, V. L., Singleton, C. S., Bailey, S. M., Bernath, P. F., Codrescu, M., Nakajima, H., & Russell, J. M. (2007). Energetic particle precipitation effects on the Southern Hemisphere stratosphere in 1992–2005. *Journal of Geophysical Research, 112*, D08308. doi:10.1029/2006JD007696.
Rayner, N. A., Parker, D. E., Horton, E. B., Folland, C. K., Alexander, L. V., Rowell, D. P., Kent, E. C., & Kaplan, A. (2003). Global analyses of sea surface temperature, sea ice, and night marine air temperature since the late nineteenth century. *Journal of Geophysical Research, 108*, D144407. doi:10.1029/2002JD002670.
Roeckner, E., Brokopf, R., Esch, M., Giorgetta, M., Hagemann, S., Kornblueh, L., Manzini, E., Schlese, U., & Schulzweida, U. (2006). Sensitivity of simulated climate to horizontal and vertical resolution in the ECHAM5 atmosphere model. *Journal of Climate, 19*, 3771. doi:10.1175/JCLI3824.1.
Rohen, G., von Savigny, C., Sinnhuber, M., Llewellyn, E. J., Kaiser, J. W., Jackman, C. H., Kallenrode, M.-B., Schröter, J., Eichmann, K.-U., Bovensmann, H., & Burrows, J. P. (2005). Ozone depletion during the solar proton events of October/November 2003 as seen by SCIAMACHY. *Journal of Geophysical Research, 110*, A09S39. doi:10.1029/2004JA010984.
Sander, R., Kerkweg, A., Jöckel, P., & Lelieveld, J. (2005). Technical note: The new comprehensive atmospheric chemistry module mecca. *Atmospheric Chemistry and Physics, 5*(2), 445–450.

doi:10.5194/acp-5-445-2005.

Sander, R., Baumgaertner, A., Gromov, S., Harder, H., Jöckel, P., Kerkweg, A., Kubistin, D., Regelin, E., Riede, H., Sandu, A., Taraborrelli, D., Tost, H., & Xie, Z.-Q. (2011). The atmospheric chemistry box model CAABA/MECCA-3.0. *Geoscientific Model Development*, *4*, 373–380. doi:10.5194/gmd-4-373-2011.

Schmidt, H., Brasseur, G. P., Charron, M., Manzini, E., Giorgetta, M. A., Diehl, T., Fomichev, V. I., Kinnison, D., Marsh, D., & Walters, S. (2006). The HAMMONIA chemistry climate model: sensitivity of the mesopause region to the 11-year solar cycle and CO2 doubling. *Journal of Climate*, *19*, 3903–3931. doi:10.1175/JCLI3829.1.

Seppälä, A., Randall, C. E., Clilverd, M. A., Rozanov, E., & Rodger, C. J. (2009). Geomagnetic activity and polar surface air temperature variability. *Journal of Geophysical Research*, *114*(a13), A10312. doi:10.1029/2008JA014029.

Solomon, S., Rusch, D. W., Gerard, J. C., Reid, G. C., & Crutzen, P. J. (1981). The effect of particle-precipitation events on the neutral and ion chemistry of the middle atmosphere, 2, Odd hydrogen. *Planetary and Space Science*, *29*(8), 885–892.

Verronen, P. T., Rodger, C. J., Clilverd, M. A., & Wang, S. (2011). First evidence of mesospheric hydroxyl response to electron precipitation from the radiation belts. *Journal of Geophysical Research. Atmospheres*, *116*(D15), D07307. doi:10.1029/2010JD014965.

Vitt, F. M., & Jackman, C. H. (1996). A comparison of sources of odd nitrogen production from 1974 through 1993 in the Earth's middle atmosphere as calculated using a two-dimensional model. *Journal of Geophysical Research*, *101*, 6729–6740. doi:10.1029/95JD03386.

Chapter 18
Solar Variability and Trend Effects in Mesospheric Ice Layers

Franz-Josef Lübken, Uwe Berger, Johannes Kiliani, Gerd Baumgarten, and Jens Fiedler

Abstract In this paper we summarize results from the SOLEIL project (SOLar variability and trend Effects in Ice Layers) which was part of the CAWSES priority program in Germany. We present results from LIMA/ICE which is a global circulation model concentrating on ice clouds (NLC, noctilucent clouds) in the summer mesopause region. LIMA/ICE adapts to ECMWF data in the lower atmosphere which produces significant short term and year-to-year variability. The mean ice cloud parameters derived from LIMA/ICE generally agree with observations. The formation, transport, and sublimation of ice particles causes a significant redistribution of water vapor ('freeze drying'). Model results are now available for all years since 1961 for various scenarios, e.g., with and without greenhouse gas increase etc. Temperatures and water vapor are affected by solar activity. In general it is warmer during solar maximum, but there is a small height region around the mesopause where it is colder. This complicates the prediction of solar cycle effects on ice layers. The magnitude of the solar cycle effect is ~1–3 K which is similar to the year-to-year variability. Therefore, only a moderate solar cycle signal is observed in temperatures and in ice layers. Temperature trends at NLC altitudes are partly caused by stratospheric trends ('shrinking effect'). Trends are generally negative, but are positive in the mesopause region. Again, this complicates a simple prediction of temperature trends on ice layers and requires a complex model like LIMA/ICE. Trends in CO_2 and stratospheric O_3 enhance mesospheric temperature trends but have comparatively small effects in the ice regime. Comparison of contemporary and historic observations of NLC altitudes leads to negligible temperature trends at NLC altitudes (~83 km). For the time period of satellite measurements (1979–2009) LIMA/ICE predicts trends in ice cloud brightness and occurrence rates, consistent with observations. Temperature trends are not uniform in time but are stronger until the mid 1990s, and weaker thereafter. This change is presumably related to stratospheric ozone recovery. The accidental coincidence of lowest temperatures and solar cycle minimum in the mid 1990s led to large NLC activity. It is important to consider the time period and the height range when studying temperature and ice cloud

F.-J. Lübken (✉) · U. Berger · J. Kiliani · G. Baumgarten · J. Fiedler
Leibniz-Institute of Atmospheric Physics, Schlossstr. 6, 18225 Kühlungsborn, Germany
e-mail: luebken@iap-kborn.de

F.-J. Lübken (ed.), *Climate and Weather of the Sun-Earth System (CAWSES)*, Springer Atmospheric Sciences,
DOI 10.1007/978-94-007-4348-9_18,

trends. In the mesosphere temperature trends can be as large as −(3–5) K/decade (in agreement with observations) or rather small, depending on the time period and height range.

18.1 Introduction

The mesosphere/lower thermosphere (MLT) is very suitable for trend studies since an increase of greenhouse gases in general affects this region much more strongly compared to the troposphere. Furthermore, some parameters are very sensitive to changes in the background atmosphere, for example ice layers. Ice layers form in the very cold summer mesopause at middle and polar latitudes and are known as noctilucent clouds (NLC) or PMC (polar mesospheric clouds). Ice layers are important for trend studies in the MLT region because they constitute the longest record of quantitative observations in the middle atmosphere starting in the 1890s [*Jesse*, 1896]. They are very sensitive to temperatures and to water vapor. As will be shown in this paper, NLC parameters show significant long term variations, but also some incredible persistences which provide a stringent constraint for any model on trends in the MLT region. Whether or not ice layers show trends is disputed in the literature [*von Zahn*, 2003; *Thomas et al.*, 2003]. In this paper we consider variations with solar cycle and on longer time scales ('trends'). We summarize results from the SOLEIL project (SOLar variability and trend Effects in Ice Layers) which concentrates on simulations of ice layers and compares with observations of NLC and PMC.

18.2 The LIMA/ICE Model

The LIMA model (Leibniz-Institute Middle Atmosphere model) is a general circulation model of the middle atmosphere which especially aims to represent the thermal structure around mesopause altitudes [*Berger*, 2008]. LIMA is a fully non-linear, global, and three-dimensional Eulerian grid-point model which extends from the ground to the lower thermosphere (0–150 km) taking into account major processes of radiation, chemistry, and transport. LIMA applies a triangular horizontal grid structure with 41804 grid points in every horizontal layer ($\Delta x \sim \Delta y \sim 110$ km) and adapts to tropospheric and lower stratospheric data from ECMWF.[1] This introduces short term and year-to-year variability which influences, amongst other parameters, ice formation. LIMA has recently been upgraded to include additional radiation parameterizations which improve simulations of solar irradiance variations, e.g., during the 11-y solar cycle. The old and new versions of LIMA are called LIMA-I and LIMA-II, respectively. One of the important applications of LIMA is modeling of NLC formation. To this extent a dust/ice particle module is added to LIMA which

[1] European Center for Medium-range Weather Forecasts.

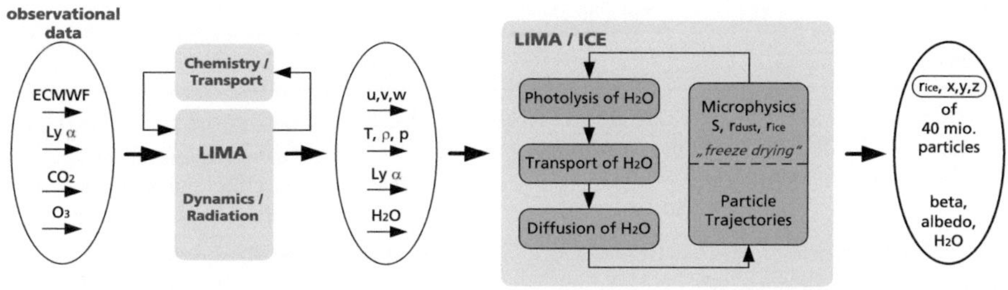

Fig. 18.1 Flow chart of LIMA and LIMA/ICE. See Table 18.1 for more details

is called LIMA/ICE. This module consists of a multiple particle system containing 40 million meteoric dust particles which act as potential condensation nuclei in the cold summer mesopause region at high latitudes. Several microphysical processes are present in LIMA/ICE, such as nucleation, growth, sublimation, sedimentation, and diffusion [see *Rapp and Thomas*, 2006, for a recent update on microphysics of ice particles]. The background water vapor interacts with ice particles which is re-distributed vertically and horizontally ('freeze drying'). It is important to note that we initialize LIMA/ICE with a fixed latitude/longitude/height distribution of H_2O at the beginning of each summer season at 20 May (15 November) in the northern (southern) hemisphere (NH/SH). This fixed water vapor distribution is used as an initial condition for all years (1961–2010), identical in both hemispheres. When the ice season starts the water vapor is exposed to photo-dissociation by varying Ly_α radiation, 3-d transport, turbulent diffusion, and redistribution by freeze drying until the season ends at 20 August (15 February) in the NH (SH). As will be discussed later, these processes induce solar cycle variations and long-term trends in water vapor in the upper mesosphere in LIMA/ICE. In Fig. 18.1 a sketch of the interplay between LIMA and LIMA/ICE is presented.

LIMA and LIMA/ICE were run for different scenarios summarized in Table 18.1. In order to account for realistic atmospheric conditions LIMA adapts several observational data sets, such as a) tropospheric and stratospheric temperatures and winds (ECMWF data) from 0 to 40 km since 1961 (every 6 h; global coverage); b) daily Lyman-alpha fluxes[2]; c) monthly carbon dioxide concentrations[3] from 0–130 km; and d) stratospheric ozone which is quantified by monthly, latitudinal, longitudinal, and altitude-dependent profiles available from SBUV/TOMS[4] satellite data for 1979–2008.[5] The uppermost level of SBUV ozone data corresponds to a pressure of 0.5 hPa ($\sim$50 km). Above this level LIMA interpolates and fits the SBUV ozone fields within one scale height to the standard ozone climatology used in LIMA.

---

[2] ftp://laspftp.colorado.edu/pub/SEE-DATA/composite-lya.

In this paper we express Ly_α radiation in units of 10^{11} photons/(cm^2·s).

[3] http://www.esrl.noaa.gov/gmd/ccgg/trends.

[4] Solar Backscatter in the Ultraviolet, Total Ozone Mapping System.

[5] http://acdb-ext.gsfc.nasa.gov/Data_services/merged/.

Table 18.1 Runs of LIMA and LIMA/ICE

	LIMA[a]	LIMA/ICE
	global, 0–150 km, $\Delta z = 1.1$ km, $\Delta x = 110$ km, ECMWF (0–35 km), dynamics, radiation, chemistry, transport, H_2O fixed	37.5°–90°N/S, $z = 78$–94 km, $\Delta z = 0.1$ km, u, v, w, T, H_2O from LIMA, 40 million dust/ice particles, cloud microphysics, H_2O: transport, photolysis, freeze drying
1	±10 % CO_2, ±10 % H_2O, ±10 % O_3[b]	Ly_α at 2005 solar conditions
2	Ly_α at solar min and max[b]	Ly_α at solar min and max
3	1961–2009 dynamics, fixed Ly_α, fixed GHG (I)	H_2O varies with Ly_α
4	1961–2009 dynamics, fixed Ly_α, fixed GHG	"
5	1961–2010 dynamics, $T(Ly_\alpha)$, GHG fixed	H_2O and temperatures vary with Ly_α
6	1961–2009 dynamics, $T(Ly_\alpha)$, CO_2⇑	"
7	1979–2009 dynamics, $T(Ly_\alpha)$, CO_2 and O_3⇑	"

[a]Run no. 3 was performed with LIMA-I, whereas all other runs were performed with LIMA-II

[b]Relative to 2005

For the period 1964–1978 we adapt annual total ozone data from ground-based measurements published in the WMO report 2011. The years 1961–1963 have been extrapolated from 1964.

Three different simulations have been performed with LIMA to assess the sensitivity of mesospheric temperature trends and their effect on ice layers: a) keeping CO_2 and O_3 constant from year to year (run no. 5), b) taking into account long term changes of CO_2 in the entire atmosphere and keeping O_3 constant from year to year (run no. 6), and c) same, but taking into account O_3 changes up to approximately 55 km from SBUV in years 1979–2008 and WMO 2011 in years 1961–1978 (run no. 7). We note that all runs except run no. 7 use the same annual standard ozone climatology constant from year to year. This global data set for January-December is a composite of a modeled ozone climatology provided by the LIMA chemistry-transport model (mesosphere/thermosphere) and an ozone climatology based on ozonesonde and satellite measurements (stratosphere) by *Fortuin and Kelder* [1998]. Furthermore, we note that global long term changes of CO_2 and O_3 in the troposphere and lower stratosphere are implicitly taken into account in all runs by the adaption of ECMWF data.

Certainly the most important background parameter controlling the morphology of ice layers is temperature. In Fig. 18.2 LIMA temperatures in the northern hemisphere ice domain are shown for run no. 7. Ice particles can grow and exist if the degree of saturation, S, is larger than unity. S increases exponentially with decreasing temperatures and linearly with enhanced water vapor concentration. For typical water vapor concentrations of 2–4 ppmv at NLC altitudes S is larger then unity for temperatures smaller than approximately 148 K. As can be seen from Fig. 18.2 NLC/PMC may exist poleward of ~50°N. As will be shown later, natural variability can extend this latitude range more southward. Note that ice particles may be present but may be too small to be detectable by remote sounding instruments.

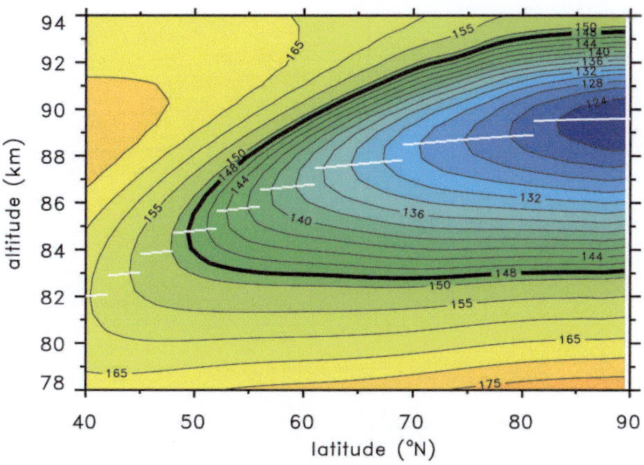

Fig. 18.2 Mean July temperatures averaged for the years 1997–2007, namely for the last solar cycle period (run no. 7). The *black isoline* highlights the approximate region where ice particles may exist. *White lines* indicate the mesopause altitude

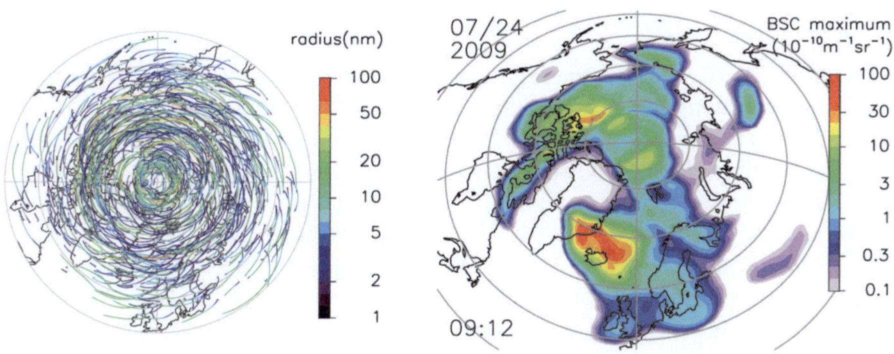

Fig. 18.3 *Left*: Trajectories of 100 ice particles (out of 40 million) for a period of 5 days. *Right*: Snapshot of the ice cloud in the northern hemisphere on 24 July 2009, 09:12 UT

Therefore, the total altitude/latitude range of super-saturation is difficult to specify experimentally. The isothermal lines at NLC heights are nearly constant in altitude from ~55°N to the pole. This suggests that also the lower edge of ice clouds should be nearly constant with latitude. Indeed, a review of lidar observations at various stations confirmed that NLC heights vary only very little with latitude [see Fig. 3 in *Lübken et al.*, 2008]. This nicely supports the background temperature fields in LIMA and imposes a prominent constraint on any model of the polar MLT. The constant NLC altitude concerns mean layer characteristics. In contrast, NLC altitudes from LIMA/ICE generally decrease with increasing brightness, again in agreement with lidar observations at ALOMAR [*Baumgarten and Fiedler*, 2008].

In Fig. 18.3 the trajectories of 100 (out of 40 million) ice particles from LIMA/ICE are shown for a period of 5 days. Particles move with the dominant wind direction which is nearly westward below the mesopause. Ice particles have a typical lifetime of a few days and only a small fraction of them grows large enough (radius larger than ~20 nm) to be visible from the ground or from satellites. In the right

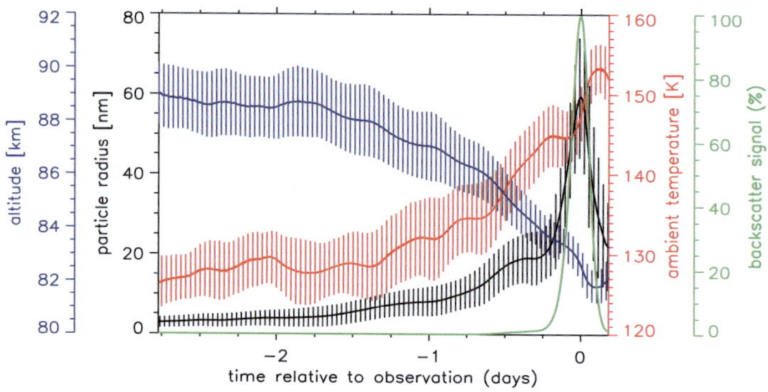

Fig. 18.4 Temporal evolution of ice particles observed at $t = 0$ at ALOMAR (69°N). Mean cloud altitudes (*blue*), backscatter signal (relative to the maximum signal at $t = 0$, *green*), particle radii (*black*), and ambient temperatures (*red*) are shown, averaged over 50 clouds which all show maximum scatter at ALOMAR. *Vertical lines* indicate the variability in these 50 clouds

panel of Fig. 18.3 we show the brightness distribution of an NLC from LIMA/ICE at a certain point in time. More precisely, peak backscatter coefficients β_{max} in units of 10^{-10}/(m·sr) at $\lambda = 532$ nm are shown.[6] This plot demonstrates that the ice cloud is heterogeneous in time and space and may occasionally extend as far southward as ~50°N. This morphology generally agrees with space borne and ground based measurements [*Russell et al.*, 2009; *Lübken et al.*, 2008].

LIMA/ICE allows to study in detail the development of ice clouds. We selected 50 ice clouds from LIMA which are detectable by our lidar at ALOMAR (69°N) and studied their genesis (Fig. 18.4). Mean values and variabilities for ice particle altitudes, radii, and ambient temperatures are shown. They agree in general with observations [see for example *von Cossart et al.*, 1999; *von Savigny et al.*, 2005]. Furthermore, the sum of the backscatter signal from all 50 ice clouds is shown, normalized to the maximum value (which is observed at ALOMAR). Ice particles initially stay in the cold region near the mesopause for an extended time period and grow very slowly. About a day before observation, particle growth and sedimentation accelerates. The main growth phase occurs within only a few hours prior to observation. This implies that the 'local' background conditions at ALOMAR determine the main features of the ice clouds as observed by lidars. After leaving ALOMAR, the ice clouds sediment quickly to warmer regions and sublimate within a few hours.

We have studied the variability of ice cloud parameters within a summer season. In Fig. 18.5 the cumulative occurrence frequency distribution as a function of backscatter coefficients β_{max} is shown taking into account all ice clouds with $\beta_{max} > 4.5$ from year 2000 at 69°N (total number of clouds: 21258). This means, for each β the relative occurrence frequency of clouds being brighter than β is shown. The distribution shows an exponential decrease with increasing β which

[6] In this paper all backscatter coefficients β are given in units of 10^{-10}/(m·sr).

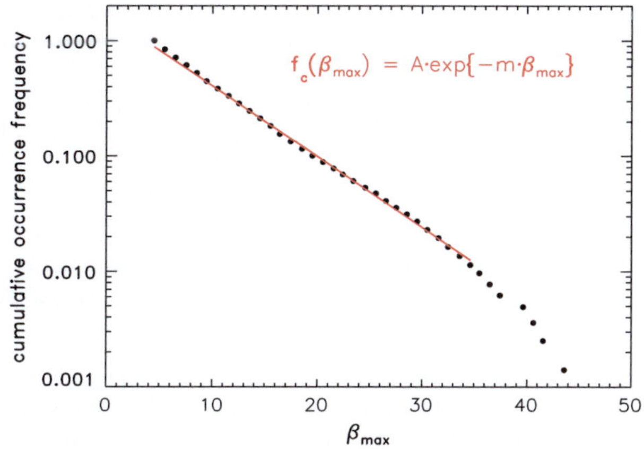

Fig. 18.5 Cumulative occurrence frequency distribution of backscatter coefficients in the summer season 2000 at 69°N (run no. 5). The occurrence is normalized to 1 at the smallest β_{max}-value ($\beta_{max} = 4.5$). All β_{max}-values are given in units of 10^{-10}/(m·sr). Fitting an exponential function $f_c(\beta_{max}) = A \cdot e^{-m \cdot \beta_{max}}$ gives $A = 1.662$ and $m = 0.1411$ (*red line*)

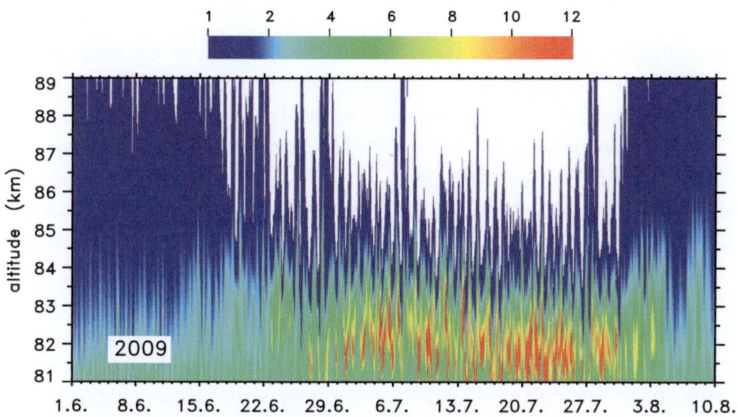

Fig. 18.6 Water vapor concentration from LIMA (run no. 5) in ppmv (see *color bar*) in the summer season 2009 at ALOMAR (69°N, 15°E). The freeze drying effect caused by ice particle formation, transport, and sublimation reduces the H_2O concentration in the upper atmosphere and leads to an enhancement around 82 km. Sporadic mixing events lead to temporary redistributions. Note the non-uniform color scale

points to randomly distributed processes forming ice [*Thomas*, 1995]. The distribution is in nice agreement with the so called 'g-distributions' deduced from satellite observations of PMC [*DeLand et al.*, 2003]. The similarity between observed and LIMA/ICE distributions suggests that the mechanisms controlling the variability of ice clouds are replicated in LIMA. In practical terms this result implies that occurrence rates from instruments with different β-thresholds can easily be compared (within limits).

Water vapor varies in LIMA/ICE due to photo-dissociation by Ly_α and due to redistribution of H_2O by ice particle formation, sedimentation, and sublimation (see Sect. 10.2). This effect is called 'freeze drying' and depends on temperatures, winds etc. As can be seen from Fig. 18.6 water vapor is redistributed, i.e., water

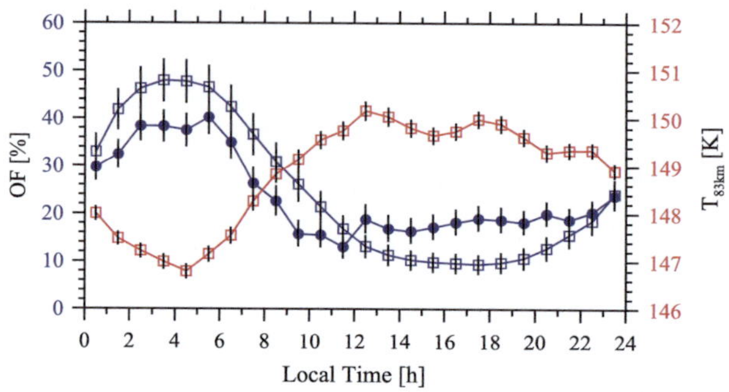

Fig. 18.7 Tides in NLC and temperatures at ALOMAR (69°N, 16°E). Occurrence rates from the RMR-Lidar at ALOMAR are shown for $\beta_{max} > 4$ using the entire data set 1997–2010 [*blue dots*, update from *Fiedler et al.*, 2011]. Temperature deviations from LIMA are shown for the same location (run no. 5, *red curve*) together with occurrence rates from LIMA/ICE (*blue squares*)

vapor concentration decreases above ~84–85 km and can be more than doubled (compared to normal conditions) in the sublimation zone at ~82 km. Indeed, satellite observations have shown strong indications for such an enhancement of water vapor at ~82–84 km [*Summers et al.*, 2001]. Significant freeze drying sets in with some delay after the first occurrence of ice layers, and terminates when ice formation disappears (beginning of August in Fig. 18.6). Ice particles can also redistribute water vapor horizontally, but mainly along latitudes since ice clouds mainly move in the zonal direction. As can be seen from Fig. 18.6 vertical winds may occasionally blow water vapor from the sublimation zone back up into the dry region ('episodic mixing') which causes a temporary humidification. The freeze drying effect has potential implications since some photochemical reactions in the MLT region involve H_2O. Furthermore, ice particles can react directly with other substances, for example with metal atoms [*Plane et al.*, 2004; *Lübken and Höffner*, 2004]. Therefore, ice layers play a more general role for the photochemistry in the MLT region which should be taken into account in general circulation models.

We have also studied tides in LIMA/ICE and compared our results with observations at ALOMAR. In Fig. 18.7 we show the diurnal variations of LIMA temperatures at 83 km and ice cloud occurrence frequencies averaged for June and July of all years from 1997–2010, as in *Fiedler et al.* [2011]. We also show occurrence frequencies for clouds with $\beta_{max} > 4$ as measured by the RMR lidar. The temperature deviations from LIMA show a nice anti-correlation with the occurrence frequency observed at ALOMAR (and modeled by LIMA/ICE) which suggests that tidal temperature variations are the main reason for tides in ice clouds.

LIMA/ICE modeling is now available for all years since 1961, both for the northern (NH) and southern hemisphere (SH). All model runs apply the nudging technique and the ice particle routine LIMA/ICE introduced above. A collection of latitudinal/longitudinal distributions of ice layers from 1961 to 2009 is presented in the literature for run no. 3 [*Lübken et al.*, 2009] and run no. 5 [*Lübken and Berger*, 2011]. In the meantime more runs with varying Ly_α and increasing greenhouse gases have been performed with LIMA-II (see Table 18.1). A total of 20 trillion

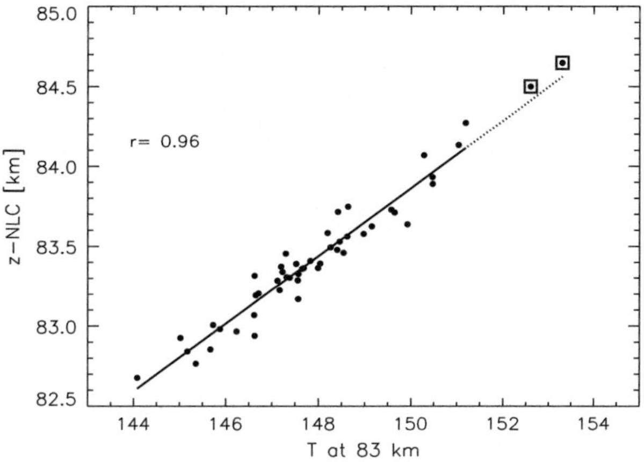

Fig. 18.8 NLC altitudes versus temperatures at 83 km (from run no. 6) showing a very high correlation ($r = 0.96$). A straight line fit ignoring the spurious years 1975/1976 (*squares*) gives a slope of 0.212 km/K

single data points are available for each run, namely 40 million particles $\times$ 24 hours $\times$ 90 days $\times$ 50 years $\times$ 3 positions $\times$ 2 parameters (radii) $\times$ 2 hemispheres $\sim 20 \times 10^{12}$.

18.3 Correlations and Sensitivity Studies

We have studied in detail various correlations between ice clouds and background parameters [see, e.g., *Lübken et al.*, 2009, where run no. 3 was used]. We have repeated these studies with LIMA-II including solar cycle variations with and without CO_2 and O_3 trends. The correlation coefficients and slopes vary somewhat for different runs, but some important results are unchanged. For example, in all runs we find a very high correlation between NLC altitudes and temperatures at 83 km (see Fig. 18.8). The correlation coefficient is close to one and is highly significant. This result allows to use historical altitude measurements of NLC heights to constrain temperature trends (see Sect. 18.5). Strong positive correlations are also found for occurrence rates versus brightness. Weaker positive correlation exists for occurrence and brightness versus water vapor concentration. More examples for (anti-)correlations are presented in the literature stated above. We will discuss the dependence of background and ice cloud parameters on Ly_α in more detail in Sect. 18.4.

We have performed sensitivity studies for NLC altitude, occurrence frequency, and brightness by varying water vapor (by ± 25 %), temperatures (by ± 1 K), and solar activity (max, min) relative to a mean state. More details are presented in *Lübken et al.* [2009]. Roughly speaking, NLC altitudes vary mostly with temperatures and are rather insensitive to H_2O change. Occurrence rates and brightness vary with all three parameters (H_2O, T, Ly_α) and none of the influences is small compared to the others. This implies that a given variation of occurrence or brightness cannot unambiguously be related to a change of one of the three background conditions.

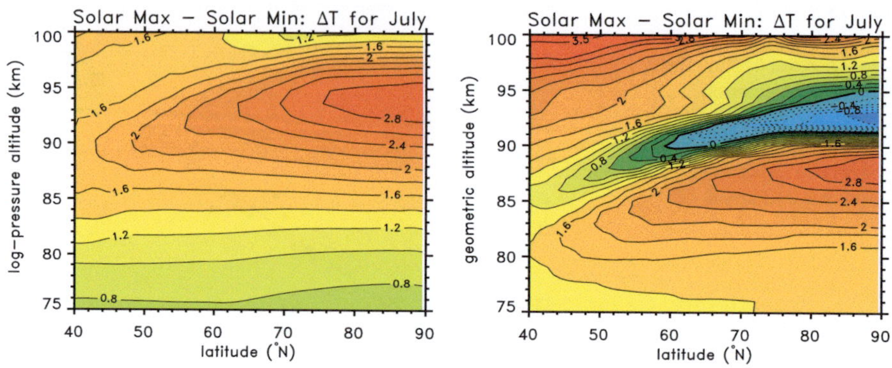

Fig. 18.9 Solar cycle effect on temperatures for run no. 2 as a function of pressure (*left panel*) and geometric altitude (*right panel*). The mean temperature difference (in Kelvin) of solar maximum minus minimum for July is shown

18.4 Solar Cycle Variations

Solar Cycle Effect on Temperatures and H_2O In Fig. 18.9 the effect of Ly_α-radiation on temperatures in July is shown (run no. 2). In pressure coordinates temperatures are generally higher during solar maximum compared to minimum by approximately 1–3 K in the MLT region. At mesopause altitudes the difference increases with latitude. The temperature increase is mainly due to increased heating caused by photo-dissociation of molecular oxygen by Ly_α. To a smaller extent variable solar activity also alters the strength of solar absorption from the near infrared to the UV. Heating generally increases with altitude up to the thermosphere. The maximum heating around 93–95 km in Fig. 18.9 (left panel) is due to very low temperatures and consequently large number densities leading to a local enhancement of optical thickness. The effect of solar radiation on the thermal structure in the polar summer MLT region is more structured in geometric altitudes (right panel of Fig. 18.9). Even negative values occur at ~90–95 km at high latitudes (i.e., colder during solar maximum). This is explained by an expansion of the atmosphere below these altitudes which leads to an increase of geopotential altitudes (inverse to the effect of increased greenhouse gases discussed later).

We have compared the solar cycle effects on temperatures in LIMA with results from general circulation and climate models such as HAMMONIA[7] and WACCM[8] which include fully coupled chemistry [*Schmidt et al.*, 2010; *Marsh et al.*, 2007; *Tsutsui et al.*, 2009]. In general we find good agreement. We note that LIMA ignores any feedback of solar irradiance on chemistry. However, fully coupled chemistry-climate models indicate that e.g. ozone changes due to solar variation are on the order of few percent only which has only little effect on the thermal structure. At

[7]HAMburg MOdel of the Neutral and Ionized Atmosphere.

[8]Whole-Atmosphere Community Climate Model.

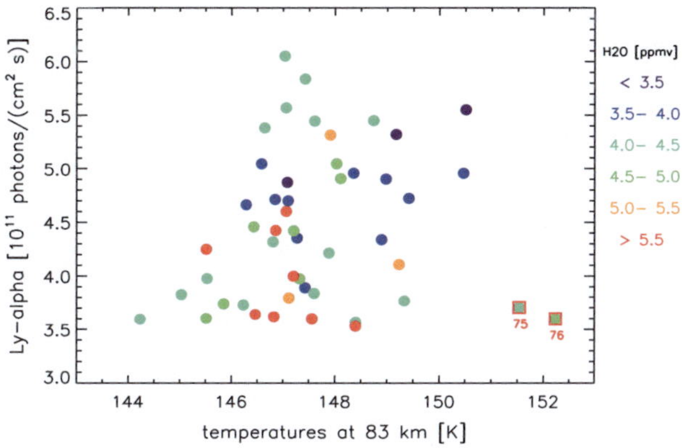

Fig. 18.10 Water vapor concentration at 83 km in ppmv (color coded, see *legend*) as a function of temperatures at 83 km and Ly_α for run no. 5. Years 1975 and 1976 are marked for reasons explained in the text

lower latitudes (44°N) and altitudes (<75 km) solar cycle temperature variations from LIMA agree with lidar observations [*Keckhut et al.*, 2005].

In summary, stronger solar activity increases MLT temperatures by a few degrees in the ice domain and cools the atmosphere just above the mesopause. The magnitude of this effect is in the same order as natural year-to-year variability. It is therefore not surprising that the correlation between temperatures at 83 km and Ly_α is poor ($r = 0.34$) because natural variability counteracts the influence of Ly_α.

Solar radiation destroys water vapor by photo-dissociation. Furthermore, H_2O is modified by freeze-drying which in turn is temperature dependent. These effects are taken into account in LIMA/ICE (note that the initial water vapor is fixed in LIMA/ICE). More precisely, enhanced photo-dissociation during solar maximum leads to less water vapor available for ice formation. Furthermore, higher temperatures reduce freeze-drying which leaves less water vapor from ice sublimation at ~83 km. In total, one would expect an anti-correlation between H_2O and Ly_α for altitudes around 83 km. Indeed, there is a tendency for high (low) water vapor concentrations to appear at low (high) Ly_α values (see Fig. 18.10). However, temperatures and Ly_α only partly control the amount of water vapor at 83 km. Other mechanisms such as natural variability, non-linear interaction in ice formation (causing freeze drying), and vertical/horizontal winds are of comparable importance. These effects destroy a clear anti-correlation of H_2O and Ly_α. For example, at low Ly_α values (3.5–4) a large range of H_2O values can appear (see Fig. 18.10).

Ly_α Effects on Ice Layers Since in LIMA/ICE NLC brightness and occurrence rates are closely correlated with water vapor, a variation of these ice cloud parameters with solar cycle and temperatures looks similar to Fig. 18.10. This means that we cannot expect a simple anti-correlation between ice cloud parameters and Ly_α. We have studied the effects of water vapor and temperatures (both modified by Ly_α) on ice cloud parameters in more detail. The effect of Ly_α on H_2O alone causes some correlation between occurrence rates and Ly_α ($r = -0.44$, run no. 4) which increases if the Ly_α effect on temperatures is added ($r = -0.59$, run no. 5). Results are similar for occurrence rates: the H_2O effect alone causes some correlation

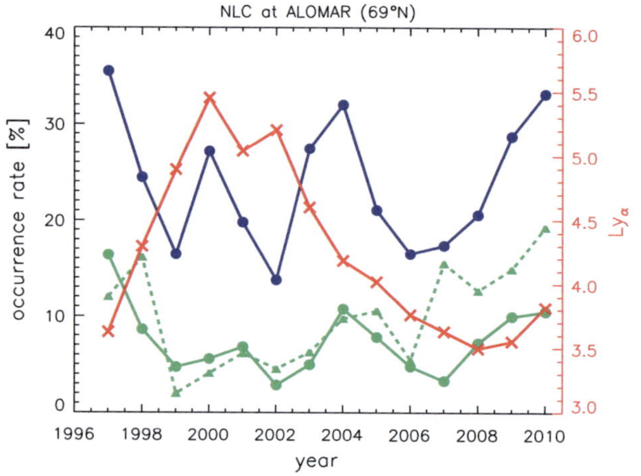

Fig. 18.11 NLC occurrence rates observed at ALOMAR (69°N) for brightness thresholds of $\beta_{max} > 13$ (*green solid line*) and $\beta_{max} > 4$ (*blue line*). The Ly_α activity is shown for comparison (*red line*, *right axis*). Results for LIMA/ICE run no. 5 are shown for strong NLC (*green dashed line*)

($r = -0.33$, run no. 4) which increases if solar effects on temperatures are included ($r = -0.51$, run no. 5). Including CO_2 increase does not significantly modify these values, as expected ($r = -0.55$ for brightness, and $r = -0.48$ for occurrence, respectively). Since NLC altitudes z_{NLC} barely depend on water vapor, the Ly_α effect on H_2O has basically no effect on z_{NLC}. Only if the temperature effect is included, we find some correlation ($r = +0.42$).

In summary, solar activity only partially influences NLC/PMC brightness and occurrence rates. This result is confirmed by observations. In Fig. 18.11 we show lidar measurements of NLC at ALOMAR (69°N). A total of 4224 hours of observations with 1023 hours of NLC contribute to this plot [see *Fiedler et al.*, 2011, for a recent update on NLC from ALOMAR]. Occurrence rates are shown for two different thresholds of NLC brightness: $\beta_{max} > 13$ ('strong') and $\beta_{max} > 4$ ('long term'). The latter represents the instrumental sensitivity when measurements started in 1997. *Fiedler et al.* [2011] studied in detail the potential influence of sampling on NLC brightness, occurrence rates, and altitudes for various β-thresholds. An important conclusion is that the main characteristics of ice clouds at ALOMAR are most likely not severely hampered by sampling issues. As can be seen from Fig. 18.11 there is a tendency for occurrence rates of strong NLC ($\beta_{max} > 13$) to be lower during solar maximum, i.e., in the years 2000 to 2003, as expected. However, occurrence rates were also very low in 2007 when solar activity was close to its record minimum. Occurrence rates with long term threshold ($\beta_{max} > 4$) also show no clear anti-correlation with Ly_α. For example, occurrence rates decreased(!) from 2004 to 2007/2008 during the declining phase of solar cycle 23. On the other hand, we observed almost maximum occurrence rates in 2010 when solar activity was still low, which is in line with expectations based on the simple scenario outlined above. The comparison between LIMA/ICE and ALOMAR in Fig. 18.11 generally shows nice agreement with some deviations, in particular in the late 2000s. We have not yet identified a specific reason for these differences and speculate that perhaps water vapor abundance was different in these years compared to LIMA/ICE (note that we assume the same May profile of H_2O in all years varying throughout the season due to Ly_α and transport).

NLC observations at mid latitudes in Kühlungsborn (54°N) show a similar non-coherent picture: in 2009 we observed record high NLC brightness and occurrence rates, whereas one year earlier (i.e., still during solar minimum) the activity was very low. Visible observations of NLC and satellite measurements of PMC also show little correlations with solar cycle. *Kirkwood et al.* [2008] analyzed 43 years of NLC data from ground based observers from the UK and from Denmark. They found correlation coefficients of typically $r = 0.45$–0.55 depending on which selection of NLC was analyzed (strong/faint, part of the season only etc.). We note that most of the year-to-year variation, namely 70–80 %, cannot(!) be explained by solar flux variability.

The longest record of PMC observations (28 years) comes from SBUV instruments on various satellites. The data set has been intensively analyzed for trends and solar cycle variations [see *DeLand et al.*, 2007; *Shettle et al.*, 2009, for some recent results]. A solar cycle modulation and an increase of PMC albedo and occurrence rates was identified. However, the correlation coefficients are on the order of $r = 0.5$ only, i.e., again the majority of the variation (75 %) cannot be explained by solar influence.

As will be shown in Sect. 18.5 the long term temperature trend at NLC altitudes led to minimum temperatures in the mid 1990s which accidentally coincides with a period of solar minimum. This combination of low temperatures and low Ly_α led to strong and frequent ice clouds detected in basically all long term records of NLC and PMC. This accidental coincidence over-emphasizes the solar signal in short time series of ice cloud observations. For example, an analysis of PMC data from HALOE from 1993 to 2005 shows comparatively large correlation between solar cycle and extinction with values up to $r \sim -0.8$ [see Fig. 4 in *Hervig and Siskind*, 2006]. But the first years of this data set is from the mid 1990s where the accidental coincidence mentioned above creates many and strong PMC. We conclude that too short time series may lead to misinterpretations of the importance of solar influence on mesospheric ice layers.

18.5 Trends

Water Vapor Trends Any temperature trend in LIMA induces trends in H_2O in LIMA/ICE due to freeze drying. More precisely the collection of water vapor by freezing ice particles is more effective for lower temperatures. Therefore, the release of water vapor around 83 km during sublimation is larger for lower temperatures. The inverse is true in the region of water vapor collection, namely above the sublimation regime. In total, a negative temperature trend causes a negative water vapor trend at ~84–92 km, and a positive trend at ~81–84 km. We note that the freeze drying efficiency depends on the degree of saturation (i.e. on temperature) which introduces a NH/SH asymmetry in water vapor trends and subsequently in ice layer

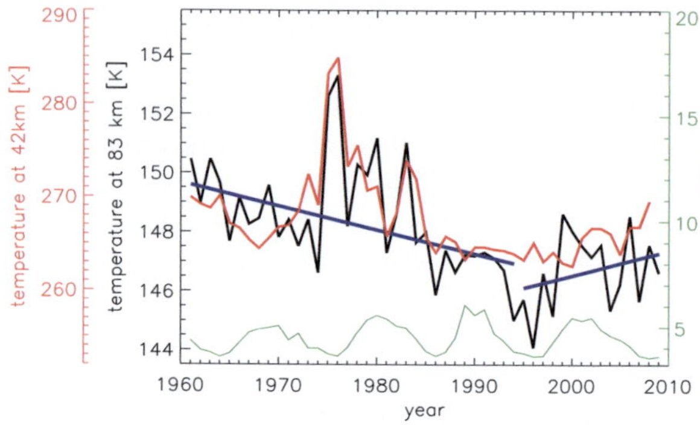

Fig. 18.12 Temperatures at 83 km (*black line*) and at 42 km (*red line*) at 69°N from run no. 6, i.e., including solar cycle effects and CO_2 increase. Note the different scales. The *blue lines* represent straight line fits. The Ly_α activity is shown for comparison (*green line, right axis*)

trends [see Fig. 10 in *Lübken and Berger*, 2011]. From observations little is known about water vapor trends in the polar summer MLT region [*von Zahn et al.*, 2004]. Some comparison of water vapor profiles from LIMA with measurements has been presented in the literature [*Lübken et al.*, 2009; *Hartogh et al.*, 2010]. In general, LIMA water vapor profiles agree with observations. In the future we intend to incorporate water vapor trends in LIMA/ICE from observations (if available), or from other models [e.g., *Grygalashvyly et al.*, 2009].

Temperature Trends and Influence of the Stratosphere In Fig. 18.12 we show seasonal (19 June–29 July) mean temperatures at 83 km for the ALOMAR station at 69°N from run no. 6, i.e., including solar cycle effects and CO_2 increase. Temperatures decrease in the period ~1961–1997 at a rate of approximately −0.80 K/decade, and increase thereafter. This implies that temperature trends in the mesosphere are not uniform in the last 50 years, but may even change sign. It is therefore important to specify the time period for which a trend is determined. This behavior is also seen in run 5 (constant CO_2 and O_3) and run 7 (trends in CO_2 and O_3) but with different magnitude. This suggests that a major source for the trend (and its reversal in the mid 1990s) is located in the stratosphere which is implicitly taken into account in LIMA in all runs. This speculation is further supported by comparing temperatures at 83 km with temperatures in the stratosphere (42 km, see Fig. 18.12). The similarity of both curves indicates that trends in the stratosphere impact mesospheric temperatures and ice layers. We note that ECMWF temperatures in 1975 and 1976 are considerably higher compared to the general trend which is due to a bias in satellite temperatures [*Gleisner et al.*, 2005]. These spurious temperatures appear both in the stratosphere and mesosphere, which highlights the coupling between stratosphere and mesosphere. We have shown in detail that shrinking of the stratosphere is an important mechanism causing mesospheric cooling and ice layer enhancement [*Lübken et al.*, 2009; *Lübken and Berger*, 2011]. Our temperature trends in the lower stratosphere at high latitudes agree with other studies based on radio sonde measurements (see Fig. 1 in *Lübken and Berger*, 2011, which is based on *Randel et al.*, 2009). This agreement is perhaps not surprising since LIMA adapts ECMWF data at these altitudes. Also, model results of mesospheric temperature trends at mid-latitudes for summer have been recently compared with

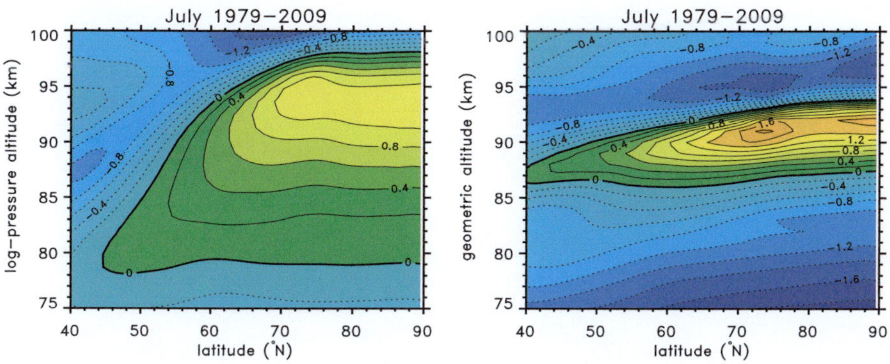

Fig. 18.13 Temperature trends (in K/decade) in the summer mesopause region for July considering increase of CO_2 and O_3 (run no. 7). Trends are determined in the period 1979–2009. *Left*: as a function of pressure, *right*: as a function of geometric altitude

lidar observations of temperatures at 44°N, phase height measurements at 51°N, and temperature trends derived from SSU[9] satellite data (channel 47X). In general there is excellent agreement between trends from LIMA and observations [*Berger and Lübken*, 2011]. We realize that little is known about the cause of long term trends in the stratosphere at polar latitudes in summer. Furthermore, for the mesosphere no direct multi-decadal measurements of temperatures by satellites are available. Presumably, greenhouse gas increase and ozone trends play an important role in this context. The latter is particularly intriguing since the curves in Fig. 18.12 show a change of trends in the mid 1990s, perhaps related to stratospheric ozone recovery where a change of trends also occurs in the mid 1990s [*Jones et al.*, 2009]. Our model results suggest that the differences of mesospheric temperature trends at an altitude of 83 km in the periods 1961–1997 and 1997–2009 originate from long term ozone changes in the upper stratosphere. A potential influence of stratospheric ozone on mesospheric ice layers was recently suggested also for the southern hemisphere [*Smith et al.*, 2010].

Temperature trends were determined from all runs listed in Table 18.1. Here we discuss results from run 7 only which includes CO_2 increase and stratospheric O_3 change. In Fig. 18.13 we show temperature trends in K/decade for the time period 1979–2009 as a function of pressure and altitude. This period was chosen because ozone trends measured by SBUV satellites are available since 1979 [*WMO*, 2011]. Furthermore, PMC measured by SBUV are also available in this period. As can be seen from this figure, temperature trends are positive in the summer mesopause region which is mainly caused by infrared radiative effects of CO_2 [see Fig. 3 in *Akmaev et al.*, 2006]. If plotted as a function of geometric altitude, temperature trends at polar latitudes are positive in a narrow region around the mesopause only (more precisely from ~86 to 92 km) with maximum values of approximately

[9]Stratospheric Sounding Units.

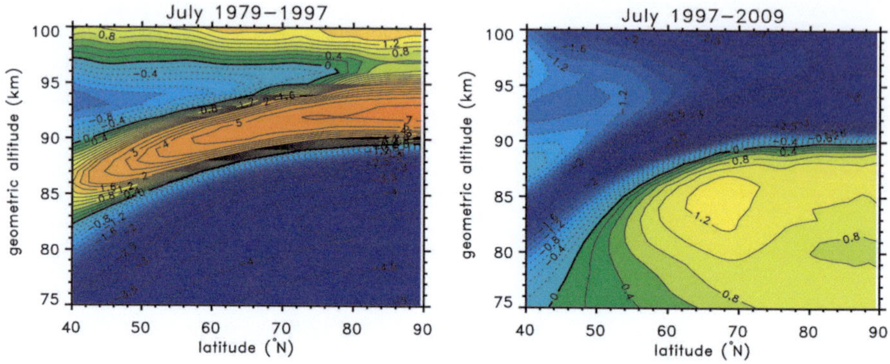

Fig. 18.14 Similar to Fig. 18.13 but for sub-periods, namely 1979–1997 (*left*) and 1997–2009 (*right panel*)

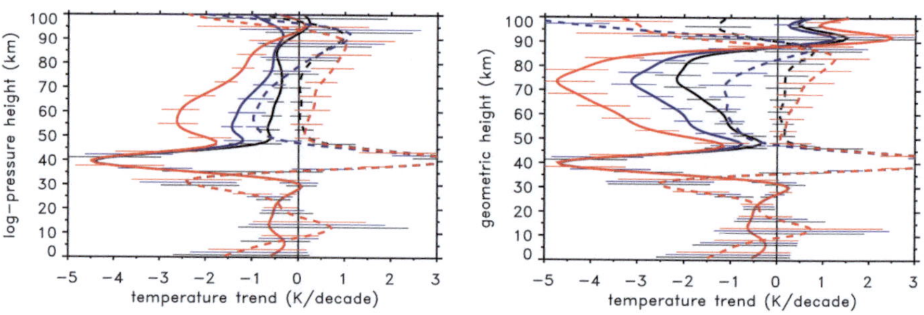

Fig. 18.15 Temperature trends from LIMA at 69°N for 3 scenarios regarding trends in mesospheric GHG: $CO_2 + O_3$ fixed (run no. 5, *black lines*), CO_2 increase (run no. 6, *blue lines*), and CO_2 plus O_3 change (run no. 7, *red lines*). Similarly to Fig. 18.14 trends are shown for two different time periods, namely 1979–1997 (*solid lines*) and 1997–2009 (*dashed lines*). Temperature trends are shown as a function of pressure (*left*) and geometric altitude (*right*). Note that trends coincide below 40 km because of adaption to ECMWF

+2–2.5 K/decade. This is in the same order of magnitude as other model studies [*Garcia et al.*, 2007]. It is important to notice that both negative and positive temperature trends occur in the ice particle regime, depending on altitude.

We noted earlier that temperature trends are not uniform (see Fig. 18.12). In order to highlight this effect further we show in Fig. 18.14 temperature trends from run 7 for sub-periods, namely 1979–1997 and 1997–2009. This facilitates comparison with lidar measurements and SBUV observations of PMC. Trends are largest around the mesopause in the 1979–1997 period with maximum values of up to +4–5 K/dec. Thereafter, trends are rather small and distributed more uniformly. It is tempting to associate this change of trends with the stratospheric ozone recovery.

In Fig. 18.15 we show temperature trends from LIMA at 69°N for 3 scenarios regarding trends of mesospheric greenhouse gases (see figure legend). Temperature trends generally increase if GHG increase is included, but the differences are

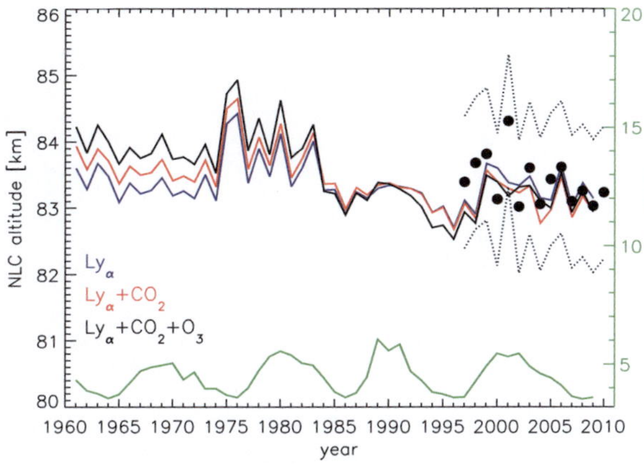

Fig. 18.16 NLC altitudes from LIMA at 69°N for runs 5 (*blue*), 6 (*red*), and 7 (*black*). For comparison, mean NLC altitudes from ALOMAR are shown for $\beta_{max} > 1$ (*dots*). A typical variability throughout the season of ±1 km is indicated by *dotted lines*. The Ly_α activity is shown for comparison (*green line*, *right axis*)

comparatively small in the ice layer domain. We note that for the 1979–1997 period trends are fairly large (up to −(3–5) K/dec) in the mesosphere, and are much less (and even positive) for the later period. We compared temperature trends from LIMA with lidar measurements at a mid latitude lidar station, namely Observatoire Haute Provence (44°N) [see Fig. 10 in *Keckhut et al.*, 2011]. For the time period analyzed in that paper, namely from 1979–1997, we find similarly large trends and good agreement in the altitude dependence of the trends.

Ice Layer Trends The altitudes of NLC from LIMA/ICE show only very little variation over the last 50 years and agree nicely with ALOMAR (Fig. 18.16). This is basically true for all long term runs, i.e., whether or not solar cycle and CO_2/O_3 trends are included. We have shown in Fig. 18.8 that there is a very strong correlation between NLC altitudes and temperatures at 83 km. For run no. 6 we arrive at a slope of 0.212 km/K corresponding to a temperature change of 4.7 K/km. The observed altitude change from the late 1890s to today is only ∼350–670 m [see *Lübken et al.*, 2009, and references therein]. Using the altitude/temperature correlation from above, this corresponds to a temperature change of only 1.6–3.2 K in nearly 120 years. This is an extremely small temperature trend and imposes a stringent constraint on all modeling of trends in the middle atmosphere. We note that the slopes given in *Lübken et al.* [2009] were slightly different (0.263 km/K) since LIMA-I was used for that study. However, this has no influence on the main conclusion outlined above. The cooling trend as inferred from NLC altitudes is somewhat smaller compared to the trends shown in Fig. 18.13 which suggests that trends were smaller in the earlier part of the 20th century.

We have determined albedo values from LIMA/ICE applying Mie scattering calculations with conditions very similar to SBUV, i.e. using the same wavelength ($\lambda = 252$ nm) and applying scattering angles which vary for each instrument. Albedos were calculated in three different latitudes bands as given in *DeLand et al.* [2007], namely 56–64°N, 64–74°N, and 74–82°N. Seasonal mean values were determined considering only those PMC which have albedos larger than thresholds of 5.5, 6.5, and 7.5 × 10^{-6}/sr for the three latitude bands given above. This is again

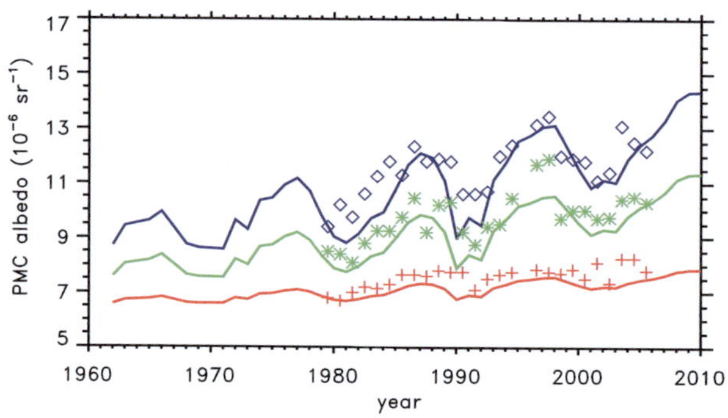

Fig. 18.17 Comparison of LIMA/ICE albedos from run no. 7 with satellite observations from SBUV. Albedos of polar mesospheric clouds (PMC) from the LIMA model (*solid lines*) for three latitude bands: 56–64°N (*red*), 64–74°N (*green*), and 74–82°N (*blue*), similar to the bins chosen for SBUV observations (symbols) presented by *DeLand et al.*, 2007. The LIMA curves represent multiple regression fits consisting of straight lines plus a solar cycle function. These fits were determined using LIMA values calculated with similar parameters as in SBUV observations, namely a wavelength of 252 nm and varying scattering angles, depending on instrument and time of observation. Zonal and seasonal mean values are determined applying albedo thresholds of 5.5, 6.5, and 7.5×10^{-6}, again similar to constraints imposed by the SBUV instruments

similar to the data analysis applied for SBUV where the sensitivity of detecting PMC decreases with increasing latitude. Note that we have determined albedos from $\beta_{\max}$ assuming that the maximum backscatter at visible wavelengths contributes most to the backscatter at UV wavelengths. As can be seen from Fig. 18.17 SBUV albedos and their variation with latitude are nicely reproduced by LIMA.

18.6 Discussion and Conclusion

We present results from the MLT region based on the LIMA model which adapts to real conditions (ECMWF) in the troposphere/lower stratosphere. A Lagrangian ice particle model (LIMA/ICE) is included which allows to study the morphology of ice clouds in the summer mesopause region. In general there is good agreement between LIMA/ICE and the characteristics of ice clouds observed from the ground (e.g., by lidars) and from satellites. This concerns, for example, cloud altitude, brightness, occurrence frequencies, variation with latitude and season, interhemispheric difference, cumulative frequency distribution etc. The formation, sedimentation, and sublimation of ice particles causes a horizontal and vertical redistribution of H_2O ('freeze drying') which significantly impacts the distribution of H_2O and thereby affects photochemistry. We note that this effect should be taken into account in GCM models. LIMA/ICE results are now available for all years since 1961 and for various scenarios: with and without Ly_α effect on temperatures, with and without trends of CO_2 and O_3, etc. (see Table 18.1). We have not yet considered trends in water vapor from the lower atmosphere since too little is known about such trends in the summer MLT region. We note that H_2O trends are observed in LIMA/ICE caused by temperature-dependent freeze drying.

We have studied interhemispheric differences in detail in *Lübken and Berger* [2007, 2011]. Generally speaking, ice layers in the southern hemisphere are weaker and less frequent compared to the northern hemisphere. Apart from the southern MLT region being slightly warmer, freeze drying is less effective, which contributes significantly to the interhemispheric differences of ice layer morphology.

Solar radiation leaves a major impact on H_2O (through photo-dissociation) and on temperatures. In general it is warmer during solar maximum, but there is a small height region around the mesopause where it is colder during solar maximum. This complicates the forecast of solar cycle effects on ice layers, since the first stage of ice particle growth (around the mesopause) is supported during solar maximum (because it's colder), whereas the later stage of growth is hampered. Therefore, a complex model like LIMA/ICE is needed to take these counteracting effects into account. The magnitude of the solar cycle effect is ~1–3 K which is on the same order of magnitude as natural year-to-year variability. Consequently, there is only a moderate solar cycle signal in temperatures and in ice layers. This is in agreement with long term observations of ice clouds where only a small fraction of variability can be accounted for by solar cycle influence.

We have studied long term effects in MLT temperatures and ice layers. Temperature trends at NLC altitudes are partly caused by stratospheric trends ('shrinking effect') without any direct effects of GHG effects in the mesosphere. Trends are generally negative, as expected, but are very small or even positive in the mesopause region. Considering the ice particle regime from roughly 80 to 95 km, trends are negative in the lower part and positive in the upper part. This complicates a prediction of the effect of temperature trends on ice layers. Again, a complex model like LIMA/ICE is needed. Adding CO_2 and O_3 trends enhances temperature trends in the mesosphere but has comparatively small effects in the ice regime. We apply the strong correlation between ice layer altitudes and temperatures at 83 km to observations made nearly 120 years ago. This gives negligible temperature trends at this altitude over this time period. For the time period of satellite measurements (1979–2009) LIMA/ICE predicts trends in ice cloud brightness and occurrence rates consistent with satellite observations.

It turns out that temperature trends are not uniform in time but are stronger in the period from 1961 to the mid 1990s, and weaker (even positive) thereafter. In a recent study *Berger and Lübken* [2011] showed that in the summer period 1979–1997 at mid-latitudes strong cooling of up to ~3–4 K/decade occurs in the middle mesosphere, in the period 1961–1979 the middle atmosphere cools significantly less, and for the period 1997–2009 they find a warming of ~1 K/decade. For the first time, modeled temperature trends confirm the extraordinarily large temperature trends observed at mid-latitudes during the period 1979–1997 derived from lidar measurements, satellite data, and phase height measurements. The differences in temperature trends in the mesosphere originate from the evolution of stratospheric ozone in the past 50 years, e.g. the observed reversal of both stratospheric and mesospheric temperature trends in the mid 1990s is caused by the recovery of stratospheric ozone [see Figs. 2.2 and 4.8 in *WMO*, 2011]. We note that lowest temperatures accidentally coincided with solar minimum in the mid 1990s which gave strong appearance of NLC/PMC. This accidental coincidence over-emphasizes the importance of solar activity on ice layers and implies that too short time series may lead to misinterpretations of the importance of solar influence on mesospheric ice layers.

Our results from the SOLEIL project show that it is important to consider the time period and the height range when studying temperature and ice clouds trends in the MLT. For example, temperature trends can be negative or positive in the ice particle regime, depending on altitude. Temperature trends can be very large (in agreement with some observations) or small, depending on the time period considered. We intend to continue our studies on trends in the MLT region including potential water vapor trends and more observations. We will also investigate potential implications for the entire atmosphere taking into account various coupling mechanisms.

Acknowledgements We appreciate the continuing financial support from the DFG for the SOLEIL project. The European Centre for Medium-Range Weather Forecasts (ECMWF) is gratefully acknowledged for providing ERA-40 and operational analysis data. Several students have spent a considerable time at the ALOMAR observatory for making measurements. FJL thanks Mrs Rosenthal for her technical support in chairing the CAWSES priority program.

References

Akmaev, R. A., Fomichev, V. I., & Zhu, X. (2006). Impact of middle-atmospheric composition changes on greenhouse cooling in the upper atmosphere. *Journal of Atmospheric and Solar-Terrestrial Physics*, *68*, 1879–1889.

Baumgarten, G., & Fiedler, J. (2008). Vertical structure of particle properties and water content in noctilucent clouds. *Geophysical Research Letters*, *35*, L10811. doi:10.1029/2007GL033084.

Berger, U. (2008). Modeling of middle atmosphere dynamics with LIMA. *Journal of Atmospheric and Solar-Terrestrial Physics*, 1170–1200. doi:10.1016/j.jastp.2008.02.004.

Berger, U., & Lübken, F.-J. (2011). Mesospheric temperature trends at mid-latitudes in summer. *Geophysical Research Letters*, *38*, L22804. doi:10.1029/2011GL049528.

DeLand, M. T., Shettle, E. P., Thomas, G. E., & Olivero, J. J. (2003). Solar backscattered ultraviolet (SBUV) observations of polar mesospheric clouds (PMCs) over two solar cycles. *Journal of Geophysical Research*, *108*(D8), 8445. doi:10.1029/2002JD002398.

DeLand, M. T., Shettle, E. P., Thomas, G. E., & Olivero, J. J. (2007). Latitude-dependent long-term variations in polar mesospheric clouds from SBUV version 3 PMC data. *Journal of Geophysical Research*, *112*(D10), D10315. doi:10.1029/2006JD007857.

Fiedler, J., Baumgarten, G., Berger, U., Hoffmann, P., Kaifler, N., & Lübken, F.-J. (2011). NLC and the background atmosphere above ALOMAR. *ACP*, 5701–5717. doi:10.5194/acp-11-5701-2011.

Fortuin, J. P., & Kelder, H. (1998). An ozone climatology based on ozonesonde and satellite measurements. *Journal of Geophysical Research*, *103*, 31709–31734.

Garcia, R. R., Marsh, D. R., Kinnson, D. E., Boville, B. A., & Sassi, F. (2007). Simulation of secular trends in the middle atmosphere, 1950–2003. *Journal of Geophysical Research*, *112*, D09301. doi:10.1029/2006JD007485.

Gleisner, H., Thejll, P., Stendel, M., Kaas, E., & Machenhauer, B. (2005). Solar signals in tropospheric re-analysis data: comparing NCEP/NCAR and ERA40. *Journal of Atmospheric and Solar-Terrestrial Physics*, *67*, 785–791. doi:10.1016/j.jastp.2005.02.001.

Grygalashvyly, M., Sonnemann, G., & Hartogh, P. (2009). Long-term behavior of the concentration of the minor constituents in the mesosphere—a model study. *Atmospheric Chemistry and Physics*, *9*, 2779–2992.

Hartogh, P., Sonnemann, G. R., Li, S., Grygalashvyly, M., Berger, U., & Lübken, F.-J. (2010). Water vapor measurements at ALOMAR over a solar cycle compared with model calculations by LIMA. *Journal of Geophysical Research*, *115*, D00I17. doi:10.1029/2009JD012364.

Hervig, M., & Siskind, D. (2006). Decadal and inter-hemispheric variability in polar mesospheric clouds, water vapor, and temperature. *Journal of Atmospheric and Solar-Terrestrial Physics*, *68*, 30–41.

Jesse, O. (1896). Die Höhe der leuchtenden Nachtwolken. *Astronomische Nachrichten*, *140*, 161–168.

Jones, O., et al. (2009). Evolution of stratospheric ozone and water vapour time series studied with satellite measurements. *Atmospheric Chemistry and Physics*, *9*, 6055–6075.

Keckhut, P., Cagnazzo, C., Chanin, M.-L., Claud, C., & Hauchecorne, A. (2005). The 11-year solar-cycle effects on the temperature in the upper-stratosphere and mesosphere: Part I—Assessment of observations. *Journal of Atmospheric and Solar-Terrestrial Physics*, *67*, 940–947.

Keckhut, P., et al. (2011). An evaluation of uncertainties in monitoring middle atmosphere temperatures with the ground-based Lidar network in support of space observations. *Journal of Atmospheric and Solar-Terrestrial Physics*. doi:10.1016/j.jastp.2011.01.003.

Kirkwood, S., Dalin, P., & Réchou, A. (2008). Noctilucent clouds observed from the UK and Denmark—trends and variations over 43 years. *Annals of Geophysics*, *26*, 1243–1254.

Lübken, F.-J., & Berger, U. (2007). Interhemispheric comparison of mesospheric ice layers from the LIMA model. *Journal of Atmospheric and Solar-Terrestrial Physics*, *69*(17–18), 2292–2308. doi:10.1016/j.jastp.2007.07.006.

Lübken, F.-J., & Berger, U. (2011). Latitudinal and interhemispheric variation of stratospheric effects on mesospheric ice layer trends. *Journal of Geophysical Research*, *116*, D00P03. doi:10.1029/2010JD015258.

Lübken, F.-J., & Höffner, J. (2004). Experimental evidence for ice particle interaction with metal atoms at the high latitude summer mesopause region. *Geophysical Research Letters*, *31*(8), L08103. doi:10.1029/2004GL019586.

Lübken, F.-J., Baumgarten, G., Fiedler, J., Gerding, M., Höffner, J., & Berger, U. (2008). Seasonal and latitudinal variation of noctilucent cloud altitudes. *Geophysical Research Letters*, *35*, L06801. doi:10.1029/2007GL032281.

Lübken, F.-J., Berger, U., & Baumgarten, G. (2009). Stratospheric and solar cycle effects on long-term variability of mesospheric ice clouds. *Journal of Geophysical Research*, *114*, D00106. doi:10.1029/2009JD012377.

Marsh, D. R., Garcia, R. R., Kinnison, D. E., Boville, B. A., Sassi, F., Solomon, S. C., & Matthes, K. (2007). Modeling the whole atmosphere response to solar cycle changes in radiative and geomagnetic forcing. *Journal of Geophysical Research*, *112*, D23306. doi:10.1029/2006JD008306.

Plane, J., Murray, B., Chu, X., & Gardner, C. (2004). Removal of meteoric iron on polar mesosphere clouds. *Science*, *304*, 426–428.

Randel, W. J., et al. (2009). An update of observed stratospheric temperature trends. *Journal of Geophysical Research*, *114*, D02107. doi:10.1029/2008JD010421.

Rapp, M., & Thomas, G. E. (2006). Modeling the microphysics of mesospheric ice particles: assessment of current capabilities and basic sensitivities. *Journal of Atmospheric and Solar-Terrestrial Physics*, *68*, 715–744.

Russell, J. M., et al. (2009). The aeronomy of ice in the mesosphere (AIM) mission: overview and early science results. *Journal of Atmospheric and Solar-Terrestrial Physics*, *71*. doi:10.1016/j.jastp.2008.08.011.

Schmidt, H., Brasseur, G. P., & Giorgetta, M. A. (2010). Solar cycle signal in a general circulation and chemistry model with internally generated quasi-biennal oscillation. *Journal of Geophysical Research*, *115*, D00114. doi:10.1029/2009JD012542.

Shettle, E. P., DeLand, M. T., Thomas, G. E., & Olivero, J. J. (2009). Long term variations in the frequency of polar mesospheric clouds in the Northern Hemispheric from SBUV. *Geophysical Research Letters*, *36*, L02803. doi:10.1029/2008GL036048.

Smith, A. K., Garcia, R. R., Marsh, D. R., Kinnison, D. E., & Richter, J. H. (2010). Simulations of the response of mesospheric circulation and temperature to the Antarctic ozone hole. *Geophysical Research Letters*, *37*, L22803. doi:10.1029/10GL045255.

Summers, M. E., Conway, R. R., Englert, C., Siskind, D. E., Stevens III, M. J. R., Gordley, L., & McHugh, M. (2001). Discovery of a layer of enhanced water vapor in the arctic summer mesosphere: implications for polar mesospheric clouds. *Geophysical Research Letters*, *28*, 3601–3604.

Thomas, G., Olivero, J., DeLand, M., & Shettle, E. (2003). Comment on 'Are noctilucent clouds truly a miner's canary for global change?'. *EOS*, *84*(36), 352–353.

Thomas, G. E. (1995). Climatology of polar mesospheric clouds: interannual variability and implications for long-term trends. *Geophysical Monograph*, *87*, 185–200.

Tsutsui, J., Nishizawa, K., & Sassi, F. (2009). Response of the middle atmosphere to the 11-year solar cycle simulated with the whole atmosphere community climate model. *Journal of Geophysical Research*, *114*, D02111. doi:10.1029/2008JD010316.

von Cossart, G., Fiedler, J., & von Zahn, U. (1999). Size distributions of NLC particles as determined from 3-color observations of NLC by ground-based lidar. *Geophysical Research Letters*, *26*, 1513–1516.

von Savigny, C., Petelina, S. V., Karlsson, B., Llewellyn, E. J., Degenstein, D. A., Lloyd, N. D., & Burrows, J. P. (2005). Vertical variation of NLC particle sizes retrieved from Odin/OSIRIS limb scattering observations. *Geophysical Research Letters*, *32*, L07806. doi:10.1029/2004GL021982.

von Zahn, U. (2003). Are noctilucent clouds truly a 'miner's canary' for global change? *EOS*, *84*(28), 261–264.

von Zahn, U., Baumgarten, G., Berger, U., Fiedler, J., & Hartogh, P. (2004). Noctilucent clouds and the mesospheric water vapour: the past decade. *Atmospheric Chemistry and Physics*, *4*, 2449–2464.

WMO (2011). *Global ozone research and monitoring project* (Report no. 52). Scientific assessment of ozone depletion: 2010, World Meteorological Organization.

Chapter 19
Charged Aerosol Effects on the Scattering of Radar Waves from the D-Region

Markus Rapp, Irina Strelnikova, Qiang Li, Norbert Engler, and Georg Teiser

Abstract Charged aerosol particles are an important contributor to the D-region charge balance and affect the scattering of radar waves. Among these particles are meteoric smoke particles (MSP) which occur at all D-region altitudes and all seasons, and mesospheric ice particles whose occurrence is confined to altitudes of ~80–90 km at polar latitudes during summer. We argue that it is the modification of electron diffusion by the heavy charged aerosol particles which is the prime effect leading to clearly detectable signatures in both incoherent and coherent radar backscatter. In the case of incoherent scatter, it is shown that the presence of charged aerosol particles modifies the incoherent scatter spectrum. Corresponding observations with the EISCAT UHF radar and the Arecibo radar have been used to detect both MSP and ice particles at D-region altitudes and characterize their radii and number densities. In the case of coherent scatter, it is argued that the modified diffusion properties of the D-region electrons lead to small scale structures at the radar Bragg wavelength due to turbulent mixing in combination with a large Schmidt number. To test this theory, calibrated echo strengths of polar mesosphere summer echoes have been measured with the EISCAT radars at Tromsø (69°N) and Svalbard (78°N) and collocated 53 MHz radars, thus covering frequencies of 53 MHz, 224 MHz, 500 MHz, and 933 MHz. Importantly, the vast majority of these observations show excellent agreement with the corresponding theoretical predictions thus providing strong support for this theory. This theory was subsequently applied to the same data sets in order to derive ice particle radii. Corresponding results are in excellent agreement with independent data sets from satellite-borne and ground-based optical observations. Finally, some suggestions for future investigations are given.

M. Rapp (✉) · I. Strelnikova · Q. Li · N. Engler · G. Teiser
Leibniz-Institute of Atmospheric Physics at the Rostock University, Schlossstr. 6, 18225 Kühlungsborn, Germany
e-mail: markus.rapp@dlr.de

Present address:
M. Rapp
Deutsches Zentrum für Luft- und Raumfahrt (DLR), Institute of Atmospheric Physics, Oberpfaffenhofen, 82234 Wessling, Germany

F.-J. Lübken (ed.), *Climate and Weather of the Sun-Earth System (CAWSES)*, Springer Atmospheric Sciences,
DOI 10.1007/978-94-007-4348-9_19,

19.1 Introduction

In recent years, the mesospheric (~50–100 km) science community has undertaken significant efforts to study two types of mesospheric aerosol particles and related phenomena. Among these are mesospheric ice particles which form under the extreme thermal conditions of the polar summer mesopause, i.e., at heights between 80–90 km and at latitudes poleward of ~55° where mean mesopause temperatures of about 130 K are regularly reached during the summer months [*Lübken*, 1999]. These ice particles reach typical radii between ~2–100 nm and can even be visually observed from the ground as so-called noctilucent clouds (NLC) when Mie-scattering by particles with radii $\gtrsim$30 nm directs photons from the twilight sun into the field of view of an observer on the post-sun set side of the Earth. Besides these visual observations, various passive and active optical techniques have been applied over the past decades and allowed us to collect a wealth of information on the properties of these clouds [e.g., *DeLand et al.*, 2006; *Baumgarten et al.*, 2008]. In addition, these ice particles also modify the ambient plasma of the D-region (i.e., the lowest part of Earth's ionosphere at ~70–90 km altitude) by acquiring a net charge such that they further lead to strong radar echoes from the HF to the UHF wavelength range [e.g., *Cho and Röttger*, 1997; *Rapp and Lübken*, 2004]. These radar echoes are today known as polar mesosphere summer echoes or PMSE. Since both NLC and PMSE are easily observed from the ground they allow us to perform precise measurements of physical processes around the mesopause which are hardly accessible otherwise. Among these processes are the effects of waves from the planetary scale to a few kilometers [e.g., *Merkel et al.*, 2003; *Hoppe and Fritts*, 1995] as well as chemical effects initiated for example by the precipitation of high energetic particles from the Sun [*von Savigny et al.*, 2007]. Most importantly, however, it was already realized in the late 1980s that the highly non-linear nature of ice microphysics should amplify minute changes of temperature and water vapor at these altitudes. Hence, it was proposed that observations of NLC brightness and occurrence frequency should be ideally suited for the detection of corresponding long term changes [*Thomas et al.*, 1989]. Indeed, global satellite observations going back to 1979 appear to indicate trends in both of these parameters [*DeLand et al.*, 2007; *Shettle et al.*, 2009], even though some issues regarding this data set (like local time dependence or the homogeneity of the data set) are still under debate [*Stevens et al.*, 2010; *von Zahn*, 2003; *Thomas et al.*, 2003].

The second type of mesospheric aerosol particles is known as meteoric smoke particles (MSP). MSP are thought to be secondary meteor ablation products [*Hunten et al.*, 1980]. I.e., it is assumed that a significant part of the ~10–100 tons/day, which are deposited in the altitude range between ~70–110 km by meteor ablation, re-condense into tiny sub-nanometer scale particles. These particles may then grow to sizes of about 1–10 nm by Brownian coagulation and are being distributed throughout the entire atmosphere by means of sedimentation and dynamical transport processes [*Hunten et al.*, 1980; *Gabrielli et al.*, 2004; *Megner et al.*, 2008]. MSP are closely related to mesospheric ice clouds since it has long been assumed that they serve as corresponding ice nuclei [see Fig. 19.1 for illustration and *Rapp*

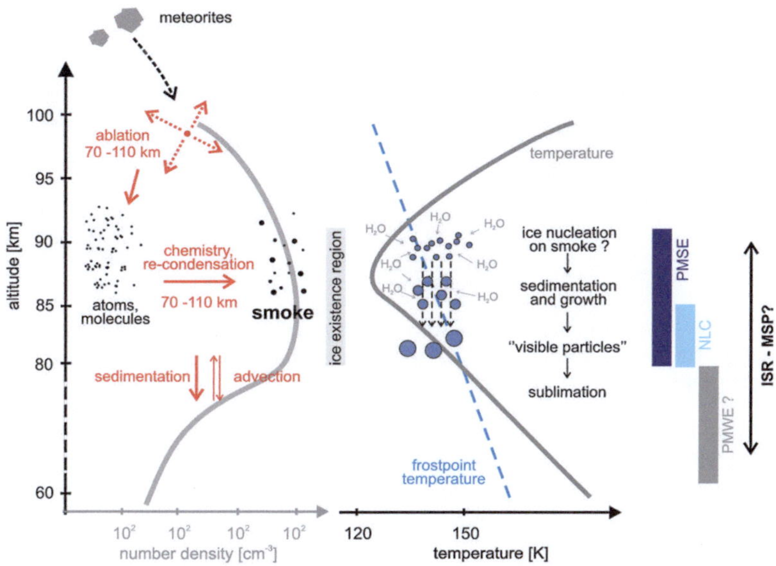

Fig. 19.1 Schematic of the most important processes leading to the formation of meteoric smoke particles (MSP, *left part*) and layers of mesospheric ice particles (*right part*). To the right, altitude ranges are indicated over which different optical and radar phenomena are observed. See text for more details

and Thomas, 2006, for a detailed discussion]. Besides their prominent role in mesospheric ice microphysics, it is further thought that MSP modify the charge balance of the D-region [e.g., *Rapp*, 2009] and subsequently cause so far unexplained radar echoes from the mid-mesosphere, i.e., so called polar mesosphere winter echoes or PMWE [*La Hoz and Havnes*, 2008]. In addition, it has been proposed that MSP are the form in which meteoric matter ultimately reaches the ground and is today found in geophysical archives like ice cores [*Gabrielli et al.*, 2004]. Finally, it has recently been proposed that MSP might also be involved in the formation of a special class of polar stratospheric clouds (so-called NAT PSC = polar stratospheric clouds consisting of nitric acid trihydrate particles) which play an important role in the formation of the Antarctic ozone hole [*Voigt et al.*, 2005].

In spite of this large scientific interest, numerous fundamental properties of these particles and/or related processes are only poorly known. In consequence, high quality observations of both mesospheric ice particles as well as MSP are needed. This book chapter reviews the major results of a 6-year project funded by the German Science foundation in the frame of the CAWSES priority program. The major aim of this project was to unravel the effect of charged mesospheric aerosol particles on the scattering of radar waves and in turn utilize such radar observations to characterize the properties of mesospheric aerosol particles. The structure of this report is as follows: In Sect. 19.2 we review our view of the current theoretical understanding of the effect of charged aerosol particles on the scattering of radar waves where we distinguish between incoherent scatter (Sect. 19.2.1) and coherent scatter (Sect. 19.2.2). Section 19.3 describes the experimental facilities used and the

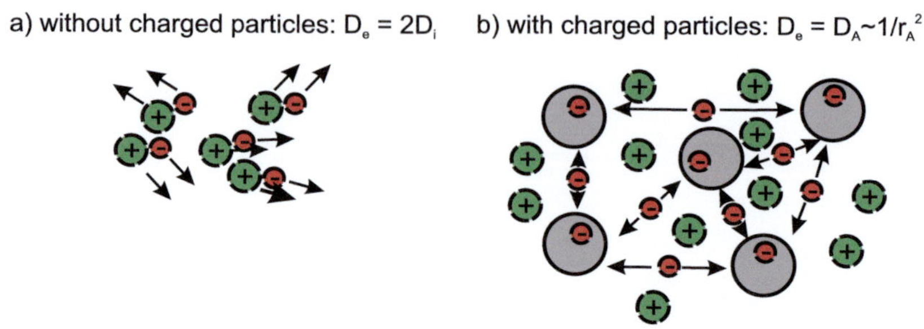

Fig. 19.2 Schematic illustrating the diffusion properties of different plasmas: *left* for the ambipolar case and *right* for the case with an additional charged constituent being much heavier than the other ionic component

data sets considered which is followed by two major parts describing our results concerning aerosol signatures in incoherent scatter data (Sect. 19.4) and coherent scatter data (Sect. 19.5). The chapter concludes with a summary of our main results and an outlook for future work (Sect. 19.6).

19.2 The Effect of Charged Aerosol Particles on the Scattering of Radar Waves

In general, radar waves are scattered from variations in the corresponding refractive index which is mainly determined by the electron density for the frequencies and the mesospheric altitude range considered in this chapter. In this subsection, we start with a short review of what we consider the prime effect of charged mesospheric aerosol particles on the free electron gas, i.e., the occurrence of a diffusion mode of the electrons which is controlled by the properties of the heavy charged aerosol particles [*Hill*, 1978; *Cho et al.*, 1992; *Rapp and Lübken*, 2003].

The general mechanism behind this process is depicted in the schematic shown in Fig. 19.2. This figure distinguishes two cases. The first (left side) illustrates the diffusion of a plasma consisting of positive ions and electrons only. In this case the plasma diffusion is characterized by the so-called ambipolar diffusion coefficient which is given by twice the diffusion coefficient of the positive ions. However, in the more complicated case (right side) with a plasma consisting of electrons, positive ions, and negatively charged aerosol particles (i.e., in our case ice particles or MSP), the Coulomb interaction between all charged species gives rise to a multipolar polarization electric field and hence leads to the occurrence of two distinct diffusion modes [see *Rapp and Lübken*, 2003, for details]. These two modes can be interpreted as being due to the Coulomb interaction of the electrons with the positive ions (which is hence close to ambipolar diffusion) and the Coulomb interaction between the electrons and the charged heavy particles. It is for the latter mode that the corresponding electron diffusion coefficient asymptotically approaches the diffusion coefficient of the heavy aerosol particles. The latter, in turn, can be described in the frame of a hard sphere collision model and is hence inversely proportional to the square of the particle radius [*Cho et al.*, 1992; *Rapp and Lübken*, 2003].

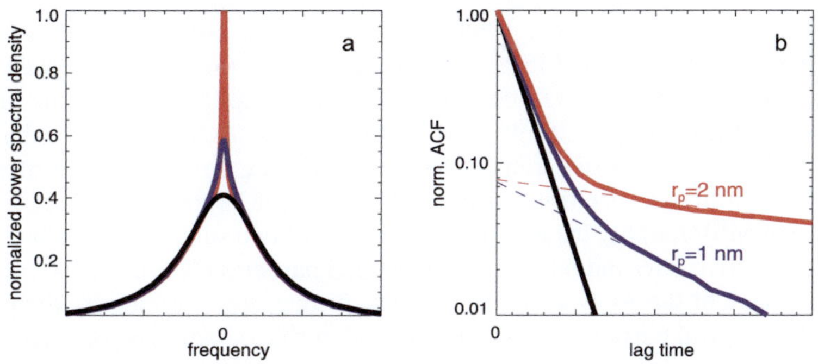

Fig. 19.3 ISR spectra (Panel **a**) and corresponding autocorrelation functions (ACF, Panel **b**) without (*black lines*) and with the presence of charged MSPs (*blue lines* for particles with $r_p = 1$ nm, red for $r_p = 2$ nm) based on the theory by *Cho et al.* [1998]. This figure is reproduced after *Strelnikova et al.* [2007]

In the following two subsections, we will briefly summarize which effect the occurrence of these two diffusion modes has on the properties of incoherent and coherent scatter, respectively.

19.2.1 Incoherent Scatter

We first consider the effect of the above described two diffusion modes on incoherent scatter. Incoherent scatter, also called Thomson scatter, was originally thought to arise from thermal irregularities in the electron gas [*Gordon*, 1958]. These are statistically uncorrelated by nature hence justifying the term 'incoherent'. However, soon after the availability of incoherent scatter radars (ISR) in the 1950s, it was found that the observed Doppler spectrum was much narrower than anticipated from pure thermal motions of the electrons but is rather determined by ion motion [*Bowles*, 1958]. It was then understood that the scatter is only then from free electrons when the radar wavelength is much smaller than the Debye-length (i.e., the shielding length of the plasma). Since the latter is in the order of centimeters, and today available incoherent scatter radars have wavelengths of tens of centimeters, the scattering is in fact from different types of plasma waves, i.e., ion acoustic waves, electron plasma waves, and ion cyclotron waves [see e.g. *Bauer*, 1973, for more details]. In the D-region, i.e., at altitudes below about 90 km, these waves are efficiently damped by the frequent collisions with neutral molecules such that the spectrum is dominated by highly damped ion acoustic waves. It is the shape of this D-region ion line of the spectrum which we will consider here in more detail.

Figure 19.3 shows calculations of ISR spectra and corresponding autocorrelation functions (ACFs) using the fluid theory-approach described in *Cho et al.* [1998]. This figure illustrates that the presence of charged MSPs leads to the occurrence of a narrow central line on top of the broad Lorentzian background of the spectrum which is due to the highly damped ion acoustic waves introduced above [*Dougherty and Farley*, 1963; *Tanenbaum*, 1968; *Mathews*, 1978]. The additional narrow line in the presence of charged MSPs is a consequence of the aforementioned second diffusion mode in the plasma due to the Coulomb coupling between the electrons and charged MSPs. Figure 19.3b shows corresponding ACFs which are derived by taking the (inverse) Fourier transforms of the spectra. In the absence of charged MSPs the Lorentzian spectrum corresponds to a single exponential decay while the presence of MSP roughly leads to a superposition of two such exponentials, i.e., one owing to positive ions and one owing to charged particles [*Strelnikova et al.*, 2007]. It is this shape of the ACF which we will use further down to identify the presence of charged aerosol particles in ISR spectra and derive parameters like their number density and radii where possible (see Sect. 19.4).

19.2.2 Coherent Scatter: The TWLS Theory

The other type of radar scattering that we consider here is termed 'coherent' scatter since it arises from refractive index variations (= electron density irregularities, see above) which are statistically correlated. From the troposphere up to the D-region, the most prominent and best studied source for such irregularities is neutral air turbulence. In this case, the mixing of electron density over a gradient leads to small scale fluctuations which may reach the radar Bragg wavelength (= half the radar wavelength for a monostatic radar, see *Tatarskii* [1961]) and hence lead to scattering of the electromagnetic wave. Consequently, such coherent radar measurements have been used for decades to perform measurements of neutral atmosphere dynamics [*Woodman and Guillen*, 1974; *Hocking*, 2011]. Noteworthy, the corresponding theory is well developed [*Tatarskii*, 1961; *Hocking*, 1985] and the volume reflectivity η (=scattering cross section per observing volume) is given by

$$\eta(k) = \frac{\pi}{2} k^4 \Phi_n(k) \tag{19.1}$$

where $k = 4\pi/\lambda_R$ is the Bragg wavenumber of the radar, λ_R is the radar wavelength, and $\Phi_n(k)$ is the power spectral density of the refractive index field n at the Bragg wavenumber k.

Hence, it is not surprising that PMSE were initially attributed to the same scattering process [*Balsley et al.*, 1983]. Closer inspection, however, reveals that this would require unrealistically large turbulent energy dissipation rates such that it was soon recognized that these echoes could not be attributed to regular turbulent scatter [see e.g., *Rapp and Lübken*, 2004, for further quantitative explanation].

This problem is also visualized in Fig. 19.4 which shows theoretical values of the volume reflectivity for various electron densities, electron density gradients, and turbulent energy dissipation rates. In this figure the solid curves show the results for pure turbulent scatter and the gray vertical bar indicates the range of volume reflectivities observed with radars operating at 50 MHz at the corresponding Bragg wavelength of 3 m. Evidently, these observed values are never reached in the purely turbulent case which is because the radar Bragg wavelength already lies in the so-called viscous subrange of turbulence, i.e., at spatial scales where any irregularity in the neutral gas is efficiently destroyed by molecular diffusion.

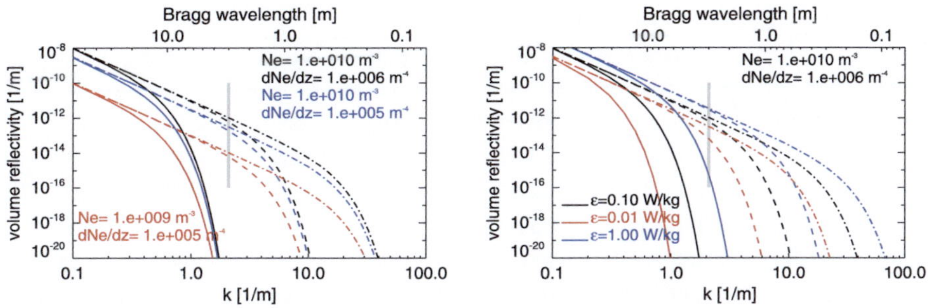

Fig. 19.4 *Left*: Calculated volume reflectivities for turbulent backscatter for a turbulent energy dissipation rate of 0.1 W/kg and electron number densities *Ne* and electron number density gradients dNe/dz indicated in the insert. *Solid*, *dashed* and *dashed-dotted lines* were calculated for $Sc = 1$, 50, and 1000, respectively. The *gray vertical bar* indicates the range of values observed by the ALWIN radar (Bragg wavelength of 3 m) in the period 28.06.–05.07.2006. *Right*: Same as *left panel*, but for different values of the turbulent energy dissipation rate ε and fixed electron number density *Ne* and electron number density gradient dNe/dz. This figure is reproduced from *Rapp et al.* [2008]

The breakthrough in our understanding of this phenomenon was subsequently achieved by *Kelley et al.* [1987] who proposed that electrons under PMSE-conditions are low diffusivity tracers, or—in other words—that the Schmidt number *Sc* (=the ratio between the kinematic viscosity of neutral air and the electron diffusion coefficient) is significantly larger than unity. It was already shown by *Batchelor* [1959] that such large Schmidt number tracers extend their power spectrum to much smaller scales than for the $Sc = 1$-case (which applies to the neutral gas). This is also depicted in Fig. 19.4 with the dashed and dashed-dotted lines which show theoretical values of the volume reflectivity for $Sc = 50$ and $Sc = 1000$, respectively. In both of these cases the theoretical values of the volume reflectivities reach the observed values such that this so-called turbulence with large Schmidt number (TWLS)-theory may indeed explain the observations—provided that *Sc* is indeed significantly enhanced beyond 1, i.e., that the electron diffusivity is significantly reduced. As explained above, this is exactly the effect of the presence of charged aerosol particles on the electrons. I.e., one of the diffusion modes of the electrons is close to the aerosol particle diffusion coefficient which in turn is much smaller than the neutral air diffusion coefficient or kinematic viscosity owing to the large particle size compared to molecular dimensions [*Cho et al.*, 1992; *Rapp and Lübken*, 2003]. In the course of the work described in this review article, the TWLS theory has both been tested and applied to observations to derive microphysical particle parameters. Corresponding results are presented below in Sect. 19.5.

19.3 Experimental Facilities and Data Sets

Measurements have been conducted with various incoherent and coherent scatter radars. Among these are the EISCAT VHF and UHF radars located at Tromsø,

Table 19.1 Radar system parameters

	EISCAT VHF	EISCAT UHF	ESR	SSR	ALWIN	Arecibo
Frequency (MHz)	224	933	500	53.5	53.5	430
Wavelength (m)	1.34	0.32	0.6	2.8	2.8	0.7
Peak power (MW)	1.2	1.2	0.8	0.004	0.036	2.0
3 dB-beam width (°)	1.2 × 1.7	0.5	1.23	5.0	6.0	0.15
Range-resolution (m)	300	300	300	300	300	150
Geogr. coord. (°)	69N, 19E	69N, 19E	78N, 16E	78N, 16E	69N, 16E	18N, 66W

Northern Norway [*Baron*, 1986; *Folkestad et al.*, 1983], the ALWIN VHF radar at Andenes, Northern Norway [*Latteck et al.*, 1999], the EISCAT Svalbard radar (ESR) and the SOUSY Svalbard radar (SSR) at Longyearbyen on Svalbard [*Wannberg et al.*, 1997; *Röttger*, 2001], and the Arecibo radar in Puerto Rico [e.g., *Janches et al.*, 2006]. Some important technical parameters of these radars are summarized in Table 19.1.

During the project period, several extended dedicated campaigns were conducted with these radars. In addition, we also analyzed already existing data where appropriate. Hence, we will here not provide a complete list with all observations. Instead we will mention observation periods and further experimental details in the appropriate result sections or we will refer to our corresponding original publications.

19.4 Aerosol Signatures in Incoherent Radar Echoes

As described above, the presence of charged aerosol particles is expected to change the shape of ISR spectra (or ACFs) and should hence allow us to obtain information on these particles. In the following two subsections, we will summarize our corresponding results. To start with, however, we shortly describe a new method to characterize the shape of ACFs which was originally introduced for the characterization of auroral backscatter and which we subsequently applied to the incoherent scatter from the D-region in order to distinguish different scattering regimes [*Jackel*, 2005; *Rapp et al.*, 2007; *Strelnikova and Rapp*, 2010].

The key idea behind this method is that D-region spectra owing to coherent, 'regular' incoherent (i.e., in the absence of charged aerosol particles), or incoherent scatter in the presence of charged particles are expected to have very different spectral shapes, i.e., they should possess a Gaussian, a Lorenzian, or a "charged dust" spectrum (see Fig. 19.3 and the corresponding text for details), respectively. After conversion to ACFs, these three different spectral types may be identified by means of one single parameter. This is because the magnitude of all these ACFs can be approximated as

$$ACF(\tau) = ACF_{\tau=0} \cdot \exp\left\{-(\tau/\tau_e)^n\right\} \tag{19.2}$$

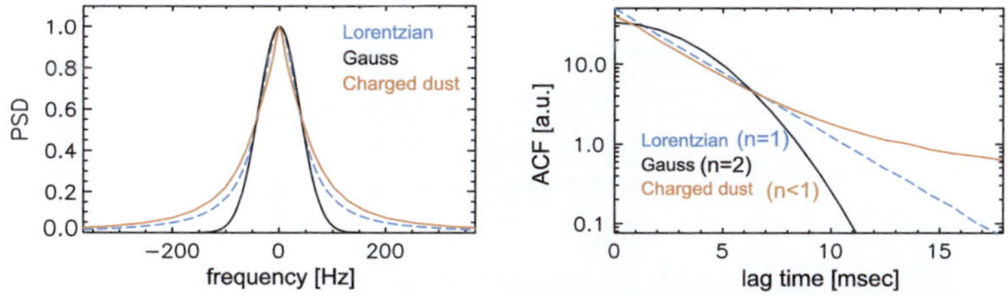

Fig. 19.5 *Left*: Different types of spectra: Gaussian (*black*), Lorentzian (*blue*), and for the presence of charged dust particles (*red*). *Right*: corresponding ACFs [*Strelnikova and Rapp*, 2010]

where τ is the time lag at which the ACF is evaluated, τ_e is a correlation time of the ACF, and the parameter n describes the shape of the ACF. Lorentzian and Gaussian shapes correspond to $n = 1$ and $n = 2$, respectively, whereas $n < 1$ reflects the presence of charged dust particles [*Rapp et al.*, 2007]. Examples of these ACFs are shown in the right panel of Fig. 19.5. We note that this method is applicable to altitudes between roughly 70–90 km. Below this altitude range, spectra might be additionally modified by the presence of negative ions, and above classical incoherent scatter theory predicts much more complicated features than a simple Lorentzian since the motion of electrons and ions is no longer collisionally dominated.

19.4.1 Meteoric Smoke Results

The above described method was first applied to observations obtained with the EISCAT UHF radar on January 15th, 2005 in search for the presence of charged MSPs. The corresponding results have been described in detail in *Rapp et al.* [2007]. These authors found altitude regions with $n < 1$ in an altitude range between ~72 km and 82 km which were indicative of charged MSP. However, at that time the authors also noted the rather poor signal-to-noise ratio of their ACFs as well as the missing independent confirmation of the presence of charged MSPs. In more recent EISCAT observations conducted in parallel to a major sounding rocket campaign in December 2010, however, much better data could be obtained. A corresponding example is presented in Fig. 19.6 which shows the height-time variation of the observed electron density (left panel) and a profile of the n-parameter for a selected time window from 11-12:45 UT on December 14th, 2010 (right panel). This reveals a clear and unambiguous region with $n < 1$ below about 85 km and hence gives evidence for the existence of charged MSPs. Importantly, this altitude range with the occurrence of charged MSPs is independently confirmed by rocket observations with a Faraday-cup type particle detector [see *Rapp et al.*, 2010, for more details] which was launched only 1 day before these particular radar observations.

Hence, the shape of ISR spectra is indeed suitable for the detection of charged MSPs. In an additional effort, further observations were also conducted with the Arecibo radar in Puerto Rico. This radar is generally deemed the most sensitive

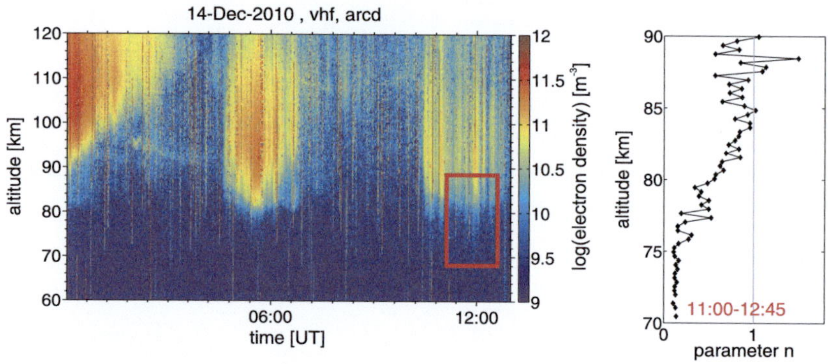

Fig. 19.6 *Left*: Height-time-intensity plot of echo power observed with the EISCAT VHF radar on December 14th, 2010. *Right*: Altitude profile of the spectral parameter n for the time and altitude range marked in *red* in the *left panel*

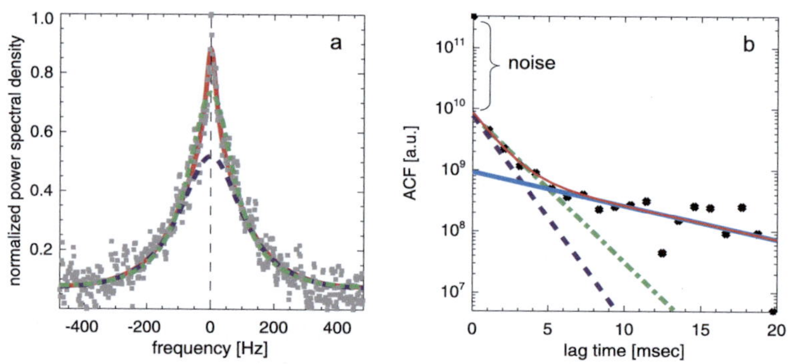

Fig. 19.7 Observed spectrum (panel **a**) and corresponding ACF (panel **b**) at an altitude of 88.2 km. *Green lines* indicate the best fit to the data for a single Lorentzian. *Red lines* show the sum of two Lorentzians (*dark* and *light blue lines*). This figure is reproduced from *Strelnikova et al.* [2007]

existing ISR and should hence be ideally suited for the detection of such subtle signals. Figure 19.7 shows a sample spectrum and corresponding ACF as observed on September 10, 2006 (see *Strelnikova et al.* [2007] for details regarding these observations). Indeed both the spectra as well as the ACF clearly revealed the features anticipated as the signatures of charged MSPs in the ISR spectrum. Owing to the excellent signal to noise ratio of these observations, *Strelnikova et al.* [2007] did not only use them to establish the presence of charged MSPs but also to infer quantitative parameters, i.e., their particle radius and charge number density. This result was obtained by fitting the observed ACF with two exponentials (see dark and light blue lines in Fig. 19.7, respectively) which was shown to be a reasonably good approximation to the full ISR theory described in *Cho et al.* [1998]. Corresponding altitude profiles of MSP radii and charge number densities are presented in Fig. 19.8. Radii are in the range between 0.8–1 nm which agrees well with predictions by microphysical models [*Hunten et al.*, 1980]. Number densities (assuming

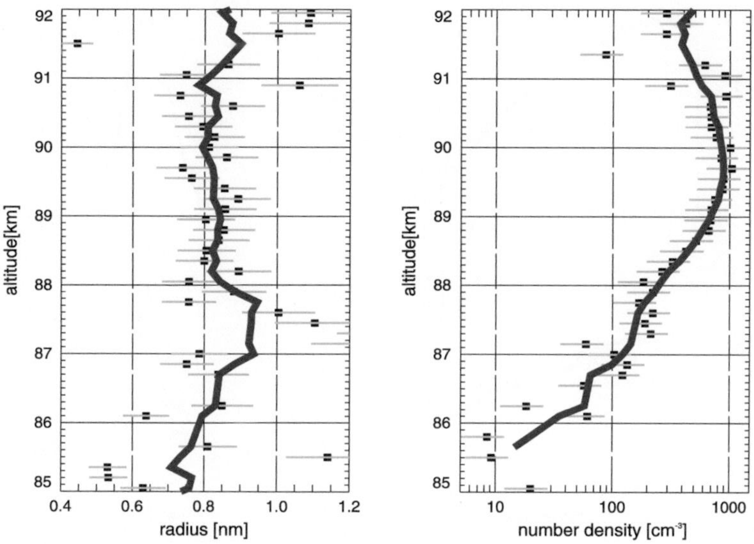

Fig. 19.8 Altitude profiles of retrieved radii (*left panel*) and number densities (*right panel*) of charged meteoric smoke particles. The *blue lines* are 3-point running means of the original data. This figure is reproduced from *Strelnikova et al.* [2007]

singly positively charged particles—see *Rapp* [2009] for a discussion of the MSP charge state) are in the range between 10–1000 cm^{-3} which is also in reasonable agreement with model predictions.

19.4.2 Mesospheric Ice Results

As a next step, it was further investigated whether similar features in the ISR spectrum could also be detected for the case of charged mesospheric ice particles. In this case, however, it must be noted that much more care in the selection of suitable data is needed since the subtle ISR-signatures that we are interested in here are likely masked by coherent contributions to the radar signal, i.e., PMSE. Hence, a case was selected for which the presence of PMSE observed with the VHF radar provided proof for the presence of charged ice particles. At the same time, however, the ISR signal which was simultaneously detected with the EISCAT UHF radar was required to show no signature of PMSE but instead revealed the narrow spectral feature illustrated in Fig. 19.3 above. We note that a combination of such observations corresponds to a situation where the coherent PMSE scatter is much larger than the incoherent scatter background at the frequency of the VHF radar but well below this background at the frequency of the UHF radar [see also Fig. 1 and the corresponding discussion in *Strelnikova and Rapp*, 2010, for more details]. The spectra and ACFs for such a case are presented in Fig. 19.9. This reveals all the characteristics anticipated for such a case: The VHF spectrum and ACF clearly show a Gaussian shape indicative of coherent scatter, i.e., PMSE. Unlike that, the UHF spectrum clearly

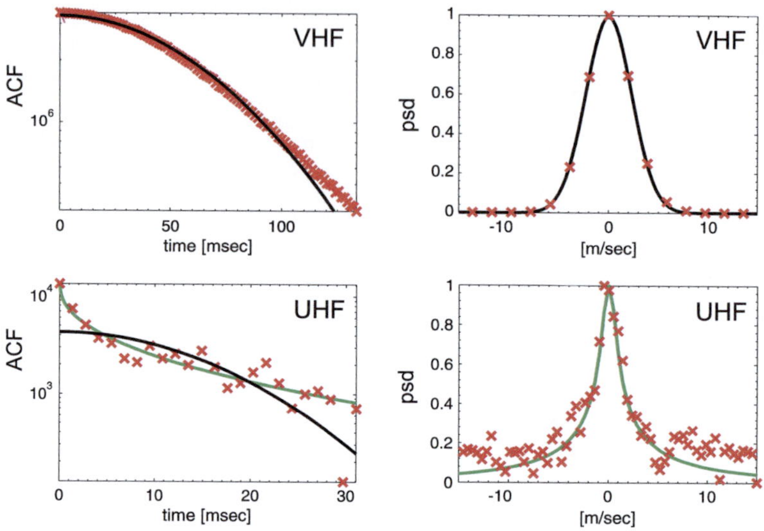

Fig. 19.9 The example of ACF (*left*) and spectra (*rights*) measured by VHF (*upper panel*) and UHF (*lower panel*). The measurements were done on 11 August 2007 at 08:00 UT and 83.7 km altitude. Reproduced from *Strelnikova and Rapp* [2010]

shows a narrow central component indicative of the presence of charged ice particles contributing to the ISR spectrum [*Strelnikova and Rapp*, 2010]. As discussed in detail in this paper, these measurements reveal the presence of charged ice particles with radii in excess of about 5 nm and underline the suitability of such spectral investigations for the detection and characterization of charged mesospheric aerosol particles.

19.5 Aerosol Signatures in Coherent Radar Echoes

We next turn to the case of coherent radar scatter due to the presence of charged aerosol particles. In the next subsections we will first summarize the most important results of a case study involving calibrated PMSE observations at three frequencies and then turn to simultaneous calibrated PMSE observations at the frequency pairs of 53 MHz and 500 MHz as well as 224 MHz and 933 MHz, respectively. These observations are used to first test the TWLS theory which is subsequently applied to the data to infer microphysical properties of the aerosol particles involved.

19.5.1 A Case Study: PMSE-Observations at 3 Frequencies

We start with a case study using the ALWIN radar in Andenes, Northern Norway and the EISCAT VHF and UHF radars in Tromsø, Northern Norway, which is located approximately 130 km to the North-East. The left panel of Fig. 19.10 shows

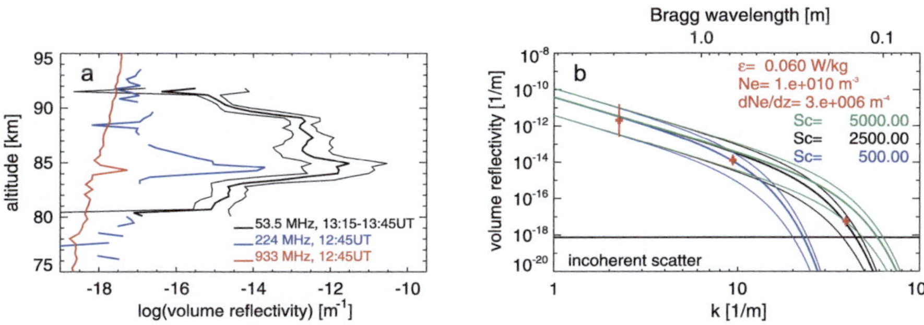

Fig. 19.10 Panel **a**: Volume reflectivities measured with the ALWIN radar (*black lines*), and the EISCAT VHF and UHF radars (*blue* and *red lines*) on July 5th, 2006. EISCAT measurements were obtained at 12:45 UT. ALWIN data are shown as mean plus/minus one standard deviation of the signal measured in the period from 13:15–13:45 UT. Panel **b**: Comparison of measured (*red symbols*) and calculated volume reflectivities for Schmidt numbers of 500, 2500, and 5000 (*blue*, *black*, and *green lines*), respectively. The corresponding *thin lines* indicate the uncertainty in the theoretical expression for the volume reflectivity owing to uncertainties in the Richardson number and the turbulent Prandtl number. The *black horizontal line* indicates the volume reflectivity due to incoherent scatter for comparison. Both panels are reproduced from *Rapp et al.* [2008]

altitude profiles of the volume reflectivities observed on July 5th, 2006, with these three radars. Note that the observations with the EISCAT radars are for 12:45 UT whereas the ALWIN data are for 13:15–13:45 UT. These different times were chosen because of the spatial separation of the two radar sites and taking into account the horizontal wind (taken from independent observations) which was assumed to passively advect the PMSE structures from one site to the other. A more detailed discussion of this issue can be found in *Rapp et al.* [2008]. For the summary of results presented here, we just add that we attempted to take the inherent uncertainty due to the spatial separation of the measurements into account by considering a rather large time window for the ALWIN observation which we subsequently characterized by the corresponding mean reflectivity and its standard deviation. Having this caveat in mind, Fig. 19.10 reveals a tremendous frequency dependence of the PMSE volume reflectivity with values in excess of $\sim 10^{-12}\ \mathrm{m}^{-1}$ at 53 MHz, $\sim 10^{-14}\ \mathrm{m}^{-1}$ at 224 MHz, and $\sim 10^{-17}\ \mathrm{m}^{-1}$ at 933 MHz.

In order to compare these observations to predictions from the TWLS theory, *Rapp et al.* [2008] derived a closed expression for the volume reflectivity as a function of turbulence-, electron density- and aerosol particle properties. According to their work, the volume reflectivity η may be expressed as

$$\eta = 8\pi^3 \cdot \frac{f_\alpha \cdot q \cdot Ri}{Pr^t} \cdot \frac{\sqrt{\varepsilon\nu}}{N^2} \cdot \widetilde{M_e}^2 \cdot r_e^2 \cdot \frac{1}{k^3} \cdot \exp\left(-\frac{2\eta_K^2}{Sc} \cdot k^2\right) \tag{19.3}$$

where ε is the turbulent energy dissipation rate, ν is the kinematic viscosity, N is the buoyancy frequency, $k = 4\pi/\lambda$ is the Bragg wavenumber of the radar, $\eta_K = (\nu^3/\varepsilon)^{\frac{1}{4}}$ is the Kolmogorov microscale, r_e is the classical electron radius, and $Sc = \nu/D_e$ is the Schmidt number introduced above. $\widetilde{M_e}$ is the reduced potential refractive index gradient, i.e., $\widetilde{M}_e = \frac{Ne\,N^2}{g} - \frac{dNe}{dz} - \frac{Ne}{H_n}$ which depends on the electron number density Ne, the buoyancy frequency, and the density scale height H_n. Finally, f_α, q, Ri and Pr^t are all 'constants' (i.e., they are constant for a given event) derived from either theory or by comparison with observations (see Appendix A in *Rapp et al.* [2008] for more details).

Importantly, except for the Schmidt number, all critical parameters like the electron density and its gradient and the turbulent energy dissipation rate can be directly derived from the ISR-observations. Hence, only one free parameter remains which may be derived from adjusting the theoretical spectrum to match the observations. Such comparisons are presented in the right panel of Fig. 19.10. This reveals that the absolute level of the volume reflectivities is surprisingly well described by the theoretical expression. Furthermore, it also reveals that a Schmidt number of about 2500–5000 is required to match theory and observations at all frequencies. Based on the physical model of the Schmidt number introduced above, it can be shown that this corresponds to particle radii in the range between 20–30 nm. This lies in the range of particle radii which may reasonably be expected to occur at such altitudes based on both modeling and independent observations [e.g., *Rapp and Thomas*, 2006; *Baumgarten et al.*, 2008].

It hence appears that this case study is indeed consistent with the TWLS theory and that the comparison of theory and calibrated observations at a minimum of two frequencies allows us to derive particle radii. However, in order to arrive at more robust statements a much better statistics of such observations—and even more important—a confirmation of derived aerosol radii by an independent method are needed. This will be presented in the next subsections.

19.5.2 PMSE-Properties at High Frequencies: Statistical Results

In order to test the TWLS theory more rigorously and see whether multi-frequency PMSE observations are indeed a suitable tool for the derivation of ice particle parameters, two major data sets were investigated: One is from observations conducted in June 2006 with the ESR and SSR (see also Table 19.1) which are both located at Longyearbyen on Svalbard [see *Li et al.*, 2010, for more details]. The other data set is from observations with the EISCAT VHF and UHF radars in Tromsø which were conducted in the years 2004 and 2005 [see *Li and Rapp*, 2011, for more details].

To start with, occurrence rates of PMSE at all four frequencies were derived and are presented in Fig. 19.11. This reveals that PMSE at 500 and 933 MHz occur much less frequently (16 % and 5 %) than PMSE at 53 MHz and 224 MHz (95 % and 85 %), respectively. In all cases, PMSE at the higher frequency occurred in the presence of PMSE at the lower frequency. In absolute numbers this amounts to about 300 min (380 min) of simultaneous PMSE observations with the SSR/ESR (EISCAT-VHF/UHF radars).

At this point we need to note, however, also a general problem with the comparison of occurrence rates. Inherently, these values also depend on radar system

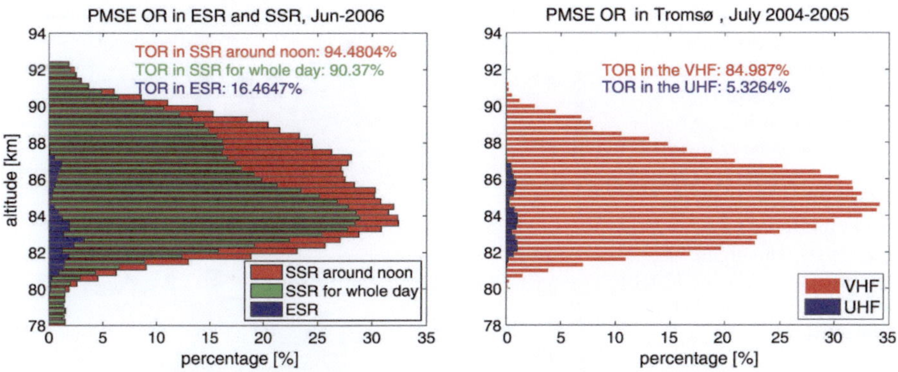

Fig. 19.11 *Left*: Comparison of PMSE occurrence rates at 500 MHz (*blue*) and 53.5 MHz (*red* and *green*) respectively derived from the ESR and SSR observations in June 2006. The *red histogram* shows the PMSE occurrence rate during 9–13 UT, i.e., during the period when the ESR was run whereas the *green histogram* is for all SSR observations. Reproduced from *Li et al.* [2010]. *Right*: Comparison of PMSE occurrence rate at 930 MHz (in *blue*) and 224 MHz (in *red*), respectively, derived from the EISCAT UHF and VHF observations in July 2004 and 2005. Reproduced from *Li and Rapp* [2011]. Total occurrence rates (TOR) are indicated in the insert

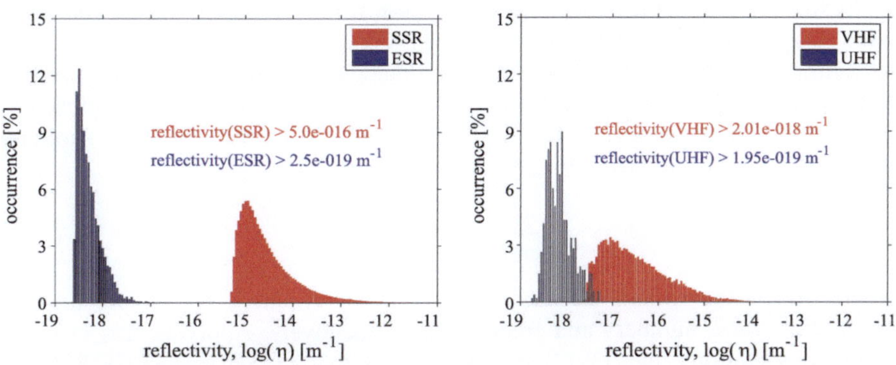

Fig. 19.12 Histograms of volume reflectivities observed with the SSR and ESR (*left panel*) and the EISCAR VHF and UHF radars (*right panel*), respectively. This figure is modified after *Li et al.* [2010] and *Li and Rapp* [2011]

parameters which are in fact very different for the different radars used in this study (see Table 19.1). Hence, beyond the qualitative measure of occurrence rates, we also considered corresponding statistical distributions of the volume reflectivities. A comparison of these distributions is presented in Fig. 19.12. This reveals values in the range 2.5×10^{-19}–1×10^{-17} m^{-1} for the ESR, 5×10^{-16}–6×10^{-12} m^{-1} for the SSR, 2×10^{-19}–7×10^{-18} m^{-1} for the EISCAT UHF radar, and finally 1.4×10^{-18}–1.6×10^{-14} m^{-1} for the EISCAT VHF radar, respectively. This indicates that PMSE not only have an impressively strong frequency dependence but also that the dynamical range of volume reflectivities at all considered frequencies

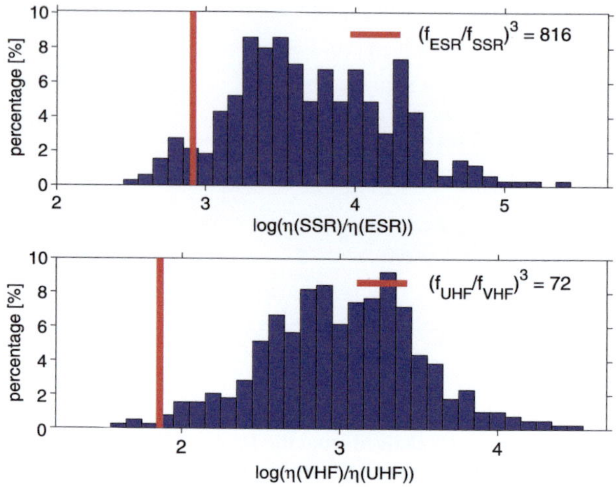

Fig. 19.13 Histograms of reflectivity ratios from simultaneous observations with the ESR/SSR (*upper panel*) and the EISCAT VHF/UHF radars (*lower panel*), respectively. The *red vertical lines* indicate the lower threshold from Eq. (19.4). This figure is modified after *Li et al.* [2010] and *Li and Rapp* [2011], respectively

is very large. This in turn implies that a meaningful comparison of PMSE observations at different frequencies must be conducted simultaneously and in a common volume.

Next, we return to the question whether these volume reflectivities are consistent with expectations from the above described TWLS theory. For this purpose, histograms of reflectivity ratios from such simultaneous and common volume observations with the SSR/ESR and EISCAT VHF/UHF were determined and are presented in Fig. 19.13. In these figures we have also marked a lower theoretical threshold for these ratios as expected from Eq. (19.3), i.e.,

$$\frac{\eta(1)}{\eta(2)} \geq \frac{k_2^3}{k_1^3} = \frac{f_2^3}{f_1^3} \tag{19.4}$$

where $\eta(i)$ is the volume reflectivity at frequency i, and k_i and f_i are the corresponding wave numbers and frequencies, respectively. Applying this criterion to our observations, the TWLS theory requires that $\eta(SSR)/\eta(ESR) \geq 816$ and $\eta(VHF)/\eta(UHF) \geq 72$. Indeed, Fig. 19.13 reveals that these criteria are met by 94 % of the SSR/ESR-data and even 99 % of the VHF/UHF-data. The few cases violating these criteria can likely be explained by slightly different observing volumes of the corresponding radars [see *Li et al.*, 2010; *Li and Rapp*, 2011, for detailed discussions].

We note that we further checked whether the simultaneously observed PMSE also complied with the additional requirement that corresponding spectral widths were identical after conversion to Doppler-velocities. This should be the case because based on the TWLS theory one would expect that the spectral widths are a measure of turbulent velocity fluctuations in the medium which should be independent of the radar frequency [e.g., *Hocking*, 1985]. Indeed, *Li et al.* [2010] and *Strelnikova and Rapp* [2011] were able to show that this criterion is fulfilled for both data sets after spectral broadening effects like wave and beam broadening (stated in order of importance) were taken into account.

In all, our studies strongly support the view that TWLS theory can explain PMSE at all considered frequencies. In the next subsection, this theory will hence be applied to the same data sets to derive aerosol parameters.

19.5.3 Microphysical Parameters from Observations at Two Frequencies

The method to derive ice particle radii from PMSE observations at two frequencies was first described in *Li et al.* [2010] and is briefly reviewed here: Using Eq. (19.3), the following expression for the Schmidt number may be derived from taking the ratio of volume reflectivities $\eta(1)/\eta(2)$ where $\eta(i)$ again denotes the volume reflectivity at frequency i:

$$Sc = \frac{2\eta_K^2(k_2^2 - k_1^2)}{\ln(\frac{\eta(1)}{\eta(2)} \cdot \frac{k_1^3}{k_2^3})}. \qquad (19.5)$$

Note that this equation only depends on known quantities or quantities which may be easily derived, i.e., the measured volume reflectivities, the radar Bragg wave numbers, and the Kolmogorov-microscale η_K. η_K can be calculated using the turbulent energy dissipation rate estimated from spectral width measurements [*Hocking*, 1985] and an estimate of the kinematic viscosity using Sutherland's formula and densities and temperatures from the MSIS-climatology [*Picone et al.*, 2002]. Taking further into account that the Schmidt number can be expressed in terms of the properties of the charged ice particles, corresponding ice particle radii can be derived from the following relation

$$r_A = \sqrt{\frac{Sc}{6.5}} \qquad (19.6)$$

where the radius r_A is in nm [*Lübken et al.*, 1998]. We note that this simple relation was derived under the assumption that the electron diffusion coefficient equals the aerosol particle diffusion coefficient. A more rigorous analysis presented in *Rapp and Lübken* [2003] shows that the electron diffusion coefficient is given by the 'slow' diffusion mode introduced above which depends on both the aerosol particle radius and the so-called Havnes-parameter $\Lambda = |Z_A|N_A/Ne$, i.e., the ratio between the charge number density of the aerosol particles $|Z_A|N_A$ and the electron density Ne. Since $|Z_A|N_A$ is usually not known (except for the case of rocket observations) we nevertheless have used Eq. (19.6) to derive radii. The corresponding error of the particle radius can be quantified by comparison of Schmidt numbers derived from the full theory described in *Rapp and Lübken* [2003] with those from the simple relation in Eq. (19.6). This reveals that for typical values of $\Lambda \leq 0.5$ [*Blix et al.*, 2003] the relative error in Sc is at most 20 % implying a relative error in the particle radius of less than 10 % from this error source alone.

In a next step, Fig. 19.14 shows contours of particle radii as a function of volume reflectivity ratios at two frequencies and the turbulent energy dissipation rate ε.

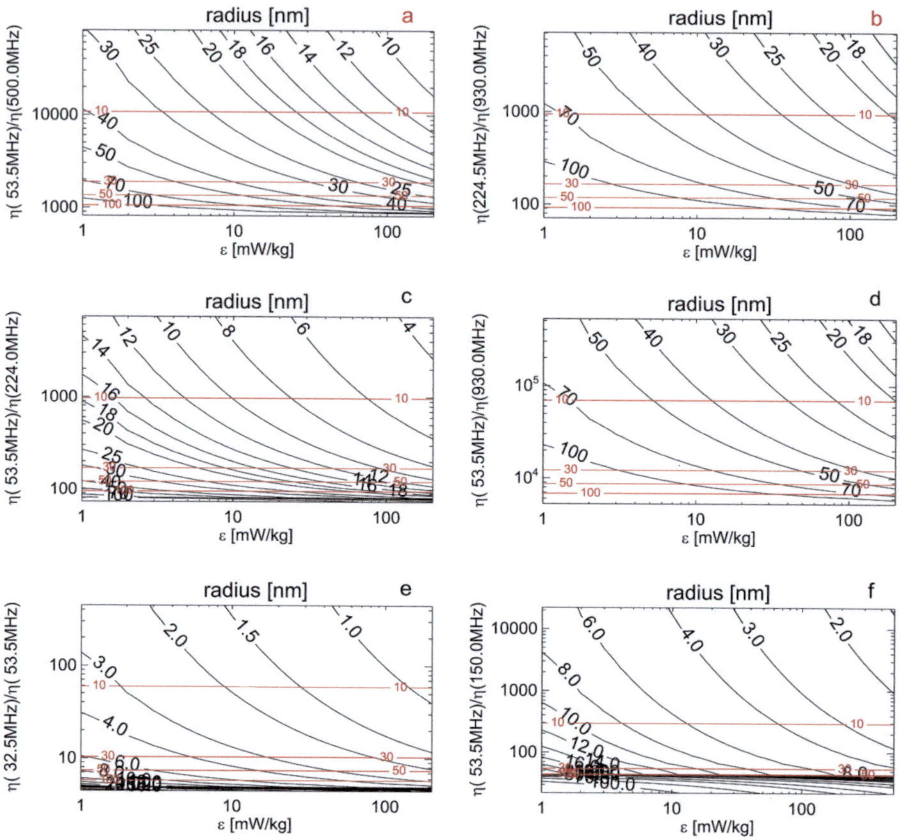

Fig. 19.14 Particle radii derived from Eqs. (19.5) and (19.6) as a function of volume reflectivity ratios for various frequencies and turbulent energy dissipation rates. The *red contours* indicate errors (in percent) due to uncertainties in the reflectivity ratios and turbulent energy dissipation rates

This reveals that different frequency combinations are sensitive to different ranges of particle radii. I.e., for the case of the two data sets presented here (namely 53 and 500 MHz as well as 224 and 933 MHz, see panels a and b) radii between 10 nm and 100 nm can be derived reliably. For even larger particle sizes, the contours converge and separation between two different values is not longer possible. Going back to Eq. (19.3) we see that this is the case when η is asymptotically described by a k^{-3}-dependence which happens in the case of large Schmidt numbers. Fig. 19.14 further shows that the determination of particle radii smaller than ~10 nm requires other frequency combinations like the combination of 53 MHz and 224 MHz (panel c) with which radii in the range of ~5–40 nm should be accessible. With the combination of 32.5 MHz (frequently used for meteor radars) and 53 MHz even much smaller particle sizes, i.e., 1–10 nm, can be reached. Figure 19.14 further shows that typical errors due to uncertainties in the determination of the reflectivity ratios and

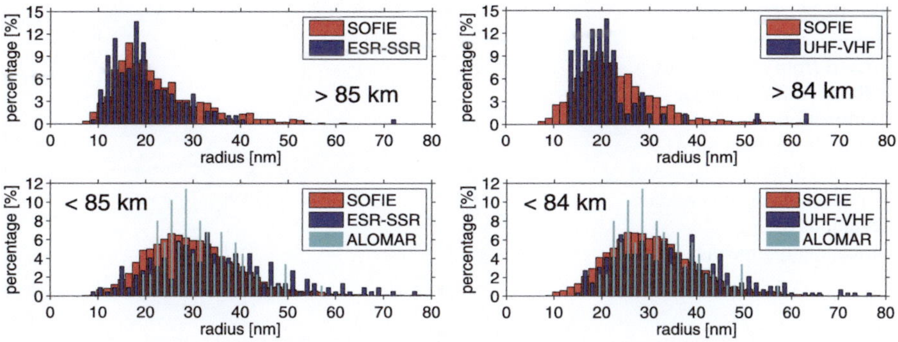

Fig. 19.15 Distribution of ice particle radii derived from radar (in *dark blue*) and optical observations from the SOFIE instrument on the AIM satellite (in *red*) and from the ALOMAR RMR lidar (in *light blue*) for altitudes below (*lower panel*) and above 85 km (*upper panel*). The *left panels* are for the SSR/ESR radar combination and are reproduced from *Li et al.* [2010]. The *right panels* are for the EISCAT VHF/UHF radar combinations

the turbulent energy dissipation rate are less than 30 % except for the region where the contour lines converge [see *Li et al.*, 2010, for a more in-depth discussion of errors].

Next we present histograms of particle radii derived from our observations with the SSR/ESR and the EISCAT VHF/UHF radars in Fig. 19.15. For both cases we have divided the data set in a population of 'low' altitude values and 'high' altitude values where the border between these two populations was chosen based on the specific altitude distribution of corresponding data [see *Li et al.*, 2010; *Li and Rapp*, 2011, for more details]. In both cases, particle sizes at the lower altitudes are larger than at the upper altitudes as expected from microphysical modeling and independent observations. More importantly, we have also compared our radar-based particle size estimates to independent optical observations from the SOFIE instrument on the AIM satellite and the ALOMAR RMR lidar [*Hervig et al.*, 2009; *Baumgarten et al.*, 2008]. Figure 19.15 reveals that these different statistical distributions agree amazingly well given the very different underlying techniques. Even more to that, we have further compared the mean altitude profiles of particle radii from both SOFIE and the two radar combinations as shown in Fig. 19.16. Again, an excellent agreement between these different data sets is found.

Finally, we stress that this new technique to derive ice particle radii cannot only be used to derive mean values of ice particle radii, but that it can also be applied to single cases and that even the time development of particle radii can be studied. One corresponding example is presented in Fig. 19.17 which shows ice particle radii with a time resolution of 1 min in the altitude range where both the EISCAT VHF and UHF radars observed PMSE. Data like this enable us to study the short term variation of mesospheric ice microphysics for example in relation to gravity waves. This will be a subject for more in-depth studies in the future.

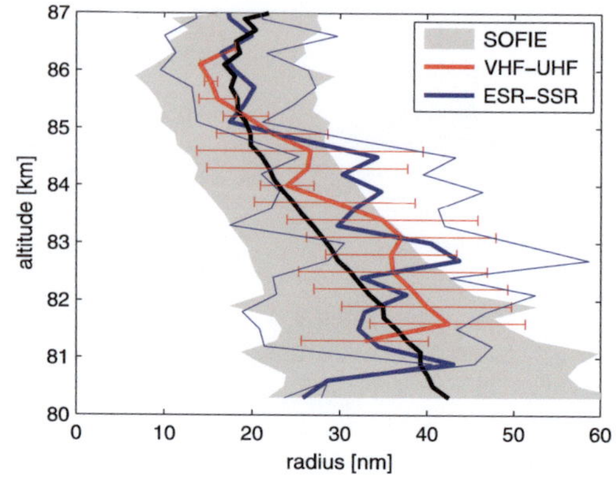

Fig. 19.16 Mean altitude profiles of particle radii derived from observations with two different radar combinations and the SOFIE-instrument on AIM. *Grey shading*, *thin lines*, and *error bars* mark the standard deviations of the corresponding means

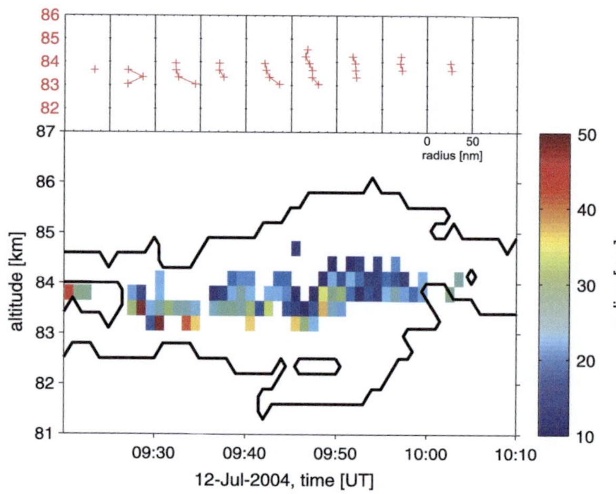

Fig. 19.17 *Lower panel*: The colored pixels show ice particle radii derived from simultaneous PMSE observations with the EISCAT VHF and UHF radars. The *black contour line* indicates the altitude range in which the EISCAT VHF radar observed PMSE. *Upper panels*: Mean ice particle radius-profiles for 5 min of observations each

19.6 Summary and Outlook

In the current manuscript we have reviewed the major results of a project dealing with the investigation of the effect of charged particles on the scattering of radar waves from the D-region. Based on the assumption that the main effect of charged aerosol particles on the ambient plasma is the generation of a 'slow' diffusion mode, consequences for both incoherent as well as coherent scattering were discussed and corresponding measurements were analyzed. The major results of this study may be listed as follows:

1. Based on the theoretical work of *Cho et al.* [1998] it was argued that the presence of charged MSPs or ice particles should lead to an additional narrow line in the ISR spectrum such that the shape of such spectra should be a suitable measure of the presence of such particles. Indeed, such features could be identified for both MSPs [*Rapp et al.*, 2007] and ice particles [*Strelnikova and Rapp*, 2010] in measurements with the EISCAT UHF radar. In addition, measurements with the

Arecibo radar were conducted which yielded an improved signal to noise ratio as compared to the EISCAT observations and allowed us to derive both MSP radii and number densities from these observations [*Strelnikova et al.*, 2007]. This method has since been further applied to additional Arecibo measurements hence yielding insight into the seasonal variation of MSPs at that particular location [*Fentzke et al.*, 2009].

2. Applying the methods for the characterization of spectral shapes of incoherent scatter data from the D-region to so-called polar mesosphere winter echoes (PMWE) with the EISCAT VHF radar it was found that the spectra revealed a clear Gaussian shape [*Lübken et al.*, 2007] and were hence not just enhanced incoherent scatter (i.e., enhanced ion acoustic waves) as proposed previously. Whether or not this also implies the involvement of charged aerosol particles in the generation of these echoes will be subject to future investigations.
3. For the case of coherent scatter, the TWLS theory was tested versus calibrated observations of simultaneous and common volume observations of PMSE at multiple frequencies. For this purpose, a case study on PMSE observations with the EISCAT VHF and UHF radars and the ALWIN radar were analyzed as well as 300 min of simultaneous PMSE observations with the SSR and ESR on Svalbard, and 380 min of simultaneous PMSE observations with the EISCAT VHF and UHF radars [*Rapp et al.*, 2008; *Li et al.*, 2010; *Li and Rapp*, 2011]. In addition, the same data sets were also used to test the theoretical prediction that spectral widths should be independent of radar frequency after conversion to Doppler velocities [*Rapp et al.*, 2008; *Li et al.*, 2010; *Strelnikova and Rapp*, 2011]. Importantly, all these tests revealed excellent agreement between theory and observations hence providing support for the applied theoretical concept.
4. Subsequently, the TWLS theory was applied to the same multi-frequency data sets to infer ice particle radii from the radar measurements. Acquired ice particle radii were in excellent agreement with independent values from optical observations with the SOFIE instrument on the AIM satellite and the ALOMAR RMR lidar [*Li et al.*, 2010, and this study]. On the one hand, this demonstrates the feasibility of the quantification of mesospheric ice parameters with radars and hence paves the way for future studies on the short term variation of mesospheric ice microphysics driven, for example, by gravity waves. On the other hand, this excellent agreement provides further very strong support for the overall correctness of the applied theory.

In summary, it hence appears that a solid understanding of the effect of charged aerosols on the scattering of radar waves from the D-region has been achieved. For the future, the most obvious point for the continuation of related research should be in our view the utilization of corresponding measurements for the characterization of the background atmosphere, e.g. for the study of waves and turbulence on all accessible spatial scales, as well as related dynamical processes. Nevertheless, the investigation of some still unexplored fundamental processes (like the generation mechanism of the aforementioned PMWE) should continue and will likely give us new and unexpected surprises and challenging research tasks for the future.

Acknowledgements This project has been supported by DFG under grants RA 1400/2-1, RA 1400/2-2, and RA 1400/2-3 in the frame of the CAWSES priority program. Fruitful discussions and collaborations with J. Röttger, S. Raizada, M. Sulzer, S. Gonzales, D. Janches, and J. Fentzke are gratefully acknowledged. EISCAT is an international association supported by research organizations in China (CRIRP), Finland (SA), France (CNRS, till end of 2006), Germany (DFG, formerly MPG), Japan (NIPR and STEL), Norway (NFR), Sweden (VR), and the United Kingdom (PPARC).

References

Balsley, B. B., Ecklund, W. L., & Fritts, D. C. (1983). VHF echoes form the high-latitude mesosphere and lower thermosphere: observations and interpretations. *Journal of the Atmospheric Sciences*, *40*, 2451–2466.

Baron, M. (1986). EISCAT progress 1983–1985. *Journal of Atmospheric and Terrestrial Physics*, *48*, 767–772.

Batchelor, G. K. (1959). Small-scale variation of convected quantities like temperature in a turbulent fluid. *Journal of Fluid Mechanics*, *5*, 113–133.

Bauer, S. J. (1973). *Physics of planetary ionospheres*. Berlin: Springer.

Baumgarten, G., Fielder, J., Lübken, F.-J., & von Cossart, G. (2008). Particle properties and water content of noctilucent clouds and their interannual variation. *Journal of Geophysical Research*, *113*, D06203. doi:10.1029/2007JD008884.

Blix, T. A., Rapp, M., & Lübken, F.-J. (2003). Relations between small scale electron number density fluctuations, radar backscatter and charged aerosol particles. *Journal of Geophysical Research*, *108*(D8), 8450. doi:10.1029/2002JD002430.

Bowles, K. L. (1958). Observations of vertical incidence scatter from the ionosphere at 41 Mc/s. *Physical Review Letters*, *1*, 454–455.

Cho, J. Y., Sulzer, M. P., & Kelley, M. C. (1998). Meteoric dust effects on D-region incoherent scatter radar spectra. *Journal of Atmospheric and Solar-Terrestrial Physics*, *60*, 349–357.

Cho, J. Y. N., & Röttger, J. (1997). An updated review of polar mesosphere summer echoes: observation, theory, and their relationship to noctilucent clouds and subvisible aerosols. *Journal of Geophysical Research*, *102*, 2001–2020.

Cho, J. Y. N., Hall, T. M., & Kelley, M. C. (1992). On the role of charged aerosols in polar mesosphere summer echoes. *Journal of Geophysical Research*, *97*, 875–886.

DeLand, M. T., Shettle, E. P., Thomas, G. E., & Olivero, J. J. (2006). A quarter-century of satellite polar mesospheric cloud observations. *Journal of Atmospheric and Solar-Terrestrial Physics*, *68*, 9–29.

DeLand, M. T., Shettle, E. P., Thomas, G. E., & Olivero, J. J. (2007). Latitude-dependent long-term variations in polar mesospheric clouds from SBUV version 3 PMC data. *Journal of Geophysical Research*, *112*, D10315. doi:10.1029/2006JD007857.

Dougherty, J. P., & Farley, D. T. (1963). A theory of incoherent scattering of radio waves by a plasma: 3. Scattering in a partly ionized gas. *Journal of Geophysical Research*, *68*, 5473–5486.

Fentzke, J., Janches, D., Strelnikova, I., & Rapp, M. (2009). Meteoric smoke particle properties derived using dual-beam Arecibo UHF observations of D-region spectra during different seasons. *Journal of Atmospheric and Solar-Terrestrial Physics*, *71*, 1982–1991.

Folkestad, K. T., Hagfors, T., & Westerlund, S. (1983). EISCAT: an updated description of technical characteristics and operational capabilities. *Radio Science*, *18*, 867–879.

Gabrielli, P., Barbante, C., Plane, J. M. C., Varga, A., Hong, S., Cozzi, G., Gasparia, V., Planchon, F. A. M., Cairns, W., Ferrari, C., Crutzen, P., Ceson, P., & Boutron, C. F. (2004). Meteoric smoke fallout over the Holocene epoch revealed by iridium and platinum in Greenland ice. *Nature*, *432*, 1011–1014.

Gordon, W. E. (1958). Incoherent scattering by free electrons with applications to space exploration by radars. *Proceedings of the IRE*, *46*, 1824–1829.

Hervig, M. E., Gordley, L. L., Stevens, J. M. R. III, Bailey, S. M., & Baumgarten, G. (2009). Interpretation of SOFIE PMC measurements: cloud identification and derivation of mass density, particle shape, and particle size. *Journal of Atmospheric and Solar-Terrestrial Physics*, *71*, 316–330.

Hill, R. J. (1978). Nonneutral and quasi-neutral diffusion of weakly ionized multiconstituent plasma. *Journal of Geophysical Research*, *83*, 989–998.

Hocking, W. (2011). A review of mesosphere-stratosphere-troposphere (MST) radar developments and studies, circa 1997–2008. *Journal of Atmospheric and Solar-Terrestrial Physics*, *73*, 848–882.

Hocking, W. K. (1985). Measurement of turbulent energy dissipation rates in the middle atmosphere by radar techniques: a review. *Radio Science*, *20*, 1403–1422.

Hoppe, U.-P., & Fritts, D. C. (1995). High resolution measurements of vertical velocity with the European incoherent scatter VHF radar 1. Motion field characteristics and measurement biases. *Journal of Geophysical Research*, *100*, 16813–16825.

Hunten, D. M., Turco, R. P., & Toon, O. B. (1980). Smoke and dust particles of meteoric origin in the mesosphere and stratosphere. *Journal of the Atmospheric Sciences*, *37*, 1342–1357.

Jackel, B. J. (2005). Characterization of auroral radar power spectra and autocorrelation functions. *Radio Science*, *35*, 1009–1032.

Janches, D., Fritts, D. C., Riggin, D. M., Sulzer, M. P., & Gonzales, S. (2006). Gravity wave and momentum fluxes in the mesosphere and lower thermosphere using 430 MHz dual-beam measurements at Arecibo: 1. Measurements, methods, and gravity waves. *Journal of Geophysical Research*, *111*, D18107. doi:10.1029/2005JD006882.

Kelley, M. C., Farley, D. T., & Röttger, J. (1987). The effect of cluster ions on anomalous VHF backscatter from the summer polar mesosphere. *Geophysical Research Letters*, *14*, 1031–1034.

La Hoz, C., & Havnes, O. (2008). Artificial modification of polar mesospheric winter echoes with an RF heater: do charged dust particles play an active role? *Journal of Geophysical Research*, *113*, D19205. doi:10.1029/2008JD010460.

Latteck, R., Singer, W., & Bardey, H. (1999). The ALWIN MST radar: technical design and performance. In *Proceedings of the 14th ESA symposium on European rocket and balloon programmes and related research*, Potsdam, Germany (ESA SP-437) (pp. 179–184).

Li, Q., & Rapp, M. (2011). PMSE-observations with the EISCAT VHF and UHF-radars: statistical results. *Journal of Atmospheric and Solar-Terrestrial Physics*, *73*, 944–956.

Li, Q., Rapp, M., Röttger, J., Latteck, R., Zecha, M., Strelnikova, I., Baumgarten, G., Hervig, M., Hall, C., & Tsutsumi, M. (2010). Microphysical parameters of mesospheric ice clouds derived from calibrated observations of polar mesosphere summer echoes at Bragg wavelengths of 2.8 m and 30 cm. *J. Geophys. Res.*, D00I13. doi:10.1029/2009JD012271.

Lübken, F.-J. (1999). Thermal structure of the Arctic summer mesosphere. *Journal of Geophysical Research*, *104*, 9135–9149.

Lübken, F.-J., Rapp, M., Blix, T., & Thrane, E. (1998). Microphysical and turbulent measurements of the Schmidt number in the vicinity of polar mesosphere summer echoes. *Geophysical Research Letters*, *25*, 893–896.

Lübken, F.-J., Singer, W., Latteck, R., & Strelnikova, I. (2007). Radar measurements of turbulence, electron densities, and absolute reflectivities during polar mesosphere winter echoes (PMWE). *Advances in Space Research*, *40*, 758–764. doi:10.1016/j.asr.2007.01.015.

Mathews, J. D. (1978). The effect of negative ions on collision-dominated Thomson scattering. *Journal of Geophysical Research*, *83*, 505–512.

Megner, L., Siskind, D. E., Rapp, M., & Gumbel, J. (2008). Global and temporal distribution of meteoric smoke: a two-dimensional simulation study. *Journal of Geophysical Research*, *113*, D03202. doi:10.1029/2007JD009054.

Merkel, A. W., THomas, G. E., Palo, S. E., & Bailey, S. (2003). Observations of the 5-day planetary wave in PMC measurements from the Student Nitric Oxide Explorer satellite. *Geophysical Research Letters*, *30*, 1196. doi:10.1029/2002GL016524.

Picone, J. M., Hedin, A. E., Drob, D. P., & Aikin, A. C. (2002). NRLMSISE-00 empirical model of the atmosphere: statistical comparison and scientific issues. *Journal of Geophysical Research*,

107(A12), 1468. doi:10.1029/2002JA009430.

Rapp, M. (2009). Charging of mesospheric aerosol particles: the role of photodetachment and photoionization from meteoric smoke and ice particles. *Annales Geophysicae*, *27*, 2417–2422.

Rapp, M., & Lübken, F.-J. (2003). On the nature of PMSE: electron diffusion in the vicinity of charged particles revisited. *Journal of Geophysical Research*, *108*(D8), 8437. doi:10.1029/2002JD002857.

Rapp, M., & Lübken, F.-J. (2004). Polar mesosphere summer echoes (PMSE): review of observations and current understanding. *Atmospheric Chemistry and Physics*, *4*, 2601–2633.

Rapp, M., & Thomas, G. E. (2006). Modeling the microphysics of mesospheric ice particles: assessment of current capabilities and basic sensitivities. *Journal of Atmospheric and Solar-Terrestrial Physics*, *68*, 715–744.

Rapp, M., Strelnikova, I., & Gumbel, J. (2007). Meteoric smoke particles: evidence from rocket and radar techniques. *Advances in Space Research*, *40*, 809–817. doi:10.1016/j.asr.2006.11.021.

Rapp, M., Strelnikova, I., Latteck, R., Hoffman, P., Hoppe, U.-P., Häggström, I., & Rietveld, M. (2008). Polar mesosphere summer echoes (PMSE) studied at Bragg wavelengths of 2.8 m, 67 cm, and 16 cm. *Journal of Atmospheric and Solar-Terrestrial Physics*, *70*, 947–961.

Rapp, M., Strelnikova, I., Strelnikov, B., Hoffmann, P., Friedrich, M., Gumbel, J., Megner, L., Hoppe, U.-P., Robertson, S., Knappmiller, S., Wolff, M., & Marsh, D. R. (2010). Rocket-borne in-situ measurements of meteor smoke: charging properties and implications for seasonal variation. *Journal of Geophysical Research*, D00I16. doi:10.1029/2009JD012377.

Röttger, J. (2001). Observations of the polar D-region and the mesosphere with the EISCAT Svalbard Radar and the SOUSY Svalbard Radar. *Memoirs of National Institute of Polar Research*, *54*, 9–20.

Shettle, E. P., DeLand, M. T., Thomas, G. E., & Olivero, J. J. (2009). Long term variations in the frequency of polar mesospheric clouds in the northern hemisphere from SBUV. *Geophysical Research Letters*, *36*, L02803. doi:10.1029/2008GL036048.

Stevens, M. H., Siskind, D. E., Eckermann, S. D., Coy, L., McCormack, J. P., Englert, C. R., Hoppel, K. W., Nielsen, K., Kochenash, A. J., Hervig, M. E., Randall, C. E., Lumpe, J., Bailey, S. M., Rapp, M., & Hoffmann, P. (2010). Tidally induced variations of polar mesospheric cloud altitudes and ice water content using a data assimilation system. *Journal of Geophysical Research*, *115*, D18209. doi:10.1029/2009JD013225.

Strelnikova, I., & Rapp, M. (2010). Studies of polar mesosphere summer echoes with the EISCAT VHF and UHF radars: Information contained in the spectral shape. *Advances in Space Research*, *45*, 247–259.

Strelnikova, I., & Rapp, M. (2011). Majority of PMSE spectral widths at UHF and VHF are compatible with a single scattering mechanism. *Journal of Atmospheric and Solar-Terrestrial Physics* doi:10.1016/j.jastp.2010.11.025.

Strelnikova, I., Rapp, M., Raizada, S., & Sulzer, M. (2007). Meteor smoke particle properties derived from Arecibo incoherent scatter radar observations. *Geophysical Research Letters*, *34*, L15815. doi:10.1029/2007GL030635.

Tanenbaum, B. S. (1968). Continuum theory of Thomson scattering. *Physical Review*, *171*, 215–221.

Tatarskii, V. I. (1961). *Wave propagation in a turbulent medium*. New York: McGraw-Hill.

Thomas, G. E., Olivero, J. J., Jensen, E. J., Schröder, W., & Toon, O. B. (1989). Relation between increasing methane and the presence of ice clouds at the mesopause. *Nature*, *338*, 490–492.

Thomas, G. E., Olivero, J. J., DeLand, M., & Shettle, E. P. (2003). A response to the article by U. von Zahn, "Are noctilucent clouds truly a miner's canary of global change? *Eos, Transactions, American Geophysical Union*, *84*(36), 352–353.

Voigt, C., Schlager, H., Luo, B. P., Dörnbrack, A., Roiger, A., Stock, P., Curtius, J., Vössing, H., Borrmann, S., Davies, S., Konopka, P., Schiller, C., Shur, G., & Peter, T. (2005). Nitric acid trihydrate (NAT) formation at low NAT supersaturation in polar stratospheric clouds (PSCs). *Atmospheric Chemistry and Physics*, *5*, 1371–1380.

von Savigny, C., Sinnhuber, M., Bovensmann, H., Burrows, J. P., Kallenrode, M.-B., & Schwartz, M. (2007). On the disappearance of noctilucent clouds during the January 2005 solar proton

events. *Geophysical Research Letters*, *34*, L02805. doi:10.1029/2006GL028106.
von Zahn, U. (2003). Are noctilucent clouds truly a "miner's canary" of global change? *Eos, Transactions, American Geophysical Union*, *84*(28), 261–268.
Wannberg, U. G., Wolf, I., Vanheinen, L. G., Koskeniemi, K., Röttger, J., et al. (1997). The EISCAT Svalbard radar: a case study in modern incoherent scatter radar design. *Radio Science*, *32*, 2283–2307.
Woodman, R. F., & Guillen, A. (1974). Radar observations of winds and turbulence in the stratosphere and mesosphere. *Journal of the Atmospheric Sciences*, *31*, 493–505.

Chapter 20
Impact of Short-Term Solar Variability on the Polar Summer Mesopause and Noctilucent Clouds

Christian von Savigny, Charles Robert, Nabiz Rahpoe, Holger Winkler, Erich Becker, Heinrich Bovensmann, John P. Burrows, and Matthew T. DeLand

Abstract The Earth's middle atmosphere is affected by short-term solar variability in a variety of ways. This chapter focuses on the investigation of two different short-term solar effects on the polar summer mesopause region and on noctilucent clouds (NLCs). First, the effect of solar proton events (SPEs) on the thermal conditions near the polar summer mesopause and consequently on NLCs is discussed. An analysis of the SBUV(/2) time series to identify examples of NLC depletion due to SPEs shows that NLCs are probably frequently affected during strong SPEs. As part of this study a physical mechanism explaining a dynamically induced warming at the polar summer mesopause during and after SPEs is investigated using model simulations with the Kühlungsborn Mechanistic General Circulation Model (KMCM). A second aspect related to the effect of SPEs on NLCs is on the SPE-induced ion-chemical conversion of H_2O to HO_x, leading to a possible sublimation of NLCs. However, this effect was found to be of minor importance compared to the dynamically induced temperature effect. Second, we discuss the recently discovered 27-day

C. von Savigny (✉) · N. Rahpoe · H. Winkler · H. Bovensmann · J.P. Burrows
Institute of Environmental Physics, University of Bremen, Otto-Hahn-Allee 1, 28359 Bremen, Germany
e-mail: csavigny@iup.physik.uni-bremen.de

Present address:
C. von Savigny
Institute of Physics, Ernst-Moritz-Arndt-University of Greifswald, Felix-Hausdorff-Str. 6, 17487 Greifswald, Germany
e-mail: csavigny@physik.uni-greifswald.de

C. Robert
Belgium Institute of Space Aeronomy (BIRA), Brussels, Belgium
e-mail: crobert@bira.be

E. Becker
Institute of Atmospheric Physics, Schlossstr. 6, 18225 Kühlungsborn, Germany
e-mail: becker@iap-kborn.de

M.T. DeLand
Science Systems and Applications, Inc., Lanham, MD, USA
e-mail: matthew.deland@ssaihq.com

F.-J. Lübken (ed.), *Climate and Weather of the Sun-Earth System (CAWSES)*, Springer Atmospheric Sciences,
DOI 10.1007/978-94-007-4348-9_20,

solar cycle signature in NLCs, which was identified in SCIAMACHY as well as SBUV satellite observations of NLCs using cross correlation analysis and the superposed epoch method. NLC occurrence rate and albedo anomalies are anti-correlated with Lyman-α anomalies with a time-lag of 1 day at most. The sensitivities of the NLC albedo anomalies to Lyman-α forcing in terms of the 27-day and the 11-year solar cycle were found to agree within their uncertainties. This finding suggests that the same underlying physical mechanism drives the 27-day as well as the 11-year solar cycle signature in NLCs. The exact mechanism is still unknown, however.

20.1 Introduction

Noctilucent clouds (NLCs) or Polar mesospheric clouds (PMCs) are optically thin H_2O ice clouds occurring at mid and polar latitudes in the summer hemisphere around summer solstice. With an altitude of about 83 km they are the highest clouds in the terrestrial atmosphere.

Despite the fact that NLCs were first reported well over one hundred years ago [e.g. *Leslie*, 1885], they are still a major focus of upper atmosphere research. This is partly because NLCs are discussed as being early indicators of global change, as their growth and existence depends sensitively on the mesopause temperature field and on the ambient amount of H_2O. It has been argued that enhanced CO_2 concentrations—caused by burning of fossil fuels—will lead to increased radiative cooling in the middle atmosphere, where CO_2 is the major radiative cooling agent [e.g. *Roble and Dickinson*, 1989]. Interestingly, the change in radiative cooling due to increasing CO_2 abundances assumes a minimum at the polar summer mesopause—due to the low temperatures—and may even by negative [*Akmaev et al.*, 2006]. Recent model simulations with the LIMA (Leibniz-Institute Middle Atmosphere) model [*Lübken et al.*, 2009] show a temperature trend of -0.8 K/decade at 83 km and a latitude of 69°N between 1961 and the mid 1990s, when considering the northern summer period from June 6 to July 21 only.

Furthermore, anthropogenically elevated levels of atmospheric CH_4 may lead to an enhanced H_2O abundance in the middle atmosphere [*Thomas et al.*, 1989]. Both processes—increased radiative cooling and increasing H_2O abundances—may lead to more favorable conditions for the formation and existence of NLC particles. Long-term satellite observations with the nadir viewing SBUV(/2) instruments indeed show a long-term increase in the NLC albedo in the UV-C spectral range [*DeLand et al.*, 2003]—apart from a pronounced 11-year solar cycle variation. The LIMA model simulations by *Lübken et al.* [2009] are able to reproduce the long-term trend in NLC occurrence rate present in the SBUV NLC time series.

Many processes affecting the formation and existence of NLCs and their long-term variation are still not well understood. This chapter deals with the investigation of the effects of short-term solar variability on the polar summer mesopause region and on NLCs in particular. Two different aspects are discussed in detail. First, the effect of solar proton events (SPEs) on NLCs and on the thermal and dynamical structure of the upper atmosphere in general. Second, the effect of the 27-day solar rotational cycle on NLCs and temperatures in the middle atmosphere.

20.2 Brief Description of NLC Data Sets

20.2.1 NLC Observations with SCIAMACHY on Envisat

SCIAMACHY, the Scanning Imaging Absorption spectroMeter for Atmospheric CHartography [*Bovensmann et al.*, 1999] onboard Envisat was launched on March 1, 2002 into a polar sun-synchronous orbit with a descending node at 10:00 local time. SCIAMACHY essentially is an 8-channel grating spectrograph performing nadir, solar/lunar occultation as well as limb-scatter observations in the 220–2380 nm spectral range. NLCs are detected in SCIAMACHY limb radiance profiles in the UV spectral range, where they produce anomalous enhancements in limb radiance near 83 km tangent height. SCIAMACHY limb observations were extensively used in recent years to study for example variability in NLCs and to retrieve NLC particle sizes [e.g. *von Savigny et al.*, 2007a, 2009; *Robert et al.*, 2009, 2010].

20.2.2 NLC Observations with the SBUV Instrument Series

SBUV, the Solar Backscatter UltraViolet instrument was first deployed on the Nimbus-7 satellite in 1978, measuring backscattered atmospheric radiance and solar irradiance at 12 discrete wavelengths between 252 nm and 340 nm. Improved versions of the instrument (SBUV/2) were since flown on a series of NOAA satellites, providing a continuous data set since 1978. Although the main purpose of the SBUV(/2) observations is the retrieval of stratospheric ozone profiles and total ozone columns, the observations can also be used to identify NLCs when exploiting the small contribution of scattering by NLC particles to the total UV backscatter signal, as first demonstrated by *Thomas et al.* [1991]. More information on the algorithm employed to produce the SBUV NLC data set used in this study can be found in *DeLand et al.* [2003].

20.3 Effect of Solar Proton Events on NLCs

Energetic solar particles, particularly protons, electrons and α-particles are ejected by the sun during coronal mass ejections (CMEs). This particle stream may be directed towards Earth. The energetic and charged particles are then guided by the Earth's magnetic field lines and they can enter the terrestrial atmosphere in the polar cap regions at magnetic latitudes exceeding about 60° in both hemispheres. Depending on their energy, highly energetic protons can penetrate down to different altitudes in the atmosphere. Protons with energies of a few MeV will deposit most of their energy in the upper mesosphere, energies of 30 MeV are sufficient to reach the stratopause and energies of several hundred MeV are required to reach the lower stratosphere. The protons lead to ionization of the atmospheric constituents and the

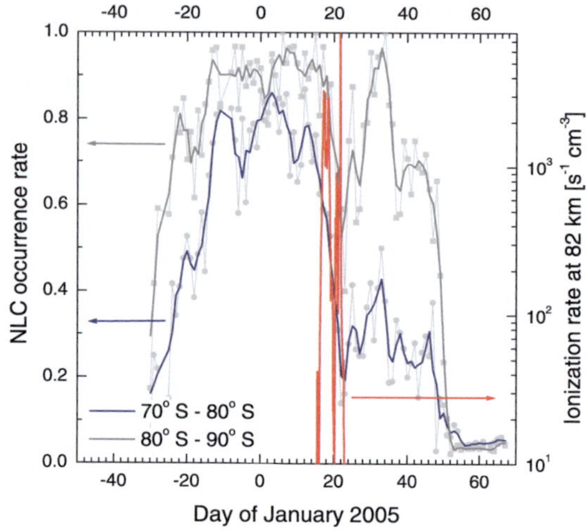

Fig. 20.1 NLC occurrence rate measured by SCIAMACHY/Envisat for two different latitude bins during the 2004/2005 NLC season in the southern hemisphere (*blue* and *gray lines*). The *red line* shows the calculated ionization rate due to precipitating solar protons at an altitude of 82 km. Figure adapted from *von Savigny et al.* [2007a]

resulting ion chemical reactions followed by neutralization reactions lead (a) to the conversion of H_2O to HO_x species [*Swider and Keneshea*, 1973; *Solomon et al.*, 1981], i.e., H and OH, and (b) to the formation of nitric oxide (NO) [*Crutzen et al.*, 1975], which is part of the NO_x (NO, NO_2) family. The formation of HO_x involves the production of intermediate H_2O cluster ions. Both families—HO_x and NO_x—participate in catalytic ozone destruction cycles, with HO_x being important above about 50 km and NO_x becoming important below about 50 km altitude. The effect of highly energetic solar particles on ozone in the middle atmosphere at polar latitudes is fairly well established, and model simulations show good qualitative and even quantitative agreement with observations [e.g. *Jackman et al.*, 2005; *Rohen et al.*, 2005].

However, the effects of solar energetic protons on the thermal and subsequently dynamical conditions of the middle atmosphere are less well understood. In this context *von Savigny et al.* [2007a] reported a rapid decrease in NLC occurrence rate based on SCIAMACHY limb scatter observations at southern polar latitudes during the January 2005 SPE (see Fig. 20.1). The NLC occurrence rate is defined as the ratio of the number of SCIAMACHY limb-scatter observations with detected NLC signatures and the total number of observations in a given latitude band. For the 70°S–80°S latitude band the NLC occurrence rate decreased from about 75 % during a period of 4 weeks before the SPE to about 15 % during the period with the enhanced ionization caused by precipitating solar protons (blue solid line in Fig. 20.1). For the 70°S–80°S latitude range the NLC occurrence rate does not reach the values before the SPE anymore. This is due to the fact that the temperature in this latitude range and around NLC altitude—despite some variability—remains at a higher level than before the SPE, as can be seen in Fig. 3 of *von Savigny et al.* [2007a].

The close temporal coincidence of the enhanced ionization and the decrease in NLC occurrence frequency suggests a causal relationship. Simultaneous observations of the polar summer mesopause temperatures carried out with the Microwave

Limb Sounder (MLS) on the Aura satellite showed an increase in zonally averaged temperatures of about 10 K during the SPE [*von Savigny et al.*, 2007a], which is sufficient to produce the observed effect in NLCs. MLS observations of H_2O in the southern polar summer mesopause region do not show any obvious signatures during the January 2005 SPE.

The obvious question is, what physical mechanism(s) may lead to the apparent increase in polar summer mesopause temperatures and the subsequent depletion of NLCs during the SPE. Two recent modeling studies [*Krivolutsky et al.*, 2006; *Jackman et al.*, 2007] suggest that dynamically induced heating—caused by reduced adiabatic cooling—in the polar summer mesopause region may lead to significant temperature perturbations during solar proton events. However, neither of these studies fully identified the underlying physical mechanism. The following hypothetical chain of effects was proposed by *Becker and von Savigny* [2010] to eventually cause the polar summer mesopause warming during solar proton events:

1. Catalytic destruction of ozone in the lower and middle polar mesosphere caused by SPE-produced HO_x. Catalytic ozone losses of up to 70 % between 50 and 70 km altitude are frequently observed during strong SPEs [e.g. *Jackman et al.*, 2005; *Rohen et al.*, 2005].
2. Reduced diabatic heating due to reduced absorption of solar radiation by ozone in the illuminated polar summer mesosphere. Diabatic heating rates associated with absorption of solar radiation by ozone in the summer middle atmosphere range from 1 K/day to about 10 K/day [*Mlynczak*, 1996].
3. The reduced diabatic heating in the polar mesosphere leads to a reduction in the meridional temperature gradient which drives the zonal wind according to the thermal wind balance.
4. The reduced zonal wind affects the filtering of gravity waves, and thereby the gravity wave driven meridional circulation, leading to reduction in the vertical motion in the polar summer mesopause region.
5. Reduced upwelling is associated with reduced adiabatic cooling, i.e., an apparent warming.

Step 1 has to be considered well understood; the same is valid for step 2. Interestingly, experimental verification of the reduced diabatic heating caused by SPE driven ozone loss is still missing. In order to investigate whether the dynamical part of the proposed chain of effects (steps 3–5) is correct, we used simulations with the Kühlungsborn Mechanistic General Circulation Model (KMCM) [*Becker*, 2009].

20.3.1 Model Simulations of the Dynamical Effects Associated with SPE Induced Ozone Depletion

In a time-slice sensitivity experiment with the KMCM, *Becker and von Savigny* [2010] have confirmed the aforementioned dynamical mechanism by which the diabatic cooling in the lower summer polar mesopause can lead to a strong dynamical warming of the mesopause region above. The sensitivity experiment was designed as follows. An arbitrary snapshot from a simulation for permanent January conditions was used as initial conditions and the model was integrated forward in time with an additional diabatic cooling of up to 2.5 K/day concentrated in the lower mesosphere above the summer pole. The SPE-induced chemical ozone loss in the mesosphere is driven by HO_x and is essentially limited to the period with enhanced ionization, because HO_x is quickly converted back to H_2O as soon as the anomalous ionization stops. Typical durations of SPEs are several days. After five model days, the diabatic cooling perturbation was switched off and the integration extended for another 10 days. The resulting 15-day time series then constituted an SPE-related perturbation simulation when compared to the corresponding 15-day time series of the unperturbed January simulation. This procedure was repeated 5 times for other arbitrarily chosen initial conditions. Averaging all day series for the unperturbed as well as the SPE cases then yielded statistically averaged composite time series.

Figure 20.2 shows the zonal and temporal mean differences between the SPE and the unperturbed case as colored contours. The black lines correspond to the unperturbed composite. Figure 20.2 clearly shows that an SPE-induced heating with a maximum around the polar summer mesopause is well simulated. Considering the temperature response together with that of the zonal wind and the gravity wave drag reveals the following mechanism: As a result of the temporary diabatic cooling in the lower polar mesosphere in the SPE composite, the averaged temperature response amounts to about −3 K in this region. The zonal wind response (Fig. 20.2b) shows the effect of this cooling by an increased vertical shear below 80 km and a resulting positive anomaly in zonal wind in the summer MLT up to 90 km. This gives rise to shorter vertical wavelengths of the eastward propagating gravity waves and induces dynamical instability at lower altitudes, as reflected by a downward shift of the gravity-wave drag (Fig. 20.2). *Becker and von Savigny* [2010] demonstrated that this downward shift is indeed mainly balanced by a downward shift of the residual circulation and that it therefore induces an anomalous large-scale dynamic warming around and below the mesopause. In this context it is also interesting to mention that radar wind observations during the July 2000 SPE show a change in zonal wind consistent with the mechanism proposed here (see Chap. 16).

20.3.2 Modeling Temperature and Ion Chemical Impacts of the January 2005 Solar Particle Event on Noctilucent Clouds

The growth and sublimation rates of icy NLC particles depend on the imbalance of the ambient water vapor pressure and the saturation pressure over the particles. The latter significantly increases with temperature. Therefore, it is reasonable to attribute the observed disappearance of NLCs after the solar particle event (SPE) in January 2005 to the observed temperature increase at the same time [*von Savigny et*

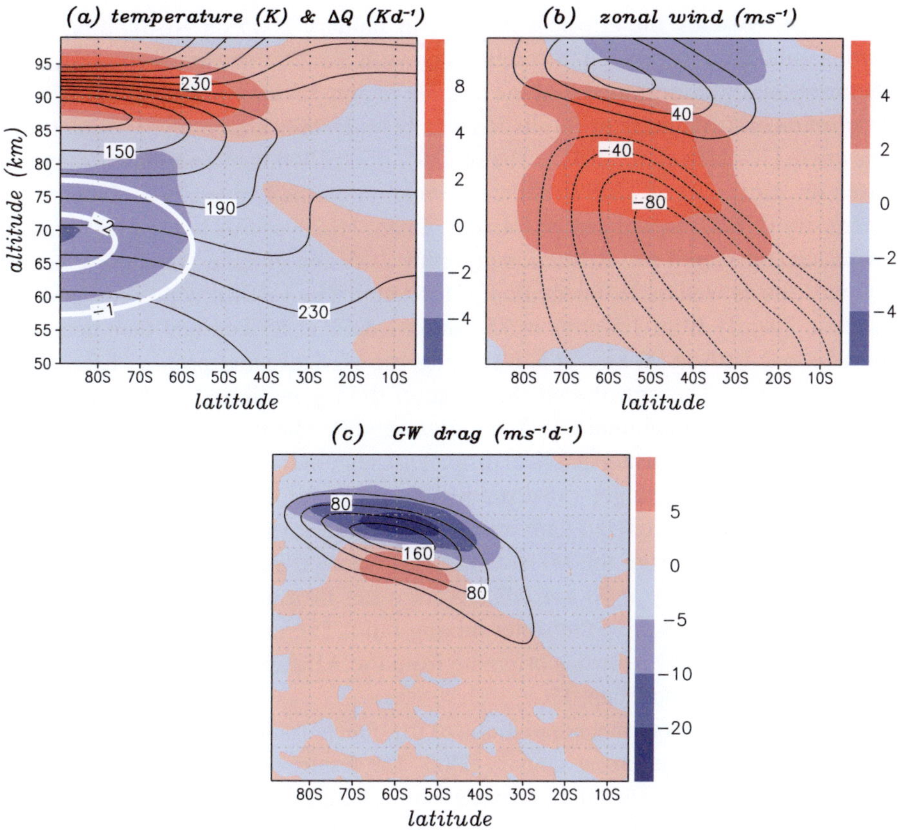

Fig. 20.2 Model response in the summer MLT (mesosphere/lower thermosphere) region to the SPE perturbation. Differences (*color shading*) in the SPE composite from the unperturbed January composite for (**a**) temperature, (**b**) zonal wind, and (**c**) gravity-wave drag. The control composite is indicated by *black contours* (interval 20 K, 20 $m\,s^{-1}$, and 40 $m\,s^{-1}d^{-1}$ with the zero contours omitted). Figure adapted from *Becker and von Savigny* [2010]

al., 2007a]. Besides the temperature impact on NLCs, there is an additional possible effect of SPEs: Reactions of positive water-cluster ions lead to a conversion of H_2O into H + OH [*Solomon et al.*, 1981]. This water depletion reduces the local water vapor pressure and in turn potentially affects the growth or sublimation of icy particles. However, while the abundance of water-cluster ions below ~80 km is large, the positive ion chemistry at higher altitudes is dominated by molecular ions such as O_2^+ and NO^+ [e.g. *Brasseur and Baets*, 1986]. The transition from cluster molecules to molecular ions takes place in the NLC altitude region [e.g. *Friedrich and Torkar*, 1988]. As a result, the SPE induced water depletion decreases drastically in that height region [*Solomon et al.*, 1981]. Therefore, an assessment of the SPE impacts on water and subsequently on NLCs requires a detailed consideration of the ion chemical processes.

In order to study the effect of the SPE in January 2005 on NLCs, a basic ice particle model has been used in combination with a one-dimensional chemistry, ion chemistry and transport model of the altitude region 75–90 km. Only a brief model description and summary of results is given here. Details are going to be published elsewhere [*Winkler et al.*, 2012]. The model's photochemistry and ion chemistry is based on *Winkler et al.* [2009], some information on the University of Bremen Ion Chemistry (UBIC) model can be found in Chap. 16.2.3 of this book. The transport processes considered are: Advection and eddy diffusion of trace gases and spherical icy particles as well as sedimentation of the latter. The vertical wind is assumed to increase linearly from 1.5 cm/s at 75 km to 4 cm/s at 85 km, and then decreases linearly to 2 cm/s at 90 km. This resembles the vertical wind profile of *Rapp and Thomas* [2006]. The growth and shrinkage of the icy particles is calculated by an approach similar to the model of *Toon et al.* [1988]. The nucleation rate of the icy particles is calculated using classical droplet nucleation theory [e.g. *Pruppacher and Klett*, 1997]. In order to get a realistic representation of the temperature fields, the model uses zonally averaged daily mean temperature measurements for the 70°S–80°S latitude range from the MLS/Aura instrument [*Schwarz et al.*, 2008]. At the lower model boundary at 70 km, MLS/Aura water measurement data [*Read et al.*, 2007] are used to prescribe the water mixing ratio. The ionization rates due to precipitating protons and electrons originate from the AIMOS model, see *Wissing and Kallenrode* [2009] and Chap. 13.

Figure 20.3 shows model simulations of the concentration of NLC particles with radii exceeding 24 nm during January 2005. The Figure demonstrates that the temperature increase after the solar particle event in January 2005 is the main driver for the depletion of NLCs. The simple particle model (dashed blue line) reproduces the observed drop in NLC abundance due to the temperature increase, although it overestimates the peak depletion. If the ion chemical depletion of water is considered (solid blue line), there is some additional reduction of NLCs. But this effect is of minor importance compared to the temperature impact on NLCs. The model simulations show that there is an immediate depletion of H_2O during the SPE at altitudes below the NLCs. At 80 km the H_2O decrease is about 1 ppmv on January 18, 2005. However, at altitudes above 80 km the positive ion chemistry is dominated by O_2^+ and NO^+ rather than the water ion clusters and the depletion of H_2O becomes inefficient. Due to the upward transport of H_2O, the water amounts at higher altitudes decrease with some delay, but compared to the impact of the temperature increase this is only a second-order effect.

Both model runs shown in Fig. 20.3 qualitatively reproduce the 2 minima also seen in the SCIAMACHY NLC occurrence rate around days 22 and 27, which appear to be related to 2 local maxima in the MLS temperature measurements. We note that the ionization rate time series at an altitude of 82 km shown in Fig. 20.1 and in *von Savigny et al.* [2007a] exhibits two local maxima. This led to the conclusion that the 2 minima in SCIAMACHY NLC occurrence rates may be related to the 2 maxima in the ionization rate time series.

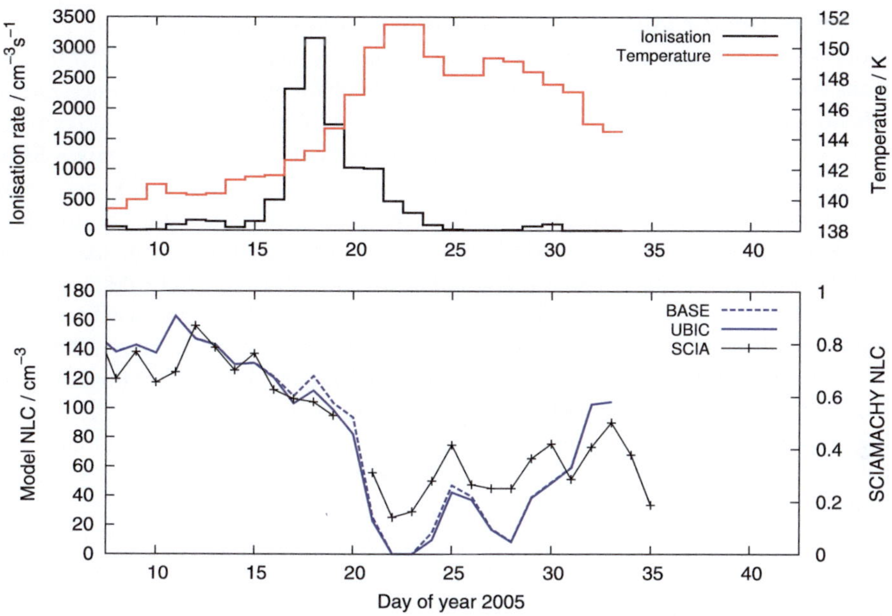

Fig. 20.3 Impact of the January 2005 solar particle event on noctilucent clouds in the southern polar region. *Upper panel*: Daily averaged mean MLS/Aura temperature for latitudes 70°–80°S and AIMOS ionization rates at 75°S. Both quantities are averaged over the 80–90 km altitude range. *Lower panel*: SCIAMACHY NLC occurrence rates (70°–80°S, *solid black line*), and modeled number density of NLC particles with radii larger than 24 nm between 80–90 km. The *dashed blue curve* shows the result of the NLC model without considering ion chemistry, and the *solid blue line* the simulation with the ion chemistry model UBIC. Figure adapted from *Winkler et al.* [2012]

20.3.3 Is NLC Depletion Frequently Observed During SPEs?

Rahpoe et al. [2011] recently investigated whether the NLC depletion observed during the January 2005 SPE is an unusual phenomenon, or whether similar effects are frequently observed during SPEs. For that purpose the SBUV(/2) NLC time series covering the period from 1978 to 2008 was used in combination with proton flux measurements with the SEM (Space Environment Monitor) instruments on the GOES (Geostationary Operational Environmental Satellites) series (GOES 5–GOES 12)—available since 1986. The NLC occurrence rate and albedo time series retrieved from SBUV(/2) measurements were correlated with the proton flux time series using SEM channels 2–7, covering proton energies between 5 MeV and about 100 MeV. These energies correspond to energy deposition altitudes ranging from about 75 km to about 30 km. For a detailed description of the analysis and the determination of the statistical significance we refer to *Rahpoe et al.* [2011]. Here we only summarize the main results of the analysis. The number of SPEs that occurred during the NLC seasons in the northern or southern hemisphere is quite limited, because (a) SPEs are rare phenomena, and (b) the core NLC seasons (here defined as ranging from day 0 to 40 with respect to solstice) cover only about 3 months per year. The analysis was limited to the core season in order to avoid spurious correlations caused by the typically rapidly changing NLC occurrence rate or albedo time series at the beginning and end of the NLC season. Since 1986 only 12 SPEs occurred during the core season in the northern hemisphere, and only 8 SPEs in the southern hemisphere. For the majority of these SPEs statistically significant negative correlations between the NLC and the proton flux time series at time lags of about 3–5 days were found, while only a few cases were found with statistically significant positive correlations (see *Rahpoe et al.* [2011] for more details). This asymmetry may be evidence—despite the high intrinsic variability in NLCs making the analysis of these effects difficult – that the effect of the January 2005 SPE on NLCs was not exceptional. We also note that for the July 2000 (Bastille day) SPE and the January 2005 SPE—which are by far the two strongest SPEs during the NLC core seasons analyzed—NLC occurrence rate reductions of about 50 % and about 70 % were observed, respectively, averaged over the 60°S–90 °S latitude range [*Rahpoe et al.*, 2011].

20.4 Evidence for a 27-Day Solar Cycle Signature in NLCs

The solar body continuously performs a differential rotation with a period of about 25 days near the poles and about 30 days at low heliospheric latitudes, leading to an effective rotational period of about 27 days. In the presence of sunspots this solar rotation can lead to measurable modulations of solar spectral irradiance, particularly in the UV spectral range. Essentially all solar activity proxies such as the Lyman-α flux, the MgII index (which is the core-to-wing ratio of the Mg^+ Fraunhofer-line at 280 nm in the solar irradiance spectrum) and the F10.7 radio flux exhibit periodic variations caused by the sun's differential rotation. Interestingly, the Lyman-α variations associated with strong 27-day cycles are of the same order of magnitude as the Lyman-α variation associated with the 11-year solar cycle.

Solar driven 27-day signatures have been identified in several middle and upper atmospheric parameters, notably in stratospheric ozone and temperatures [e.g. *Hood*, 1986; Dikty et al., 2010], thermospheric density [*Rhoden et al.*, 2000], and total electron content in the ionosphere [*Oinats et al.*, 2008]. Several studies also showed evidence for solar 27-day signatures in mesospheric and mesopause temperatures [*Keating et al.*, 1987; *Hood*, 1991]. *Höppner and Bittner* [2007] recently reported on a 27-day modulation of the fluctuations of OH*(3–1) rotational temperature measurements. An 11-year solar cycle signature in NLCs [*Thomas et al.*, 1991; *DeLand et al.*, 2003] and mesopause temperature [*Hervig and Siskind*, 2006]—not only at polar latitudes during summer [*Beig et al.*, 2008]—is well established, although the mechanisms leading to these signature are not well known.

During the DFG-CAWSES projects WAVE-NLC I and II, a 27-day signature in NLCs was identified for the first time, and was successfully demonstrated to be a consequence of a similar signature in polar summer mesopause temperatures

[*Robert et al.*, 2010]. In the following the main results of these investigations are summarized and new results regarding the sensitivity of NLC albedo changes to solar Lyman-α changes are presented.

The occurrence of a 27-day solar cycle signature in NLCs was found in SCIAMACHY NLC occurrence rate time series, as well as in SBUV(/2) measurements of NLC albedo and occurrence rate. The statistical analysis techniques applied to identify the solar signature included cross-correlation of NLC (albedo or occurrence rate) anomalies and Lyman-α anomalies, as well as the more sophisticated superposed epoch analysis method. The anomaly time series were determined by subtracting a 35-day running mean from the original time series. In the following only the results for the superposed epoch analysis will be summarized. More detailed information can be found in *Robert* [2009] and *Robert et al.* [2010].

20.4.1 Superposed Epoch Analysis

The superposed epoch analysis method first introduced by *Chree* [1912] is highly suited to identify weak signatures in time series affected by different sources of variability. The method requires the definition of so-called epochs in the response time series based on events in the forcing time series. In this application the Lyman-α anomaly time series is the forcing time series and the NLC occurrence rate or albedo anomaly time series are the response time series. The epochs are defined as 61-day periods around the maxima (or minima) of the Lyman-α anomaly time series, i.e., ranging from 30 days before until 30 days after the extremum in the Lyman-α anomaly time series. The basic idea of the technique is to average the corresponding data points of the response time series (e.g., day 2 with respect to the presumed forcing) for all available epochs. If a 27-day solar signature is indeed present in the NLC time series, it should emerge after averaging a sufficient number of epochs, while all other effects may be expected to be of random nature and will generally disappear. The analysis was applied to NLC occurrence rate time series measured with SCIAMACHY as well as to NLC albedo and occurrence rate times series observed with the SBUV(/2) instrument series, as described in *Robert et al.* [2010]. Figure 20.4 shows as an example the superposed epoch analysis result for the SBUV occurrence rate time series in the two hemispheres for both maxima and minima in the Lyman-α anomaly time series as forcing dates. The analysis is based on 66 epochs of northern hemispheric data and 62 epochs of southern hemispheric data. Apparently the averaged NLC anomalies show a pronounced minimum near a time lag of 0 days for the Lyman-α maximum analysis, as well as minima near $\pm$ 27 days. Furthermore, maxima in Lyman-α correspond to minima in the NLC anomaly—and vice versa—with a time lag of a few days at most. These findings are clear evidence for the existence of a 27-day signature in NLC occurrence rate, which is apparently driven by solar variability. Similar results were obtained for SCIAMACHY NLC occurrence rate as well as for SBUV NLC albedo time series.

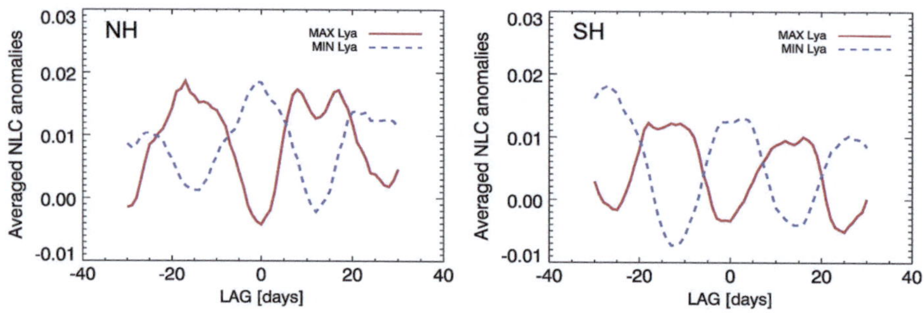

Fig. 20.4 Superposed epoch analysis of the SBUV NLC occurrence rate response to the Lyman-α forcing for the northern (*left panel*) and southern (*right panel*) hemisphere. NLC anomalies were calculated using zonally averaged NLC occurrence rates for the 60° to 80° latitude range. The epochs are centered around maxima (*red solid line*) and minima (*blue dashed line*) of the Lyman-α anomaly time series. Figure adapted from *Robert et al.* [2010]

The statistical significance of the results was tested with the modified bootstrapping method developed by *Ebisuzaki* [1997], which uses a statistical ensemble produced by adding a random phase in Fourier space followed by a Fourier transform back to physical space. The advantage of the method is that the randomly created representations of the time series have the same auto-correlation as the original time series, because adding a random phase does neither affect the power spectrum nor the auto-correlation function, according to the Wiener-Khintchin-Theorem. As shown in Table 20.1, the 27-day signatures identified in the different NLC time series are all highly significant, except for the SCIAMACHY NLC occurrence rate times series in the southern hemisphere, which is only significant at the 82 % confidence level.

The obvious question is what process is responsible for the 27-day solar cycle signature in NLC properties. Such a signature may be caused by a similar signature in polar summer mesopause temperatures or in the abundance of H_2O or by a combination of both. In order to investigate the causes further, *Robert et al.* [2010] used mesospheric temperature and H_2O profile measurements with the Microwave Limb Sounder (MLS) on NASA's Aura satellite. Both, temperature and H_2O abundance show a 27-day signature in the mesosphere, and the effects are particularly pronounced in the polar summer mesopause region [*Robert*, 2009]. However, the amplitude found in H_2O was only about 0.1 ppm at most, which is most likely

Table 20.1 Level of significance of the NLC response to the epochs around maxima and minima in the 27-day Lyman-α variation for different NLC data sets and both hemispheres

Data set	Northern hemisphere		Southern hemisphere	
	Min	Max	Min	Max
SCIAMACHY occurrence rate	97 %	96 %	82 %	98 %
SBUV occurrence rate	99.9 %	99.2 %	98.7 %	99.3 %
SBUV albedo	99.6 %	97.3 %	98.7 %	99.3 %

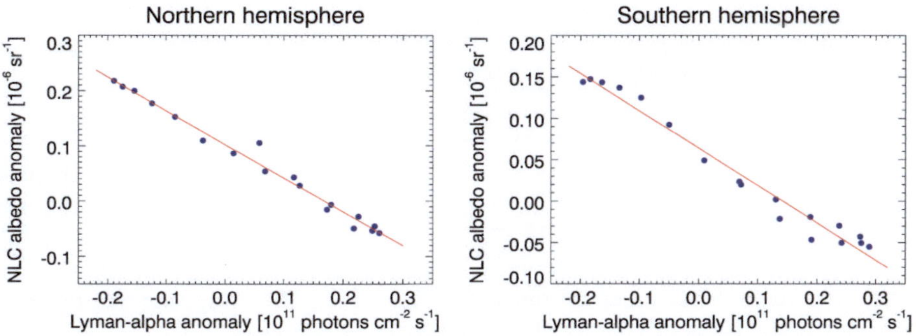

Fig. 20.5 Scatter plots of the SBUV albedo responses determined from the superposed epoch analysis as a function of the mean Lyman-α anomaly for the northern (*left panel*) and the southern (*right panel*) hemisphere. The correlation coefficients between the albedo and the Lyman-α anomalies are -0.992 for the northern and -0.974 for the southern hemisphere

not sufficient to explain the signature in NLC properties. The 27-day signature in temperature reaches amplitudes of up to 1 K, strongly indicating that the 27-day signature in summer mesopause temperatures is the main driver for the 27-day solar cycle signature in NLCs. MLS observations of temperature fluctuations near the polar summer mesopause on the order of a few K caused by the quasi 5-day wave were shown to cause significant variations in NLC occurrence rates observed with SCIAMACHY [*von Savigny et al.*, 2007b].

20.4.2 Sensitivity of NLC Anomalies to Solar Forcing

The average response of the NLC time series to the solar 27-day forcing can be investigated quantitatively using the results of the superposed epoch analysis presented above. Figure 20.5 shows scatter plots of the NLC albedo anomaly derived from SBUV observations and the Lyman-α anomaly for the time period of $\pm$ 8 days around the local maxima in Lyman-α. Apparently, the two anomalies are highly anticorrelated, which is not surprising given the result of the superposed epoch analysis shown in Fig. 20.4. The sensitivity of the NLC time series to the solar forcing is then determined as the slope of a linear function fitted to the points in Fig. 20.5, and it corresponds to the mean change in the NLC anomaly for a certain change in Lyman-α anomaly. The derived sensitivities for the SBUV NLC albedo time series are presented in Table 20.2 together with similar sensitivity parameters derived by *DeLand et al.* [2003] for the 11-year solar cycle. Surprisingly, the sensitivities of the NLC albedo for the 27-day and the 11-year solar cycle agree within the combined uncertainties. This remarkable finding may suggest that the same underlying physical mechanisms are responsible for the 27-day and the 11-year signature observed in NLCs.

Considering the similar sensitivities of NLC albedo to the 27-day and 11-year solar cycle variations in combination with the essentially negligible 27-day signature

Table 20.2 Sensitivity of the SBUV albedo to the 27-day and 11-year solar variation. Values of the sensitivities of the 11-year solar cycle were taken from *DeLand et al.* [2003] for corresponding latitudes. Sensitivity unit: 10^{-6} sr^{-1}/(10^{11} photons cm^{-2} s^{-1})

Hemisphere	27-day	11-year
North	-0.61 ± 0.11	-0.71 ± 0.02
South	-0.46 ± 0.12	-0.41 ± 0.02

in H_2O at the polar summer mesopause [*Robert et al.*, 2010] suggests that the temperature effect is the driving force for both solar signatures in NLCs.

Although the exact connections between the upper tropical stratosphere and the polar summer mesopause are not clear, *Fioletov* [2009] recently found that the sensitivities of tropical ozone in the upper stratosphere with respect to the 27-day and the 11-year solar cycle signature are similar as well. *Fioletov* [2009] used the ozone sensitivity for the 27-day solar cycle to model the amplitude of the 11-year solar cycle signature in tropical upper stratospheric ozone and found good agreement with experimental results based on SBUV(/2) stratospheric ozone profile retrievals. It is conceivable that the 27-day solar cycle signature in polar summer mesopause temperature is dynamically driven by the diabatic heating changes caused by the 27-day solar cycle signature in middle atmospheric ozone. A potential mechanism may be analogous to the dynamically induced warming of the polar summer mesopause during SPEs explained above: Ozone changes cause changes in the meridional temperature gradient, affecting the zonal wind according to the thermal wind balance; this in turn affects the propagation of gravity waves and ultimately the upper mesospheric meridional circulation along with the adiabatic cooling near the polar summer mesopause. The validity of this concept should be tested in the future. Preliminary results show that the sensitivity of middle atmospheric temperatures to Lyman-α forcing maximizes near the polar summer mesopause which may be an indication for a dynamically driven mechanism. In this context the superposed epoch analysis was also applied to the MLS temperature data set. Figure 20.6 shows the results of the superposed epoch analysis of the available MLS temperature measurements in the northern hemisphere during the northern hemisphere NLC seasons (left panel) together with the derived sensitivities (right panel) in $K/(10^{11}$ photons cm^{-2} s$^{-1})$. Evidently there is a clear correlation between the temperature and the Lyman-α anomalies with a time-lag near zero days in the upper mesosphere. The amplitude of the temperature variation is about 0.5 K. The sensitivity reaches its maximum values at the highest latitudes and altitudes between 80 and 90 km.

Other possible mechanisms contributing to a solar driven 27-day cycle in upper mesospheric temperatures may in principle be changes in diabatic solar heating due to absorption by ozone and O_2 in the upper mesosphere. However, in terms of O_3, *Hood et al.* [1991] found that ozone was negatively correlated to solar irradiance at 205 nm around 70 km altitude (essentially at all latitudes), while temperature was positively correlated to the irradiance at 205 nm. This implies that near 70 km the changes in diabatic heating due to the 27-day signature in ozone are unlikely the cause for the 27-day signature in temperature. Furthermore, heating by exothermic

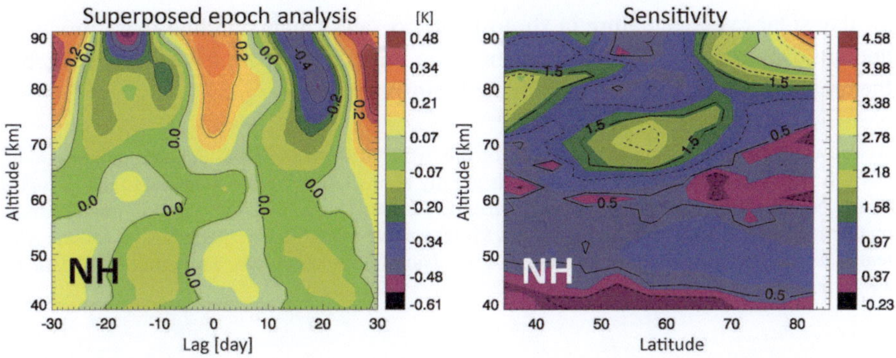

Fig. 20.6 Superposed epoch analysis of MLS temperature measurements in the northern hemisphere. *Left panel*: altitude-lag cross section of the temperature response to the solar forcing for latitudes between 60°N–80°N. *Right panel*: Sensitivity of the temperature response to the mean Lyman-α anomaly variation (unit: $K/(10^{11}\ \text{photons}\,\text{cm}^{-2}\,\text{s}^{-1})$)

chemical reactions (e.g. $H + O_3 \longrightarrow OH^* + O_2$) whose reactants show a 27-day signature may also contribute to a 27-day signature in temperature. However, particularly near the polar mesopause small changes in the mesospheric meridional circulation may be the main driver for the a 27-day solar cycle signature in temperature. Further research is required to improve the understanding of the relevant processes.

20.5 Conclusions

Two aspects related to the impact of short-term solar variability on Earth's upper mesosphere region and NLCs in particular were the main foci of this contribution. First, the effect of solar proton events on middle atmospheric temperature and dynamics in the summer hemisphere—and subsequently on NLCs—were investigated. A hypothetical chain of effects leading from reduced diabatic heating in the polar middle mesosphere to a dynamically induced warming near the polar summer mesopause was tested with the Kühlungsborn Mechanistic General Circulation Model (KMCM). The ion-chemical removal of H_2O at NLC altitude following the ionization by highly energetic solar protons during the SPE was found to have only a minor effect on NLCs. The second part of the study was devoted to a solar driven 27-day signature in NLCs, which is most likely caused by a similar signature in mesopause temperature. Interestingly, the sensitivities of the NLC albedo response (derived from SBUV observations) to the Lyman-α forcing have similar values for the 27-day and the 11-year solar cycle, suggesting similar mechanisms. However, the underlying mechanism leading to the 27-day signatures in polar summer mesopause temperatures is not yet established. The fact that temperature sensitivity with respect to Lyman-α forcing maximizes near the polar summer mesopause may suggest a dynamically driven effect which may be related to the well established

27-day solar cycle signature in stratospheric ozone. Understanding these processes and atmospheric coupling mechanisms is an important aspect of further research in the area of solar impacts on the middle atmosphere.

Acknowledgements Financial support for this work through the DFG Schwerpunktprogramm CAWSES (Projects WAVE-NLC I and II) is greatly appreciated. We are indebted to the European Space Agency (ESA) for providing SCIAMACHY Level 1 data. SCIAMACHY is jointly funded by Germany, the Netherlands and Belgium. Moreover, we would like to thank NASA and the MLS team—in particular Michael Schwartz (Jet Propulsion Laboratory)—for providing MLS data and assistance with the use of this data set.

References

Akmaev, R. A., Fomichev, V. I., & Zhu, X. (2006). Impact of middle-atmospheric composition changes on greenhouse cooling in the upper atmosphere. *Journal of Atmospheric and Solar-Terrestrial Physics*, *68*, 1879–1889.

Becker, E. (2009). Sensitivity of the upper mesosphere to the Lorenz energy cycle of the troposphere. *Journal of the Atmospheric Sciences*, *66*, 647–666.

Becker, E., & von Savigny, C. (2010). Dynamical heating of the polar summer mesopause induced by solar proton events. *Journal of Geophysical Research*, *115*, D00I18. doi:10.1029/2009JD012561.

Beig, G., Scheer, J., Mlynczak, M. G., & Keckhut, P. (2008). Overview of the temperature response in the mesosphere and lower thermosphere to solar activity. *Reviews of Geophysics*, *46*, RG3002. doi:10.1029/2007RG000236.

Bovensmann, H., Burrows, J. P., Buchwitz, M., Frerick, J., Noël, S., Rozanov, V. V., Chance, K. V., & Goede, A. P. H. (1999). Sciamachy: mission objectives and measurement modes. *Journal of the Atmospheric Sciences*, *56*, 127–150.

Brasseur, G., & Baets, P. D. (1986). Ions in the mesosphere and lower thermosphere: a two-dimensional model. *Journal of Geophysical Research*, *91*, 4025–4046.

Chree, C. (1912). Some phenomena of sunspots and of terrestrial magnetism at kew observatory. *Philosophical Transactions of the Royal Society of London*, *212*, 75–116.

Crutzen, J. P., Isaksen, I. S. A., & Reid, G. C. (1975). Solar proton events: stratospheric sources of nitric oxide. *Science*, *189*, 3179–3186.

DeLand, M. T., Shettle, E., Thomas, G., & Olivero, J. (2003). Solar backscattered ultraviolet (sbuv) observations of polar mesospheric clouds (pmcs) over two solar cycles. *Journal of Geophysical Research*, *108*, 8445. doi:10.1029/2002JD002398.

Dikty, S., Weber, M., von Savigny, C., Sonkaew, T., Rozanov, A., & Burrows, J. P. (2010). Modulations of the 27 day solar rotation signal in stratospheric ozone from scanning imaging absorption spectrometer for atmospheric cartography (sciamachy) (2003–2008). *Journal of Geophysical Research*, *115*, D00I15. doi:10.1029/2009JD012379.

Ebisuzaki, W. (1997). A method to estimate the statistical significance of a correlation when the data are serially correlated. *Journal of Climate*, *10*, 2147–2153.

Fioletov, V. E. (2009). Estimating the 27-day and 11-year solar cycle variations in tropical upper stratospheric ozone. *Journal of Geophysical Research*, *114*, D02302. doi:10.1029/2008JD010499.

Friedrich, M., & Torkar, K. M. (1988). Empirical transition heights of cluster ions. *Advances in Space Research*, *8*, 4235–4238.

Hervig, M., & Siskind, D. (2006). Decadal and inter-hemispheric variability in polar mesospheric clouds, water vapor, and temperature. *Journal of Atmospheric and Solar-Terrestrial Physics*, *68*, 30–41.

Hood, L. L. (1986). Coupled stratospheric ozone and temperature response to short-term changes in solar ultraviolet flux: an analysis of nimbus 7 sbuv and sams data. *Journal of Geophysical Research*, *91*, 5264–5276.

Hood, L. L., Huang, Z., & Bougher, S. W. (1991). Mesospheric effects of solar ultraviolet variations: further analysis of sme ir ozone and nimbus 7 sams temperature data. *Journal of Geophysical Research*, *96*, 12989–13002.

Höppner, K., & Bittner, M. (2007). Evidence for solar signals in the mesopause temperature variability? *Journal of Atmospheric and Solar-Terrestrial Physics*, *69*, 431–448.

Jackman, C. H., DeLand, M. T., Labow, G. J., Fleming, E. L., Weisenstein, D. K., Ko, M. K. W., Sinnhuber, M., & Russell, J. M. (2005). Neutral atmospheric influences of the solar proton events in October–November 2003. *Journal of Geophysical Research*, *110*, A09S27. doi:10.1029/2004JA010888.

Jackman, C. H., Roble, R. G., & Fleming, E. L. (2007). Mesospheric dynamical changes induced by solar proton events in October–November 2003. *Geophysical Research Letters*, *34*, L04812. doi:10.1029/2006GL028328.

Keating, G. M., Pitts, M. C., Brasseur, G., & de Rudder, A. (1987). Response of middle atmosphere to short-term solar ultraviolet variations: 1. Observations. *Journal of Geophysical Research*, *92*, 889–902.

Krivolutsky, A. A., Klyuchnikova, A. V., Zakharov, G. R., Vyushkova, T. Y., & Kuminov, A. A. (2006). Dynamical response of the middle atmosphere to solar proton event of July 2000: three-dimensional model simulations. *Advances in Space Research*, *37*, 1602–1613.

Leslie, R. C. (1885). Sky glows. *Nature*, *32*, 245.

Lübken, F.-J., Berger, U., & Baumgarten, G. (2009). Stratospheric and solar cycle effects on long-term variability of mesospheric ice clouds. *Journal of Geophysical Research*, *114*, D00106. doi:10.1029/2009JD012377.

Mlynczak, M. G. (1996). Energetics of the middle atmosphere: theory and observation requirements. *Advances in Space Research*, *17*, 117–126.

Oinats, A., Ratovsky, K., & Kotovich, G. (2008). Influence of the 27-day solar flux variations on the ionosphere parameters measured at Irkutsk in 2003–2005. *Advances in Space Research*, *42*, 639–644.

Pruppacher, H. R., & Klett, J. D. (1997). *Microphysics of clouds and precipitation*. Berlin: Springer.

Rahpoe, N., von Savigny, C., Robert, C. E., Deland, M. T., & Burrows, J. P. (2011). Impact of solar proton events on noctilucent clouds. *Journal of Atmospheric and Solar-Terrestrial Physics*, *73*, 2073–2081.

Rapp, M., & Thomas, G. E. (2006). Modeling the microphysics of mesospheric ice particles: assessment of current capabilities and basic sensitivities. *Journal of Atmospheric and Solar-Terrestrial Physics*, *68*, 715–744.

Read, W. G., et al. (2007). Aura microwave limb sounder upper tropospheric and lower stratospheric h2o and relative humidity with respect to ice validation *Journal of Geophysical Research*, *112*, D24S35. doi:10.1029/2007JD008752.

Rhoden, E., Forbes, J., & Marcos, F. (2000). The influence of geomagnetic and solar variabilities on lower thermosphere density. *Journal of Atmospheric and Solar-Terrestrial Physics*, *62*, 999–1013.

Robert, C. (2009). *Investigation of noctilucent cloud properties and their connection with solar activity*. Ph.D. thesis, Univ. Bremen.

Robert, C. E., von Savigny, C., Burrows, J. P., & Baumgarten, G. (2009). Climatology of noctilucent cloud radii and occurrence frequency using sciamachy. *Journal of Atmospheric and Solar-Terrestrial Physics*, *71*, 408–423.

Robert, C. E., von Savigny, C., Rahpoe, N., Bovensmann, H., Burrows, J. P., DeLand, M. T., & Schwartz, M. J. (2010). First evidence of a 27 day solar signature in noctilucent cloud occurrence frequency. *Journal of Geophysical Research*, *115*, D00I12. doi:10.1029/2009JD012359.

Roble, R. G., & Dickinson, R. E. (1989). How will changes in carbon dioxide and methane modify the mean structure of the mesosphere and thermosphere. *Geophysical Research Letters*, *16*,

1441. doi:10.1029/GL016i012p01441.

Rohen, G. J., von Savigny, C., Sinnhuber, M., Eichmann, K.-U., Kaiser, J. W., Llewellyn, E. J., Rozanov, A., Bovensmann, H., & Burrows, J. P. (2005). Impact of the October/November 2003 solar proton events on mesospheric ozone: sciamachy measurement and model results. *Journal of Geophysical Research*, *110*, A09S39. doi:10.1029/2004JA010984.

Schwartz, M., Lambert, A., Manney, G., Read, W., Livesey, N., Froidevaux, L., Ao, C., Bernath, P., Boone, C., Cofield, R., Daffer, W., Drouin, B., Fetzer, E., Fuller, R., Jarnot, R., Jiang, J., Jiang, Y., Knosp, B., Krüger, K., Li, J.-L., Mlynczak, M., Pawson, S., Russell III, J., Santee, M., Snyder, W., Stek, P., Thurstans, R., Tompkins, A., Wagner, P., Walker, K., Waters, J., & Wu, D. (2008). Validation of the aura microwave limb sounder temperature and geopotential height measurements. *Journal of Geophysical Research*, *113*, D15S11.

Solomon, S., Rusch, D. W., Gerard, J.-C., Reid, G. C., & Crutzen, P. J. (1981). The effect of particle precipitation events on the neutral and ion chemistry of the middle atmosphere ii: odd hydrogen. *Planetary and Space Science*, *29*, 885–892.

Swider, W., & Keneshea, T. J. (1973). Decrease of ozone and atomic oxygen in the lower mesosphere during a pca event. *Planetary and Space Science*, *21*, 1969.

Thomas, G. E., Olivero, J. J., Jensen, E. J., Schröder, W., & Toon, O. B. (1989). Relation between increasing methane and the presence of ice clouds at the mesopause. *Nature*, *338*, 490–492.

Thomas, G. E., McPeters, R. D., & Jensen, E. J. (1991). Satellite observations of polar mesospheric clouds by the solar backscattered ultraviolet spectral radiometer: evidence of a solar cycle dependence. *Journal of Geophysical Research*, *96*, 927–939.

Toon, O. B., Turco, R. P., Westphal, D., Malone, R., & Liu, M. S. (1988). A multidimensional model for aerosols: description of computational analogs. *Journal of the Atmospheric Sciences*, *45*, 2123–2143.

von Savigny, C., Sinnhuber, M., Bovensmann, H., Burrows, J., Kallenrode, M.-B., & Schwartz, M. (2007a). On the disappearance of noctilucent clouds during the January 2005 solar proton events. *Geophysical Research Letters*, *34*, L02805. doi:10.1029/2006GL028106.

von Savigny, C., Robert, C., Bovensmann, H., Burrows, J. P., & Schwartz, M. (2007b). Satellite observations of the quasi 5-day wave in noctilucent clouds and mesopause temperatures. *Geophysical Research Letters*, *34*, L24808. doi:10.1029/2007GL030987.

von Savigny, C., Robert, C., Baumgarten, G., Bovensmann, H., & Burrows, J. P. (2009). Comparison of nlc particle sizes derived from sciamachy/envisat observations with ground-based lidar measurements at alomar (69 n). *Atmospheric Measurement Techniques*, *2*, 523–531.

Winkler, H., Kazeminejad, S., Sinnhuber, M., Kallenrode, M.-B., & Notholt, J. (2009). The conversion of mesospheric hcl into active chlorine during the solar proton event in July 2000 in the northern polar region. *Journal of Geophysical Research*, *114*, D00I03. doi:10.1029/2008JD011587.

Winkler, H., von Savigny, C., Burrows, J. P., Wissing, J. M., Schwartz, M. J., Lambert, A., & Garcia-Comas, M. (2012). Impacts of the January 2005 solar particle event on noctilucent clouds and water at the polar summer mesopause. *Atmospheric Chemistry and Physics*, *12*, 5633–5646. doi:10.5194/acp-12-5633-2012.

Wissing, J.-M., & Kallenrode, M.-B. (2009). Atmospheric ionisation module osnabrück (aimos): a 3-d model to determine atmospheric ionization by energetic charged particles from different populations. *Journal of Geophysical Research*, *114*, A06104. doi:10.1029/2008JA013884.

Chapter 21
Observations and Ray Tracing of Gravity Waves: Implications for Global Modeling

Manfred Ern, Christina Arras, Antonia Faber, Kristina Fröhlich, Christoph Jacobi, Silvio Kalisch, Marc Krebsbach, Peter Preusse, Torsten Schmidt, and Jens Wickert

Abstract Vertical coupling by atmospheric waves is essential for the wind and temperature structure of the middle atmosphere. In particular, momentum carried by atmospheric gravity waves (GWs) governs the global circulation in the mesosphere and is for instance the reason for the cold summer mesopause. However, the small horizontal scales of GWs (tens to thousands of km) are challenging both global modeling and observations from satellite. Further, due to the small scales involved, there is a severe lack of understanding about GWs themselves, as well as dynamical phenomena involving GWs. Until recently, global observations of GWs were sparse and little was known about the global distribution of GWs, as well as their seasonal variation. Therefore, several projects in the priority program Climate And Weather of the Sun-Earth System (CAWSES) of the Deutsche Forschungsgemeinschaft (DFG) have addressed a number of the most pressing problems. Global distributions of GW activity and momentum fluxes have been derived from observations with number of satellite instruments, resulting in the first multi-year global data sets of GW parameters, covering time scales from seasonal variations up to the duration of almost a full 11-year solar cycle. In addition, seasonal and tidal variations of sporadic E layers in the ionosphere were studied in Global Positioning System (GPS) radio occultation data. Satellite observations of GWs and sporadic E layers were complemented by

M. Ern (✉) · S. Kalisch · P. Preusse
Institute of Energy and Climate Research (IEK-7), Forschungszentrum Jülich, 52425 Jülich, Germany
e-mail: m.ern@fz-juelich.de

C. Arras · A. Faber · T. Schmidt · J. Wickert
Dept. Geodesy and Remote Sensing, German Research Centre for Geosciences GFZ, Potsdam, 14473 Potsdam, Germany

K. Fröhlich
Dept. Climate and Environment, Deutscher Wetterdienst, 63067 Offenbach, Germany

C. Jacobi
Institute for Meteorology, University of Leipzig, 04103 Leipzig, Germany

M. Krebsbach
Physics Department, University of Wuppertal, 42119 Wuppertal, Germany

F.-J. Lübken (ed.), *Climate and Weather of the Sun-Earth System (CAWSES)*, Springer Atmospheric Sciences,
DOI 10.1007/978-94-007-4348-9_21,

ground-based observations (radar and low-frequency (LF) drift measurements). All these observations, as well as accompanying modeling activities provided important constraints for GW parameterizations. Further activities addressed important aspects of GW propagation usually neglected in global modeling: GW ray tracing studies revealed the importance of non-vertical propagation of GWs and first steps were undertaken to develop an improved GW parameterization based on GW ray tracing techniques.

21.1 Introduction

Vertical coupling by atmospheric waves is one of the main ingredients in atmospheric dynamics. Fundamental understanding of this coupling is required for global modeling of the atmospheric circulation and other atmospheric processes. Global scale waves are resolved by the model grid of general circulation models (GCMs), and their effect can therefore explicitly be modeled. Gravity waves (GWs), on the other hand, are mesoscale waves with horizontal scales between tens and several thousand kilometers, i.e., scales that are only partly resolved by most GCMs. Therefore the effect of GWs on the global circulation has to be parameterized by gravity wave drag (GWD) schemes. Usually at least two different schemes are used. One covers the effect of GWs generated by large-scale flow over mountains (orographic GWD scheme), and another one covers the effect of GWs generated by all other source processes (nonorographic GWD scheme).

It is generally assumed that most GW sources are located in the troposphere, GWs propagate upward, and deposit energy and momentum at higher atmospheric levels where they break and interact with the background flow. This is why GWs are an important vertical coupling mechanism, redistributing energy and momentum between different altitude levels and locations in the atmosphere.

GWs play a significant role in atmospheric dynamics in the stratosphere, mesosphere and also at even higher altitudes. For example, the quasi-biennial oscillation (QBO) of the zonal wind in the tropical stratosphere is mainly driven by GWs [e.g., *Dunkerton*, 1997; *Ern and Preusse*, 2009a, 2009b]. The QBO influences the stratospheric polar night jets [*Holton and Tan*, 1980] and by downward coupling has also effect on the weather in the troposphere [e.g., *Baldwin and Dunkerton*, 2001; *Marshall and Scaife*, 2009]. GWs are also important for the Brewer Dobson (BD) circulation in the stratosphere, which is directed upward in the tropics, towards the poles at midlatitudes, and downward over the poles. Recent simulations with a number of GCMs predict an acceleration of the BD circulation in a warming climate [e.g., *Garcia and Randel*, 2008; *Li et al.*, 2008]. There are indications that large part of this possible acceleration is related to mountain waves (i.e., GWs excited by large-scale flow over mountains) [*McLandress and Shepherd*, 2009]. This implies that realistic GW source distributions are needed in climate models, including both spatial and temporal variations, to enable GCMs to provide more reliable climate predictions, potentially even on regional scales.

In the mesosphere GWs are even more important. They drive the residual mesospheric circulation from summer to winter pole, strongly influence the temperature structure in the mesosphere and, via gradient wind balance, the mean zonal wind. In particular, they are responsible for the cold summer mesopause. See also, for example, *Holton* [1982, 1983].

Both global-scale waves and GWs are possible sources for the formation of ionospheric irregularities. For example, the importance of GWs [e.g., *Hocke and Tsuda*, 2001], as well as semidiurnal tides [*Arras et al.*, 2009], has been pointed out for the formation of sporadic E layers around 100 km altitude. These irregularities are one of the error sources of GPS positioning systems. Therefore also the study of sporadic E layers is of great interest.

The key quantity relevant for the interaction of GWs with the atmospheric background wind is the vertical flux of horizontal wave momentum due to GWs:

$$(F_{px}, F_{py}) = \rho(\overline{u'w'}, \overline{v'w'}). \tag{21.1}$$

F_{px} and F_{py} are the vertical flux of zonal and meridional momentum, ρ is the atmospheric density, (u, v, w) the atmospheric background wind vector with u the zonal, v the meridional and w the vertical component, and (u', v', w') the vector of wind perturbations due to GWs. The overbar indicates temporal and/or spatial averaging. The acceleration of the background flow due to GWs is given by the vertical gradient of this momentum flux:

$$(X, Y) = -\frac{1}{\rho}\frac{\partial(F_{px}, F_{py})}{\partial z} \tag{21.2}$$

with X the zonal and Y the meridional acceleration. A more detailed discussion is given, for example, in *Fritts and Alexander* [2003].

Since the 1980s most middle atmosphere GCMs agree on the strength of GWD in the mesosphere/lower thermosphere (MLT) and also on the strength of the GW-driven summer-to-winter-pole circulation in the mesosphere (leading, for example, to reasonable mean temperature, zonal wind, and traveling planetary waves). This was achieved by combining the theory by *Lindzen* [1981] with the TEM picture of the general circulation and the data sets combined in the CIRA climatology [*Fleming et al.*, 1990]. In addition, observations at single measurement stations provided first direct constraints for the representation of GWs in GCMs [e.g., *Lübken*, 1997; *Rauthe et al.*, 2006; *Hoffmann et al.*, 2010]. Further improvements in the representation of GWs in GCMs can be expected if direct observations of the global distribution of GWs from satellite are also considered.

Nonorographic GWD schemes in GCMs [e.g., *Hines*, 1997a, 1997b; *Warner and McIntyre*, 2001] usually launch GW momentum flux from a given fixed source level. At this level source parameters have to be prescribed, for example, the momentum flux amplitude, characteristic horizontal and/or vertical wavelengths, and the shape of the GW spectrum. However, there are only very few global observations that provide useful constraints for GWD schemes. In particular, until recently, global observations of GW momentum fluxes that can directly be compared to fluxes produced by GWD schemes were completely missing. Therefore, source parameters of GWD schemes often are guessed or tuned for the desired model response, the same GW source parameters are used for the whole globe (independently of geolocation and direction of the source momentum flux), and seasonal variations of the GW sources are neglected. Recently, this situation has however started to improve: Some attempts have been made to couple nonorographic GW momentum fluxes at the source altitude to physical processes, like GW generation by fronts [e.g., *Charron and Manzini* 2002; *Richter et al.*, 2010], or GW excitation by convection in the tropics and subtropics [e.g., *Song and Chun*, 2008; *Richter et al.*, 2010].

Not all aspects of GW propagation are covered by standard GWD schemes. For example, in most approaches GW propagation is considered only in a vertical column. Thereby the horizontal redistribution of GW momentum with altitude, as well as the horizontal refraction of waves, are neglected. Another assumption is that GWs propagate quasi-instantaneously from their sources to the regions where they dissipate. These idealizations could be relaxed, for example, by ray tracing of GWs.

To provide global experimental data sets of GW parameters and to investigate the effect of simplifications made in standard GWD schemes, the three projects GW-CODE, GW-EXCITES and GRAPES were proposed in the priority program Climate And Weather of the Sun-Earth System (CAWSES) of the Deutsche Forschungsgemeinschaft (DFG).

GW-CODE:

The project GW-CODE was a cooperation of the three institutes Forschungszentrum Jülich (FZJ), Geoforschungszentrum Potsdam (GFZ), and the Institute of Meteorology at the University of Leipzig (LIM).

One part of the project GW-CODE provided global long-term data sets of GW potential energies from Global Positioning System (GPS) radio occultations in the stratosphere, as well as global climatologies of sporadic E layers in the altitude region around 100 km (GW-CODE/GFZ). In addition, GW temperature variances were derived from Sounding of the Atmosphere using Broadband Emission Radiometry (SABER) satellite data and screened for seasonal variations (GW-CODE/FZJ). Further, a long-term data set of GW wind variances was derived in the MLT region from low-frequency (LF) electron drift measurements and winds and temperatures were derived from meteor radar observations (GW-CODE/LIM).

Another part of the project GW-CODE was focused on GW modeling. Based on global satellite observations of GW momentum fluxes derived from Cryogenic Infrared Spectrometers and Telescopes for the Atmosphere (CRISTA) temperature data, constraints were deduced for the source distribution of the *Warner and McIntyre* [2001] (WM01) GWD scheme (GW-CODE/FZJ). The WM01 GWD scheme was implemented in the simple GCM COMMA-LIM (GW-CODE/LIM). Further, in COMMA-LIM simulations for solar maximum and solar minimum conditions of the 11-year solar cycle, possible differences in the propagation of GWs and planetary-scale waves were investigated (GW-CODE/LIM).

In a third part of the project, ray tracing simulations were carried out to model the global distribution of GWs, based on the experiences with the

WM01 GWD scheme and global SABER temperature observations. A global GW launch distribution was derived, and weaknesses of conventional GWD schemes were studied by GW ray tracing (GW-CODE/FZJ).

GRAPES:

The GW ray tracing study was continued in the DFG project GRAPES. In this project the GW ray tracing approach is used to provide GW drag for comparison with observations and GCM results.

GW-EXCITES:

The work to provide experimental constraints for GWD schemes was continued. In the DFG project GW-EXCITES, GW momentum fluxes were derived from the newly available three years (2005–2007) of High Resolution Dynamics Limb Sounder (HIRDLS) high spatial resolution satellite data. During the project it turned out that this was also possible from the nine years of SABER data (2002–2010) available to date. These data sets now provide information about seasonal and longer-term variations of GW parameters, in particular GW momentum fluxes, up to time scales of almost a full 11-year solar cycle in both strato- and mesosphere.

The remainder of this paper is structured as follows. In Sect. 21.2 we briefly introduce the experimental data sets used in our studies. In Sect. 21.3 we discuss observed seasonal and in Sect. 21.4 observed longer-term variations. Constraints for standard GWD schemes derived from the observations are discussed in Sect. 21.5. Section 21.6 summarizes the results obtained from global modeling with the GCM COMMA-LIM. GW ray tracing simulations showing the importance of effects of GW propagation that are neglected in standard GWD schemes, as well as first steps toward the development of a new GWD scheme based on GW ray tracing techniques are presented in Sect. 21.7. Finally, a summary is given in Sect. 21.8.

21.2 Experimental Datasets

For comparison with global modeling by GCMs it is required to provide global observations of GWs over a large altitude range. Coverage is needed from the troposphere, where most GWs sources are located, up to the mesosphere and lower thermosphere, where GWs are the main driver of the global circulation. Further, temporal variations from seasonal changes up to timescales of the 11-year cycle of solar activity have to be covered to obtain information about GW source variations, as well as variations in the interaction between GWs and the mean atmospheric circulation. Observations should ideally cover a broad range of the GW spectrum, i.e. ranges of horizontal and vertical wavelengths as broad as possible. Of course, no single observing system is capable to fulfill all these requirements [e.g., *Alexander et al.*, 2010]. Therefore a combination of both ground- and space-based measurement techniques is required. The latter must consist of different instruments. The data sets used in our studies are briefly discussed in the following.

21.2.1 GPS Radio Occultations

High precision altitude profiles of atmospheric temperatures were derived from data of GPS receivers onboard satellites in low Earth orbit (GW-CODE/GFZ). These altitude profiles are measured when one of the transmitter satellites of the GPS system is rising or setting at the Earth's horizon as seen from a receiver satellite. This measurement method is called GPS radio occultation. Temperatures are derived in the altitude range between about 5 and 35 km with an excellent vertical resolution of about 1 km. GPS radio occultations are not affected by clouds and therefore they are one of the few data sets providing global temperature observations in the troposphere under all weather conditions. In addition, ionospheric irregularities (layers of enhanced ion density, so called sporadic E layers) can be detected at altitudes 80–120 km.

Radio occultations of the German Challenging Minisatellite Payload (CHAMP) satellite are the first long-term GPS radio occultation data set, starting from May 2001 with about 150–200 measured altitude profiles per day. The CHAMP mission ended on 19 September 2010 when the satellite re-entered the Earth's atmosphere. Also data from other GPS satellites were used in our studies. In particular, since April 2006 the six-satellite constellation COSMIC provides a much higher number of about 2000 radio occultations per day. For more details see also *Wickert et al.* [2006, 2009] (GW-CODE/GFZ).

21.2.2 Infrared Limb Sounding

Another space-based measurement technique is infrared (IR) limb sounding from satellites in low Earth orbit. High vertical resolution temperature altitude profiles can be derived from atmospheric IR emissions. In our studies we use temperature observations of three IR limb sounders: CRISTA, HIRDLS and SABER.

During two one-week missions in November 1994 and in August 1997 CRISTA measured high spatial resolution global temperature distributions in the stratosphere using three viewing directions simultaneously [e.g., *Riese et al.*, 1999]. The HIRDLS instrument onboard the EOS-Aura satellite measured global temperature distributions in the stratosphere with very short distances of about 100 km between subsequent altitude profiles along the measurement track. These short along-track distances are very useful for the study of mesoscale atmospheric structures like GWs. HIRDLS data are available from January 2005 until December 2007. The SABER instrument onboard the TIMED satellite provides global temperature measurements over a very large altitude range of about 15–125 km, i.e., in the stratosphere, mesosphere and lower thermosphere. SABER observations start in January 2002 and are still ongoing at the time of writing. The vertical resolution of measured temperature altitude profiles is $\sim$1 km for HIRDLS, $\sim$2 km for SABER, and $\sim$3 km for CRISTA.

All three instruments provide observations with short distances between subsequent altitude profiles along the measurement track. From single altitude profiles only GW amplitudes and vertical wavelengths can be determined. But, given the short-distance along-track sampling of all three instruments, also horizontal wavelengths and GW momentum fluxes can be estimated for pairs of altitude profiles [*Ern et al.*, 2004]. There is however no information about the true propagation direction of an observed GW, and derived momentum fluxes are only absolute values, i.e., no directional information is available [see also *Preusse et al.*, 2009a]. In particular, GW amplitudes and vertical wavelengths were derived from SABER data in GW-CODE/FZJ and GW amplitudes, vertical and horizontal wavelengths, as well as GW momentum fluxes from SABER and HIRDLS data in GW-EXCITES (see also below).

21.2.3 Ground-Based Observations

No satellite data set suited for the analysis of GWs is long enough to cover more than about one 11-year solar cycle. Longer data sets are only available from ground-based observations. One of those longer data sets are low-frequency drift measurements at Collm (51.3°N, 13.0°E), covering more than two 11-year solar cycles. In the altitude range 80–95 km zonal and meridional winds as well as wind fluctuations due to GWs are derived from daily radio drift and reflection height measurements using the ionospheric reflection of the sky wave of commercial radio transmitters (location of the reflection point: 52°N, 15°E).

For more details see *Jacobi et al.* [2006] (GW-CODE/LIM). At the same measurement location, wind and temperatures in the altitude range of about 80–100 km are also obtained from meteor radar observations [*Stober et al.*, 2008 (GW-CODE/LIM)].

21.3 Observed Seasonal Variations of Gravity Wave Activity

Until recently, long-term data sets providing information about the global distribution of GWs were not available. Therefore one of the main aims in our studies was to deduce seasonal variations of the global GW distribution. These observations could be used in the future to improve the source distributions of GWD schemes used in GCMs. The main findings are summarized in the following.

21.3.1 GPS Observations of Sporadic E Layers

Both global-scale waves and GWs are believed to play an important role in the formation of sporadic E layers. Therefore, seasonal variations in sporadic E occurrence frequencies might provide information about differences in the atmospheric

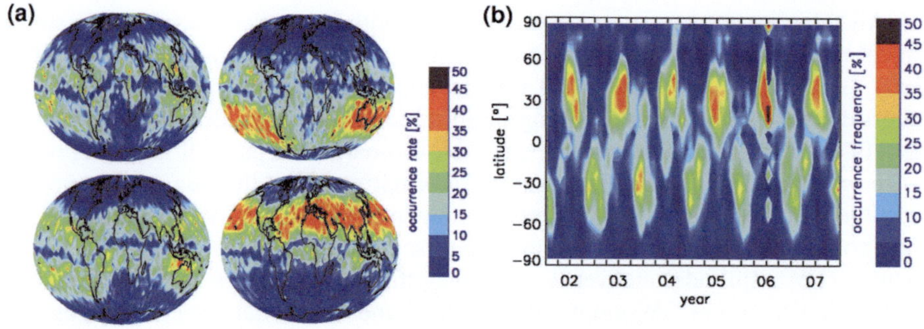

Fig. 21.1 Sporadic E occurrence frequencies derived from GPS radio occultations. (**a**) Occurrence frequencies derived from CHAMP, GRACE and COSMIC radio occultations for the months Sep.–Nov. 2006 (*upper left*), Dec.–Feb. 2006/2007 (*upper right*), March–May 2007 (*lower left*), and June–Aug. 2007 (*lower right*). (**b**) Latitude-time cross section of sporadic E zonally averaged occurrence frequencies measured by CHAMP from January 2002 until December 2007. Fig. 21.1a was adapted from Fig. 3 in *Arras et al.* [2008], Fig. 21.1b from Fig. 2 in *Arras et al.* [2008] (GW-CODE/GFZ). Reproduced/modified from "Arras, C., J. Wickert, G. Beyerle, S. Heise, T. Schmidt, and C. Jacobi, A global climatology of ionospheric irregularities derived from GPS radio occultation, Geophys. Res. Lett., 35, L14809, 2008." Copyright [2008] American Geophysical Union. Reproduced/modified by permission of American Geophysical Union

propagation conditions for GWs. From CHAMP and COSMIC radio occultations it has been found that sporadic E layers exhibit a pronounced seasonal cycle with high occurrence frequencies in the summer hemisphere [*Arras et al.*, 2008 (GW-CODE/GFZ)]. This is illustrated in Fig. 21.1. Particularly high occurrence frequencies are found in the northern hemisphere during June–August and in the southern hemisphere during December–February (see Fig. 21.1a). In a follow-up study it has been found that the formation of sporadic E layers is governed by atmospheric tides rather than by GWs [*Arras et al.*, 2009 (GW-CODE/GFZ)]. Further studies about the role of GWs in the formation of sporadic E layers are however required.

21.3.2 Gravity Wave Potential Energies and Momentum Fluxes Derived from GPS Radio Occultations

Observations of GWs in the lower stratosphere close to levels where they are excited are very important for constraining GW parameterizations. For the identification of GW-related temperature variations in temperature altitude profiles usually temperature variations over an altitude range (e.g. 10 km) are investigated. In the tropopause region, temperature altitude profiles are however dominated by the strong temperature variations related to the tropopause, and the separation of GW-related temperature variations from background temperature profiles is complicated. Therefore several studies were carried out: A climatology of multiple tropopauses was derived by *Schmidt et al.* [2006] (GW-CODE/GFZ), and potential errors in GW potential energies observed by GPS radio occultation were studied by *de la Torre et al.* [2010] (GW-CODE/GFZ). In addition, *Schmidt et al.* [2008] showed that it is possible to

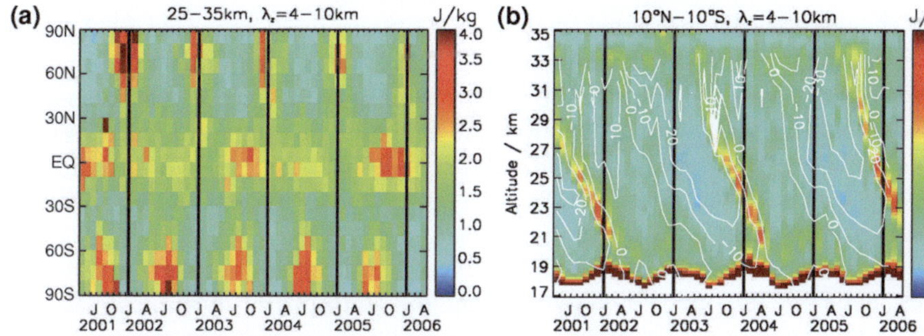

Fig. 21.2 GW potential energies derived from GPS radio occultations for vertical wavelengths of 4–10 km. (**a**) Latitude/time cross section of potential energies in the altitude interval 25–35 km, (**b**) altitude/time cross section of GW potential energies averaged over the latitude band 10°S–10°N. *White contour lines* in (**b**) indicate the zonally averaged zonal wind in m/s obtained in the latitude band 10°S–10°N from the global radiosonde network. Figure 21.2a was adapted from Fig. 1f, and Fig. 21.2b from Fig. 2a in *de la Torre et al.* [2006a] (GW-CODE/GFZ). Reproduced/modified from "de la Torre, A., T. Schmidt, and J. Wickert, A global analysis of wave potential energy in the lower stratosphere derived from 5 years of GPS radio occultation data with CHAMP, Geophys. Res. Lett., 33, L24809, 2006." Copyright [2006] American Geophysical Union. Reproduced/modified by permission of American Geophysical Union

derive GW potential energies from GPS radio occultations even in the tropopause region.

The first multi-year data set of GW potential energies from GPS radio occultations was derived by *de la Torre et al.* [2006a, 2006b (GW-CODE/GFZ)]. Figure 21.2a shows a latitude/time cross section of GW potential energies derived from CHAMP radio occultations for the altitude range 25–35 km and GWs with vertical wavelengths between 4 and 10 km. High potential energies are found at high latitudes during the winter season of the respective hemisphere. In addition, high potential energies are found in the tropics. The potential energies in the tropics exhibit a quasi-biennial variation rather than a strong seasonality.

It is known for a long time that the QBO of the zonal wind in the equatorial region is a wave-driven circulation [*Lindzen and Holton*, 1968; *Holton and Lindzen*, 1972] and that GWs are one of the main drivers [*Dunkerton*, 1997]. In Fig. 21.2a the observed variations of GW potential energies in the tropics reflect this strong interaction of GWs with the QBO. This is further illustrated in Fig. 21.2b in an altitude-time cross section of GW potential energies averaged over the latitude band 10°S–10°N. The white zonal wind contour lines highlight the close relationship between GWs and equatorial zonal wind: Enhanced GW potential energies are observed particularly during the QBO transition from westward to eastward winds, and not during the reverse transition. This might be related to the fact that zonal wind vertical gradients are much stronger during the westward-eastward transition, causing large part of the GW spectrum to dissipate at about the same altitude. In addition, differences in the phase speed spectrum between eastward and westward propagating GWs could also play an important role.

It should however be noted that these GPS analyses use a vertical wavelength separation (vertical wavenumber high-pass filtering) in order to distinguish between global-scale waves and GWs. Therefore in the tropics potential energies derived from CHAMP radio occultations are a mixture of both mesoscale GWs and global-scale waves [*de la Torre et al.*, 2006a (GW-CODE/GFZ)]. This could also partly contribute to the enhanced GW potential energies during the QBO westward to eastward transition.

The previous findings are based on GW potential energies. GW momentum fluxes, however, are more directly relevant for atmospheric dynamics (see also Sect. 21.1). In addition, GW momentum flux is one of the main parameters directly simulated by GWD schemes. Therefore, GW momentum flux distributions are especially suited for a comparison of observations with GWD schemes. In particular, since global distributions of GW momentum flux are very different from global distributions of, for example, GW variances or potential energies. This is the case because also GW horizontal and vertical wavelengths vary significantly depending on geolocation and season, and are also relevant for the momentum flux distribution [e.g., *Ern et al.*, 2004, 2011].

Therefore, the attempt was made to derive, for a first time, seasonal variations of GW momentum flux absolute values from GW potential energies observed by CHAMP and SAC-C [*Fröhlich et al.*, 2007 (GW-CODE/LIM)]. For this purpose the global distribution of GW horizontal wavelengths, which could not be obtained from the GPS observations alone, was estimated from the two one-week snapshots of global GW distributions derived from CRISTA observations [see also *Ern et al.*, 2004, 2006; *Preusse et al.*, 2006 (GW-CODE/FZJ)]. Main findings are especially high values of GW momentum flux in the polar night jets of the winter hemisphere, and indications of enhanced GW momentum fluxes over regions with deep convection in the subtropics of the summer hemisphere. Of course, because of the assumptions made, the distributions of GW momentum fluxes derived by *Fröhlich et al.* [2007] are subject to large uncertainties and need to be verified by other satellite observations (see also the next subsection).

21.3.3 Infrared Limb Sounding

While GPS radio occultations have a more irregular sampling pattern, limb sounding of infrared (IR) emissions from satellite provides a regular sampling pattern with short distances between subsequent altitude profiles along the measurement track. If along-track distances are shorter than about 300 km, horizontal wavelengths of observed GWs can be derived from pairs of subsequent altitude profiles. This technique introduced by *Ern et al.* [2004] resulted in the first global distributions of GW momentum fluxes observed from satellite. However, since the propagation direction of an observed wave cannot be deduced from a single measurement track, the derived values of GW momentum flux are only absolute values, i.e., no information about the direction of momentum flux is provided.

These first observed GW momentum flux distributions were derived from only two one-week global snapshots of CRISTA measurements. Therefore, studies of GW seasonal variations from IR limb sounding were initially carried out for GW squared temperature amplitudes and not GW momentum fluxes. Seasonal variations of observed SABER GW squared amplitudes were investigated in detail by *Krebsbach and Preusse* [2007], and *Preusse et al.* [2009b] (both GW-CODE/FZJ). Later, following first case studies based on HIRDLS GW momentum fluxes [e.g. *Alexander et al.*, 2008; *Wright et al.*, 2010], the method by *Ern et al.* [2004] has been applied systematically to the whole HIRDLS and SABER temperature data sets [*Ern et al.*, 2011 (GW-EXCITES)], and seasonal variations of GW momentum fluxes were derived. Further, global multi-year climatologies of GW amplitudes, vertical and horizontal wavelengths were obtained. In addition, a first attempt was made to estimate GWD directly from the observed momentum fluxes [*Ern et al.*, 2011 (GW-EXCITES)].

In *Ern et al.* [2011] an improved removal of the global-scale temperature background is used, replacing the methods used before (for example, the Kalman filter method used in *Ern et al.* [2004] or the S-transform used in *Alexander et al.* [2008]). The new method developed by *Ern et al.* [2011] is based on a two-dimensional Fourier decomposition in longitude and time and can be applied to the more regular sampling pattern of IR limb sounders (e.g., HIRDLS and SABER). Planetary-scale waves with zonal wavenumbers as high as 6 and wave periods as short as 1–2 days can be identified and removed as global temperature background. Since this method is capable to remove even short-period global-scale waves, reliable estimates of residual temperatures due to GWs, as well as GW momentum fluxes could even be derived in the mesosphere for the first time [*Ern et al.*, 2011 (GW-EXCITES)].

Figure 21.3 shows global distributions of GW momentum flux derived from SABER temperatures at 25 km altitude, averaged over the years 2002–2006 for the months of (a) January, (b) April, (c) July and (d) October. For a better comparison with the previous estimates by *Fröhlich et al.* [2007], a vertical wavenumber high-pass filter was applied to each SABER altitude profile such that only GWs with vertical wavelengths between about 4 and 10 km are contained in the data. The time period 2002–2006 of the averages shown in Fig. 21.3 is also about the same as for Fig. 2 in *Fröhlich et al.* [2007]. It should however be mentioned that, unlike in the GPS data used in *de la Torre et al.* [2006a, 2006b] and *Fröhlich et al.* [2007], the global-scale temperature background has been removed in the SABER data prior to the vertical wavenumber filtering. In Fig. 21.3 the characteristic seasonal variations of the global distribution of GW momentum flux, indicated in *Fröhlich et al.* [2007] can now be clearly identified. This shows that the assumption made in *Fröhlich et al.* [2007] of a zonal average global GW horizontal wavelength distribution, common for all months, is too simple.

The GW momentum flux distributions in Fig. 21.3 reveal a number of characteristic seasonal features. High values of GW momentum flux can be found in the polar night jets during January (Fig. 21.3a) and July (Fig. 21.3c). In addition, in January and July, several well localized maxima of GW momentum flux are found in the subtropics of the respective summer hemisphere. In the northern hemisphere enhanced GW activity is found above the Gulf of Mexico, North Africa, and Southeast

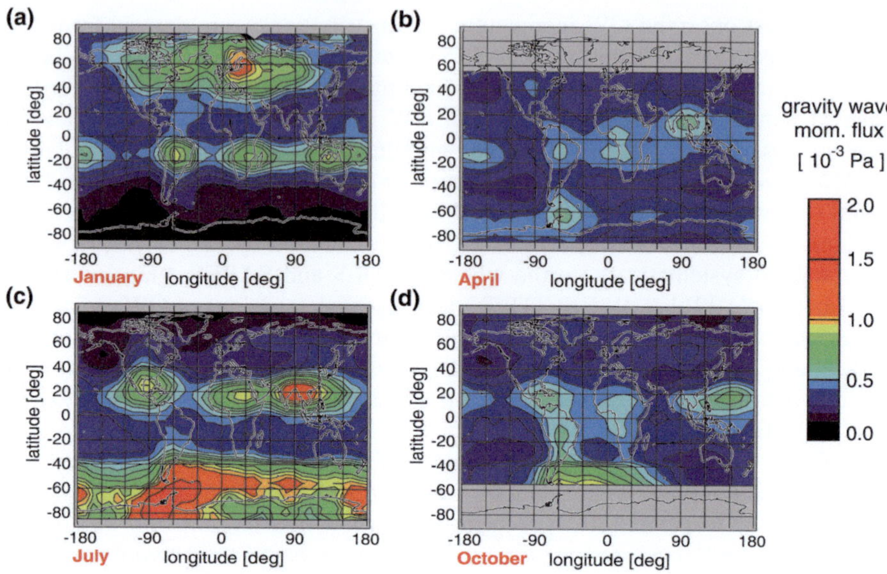

Fig. 21.3 Global distributions of GW momentum flux derived from SABER temperature at 25 km altitude, averaged over the years 2002–2006 for the months of (**a**) January, (**b**) April, (**c**) July and (**d**) October. Values represent the GW vertical wavelength range of about 4–10 km (GW-EXCITES)

Asia. In the southern hemisphere high GW activity is found above South America, South Africa, and Australia and the Pacific. In the intermediate seasons in April (Fig. 21.3b) and October (Fig. 21.3d) GW momentum fluxes are strongly reduced and the global distribution is less structured.

As has been pointed out before [e.g., *Jiang et al.*, 2004a, 2004b], GWs seen at different geolocations can often be attributed to different source processes. For example, the high values of GW activity observed in the summer hemisphere subtropics are likely due to GWs generated by deep convection. On the other hand, GWs observed above mountain ridges during periods of strong zonal wind jets are likely mountain waves. Some hot spots known for enhanced mountain wave activity are, for example, the Southern Andes and the Antarctic Peninsula in the southern hemisphere, the Rocky Mountains, Greenland, and Scandinavia in the northern hemisphere. Especially over the Southern Andes and the Antarctic Peninsula high values of GW momentum flux can be seen in Fig. 21.3 for all months except January.

21.3.4 Ground-Based Observations

Low-frequency electron drift measurements at the Collm observatory (52.1°N, 13.2°E) provide wind data in an altitude range of about 80–95 km. Over the 20-year period 1984–2003 observed wind variances due to GWs show a clear annual cycle with maximum GW wind variances in spring/summer [*Jacobi et al.*, 2006 (GW-CODE/LIM)]. There are also measurement locations where the seasonality of derived GW variances is dominated by semiannual variations [e.g., *Gavrilov et al.*, 1995]. This change in the observed seasonality could be related to changes in the zonal wind depending on geolocation, as well as differences in the parts of the whole GW spectrum seen by the different instruments. A recent study about the seasonal cycle observed in the mesosphere/lower thermosphere at northern mid and high latitudes is given, for example, by *Hoffmann et al.* [2010].

These findings show that global observations of GW variances and momentum fluxes in the mesosphere are very important to provide a global picture and they can help to explain these observed differences. One data set which could be useful in future studies are the GW distributions derived in the mesosphere by *Ern et al.* [2011] (GW-EXCITES), which are based on SABER satellite data.

The meteor radar at the Collm observatory has started its operation in August 2004. At first, the observed temperatures in the mesopause region were validated by comparing with temperatures derived from SABER satellite data, as well as temperatures derived from OH airglow observations at Wuppertal University (51.3°N, 7°E). This comparison shows very good agreement over the whole year 2005. In addition, first analyses of the dominant atmospheric tides were started [*Stober et al.*, 2008 (GW-CODE/LIM)]. These results indicate that the data obtained from the meteor radar at Collm are also well suited for the analysis of GWs and first analyses have been carried out [e.g., *Placke et al.*, 2011]. However, at the time when the DFG project GW-CODE finished, it was too early to start these investigations.

21.4 Observed Longer-Term Variations of Gravity Wave Activity

Until recently, there was only little information about observed seasonal variations of the global distribution of GWs. There are even fewer observations addressing longer-term variations of global GW activity. In our studies we investigated variations of GW activity related to the quasi-biennial oscillation (QBO) in the tropics, as well as variations potentially related to the 11-year solar cycle.

21.4.1 Variations Related to the Quasi-Biennial Oscillation (QBO)

The importance of GWs for driving the QBO in the tropics was discussed in Sect. 21.3.2 [see also *de la Torre et al.*, 2006a (GW-CODE/GFZ)]. A more detailed study by *Krebsbach and Preusse* [2007] (GW-CODE/FZJ) based on four years of SABER GW squared temperature amplitudes addressed annual, semi-annual, and also quasi-biennial variations. This study shows that QBO-related variations of the global distribution of GWs are found not only in the tropical stratosphere. There are also indications for QBO-related variations of GW amplitudes in the mesosphere above about 60 km.

Figure 21.4 shows a latitude/time cross section of zonally averaged GW momentum flux absolute values at 25 km altitude derived from SABER temperatures. The most striking features governing the global distribution are the enhanced values in

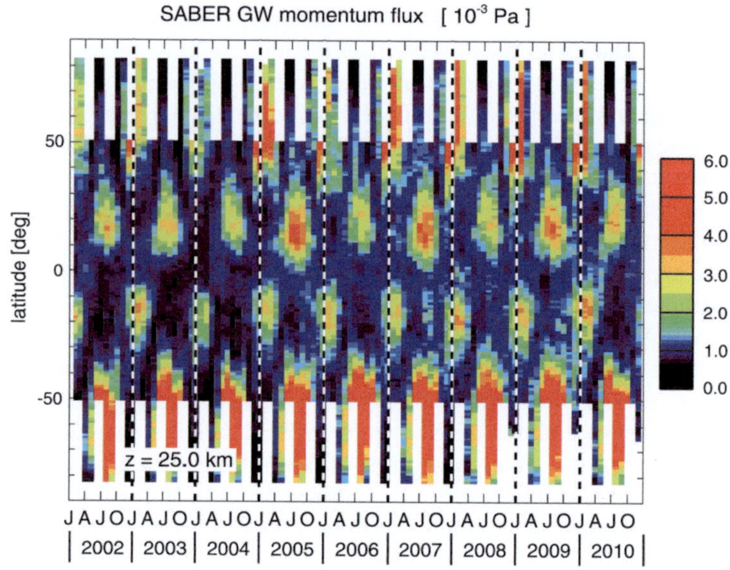

Fig. 21.4 Latitude-time cross section of zonal average GW momentum fluxes for the GW vertical wavelength range 4–25 km derived from SABER temperatures at 25 km altitude (GW-EXCITES)

the polar night jets on the winter hemisphere, as well as the enhanced momentum fluxes in the subtropics of the summer hemisphere, centered at about 20°S during austral summer and 20°N during boreal summer. Especially in the subtropical momentum flux maxima a quasi-biennial variation is clearly visible [see also *Ern et al.*, 2011 (GW-EXCITES)].

Other observations show significant QBO-related variations even in the exosphere at altitudes as high as 400 km [e.g., *Oberheide et al.*, 2009]. GWs could play an important role in conveying those QBO-related variations to higher altitudes, which underlines the importance of the observations made by *de la Torre et al.* [2006a], *Krebsbach and Preusse* [2007], and *Ern et al.* [2011].

21.4.2 Variations Potentially Related to the 11-Year Solar Cycle

Evidence for variations of GW activity on time scales of the 11-year solar cycle is quite contradictory. For example, observations in the MLT region by *Gavrilov et al.* [1995, 2002, 2004] indicate both positive and negative correlations between GW activity and solar activity during an 11-year solar cycle. In particular, 20 years (1984–2003) of LF drift observations by *Jacobi et al.* [2006] indicate a positive correlation between GW wind variances and solar activity, consistent with a decadal change of mesospheric GW filtering through the zonal wind jet (GW-CODE/LIM). Possible reasons for those contradictory observations could be related to differences in the local meteorological conditions at different stations that might influence the observed sign of the correlation.

This shows the need for longer-term global observations from satellites to obtain also a global view of the decadal variations of GW activity. Long-term variations on a decadal time scale are also visible in Fig. 21.4 and might indicate an anti-correlation between solar activity and GW momentum fluxes. However, the SABER data set is still too short to give reliable evidence [see also *Ern et al.*, 2011 (GW-EXCITES)].

21.5 Observational Constraints for Standard Nonorographic Gravity Wave Drag Schemes

In standard nonorographic GWD schemes usually a number of simplifying assumptions are made for the GW source distribution. One assumption is a globally constant GW momentum flux or GW potential energy source amplitude. Further, the GW source distribution is assumed to be isotropic, i.e., the same amount of momentum flux is launched in all directions used (usually 4 or 8, depending on implementation). In addition, assumptions are made about the spectral shape of the GW source spectrum. This very simplified treatment of GWs in GCMs is widely used. Therefore it makes sense to optimize the launch parameters of such a simplified global GW source distribution by comparison with observations. In particular, results of this kind can then be implemented in existing GWD schemes.

This is why *Ern et al.* [2006] (GW-CODE/FZJ) tried to constrain the launch parameters of the WM01 GWD scheme by comparison with the two one-week average global distributions of CRISTA GW momentum fluxes [see also *Ern et al.*, 2004]. The criteria were to obtain good agreement between observed and modeled GW momentum flux, both for the horizontal global structures, as well as for absolute values. For this comparison of modeled and observed GW momentum flux distributions it is very important that the observational filter of the satellite instrument is applied to the modeled GW momentum flux spectrum as has been shown by *Ern et al.* [2005]. The main findings by *Ern et al.* [2006] (GW-CODE/FZJ) are:

1. Due to wind filtering of the GW source spectrum in the upper troposphere and lower stratosphere good agreement of modeled and observed global GW momentum flux horizontal distributions can be achieved (even for a globally constant and isotropic GW source distribution) if a low GW source level is selected (at altitudes below about 8 km).
2. The characteristic vertical wavelength where the maximum of the GW vertical wavenumber spectrum is attained should be in the range 2–4 km at the source level.
3. The spectral slope of the vertical wavenumber spectrum at low vertical wavenumbers should be about 2.

Further, it has been found that simulated GW momentum flux values are much too strong in the summer hemisphere and are not in agreement with the observations.

This shows that a more sophisticated approach is required in the future to provide more realistic results.

Recent studies by *Ern et al.* [2011] (GW-EXCITES) indicate that poleward propagation of GWs generated in the summer hemisphere subtropics is an important effect and might contribute to the poleward tilt of the mesospheric westward wind jet in the summer hemisphere, as well as to the rather semiannual than annual variations of GW activity observed at some midlatitude measurement stations (see also Sect. 21.3.4). Another finding by *Ern et al.* [2011] is a stronger than expected decrease of observed GW momentum fluxes in the stratosphere, suggesting higher than expected values of GWD already in the stratosphere, in particular in the polar night jet in the southern hemisphere. This finding is currently still not understood and might indicate either another shortcoming of current GWD schemes or aspects of GW propagation still not understood.

The results obtained by *Ern et al.* [2006] have proven to be helpful for the setup of nonorographic GWD schemes in GCMs. For example, the observational evidence that a low launch level is more favorable for GWD schemes confirms earlier GCM setups [e.g., *Fomichev et al.*, 2002; *Garcia et al.*, 2007]. Already these setups used a low launch level for their nonorographic GWD schemes and benefited from wind filtering at low levels, which made sure that a more realistic global momentum flux distribution entered the stratosphere. This approach has also been used when a nonorographic GWD scheme was included in the European Centre for Medium-Range Weather Forecasts (ECMWF) global weather forecasting model to provide an improved global circulation in the stratosphere [*Orr et al.*, 2010].

21.6 GCM Modeling with COMMA-LIM

21.6.1 Testing a State-of-the-Art GWD Scheme with Launch Conditions Adapted from Observations

The example of *Orr et al.* [2010] shows that observations can help to constrain the launch distribution of a GW parameterization. However, GW dissipation is treated differently in different GWD schemes [e.g., *McLandress and Scinocca*, 2005]. For instance, the frequently used Hines scheme tends to produce a very sharp wind tendency peak in the region of vertical wind shear. On the other hand, the semi-empirical *Warner and McIntyre* [2001] (WM01) scheme generates a smoother vertical structure of the wind tendencies and deposits notable momentum already in the stratosphere and lower mesosphere. This is more consistent with the unexpectedly strong vertical gradients of GW momentum fluxes found in satellite observations already in the stratosphere, especially in the southern polar night jet [*Ern et al.*, 2011].

Therefore, the aim of the project GW-CODE/LIM was to study the performance of a current GWD scheme, using the forcing of observationally derived launch spectra. For this purpose a spectral GWD scheme (WM01) has been implemented into

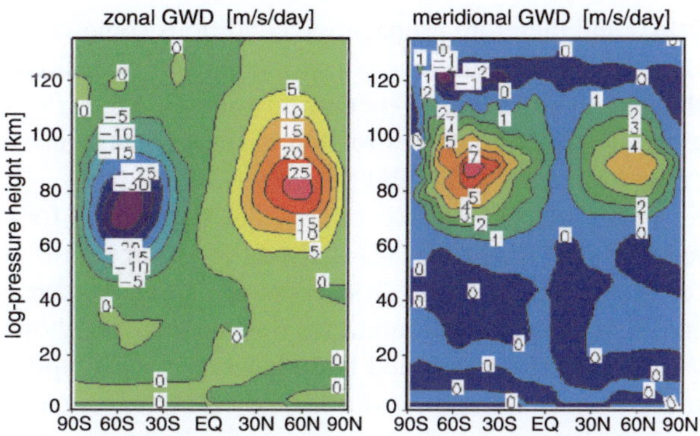

Fig. 21.5 Zonal average distributions of GW drag of the WM01 GWD scheme implemented in COMMA-LIM for a stable model run under July conditions using the following GWD scheme settings: launch level at 7 km, characteristic vertical wavelength of 2.3 km at launch level, spectral slope of 2 for the unsaturated part of the GW launch spectrum (GW-CODE/LIM)

the GCM COMMA-LIM [*Fröhlich et al.*, 2006 (GW-CODE/LIM)] and the GW launch spectrum has been adapted following *Ern et al.* [2006]. In particular, to match the observations, it was assumed that already the GW launch spectrum is saturated at high vertical wavenumbers. A similar study carried out before by *McLandress and Scinocca* [2005] resulted in GW momentum deposition over a larger altitude range. This is qualitatively in good agreement with the results obtained in our study (see Fig. 21.5), but different from standard setups in GCMs showing sharper peaks of GWD in the mesopause region. In the study by *McLandress and Scinocca* [2005] agreement with the standard setups could only be achieved by using very small launch amplitudes, not in agreement with observations. This suggests that the vertical structure of momentum deposition and resulting wind tendencies merit further investigation.

21.6.2 Differences in Wave Propagation Conditions Between Solar Maximum and Solar Minimum

Further simulations addressed differences in the vertical propagation conditions of GWs and planetary-scale waves for solar maximum and solar minimum conditions. For this purpose the GCM COMMA-LIM in its original configuration (i.e., using a Lindzen-type GWD scheme) was nudged to an NCEP data set of 11-year averaged (1992–2002) zonal mean temperature and geopotential fields for each year of the month to prescribe realistic atmospheric conditions below about 30 km altitude [see also *Pogoreltsev et al.*, 2007 (GW-CODE/LIM)]

More enhanced GW momentum flux was found in the MLT region during solar maximum conditions [*Jacobi et al.*, 2006 (GW-CODE/LIM)]. This is the case because during solar maximum the mesospheric jets are somewhat stronger [e.g.,

Kodera et al., 2003] and therefore propagation conditions somewhat better for GWs propagating in opposite direction to the prevailing background wind. It should however be noted that these simulations were carried out with fixed GW source strength and also variations of the GW sources related to the 11-year solar cycle might be important (see also *Ern et al.* [2011] and Sect. 21.4.2).

Figure 21.6a shows planetary wave zonal wind amplitudes for different wave modes simulated with the GCM COMMA-LIM during solar maximum atmospheric background conditions. In Fig. 21.6b differences of planetary wave amplitudes between solar maximum and solar minimum for the wave modes shown in Fig. 21.6a are given (GW-CODE/LIM). It is found that planetary wave amplitudes are somewhat higher during solar minimum conditions, however, this is only a small effect. The finding of somewhat enhanced planetary wave amplitudes during solar minimum conditions is qualitatively in good agreement with the results of a recent study by *Powell and Xu* [2011], who analyzed planetary wave amplitudes in NCEP and ERA-40 reanalysis data, as well as satellite observations.

21.7 Gravity Wave Ray Tracing

Not all aspects of GW propagation are included in standard GWD schemes. In particular, only purely vertical propagation of GWs is considered. Simulations based on ray tracing of GWs make use of a much more general approach and do not rely on this strong assumption. Figure 21.7 shows the result of a GW ray tracing simulation for 15 July 2003. GWs were launched at 0° longitude in eight directions from a launch altitude at 5 km. The launch locations are indicated by gray asterisks. The propagation paths of the different GWs are indicated by the colored lines. The color indicates the altitude that a GW has reached during its upward propagation from the source level. This simulation shows that GWs can travel large horizontal distances before reaching an altitude of 100 km and that the assumption of purely vertical propagation of GWs is too simple [*Preusse et al.*, 2009b (GW-CODE/FZJ)].

Since GW ray tracing uses a more general physical approach, this technique can be promising for obtaining more realistic global distributions of GWs and achieve better agreement with observations than standard GWD schemes. In a first approach *Preusse et al.* [2006] (GW-CODE/FZJ) derived an empirical source distribution consisting of a limited number of single wave components launched at 5 km source altitude. The source amplitude of the single wave components is the same for the whole globe, and the same distribution of waves is launched in each of the eight directions used. This means that, similar to the standard nonorographic GWD schemes, a global homogeneous and isotropic GW source distribution is assumed. With this approach it was possible to reproduce the global distribution of GW squared temperature amplitudes observed by the SABER instrument in August 2003.

This approach has been more refined in a follow-up study by *Preusse et al.* [2009b] (GW-CODE/FZJ, GRAPES). Instead of an empirical set of single wave components, amplitudes and intermittency of wave components used for the launch

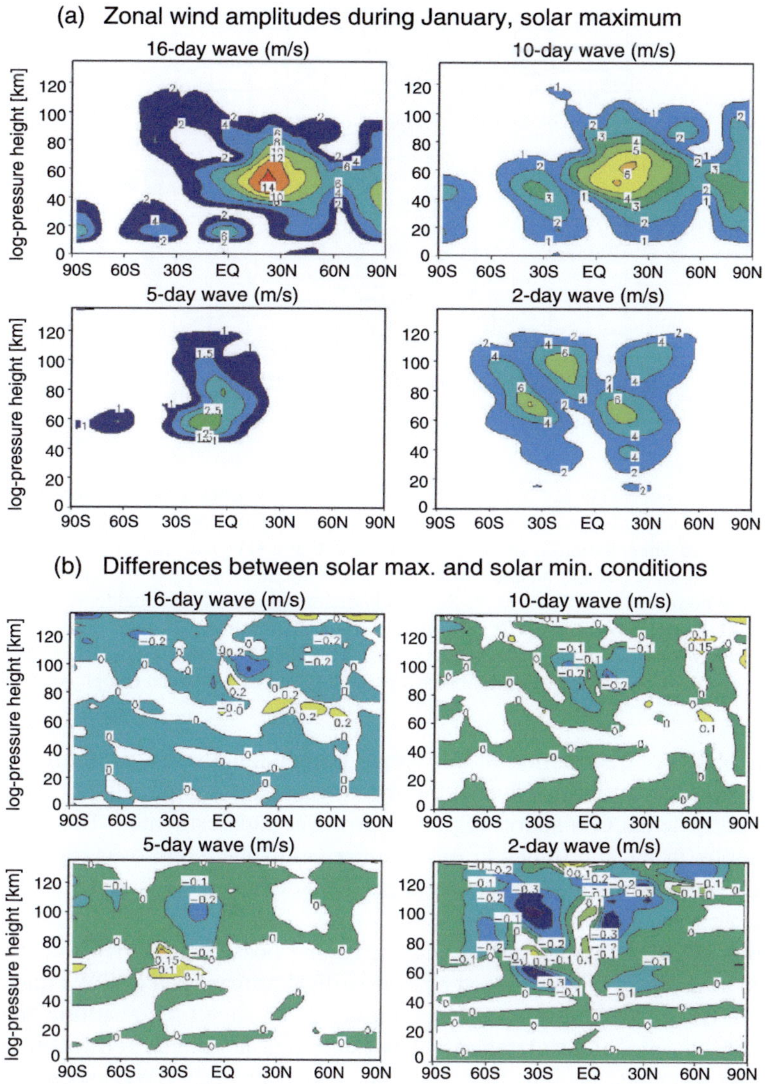

Fig. 21.6 Sensitivity of planetary wave amplitudes to the 11-year solar cycle studied in simulations with the GCM COMMA-LIM. Shown are planetary wave zonal wind amplitudes for (**a**) different wave modes during solar maximum atmospheric background conditions and (**b**) differences of planetary wave amplitudes between solar maximum and solar minimum conditions for the wave modes shown in (**a**) (GW-CODE/LIM)

distribution were now determined by a least-squares approach minimizing the differences between simulated and observed zonal average SABER GW squared amplitudes for the single month of July. Even though the source distribution was adapted to only one single month of observations, it was possible to reproduce not

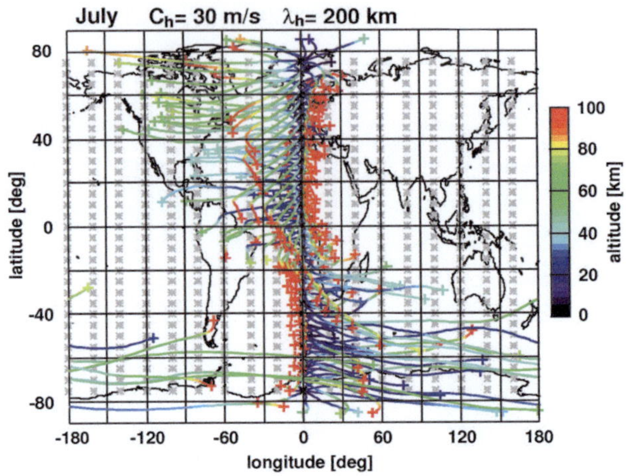

Fig. 21.7 Results of a GW ray tracing simulation for 15 July 2003. GWs are launched at 5 km altitude and 0° longitude into eight directions. Propagation paths of the single waves are indicated by *colored lines*, the color indicates the altitude that a GW has reached during its upward propagation. Figure 21.7 was adapted from Fig. 1 in *Preusse et al.* [2009b], (GW-CODE/FZJ). Reproduced/modified from "Preusse, P., S. D. Eckermann, M. Ern, J. Oberheide, R. H. Picard, R. G. Roble, M. Riese, J. M. Russell III, and M. G. Mlynczak, Global ray tracing simulations of the SABER gravity wave climatology, J. Geophys. Res., 114, D08126, 2009." Copyright [2009] American Geophysical Union. Reproduced/modified by permission of American Geophysical Union

only the observed distribution in this single month, but also to qualitatively reproduce the seasonal variations during the whole year over a large altitude range. GW activity in particular GW source regions is however underestimated. This is the case, for example, in the subtropics in the summer hemisphere and over regions where mountain wave activity is observed (e.g., the Southern Andes and the Antarctic Peninsula).

By this GW ray tracing approach it was further possible to derive values of GW drag and to quantify effects caused by non-vertical propagation of GWs that are neglected in the standard GWD schemes. Figure 21.8a shows the resulting zonal and Fig. 21.8b meridional accelerations. It is found that horizontal refraction of GWs into strong wind jets, an effect that is not included in standard GWD schemes, is very important for meridional accelerations of the background wind by GWs (GRAPES). It is also found that meridional propagation of GWs cannot be neglected. This effect is also not included in standard GWD schemes. This finding is further illustrated in Figs. 21.8c and 21.8d, showing the average latitudinal propagation of GWs weighted by the momentum flux carried by a GW (Fig. 21.8c), as well as by the acceleration exerted by a GW (Fig. 21.8d). These results show that especially poleward propagation of GWs from the subtropics in the summer hemisphere toward midlatitudes is an important effect and poleward propagation of those GWs of about 15° latitude between about 30 and 70 km altitude is indicated (GW-CODE/FZJ). This result is in good agreement with the global GW momentum flux distribution derived from SABER observations [*Ern et al.*, 2011 (GW-EXCITES)].

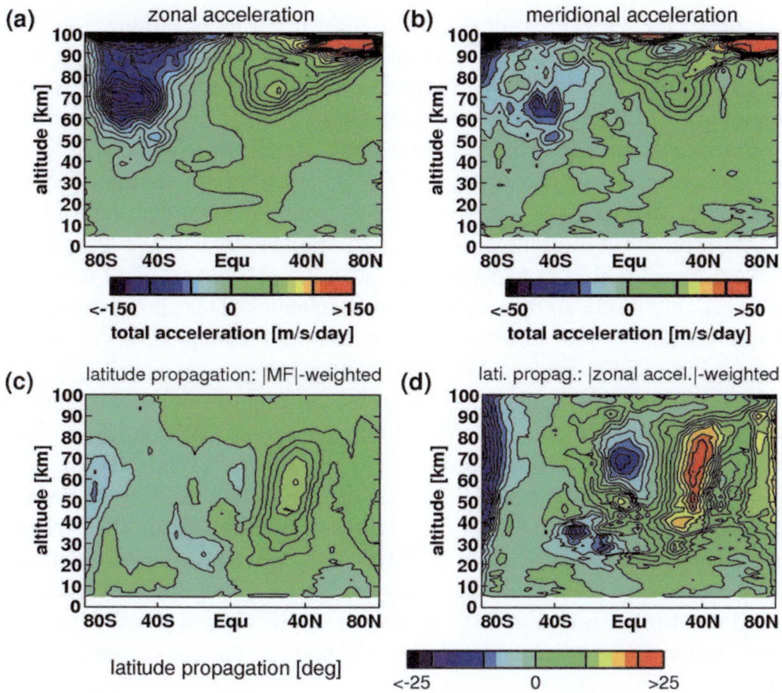

Fig. 21.8 In a global GW ray tracing simulation accelerations of the background wind have been derived. Shown are: (**a**) zonal, (**b**) meridional accelerations in m/s/day. In addition, the latitudinal shift of GWs can be deduced by GW ray tracing (**c**, **d**): Average latitudinal shift of GW rays weighted by the (**c**) momentum flux carried, (**d**) wind acceleration exerted. Figures 21.8a and 21.8b were adapted from Figs. 12a and 12b in *Preusse et al.* [2009b] and Figs. 21.8c and 21.8d were adapted from Figs. 13b and 13c in *Preusse et al.* [2009b] (GW-CODE/FZJ, GRAPES). Reproduced/modified from "Preusse, P., S. D. Eckermann, M. Ern, J. Oberheide, R. H. Picard, R. G. Roble, M. Riese, J. M. Russell III, and M. G. Mlynczak, Global ray tracing simulations of the SABER gravity wave climatology, J. Geophys. Res., 114, D08126, 2009." Copyright [2009] American Geophysical Union. Reproduced/modified by permission of American Geophysical Union

This shows that many assumptions made for standard GWD schemes are too simple and large benefits could be expected from a GWD scheme based on GW ray tracing techniques. Therefore in the DFG project GRAPES the attempt is made to develop a GWD scheme based on GW ray tracing techniques and to include this scheme in the GCM HAMMONIA.

21.8 Summary

Different global and ground-based data sets of GW observations created in the framework of the German DFG priority program CAWSES produced a number of important scientific results and they also provide valuable constraints for GWD schemes used in GCMs. These data sets cover time scales from seasonal, over biennial, up to the 11-year solar cycle. In addition, a large altitude range is covered, starting in the tropopause region and reaching up to above 100 km altitude.

Satellite observations and GW ray tracing have revealed significant weaknesses of standard GWD schemes. There are indications that under certain conditions the GWD influence in the stratosphere is more important than expected. In addition, it is found that non-vertical propagation of GWs, an effect not contained in standard GWD schemes, is important to derive realistic zonal average distributions of zonal GWD. Distributions of meridional GWD are dominated by non-vertical propagation of GWs and by refraction of GWs into the strong mesospheric wind jets and likely cannot be reliably modeled by standard GWD schemes. This shows the importance to develop new GWD schemes based on ray tracing techniques, which has already been started in the DFG-project GRAPES.

The relevance of the results obtained in the different projects summarized here is reflected by the facts that constraints of nonorographic GWD schemes derived in the project GW-CODE are now used in many GCMs and even in the ECMWF operational weather forecasting model. Global distributions of GW momentum fluxes derived in the project GW-EXCITES are currently being used in the Stratospheric Processes And their Role in Climate (SPARC) gravity wave initiative and in an international gravity wave project organized by the International Space Science Institute (ISSI)/Bern to further improve GWD schemes in GCMs and to validate high horizontal resolution GCMs that explicitly resolve parts of the global GW distribution.

Acknowledgements Very helpful comments of two anonymous reviewers are gratefully acknowledged. The projects GW-CODE/FZJ, GW-CODE/GFZ, GW-CODE/LIM, GRAPES, and GW-EXCITES were funded by Deutsche Forschungsgemeinschaft (DFG) grants no. ER474/1-1, WI2634/2-1, JA836/21-1, PR919/2-1, and ER474/2-1 within the DFG priority program CAWSES (SPP1176).

References

Alexander, M. J., Gille, J. C., Cavanaugh, C., Coffey, M., Craig, C., Eden, T., Francis, G., Halvorson, C., Hannigan, J., Khosravi, R., Kinnison, D., Lee, H., Massie, S., Nardi, B., Barnett, J. J., Hepplewhite, C., Lambert, A., & Dean, V. (2008). Global estimates of gravity wave momentum flux from High Resolution Dynamics Limb Sounder observations. *Journal of Geophysical Research*, *113*, D15S18. doi:10.1029/2007JD008807.

Alexander, M. J., Geller, M., McLandress, C., Polavarapu, S., Preusse, P., Sassi, F., Sato, K., Eckermann, S., Ern, M., Hertzog, A., Kawatani, Y., Pulido, M., Shaw, T. A., Sigmond, M., Vincent, R., & Watanabe, S. (2010). Recent developments in gravity-wave effects in climate models and the global distribution of gravity-wave momentum flux from observations and models. *Quarterly Journal of the Royal Meteorological Society*, *136*, 1103–1124. doi:10.1002/qj.637.

Arras, C., Wickert, J., Beyerle, G., Heise, S., Schmidt, T., & Jacobi, C. (2008). A global climatology of ionospheric irregularities derived from GPS radio occultation. *Geophysical Research Letters*, *35*, L14809. doi:10.1029/2008GL034158.

Arras, C., Jacobi, C., & Wickert, J. (2009). Semidiurnal tidal signature in sporadic E occurrence rates derived from GPS radio occultation measurements at higher midlatitudes. *Annales Geophysicae*, *27*, 2555–2563.

Baldwin, M. P., & Dunkerton, T. J. (2001). Stratospheric harbingers of anomalous weather regimes. *Science*, *294*, 581–584.

Charron, M., & Manzini, E. (2002). Gravity waves from fronts: parameterization and middle atmosphere response in a general circulation model. *Journal of the Atmospheric Sciences, 59*, 923–941.

de la Torre, A., Schmidt, T., & Wickert, J. (2006a). A global analysis of wave potential energy in the lower stratosphere derived from 5 years of GPS radio occultation data with CHAMP. *Geophysical Research Letters, 33*, L24809. doi:10.1029/2006GL027696.

de la Torre, A., Alexander, P., Llamedo, P., Menéndez, C., Schmidt, T., & Wickert, J. (2006b). Gravity waves above the Andes detected from GPS radio occultation temperature profiles: jet mechanism? *Geophysical Research Letters, 33*, L24810. doi:10.1029/2006GL027343.

de la Torre, A., Llamedo, P., Alexander, P., Schmidt, T., & Wickert, J. (2010). Estimated errors in a global gravity wave climatology from GPS radio occultation temperature profiles. *Advances in Space Research, 46*, 174–179. doi:10.1016/j.asr.2010.02.033.

Dunkerton, T. J. (1997). The role of gravity waves in the quasi-biennial oscillation. *Journal of Geophysical Research, 102*, 26053–26076.

Ern, M., & Preusse, P. (2009a). Wave fluxes of equatorial Kelvin waves and QBO zonal wind forcing derived from SABER and ECMWF temperature space-time spectra. *Atmospheric Chemistry and Physics, 9*, 3957–3986. doi:10.5194/acp-9-3957-2009.

Ern, M., & Preusse, P. (2009b). Quantification of the contribution of equatorial Kelvin waves to the QBO wind reversal in the stratosphere. *Geophysical Research Letters, 36*, L21801. doi:10.1029/2009GL040493.

Ern, M., Preusse, P., Alexander, M. J., & Warner, C. D. (2004). Absolute values of gravity wave momentum flux derived from satellite data. *Journal of Geophysical Research, 109*, D20103. doi:10.1029/2004JD004752.

Ern, M., Preusse, P., & Warner, C. D. (2005). A comparison between CRISTA satellite data and Warner and McIntyre gravity wave parameterization scheme: Horizontal and vertical wavelength filtering of gravity wave momentum flux. *Advances in Space Research, 35*, 2017–2023. doi:10.1016/j.asr.2005.04.109.

Ern, M., Preusse, P., & Warner, C. D. (2006). Some experimental constraints for spectral parameters used in the Warner and McIntyre gravity wave parameterization scheme. *Atmospheric Chemistry and Physics, 6*, 4361–4381. doi:10.5194/acp-6-4361-2006.

Ern, M., Preusse, P., Gille, J. C., Hepplewhite, C. L., Mlynczak, M. G., Russell III, J. M., & Riese, M. (2011). Implications for atmospheric dynamics derived from global observations of gravity wave momentum flux in stratosphere and mesosphere. *Journal of Geophysical Research, 116*, D19107. doi:10.1029/2011JD015821.

Fleming, E. L., Chandra, S., Barnett, J. J., & Corney, M. (1990). Zonal mean temperature, pressure, zonal wind and geopotential height as functions of latitude. *Advances in Space Research, 12*, 1211–1259.

Fomichev, V. I., Ward, W. E., Beagley, S. R., McLandress, C., McConnell, J. C., McFarlane, N. A., & Shepherd, T. G. (2002). Extended Canadian Middle Atmosphere model: zonal-mean climatology and physical parameterizations. *Journal of Geophysical Research, 107*, 4087. doi:10.1029/2001JD000479.

Fritts, D. C., & Alexander, M. J. (2003). Gravity wave dynamics and effects in the middle atmosphere. *Reviews of Geophysics, 41*, 1003. doi:10.1029/2001RG000106.

Fröhlich, K., Ern, M., Jacobi, C., & Preusse, P. (2006). Implementation of the Warner-McIntyre scheme of gravity wave parametrization into COMMA-LIM, Part I: code transfer. *Reports of the Institute of Meteorology, University of Leipzig, 37*, 11–16.

Fröhlich, K., Schmidt, T., Ern, M., Preusse, P., de la Torre, A., Wickert, J., & Jacobi, C. (2007). The global distribution of gravity wave energy in the lower stratosphere derived from GPS data and gravity wave modelling: Attempt and challenges. *Journal of Atmospheric and Solar-Terrestrial Physics, 69*, 2238–2248. doi:10.1016/j.jastp.2007.07.005.

Garcia, R. R., & Randel, W. J. (2008). Acceleration of the Brewer Dobson circulation due to increases in greenhouse gases. *Journal of the Atmospheric Sciences, 65*, 2731–2739.

Garcia, R. R., Marsh, D. E., Kinnison, D. E., Boville, B. A., & Sassi, F. (2007). Simulation of secular trends in the middle atmosphere, 1950–2003. *Journal of Geophysical Research, 112*,

D09301. doi:10.1029/2006JD007485.
Gavrilov, N. M., Manson, A. H., & Meek, C. E. (1995). Climatological monthly characteristics of middle atmosphere gravity waves (10 min–10 h) during 1979–1993 at Saskatoon. *Annales Geophysicae*, *13*, 285–295.
Gavrilov, N. M., Fukao, S., Nakamura, T., Jacobi, C., Kürschner, D., Manson, A. H., & Meek, C. E. (2002). Comparative study of interannual changes of the mean winds and gravity wave activity in the middle atmosphere over Japan, Central Europe and Canada. *Journal of Atmospheric and Solar-Terrestrial Physics*, *64*, 1003–1010.
Gavrilov, N. M., Riggin, D. M., & Fritts, D. C. (2004). Interannual variations of the mean wind and gravity wave variances in the middle atmosphere over Hawaii. *Journal of Atmospheric and Solar-Terrestrial Physics*, *66*, 637–645.
Hines, C. O. (1997a). Doppler-spread parameterization of gravity-wave momentum deposition in the middle atmosphere. Part 1: basic formulation. *Journal of Atmospheric and Solar-Terrestrial Physics*, *59*, 371–386.
Hines, C. O. (1997b). Doppler-spread parameterization of gravity-wave momentum deposition in the middle atmosphere. Part 2: broad and quasi monochromatic spectra and implementation. *Journal of Atmospheric and Solar-Terrestrial Physics*, *59*, 387–400.
Hocke, K., & Tsuda, T. (2001). Gravity waves and ionospheric irregularities over tropical convection zones observed by GPS/MET radio occultation. *Geophysical Research Letters*, *28*, 2815–2818.
Hoffmann, P., Becker, E., Singer, W., & Placke, M. (2010). Seasonal variation of mesospheric waves at northern middle and high latitudes. *Journal of Atmospheric and Solar-Terrestrial Physics*, *72*, 1068–1079. doi:10.1016/j.jastp.2010.07.002.
Holton, J. R. (1982). The role of gravity wave induced drag and diffusion in the momentum budget of the mesosphere. *Journal of the Atmospheric Sciences*, *39*, 791–799.
Holton, J. R. (1983). The influence of gravity wave breaking on the general circulation of the middle atmosphere. *Journal of the Atmospheric Sciences*, *40*, 2497–2507.
Holton, J. R., & Lindzen, R. S. (1972). An updated theory for the quasi-biennial cycle of the tropical stratosphere. *Journal of the Atmospheric Sciences*, *29*, 1076–1080.
Holton, J. R., & Tan, H.-C. (1980). The influence of the equatorial quasi-biennial oscillation on the global circulation at 50 mb. *Journal of the Atmospheric Sciences*, *37*, 2200–2208.
Jacobi, C., Gavrilov, N. M., Kürschner, D., & Fröhlich, K. (2006). Gravity wave climatology and trends in the mesosphere/lower thermosphere region deduced from low-frequency drift measurements 1984–2003 (52.1°N, 13.2°E). *Journal of Atmospheric and Solar-Terrestrial Physics*, *68*, 1913–1923. doi:10.1016/j.jastp.2005.12.007.
Jiang, J. H., Eckermann, S. D., Wu, D. L., & Ma, J. (2004a). A search for mountain waves in MLS stratospheric limb radiances from the winter northern hemisphere: Data analysis and global mountain wave modeling. *Journal of Geophysical Research*, *109*, D03107. doi:10.1029/2003JD003974.
Jiang, J. H., Wang, B., Goya, K., Hocke, K., Eckermann, S. D., Ma, J., Wu, D. L., & Read, W. G. (2004b). Geographical distribution and interseasonal variability of tropical deep convection: UARS MLS observations and analyses. *Journal of Geophysical Research*, *109*, D03111. doi:10.1029/2003JD003756.
Kodera, K., Matthes, K., Shibata, K., Langematz, U., & Kuroda, Y. (2003). Solar impact on the lower mesospheric subtropical jet: a comparative study with general circulation model simulations. *Geophysical Research Letters*, *30*, 1315. doi:10.1029/2002GL016124.
Krebsbach, M., & Preusse, P. (2007). Spectral analysis of gravity wave activity in SABER temperature data. *Geophysical Research Letters*, *34*, L03814. doi:10.1029/2006GL028040.
Li, F., Austin, J., & Wilson, J. (2008). The strength of the Brewer Dobson circulation in a changing climate: coupled chemistry climate model simulations. *Journal of Climate*, *21*, 40–57. doi:10.1175/2007JCLI1663.1.
Lindzen, R. S. (1981). Turbulence and stress owing to gravity wave and tidal breakdown. *Journal of Geophysical Research*, *86*, 9707–9714.

Lindzen, R. S., & Holton, J. R. (1968). A theory of the quasi-biennial oscillation. *Journal of the Atmospheric Sciences*, *25*, 1095–1107.

Lübken, F.-J. (1997). Seasonal variation of turbulent energy dissipation rates at high latitudes as determined by in situ measurements of neutral density fluctuations. *Journal of Geophysical Research*, *102*, 13441–13456.

Marshall, A. G., & Scaife, A. A. (2009). Impact of the QBO on surface winter climate. *Journal of Geophysical Research*, *114*, D18110. doi:10.1029/2009JD011737.

McLandress, C., & Scinocca, J. F. (2005). The GCM response to current parameterizations of nonorographic gravity wave drag. *Journal of the Atmospheric Sciences*, *62*, 2394–2413.

McLandress, C., & Shepherd, T. G. (2009). Simulated anthropogenic changes in the Brewer Dobson circulation, including its extension to high latitudes. *Journal of Climate*, *22*, 1516–1540. doi:10.1175/2008JCLI2679.1.

Oberheide, J., Forbes, J. M., Häusler, K., Wu, Q., & Bruinsma, S. L. (2009). Tropospheric tides from 80 to 400 km: propagation, interannual variability, and solar cycle effects. *Journal of Geophysical Research*, *114*, D00I05. doi:10.1029/2009JD012388.

Orr, A., Bechtold, P., Scinocca, J. F., Ern, M., & Janiskova, M. (2010). Improved middle atmosphere climate and forecasts in the ECMWF model through a non-orographic gravity wave drag parametrization. *Journal of Climate*, *23*, 5905–5926. doi:10.1175/2010JCLI3490.1.

Placke, M., Stober, G., & Jacobi, C. (2011). Gravity wave momentum fluxes in the MLT—Part I: seasonal variation at Collm (51.3°N, 13.0°E). *Journal of Atmospheric and Solar-Terrestrial Physics*, *73*, 904–910. doi:10.1016/j.jastp.2010.07.012.

Pogoreltsev, A. I., Vlasov, A. A., Fröhlich, K., & Jacobi, C. (2007). Planetary waves in coupling the lower and upper atmosphere. *Journal of Atmospheric and Solar-Terrestrial Physics*, *69*, 2083–2101. doi:10.1016/j.jastp.2007.05.014.

Powell, A. M., & Xu, J. (2011). Possible solar forcing of interannual and decadal stratospheric planetary wave variability in the northern hemisphere: an observational study. *Journal of Atmospheric and Solar-Terrestrial Physics*, *73*, 825–838. doi:10.1016/j.jastp.2011.02.001.

Preusse, P., Ern, M., Eckermann, S. D., Warner, C. D., Picard, R. H., Knieling, P., Krebsbach, M., Russell III, J. M., Mlynczak, M. G., Mertens, C. J., & Riese, M. (2006). Tropopause to mesopause gravity waves in August: measurement and modeling. *Journal of Atmospheric and Solar-Terrestrial Physics*, *68*, 1730–1751. doi:10.1016/j.jastp.2005.10.019.

Preusse, P., Schroeder, S., Hoffmann, L., Ern, M., Friedl-Vallon, F., Oelhaf, H., Fischer, H., & Riese, M. (2009a). New perspectives on gravity wave remote sensing by spaceborne infrared limb imaging. *Atmospheric Measurement Techniques*, *2*, 299–311.

Preusse, P., Eckermann, S. D., Ern, M., Oberheide, J., Picard, R. H., Roble, R. G., Riese, M., Russell III, J. M., & Mlynczak, M. G. (2009b). Global ray tracing simulations of the SABER gravity wave climatology. *Journal of Geophysical Research*, *114*, D08126. doi:10.1029/2008JD011214.

Rauthe, M., Gerding, M., Höffner, J., & Lübken, F.-J. (2006). Lidar temperature measurements of gravity waves over Kühlungsborn (54°N) from 1 to 105 km: a winter-summer comparison. *Journal of Geophysical Research*, *111*, D24108. doi:10.1029/2006JD007354.

Richter, J. H., Sassi, F., & Garcia, R. R. (2010). Toward a physically based gravity wave source parameterization in a general circulation model. *Journal of the Atmospheric Sciences*, *67*, 136–156.

Riese, M., Spang, R., Preusse, P., Ern, M., Jarisch, M., Offermann, D., & Grossmann, K. U. (1999). Cryogenic infrared spectrometers and telescopes for the atmosphere (CRISTA) data processing and atmospheric temperature and trace gas retrieval. *Journal of Geophysical Research*, *104*, 16349–16367.

Schmidt, T., Beyerle, G., Heise, S., Wickert, J., & Rothacher, M. (2006). A climatology of multiple tropopauses derived from GPS radio occultations with CHAMP and SAC-C. *Geophysical Research Letters*, *33*, L04808. doi:10.1029/2005GL024600.

Schmidt, T., de la Torre, A., & Wickert, J. (2008). Global gravity wave activity in the tropopause region from CHAMP radio occultation data. *Geophysical Research Letters*, *35*, L16807. doi:10.1029/2008GL034986.

Song, I.-S., & Chun, H.-Y. (2008). A Lagrangian spectral parameterization of gravity wave drag induced by Cumulus convection. *Journal of the Atmospheric Sciences*, *65*, 1204–1224.

Stober, G., Jacobi, C., Fröhlich, K., & Oberheide, J. (2008). Meteor radar temperatures over Collm (51.3°N, 13°E). *Advances in Space Research*, *42*, 1253–1258. doi:10.1016/j.asr.2007.10.018.

Warner, C. D., & McIntyre, M. E. (2001). An ultrasimple spectral parameterization for nonorographic gravity waves. *Journal of the Atmospheric Sciences*, *58*, 1837–1857.

Wickert, J., Schmidt, T., Beyerle, G., Heise, S., & Reigber, C. (2006). Global atmospheric sounding with GPS radio occultation aboard CHAMP. In J. Flury, R. Rummel, C. Reigber, M. Rothacher, G. Boedecker & U. Schreiber (Eds.), *Observation of the Earth system from space* (pp. 55–67). Berlin: Springer.

Wickert, J., Michalak, G., Schmidt, T., Beyerle, G., Cheng, C.-Z., Healy, S. B., Heise, S., Huang, C.-Y., Jakowski, N., Köhler, W., Mayer, C., Offiler, D., Ozawa, E., Pavelyev, A. G., Rothacher, M., Tapley, B., & Arras, C. (2009). GPS radio occultation: results from CHAMP, GRACE and FORMOSAT-3/COSMIC. *Terrestrial Atmospheric and Oceanic Sciences*, *20*, 35–50. doi:10.3319/TAO.2007.12.26.01(F3C).

Wright, C. J., Osprey, S. M., Barnett, J. J., Gray, L. J., & Gille, J. C. (2010). High resolution dynamics limb sounder measurements of gravity wave activity in the 2006 Arctic stratosphere. *Journal of Geophysical Research*, *115*, D02105. doi:10.1029/2009JD011858.

Chapter 22
Atmospheric Coupling by Gravity Waves: Climatology of Gravity Wave Activity, Mesospheric Turbulence and Their Relations to Solar Activity

Werner Singer, Peter Hoffmann, G. Kishore Kumar, Nicholas J. Mitchell, and Vivien Matthias

Abstract Gravity waves (GW) are important for the coupling between the different regions of the middle atmosphere. They are normally generated in the troposphere, are filtered by the wind field in the stratosphere and lower mesosphere and dissipate at least partly in upper mesosphere and lower thermosphere (MLT). The activity of gravity waves, their filtering by the mean circulation, and the variation of GW activity with solar activity have been studied using long-term wind measurements with Medium Frequency (MF) radars and meteor radars at high and middle northern latitudes. The GW activity is characterized by a semi-annual variation with a stronger maximum in winter and a weaker in summer consistent with the selective filtering of westward and eastward propagating GWs by the mean zonal wind. The latitudinal variation of GW activity shows the largest values in summer at mid-latitudes between 65 km and 85 km accompanied with an upward shift of the height of wind reversal towards the pole. Long-term observations of the MLT winds at mid latitudes indicate a stable increase of westward directed winds below about 85 km and an increase of eastward directed winds above 85 km especially during summer. The observed long-term trend of zonal wind at about 75 km goes along with an enhanced activity of GWs with periods of 3 to 6 hours at altitudes between 80 km and 88 km. In addition, the mesosphere responds to severe solar proton events (SPE) with increased eastward directed winds above about 85 km. The vertical coupling from the troposphere up to the lower thermosphere due to gravity waves and planetary waves is discussed for major sudden stratospheric warmings (SSW) for the winters 2006 and 2009.

W. Singer (✉) · P. Hoffmann · G. Kishore Kumar · V. Matthias
Leibniz-Institute of Atmospheric Physics at the Rostock University, Schlossstr. 6, 18225 Kühlungsborn, Germany
e-mail: singer@iap-kborn.de

P. Hoffmann
e-mail: hoffmann@iap-kborn.de

N.J. Mitchell
Centre for Space, Atmospheric and Oceanic Science, Department of Electronic and Electrical Engineering, University of Bath, Claverton Down, Bath BA2 7AY UK
e-mail: N.J.Mitchell@bath.ac.uk

F.-J. Lübken (ed.), *Climate and Weather of the Sun-Earth System (CAWSES)*, Springer Atmospheric Sciences,
DOI 10.1007/978-94-007-4348-9_22,

22.1 Introduction

During recent years the middle atmosphere (10 to 110 km) has been recognized as an essential part of the climate system. The natural internal variability of the climate and its sensitivity to perturbations can only be understood when the middle atmosphere is taken into account. The mesosphere (50 to 100 km) is of particular importance because both short-term and long-term variations exceed those in the troposphere (0–10 km) by an order of magnitude. To understand the general structure and dynamics of the atmosphere and its variability it is important to know the coupling between different atmospheric regions from the troposphere up to the thermosphere. Often changes caused by solar or non-solar sources are observed at height regions which could be different from the origin of this effect. Important coupling processes are radiative exchanges, transport of chemically active minor constituents and dynamical processes. The dynamical coupling mainly includes forcing of different atmospheric waves (GW, tides, and planetary waves), their propagation through the atmosphere, the interaction between different waves and their impact upon the mean circulation. The dynamical coupling is considered to be the most important aspect of atmospheric coupling. Here especially the gravity waves play an essential role.

Starting with the basic study of *Hines* [1960], many experimental and theoretical investigations have been carried out concerning different aspects of GW sources, propagation, interaction with other waves and the mean circulation, and their dissipation and the creation of turbulence. Excellent reviews of the most important results have been given by *Fritts* [1984] and *Fritts and Alexander* [2003]. As most important GW sources have been detected: topographic generation [e.g., *Nastrom and Fritts*, 1992], convection events [mainly in the tropics, e.g., *Vincent and Alexander*, 2000], frontal systems and jet streams [*Fritts and Nastrom*, 1992]. A survey on the characteristics and influences arising from various GW sources was given by *Fritts et al.* [2006]. The propagation of GW is mainly controlled by the atmospheric wind field. For the normal case of tropospheric sources the GW spectrum in the mesopause region depends markedly on the filtering effect of the wind field in the strato- and mesosphere [*Lindzen*, 1981] due to reflection and absorption of GW at critical levels. Therefore, the observed GW signatures observed in the MLT region contain information about their sources and also about the wind field between the source and the observing region.

The chapter is organized as follows: In Sect. 22.2 our data base and methods are described. Long-term variations of winds and GWs are presented in Sect. 22.3. Our results of vertical coupling processes during SSWs are a summarized in Sect. 22.4, whereas Sect. 22.5 presents the responses of the MLT wind field on SPE. Finally, the main results are summarized in Sect. 22.6.

22.2 Experimental Techniques

Continuous measurements of winds in the mesosphere and lower thermosphere as well as temperatures around the mesopause have been carried out with MF radars

Table 22.1 Technical details of the MF and meteor radars at Juliusruh and Andenes

	Juliusruh 55°N, 13°E		Andenes 69°N, 16°E		
Radar	MF	Meteor	MF	Meteor	MF Doppler
Frequency	3.18 MHz	32.55 MHz	1.98 MHz	32.55 MHz	3.17 MHz
Beam width	$\sim$15°	All-sky <60°	$\sim$60°	All-sky <60°	6.4°
Pulse length	27 μs	13 μs	27 μs	13 μs	10 μs
Sampling res.	2 km	2 km	2 km	2 km	1 km
Height range	70–94 km	80–100 km	70–94 km	80–100 km	60–94 km
Parameters	winds (FCA)	winds (Doppler)	winds (FCA)	winds (Doppler)	electron density, Doppler winds, turbulence
Observations since	1990–	1999–2001 2007–	1998–	2001– Kiruna: 1999–	2003–

and meteor radars at Andenes (69°N, 16°E), and Juliusruh (55°N, 13°E). The basic parameters of the radars are summarized in Table 22.1. Winds from the MF radars at Juliusruh (3.18 MHz) and Andenes (1.98 MHz) are determined by the spaced-antenna method [e.g., *Briggs*, 1984], details about these radars are given by *Singer et al.* [1997], *Keuer et al.* [2007] and *Hoffmann et al.* [2011].

The all-sky meteor radars at Andenes and Juliusruh are classical meteor radars [e.g., *Hocking et al.*, 2001; *Singer et al.*, 2003]. From each meteor the radial velocity of the meteor trail due to its movement with the background wind is estimated. The data are binned in height intervals of 3 km and time intervals of one hour to determine horizontal winds between 82 and 98 km. Furthermore, at the peak of the meteor layer at about 90 km (for 32.55 MHz) daily mean temperatures were derived from the height variation of the meteor decay time in combination with an empirical model of the mean temperature gradient at the peak altitude [for details see *Singer et al.*, 2004; *Hocking et al.*, 2004].

The acquired time series from the MF and meteor radars with lengths up to 20 years allow studies of possible trends and solar cycle variations of the wind field in the MLT region. Prevailing winds as well as diurnal and semidiurnal tides are obtained from harmonic analysis of a sliding 4-day composite of hourly winds for each radar [e.g. *Singer et al.*, 2005]. If available, we combine the winds obtained by the MF radars from 70–82 km and meteor radars from 82–95 km to get reliable winds over an extended altitude range from 70–95 km at Andenes and Juliusruh [see also *Hoffmann et al.*, 2007, 2010]. The advantages of this method are obvious when comparing both panels in Fig. 22.1, showing an excellent agreement between both data sets at 82–84 km, but also indicating differences at altitudes above 85 km where the MF radar winds tend to be smaller than those recorded by the meteor radars. This is in agreement with earlier works showing MF radar winds above about 85 km were weak compared with winds measured by the UARS satellite [e.g. *Burrage et al.*, 1996]. Monthly medians of prevailing winds and tides are used for time series

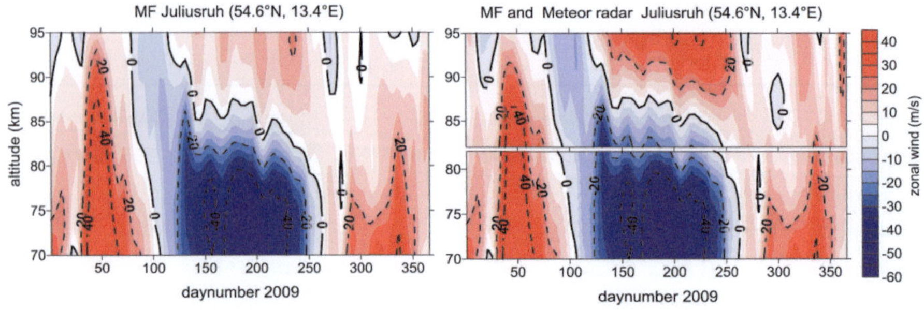

Fig. 22.1 Mean zonal winds observed at Juliusruh in 2009. *Left panel*: MF radar winds (70–95 km); *Right panel*: combined MF radar (70–82 km) and meteor radar (82–95 km) winds

analysis, as medians are outlier-resistant [e.g. *Weatherhead et al.*, 1998]. Ground-based and satellite observations of tidal signatures in the MLT region during the CAWSES tidal campaigns showed that satellite and ground-based tidal characteristics are in good agreement and it could be proved for the first time that both types of observations are consistent with each other [e.g., *Ward et al.*, 2010].

Additional parameters, e.g. turbulence and electron density used in Sects. 22.4 and 22.5, were derived with the narrow beam Saura MF Doppler radar [*Latteck et al.*, 2005; *Singer et al.*, 2008, 2011].

22.3 Climatology of Winds and Gravity Wave Activity

22.3.1 Trends and Solar Cycle Variation of Mean Winds

Ten years of meteor wind observations at Andenes cover nearly a complete solar cycle. The data base could be extended to a full solar cycle adding the meteor radar observations from Kiruna, Sweden [*Mitchell et al.*, 2002] for the period August 1999 to August 2001. Both systems (270 km apart) use identical radar technology and a comparison of the wind measurements demonstrated that both sites give representative information about the MLT dynamics over northern Scandinavia [*Pancheva et al.*, 2007]. The extended data set has been used to get an initial estimate of possible trends and solar activity variations of the MLT wind field although longer time series would be preferable for a reliable separation of solar activity influence and trend. The seasonal variation of the zonal mean wind for the period 1999 to 2010 is dominated by the persistent feature of a strong westward directed jet below about 88 km and an eastward directed jet above that altitude in summer (Fig. 22.2).

The wind field is subject to strong variability in winter due to planetary waves and associated sudden stratospheric warmings (see also Sect. 22.4). A solar activity related wind variation is obvious for altitudes above 90 km in summer with larger winds at high solar activity. The seasonal variation of zonal wind at 88 km is shown

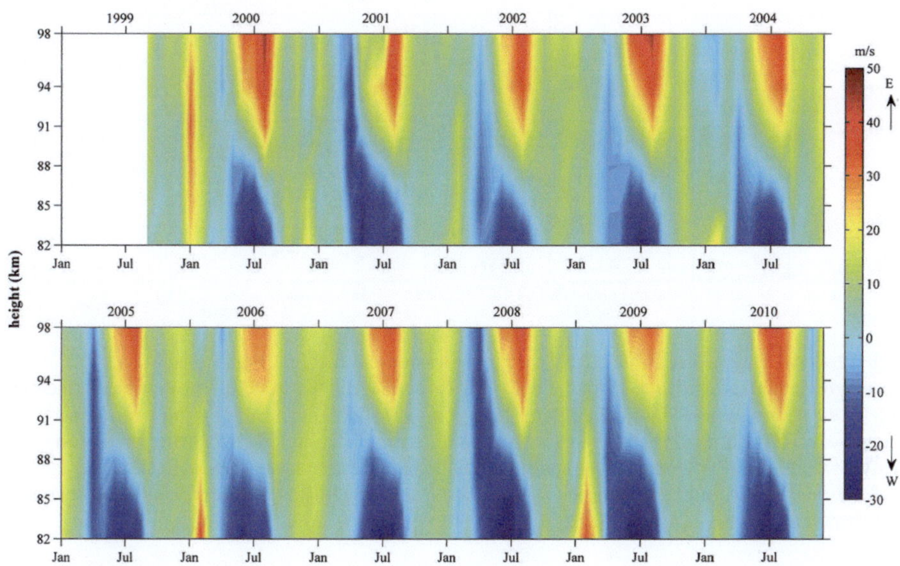

Fig. 22.2 Height-time cross section of zonal winds derived from Meteor radar measurements at Andenes and Kiruna between August 1999 and December 2010

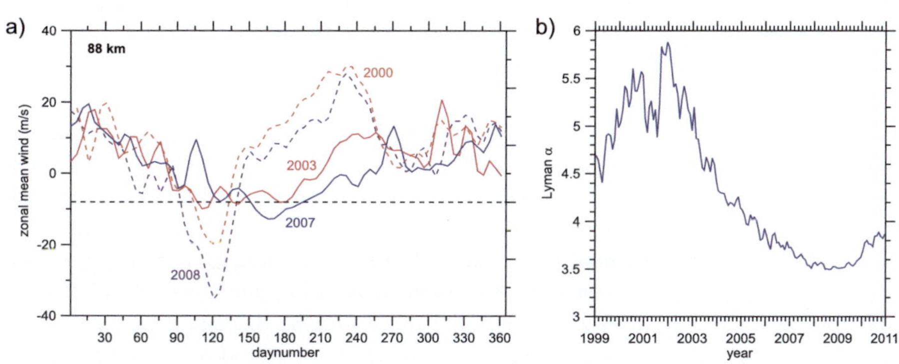

Fig. 22.3 (**a**) Seasonal variation of zonal winds at Andenes (69°N, *full lines*) for 2003 (enhanced solar activity) and 2007 (low solar activity) and at Juliusruh (55°N, *dashed lines*) for 2000 (solar maximum) and 2008 (solar minimum) at 88 km. Mean zonal wind from 10-day composite analyses are depicted. (**b**) Variation of the solar Lyman α radiation (units: 10^{11} ph. cm^{-2} s^{-1}) between 1999 and 2010 basing on monthly mean values

for a year of enhanced solar activity (2003) and a year of low solar activity (2008) in Fig. 22.3a. Enhanced eastward winds are evident in summer 2003 (June, July, August) compared to summer at low solar activity whereas no clear signatures can be seen in winter. The solar activity varied during cycle 23 by a factor 1.7 between solar maximum and minimum as shown by the Lyman α radiation in Fig. 22.3b.

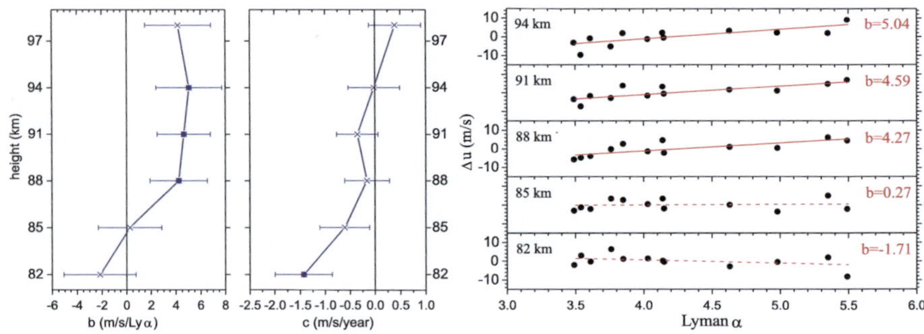

Fig. 22.4 (**a**) Height variation of the partial regression coefficients b and c (see Eq. (22.1)) between zonal wind at Andenes and solar activity (*left panel*) and the trend of zonal wind at Andenes (*right panel*) for summer (July, August). The significance level is marked by *full dots*, nonsignificant coefficients are marked by *crosses*. (**b**) Dependence of detrended zonal wind values on solar activity at altitudes between 82 km and 94 km. The significance level ≥95 % is characterized by *red full lines*, nonsignificant levels by *red dashed lines*. In addition, the regression coefficients b (corresponding to Fig. 22.4a) are added

Solar activity variations and trends of the MLT wind field are studied using robust regression [e.g. *Holland and Welsch*, 1977; *MATLAB*, 2011]. The robust fitting method is less sensitive than ordinary least squares to large changes in small parts of the data. The influence of the solar activity on the wind field and trends in the wind field are analyzed using the following twofold regression equation

$$u = a + b \times Ly\alpha + c \times year. \tag{22.1}$$

Regression analyses have been done separately for the zonal and meridional mean winds as well as for the tidal components at heights between 82 km and 98 km on the basis of seasonal median values. Summer median values are estimated from the results of harmonic analyses of 4-day composites shifted by one day covering July and August in total of 62 days. The most significant partial regression coefficients have been obtained for the zonal mean wind and summer conditions. The solar activity variation and a possible trend will be discussed in the following. Significantly increasing eastward directed winds are found between 88 km and 94 km with positive wind changes in the order of 4 to 5 m/s per Lyman α flux unit (Fig. 22.4a).

This value is in good agreement with the wind variations of about 5 to 10 m/s presented in Fig. 22.3a for summer 2003/2007 if we consider the corresponding solar activity variation by a factor of 1.3. The wind changes as a function of Lyman α flux are presented in Fig. 22.4b. A significant but small negative trend in zonal wind has been found only at 82 km with values around -1 m/s per year. This value agrees well with the results obtained at mid-latitudes from MF radar wind measurements at 3.18 MHz [Fig. 14 in *Keuer et al.*, 2007]. Increasing eastward directed winds with increasing solar activity are found at high latitudes between 88 km and 94 km in comparison with decreasing eastward directed winds at mid-latitudes at 90 km and below [Fig. 7 in *Keuer et al.*, 2007]. This difference is possibly related to the

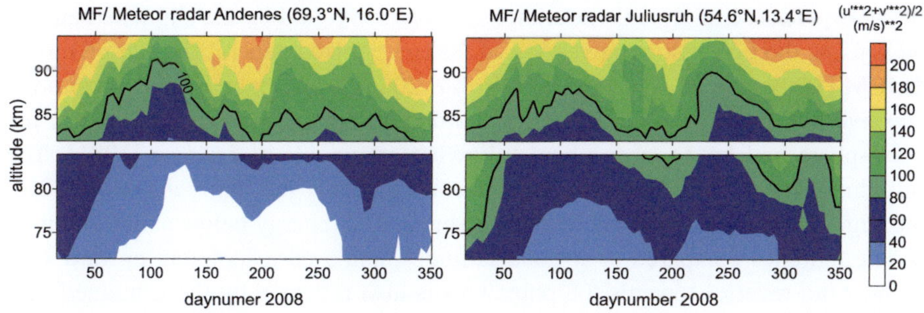

Fig. 22.5 Mean seasonal variations of GW kinetic energy derived from wind variances periods 3–9 hours from MF and meteor radar observations at Andenes (69°N) and Juliusruh (55°N) during 2008. Adopted in a modified form from *Hoffmann et al.* [2010]

different measurement techniques, as MF radars estimate smaller wind magnitudes above about 86 km–90 km compared to other techniques. Similar results are known from other investigations [for details see also *Hocking and Thayaparan*, 1997; *Manson et al.*, 2004; *Jacobi et al.*, 2009; *Engler et al.*, 2008]. Meteor radar observations at Juliusruh during solar maximum (2000) and solar minimum (2008) showed a smaller increase of zonal winds with solar activity compared to the high-latitudes but the magnitudes at mid-latitudes are greater by a factor of about two (Fig. 22.3a). Nonsignificant solar activity variations and trends were found in the MF radar winds at 1.98 MHz at Andenes which are possibly related to the general weaker winds at high-latitudes and additionally influenced by the reduced MF radar winds above 84 km compared to other methods.

22.3.2 Annual and Latitudinal Variations of Gravity Waves

Continuous observations of mesospheric MF and Meteor radar winds allow the investigation of GW, first shown by *Manson and Meek* [1986]. In the following we use the kinetic energy derived by the sum of zonal and meridional wind variances divided by two as a measure of the GW activity. First, mean winds and tidal variations averaged over 7 days are removed from the hourly wind values to reduce the effects of background winds and tidal amplitudes on the GW activity. Then the variances for defined period bands are derived by wavelet transforms applying the Morlet mother wavelet of the 6-th order [for details see *Torrence and Compo*, 1998; *Serafimovich et al.*, 2005]. A detailed description of the method applied to the data of one particular year is given in *Hoffmann et al.* [2010]. As an example, Fig. 22.5 shows the seasonal variation of the GW activity during 2008, as evaluated using the combined MF and meteor radar winds at Andenes and Juliusruh.

At both latitudes, the observations show a semi-annual variation with strongest GW energy during winter and a secondary maximum during summer. Minima of the GW activity appeared during the equinoxes at altitudes of about 90 km [*Hoffmann et al.*, 2010]. This semi-annual variation has been also found by the simulated annual cycle using the GW resolving version of the mechanistic general circulation model KMCM [for details of the model see *Becker*, 2009], and can be explained with the selective filtering of westward and eastward GWs by the mean zonal wind in the stratosphere and lower mesosphere following arguments of *Lindzen* [1981]. The latitudinal dependence during summer [for more details see *Hoffmann et al.*, 2010, and Fig. 13 therein] is characterized by stronger GW energy below about 85 km at middle latitudes compared to at polar latitudes, and a corresponding upward shift of the wind reversal towards the pole. This is also reflected by the simulated GW drag. A possible explanation is the reduced westward flow in the polar summer stratosphere and lower mesosphere relative to middle latitudes, which results in a reduced filtering of westward propagating GWs and a stronger damping of eastward propagating waves.

22.3.3 Trends of Gravity Wave Activity

The seasonal variation of the GW activity as investigated in detail in [*Hoffmann et al.* [2010] is based on radar data from one particular year and simulations with the KMCM. However, using all data derived from observations with the MF radar in Juliusruh from 1990–2010, the averaged annual cycle of GW activity [see *Hoffmann et al.*, 2011, and Fig. 1b therein] confirms the semiannual variation with minima during the equinoxes, maxima during winter and slightly weaker during July/August. As mentioned above, the variability of the GW activity is generally assumed to be determined by the filtering due to the background winds in the stratosphere and lower mesosphere. Therefore, in *Hoffmann et al.* [2011] we investigated the following hypothesis: If there are trends in the background winds as e.g. shown by *Keuer et al.* [2007] using the Juliusruh-MF radar winds from 1990–2005, then these trends are considered to be responsible for long-term variations of gravity waves due to the filtering effect.

At first, we extended the trend analysis to the years 1990–2010. We found, as shown in Fig. 22.6a, stable negative trends of the zonal winds during summer, i.e., an enhancement of the westward directed summer jet, which dominates the seasonal variation during all years (see also Fig. 22.1). Using monthly mean values for July we found that the observed zonal wind trend at about 75 km goes along with an enhanced activity of GW with periods between 3–6 hours at altitudes above 80 km (Fig. 22.6b).

Indeed, also the year-to-year variation of maxima of the observed westward directed winds at altitudes near 75 km and the GW activity at about 80 km are significantly correlated (Fig. 22.6c) thus stimulating the further study of long-term wind changes and corresponding GW trends.

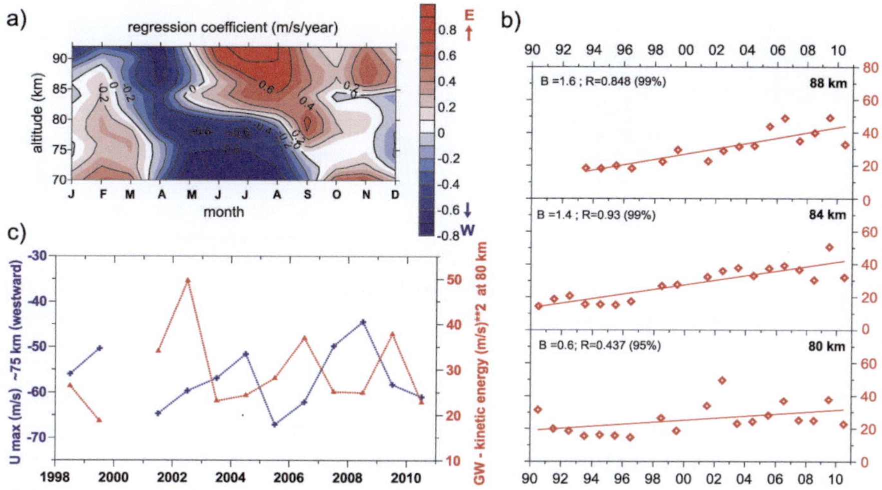

Fig. 22.6 (**a**) Mean height-time cross section of the trend in the zonal wind at Juliusruh from 1990–2010. (**b**) Trend of kinetic energy of GW (3–6 h) at Juliusruh at 80, 84 and 88 km for July. (**c**) Variation of GW—activity (kinetic energy, *red line*) at 80 km and westward directed jet strength (~75 km, *blue line*) during July at 55°N from 1998–2010 (no data during July 2000). Modified version of Figs. 2, 4, and 5 in *Hoffmann et al.* [2011]

22.4 Vertical Coupling During Sudden Stratospheric Warmings

22.4.1 Sudden Stratospheric Warmings in Winters 2006 and 2009

Continuous MF and meteor radar observations have been used for detailed studies of winds and temperatures in the upper mesosphere and lower thermosphere to understand the strong variability in winter 2006/07 due to enhanced planetary wave activity and related stratospheric warming events [*Hoffmann et al.*, 2007]. Such events are considered as an exceptionally striking vertical coupling process between lower, middle and upper atmosphere. It has been shown, that the strength of the observed zonal wind reversal in the mesosphere during SSW is decreasing with latitude. There are only weak longitudinal variations of zonal winds and temperatures at mesopause heights during major warmings but stronger longitudinal variations of meridional winds, as confirmed by comparisons to similar observations at Resolute Bay (75°N, 95°W) and Poker Flat (65°N, 147°W). The results during winter 1998/1999 and 2006/07 indicate the occurrence of a planetary wave 1 structure in the mesosphere. The short term reversals of the mesospheric winds during SSWs are followed by periods of intensified westerly winds connected with enhanced turbulent energy dissipation rates below 85 km, as estimated with the narrow beam Saura MF radar, and an increase of the gravity wave activity in the altitude range 70–85 km [more details about the winter 2006/2007 are given by *Hoffmann et al.*, 2007]. The seasonal variation of the gravity wave activity derived from the horizontal wind variability for periods between 3–9 hours and the turbulent energy dissipation rates in 2006 are

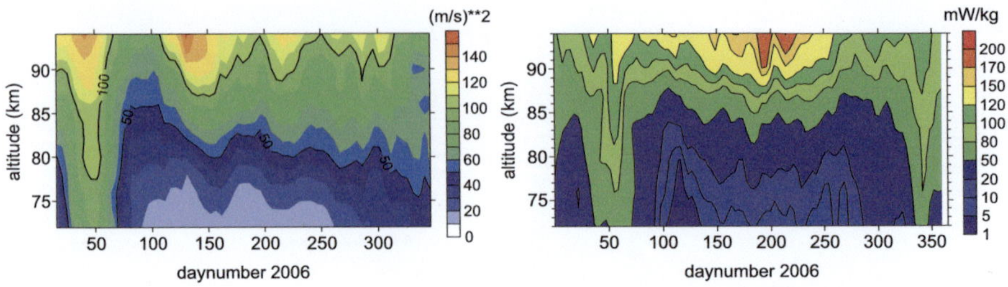

Fig. 22.7 Seasonal variation of horizontal wind variability for periods 3–9 hours (*left panel*) and seasonal variation of turbulent kinetic energy dissipation rate (*right panel*) over Andenes in 2006

shown in Fig. 22.7. Enhanced turbulent energy dissipation rates in February and in December go along with enhanced GW activity indicating that enhanced turbulence at that time is related to GW dissipation. In addition, the seasonal evolution and turbulence strength observed by the Saura MF radar is in general agreement with in situ measurements at Andenes [*Lübken*, 1997]. In February, the enhanced gravity wave activity and the intensified westerly winds are associated also with a period of reduced planetary wave 1 activity in the stratosphere [*Hoffmann et al.*, 2007].

22.4.2 Major Stratospheric Warming 2009—Downwelling of NO

The major stratospheric warming in January 2009 is accompanied by a jump of the stratopause and its reformation at about 80 km in the first days of February as shown in (Fig. 22.8) by zonal averaged temperatures (0°E–30°E) over Andenes from the Microwave Limb Sounder (MLS) instrument on the NASA-EOS Aura satellite [*Livesey et al.*, 2007].

The stratopause returns to its normal level over a period of about six weeks. The observed cooling of the lower mesosphere is associated with forced propagation of eastward gravity waves during the enhanced zonal wind after wind reversal (see also Fig. 22.10). The major stratospheric warmings in winters 2004 and 2006 showed a similar behavior comparable to the descent of the stratopause in February 2009. Also *von Zahn et al.* [1998] observed a descent of an elevated stratopause by about 30 km in January/February 1998 with lidar observations at Andenes.

The descent of the stratopause to its normal level brings down minor constituents in the winter middle atmosphere which are present in the upper mesosphere/lower thermosphere. One important trace species of the lower thermosphere is nitric oxide NO which has a long life time in absence of sun light during the polar night. NO will be immediately ionized by solar UV at solar zenith distances less than about 98° under production of free electrons. The downwelling of nitric oxide has been studied using electron density as proxy measured with the Saura MF radar [*Singer et al.*, 2011] as local ionization of nitric oxide by precipitating energetic particles should not happen under solar minimum conditions. Time series of electron density

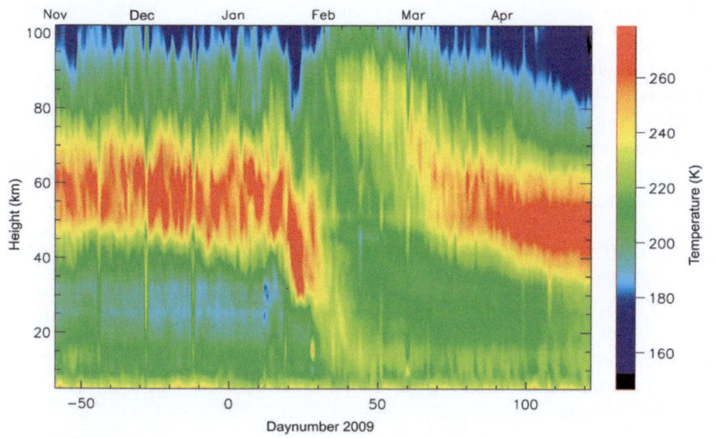

Fig. 22.8 MLS temperatures over Andenes (69°N, 16°E) around the major stratospheric warming in January 2009, a zonal average for the longitudes between 0°E and 30°E is shown. The stratospheric warming is accompanied by a jump of the stratopause and its reformation at about 80 km

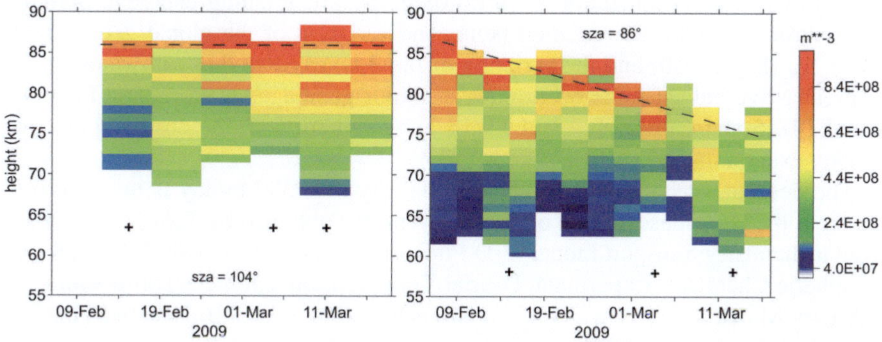

Fig. 22.9 Height-time cross-section of electron densities over Andenes at constant solar zenith angles (sza) 86° and 104° in February/March 2009 during the reformation phase of the stratopause after the stratospheric warming. The electron densities under sunlit conditions (sza = 86°) are descending with a rate of about 300 m/day whereas nearly constant electron densities are observed during night (sza = 104°) as indicated by *dashed lines*. The *crosses* mark geomagnetically disturbed periods with possibly direct production of NO_X by energetic particles (for details see text)

at constant solar zenith angles of 86° (sunlit conditions) and 104° (absence of sun light) were analyzed over periods of three days and six days, respectively due to the high background noise during polar night. Hight-time cross-sections of the observed electron densities for the period February 6 to March 17 are depicted in Fig. 22.9.

Under sunlit conditions (right panel) a descent of the electron peak density was found with a mean descent rate of about 300 m/day. During polar night a more or less constant height of peak electron density was estimated (Fig. 22.9, left panel). In both cases excursions are observed which are marked by crosses which are possibly

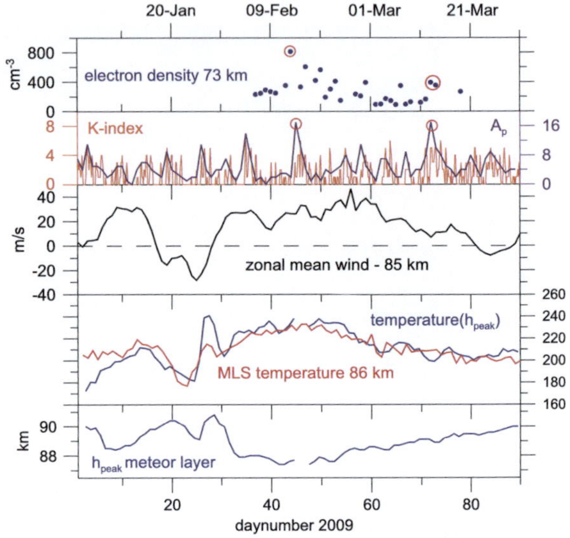

Fig. 22.10 Mesospheric electron density at 73 km during local noon, global geomagnetic activity A_p and 3-hourly K-index from Tromsø, mean zonal wind at 85 km prior, during, and after the stratospheric warming event, mesopause temperatures from MLS measurements and meteor decay times, and peak height of the meteor layer for the period January to March 2009. The *red circles* indicate minor geomagnetic storms

related to the direct ionization of nitric oxide by precipitating energetic particles from aurora or from the radiation belts. The variation of the global geomagnetic index A_p (Fig. 22.10) and the variation of the 3-hourly K index with values equal or greater five indicates the occurrence of minor geomagnetic storms on February 4 and 14/15 and on March 4 and 13/14 in agreement with the observed electron density excursions.

The descent rate of day-time electron density of ~300 m/day is not so different from the NO_X descent rate of about 700 m/day obtained by *Salmi et al.* [2011] from a chemistry transport model (3-D FinRose) in connection with ACE-FTS (Atmospheric Chemistry Experiment Fourier Transform Spectrometer) observations in February/March 2009. WACCM simulations by *Smith et al.* [2011] showed that the wave driven mean circulation is the dominating transport process bring down trace species in the winter middle atmosphere. The downwelling phase in February is accompanied with a cooling and shrinking of the middle atmosphere which is also reflected by a lowering of the peak height of the meteor layer during that time as the meteoroids descending on the upper atmosphere with the same entry velocities burn up at the same pressure level (lowest panel of Fig. 22.10). In addition, the neutral temperatures derived from meteor decay times at the peak height of the meteor layer agree well with the temperatures from MLS measurements at 86 km in magnitude and temporal evolution.

22.5 Mesospheric Response on Solar Activity Storms

During solar proton events (SPE) large fluxes of energetic protons impacted the mesosphere and stratosphere causing ionization, dissociation, and excitation. Drastic changes of the ionospheric plasma with excessive ionization enhancements down

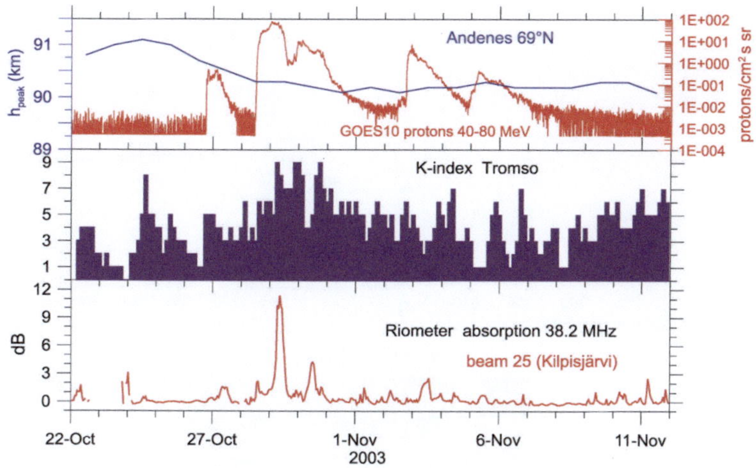

Fig. 22.11 Solar proton events in October/November 2003: solar proton fluxes with energies 40–80 MeV from GOES 10 observations and variation of the peak height of the meteor layer, geomagnetic activity expressed by the 3-hourly K-index from Tromso indicating severe geomagnetic storms (K = 9) after October 28, and ionospheric radio wave absorption of the imaging riometer at Kilpisjärvi (beam 25 sampled the ionosphere over Andenes)

to altitudes of 60 km are well known against it measurements of dynamical and structural changes during these perturbations are rare. The MF radar and meteor radar observations at Andenes (69°N) and Juliusruh (55°N) have been used to study the influence of the strongest four SPEs on the MLT dynamics occurring during the last decade. These SPE events appeared in July 2000 (24000 pfu), October 2003 (29500 pfu), January 2005 (5040 pfu), and December 2006 (1980 pfu), the values in brackets characterize the strength of the event through the proton flux with energies greater than 10 MeV: 1 pfu = 1 protons/$cm^2\,s^{-1}$ sr. Special emphasis is given to the observations during the strongest SPE in October 2003 as *Jackman et al.* [2007] used the Thermosphere Ionosphere Mesosphere Electrodynamic General Circulation Model (TIME-GCM) to study the mesospheric changes induced by the October 2003 event.

The interactions of protons with air molecules and trace species resulted in the production of odd hydrogen and odd nitrogen. The enhancements of these constituents lead to a destruction of ozone via catalytic cycles resulting in a cooling of the lower mesosphere below about 80 km due to reduced ozone heating. A weak temperature increase was found above 80 km peaking around 95 km. The Joule heating rates were much less than the solar/chemical heating rates. Temperature changes up to ±2.6 K and wind changes up to 20–25 % were obtained whereas the largest response was located in the sunlit southern hemisphere.

MF and meteor radar winds cover the height range between 70 and 94 km during the October 2003 SPE on the northern winter hemisphere. The proton event started on October 28 and reached its maximum on October 29, the geomagnetic field over northern Scandinavia was severely disturbed from October 29 onward

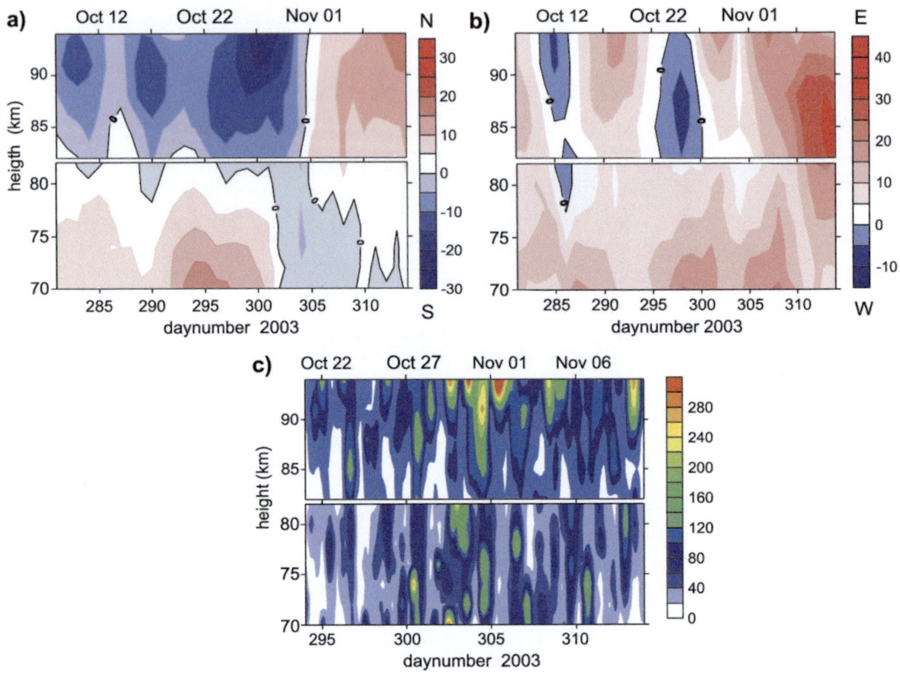

Fig. 22.12 MF/meteor radar observations during the SPEs in October/November 2003: (**a**) mean meridional winds, (**b**) mean zonal winds, (**c**) variances of horizontal wind disturbances for periods 3–9 hours

causing additional ionization due to precipitation energetic particles from the radiation belts indicated by the excessive riometer absorption over Andenes (Fig. 22.11). In addition, the peak altitude of the meteor layer was decreasing during the SPE by up to about 1 km indicating a cooling and shrinking of the lower mesosphere below. At the maximum of the event on October 29 (day 302) and a few days later a reversal of the meridional winds from equatorwards to polewards was observed above 82 km and the opposite situation below that altitude. At same time the zonal winds are increasing above about 80 km and below 78 km (Fig. 22.12).

The increase of eastward directed winds is associated with enhanced GW activity at these altitudes and above as eastward propagating GWs can dissipate at lower altitudes (Fig. 22.12c).

An increase of the eastward directed zonal mean winds has been observed above 85 km for all SPEs studied here (Fig. 22.13). Below 85 km a positive zonal wind response was found only in winter whereas no significant differences (less than 2–5 m/s) between undisturbed and disturbed wind fields appeared in summer and autumn. *Becker and von Savigny* [2010] studied the mesospheric response on an SPE during summer using a mechanistic general circulation model. They found that the temporary ozone reduction during an SPE results in an positive zonal wind response between 70 km and about 95 km in the order of 2–8 m/s what is in general agreement with our observations during the July 2000 event.

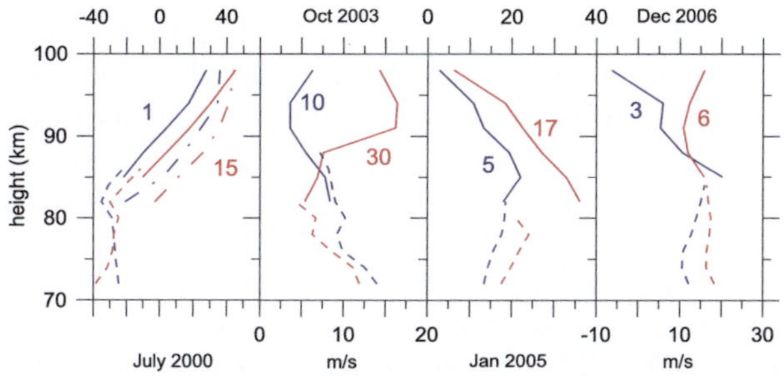

Fig. 22.13 Zonal mean winds (m/s) before the solar proton events (*blue lines*) and at the maximum phase (*red lines*) of the events in July 2000, October 2003, January 2005, and December 2006. Meteor radar observations at Andenes/Kiruna (*full lines*) and Juliusruh (*dashed-dotted lines*) as well as MF radar (*dashed lines*) observations are depicted (the figures indicate the corresponding days, note the different scales for summer and winter)

22.6 Summary

Long-term wind measurements by MF and meteor radars at high and middle latitudes from 1990 to 2011 provide reliable data to study trends and apparent solar activity related variations of the wind field between about 70 km and 95 km during the measurement period. Significant trends with time and with solar activity are found for the zonal wind preferred in summer. At mid-latitudes a negative zonal wind response (decreasing strength of eastward winds) with time was found below about 82 km from MF radar observations whereas at high latitudes a negative zonal wind response was determined in a height of 82 km from meteor radar observations in the same order of about -1 m/s per year. A significant variation with solar activity was observed at high latitudes in summer (July and August of the years 2000 to 2011) at altitudes above 85 km with a positive zonal wind response (increasing strength of eastward winds) in the order of 4 to 5 m/s per Lyman α flux unit. That is a first indication of an apparent solar activity induced wind variation but longer time series are required to verify the finding. *Keuer et al.* [2007] obtained from MF radar measurements at mid-latitudes for summer (April to September, 1990 to 2005) a negative zonal wind response in the order of about 5 m/s per Lyman α flux unit for altitudes below 90 km. In addition, a general positive zonal wind response was found during the strongest solar proton events occurred between 2000 and 2006 for altitudes above 85 km. No significant changes between disturbed and undisturbed conditions have been seen below 85 km. The strong response of the upper mesosphere on severe solar activity storms during all seasons indicates that also high levels of solar UV radiation can possibly induce changes of the wind field above 85 km as found at high latitudes for the period 2000 to 2011. A possible mechanism effecting the wind field at altitudes below 85 km can be related to the influence of the global warming effect on the mesospheric circulation. Changes of the zonal wind field obtained with the Hamburg Model of the Neutral and Ionized Atmosphere (HAMMONIA) for doubling the CO2 content show qualitatively the same behavior of a decreasing strength of eastward winds below 83 km in summer and increasing eastward winds above 83 km [*Schmidt et al.*, 2006]. The analysis of 12 years of observations provides some evidence of possible trends and solar activity related variations of the mesospheric wind field but at least 10 years more observations are needed to verify the result.

The wind measurements have been used to study the seasonal and inter-annual variations of GW at high and middle northern latitudes. The annual cycle of the GW activity is characterized by a semi-annual variation with a stronger maximum in winter and a weaker in summer consistent with the selective filtering of westward and eastward propagating GWs by the mean zonal wind. Long-term changes of the background winds influence the activity of upward propagating GWs at mid-latitudes. The most significant long-term trend of zonal wind at about 75 km during July goes along with an enhanced activity of GWs with periods of 3 to 6 hours at altitudes between 80 km and 88 km. The direct relation between the maxima of the westward directed winds at altitudes near 75 km and the enhanced GW activity at about 80 km shows a significant anticorrelation and stimulates further studies of long-term wind changes and corresponding GW trends.

The mesospheric response on major sudden stratospheric warmings and on solar proton events looks quite similar. After SSWs and SPEs in winter a general increase of the eastward directed winds associated with mesospheric coolings and an increase of turbulence are observed but the reasons are different. Gravity waves play an important role in the coupling during SSWs as the mesospheric coolings associated with SSWs are intensified by improved propagation of eastward directed gravity waves due to enlarged eastward background winds. The mesospheric coolings and related wind changes after SPEs result from reduced ozone heating due to ozone destruction by energetic protons.

Acknowledgements The authors are grateful to Erich Becker and Markus Rapp for their support and helpful discussions. We also thank Ralph Latteck and Dieter Keuer for their support running the radars at Andenes and Juliusruh. This work has been supported by DFG in the frame of the CAWSES priority program SPP 1176 under grants SI 501/5-1 and SI 501/5-2. We thank the Jet Propulsion Laboratory/NASA for providing data access to the Aura/MLS level 2.2 retrieval product. The data originated from the Imaging Riometer for Ionospheric Studies (IRIS), operated by the Space Plasma Environment and Radio Science (SPEARS) group, Department of Physics, Lancaster University (UK) in collaboration with the Sodankylä Geophysical Observatory.

References

Becker, E. (2009). Sensitivity of the upper mesosphere to the Lorenz energy cycle of the troposphere. *Journal of the Atmospheric Sciences*, *66*, 647–666.

Becker, E., & von Savigny, C. (2010). Dynamical heating of the polar summer mesopause by solar proton events. *Journal of Geophysical Research*, *115*, D00I18. doi:10.1029/2009JD012,561.

Briggs, B. (1984). The analysis of spaced sensor records by correlation techniques. In R. Vincent (Ed.), *Handbook for MAP: Vol. 13. Middle atmosphere program* (pp. 166–186). SCOSTEP.

Burrage, M. D., Skinner, W. R., Gell, D. A., Hays, P. B., Marshall, A. R., Ortland, D. A., Manson, A. H., Franke, S. J., Fritts, D. C., Hoffmann, P., McLandress, C., Niciejewski, R., Schmidlin, F. J., Shepherd, G. G., Singer, W., Tsuda, T., & Vincent, R. A. (1996). Validation of mesosphere and lower thermosphere winds from the high resolution Doppler imager on UARS. *Journal of Geophysical Research*, *101*(D6), 10365–10392. doi:10.1029/95JD01700.

Engler, N., Singer, W., Latteck, R., & Strelnikov, B. (2008). Comparison of wind measurements in the troposphere and mesosphere by VHF/MF radars and in-situ techniques. *Annales Geophysicae*, *26*, 3693–3705.

Fritts, D. C. (1984). Gravity wave saturation in the middle atmosphere: a review of theory and observations. *Reviews of Geophysics and Space Physics*, *22*, 275–308.

Fritts, D. C., & Alexander, M. J. (2003). Gravity wave dynamics and effects in the middle atmosphere. *Reviews of Geophysics*, *41*(1), 1003. doi:10.1029/2001RG000106.

Fritts, D. C., & Nastrom, G. D. (1992). Sources of mesoscale variability of gravity waves. Part II: frontal, convective, and jet stream excitation. *Journal of the Atmospheric Sciences*, *49*, 111–127.

Fritts, D. C., Vadas, S. L., Wan, K., & Werne, J. A. (2006). Mean and variable forcing of the middle atmosphere by gravity waves. *Journal of Atmospheric and Solar-Terrestrial Physics*, *68*, 247–265.

Hines, C. O. (1960). Internal atmospheric gravity waves at ionospheric heights. *Canadian Journal of Physics*, *38*, 1441–1481.

Hocking, W. K., & Thayaparan, T. (1997). Simultaneous and colocated observation of winds and tides by MF and meteor radars over London, Canada (43°N, 81°W), during 1994–1996. *Radio Science*, *32*(2), 833–865. doi:10.1029/96RS03467.

Hocking, W. K., Fuller, B., & Vandepeer, B. (2001). Real-time determination of meteor-related parameters utilizing modern digital technology. *Journal of Atmospheric and Solar-Terrestrial Physics*, *63*, 155–169.

Hocking, W. K., Singer, W., Bremer, J., Mitchell, N. J., Batista, P., Clemensha, B., & Donner, M. (2004). Meteor radar temperatures at multiple sites derived with SKiMET radars and compared to OH, rocket and lidar measurements. *Journal of Atmospheric and Solar-Terrestrial Physics*, *66*, 585–593.

Hoffmann, P., Singer, W., Keuer, D., Hocking, W. K., Kunze, M., & Murayama, Y. (2007). Latitudinal and longitudinal variability of mesospheric winds and temperatures during stratospheric warming events. *Journal of Atmospheric and Solar-Terrestrial Physics*, *69*(17–18), 2355–2366. doi:10.1016/j.jastp.2007.06.010.

Hoffmann, P., Becker, E., Singer, W., & Placke, M. (2010). Seasonal variation of mesospheric waves at northern middle and high latitudes. *Journal of Atmospheric and Solar-Terrestrial Physics*, *72*(14–15), 1068–1079. doi:10.1016/j.jastp.2010.07.002.

Hoffmann, P., Rapp, M., Singer, W., & Keuer, D. (2011). Trends of mesospheric gravity waves at northern middle latitudes during summer. *Journal of Geophysical Research*, *116*, D00P08. doi:10.1029/2011JD015717.

Holland, P. W., & Welsch, R. (1977). Robust regression using iteratively reweighted least squares. *Communications in Statistics. Theory and Methods*, *A6*, 813–827.

Jackman, C. H., Roble, R. G., & Fleming, E. L. (2007). Mesospheric dynamical changes induced by the solar proton events in October–November 2003. *Geophysical Research Letters*, *34*, L04812. doi:10.1029/2006GL028328.

Jacobi, C., Arras, C., Kürschner, D., Singer, W., Hoffmann, P., & Keuer, D. (2009). Comparison of mesopause region meteor radar winds, medium frequency radar winds and low frequency drifts over Germany. *Advances in Space Research*, *43*, 247–252. doi:10.1016/j.asr.2008.05.009.

Keuer, D., Hoffmann, P., Singer, W., & Bremer, J. (2007). Long-term variations of the mesospheric wind field at mid-latitudes. *Annales Geophysicae*, *25*(8), 1779–1790.

Latteck, R., Singer, W., & Hocking, W. K. (2005). Measurement of turbulent kinetic energy dissipation rates in the mesosphere by a 3 MHz Doppler radar. *Advances in Space Research*, *35*(11), 1905–1910.

Lindzen, R. S. (1981). Turbulence and stress owing to gravity wave and tidal breakdown. *Journal of Geophysical Research*, *86*, 9707–9714.

Livesey, N. J., Read, W. G., Lambert, A., Cofield, R. E., Cuddy, D. T., Froidevaux, L., Fuller, R. A., Jarnot, R. F., Jiang, J. H., Jiang, Y. B., Knosp, B. W., Kovalenko, L. J., Pickett, H. M.,

Pumphrey, H. C., Santee, M. L., Schwartz, M. J., Stek, P. C., Wagner, P. A., Waters, J. W., & Wu, D. L. (2007). *EOS MLS version 2.2 level 2 data quality and description document* (Technical Report, Version 2.2 D-33509). Jet Propulsion Lab., California Institute of Technology, Pasadena, California, 91198-8099.

Lübken, F.-J. (1997). Seasonal variation of turbulent energy dissipation rates at high latitudes as determined by in situ measurements of neutral density fluctuations. *Journal of Geophysical Research, 102*, 13441–13456.

Manson, A. H., & Meek, C. E. (1986). Dynamics of the middle atmosphere at Saskatoon (52°N, 107°W): a spectral study during 1981, 1982. *Journal of Atmospheric and Solar-Terrestrial Physics, 48*, 1039–1055.

Manson, A. H., Meek, C. E., Hall, C. M., Nozawa, S., Mitchell, N. J., Pancheva, D., Singer, W., & Hoffmann, P. (2004). Mesopause dynamics from the Scandinavian triangle of radars within the PSMOS-DATAR project. *Annales Geophysicae, 22*, 367–386.

MATLAB (2011). Toolbox statistics. www.mathworks.de.

Mitchell, N., Pancheva, D., Middleton, H., & Hagan, M. (2002). Mean winds and tides in the arctic mesosphere and lower thermosphere. *J. Geophys. Res., 107*. doi:10.1029/2001JA900127.

Nastrom, G. D., & Fritts, D. C. (1992). Sources of mesoscale variability of gravity waves. I: topographic excitation. *Journal of the Atmospheric Sciences, 49*, 101–110.

Pancheva, D., Singer, W., & Mukhatarov, P. (2007). Regional response of the mesosphere-lower thermosphere dynamics over Scandinavia to solar proton events and geomagnetic storms in late October 2003. *Journal of Atmospheric and Solar-Terrestrial Physics, 69*, 1075–1094. doi:10.1016/j.jastp.2007.04.005.

Salmi, S.-M., Verronen, P. T., Thölix, L., Kyrölä, E., Backman, L., Karpechko, A. Y., & Seppälä, A. (2011). Mesosphere-to-stratosphere descent of odd nitrogen in early 2009 after a major stratospheric warming. In *3rd workshop on high energy particle precipitation in the atmosphere*, Granada, Spain, 9–11 May.

Schmidt, H., Brasseur, G. P., Charron, M., Manzini, E., Giorgetta, M. A., Diehl, T., Fomichev, V. I., Kinnison, D., Marsh, D., & Walters, S. (2006). The HAMMONIA chemistry climate model: sensitivity of the mesopause region to the 11-year solar cycle and CO_2 doubling. *Journal of Climate, 19*(16), 3903–3931. doi:10.1175/JCLI3829.1.

Serafimovich, A., Hoffmann, P., Peters, D., & Lehmann, V. (2005). Investigation of inertia-gravity waves in the upper troposphere/lower stratosphere over Northern Germany observed with collocated VHF/UHF radars. *Atmospheric Chemistry and Physics, 5*, 295–310.

Singer, W., Keuer, D., & Eriksen, W. (1997). The ALOMAR MF radar: technical design and first results. In B. Kaldeich-Schürmann (Ed.), *Proceedings 13th ESA symposium on European rocket and balloon programmes and related research*, Oeland, Sweden, 26–29 May 1997 (ESA SP-397) (pp. 101–103). ESA Publications Division.

Singer, W., Bremer, J., Hocking, W. K., Weiss, J., Latteck, R., & Zecha, M. (2003). Temperature and wind tides around the summer mesopause at middle and arctic latitudes. *Advances in Space Research, 31*(9), 2055–2060.

Singer, W., Bremer, J., Weiss, J., Hocking, W. K., Höffner, J., Donner, M., & Espy, P. (2004). Meteor radar observations at middle and arctic latitudes, part 1: mean temperatures. *Journal of Atmospheric and Solar-Terrestrial Physics, 66*, 607–616.

Singer, W., Latteck, R., Hoffmann, P., Williams, B. P., Fritts, D. C., Murayama, Y., & Sakanoi, K. (2005). Tides near the Arctic summer mesopause during the MaCWAVE/MIDAS summer program. *Geophysical Research Letters, 32*, L07S90. doi:10.1029/2004GL021607.

Singer, W., Latteck, R., & Holdsworth, D. (2008). A new narrow beam Doppler Radar at 3 MHz for studies of the high-latitude middle atmosphere. *Advances in Space Research, 41*, 1487–1493. doi:10.1016/j.asr.2007.10.006.

Singer, W., Latteck, R., Friedrich, M., Wakabayashi, M., & Rapp, M. (2011). Seasonal and solar activity variability of D-region electron density at 69°N. *Journal of Atmospheric and Solar-Terrestrial Physics, 73*, 925–935.

Smith, A. K., Garcia, R., Marsh, D., & Richter, J. (2011). WACCM simulations of the mean circulation and trace species transport in the winter mesosphere. *Journal of Geophysical Research,*

116, D20115. doi:10.1029/2011JD016083.

Torrence, C., & Compo, G. P. (1998). A practical guide to wavelet analysis. *Bulletin of the American Meteorological Society*, *79*(1), 61–78.

Vincent, R. A., & Alexander, M. J. (2000). Gravity waves in the tropical lower stratosphere: an observational study of seasonal and interannual variability. *Journal of Geophysical Research*, *105*, 17971–17982.

von Zahn, U., Fiedler, J., Naujokat, B., Langematz, U., & Krüger, K. (1998). A note on record–high temperatures at the northern polar stratopause in winter 1997/98. *Geophysical Research Letters*, *25*, 4169–4172.

Ward, W. E., Oberheide, J., Goncharenko, L. P., Nakamura, T., Hoffmann, P., Singer, W., Chang, L. C., Du, J., Wang, D.-Y., Batista, P., Clemesha, B., Manson, A. H., Riggin, D. M., She, C.-Y., Tsuda, T., & Yuan, T. (2010). On the consistency of model, ground-based, and satellite observations of tidal signatures: initial results from the CAWSES tidal campaigns. *Journal of Geophysical Research*, *D115*, D07107. doi:10.1029/2009JD012593.

Weatherhead, E. C., Reinsel, G., Tiao, G., Meng, X.-L., Choi, D., Cheang, W.-K., Keller, T., DeLuisi, J., Wuebbles, D., Kerr, J., Miller, A., Oltmans, S., & Frederick, J. (1998). Factors affecting the detection of trends: statistical considerations and applications to environmental data. *Journal of Geophysical Research*, *103*(D14), 17149–17161.

Chapter 23
Infra-red Radiative Cooling/Heating of the Mesosphere and Lower Thermosphere Due to the Small-Scale Temperature Fluctuations Associated with Gravity Waves

Alexander A. Kutepov, Artem G. Feofilov, Alexander S. Medvedev, Uwe Berger, Martin Kaufmann, and Adalbert W.A. Pauldrach

Abstract We address the effect of an additional infrared radiative cooling/heating of the mesosphere and lower thermosphere (MLT) in the infrared bands of CO_2, O_3 and H_2O due to small-scale irregular temperature fluctuations associated with gravity waves (GWs). These disturbances are not well resolved by present general circulation models (GCMs), but they alter the radiative transfer and cooling rates significantly. A statistical model of gravity wave-induced temperature variations was applied to large-scale temperature profiles, and the corresponding direct radiative calculations were performed with accounting for the breakdown of the local

A.A. Kutepov (✉)
Department of Physics, The Catholic University of America/NASA Goddard Space Flight Center, 8800 Greenbelt Rd, Greenbelt, MD 20771-2400, USA
e-mail: alexander.a.kutepov@nasa.gov

A.G. Feofilov
Laboratory of Dynamical Meteorology, École Polytechnique, 91128 Palaiseau Cedex, France
e-mail: artem.feofilov@lmd.polytechnique.fr

A.S. Medvedev
Max Planck Institute for Solar System Research, Max-Planck-Strasse 2, 37191 Katlenburg-Lindau, Germany
e-mail: medvedev@mps.mpg.de

U. Berger
Leibniz-Institute of Atmospheric Physics, Schlossstr. 6, 18225 Kühlungsborn, Germany
e-mail: berger@iap-kborn.de

M. Kaufmann
Institute for Chemistry and Dynamics of Geosphere, Forschungszentrum Jülich GmbH, 52425 Jülich, Germany
e-mail: m.kaufmann@fz-juelich.de

A.W.A. Pauldrach
University Observatory Munich/Wendelstein Observatory, Scheinerstr. 1, 81679 Munich, Germany
e-mail: uh10107@usm.uni-muenchen.de

F.-J. Lübken (ed.), *Climate and Weather of the Sun-Earth System (CAWSES)*, Springer Atmospheric Sciences,
DOI 10.1007/978-94-007-4348-9_23,

thermodynamic equilibrium (non-LTE). We show that temperature fluctuations can cause an additional cooling of up to 4 $K\,day^{-1}$ near the mesopause. The effect is produced mainly by the fundamental 15 μm band of the main CO_2 isotope $^{12}C^{16}O_2$ (626). A simple parametrization has been derived that computes corrections depending on the temperature fluctuations variance, which need to be added in radiative calculations to the mean temperature and the volume mixing ratios (VMRs) of CO_2 and $O(^3P)$ to account for additional cooling/heating caused by the unresolved disturbances. Implementation of this scheme into the LIMA model resulted in a colder and broader simulated summer mesopause in agreement with recent lidar measurements at Spitsbergen.

23.1 Introduction

Light absorption and scattering in the atmosphere strongly depend on distributions of temperature and radiatively active constituents. In the middle atmosphere, instantaneous profiles are highly irregular due to disturbances associated with small-scale gravity waves (GWs). These waves with vertical wavelengths $\lambda_z \ll 2\pi H$, H being a scale height, are dynamically extremely important in the mesosphere and lower thermosphere (MLT). Because they are not well resolved by contemporary general circulation models (GCMs), their effects in the atmosphere are usually accounted for in form of GW parametrization [*Fritts and Alexander*, 2003, and references therein]. The influence of these subgrid-scale GW disturbances on radiative transfer and the related heating/cooling is always missing in GCMs mainly because of the two reasons: 1) These GWs are not well resolved. If δz is the vertical grid step, then $2\delta z$-harmonics are not reproduced, 3 to $4\delta z$-waves are resolved poorly because of the numerical and physical diffusion at small scales, and only harmonics with $\lambda_z \geq 5\delta z$ can be considered as resolved by GCMs; 2) Parametrization of radiative heating and cooling normally perform calculations at grids even sparser than the model resolution, followed by an interpolation onto the model grid.

In this study we estimate the effect of the subgrid temperature fluctuations T' on the infrared radiative cooling/heating of the MLT. Section 23.2 describes the model of GW induced temperature and volume mixing ratio fluctuations. Infra-red radiative cooling/heating of the MLT and its calculation are described in Sect. 23.3. Section 23.4 discusses the computational approach and inputs used in this study. The additional CO_2 cooling/heating caused by the subgrid GW-induced T' as well as its parametrization are discussed in Sect. 23.5 and Sect. 23.6, respectively. In Sect. 23.7 the effects of subgrid GWs on the O_3 and H_2O cooling and their parameterizations are presented. The accounting for the subgrid GW CO_2 cooling in the LIMA model and its impact on calculated temperatures are discussed in Sect. 23.8 The Conclusion summarizes results and provides recommendations for implementation of obtained parameterizations in GCMs.

23.2 Stochastic Model of Gravity Waves

Temperature fluctuations associated with GWs are highly variable and irregular. In many observational studies their properties are usually summarized in the form of averaged spectra [*Fritts and Alexander*, 2003]. Here we applied a simplified model described below which randomly generates individual temperature profiles reminiscent of instantaneous measurements during periods of strong GW activity, whose statistical properties agree with observed GW spectra. Since individual profiles cannot be reconstructed from observed spectra, the model required additional plausible assumptions.

A vertical temperature distribution created by GW can be represented as a sum of individual harmonics:

$$T'(z) = \sum_k T_k(z) = \sum_k \hat{T}_k(z_0) \exp\left(i \int_{z_0}^{z} m_k(z')\, dz' + \phi_k \right), \tag{23.1}$$

where $\hat{T}_k(z_0)$ is the amplitude of the k-th harmonic at the source level z_0, $m_k(z) = m_k^{(R)} + i m_k^{(I)}$ is its complex vertical wavenumber, ϕ_k is the phase. This yields for each k

$$dT_k/dz = i m_k T_k. \tag{23.2}$$

If $T_k(z_0)$, $m_k(z_0)$, and ϕ_k are known at the source level, (23.2) can be integrated vertically for each harmonic k to obtain the profile $T'(z)$. Real parts of the vertical wavenumbers $m_k^{(R)}$ for GW harmonics are, generally, functions of z as they depend on the refractive properties of the atmosphere determined by the background wind and temperature. However, $T_k(z)$ is not an individual plane wave, but rather a sum of GW harmonics with different phase velocities traveling at different angles to the mean wind. Here we assume no vertical dependence for $m_k^{(R)}$ since there exists no rigorous requirement for the opposite in the given formulation. This also helps to make the calculations tractable. The imaginary part of the vertical wavenumber, $m_k^{(I)}$, should include the effects of the amplification due to the air density decrease, damping by the molecular viscosity, and the effect of amplitude saturation which limits the instantaneous temperature gradients by the value of the adiabatic lapse rate, i.e., $\bar{T}_z + T'_z \geq -g c_p^{-1}$, c_p being the specific heat capacity, g is the acceleration of gravity. The first two effects enter (23.2) in the form [see e.g., *Vadas and Fritts*, 2005]

$$m_k^{(I)} = -1/2H(z) + \mu m_k^{(R)3}/(\rho\omega), \tag{23.3}$$

where H is the scale height; $\mu = 0.017\ \mathrm{gm^{-1}\,s^{-1}}$ is the molecular viscosity, ρ is the density, and ω is the assumed frequency of the harmonic. In our calculations, ω served as a tuning parameter to ensure that all wave disturbances dissipate above the turbopause. A "convective adjustment scheme" turns on to modify the temperature $T'(z)$ when the net vertical gradient of the potential temperature becomes negative.

We use the canonical "modified DeSaubies" vertical wavenumber spectrum to prescribe the wave field at the source level $z_0 = 15$ km: $S(m/m_*) = A_0(m/m_*)^s \times [1 + (m/m_*)^{s+3}]^{-1}$, where S is the wave spectral density, and the values for the following parameters were borrowed from [*Medvedev and Klaassen*, 2000]: $s = 0$ or 1,

$m_* = 0.006\ \mathrm{m}^{-1}$, $A_0 = 10$ to 100. The algorithm of the temperature profile generation is as follows. First, several vertical wavenumbers $2\pi\lambda_{max}^{-1} \le m_k^{(R)} \le 2\pi\lambda_{min}^{-1}$ and the corresponding phases ϕ_k are randomly generated, where λ_{max} and λ_{min} are the longest and shortest vertical wavelengths, respectively. The amplitudes of the harmonics are assigned according to the DeSaubies spectrum. Then, (23.2) is integrated upward to produce an individual profile $T'(z)$. There is ample evidence [e.g., *Fritts and Alexander*, 2003, Sect. 4.3] that GW spectra in the atmosphere are composed of only a few harmonics rather than represent a broad mix of waves. *Sica and Russell* [1999] argued that, in observations, the spectra most often are dominated by 1 to 3 waves. We use 2 GW harmonics in our simulations.

The following parameters have been used in the calculations presented below: the vertical grid step $\Delta z = 300$ m, $A_0 = 40$, $s = 0$, $\lambda_{max} = 11$ km, $\lambda_{min} = 2.5$ km. As there is no data on the frequency of occurrence of waves with different vertical wavenumbers in observations, we assumed the same probability for harmonics with all wavelengths.

Kutepov et al. [2007] applied the model described above in this section for investigating the additional CO_2, O_3 and H_2O coolings of the MLT due to the small-scale T' alone. In addition to [*Kutepov et al.*, 2007], we took here into account the effect of GW induced vertical motions of $O(^3P)$ and CO_2. The lifetime of these species is rather long in comparison to the typical period of GWs so that these species can be considered as conservative tracers. Their vertical displacement due to GWs is calculated adiabatically from GW induced temperature fluctuations weighted by static stability [*Smith et al.*, 2010]

$$c'_{\mathrm{M}} = \varepsilon_{\mathrm{M}} T', \quad \varepsilon_{\mathrm{M}} = \frac{dc_{\mathrm{M}}}{dz} \Big/ \left(\frac{dT}{dz} + \frac{RT}{c_p H} \right), \tag{23.4}$$

where M = CO_2, $O(^3P)$, c_{M} and T are the mean VMR and temperature profiles, respectively, R is the gas constant, and H is the density scale height.

23.3 Infra-red Radiative Cooling/Heating of MLT

The radiative cooling of the MLT, both at day and night, is affected mainly by the vibration-rotational bands of CO_2 around 15 μm. The 9.6 μm O_3 band is next in importance contributing up to 25 % of the total cooling at and around the stratopause. The water vapor cooling (dominated by the H_2O rotational band) plays a tertiary role contributing about 5–7 % of the total IR cooling of the MLT. At daytime, the infra-red heating of MLT is caused mostly by the CO_2 absorption of the Solar radiation in the near infrared spectral region of 1–4.3 μm.

The radiative cooling/heating of the atmosphere h is described by the radiative flux divergence taken with the minus sign. Integrating the radiative transfer equation over the total solid angle Ω and frequency ν, one obtains

$$h = -\int d\Omega \int d\nu \, \frac{dI_{\Omega\nu}}{ds} = \chi(J - S), \tag{23.5}$$

where $dI_{\Omega\nu}$ is the change of radiative intensity $I_{\Omega\nu}$ when traveling a pass ds across the medium, $J = (1/4\pi\chi)\int d\Omega \int d\nu\, \chi(\nu) I_{\Omega\nu}$ is the mean radiative intensity, $S = \eta(\nu)/\chi(\nu)$ is the source function defined as a ratio of the atmospheric emissivity $\eta(\nu)$ and opacity $\chi(\nu)$, and $\chi = \int \chi(\nu)\, d\nu$ is the mean atmospheric opacity in the spectral region under consideration. χS in (23.5) describes the radiative loss of energy by a unit volume, and χJ describes the absorption by the same unit volume of the radiation coming from other atmospheric layers, the upwelling radiation from the surface, and the solar radiation.

In the lower atmosphere, inelastic molecular collisions determine the population of molecular levels. As a result local thermodynamic equilibrium (LTE) exists, where the populations n_v of molecular vibrational levels v obey the Boltzmann law with the local kinetic temperature T, and S in (23.5) is equal to the Planck function $B_\nu(T)$. In the middle and upper atmosphere, the frequency of collisions is low, and other processes which populate and depopulate molecular levels (such as an absorption and emission of the radiation in molecular bands, redistribution of excitation between colliding molecules, chemical excitation, etc.) must be taken into account. Under these non-LTE conditions n_v deviate from the Boltzmann distribution for the local T, and S is no longer equal to $B_\nu(T)$. Non-LTE n_v are found from a set of rate equations expressing a balance of all processes, which populate and depopulate vibrational levels v, and the radiative transfer equation in the ro-vibrational bands.

23.4 Calculation Procedure and Inputs

We used the ALI-ARMS (for Accelerated Lambda Iteration for Atmospheric Radiation and Molecular Spectra) code [*Kutepov et al.*, 1998; *Gusev and Kutepov*, 2003] for solving the non-LTE problem and calculating the radiative cooling/heating in the CO_2, O_3 and H_2O bands. 60 vibrational levels and 150 vibrational-rotational bands of four CO_2 isotopes were taken into account. They included a large number of weak combinational bands that absorb the near-infrared solar radiation during day time. The set of collisionally induced vibrational-translational (V-T) and vibrational-vibrational (V-V) energy exchange processes and rate coefficients correspond to those described by *Shved et al.* [1998].

We examined five atmospheric models corresponding to the summer solstice in the Northern hemisphere: at −70.0° for the subarctic winter (SAW), −40.0° for the midlatitude winter (MLW), 0.0° for the tropics (TROP), 40.0° for the midlatitude summer (MLS), and 70.0° for the subarctic summer (SAS). The local noons were set for each model with the solar zenith angles equal to 93.5°, 63.5°, 23.5°, 16.5°, and 46.5°, respectively. Temperature T, pressure p, [N_2], [O_2] and [$O(^3P)$] were taken from the MSISE-90 model [*Hedin*, 1991].[1] The CO_2 VMR corresponds to the mean profile retrieved in the CRISTA experiments [*Kaufmann et al.*, 2002]. It deviates from well mixed values between 70 and 80 km, and is significantly smaller

[1] http://omniweb.gsfc.nasa.gov/vitmo/msis_vitmo.html.

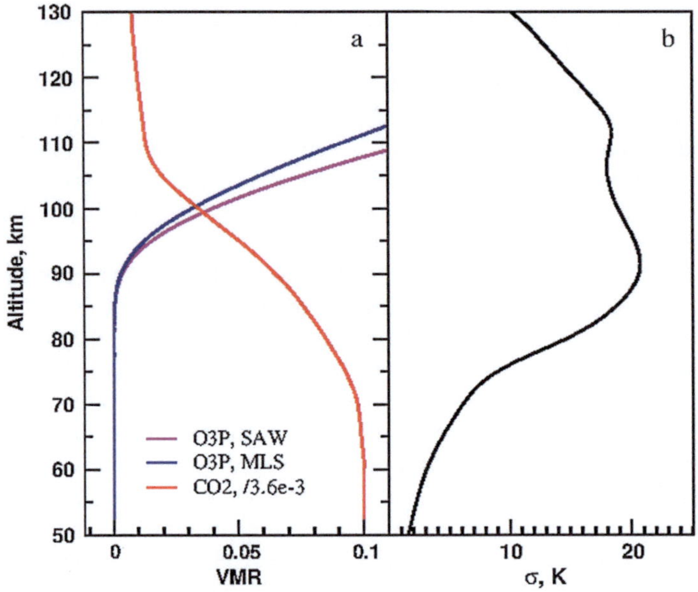

Fig. 23.1 (**a**) VMRs of $O(^3P)$ and CO_2 used in the this study (see text); (**b**) the temperature variance σ_T for MLW model

above these altitudes than the profile obtained from earlier rocket-borne mass spectrometer experiments, in which c_{CO_2} is constant up to 90 km. For each temperature model $T(z)$, 1000 individual $T'(z)$ profiles were randomly generated as described in Sect. 23.2. The non-LTE problem was solved and radiative cooling/heating $h(z)$ computed for each profile $T(z) + T'(z)$ and corresponding profiles $c_M + c'_M$ for M = $O(^3P)$, CO_2, where c'_M were calculated following (23.4). The mean cooling $\overline{h(z)}$ was then obtained by averaging $h(z)$ at each altitude z. It was compared with the cooling $h_T(z)$ for the mean temperature and the $O(^3P)$ and CO_2 VMR profiles.

23.5 Results and Discussion

In Fig. 23.1(a) we show the employed VMRs of $O(^3P)$ (for SAW and MLW) and CO_2 as well as (b) the temperature variance profile $\sigma_T(z) = \{\overline{[T'(z)]^2}\}^{1/2}$ (for MLW) computed from 1000 distributions of $T'(z)$. Panels (a) and (b) of Fig. 23.2 show $\Delta h(z) = \overline{h(z)} - h_T(z)$ for SAW and MLW, obtained for Case 1, discussed in detail by *Kutepov et al.* [2007], with only T' disturbances (solid green lines), for Case 2 with both T' and $[O(^3P)]'$ disturbances (solid blue lines), and for Case 3 with T', $[O(^3P)]'$ and $[CO_2]'$ (solid red lines) accounted for. Accounting for $[O(^3P)]'$ increases additional cooling for SAW at 85–100 km (to about 1 K day^{-1} at 90 km). This increase is obviously related to the large positive gradient of the $O(^3P)$

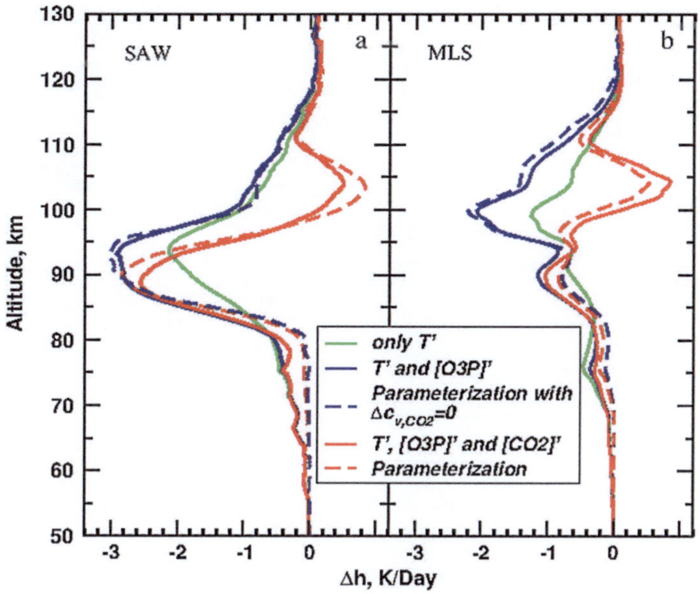

Fig. 23.2 (**a**) Δh for SAW model (**a**) and MLS model (**b**). Parameterization calculations corresponds to the usage of Eqs. (23.9), see text for more details

VMR at these altitudes (see Fig. 23.1(a)). According to (23.4), it causes intensive variations of $[O(^3P)]'$. On the other hand, additional taking into account $[CO_2]'$ for this model (Case 3) leads to an opposite effect: the cooling decreases between 88 and 110 km compared to the Case 2 to $\sim$1.7 K day^{-1} at 95 km. The latter is related to the rapid fall of the CO_2 VMR at these heights (Fig. 23.1(a)) that causes significant negative gradient. Similar effects are seen for MLS (panel b): accounting for $[O(^3P)]'$ increases the cooling effect to about 1 K day^{-1} at 100 km, whereas the additional accounting for $[CO_2]'$ causes the warming up to about 2 K day^{-1} in the layer between 100 and 105 km. In the Case 3, the total effect changes its sign at about 100 km: the cooling reaching about 1 K day^{-1} at 89 km is accompanied by the warming of up to 1 K/day at 105 km.

In Fig. 23.3, additional cooling/heating is shown for the Cases 1 and 3 for three other temperature models, for which it was not shown in Fig. 23.2. One may see in Fig. 23.2 and Fig. 23.3 that compared to Case 1, accounting for $[O(^3P)]'$ and $[CO^2]'$ simultaneously with T' makes the total effect significantly more variable depending on peculiarities of temperature and VMR profiles. For instance, for MLW, SAW, TROP and SAS models the cooling is stronger (up to 0.7 K day^{-1} for SAW), however, the corresponding cooling maxima are now located at the altitudes which are few km lower than those when only variations of T' were taken into account. Additionally, a weak warming appeared for all considered models at 100–108 km reaching 0.8 K day^{-1} for MLS.

The observed behavior of GW-caused additional cooling/heating is clarified below in the next section.

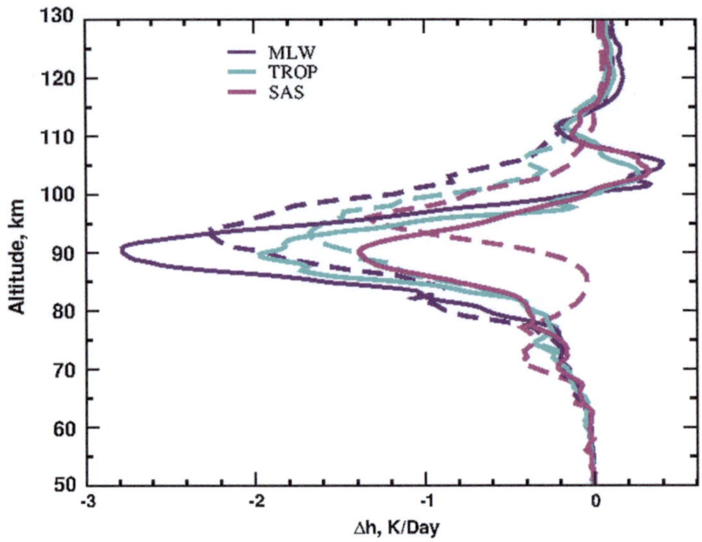

Fig. 23.3 Δh for 3 atmospheric models: *solid lines*—this study; *dashed lines*—results of *Kutepov et al.* [2007]

23.6 Parametrization

Equation (6) in [*Kutepov et al.*, 2007] defines the non-LTE radiative flux divergence for the fundamental transition of main CO_2 isotope 626, which is the main contributor to the effect discussed above, as

$$h_{T,626FB}(z) = \chi \frac{C}{A+C}(J-B), \tag{23.6}$$

where A and C are the Einstein coefficient for the spontaneous emission in the band and the total collisional quenching rate of the upper vibrational level, respectively; J is the mean radiative intensity in the band, B is the Planck function for the band center frequency ν_0, and

$$\chi = \int \chi(\nu)\, d\nu = \frac{h\nu_0}{4\pi} B_{01} n_0 \sim B_{01} n_a c_{CO_2} \tag{23.7}$$

is the mean atmospheric opacity in the spectral band of fundamental transition (with neglect of the stimulated emission). In the latter equation, B_{01} is the Einstein absorption coefficient for the band, and n_0 is the population of the CO_2 vibrational ground level. With taking into account the low degree of the CO_2 vibrational excitation in the atmosphere, it is usually assumed that n_0 is approximately equal to the total CO_2 density $n_{CO_2} = n_a c_{CO_2}$, where n_a is the total atmospheric density, and c_{CO_2} is the CO_2 VMR. In (23.6) the collisional quenching rate $C = kn_a$, where k is the total quenching rate coefficient, see (23.10) below.

Following the approach of *Kutepov et al.* [2007] we concentrate here on the analysis of the second term in (23.6) that describes the non-LTE emission of the volume unit. According to the analysis above, this term is proportional to $E = c_{CO_2} C B(T)$. In this expression variations T' influence $B(T)$ as well as induce fluctuations c'_{CO_2} and $c'_{O(^3P)}$, which are estimated according to (23.4). Averaging E over GW-induced fluctuations, one may obtain that

$$\overline{E} \approx [c_{CO_2} + \Delta c_{CO_2}][C + \Delta C] B(T + \Delta T), \tag{23.8}$$

where Δc_{CO_2}, ΔC, and ΔT depend on the temperature variance and gradients of c_{CO_2} and $c_{O(^3P)}$

$$\begin{aligned} \Delta c_{CO_2} &= \frac{1.3 \times 10^3 \varepsilon_{CO_2} \sigma_T^2}{T^2}, \qquad \Delta C = n_a k_{O(^3P)} \Delta c_{O(^3P)}, \\ \text{where } \Delta c_{O(^3P)} &= \frac{8.9 \times 10^2 \varepsilon_{O(^3P)} \sigma_T^2}{T^2}, \\ \Delta T &= \left[4.8 \times 10^2 \frac{1\text{–}2.1 \times 10^{-3} T}{T^2} + 7 \times 10^{-2} \varepsilon_{O(^3P)} \frac{k_{O(^3P)}}{k} \right] \sigma_T^2. \end{aligned} \tag{23.9}$$

In these expressions

$$\begin{aligned} \sigma_T^2 &= \overline{(T')^2}, \\ k &= c_{O(^3P)} k_{O(^3P)} + c_{N_2} k_{N_2} + c_{O_2} k_{O_2}, \\ k_{O(^3P)} &= 6 \times 10^{-12} \sqrt{\left(\frac{T}{300}\right)}, \\ k_{N_2} &= 5.5 \times 10^{-17} \sqrt{T} + 6.7 \times 10^{-10} \exp(-83.8\, T^{-1/3}), \\ k_{O_2} &= 10^{-15} \exp(23.37 - 230.9\, T^{-1/3} + 564.0\, T^{-2/3}). \end{aligned} \tag{23.10}$$

Here σ_T is in K, all rate coefficients k are in $cm^3\,c^{-1}$ and $k_{O(^3P)}$ corresponds to that of *Sharma and Wintersteiner* [1990], whereas k_{N_2} and k_{O_2} are taken from [*Shved et al.*, 1998]. Detailed derivation of expressions (23.9) is given in [*Kutepov et al.*, 2012]. Note that with $\varepsilon_{CO_2}, \varepsilon_{O(^3P)} = 0$ (23.9) reduces to

$$\Delta T(z) = \frac{1.44 \nu_0}{2T^2(z)} \left[1 - \frac{2T(z)}{1.44 \nu_0} \right] \overline{[T'(z)]^2} \tag{23.11}$$

for $\Delta T(z)$ of *Kutepov et al.* [2007] with $\nu_0 = 667.4\ cm^{-1}$ for fundamental band of the CO_2 isotope 626.

In Fig. 23.2(a) and (b) we compare the rigorously computed Δh (solid blue and red curves) for SAW and MLS models with those obtained with T, c_{CO_2} and $c_{O(^3P)}$ replaced with $T + \Delta T$, $c_{CO_2} + \Delta c_{CO_2}$ and $c_{O(^3P)} + \Delta c_{O(^3P)}$, respectively, using the parametrization (23.9). One can see in this figure that the expressions (23.9) allow for very accurate accounting of extra cooling/heating associated with the additional influence of $[O(^3P)]'$ (blue dashed curves) as well with both $[O(^3P)]'$ and $[CO^2]'$ variations included (red dashed curves). We tested (23.9) for the 5 temperature models, and found that the absolute error of applying these expressions does not exceed 0.4 K day^{-1}. The parametrization (23.9) demonstrates a high flexibility and allows accounting for a variety of atmospheric conditions. As in the case of pure temperature fluctuations studied in [*Kutepov et al.*, 2007], the effect is related to the strong temperature dependence of the Planck function B. In [*Kutepov et al.*, 2007], however, the term in the power expansion of B, which is proportional T', canceled out with averaging. In this study this term correlates with $[O(^3P)]'$ and $[CO_2]'$, which both are proportional T'. As a result, the combined effect of subgrid variations turn out to be stronger than that of pure temperature fluctuations and significantly more variable depending on the gradients of T, $O(^3P)$ and CO_2.

Above $\approx$85 km, the $O(^3P)$ atoms begin to dominate the quenching of the CO_2 ν_2 vibrations [*Kutepov et al.*, 2006]. As the result, the CO_2 radiative cooling in this region strongly depends on the product $[O(^3P)] \times k_{O(^3P)}$, which provides the atomic oxygen contribution to the total quenching rate C. The values of Δh discussed above were obtained for the high value of quenching rate coefficient $k_{O(^3P)} = 6 \times 10^{-12}$ cm^3 s^{-1} (for $T = 300$ °K), retrieved from space observations of 15 μm emissions [*Sharma and Wintersteiner*, 1990]. Application of the lower $k_{O(^3P)} = 1.5 \times 10^{-12}$ cm^3 s^{-1}, which is more consistent with recent laboratory measurements [*Castle et al.*, 2006], reduces the maximum values of the additional cooling shown in Figs. 23.2 and 23.3 by a factor of about 2 following the drop of the ratio $C/(C + A)$ in (23.6) for these altitudes when $k_{O(^3P)}$ is replaced with $k_{O(^3P)}/4$.

The expressions (23.9) explain why the accounting for $[O(^3P)]'$ and $[CO_2]'$ causes additional cooling and heating, respectively, which are seen in Figs. 23.2 and 23.3. The correlation between T' and $[O(^3P)]'$ due to the positive $O(^3P)$ gradient leads to the positive $\Delta c_{O(^3P)}$. In accordance with (23.8) this increases the second term in (23.6) responsible for radiative cooling. Opposite to this the anti-correlation between T' and $[CO_2]'$ due to the negative CO_2 gradient results in negative $\Delta c_{O(^3P)}$, which decreases this cooling term.

23.7 Effects of Subgrid GWs on the O_3 and H_2O Cooling

Following the calculation procedure described in Sect. 23.4 we also applied the ALI-ARMS code for 5 selected atmospheric models to estimate the effect of the subgrid T' on the cooling of the MLT in the O_3 9.6 μm and the H_2O rotational bands. $[O_3]$ for these calculations were taken from the SABER V1.07 data product. $[H_2O]$ were those from the HALOE V19 data product. The H_2O rotational band is in LTE throughout the MLT, whereas the O_3 9.6 μm band departs from LTE above an altitude of 65–70 km. We utilized the model of non-LTE in O_3 developed by *Manuilova et al.* [1998]. It was found that the O_3 9.6 μm additional cooling caused by subgrid T' reached between 90 and 100 km 0.3, 0.24, 0.15, 0.1 and 0.05 K day^{-1} for the SAW, MLW, TROP, MLS, and SAS models, respectively. The same effect for the H_2O rotational band did not exceed 0.05 K/day for all the models considered. We also found that the additional cooling can be well accounted for by correcting the input temperature by means of (23.11) with $\nu_0 = 1043$ and 500 cm^{-1} for O_3 and H_2O, respectively.

23.8 Application in the LIMA Model

To estimate the effects of the GW-induced CO_2 cooling, we implemented the parametrization (23.9) into the Leibniz Institute Middle Atmosphere (LIMA) GCM, which was described in detail in [*Berger*, 2008]. The peculiarity of the model is that its dynamical core employs a triangle grid with more than 42000 edge points over the globe with a distance between grids of about 110 km. Together with a relatively small vertical grid step of about 1.1 km this allows to resolve a significant portion of GWs and avoid using GW parametrization. Radiative calculations in LIMA are performed, however, on a much coarser grid (16 × 36 in the longitude and latitude directions respectively) with vertical grid step of about 1.8 km. Typically, about 20 or more triangles are located inside such a latitude-longitude segment. The background fields required for the radiative CO_2 scheme are the averaged over each bin with 20 and more points. We attributed the deviations from the mean fields within a bin to GWs, and, therefore, were able to calculate the temperature variance $\sigma_T = (\overline{T'^2})^{1/2}$ corresponding to the bin. This variance, however, corresponds to GWs resolved by LIMA, $\sigma_{T,res}$ whereas our aim is to estimate the variance of subgrid-scale harmonics unresolved by the model, $\sigma_{T,unres}$. To relate $\sigma_{T,unres}$ to $\sigma_{T,res}$, we performed calculations with our stochastic model of GW-like disturbances, described in Sect. 23.2, and estimated for 5 atmospheric models variances produced by 1000 harmonics with vertical wavelengths between 2 and 11 km, and those with wavelengths between 6 and 11 km. The former set is referred to as "all GW spectrum", whereas the second set as the "resolved GW spectrum". This implies that waves shorter than 6 km are considered as unresolved by LIMA due to the reasons discussed in the Introduction. Since the temperature variance is the sum of variances of the resolved and unresolved parts of GW spectra, we thus were able to calculate $\sigma_{T,unres}$ from these two variances. The mean for 5 atmospheric models ratio of the "unresolved" and "resolved" rms temperature fluctuations is given in Fig. 23.4. It is seen that between 80 and 100 km, the height interval where the effect of additional CO_2 cooling is most considerable, this ratio varies between 5 and slightly less than 2.

In our LIMA simulations we used in the parametrization (23.9) $\sigma_T = \sigma_{T,unres} = 2 \times \sigma_{T,res}$ as a modest estimate of the subgrid GW-induced rms temperature fluctuations, where $\sigma_{T,res}$ is the variance of the LIMA resolved GWs calculated as described above. The calculations were performed with $k_{O(^3P)} = 6 \times 10^{-12}$ cm^3 s^{-1} (for $T = 300$ °K) of *Sharma and Wintersteiner* [1990].

The effects of additional cooling/heating on LIMA temperatures are presented in Fig. 23.5 for the monthly and zonal mean temperatures for July 2005. It is seen in that subgrid GWs cause warming above about 105 km and cooling below at latitudes between 90°S and 40°N. In the winter hemisphere the effects are larger due to the larger GW activity. The simulated polar summer mesopause region turns out to be broader and colder when the additional cooling is accounted for. The difference reaches appreciable values of about of 5–6 K. Thus simulated temperature in the region as well as in the transition to the lower thermosphere layer is more consistent with the recent lidar measurements at Spitsbergen [*Höfner and Lübken*, 2007].

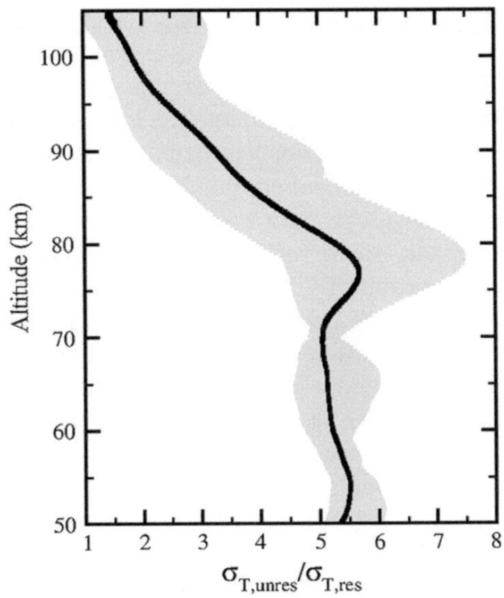

Fig. 23.4 The ratios of the "unresolved" and "resolved" rms temperature fluctuations. *Gray patter*: corridor of ratio values for 5 atmospheric models used in this study (see text for details). *Black solid line*: mean ratio for 5 atmospheric models

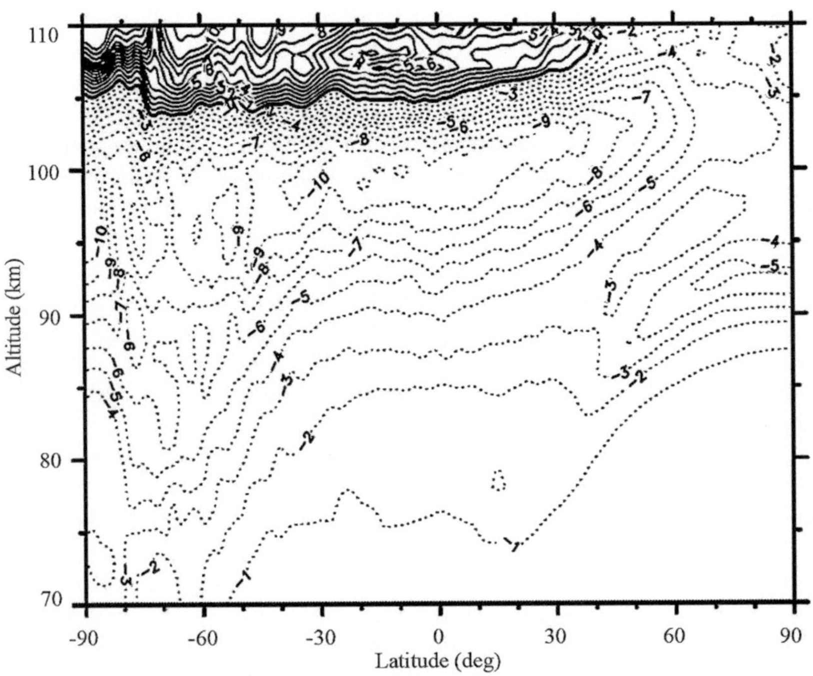

Fig. 23.5 Differences of simulated monthly and zonal mean temperatures for July 2005 obtained with and without accounting for the additional IR CO_2

23.9 Conclusion

In this study we analyzed additional cooling/heating of MLT in the infra-red bands of CO_2 O_3 and H_2O caused by the subgrid GWs. We applied a statistical model of GWs to perform direct calculations of the infra-red cooling/heating taking into account the non-LTE. The results show that additional cooling/heating is very variable depending on peculiarities of temperature and VMR profiles. In the periods of strong GW activity it can produce an additional cooling up to 4 K day^{-1} near the mesopause. A simple parametrization has been derived that computes corrections depending on the temperature fluctuations variance of subgrid GWs, which should be added in radiative calculations to the mean temperature, CO_2 and $O(^3P)$ VMR profiles to account for the additional cooling associated with the unresolved disturbances. Implementation of this scheme into the LIMA model resulted in a colder and broader simulated summer mesopause in agreement with recent lidar measurements at Spitsbergen.

Finally we note, that corrections (23.9) depend on the mean variance of temperature perturbations, $\overline{[T'(z)]^2}$. For applications in GCMs, which utilize GW parameterizations, this quantity can be obtained from these parameterizations which normally calculate the wind variance for GW fluctuations, $\overline{[u(z)']^2}$. The variance $\overline{[T'(z)]^2}$ can readily be computed from the variance $\overline{[u'(z)]^2}$ with the help of the polarization relations for GW.

References

Berger, U. (2008). Modeling of middle atmosphere dynamics with LIMA. *Journal of Atmospheric and Solar-Terrestrial Physics*, *70*, 1170–1200. doi:10.1016/j.jastp.2008.02.004.

Castle, K. J., Kleissas, K. M., Rhinehart, J. M., Hwang, E. S., & Dodd, J. A. (2006). Vibrational relaxation of $CO_2(\nu_2)$ by atomic oxygen. *Journal of Geophysical Research*, *111*(A10), A09303. doi:10.1029/2006JA011736.

Fritts, D. C., & Alexander, M. J. (2003). Gravity wave dynamics and effects in the middle atmosphere. *Annals of Geophysics*, *41*(1), 1003–1066. doi:10.1029/2001RG000106.

Gusev, O. A., & Kutepov, A. A. (2003). Non-LTE gas in planetary atmospheres. In K. W. I. Hubeny & D. Mihalas (Eds.), *Astronomical society of the pacific conference series: Vol. 288. Stellar atmosphere modeling* (pp. 318–330).

Hedin, A. E. (1991). Extension of the msis thermospheric model into the middle and lower atmosphere. *Journal of Geophysical Research*, *96*, 1159.

Höffner, J., & Lübken, F.-J. (2007). Potassium lidar temperatures and densities in the mesopause region at Spitsbergen (78_{deg}n). *Journal of Geophysical Research*, *112*(D11), D20114. doi:10.1029/2007JD008612.

Kaufmann, M., Gusev, O. A., Grossmann, K. U., Roble, R. G., Hagan, M. E., Hartsough, C., & Kutepov, A. A. (2002). The vertical and horizontal distribution of CO_2 densities in the upper mesosphere and lower thermosphere as measured by crista. *Journal of Geophysical Research*, *107*, 8182. doi:10.1029/2001JD000704.

Kutepov, A. A., Gusev, O. A., & Ogibalov, V. P. (1998). Solution of the non-LTE problem for molecular gas in planetary atmospheres: superiority of accelerated lambda iteration. *Journal of Quantitative Spectroscopy & Radiative Transfer*, *60*. doi:10.1016/S0022-4073(97)00167-2.

Kutepov, A. A., Feofilov, A. G., Marshall, B. T., Gordley, L. L., Pesnell, W. D., Goldberg, R. A., & Russell, J. M. (2006). Saber temperature observations in the summer polar mesosphere and lower thermosphere: importance of accounting for the CO_2 ν_2 quanta v-v exchange. *Geophysical Research Letters*, *33*, L21809. doi:10.1029/2006GL026591.

Kutepov, A. A., Feofilov, A. G., Medvedev, A. S., Pauldrach, A. W. A., & Hartogh, P. (2007). Small-scale temperature fluctuations associated with gravity waves cause additional radiative cooling of the mesopause region. *Geophysical Research Letters*, *34*, L24807. doi:10.1029/2007GL032392.

Kutepov, A. A., Feofilov, A. G., Medvedev, A. S., Berger, U., & Kaufmann, M. (2012). Radiative cooling of the mesopause region due to small-scale fluctuations of temperature, [$O(^3P)$] and [CO_2] associated with gravity waves. Submitted to *Geophysical Research Letters*.

Manuilova, R. O., Gusev, O. A., Kutepov, A. A., von Clarmann, T., Oelhaf, H., Stiller, G. P., Wegner, A., López Puertas, M., Martín-Torres, F. J., Zaragoza, G., & Flaud, J. M. (1998). Modelling of non-LTE limb radiance spectra of IR ozone bands for the MIPAS space experiment. *Journal of Quantitative Spectroscopy & Radiative Transfer*, *59*, 405–422. doi:10.1016/S0022-4073(97)00120-9.

Medvedev, A. S., & Klaassen, G. P. (2000). Parameterization of gravity wave momentum deposition based on nonlinear wave interactions: basic formulation and sensitivity tests. *Journal of Atmospheric and Solar-Terrestrial Physics*, *62*, 1015–1033. doi:10.1016/S1364-6826(00)00067-5.

Sharma, R. D., & Wintersteiner, P. P. (1990). Role of carbon dioxide in cooling planetary thermospheres. *Geophysical Research Letters*, *17*, 2201–2204. doi:10.1029/GL017i012p02201.

Shved, G. M., Kutepov, A. A., & Ogibalov, V. P. (1998). Non-local thermodynamic equilibrium in CO_2 in the middle atmosphere. I. Input data and populations of the ν_3 mode manifold states. *JASTP*, *60*, 289–314. doi:10.1016/S1364-6826(97)00076-X.

Sica, R. J., & Russell, A. T. (1999). How many waves are in the gravity wave spectrum? *Geophysical Research Letters*, *24*, 3617–3620. doi:10.1029/1999GL003683.

Smith, A. K., Marsh, D. R., Mlynczak, M. G., & Mast, J. C. (2010). Temporal variations of atomic oxygen in the upper mesosphere from saber. *Journal of Geophysical Research*, *115*, D18309. doi:10.1029/2009JD013434.

Vadas, S. L., & Fritts, D. C. (2005). Thermospheric responses to gravity waves: Influences of increasing viscosity and thermal diffusivity. *Journal of Geophysical Research*, *110*, D15103. doi:10.1029/2004JD005574.

Chapter 24
The Influence of Zonally Asymmetric Stratospheric Ozone on the Coupling of Atmospheric Layers

Axel Gabriel, Ines Höschel, Dieter H.W. Peters, Ingo Kirchner, and Hans-F. Graf

Abstract The radiation perturbation due to zonal asymmetries in stratospheric ozone (O_3^*) is an important factor in the atmosphere-ocean circulation system, affecting the coupling processes between the middle atmosphere, the troposphere and the ocean. We investigate the effects of observed O_3^* in general circulation model simulations without and with an interactively coupled ocean (GCM, AOGCM). Observations show a pronounced wave one pattern in O_3^* during northern winter, with amplitudes of 10–20 % of zonal mean values. Based on sensitivity studies with the GCM MAECHAM5, and accompanying linear solutions of forced Rossby waves, we demonstrate that the radiative forcing due to O_3^* affects the stratospheric circulation by imposing a wave perturbation in geopotential height (in the order of 10–20 % of the zonal mean values) and the associated large-scale flow. Subsequent changes due to non-linear stratosphere-troposphere coupling indicate an intensification of the tropospheric Aleutian low but a weakening of the Iceland low, i.e. a shift of tropospheric circulation towards negative phase of North Atlantic Oscillation (NAO). Multi-decadal simulations with the AOGCM COSMOS show that this signal is more pronounced and spatially southward shifted due to additional interactions between

A. Gabriel (✉) · D.H.W. Peters
Leibniz-Institute of Atmospheric Physics at the University Rostock, Schlossstr. 6, 18225 Kühlungsborn, Germany
e-mail: gabriel@iap-kborn.de

D.H.W. Peters
e-mail: peters@iap-kborn.de

I. Kirchner · I. Höschel
Institut für Meteorologie, Carl-Heinrich-Becker-Weg. 6-10, 12165 Berlin, Germany

I. Kirchner
e-mail: ingo.kirchner@met.fu-berlin.de

I. Höschel
e-mail: ines.hoeschel@met.fu-berlin.de

H.-F. Graf
Centre for Atmospheric Science, Cambridge University, Downing Place, Cambridge CB2 3EN, UK
e-mail: hans.graf@geog.cam.ac.uk

F.-J. Lübken (ed.), *Climate and Weather of the Sun-Earth System (CAWSES)*, Springer Atmospheric Sciences,
DOI 10.1007/978-94-007-4348-9_24,

the troposphere and the ocean. In particular, we identify significant changes in surface temperatures, precipitation, wind-driven ocean currents and sea ice thickness induced by the wave one pattern in O_3^*. The results suggest that including the effects of zonal asymmetries in stratospheric ozone and other important absorbers may help to improve current general circulation models and climate change studies.

24.1 Introduction

In the middle atmosphere (10–100 km) quasi-stationary planetary wave patterns in temperature and geopotential height, usually occurring during winter with a pronounced zonal wave number one in the wave spectrum, are well known phenomena [e.g., *Andrews et al.*, 1987]. On the other hand, stationary wave patterns or zonal asymmetries in middle atmospheric ozone, water vapour and other important absorbers are examined only very sparsely up to now, although they may play an important role in the atmospheric circulation system [*Kirchner and Peters*, 2003; *Sassi et al.*, 2005; *Gabriel et al.*, 2007; *Crook et al.*, 2008; *Gillet et al.*, 2009; *Waugh et al.*, 2009]. The aim of our paper is to diagnose the zonal asymmetries in stratospheric ozone based on observational data (assimilations and satellite data), and to investigate their influence on the atmospheric circulation via the related radiation perturbations based on general circulation model simulations without and with an interactively coupled ocean (GCM, AOGCM). The paper is related to the project SORACAL (The Influence of Solar Radiation Perturbations on the Coupling of Atmospheric Layers) in the framework of the DFG CAWSES program.

Stratospheric ozone is produced mainly in the tropics and transported poleward by the zonal mean meridional mass circulation, the so-called Brewer-Dobson circulation [*Andrews et al.*, 1987]. Observed planetary wave patterns in stratospheric ozone, as indicated by the zonally asymmetric component $O_3^* = O_3 - [O_3]$ ([ ]: zonal mean), were found to be largely related to ozone transports due to ultra-long planetary waves [*Hood and Zaff*, 1995; *Peters et al.*, 1996, 2008; *Peters and Entzian*, 1999; *Knudsen and Andersen*, 2001; *Gabriel et al.*, 2011a]. Planetary waves lead to temperature-dependent chemical ozone depletions in the lower stratosphere [e.g., *Solomon et al.*, 1998]. In a similar way, heterogeneous ozone depletions within the cold winter polar vortex, which usually shows a zonally asymmetric configuration [*Waugh and Randel*, 1999; *Karpetchko et al.*, 2005; *WMO*, 2007], contribute to zonal asymmetries in stratospheric ozone. However, a detailed examination of the zonal asymmetries in the Brewer-Dobson circulation and their effects on the distribution of ozone and other important absorbers are an outstanding issue.

Recently the attention has been drawn to the possible influence of O_3^* on the atmospheric circulation. Theoretical examinations of ozone-dynamic feedbacks suggest an induced change in zonal mean wave activity via wave-induced ozone heating [*Nathan and Cordero*, 2007]. For the Northern Hemisphere, previous simulations with general circulation models have demonstrated that O_3^* and its long-term changes can principally affect wave variability and circulation in the troposphere as well as the planetary wave patterns in the stratosphere and mesosphere [e.g.,

Kirchner and Peters, 2003; *Gabriel et al.*, 2007]. In the southern hemisphere, during southern winter polar vortex break-up, the amplitude in O_3^* becomes particularly strong and O_3^* induces significant changes in tropospheric circulation and surface temperatures [*Crook et al.*, 2008]. These results were reproduced by different general circulation models with interactive chemistry, here referred to CCMs [*Gillet et al.*, 2009; *Waugh et al.*, 2009]. However, there are still large uncertainties regarding how well current CCMs describe the observed stationary wave patterns found in atmospheric circulation and in stratospheric ozone.

Generally the discussion on the individual coupling processes between troposphere, stratosphere and mesosphere intensifies. For example, simplified model simulations have demonstrated that the stratospheric westerly jet, which is related to the configuration of the winter stratospheric polar vortex, can have an impact on tropospheric baroclinic wave development [*Polvani and Kushner*, 2002; *Wittman et al.*, 2004; *Song and Robinson*, 2004]. Diagnosis of observations and general circulation model simulations have shown that the long-term variations of the strength of the polar vortex or of the stratospheric jet can have an impact on the variations in the tropospheric circulation pattern over the North Atlantic [*Walter and Graf*, 2005; *Scaife et al.*, 2005]. A detailed diagnosis of sudden stratospheric warming events (SSWs) in the ERA-40 data revealed a change in surface geopotential during major SSWs [*Charlton and Polvani*, 2007; *Charlton et al.*, 2007]. On the other hand, observed temperature trends in stratosphere and mesosphere indicate vertical large-scale wave structures [e.g., *Kubicki et al.*, 2006] which are presumably related to changes in vertical planetary wave propagation forced by tropospheric wave dynamics. Stratospheric ozone is an important factor in configuring stratospheric dynamics. Therefore diagnosing the origin and effects of zonally asymmetric ozone may help to understand these coupling processes and to improve current state-of-the-art general circulation models used for climate predictions.

In Sect. 24.2, we describe the used data, including the European Centre of Medium Weather Forecasts (ECMWF) Reanalysis data and model-independent satellite data, as well as the used general circulation models without and with an interactively coupled ocean (GCM MAECHAM5, AOGCM COSMOS). In Sect. 24.3 we present observed wave patterns of O_3^* and the corresponding radiation perturbation introduced in the models. In Sect. 24.4, the effects of O_3^* on the atmospheric circulation and on important climate parameters are demonstrated based on the model simulations with prescribed O_3^*. Simplified linear solutions help to understand the involved stratosphere-troposphere coupling processes. Section 24.5 concludes with summary and discussion.

24.2 Data and Model Setup

We derived monthly means of zonally asymmetric stratospheric ozone $O_3^* = O_3 - [O_3]$ from the ozone data produced by the assimilation model of the European Centre of Medium Weather Forecast (ECMWF), which provide one of the largest data sets of meteorologically consistent three-dimensional (3D) ozone fields. We used the ECMWF Reanalysis data of ERA-40 for the time period 1957–2002 [*Uppala et al.*, 2005; *Dethof and Hólm*, 2004] and ERA-Interim beginning with 1989 and ongoing up to recent years [*Dee et al.*, 2011; *Dragani*, 2010a, 2010b]. As described in more detail by *Dethof and Hólm* [2004], the ozone fields of ERA-40 are produced by a tracer transport scheme including parameterisations of photochemical sources and sinks [an updated version of *Cariolle and Déqué*, 1986]. The ERA-40 assimilations include ozone observations of the Total Ozone Mapping Spectrometer (TOMS) and Solar Backscattered UltraViolet (SBUV) instruments on various satellites from the end of 1978 onwards [*Dethof and Hólm*, 2004]. Therefore the ERA-40 ozone of earlier years is a model variable as a function of transport and ozone chemistry parameterisation alone and might be somewhat uncertain for the pre-satellite period. Note that the assimilated ozone fields are not used in the radiation scheme of the assimilation model but a zonal mean ozone climatology instead [*Dethof and Hólm*, 2004]. For the ERA-Interim ozone, which begins with the year 1989, an upgraded ozone parameterisation is used [*Cariolle and Teyssedre*, 2007]. Also much more ozone satellite measurements are included in the assimilation procedure than for the production of the ERA-40 ozone, i.e., in addition to the TOMS and SBUV measurements, ozone observations of the SCIAMACHY and MIPAS measurements on Envisat, GOME measurements on ERS-2, and OMI and MLS measurements on EOS-Aura are included [*Dragani et al.*, 2010a, 2010b]. The ERA assimilation data are very suitable for incorporating in the model sensitivity experiments described in the next sections, because of the high spatial and temporal resolution on a homogeneous grid. However, because of a number of uncertainties which are unavoidable when using an assimilation model we verify the derived fields by model-independent satellite data.

For this purpose we have used ozone profiles of the SAGE II[1] and the EOS Aura MLS[2] satellite measurements. For both data sets we calculate monthly mean values of the irregularly spaced ozone profiles for a $10° \times 10°$ grid in longitude and latitude. Additionally the ozone fields were spatially smoothed by a 9×9 grid point filter to avoid small-scale perturbations. Note here that the limitations in horizontal coverage of the earlier SAGE II profiles (maximum $\approx$ 900 irregularly spaced profiles per month) can lead locally to some gaps or inaccuracies in the wave patterns for a specific month. On the other hand, the more recent Aura MLS data provide a much better resolution in time and space (maximum $\approx$ 3500 irregularly spaced profiles per day) leading to a much better description of the amplitude and phase of O_3^* based on the used sampling procedure.

Firstly the effects of O_3^* were investigated based on simulations with the GCM MAECHAM5. The MAECHAM5 is described in detail by *Roeckner et al.* [2006] and *Manzini et al.* [2006]. We use a model version with spectral resolution of T42

[1]Data are kindly provided by NASA at http://www-sage2.larc.nasa.gov.

[2]Data are also kindly provided by NASA at http://mls.jpl.nasa.gov/index-eos-mls.php.

and 39 levels up to 0.01 hPa. The standard version of MAECHAM5 includes a zonal and monthly mean ozone climatology according to *Fortuin and Kelder* [1998] which is prescribed for the radiation calculations. As a control, we have performed a model simulation over ten years forced by AMIP sea surface temperatures (SST) 1989–1999 [*Gates et al.*, 1999]. As a model experiment, we recalculated the ten winter periods by adding the January mean O_3^* of the 1990s to the zonal mean ozone climatology, for altitudes between 500 hPa and 1 hPa and latitudes $\varphi \geq 30°$N. Note that January mean O_3^* is a representative for winter conditions since phase and amplitude of decadal mean O_3^* do not vary strongly during winter months. The ten simulations of the experiment with O_3^* start in October of each year's control simulation, i.e. we include a spin-up period of 2 months up to winter months. Overall, we yield a data set of ten pairs of winter periods with and without O_3^*. A similar model setup was used by *Gabriel et al.* [2007] to investigate the effects of O_3^* on stratospheric and mesospheric temperature and planetary wave propagation. In the present paper, we focus on the stratosphere-troposphere coupling processes.

For the experiments with a coupled ocean we use the AOGCM COSMOS and a very similar model setup as for the MAECHAM5 simulations. For the AOGCM COSMOS simulations, the model resolution of the atmospheric part, which is also ECHAM5 [*Roeckner et al.*, 2006; *Jungclaus et al.*, 2006], is T31 with 90 levels from the surface to about 80 km height. Thus the model simulates stratospheric dynamics including a freely developing QBO with a frequency of 27–29 months. The ocean model MPI-OM [*Marsland et al.*, 2003] includes a dynamic/thermodynamic sea-ice model and has a 3 degree resolution with 40 vertical levels to 5.7 km depth. After a spin-up phase of 200 years under present day conditions two simulations are launched. In addition to the 80 year reference experiment an 80 year sensitivity experiment is performed. In extension of the 10 separate winter simulations with O_3^* performed with the MAECHAM5, the sensitivity experiment with the COSMOS includes a full annual cycle of the monthly mean O_3^* of thc 1990s decade from 500 hPa level to 2 hPa northward of 30°N, as derived from ERA-40, in addition to the standard zonal mean ozone. Note that the differences between the two simulations therefore include much more interannual variability. For the analysis we use the last 60 years of the model simulations.

24.3 Zonally Asymmetric Ozone O_3^* and Radiation Perturbation

Figure 24.1 shows the zonally asymmetric ozone O_3^* of the Northern Hemisphere at a level of 10 hPa (≈30 km) for mean January of the 1990s decade, derived from SAGE II (left), ERA-40 [middle, which is—for comparison—from *Gabriel et al.*, 2007], and ERA-Interim (right). There is a pronounced spiral-shaped ultra-long wave one pattern with maximum over Aleutians and minimum over Northern Europe, with an amplitude of about 0.6–0.8 mg kg^{-1} as indicated by SAGE II and by ERA-40, but with an amplitude of about 1.5–1.6 mg kg^{-1} as indicated by ERA-Interim. Note that this wave pattern intensifies during autumn, maintains its strength

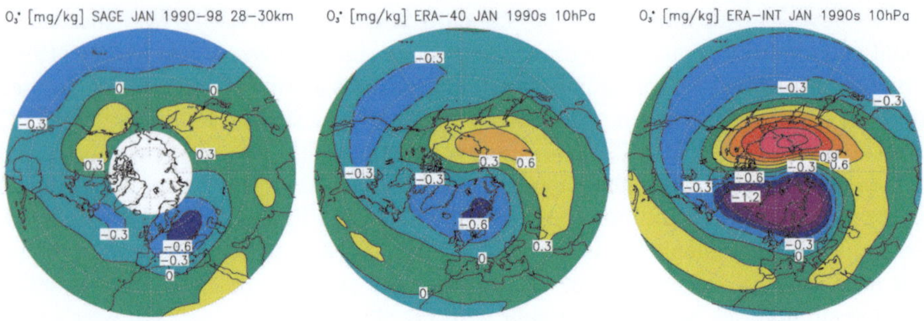

Fig. 24.1 Decadal mean zonally asymmetric ozone O_3^* for January of the 1990s at 10 hPa (≈30 km), derived from (*left*) SAGE II 1990–1998, (*middle*) ERA-40 1990–1999 and (*right*) ERA-Interim 1990–1999; distance of isolines: 0.3 mg kg^{-1}

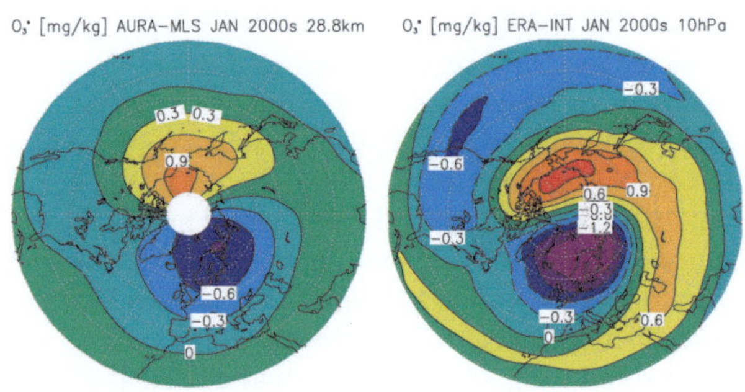

Fig. 24.2 As for Fig. 24.1, but with O_3^* derived from (*left*) Aura MLS 2005–2010 and (*right*) ERA-Interim 2001–2008; distance of isolines: 0.3 mg kg^{-1}

during winter, and decays during spring [e.g., *Gabriel et al.*, 2011a]. The amplitude of the wave one pattern derived from SAGE II might be somewhat underestimated because of the coarse resolution of the data in time and space as mentioned in the previous section. On the other hand, for the 1990s decade, the ERA-Interim data include more satellite observations than the ERA-40 data (e.g., NOAA-14 SBUV of the early 1990s decade and NOAA-14 SBUV and ERS-2 GOME of the late 1990s decade, as described in detail by *Dragani* [2010a]), which might be responsible for the stronger amplitude in the wave one pattern.

For more recent years we compare O_3^* derived from ERA-Interim and from Aura MLS data (Fig. 24.2). As shown by Fig. 24.2, the ERA-Interim data capture the observed features of O_3^* quite well. The time mean Aura MLS data are derived for a horizontal 10° × 10° grid which leads to a more smoothed picture of the O_3^* patterns and a somewhat smaller amplitude than those of ERA-Interim provided for a horizontal 1.5° × 1.5° grid. Note here that this feature holds also when comparing specific years of the overlapping time period 2005–2008 of the used data. Overall we conclude that the spatial structure of the wave one pattern is captured by both ERA-40 and ERA-Interim, but with an off-set in the amplitude of the wave one patterns. In the following we use O_3^* of ERA-40 of the 1990s in the model

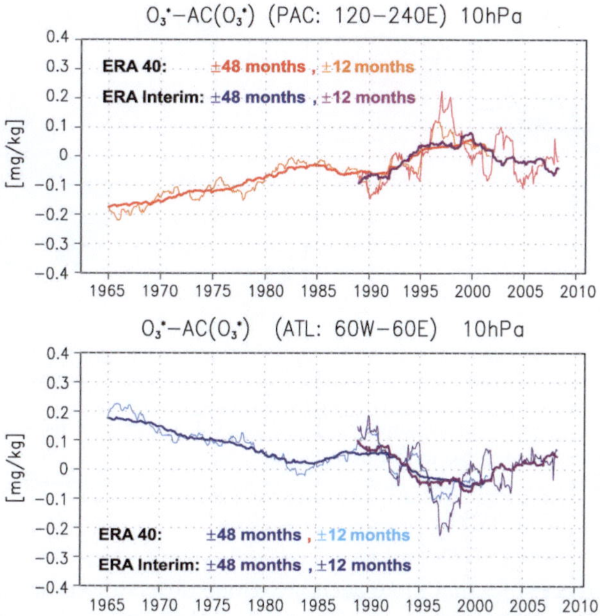

Fig. 24.3 Time series of the maximum and the minimum of monthly mean O_3^* derived from ERA-40 and ERA-Interim, smoothed by running means over (*thin lines*) ±12 months and (*thick lines*) ±48 months; O_3^* is averaged from 55°N to 75°N, and (*upper panel*) at the Pacific side of the hemisphere from 150°E to 210°E and (*lower panel*) at the Atlantic/European side of the hemisphere from 30°W to 30°E; a mean annual cycle of 1989–2002 is removed from both ERA-40 and ERA-Interim to focus on the interannual and decadal changes of O_3^*; note that satellite measurements are assimilated from the year 1978 onwards

simulations allowing a comparison with previous work. The use of ERA-40 data with lower wave amplitudes therefore represents a lower limit on the radiative effect of O_3^*.

Figure 24.3 shows the time series of mean January values of O_3^* at 10 hPa derived from ERA-40 and ERA-Interim, averaged over a region covering the maximum in O_3^* over the North Pacific and the minimum in O_3^* over the North Atlantic/Northern Europe. The values are smoothed by a running mean over ±48 months to illustrate the long-term change over decades, and over ±12 months to illustrate the interannual variability. Note that a mean annual cycle of O_3^* for the overlapping time period 1989–2002 is removed, i.e. the off-set in the amplitude of the wave one pattern described above is excluded in order to focus on the interannual and decadal changes in the amplitude of O_3^*.

Figure 24.3 suggests that the amplitude of O_3^* increased nearly linearly during the last decades, modulated by decadal and interannual variations which are more pronounced in ERA-Interim than in ERA-40. On the one hand, these changes are primarily related to transport processes due to ultra-long planetary waves. On the other hand, chemical depletion of ozone in the stratospheric winter polar vortex, which is usually zonal asymmetric, contribute additionally to the amplitude in O_3^*.

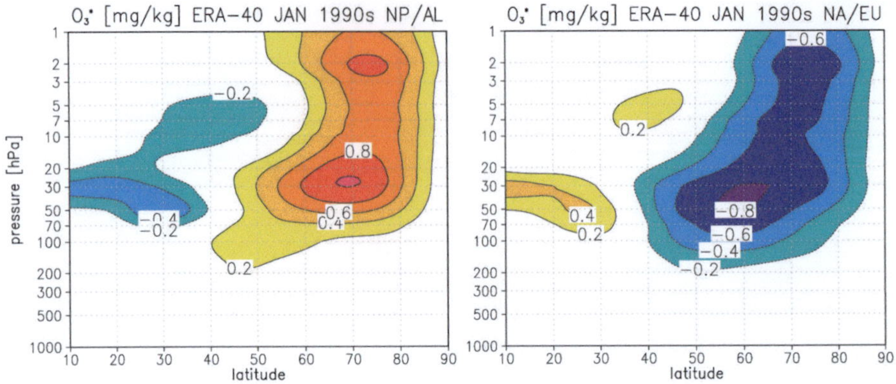

Fig. 24.4 Latitude-height cross section of January mean values of O_3^* for the 1990s decade, (*left*) longitudinally averaged over North Pacific/Aleutians (NP/AL) between 150°E and 150°W and (*right*) longitudinally averaged over North Atlantic/Europe (NA/EU) between 30°W and 30°E; distance of isolines is 0.2 mg kg^{-1}

We have also to consider that more accurate ozone observations are included in the assimilations since 1979, which affect the picture of variations. However, the nearly linear trend in O_3^* seems to be a robust feature of the time series covering both the pre-satellite and the satellite periods. Note here that the amplitude of O_3^* during summer is weaker than during winter by more than one order, because of weaker planetary wave activity. Therefore both the decadal and interannual variations of the winter months are somewhat smoothed when averaging over the full annual cycle as done in Fig. 24.3.

Figure 24.4 gives an impression of the vertical structure of O_3^* by showing latitude-height cross sections for January of the 1990s. The two cross-sections are averaged over two representative regions of the wave one pattern. At mid- and polar latitudes there are positive anomalies over North Pacific/Aleutians (Fig. 24.4, left) and negative anomalies over North Atlantic/Northern Europe (Fig. 24.4, right) throughout the whole stratosphere. Slightly enhanced values of O_3^* can be found in the lower stratosphere at levels around 50–30 hPa and in the upper stratosphere at levels around 2–3 hPa. There are also some weaker anomalies of opposite sign at lower mid- and subtropical latitudes, which are related to the spiral-shaped horizontal structure of O_3^* shown in Fig. 24.1.

Stratospheric ozone is one of the most important absorbers for both incoming solar radiation and outgoing thermal radiation. Therefore the wave one pattern in stratospheric ozone introduces a corresponding radiation perturbation in the Earth's radiation budget. The instantaneous radiation perturbation induced by O_3^* was calculated with the radiation scheme of MAECHAM5. Figure 24.5 shows the solar and thermal radiative heating rates due to January mean O_3^* of the 1990s, averaged over the same regions as for O_3^* in Fig. 24.4. Over North Pacific/Aleutians, in regions of positive O_3^*, we find positive solar heating perturbations of 0.05 to 0.1 K day^{-1} at high mid-latitudes because more solar radiation is absorbed, and

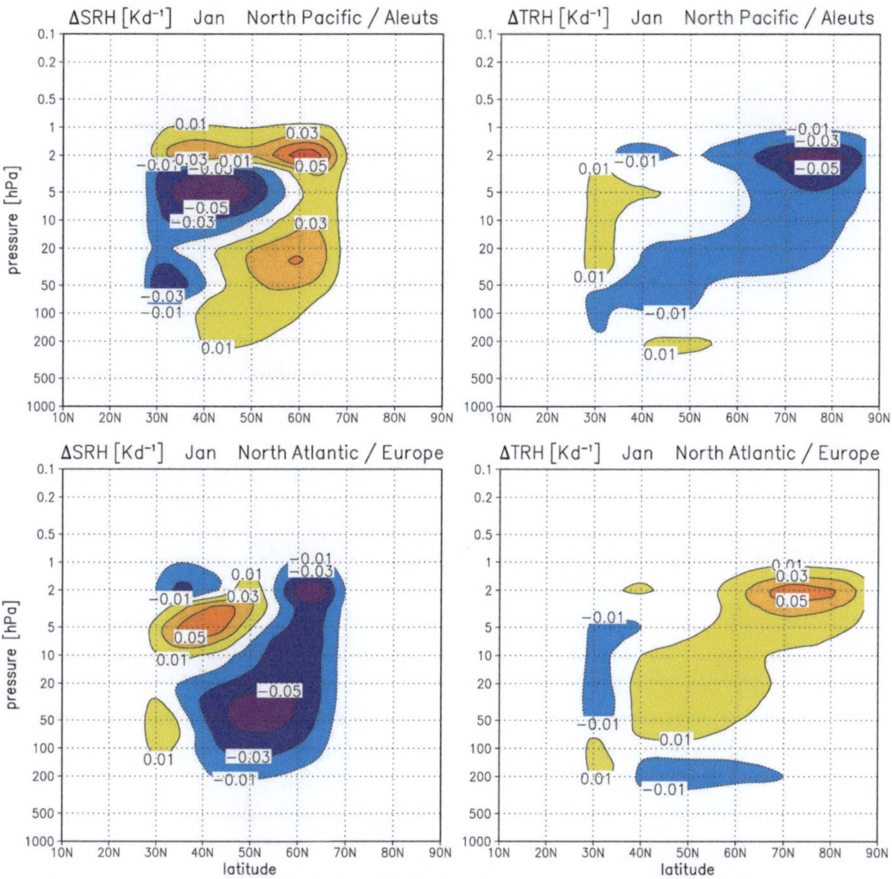

Fig. 24.5 (*Left panels*) solar and (*right panels*) thermal radiative heating rates due to O_3^* for extra-tropics ($\varphi \geq 30°N$), averaged over (*upper panels*) North Pacific/Aleutians between 150°E and 150°W and (*lower panels*) North Atlantic/Europe between 30°W and 30°E; distance of isolines is 0.02 K day^{-1}

negative thermal heating perturbations of -0.05 to -0.1 K day^{-1} at polar upper stratosphere because more long-wave radiation from the Earth's surface is emitted into space (Fig. 24.5, left panels); vice versa, over North Atlantic/Europe, in regions of negative O_3^*, we found negative solar heating perturbations of -0.05 to -0.1 K day^{-1} at high mid-latitudes and positive thermal heating perturbations of 0.05 to 0.1 K day^{-1} in the polar upper stratosphere (Fig. 24.5, right panels). Note that the radiation perturbations are changing during winter months because of the seasonal cycle in solar irradiance. Note also that—consistent with the relative amplitude of the wave one pattern—the relative contribution of radiative forcing due to O_3^* is approximately 10 % of those due to zonal mean ozone.

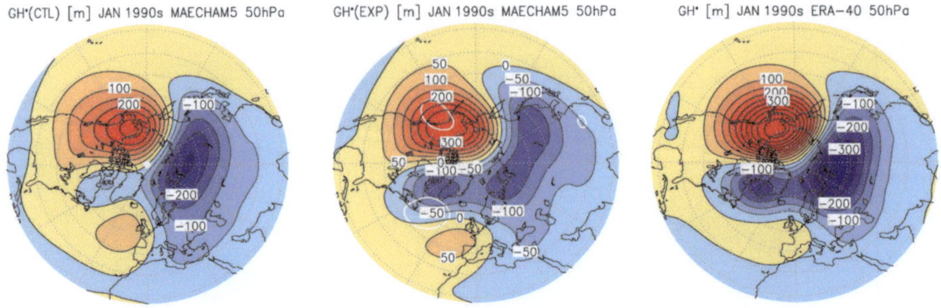

Fig. 24.6 Decadal mean of zonally asymmetric geopotential height $\Phi^* = \Phi - [\Phi]$ (denoted also as GH*) at 50 hPa for mean January of the 1990s, (*left*) for the model without O_3^*, (*middle*) for the model with O_3^* and (*right*) for ERA-40; distance of isolines is 50 m

24.4 Influence of O_3^* on the Atmospheric Coupling Processes

24.4.1 Effects of O_3^* in a GCM Sensitivity Study

The influence of the radiation perturbation due to O_3^* on the wave one pattern in geopotential height Φ at lower stratospheric altitudes (50 hPa) is shown in Fig. 24.6, as indicated by the deviation from the zonal mean $\Phi^* = \Phi - [\Phi]$, in comparison to the observed pattern derived from ERA-40. As mentioned in the introduction, we find usually a pronounced stationary wave one component in Φ^* including a positive anomaly over the northern North Pacific/Aleutians and a negative anomaly over Siberia/Northern Europe with westward elongation towards Labrador/Greenland, which is forced by tropospheric wave activity. The negative anomaly in Φ^* is related to the centre of the polar vortex which is usually found over Northern Europe/West Siberia [*Waugh and Randel*, 1999; *Karpetchko et al.*, 2005].

Figure 24.6 reveals a change in the January mean Φ^* when introducing O_3^* in the model. Both model simulations generate a wave one anomaly in Φ^* similar to those of ERA-40, but the strength of the Aleutian high anomaly and the spatial structure of the polar low anomaly towards Labrador/Greenland are much better captured by the model simulation with O_3^*. In particular, we find significant changes of about 60–80 m at the eastern side of the Aleutian high over northern North America and of 120–160 m at the western side of the polar low over Labrador/Greenland. Comparing the pictures, we conclude that the inclusion of O_3^* leads to an improvement of the model in capturing the observed spatial structure in the wave structure of Φ^*, although, as usually found in GCMs, the amplitude remains too weak in comparison to observations.

The pathway of how O_3^* affects the circulation becomes evident applying a linear solution of forced Rossby waves, similar to *Charney and Drazin* [1961]. This gives an impression of the instantaneous excitation of a planetary wave perturbation in geopotential height $\Phi^*(O_3^*)$ or horizontal stream function $\Psi^*(O_3^*)$ due to the heating rates $Q^*(O_3^*)$ (for the stream function we assume geostrophic balance:

$\phi^* \approx f_0 \Psi^*$, where ϕ is geopotential with $\Phi = \phi/g$, g: acceleration of the Earth's gravity, and f_0: Coriolis parameter at a specific latitude). We use the tendency of the zonally asymmetric quasi-geostrophic potential vorticity perturbation $q^* = q - q_0$ for a zonal mean basic state q_0 on a β-plane and steady state conditions ($\frac{\partial}{\partial t} = 0$), forced by a source $s^* = s^*(Q^*)$:

$$\frac{dq^*}{dt} \approx u_0 q_x^* + v^* q_{0y} = s^*(Q^*), \quad \text{with } q^* = \nabla^2 \Psi^* + \frac{f_0^2}{N_0^2} \Psi_{zz}^* \tag{24.1}$$

Here, u_0 is the zonal wind of the basic state, $v^* = \Psi_x^*$, $q_{0y} \approx \beta$ with $\beta = 2\frac{\Omega}{r}\cos\varphi$, $s^* = \frac{f_0}{N_0^2}(\frac{g}{\theta_0} Q^*)_z$, $\theta_0 = \theta_0(z)$ is the mean global potential temperature, N_0^2 is the Brunt-Väisälä frequency, ∇^2 is the horizontal Laplace operator, and the subscripts x, y and z denote the derivation in longitude, latitude and height. We can formulate Eq. (24.1) as a Poisson-type differential equation for the horizontal gradient Ψ_x^* (or the corresponding wind field $v^* = \Psi_x^*$):

$$\nabla^2 \Psi_x^* + \frac{f_0^2}{N_0^2} (\Psi_x^*)_{zz} + \frac{\beta}{u_0} \Psi_x^* = \frac{s^*}{u_0} \tag{24.2}$$

Assuming that the wave one pattern in $Q^*(O_3^*)$ is located in the extra-tropical stratosphere far from the surface and the top of atmosphere, and far from the tropics and the North Pole, we can derive a simple solution in case of a zonal mean westerly wind $u_0 > 0$ (which is usually fulfilled at pressure levels $p < 500$ hPa during winter) by applying a Fourier decomposition for both sides of Eq. (24.2). Then a relation of the individual Fourier components $FV_k^{lm}(\chi^*)$ (with $v^* := \chi_z^*$) and $FQ_k^{lm}(Q^*)$ for ultra-long Rossby waves (zonal wave number: $k = 1$, meridional wave number: $l = 1$–3 for 0°–90°N, vertical wave number $m = 1$–3 for $H_{surface}$–H_{top}) is given by

$$\left(-\frac{N_0^2}{f_0^2}\left(k^2 + l^2 - \frac{\beta}{u_0}\right) - m^2\right) FV_k^{lm} = FQ_k^{lm} \tag{24.3}$$

Together with the wave parameters and the basic state, the coefficients FQ_k^{lm} determine the coefficients FV_k^{lm} providing a re-transformation to a linear solution v_{LIN}^*. Then Ψ_{LIN}^* can be derived easily via $v_{LIN}^* = (\Psi_{LIN}^*)_x$, presuming a zonal wave number one for each latitude, and the perturbation in geopotential height is given by $\Phi_{LIN}^* \approx f_0 \Psi_{LIN}/g$. For the mean basic state (u_0, θ_0, N_0^2) we use the results of the model simulation without O_3^* for mean January.

As illustrated in Fig. 24.7 (upper, left), the linear solution Φ_{LIN}^* indicates that the radiation perturbation due to O_3^* forces a planetary Rossby wave one pattern in lower stratospheric geopotential height with an amplitude of about 15–20 m. This perturbation becomes evident when considering that radiative heating and cooling induce a flow perturbation from colder regions where $O_3^* < 0$ to warmer regions where $O_3^* > 0$, at least leading to the clockwise rotation of the cyclonic perturbation over North America and the anti-clockwise perturbation of the anti-cyclonic perturbation over North Siberia.

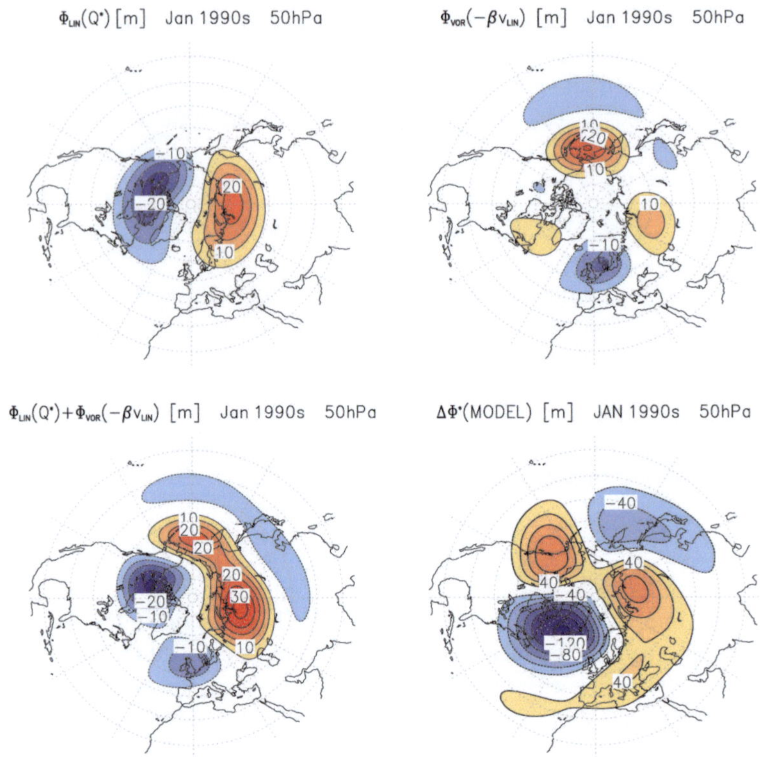

Fig. 24.7 Comparison of simplified linear solutions of the induced perturbations in geopotential height (*upper, left*) Φ^*_{LIN}, (*upper right*) Φ^*_{VOR} and (*lower, left*) $\Phi^*_{LIN} + \Phi^*_{VOR}$ with the perturbation of the model experiment $\Delta\Phi^*$ at 50 hPa (isolines for Φ^*_{LIN}, Φ^*_{VOR} and $\Phi^*_{LIN} + \Phi^*_{VOR}$: 5 m, isolines for $\Delta\Phi^*$: 20 m; for details of the solutions see text)

The induced poleward flow over the northern North Atlantic and the southward flow over the North Pacific/Aleutians can stimulate further and much stronger perturbations in the stratospheric circulation. This becomes more evident when considering that the northward and southward advection of planetary vorticity $\eta = f + \zeta$ changes efficiently the local development of vorticity because of the mean meridional gradient in planetary vorticity $\frac{\partial\eta}{r\partial\varphi} \approx \frac{\partial f}{r\partial\varphi} \approx \beta > 0$, i.e. because of the β-effect [*Pedlosky*, 1979]. Here, $f = 2\Omega \sin\varphi$ is the Coriolis parameter, ζ is the relative vorticity, $\beta = 2\Omega r^{-1} \cos\varphi$, and r is the radius of the Earth. In a simplified way we can demonstrate this effect when introducing the linear wind perturbation v^*_{LIN} in the barotropic vorticity equation, together with a simple relaxation term for the total tendency of an induced perturbation in relative vorticity ζ^*_{VOR} ($\frac{d\zeta^*}{dt} \approx -\alpha\zeta^*_{VOR}$, $\alpha = 1/\tau$, τ: relaxation time), i.e. we obtain:

$$\frac{d\zeta}{dt} = -\beta v \quad \rightarrow \quad -\beta v^*_{LIN} \approx -\alpha\zeta^*_{VOR} \tag{24.4}$$

Assuming geostrophic balance for the extra-tropics and a corresponding perturbation with zonal wave number one, i.e. $f_0 \zeta^*_{VOR} \approx -k^2 \phi^*_{VOR}$ ($k = 2\pi/L$, $L = 2\pi r \cos\phi$), we obtain an estimation of the perturbation in geopotential height $\Phi^*_{VOR} = \phi^*_{VOR}/g$ due to the local change in relative vorticity ζ^*_{VOR} induced by the linear flow perturbation v^*_{LIN}:

$$\Phi^*_{VOR} \approx -\tau \left(g \frac{f_0}{k^2} \right) \beta v^*_{LIN} \tag{24.5}$$

Figure 24.7 (upper, right) shows the induced perturbation Φ^*_{VOR} at 50 hPa for a relaxation time of $\tau = 7$ days usually used for relaxation in the stratosphere [e.g., *Dunkerton*, 1991]. The flow perturbation v^*_{LIN} induces a pronounced cyclonic perturbation ($\Phi^*_{VOR} < 0$) over the northern North Atlantic and an anticyclonic perturbation ($\Phi^*_{VOR} > 0$) over the North Pacific/Aleutians, i.e. a wave one pattern in Φ^*_{VOR} with an amplitude of about 20–30 m. Overall, the sum of the linear perturbations $\Phi^*_{LIN} + \Phi^*_{VOR}$ (Fig. 24.7, lower, left) elucidates the perturbation in the large-scale flow at 50 hPa which is imposed by the radiation perturbation $Q^*(O_3^*)$ without non-linear interaction with the time mean flow and vertical planetary wave propagation.

For comparison, Fig. 24.7 (lower, right) depicts the differences in zonally asymmetric geopotential height $\Delta\Phi^*$ at 50 hPa between the model simulations with and without O_3^*, which illustrates that the imposed perturbation excites further non-linear processes strongly enhancing and modifying the linear response to the radiation perturbation $Q^*(O_3^*)$. In particular, in comparison to Fig. 24.7 (lower, left), the centres of the perturbation pattern are slightly shifted eastward due to non-geostrophic vorticity advection which is not included in the linear approach. The imposed cyclonic perturbation is strongly amplified and spatially focused over the northern North Atlantic between Labrador and Scandinavia, whereas the imposed anti-cyclonic perturbation over the northern North Pacific/Aleutians is spatially more expanded indicating enhanced wave dissipation processes in these regions in the model with O_3^*.

In the model simulation the perturbations are strongly modified by non-linear interactions via horizontal and vertical advection of vorticity and by induced changes in the coupling of the troposphere and stratosphere via vertical wave propagation. These changes become more evident in Fig. 24.8 depicting the three-dimensional wave activity flux vectors **F** and flux divergences $\nabla \cdot \mathbf{F}$ as they follow from the wave activity equation for quasi-stationary waves [*Plumb*, 1985]. This illustrates the sources, sinks and propagation characteristics of quasi-stationary wave trains. For orientation, Fig. 24.8 (left panels) shows the January mean values of **F** and $\nabla \cdot \mathbf{F}$ of the model simulations without O_3^*, averaged from 90°E to 90°W, here called the Pacific half of the Northern hemisphere, and from 90°W to 90°E which is here called the Atlantic half of the Northern hemisphere.

For the Pacific half of hemisphere, Fig. 24.8 (upper left) indicates a strong source of wave activity in the middle troposphere at mid-latitudes and an up- and northward propagation of the wave trains into the stratosphere, but also somewhat weaker southward fluxes in the subtropical upper troposphere. For the Atlantic half of hemisphere (Fig. 24.8, lower left), the source of wave activity and the up- and northward

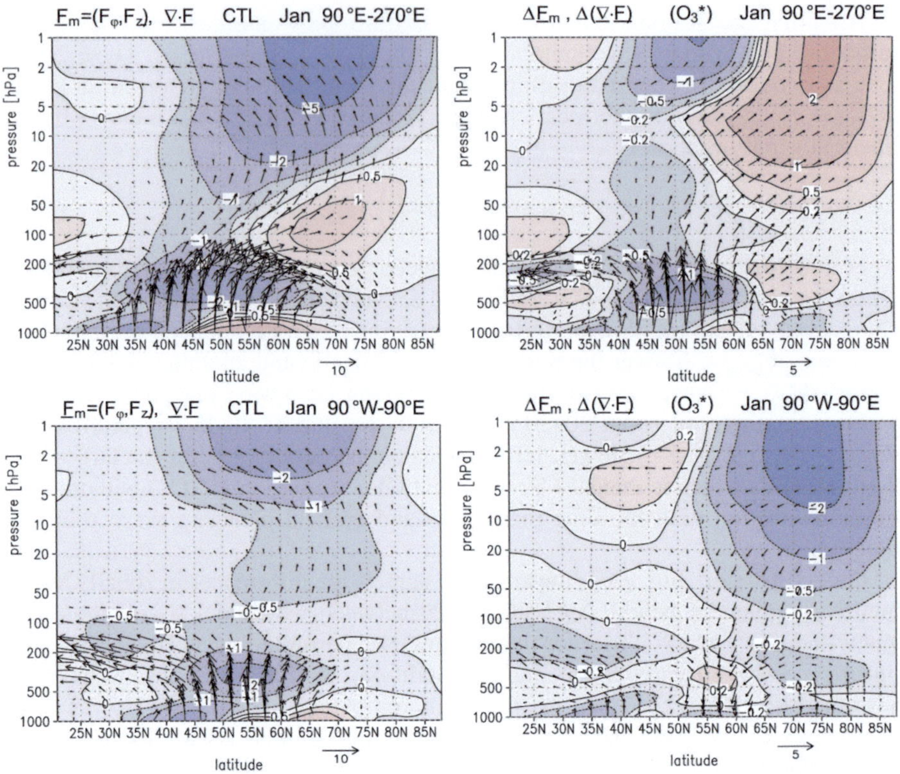

Fig. 24.8 Meridional component $\mathbf{F}_m = (F_\varphi, F_z)$ and divergence $\nabla \cdot \mathbf{F}$ of three-dimensional wave flux vectors $\mathbf{F}$ for quasi-stationary waves, (*left panels*) for the control run without O_3^* and (*right panels*) for the O_3^*-induced differences, (*upper panels*) longitudinally averaged over the Pacific-side half of hemisphere 90°E–90°W (i.e. 90°E to 270°E) and (*lower panels*) longitudinally averaged over the Atlantic-side half of hemisphere (90°W to 90°E); isolines of $\nabla \cdot \mathbf{F}$ in ms^{-1} per day, vector units at bottom in $m^2\,s^{-2}$, vectors are scaled by $(F_\varphi, F_z) \rightarrow (p/p_0)^{-1/2}(F_\varphi, 100 \cdot F_z)$

fluxes are weaker, whereas the up- and southward fluxes towards the subtropical upper troposphere are somewhat stronger than over the North Pacific. Note here that, in the troposphere, the wave flux patterns are mainly related to the pronounced regions of baroclinic Rossby wave developments over the ocean basins, which are different over the North Pacific and over the North Atlantic [e.g., *Gabriel and Peters*, 2008]. Note also that the regional distribution of upward propagating wave trains might be related to the wave one pattern in the stratosphere, i.e. the centre of the winter polar vortex occurs most frequently over Northern Europe/West Siberia. This leads to stronger damping of vertical propagating waves in the Atlantic half of hemisphere.

Figure 24.8 (right panels) shows how the change in the wave one pattern affects the wave activity fluxes by the differences $\Delta\mathbf{F}$ and $\Delta(\nabla \cdot \mathbf{F})$ between the simulations with and without O_3^*. In the Pacific half of hemisphere (Fig. 24.8, upper

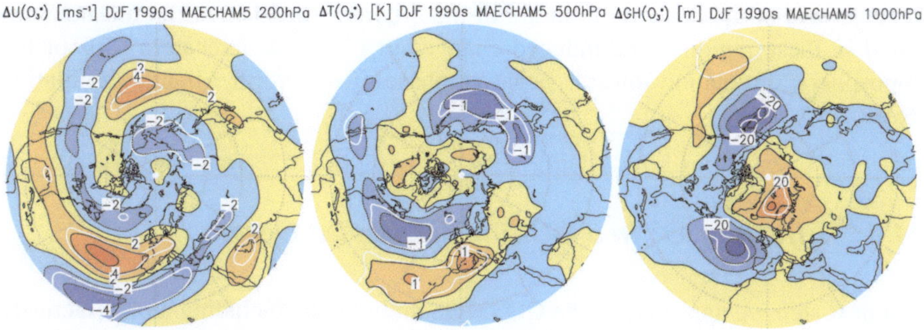

Fig. 24.9 Differences between the simulations with and without O_3^*, for zonal wind ΔU at 200 hPa (isolines in $\mathrm{m\,s^{-1}}$) (*left*), temperature ΔT at 500 hPa (isolines in K) (*middle*), and geopotential height $\Delta\Phi$ at 1000 hPa (isolines in m); *white line* indicate areas of significant ($> 90\,\%$) differences

right), both the strength of the mid-troposphere sources and the north- and southward flux vectors become stronger when introducing O_3^* (maximum changes in the order of 25 %). In the Atlantic half of hemisphere (Fig. 24.8, lower right), the strength of both the mid-tropospheric sources and the up- and northward fluxes become somewhat weaker but the southward fluxes in the sub-tropics are slightly stronger when introducing O_3^* (maximum changes in the order of 10 %). We conclude that the imposed flow perturbation in the stratosphere leads to a change in tropospheric wave activity and that these changes feed back to the stratosphere by upward propagating wave trains. These feedbacks contribute to the non-linear changes in the wave one pattern discussed above when comparing Fig. 24.7 (lower, left) and Fig. 24.7 (lower, right). The meridional wave flux component is related to the frequency of Rossby wave breaking events occurring at the northern and southern flank of the westerly jet producing cut-off anticyclones or cut-off cyclones in the upper troposphere [*Peters and Waugh*, 1996, 2003; *Esler and Haynes*, 1999; *Gabriel and Peters*, 2008]. Such breaking events affect the mean configuration of the tropospheric westerlies and associated climate-relevant patterns in the atmospheric circulation.

As an example, Fig. 24.9 shows the total differences in zonal wind ΔU at 200 hPa (left), middle tropospheric temperature ΔT at 500 hPa (middle) and geopotential height $\Delta\Phi$ at mean sea level pressure (right) between the model simulations with and without O_3^*. The differences ΔU indicate significant changes of up to 4–6 $\mathrm{m\,s^{-1}}$ in the westerly jet particularly over the ocean basins. Correspondingly we find an enhanced transport of warm air from subtropical latitudes over the North Atlantic towards the Mediterranean at the southward flank of ΔU (leading to the change in ΔT of about 1 K in this region), and an enhanced transport of cold air from Eastern Siberia towards northern North Pacific at the southward flank of ΔU (leading to a change in ΔT of about -1 K in this region). In high mid- and polar latitudes, over the northern North Pacific/Aleutians, we find a cyclonic perturbation ($\Delta\Phi < 0$) near the Aleutian low usually found in this region. But over the northern North Atlantic an anticyclonic perturbation ($\Delta\Phi > 0$) develops near the Iceland low. The changes in ($\Delta\Phi < 0$) of up to about 30 m are in the order of 20–30 % of the mean values found in these regions. The induced changes in U and Φ indicate a shift of the tropospheric circulation towards negative phase of the North Atlantic Oscillation (NAO).

24.4.2 *Effects of* O_3^* *in a GCM with Coupled Ocean*

In a next step we apply O_3^* in the COSMOS simulations including an interactively coupled ocean, providing further evidence of how O_3^* influences the climate system. Figure 24.10 (left panels) depicts the differences in geopotential height in the stratosphere at 50 hPa and in the middle troposphere at 500 hPa. For the stratosphere we find a similar but horizontally shifted structure in the signal as described in the MAECHAM5 experiment (compare with Fig. 24.7, lower, right), i.e. a negative anomaly over Northern Europe and a positive anomaly over North Canada/Labrador. In the troposphere, the structure of the response indicates very clearly a shift of the tropospheric circulation towards a negative phase of the NAO. The Island low is weakened and the North Pacific low is intensified during northern winter.

In comparison to the simulations with the GCM MAECHAM5 the signal at 500 hPa is more pronounced and shifted southward. Here we only note that also other fields, e.g. the zonal westerlies in the upper troposphere at 200 hPa, the temperature in middle troposphere at 500 hPa and the surface geopotential height at 1000 hPa show also a similar but southward shifted pattern as in the MAECHAM5 sensitivity study. The differences in the response to O_3^* between the model simulations without and with coupled ocean indicate an imposed change in the processes coupling the troposphere and the ocean, i.e. the induced changes in the wind pattern leads to change in the wind-driven ocean circulation and the induced changes in the ocean circulation feed back to tropospheric wave developments via the related changes in the advection of warm or cold water.

In a further step we have examined the interannual variability of the 60 model years in more detail. The 60 years of each AOGCM experiment are sorted by the zonal winter means of the zonal wind at 10 hPa and 60°N (UM10) and are separated into two parts (strong and moderate) with 30 members each. In the strong vortex part for the reference and the sensitivity experiment years with UM10 of more than 31 $m\,s^{-1}$ and 32 $m\,s^{-1}$ are included, respectively. It turned out that the 30 years with strong vortex are characterised by a very weak zonally asymmetric component, whereas the 30 years with a moderate polar vortex are characterised by a stronger wave one pattern that is more related to the observed field of the 1990s decade. Here we have to consider that the strength of the polar vortex is an important factor configuring the tropospheric circulation [*Walter and Graf*, 2005]. Therefore it is reasonable to focus on the 30 years of the simulations with a moderate polar vortex, which are more related to the observed wave pattern in O_3^* of the 1990s decade. For comparison, the results of the moderate polar vortex case are also shown in

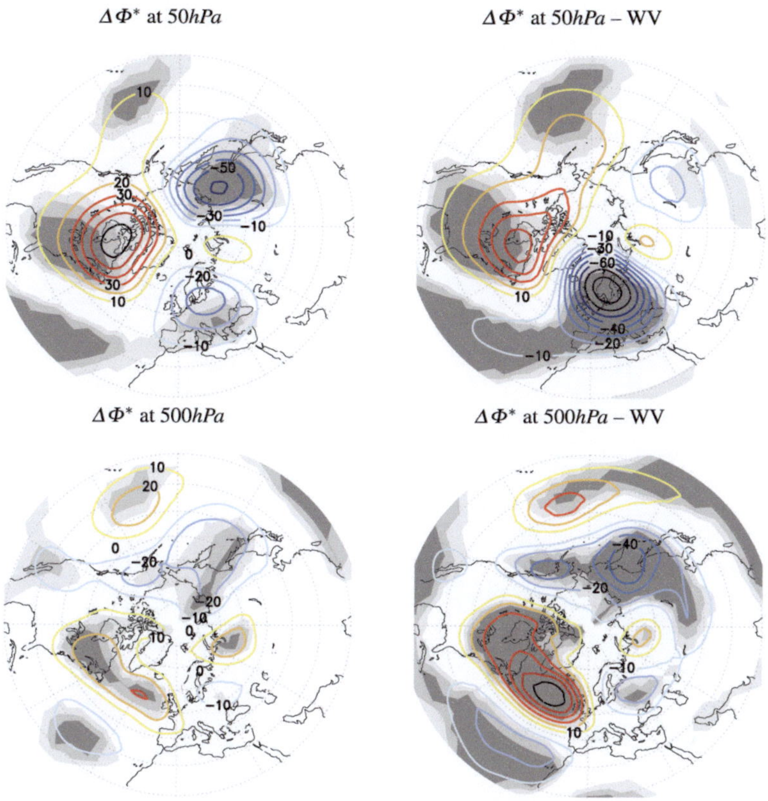

Fig. 24.10 Differences in zonally asymmetric geopotential height Φ^* in the model simulation with coupled ocean, for the overall 60 years (*left panels*) and 30 years excluding winter with a very strong polar vortex (*right panels*), where distances of isolines are 10 m. *Gray shaded areas* indicate 80, 90 and 95 % significance of changes

Fig. 24.10 (right panels) indicating a more pronounced wave one perturbation with a cyclonic perturbation over the northern North Atlantic and an anti-cyclonic perturbation over the northern North Pacific. The pattern is much more significant because of the consistent configuration of the prescribed wave pattern in O_3^* and the polar vortex. In the following we focus on the 30 years with moderate polar vortex conditions.

Figure 24.11 illustrates the influence of O_3^* on some important climate relevant parameters. In relation to the change towards the negative phase of NAO, we find a band of enhanced convective precipitation over the middle North Atlantic but less convective precipitation over the northern North Atlantic. The changes in precipitation over the North Pacific are less significant probably due to the somewhat stronger wave activity in this region. On the other hand, we find less snowfall in a band between northern North America and England, but more snowfall over Labrador/Greenland. The induced changes in near surface temperature show

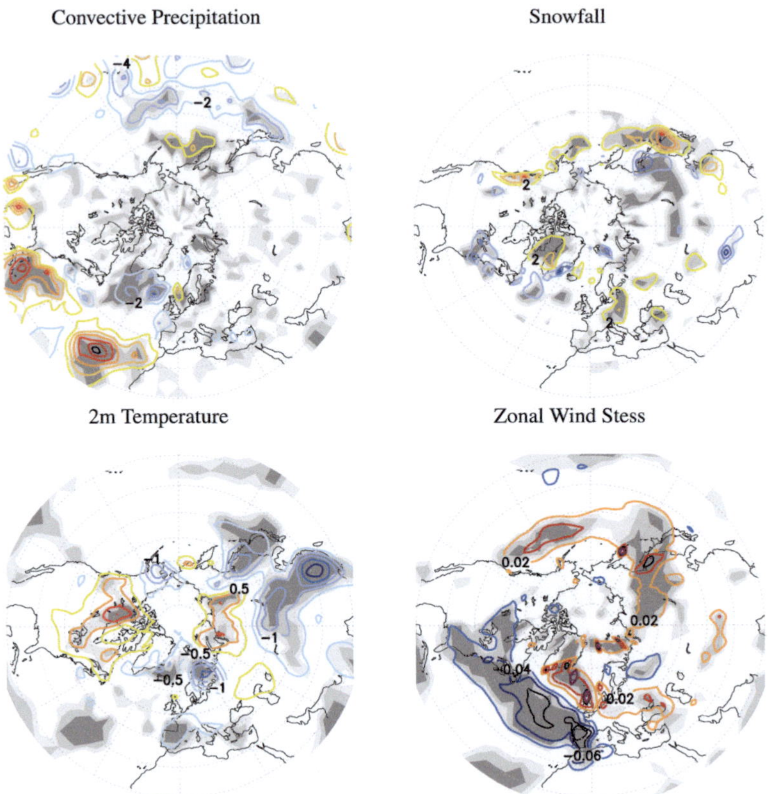

Fig. 24.11 Changes in climate parameters due to O_3^* for 30 years excluding winter with a very strong polar vortex. For convective precipitation (*upper panel left*) and snowfall (*upper right*) isoline intervals are $2 \cdot 10^{-6}$ kg m^{-1} s^{-1} for near surface temperature at 2 m in 0.5 K (*lower left*) and zonal wind stress at the Earth's surface (*lower right*) they are 0.02 Pa. Gray shaded areas indicate 80, 90 and 95 % significance of changes

warming of about +1 K over northern North America and cooling of about −1 K over south-east Asia. These changes become more evident in relation to the induced changes in other important parameters. For example, a very pronounced signal is found in the zonal wind stress with an increase over northern North Pacific and eastern Siberia and a decrease over the northern North Atlantic (note here that local changes of 0.02–0.04 Pa are in the order of 10–20 % of the mean values). The wind stress is one of the most important factors driving the near-surface ocean currents, as given approximately by the Sverdrup-Balance [e.g. *Pedlosky*, 1979].

Figure 24.12 (left) shows the induced changes in the wind-driven ocean currents by the difference of the ocean's barotropic stream function. Related to the differences in the wind stress we find a cyclonic perturbation (negative values) in the middle North Atlantic and an anti-cyclonic perturbation (positive values) in the northern North Atlantic, i.e. a decrease in the strength of the Gulf-stream (note here that local

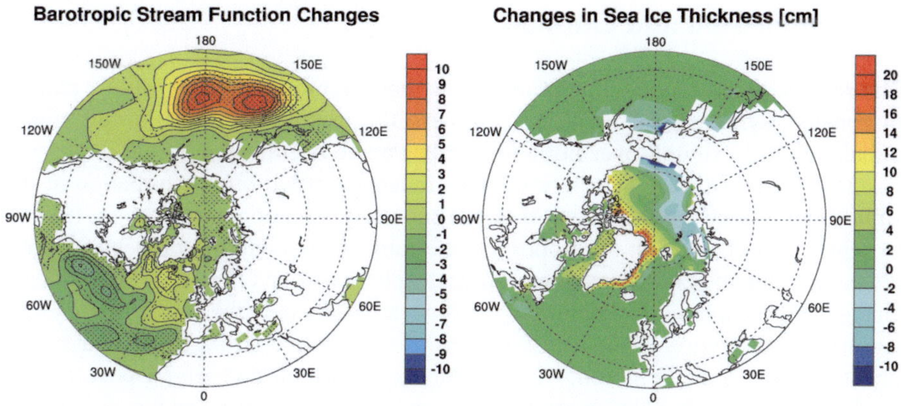

Fig. 24.12 Changes in ocean barotropic stream function [10^6 $m^3\,s^{-1}$] (*left*) and sea-ice thickness [cm] (*right*) for the 30 years excluding winter with a very strong polar vortex. Stippling denotes areas with significant changes (t-test, 90 %)

changes of $3\text{–}10 \cdot 10^6$ $m^3\,s^{-1}$ are in the order of 2–5 % of the mean values). We also find a poleward flow anomaly towards Labrador which could influence temperatures and snowfall over northern North America/Labrador. Vice versa, the anti-cyclonic perturbation over the middle North Pacific enhances the north-eastward flow of the Kuroshio-stream.

The change in the ocean currents influences the meridional transport of warm water, which has an impact on local temperatures, precipitation patterns and sea-ice cover. The induced changes in sea-ice thickness are shown in Fig. 24.12 (right). We find an increase of 10–20 cm in a region covering Labrador/Greenland, and a decrease of 5–10 cm in a region extending from the Barents-Sea to the Aleutians. Subsequently, a positive temperature—sea ice melting—albedo feedback also contribute to the pattern of significant changes in 2 m temperature, like the increase in 2 m temperature over north-eastern Siberia.

A comparison of the results without and with a coupled ocean model provide some more evidence on the influence of O_3^* on the coupling processes between troposphere and the ocean. For example, the decrease in the Gulf-stream and associated transport of warm water towards northern North Atlantic lead to a damping of baroclinic wave development, less northward transport of latent heat and colder air masses in this region. Therefore the mean Iceland low weakens over a more extended and southward shifted region than predicted by the model simulations without coupled ocean. In a similar way but with opposite sign, the imposed intensification of the Kuroshio leads to more pronounced and somewhat more northward shifted patterns in the simulations with ocean than in the simulations without ocean. Subsequently the related changes in precipitation and temperatures introduce an additional temperature—ice—albedo feedback, contributing additionally to the wave one pattern in the differences of sea-ice thickness.

24.5 Discussion and Outlook

The results of our model simulations show that the pronounced wave one pattern in zonally asymmetric stratospheric ozone (O_3^*), usually observed during winter months with maximum over northern North Pacific/Aleutians and minimum over northern North Atlantic/Northern Europe, induces a wave one perturbation in the stratospheric large-scale flow via the corresponding radiation perturbation, leading to subsequent changes in the tropospheric-stratospheric coupling processes. Overall we find that the zonally asymmetric stratospheric ozone O_3^* impose a significant change in tropospheric circulation towards negative phase of the North Atlantic Oscillation (NAO) via the imposed flow perturbation in the stratosphere and subsequent changes in precipitation, wind-driven ocean currents and sea-ice covering.

Based on simplified linear solutions of forced Rossby waves and barotropic vorticity development we demonstrate the pathway of how O_3^* affects the atmospheric circulation. Investigating the effects of stratospheric O_3^* with a numerical model includes a lot of uncertainties due to unavoidable restrictions in the discrete formulations of the model equations and physical parameterisations, therefore the simplified solutions are also very helpful in identifying the primary processes and the amplification of the perturbation in the atmospheric circulation in comparison to the numerical simulations. We find that, as a primary process, the flow perturbation in the stratosphere can be understood as a result of northward and southward transport of barotropic vorticity imposed by the planetary Rossby wave perturbation forced by O_3^*. This imposed change in the stratosphere towards a mean cyclonic perturbation over the northern North Atlantic and a mean anti-cyclonic perturbation over the northern North Pacific leads to the change in tropospheric circulation towards the negative phase of the NAO, which is associated with corresponding changes of about 10–20 % in the upper tropospheric westerlies, tropospheric heat and moisture transports, and upward propagating wave trains modulating the primary perturbation in the stratosphere.

Additional feedbacks via troposphere-ocean coupling lead to an intensification and spatial shift of the signal, including an imposed change in wind-driven ocean currents with a weakening of the Gulf-Stream and an intensification of the Kuroshio stream (in the order of 2–5 % of the mean values). The signal of O_3^* in the troposphere includes significant changes in climate-relevant parameters like convective precipitation, snowfall, surface temperatures and sea-ice covering, which indicate an imprint of the imposed wave one pattern in the stratosphere-troposphere circulation.

The long-term change in the amplitude of observed O_3^* shows a nearly linear positive trend over the last decades. As mentioned in the introduction, a large fraction ($\approx$50–60 %) of the trend patterns in the zonally asymmetric component of total ozone might be due to decadal changes in ozone transport by geostrophic winds as determined by the stationary ultra-long planetary waves in geopotential height [e.g., *Hood and Zaff*, 1995; *Peters and Entzian*, 1999; *Peters et al.*, 2008]. However, much of the processes configuring the amplitude, the phase, the vertical structure and the long-term changes of the wave one pattern in ozone and other important absorbers (e.g., stratospheric water and stratospheric aerosol) remain unclear. Further investigation is needed including a detailed assessment of zonal asymmetries in the Brewer-Dobson circulation and of the involved photo-chemical reaction cycles affecting the configuration of the wave patterns and its long-term changes.

Our results suggest that the observed trend in O_3^* might contribute to the observed trends in temperature, planetary wave activity and circulation patterns from troposphere to upper mesosphere. However, a more detailed examination of the influence of O_3^* on the observed long-term changes in the stratosphere and mesosphere is an outstanding issue. We can also assume that long-term variations in the intensity of radiation components absorbed by ozone can modify the atmospheric circulation by modulating the effect of O_3^*. In particular, ozone strongly absorbs the ultra-violet components of incoming solar irradiation, and therefore the effects of O_3^* might be somewhat stronger during years around the maximum and somewhat weaker during years around the minimum of the 11-year solar cycle. Based on long-term equilibrium simulations with a general circulation model with coupled chemistry for solar maximum and minimum conditions and ensemble means derived from ERA-40, we have recently demonstrated that the effects of O_3^* are indeed modulated by the 11-year cycle in solar irradiation [*Gabriel et al.*, 2011b].

Our results suggest that diagnosing the configuration and effects of zonal asymmetries in stratospheric ozone and other important absorbers may help to improve current general circulation models. For example, in the climate models used for the IPCC scenarios, which are based on GCMs or AOGCMs, only the zonally symmetric part of the ozone distribution is implemented [e.g., *IPCC*, 2007]. Current GCMs and AOGCMs as well as chemistry-climate models including interactively coupled ozone chemistry (CCMs) usually do not reproduce the amplitude and spatial structure of the observed stratospheric stationary wave patterns correctly [e.g., *SPARC-CCMVal*, 2010]. Our model simulations demonstrate that including zonally asymmetric ozone can lead to an improvement of the stratospheric wave one pattern imposing subsequent changes in tropospheric circulation and wind-driven ocean currents via top-down and bottom-up stratosphere-troposphere coupling processes.

Acknowledgements We thank the ECMWF who provided the ERA-40 and the ERA-Interim data. Thanks also to the NASA who provided SAGE II and Aura MLS data. We thank the Max-Planck-Institute for Meteorology (MPI-Met, Hamburg) for providing the GCM MAECHAM5 and the Deutsches Klimarechenzentrum (DKRZ, Hamburg) for providing computer resources. This project was funded by the Deutsche Forschungsgemeinschaft.

References

Andrews, D., Holton, J., & Leovy, C. (1987). *International geophysical series: Vol. 40. Middle atmosphere dynamics*. New York: Academic Press. 489 pp.

Cariolle, D., & Déqué, M. (1986). Southern hemisphere medium-scale waves and total ozone disturbances in a spectral general circulation model. *Journal of Geophysical Research*, *91*(D10), 10825–10846.

Cariolle, D., & Teyssedre, H. (2007). A revised linear ozone photochemistry parameterization for use in transport and general circulation models: multi-annual simulations. *Atmospheric Chemistry and Physics*, *7*(9), 2183–2196.
Charlton, A., & Polvani, L. (2007). A new look at stratospheric sudden warmings. part i: Climatology and modeling benchmarks. *Journal of Climate*, *20*(3), 449–469.
Charlton, A., Polvani, L., Perlwitz, J., Sassi, F., Manzini, E., Shibata, K., Pawson, S., Nielsen, J., & Rind, D. (2007). A new look at stratospheric sudden warmings. part ii: Evaluation of numerical model simulations. *Journal of Climate*, *20*(3), 470–488.
Charney, J., & Drazin, P. (1961). Propagation of planetary scale disturbances from the lower atmosphere into the upper atmosphere. *Journal of Geophysical Research*, *66*, 83–109.
Crook, J., Gillett, N., & Keeley, S. (2008). Sensitivity of southern hemisphere climate to zonal asymmetry in ozone. *Geophysical Research Letters*, *35*. doi:10.1029/2007GL032698.
Dee, D., Uppala, S., Simmons, A., Berrisford, P., Poli, P., Kobayashi, S., Andrae, U., Balmaseda, M., Balsamo, G., Bauer, P., et al. (2011). The era-interim reanalysis: configuration and performance of the data assimilation system. *Quarterly Journal of the Royal Meteorological Society*, *137*(656), 553–597.
Dethof, A., & Hólm, E. (2004). Ozone assimilation in the erqa-40 reanalyses project. *Quarterly Journal of the Royal Meteorological Society*, *130*(603), 2851–2872.
Dragini, R. (2010a). *On the quality of the era-interim ozone reanalyses. Part i: Comparisons with in situ measurements* (Tech. Rep. 2). European Centre for Medium-Range Weather Forecasts, Reading, UK.
Dragini, R. (2010b). *On the quality of the era-interim ozone reanalyses. Part ii: Comparisons with satellite data* (Tech. Rep. 3). European Centre for Medium-Range Weather Forecasts, Reading, UK.
Dunkerton, T. (1991). Nonlinear propagation of zonal winds in an atmosphere with Newtonian cooling and equatorial wave driving. *Journal of the Atmospheric Sciences*, *48*, 236–263.
Esler, J., & Haynes, P. (1999). Mechanisms of wave packet formation in a quasi-geostrophic two-layer model. *Journal of the Atmospheric Sciences*, *56*, 2457–2490.
Fortuin, J. P. F., & Kelder, H. (1998). An ozone climatology based on ozonesonde and satellite measurements. *Journal of Geophysical Research*, *103*(D24), 31709–31734.
Gabriel, A., & Peters, D. H. W. (2008). A diagnostic study of different types of Rossby wave breaking events in the Northern extratropics. *Journal of the Meteorological Society of Japan*, *86*(5), 613–631.
Gabriel, A., Peters, D. H. W., Kirchner, I., & Graf, H. (2007). Effect of zonally asymmetric ozone on stratospheric temperature and planetary wave propagation. *Geophysical Research Letters*, *34*(6), L06807. doi:10.1029/2006GL028998.
Gabriel, A., Körnich, H., Lossow, S., Peters, D. H. W., Urban, J., & Murtagh, D. (2011a). Zonal asymmetries in middle atmospheric ozone and water vapour derived from odin satellite data 2001–2010. *Atmospheric Chemistry and Physics*, *11*, 9865–9885. doi:10.5194/acpd-11-9865-2011.
Gabriel, A., Schmidt, H., & Peters, D. H. W. (2011b). Effects of the 11-year solar cycle on middle atmospheric stationary wave patterns in temperature, ozone and water vapor. *Journal of Geophysical Research*, *116*, D23301. doi:10.1029/2011JD015825.
Gates, W., & L. L. N. L. P. for Climate Model Diagnosis, and Intercomparison (1999). An overview of the results of the atmospheric model intercomparison project (amip i). *Bulletin of the American Meteorological Society*, *80*, 29–56.
Gillett, N., Scinocca, J., Plummer, D., & Reader, M. (2009). Sensitivity of climate to dynamically-consistent zonal asymmetries in ozone. *Geophysical Research Letters*, *36*, L10809. doi:1029/2009GL037246.
Hood, L., & Zaff, D. (1995). Lower stratospheric stationary waves and the longitude dependence of ozone trends in winter. *Journal of Geophysical Research*, *100*(D12), 25791–25800.
IPCC (2007). In S. Solomon, D. Qin, M. Manning, Z. Chen, M. Marquis, K. B. Averyt, M. Tignor, & H. L. Miller (Eds.), *Contribution of working group I to the fourth assessment report of the intergovernmental panel on climate change*. Cambridge: Cambridge University Press.

Jungclaus, J. H., Keenlyside, N., Botzet, M., Haak, H., Luo, J. J., Latif, M., Marotzke, J., Mico-lajewicz, U., & Roeckner, R. (2006). Ocean circulation and tropical variability in the coupled model echam5/mpi-om. *Journal of Climate*, *19*(16), 3952–3972.

Karpetchko, A., Kyrö, E., & Knudsen, B. (2005). Arctic and antarctic polar vortices 1957–2002 as seen from the era-40 reanalyses. *Journal of Geophysical Research*, *110*, D21109. doi:10.1029/2005JD006113.

Kirchner, I., & Peters, D. (2003). Modelling the wintertime response to upper tropospheric and lower stratospheric ozone anomalies over the North Atlantic and Europe. *Annales Geophysicae*, *21*, 2107–2118.

Knudsen, B., & Andersen, S. (2001). Geophysics: longitudinal variation in springtime ozone trends. *Nature*, *413*(6857), 699–700.

Kubicki, A., Keckhut, P., Chanin, M., Hauchecorne, A., Lysenko, E., & Golitsyn, G. (2006). Temperature trends in the middle atmosphere as seen by historical Russian rocket launches: Part 1, Volgograd (48.68°n, 44.35°e). *Journal of Atmospheric and Solar-Terrestrial Physics*, *68*(10), 1075–1086.

Manzini, E., Giorgetta, M., Esch, M., Kornblueh, L., & Roeckner, E. (2006). The influence of sea surface temperatures on the northern winter stratosphere: ensemble simulations with the maecham5 model. *Journal of Climate*, *19*(16), 3863–3881.

Marsland, S. J., Haak, H., Jungclaus, J. H., Latif, M., & Roeske, F. (2003). The Max-Planck-Institute global ocean/sea-ice model with orthogonal curvilinear coordinates. *Ocean Modelling*, *5*, 91–127.

Nathan, T., & Cordero, E. (2007). An ozone-modified refractive index for vertically propagating planetary waves. *Journal of Geophysical Research*, *112*, D02105. doi:10.1029/2006/D007357.

Pedlosky, J. (1979). *Geophysical fluid dynamics*. Berlin: Springer. 624 pp.

Peters, D., & Entzian, G. (1999). Longitude-dependent decadal changes of total ozone in boreal winter months during 1979–92. *Journal of Climate*, *12*(4), 1038–1048. doi:10.1175/1520-0442(1999)012<1038:LDDCOT>2.0.CO;2.

Peters, D., & Waugh, D. (1996). Influence of barotropic shear on the poleward advection of upper-tropospheric air. *Journal of the Atmospheric Sciences*, *53*, 3013–3031.

Peters, D., & Waugh, D. (2003). Rossby wave breaking in the southern hemisphere wintertime upper troposphere. *Monthly Weather Review*, *131*(11), 2623–2634.

Peters, D., Entzian, G., & Schmitz, G. (1996). Ozone anomalies over the North Atlantic-European region during January 1979–1992-linear modelling of horizontal and vertical ozone transport by ultra-long waves. *Contributions to Atmospheric Physics*, *69*(4), 477–489.

Peters, D., Gabriel, A., & Entzian, G. (2008). Longitude-dependent decadal ozone changes and ozone trends in boreal winter months during 1960–2000. *Annales Geophysicae*, *26*, 1275–1286.

Plumb, R. (1985). On the three-dimensional propagation of stationary waves. *Journal of the Atmospheric Sciences*, *42*, 217–229.

Polvani, L., & Kushner, P. (2002). Tropospheric response to stratospheric perturbations in a relatively simple general circulation model. *Geophysical Research Letters*, *29*(7), 1114. doi:10.0129/2001GL14284.

Roeckner, E., Brokopf, R., Esch, M., Giorgetta, M., Hagemann, S., Kornblueh, L., Manzini, E., Schlese, U., & Schulzweida, U. (2006). Sensitivity of simulated climate to horizontal and vertical resolution in the echam5 atmosphere model. *Journal of Climate*, *19*(16), 3771–3791.

Sassi, F., Boville, B., Kinnison, D., & Garcia, R. (2005). The effects of interactive ozone chemistry on simulations of the middle atmosphere. *Geophysical Research Letters*, *32*, L07811. doi:10.1029/2004GL022131.

Scaife, A., Knight, J., Vallis, G., & Folland, C. (2005). A stratospheric influence on the winter NAO and North Atlantic surface climate. *Geophysical Research Letters*, *32*(18), L18715. doi:10.1029/2005GL023226.

Solomon, S., Portmann, R., Garcia, R., Randel, W., Wu, F., Nagatani, R., Gleason, J., Thomason, L., Poole, L., & McCormick, M. (1998). Ozone depletion at mid-latitudes: coupling of volcanic aerosols and temperature variability to anthropogenic chlorine. *Geophysical Research Letters*, *25*(11), 1871–1874.

Song, Y., & Robinson, W. A. (2004). Dynamical mechanisms for stratospheric influences on the troposphere. *Journal of the Atmospheric Sciences*, *61*(14), 1711–1725. doi:10.1175/1520-0469(2004)061.

SPARC-CCMVal, A. (2010). In V. Eyring, T. G. Shepherd & D. W. Waugh (Eds.), *SPARC report on the evaluation of chemistry-climate models* (SPARC Report No. 5). WCRP-132, WMO/TD-No. 1526.

Uppala, S., Kållberg, P., Simmons, A., Andrae, U., Bechtold, V., Fiorino, M., Gibson, J., Haseler, J., Hernandez, A., Kelly, G., et al. (2005). The era-40 re-analysis. *Quarterly Journal of the Royal Meteorological Society*, *131*(612), 2961–3012.

Walter, K., & Graf, H. (2005). The North Atlantic variability structure, storm tracks, and precipitation depending on the polar vortex strength. *Atmospheric Chemistry and Physics*, *5*(1), 239–248.

Waugh, D., & Randel, W. (1999). Climatology of Arctic and Antarctic polar vortices using elliptical diagnostics. *Journal of the Atmospheric Sciences*, *56*, 1594–1613.

Waugh, D., Oman, L., Newman, P., Stolarski, R., Pawson, S., Nielsen, J., & Perlwitz, J. (2009). Effects of zonal asymmetries in stratospheric ozone on simulated southern hemisphere climate trends. *Geophysical Research Letters*, *36*, L18701. doi:10.1029/2009GL040419.

Wittman, M., Polvani, L., Scott, R., & Charlton, A. (2004). Stratospheric influence on baroclinic lifecycles and its connection to the arctic oscillation. *Geophysical Research Letters*, *31*, L16113. doi:10.1029/2004GL020503.

WMO (2007). *Scientific assessment of ozone depletion: 2006* (Tech. rep.). World Meteorological Organisation.

Chapter 25
Extending the Parameterization of Gravity Waves into the Thermosphere and Modeling Their Effects

Erdal Yiğit and Alexander S. Medvedev

Abstract Vertical coupling by gravity waves (GWs) between the lower atmosphere and thermosphere is studied with a general circulation model (GCM) extending from the tropopause to the upper atmosphere. A newly developed nonlinear spectral GW parameterization, which accounts for wave propagation in the highly dissipative thermosphere, has been implemented into the Coupled Middle Atmosphere-Thermosphere-2 (CMAT2) GCM. In addition to the nonlinear saturation, the extended scheme considers wave dissipation suitable for the thermosphere-ionosphere, such as molecular viscosity and thermal conduction, ion drag, eddy diffusivity, and radiative damping. The results of simulations for June solstice show that the dynamical and thermal effects of GWs are strong and cannot be neglected in the thermosphere. At F region heights, GW momentum deposition is comparable to the ion drag and GW-induced heating/cooling competes with the high-latitude Joule heating.

25.1 Introduction

Small-scale internal gravity waves (GWs) with intrinsic frequencies $\hat{\omega}$ smaller than the Brunt-Väisälä frequency N but much larger than the inertial one, $f \ll \hat{\omega} < N$, are plentiful in the lower atmosphere. They are generated owing to a variety of processes like convection, flow instabilities, latent heat release, etc., and can propagate

E. Yiğit (✉)
Department of Atmospheric Oceanic and Space Sciences, 2455 Hayward Street, Ann Arbor, MI 48109-2143, USA
e-mail: erdal@umich.edu

Present address:
E. Yiğit
Space Sciences Laboratory, UC Berkeley, Berkeley, CA, USA
e-mail: erdal@ssl.berkeley.edu

A.S. Medvedev
Max Planck Institute for Solar System Research, Max-Planck-Str. 2, 37191, Katlenburg-Lindau, Germany
e-mail: Medvedev@mps.mpg.de

F.-J. Lübken (ed.), *Climate and Weather of the Sun-Earth System (CAWSES)*, Springer Atmospheric Sciences,
DOI 10.1007/978-94-007-4348-9_25,

upwards growing in amplitude in response to exponentially decreasing background neutral density ρ. In the middle atmosphere and above, they intensively attenuate, thus depositing their momentum and energy into the mean flow. It is well established that GWs are extremely important in the mesosphere and lower thermosphere (MLT): they are responsible for the mean zonal wind reversal and the associated reversal of the meridional temperature gradient ("mesopause anomaly"), turbulent mixing, and energy deposition [*Fritts and Alexander*, 2003, and references therein]. GW effects lead to a significant departure of the middle atmosphere from radiative equilibrium. Due to limitations in spatial resolution, atmospheric general circulation models (GCMs) require appropriate GW parameterizations in order to account for GW effects and produce more realistic circulation of the middle atmosphere with respect to observations. A number of GW schemes have been developed to date to incorporate the effects of these waves into GCMs [e.g., *Lindzen*, 1981; *Holton*, 1982; *Matsuno*, 1982; *Fritts and Lu*, 1993; *Medvedev and Klaassen*, 1995; *Alexander and Dunkerton*, 1999; *Warner and McIntyre*, 2001].

However, in the thermosphere above the turbopause ($\sim$105 km), dynamical and thermal effects of GWs have been studied to a much lesser extent. Among others, there were two major reasons for this lack of development: (1) Most lower atmosphere GCMs either did not extend well into the thermosphere, [e.g., *Boville and Randel*, 1992; *Beagley et al.*, 1997; *Manzini et al.*, 2006], or lower boundaries of thermosphere models were set too high to capture GW propagation from the troposphere (typically, at 80–100 km) [e.g., *Roble et al.*, 1988; *Richmond et al.*, 1992]; (2) Existing GW parameterizations were not designed to account for wave propagation in highly dissipative atmosphere above the turbopause. For instance, in some studies, GW schemes were adapted from middle atmosphere GCMs by simply imposing an unphysical exponential decay of wave activity above the MLT region [e.g., *Fomichev et al.*, 2002]. With the arrival of so-called "Whole Atmosphere GCMs" that cover the atmosphere from the surface to the thermosphere and even to the exosphere [*Schmidt et al.*, 2006; *Liu et al.*, 2010], the task of developing suitable GW parameterizations becomes urgent.

In this article, we report on the recent progress with developing such a parameterization, and with modeling the effects of small-scale internal GWs of lower atmospheric origin in the thermosphere above the turbopause. The primary focus of this paper is to outline our recently developed nonlinear spectral GW scheme (an "extended scheme" hereafter) [*Yiğit et al.*, 2008], and to discuss the first modeling results obtained with its implementation into a comprehensive GCM extending from the tropopause to the middle thermosphere [*Yiğit et al.*, 2009].

Some observations of GWs in the thermosphere-ionosphere (TI) are briefly reviewed in the next section. The extended nonlinear GW scheme [*Yiğit et al.*, 2008] along with the analytical formulae parameterizing the influence of various dissipative processes on GW harmonics are outlined in Sect. 25.3. Section 25.4 describes the GCM used in this study, specification of GW spectra at the source level, and the implementation of the scheme into the model. Sections 25.5 and 25.6 present the dynamical and thermal effects of GWs on the thermosphere inferred from GCM simulations. Summary and future perspectives are given in Sect. 25.7.

25.2 Gravity Wave Observations in the Thermosphere

Observations of GWs in the upper atmosphere are numerous. The first were conducted more than half a century ago, and revealed wave-like disturbances in the F-region ionization [*Munro*, 1950]. They were termed traveling ionospheric disturbances (TIDs), however their link to internal GWs were then not entirely established. *Hines* [1960] was one of the first that related TIDs to GWs. Other early observations in the TI showed large-scale GW-like disturbances in the plasma [*Hunsucker*, 1982]. Since those times, internal GWs were routinely observed in the TI over decades, and our knowledge of their characteristics have rapidly grown. It becomes more evident that GWs are continuously present in the upper atmosphere. More recently, incoherent scatter radar data have presented GW signatures in the thermosphere whose sources are increasingly interpreted as being in the lower atmosphere. Based on a number of measurements, wave characteristics, such as dispersion relation and propagation directions, have been deduced, and the amount of variability produced by GWs were estimated [*Oliver et al.*, 1997]. Independent of solar illumination and geomagnetic activity, GW signatures represent an intrinsic part of the thermosphere dynamics [*Djuth et al.*, 2004; *Livneh et al.*, 2007].

25.3 Parameterization of Gravity Wave Effects in the Whole Atmosphere System

25.3.1 Vertical Evolution of the GW Horizontal Momentum Flux

Under the assumptions of vertical propagation, no horizontal wind shear, and mid-frequency GWs, the equation governing the evolution of the horizontal momentum flux (per unit mass), $\overline{u'w'}_i(z)$, associated with a harmonic "i" has the form:

$$\frac{d\overline{u'w'}_i}{dz} = \left(-\frac{\rho_z}{\rho} - \beta^i_{tot}\right)\overline{u'w'}_i, \tag{25.1}$$

where ρ is the background neutral density, β^i_{tot} is the total vertical damping rate affecting the given harmonic i, and the index "z" denotes vertical derivatives. The solution of (25.1) can be expressed as [*Medvedev and Klaassen*, 1995; *Vadas and Fritts*, 2005; *Yiğit et al.*, 2008]

$$\overline{u'w'}_i(z) = \overline{u'w'}_i(z_0)\frac{\rho(z_0)}{\rho(z)}\tau_i(z), \tag{25.2}$$

where $\overline{u'w'}_i(z_0)$ is the GW momentum flux at the source level z_0, and τ_i is the so-called *transmissivity* of the atmosphere for the given harmonic. Equation (25.2) describes how wave dissipation counteracts the exponential growth of the wave fluxes due to the density stratification. For conservative wave propagation, $\tau = 1$, and no momentum is deposited into the mean flow. If τ varies with altitude, then waves either draw their momentum from the mean flow ($\tau_i > 0$), or deposit it ($\tau_i < 0$). For the dissipative atmosphere, $\tau_i(z)$ for a test harmonic "i" can be written as [*Yiğit et al.*, 2008, Eq. (2)]

$$\tau_i(z) = \exp\left[-\int_{z_0}^{z} \sum_d \beta_d^i(z')\, dz' \right]. \tag{25.3}$$

In (25.3), the vertical damping rates $\beta_d^i(z')$ are attributed to various attenuation processes denoted by "d" that simultaneously affect the harmonic in the real atmosphere. The analytical expressions for the different dissipation mechanisms are given in Sect. 25.3.2. The total vertical damping rate β_{tot}^i is the sum of all the individual contributions, as implied in Eq. (25.3).

The total GW momentum deposition $a(z)$ (or the GW drag) is the density scaled divergence of the total momentum flux, $a = \rho^{-1} d(\rho\overline{u'w'})/dz$, calculated as a sum over all the contributions from all of M individual harmonics in the GW spectrum:

$$a(z) = \sum_i^M a_i(z) = \sum_i^M \sum_d \beta_d^i \overline{u'w'}_i(z) \quad \left[\mathrm{m\,s^{-2}}\right]. \tag{25.4}$$

The resulting wave heating/cooling ε_i associated with a harmonic is composed of two terms, the irreversible heating E_i and the differential heating/cooling Q_i, $\varepsilon_i = E_i + Q_i$, [*Medvedev and Klaassen*, 2003; *Becker*, 2004; *Yiğit and Medvedev*, 2009]:

$$E_i = \frac{a_i(c_i - \bar{u})}{c_p}, \qquad Q_i = \frac{H}{2\rho R}\frac{\partial}{\partial z}\left[\rho a_i(c_i - \bar{u})\right] \quad \left[\mathrm{K\,s^{-1}}\right]. \tag{25.5}$$

In (25.5) and hereafter, c_i is the observed horizontal phase velocity for the i-th harmonic, $\bar{u}$ is the projection of the local background wind onto the direction of propagation of the harmonic, H is the density scale height, $H = -\rho/\rho_z$, c_p is the specific heat at constant pressure, and R is the gas constant. The total heating/cooling is the sum of the contributions of all individual harmonics: $\varepsilon = \sum_1^M (E_i + Q_i)$.

25.3.2 Gravity Wave Dissipation

Conservative propagation of a harmonic is always described by a dispersion relation with a real intrinsic frequency $\hat{\omega}$ and wave vector $\mathbf{k} = (k, l, m) = (2\pi)/(\lambda_x, \lambda_y, \lambda_z)$, where λ's are the wavelengths yielding the corresponding wavenumbers. Under the mid-frequency approximation, $f \ll \hat{\omega} \ll N$, the dispersion relation for GWs is

$$\hat{\omega} = N\left|\frac{k_H}{m}\right|, \tag{25.6}$$

with $k_H = |\mathbf{k_H}| = \sqrt{k^2 + l^2}$ being the horizontal wavenumber, $\lambda_H = 2\pi/k_H$.

Dissipation modifies the dispersion relation by introducing the imaginary component to the intrinsic frequency, $\hat{\omega} = \hat{\omega}^R + i\hat{\omega}^I$, $\hat{\omega}^R$ and $\hat{\omega}^I$ being the real and imaginary parts, correspondingly, i is the imaginary unit. $\hat{\omega}^I$ describes a temporal decay of wave fluctuations with given $\hat{\omega}^R$ and $\mathbf{k}$. If the approximation of steady waves is applied to the complex dispersion relation, the temporal decay converts to a vertical damping. Thus, dissipation alters the vertical wavenumber m by introducing the corresponding real and imaginary parts, m^R and m^I:

$$m = m^R - im^I. \tag{25.7}$$

Since all perturbation variables in the harmonic are proportional to $\exp(-imz)$, the imaginary part of the vertical wavenumber describes the vertical attenuation. m^I enters GW parameterizations because the momentum flux $\overline{u'w'} \propto \exp(-2m^I z)$.

Small-scale GWs are continuously affected by various dissipation mechanisms as they propagate upward from the place of their excitation in the lower atmosphere. In the middle atmosphere, nonlinear interactions, which limit the wave amplitude when it becomes large due to the exponential growth with height, are the most important mechanism of attenuation. Above the turbopause, other processes add to, and, ultimately, compete with the nonlinear interactions. They are: molecular viscosity and heat conduction [*Vadas and Fritts*, 2005], ion friction [*Gossard and Hooke*, 1975], radiative damping in form of Newtonian cooling [*Holton*, 1982], and eddy diffusion. The following subsections briefly summarize the way of parameterizing m^I for the different dissipation mechanisms, and, therefore, how they are included in (25.3). Further details can be found in the work by *Yiğit et al.* [2008, Sect. 3].

25.3.3 Nonlinear Gravity Wave Saturation

The expression for the nonlinear dissipation term β_{non} follows from the works by *Weinstock* [1982], and *Medvedev and Klaassen* [1995, 2000]. It accounts for wave-wave interactions within a broad spectrum of GWs. The interactions are scale-dependent, i.e., larger-scale harmonics act as an additional background for the smaller-scale waves, and smaller-scale harmonics impinge on the larger-scale ones and erode them. Therefore, β^i_{non} describes the dissipation of the harmonic i caused by nonlinear interactions with other waves in the spectrum. This dissipation is the result of instabilities on scales shorter than the vertical wavelength of a given harmonic. If the spectrum consists of a single harmonic, the wave interacts nonlinearly with itself ("self-interaction"), and the corresponding β_{non} describes the well-known Hodges-Lindzen breaking due to convective instability but at $\sim 70\,\%$ $(1/\sqrt{2})$ lower amplitudes. For a spectrum with multiple waves, β_{non} takes the form [*Medvedev and Klaassen*, 1995, 2000]:

$$\beta^i_{non} = \frac{\sqrt{2\pi}N}{\sigma_i} \exp\left(-\frac{|c_i - \bar{u}|^2}{2\sigma_i^2}\right), \tag{25.8}$$

where the variance σ_i^2 is produced by velocity fluctuations due to harmonics with intrinsic phase velocities $|c_j - \bar{u}|$ slower than the reference velocity $|c_i - \bar{u}|$, or $\sigma_i^2 = \sum_{c_j < c_i} \overline{u'^2_j}$. Unlike in the Hodges-Lindzen case, β^i_{non} is a continuous function that grows when amplitudes of the harmonics grow. Therefore, by contrast to linear schemes, no artificial factors (intermittency) are needed to scale GW effects. The expressions for Froude numbers for a monochromatic, Fr_m, and broad spectra, Fr_s, illustrate the difference between the Lindzen-type linear scheme and the nonlinear scheme [*Yiğit et al.*, 2008, Eq. (11)]:

$$Fr_m = \frac{|m_i^R u'|}{N} = \left|\frac{u'}{c_i - \bar{u}(z)}\right|, \qquad Fr_s = \frac{|m_i^R \sigma_i|}{N} = \left|\frac{\sigma_i}{c_i - \bar{u}(z)}\right|. \tag{25.9}$$

Since σ_i contains contributions from other harmonics from the spectrum in addition to the fluctuations of the harmonic i itself, $\sigma_i \geqslant u'$. As the nonlinear wave saturation takes place at $Fr_s \to 1$, and the linear one at $Fr_m \to 1$, the nonlinear saturation/overturning is likely to precede the linear one. Thus, accounting for nonlinear interactions with other harmonics leads to lower breaking levels in the atmosphere.

25.3.4 Molecular and Eddy Viscosity and Thermal Conduction

The effects of molecular viscosity become increasingly important above ~80 km owing to their exponential growth with height. The kinematic molecular viscosity coefficient ν_{mol} is related to the thermal conduction κ by the Eucken formula derived from kinetic theory: $\kappa = 0.25[9c_p - 5(c_p - R)]\nu_{mol}$. After *Banks and Kockarts* [1973], this yields for the Earth atmosphere

$$\nu_{mol} = 3.563 \times 10^{-7} T^{0.69}/\rho \quad \left[\mathrm{m^2\,s^{-1}}\right] \tag{25.10}$$

where T is the temperature. The expressions for the corresponding vertical damping rate β^i_{mol} for certain Prandtl numbers, Pr, are adopted from [*Vadas and Fritts*, 2005, Eqs. (41) and (47)]

$$\begin{aligned} \beta^i_{mol}(z) &= \frac{\nu_{mol} N^3}{k_h[c_i - \bar{u}(z)]^4} \qquad (Pr = \infty), \\ \beta^i_{mol}(z) &= \frac{2\nu_{mol} N^3}{k_h[c_i - \bar{u}(z)]^4} \qquad (Pr = 1). \end{aligned} \tag{25.11}$$

Wave dissipation due to the background eddy viscosity can be accounted for similarly with ν_{mol} replaced by the eddy diffusion coefficient, D_{eddy}.

25.3.5 Ion Drag

In the ionosphere above 80 km, charged and neutral particles have different mobilities. By contrast to the neutrals, the ionized particle motion ($\mathbf{E} \times \mathbf{B}$ drift) is confined to the ambient magnetic field $\mathbf{B}$. With their much larger velocities, ions can exert a frictional torque on the neutrals that is proportional to the relative differences between their speeds, and to the effective collision frequency for momentum transfer between neutrals and ions, ν_{ni}: $\nu_{ni}(\mathbf{v}_i - \mathbf{u})$. This so-called "ion friction" is significant at F_2 layer heights, and tends to damp wave fluctuations. The expression for the corresponding vertical damping rate is given by [*Gossard and Hooke*, 1975]:

$$\beta^i_{ion} = \frac{2\nu_{ni} N}{k_H [c_i - \bar{u}(z)]^2}. \tag{25.12}$$

In simulations, the collision frequency was taken to be proportional to the ion number density n_i and depend on temperature T, following *Klostermeyer* [1972]:

$$\nu_{ni} = 7.22 \times 10^{17} T^{0.37} n_i. \tag{25.13}$$

25.3.6 Radiative Cooling

Damping of GWs by radiative processes is represented in form of Newtonian cooling. The associated cooling coefficients that are required for the evaluation of β_{rad} have been adapted from *Holton* [1982, Eq. (41)]. β_{rad} is then similar to (25.12), but with ν_{ni} replaced with the Newtonian cooling coefficient α_{newt}.

25.4 Model Description and Configuration of Runs

The Coupled Middle Atmosphere Thermosphere Model-2 (CMAT2) [*Yiğit*, 2009] is a 3-D finite difference model developed at the University College London from its first generation [*Harris*, 2001; *Dobbin*, 2005]. In a pressure vertical coordinate system, it extends from 100 hPa (~15 km) up to 1.43×10^{-8} hPa that corresponds to 250 to 600 km, depending on the solar activity described by the $F_{10.7}$ cm radio flux. Simulations are performed on a 2° × 18° latitude-longitude grid. As the model does not have a troposphere, National Centers for Environmental Prediction and Global Scale Wave Model data were used to force planetary waves and tides at the lower boundary. Further details can be found in the work by *Yiğit et al.* [2009, and references therein]. All simulations presented in this article have been performed at constant low geomagnetic and solar activities ($K_p = 2^+$; $F_{10.7} = 80 \times 10^{-22}\ \mathrm{W\,m^{-2}\,Hz^{-1}}$).

25.4.1 Source-Level Gravity Wave Activity

All GW parameterizations require a specification of the GW spectrum at the lower boundary taken as a source level z_0 (cf. Eqs. (25.1) and/or (25.2)). In a number of GCMs extending from the surface upwards, GW source is assumed at or around the tropopause [*Medvedev et al.*, 1998], presuming the tropopause to be the predominant region of wave generation. In the presented simulations, GWs were launched at the

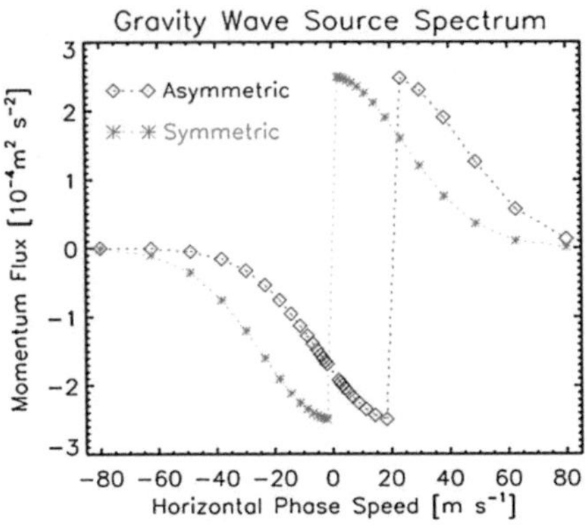

Fig. 25.1 Source-level GW spectrum of momentum fluxes (per unit mass) $(\overline{u'w'}_i)$ as a function of horizontal phase speeds c_i. The source amplitude is $\overline{u'w'}_{max} = 2.5 \times 10^{-4}\ \mathrm{m}^{-2}\,\mathrm{s}^{-2}$, the spectral width is $c_w = 35\ \mathrm{m\,s}^{-1}$. *Asterix* and *diamonds* denote the symmetric and asymmetric spectra, respectively. The fastest and slowest harmonics are $|c_{max}| = 80\ \mathrm{m\,s}^{-1}$ and $2\ \mathrm{m\,s}^{-1}$, correspondingly

bottom of the model at ~15 km. The adopted source spectrum is Gaussian [*Yiğit et al.*, 2008, Eq. (17)]:

$$\overline{u'w'}_i(z_0) = \mathrm{sgn}(c_i - \bar{u}_0)\overline{u'w'}_{max} \exp\left[-(c_i - \bar{u}_0)^2/c_w^2\right], \qquad (25.14)$$

where $\overline{u'w'}_{max}$ is the magnitude of the flux, c_w is the characteristic half-width of the spectrum, and $\bar{u}_0 = \bar{u}(z = z_0)$ is the local wind at the source level. The shape and strength of this model spectrum are in a very good agreement with the balloon measurements in the upper troposphere [*Hertzog et al.*, 2008, Fig. 6]. A representative source-level spectrum (25.14) along with its discretization are plotted in Fig. 25.1. It is seen that the spectrum is symmetric with respect to phase velocities c_i only if $\bar{u}_0 = 0$. If the local wind at the launch level is non-zero, the spectrum becomes clearly asymmetric. The rationale for such anisotropic setup has been discussed in detail by *Medvedev et al.* [1998]. In practical applications, it produces the best agreement between simulated fields and observations in the middle atmosphere. The direction of propagation of GW harmonics coincides with that of local wind for $c_i > 0$, and is against it for $c_i < 0$. Therefore, despite the assumed constant parameters of (25.14), the wave source varies in time and space being modulated by the local wind $\bar{u}_0$.

25.4.2 Implementation Algorithm

A GW spectrum (25.14) approximated with 30 harmonics was launched at the source level at ~15 km in each grid point at every time step. The vertical evolution of fluxes along with the damping rates β_d^i were then computed according to (25.2) and as described in Sects. 25.3.3–25.3.6. The GW dynamical and thermal effects were calculated at each point according to (25.4) and (25.5), respectively, and added to the corresponding tendencies in the momentum and thermodynamic equations for the resolved motions in the GCM.

25.5 Dynamical Effects of Gravity Waves in the Thermosphere

How large is the influence of GWs on the large-scale dynamics of the middle and upper atmosphere? In order to evaluate it, we performed two simulations: one with the fully implemented parameterization (EXP2), and one with the GW forcing artificially set to zero above the turbopause (EXP1). The latter run completely eliminates the direct GW effects above ~105 km. Below the turbopause, GW contribution is required to reproduce the observed solstitial jets in the MLT. Figure 25.2a,b presents the zonal 21-day mean drag (in $\mathrm{m\,s^{-1}\,day^{-1}}$) produced by dissipating GWs during the June solstice. GWs not only penetrate into the thermosphere, but create there an appreciable forcing, which is at least as large as the GW drag in the MLT. For comparison, the ion drag for the two simulations is plotted in Fig. 25.2c,d. The deposited GW momentum is predominantly directed against the mean wind, much like the midlatitude ion drag. Magnitudes of the GW drag (hundreds of $\mathrm{m\,s^{-1}\,day^{-1}}$) are comparable to those of the ion drag at least up to F region altitudes (180–200 km). The effects of GWs are particularly strong in the winter (southern) hemisphere. The imposed westward GW momentum significantly (from ~100 to 30 $\mathrm{m\,s^{-1}}$ in the mean sense) weakens the westerlies, and enhances the high-latitude easterlies (from -10 to -30 $\mathrm{m\,s^{-1}}$) [*Yiğit et al.*, 2009, Fig. 2]. Smaller compared to EXP1 values of ion drag (Fig. 25.2c,d) reflect these changes in the mean wind. For instance, the ion drag is almost twice weaker in midlatitudes at 180 km in the run with GWs accounted for. This indicates that far from the high-latitude convection electric fields, neutrals constitute a momentum source for ions, and GW drag clearly modulates the exchange of momentum between them. Overall, accounting for the GW momentum deposition noticeably improves the simulated thermospheric mean winds with respect to the Horizontal Wind Model [*Yiğit et al.*, 2009].

25.6 Thermal Effects of Gravity Waves in the Thermosphere

Figure 25.3 displays the calculated GW heating and cooling rates, and compares them with other diabatic effects. The irreversible heating due to a conversion of the wave mechanical energy into heat, $\sum_i E_i$ in (25.5), is large in the high-latitudes of both hemispheres (up to 90–100 $\mathrm{K\,d^{-1}}$ at 200–210 km in Fig. 25.3a). This heating is created by the dissipating harmonics with fast phase speeds traveling against the mean wind and responsible for the high-latitude maxima of drag in Fig. 25.2b. As the zonally averaged wind, $\bar{u}$, decreases towards the poles, the quantity $a_i(c_i - \bar{u})$ for a wave c_i increases. Therefore, although the dissipating harmonics deposit approximately equal momentum a_i in middle- and high-latitudes (Fig. 25.3b), the GW heating is much stronger in the high-latitudes. The role of this heating is clearly

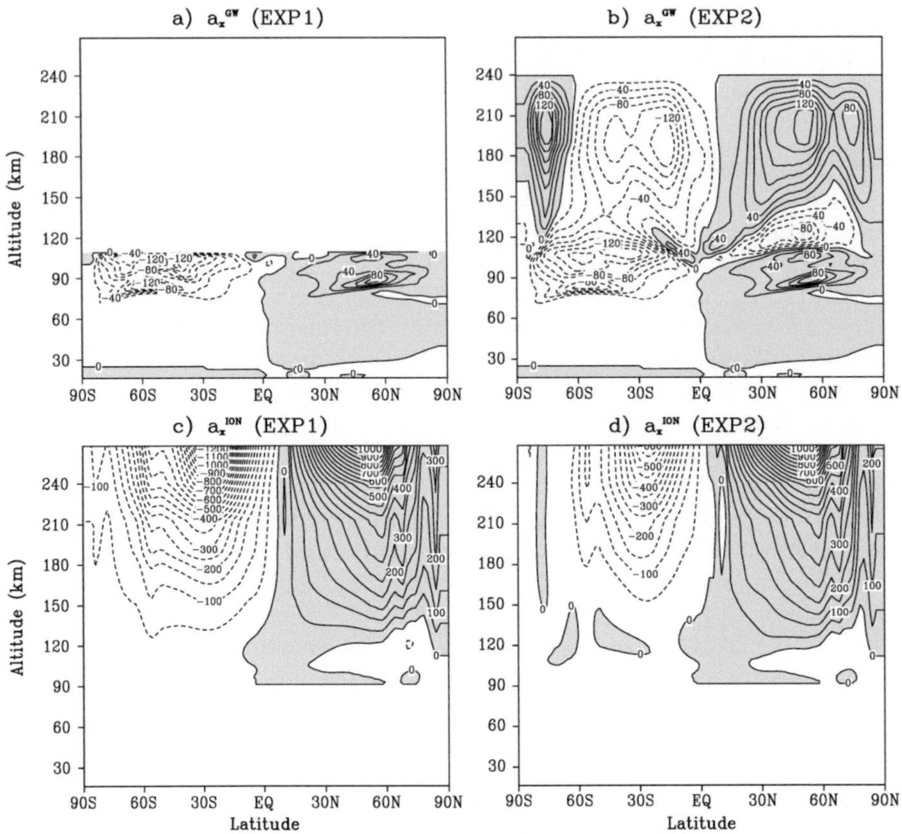

Fig. 25.2 Altitude-latitude cross sections of the zonal mean zonal torque created by parameterized GWs (in $\mathrm{m\,s^{-1}\,day^{-1}}$) over the June solstice in the (**a**) cut-off drag and (**b**) full-drag simulations. (**c**) and (**d**) are the mean zonal ion drag for the cut off and full-drag runs, correspondingly. *Grey shadings with solid line* indicate regions with eastward drag, *dotted contours* denote the westward drag (adopted from [*Yiğit et al.*, 2009, Fig. 4])

seen when compared with the Joule heating (Fig. 25.3c), which is an important heat source, resulting from ion-neutral collisional dissipation. The Joule heating is the largest in the high-latitudes (up to 250 and 400 $\mathrm{K\,d^{-1}}$ in the Southern and Northern hemispheres, respectively), where GW heating is between 20 and 40 % of the frictional heating.

Despite the strong irreversible heating, the net thermal effect of GWs is cooling (Fig. 25.3b). It is created by the downward transport of the background potential temperature by dissipating waves. The total cooling rates are up to -150 and -180 $\mathrm{K\,d^{-1}}$ at 210 km in the winter and summer hemispheres, respectively. For comparison, the cooling rates due the main cooling mechanism in the thermosphere—the molecular thermal conduction—are plotted in Fig. 25.3d. They are up to -1100 and -2300 $\mathrm{K\,d^{-1}}$ in the Southern and Northern hemispheres,

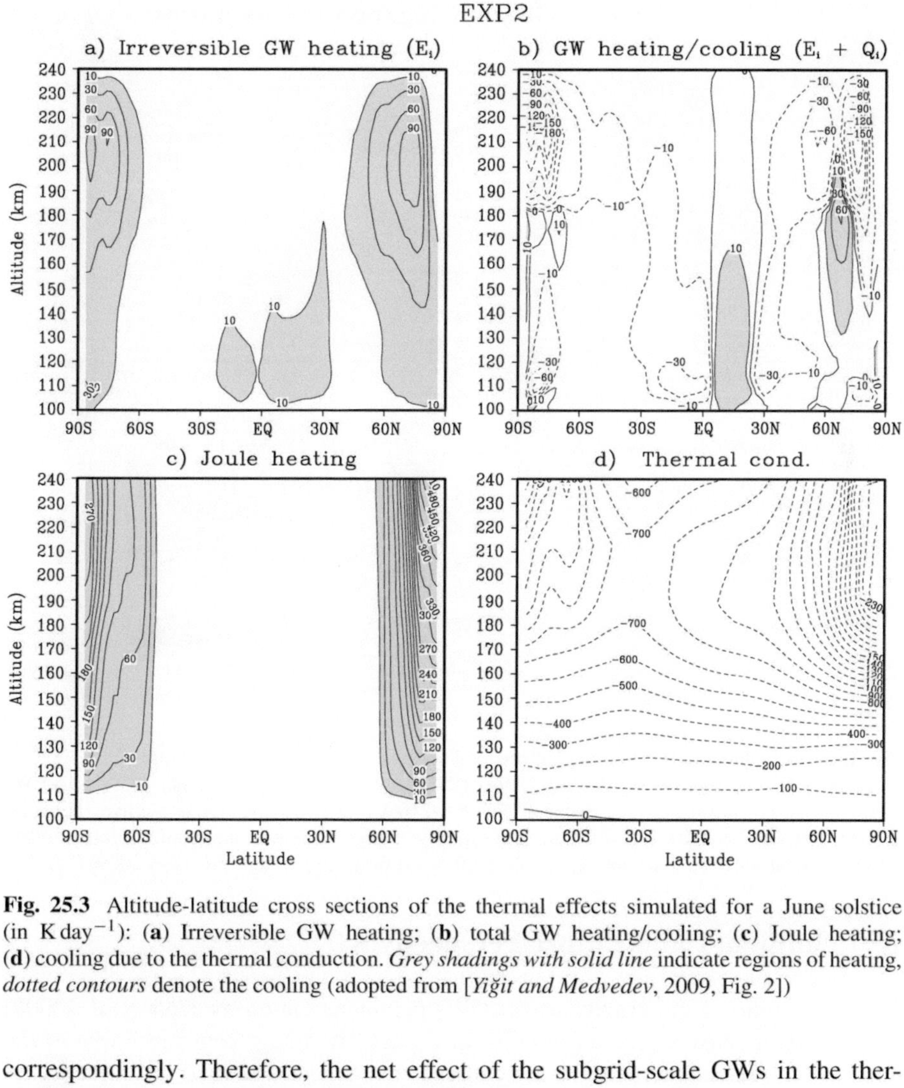

Fig. 25.3 Altitude-latitude cross sections of the thermal effects simulated for a June solstice (in $\mathrm{K\,day^{-1}}$): (**a**) Irreversible GW heating; (**b**) total GW heating/cooling; (**c**) Joule heating; (**d**) cooling due to the thermal conduction. *Grey shadings with solid line* indicate regions of heating, *dotted contours* denote the cooling (adopted from [*Yiğit and Medvedev*, 2009, Fig. 2])

correspondingly. Therefore, the net effect of the subgrid-scale GWs in the thermosphere is to enhance cooling by 10 to 20 %.

Instantaneous distributions illustrating the extent of variability of GW heating/cooling are shown in Fig. 25.4. A dipole pattern of strong heating (up to $500\ \mathrm{K\,d^{-1}}$) at 170–180 km, and cooling of more than $-1000\ \mathrm{K\,d^{-1}}$ near ~210 km is seen between the local midnight and noon (panel *a*). A reversed but weaker pattern occurs during the other half of the day. These variations of heating rates are related to the influence of tides on GW propagation. Most of the modulation in the model results is due to sun-synchronous tides, as follows from the two distributions in Fig. 25.4c,d, which are 12 hours apart. Locally, thermal effects of GWs can significantly exceed the average values, as, for instance, a striking cooling of more than $3000\ \mathrm{K\,d^{-1}}$ at 150°W in the Southern Hemisphere.

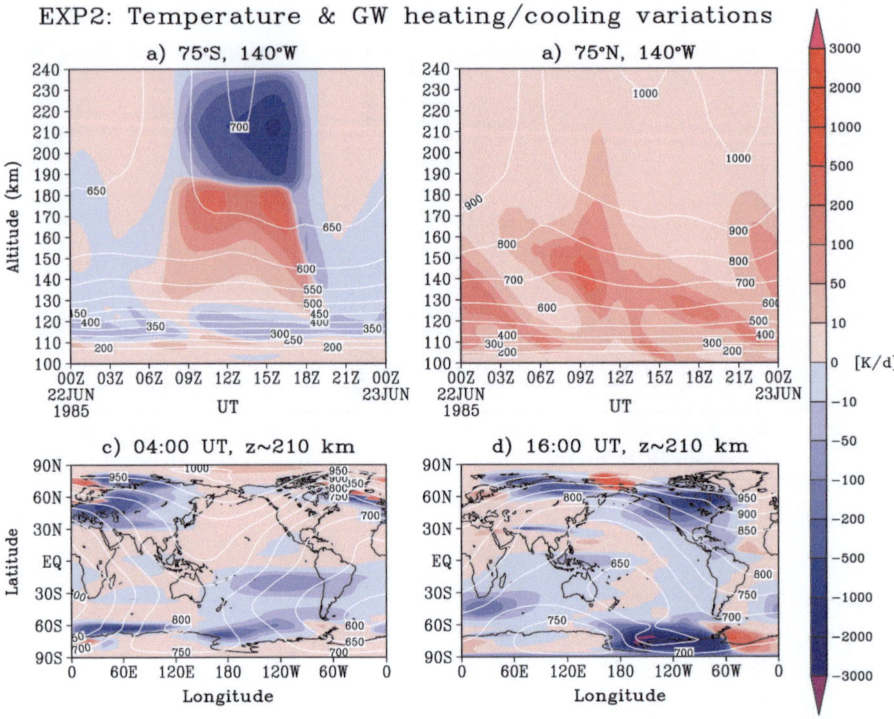

Fig. 25.4 Instantaneous distributions of the total GW-induced heating/cooling rates (in K day^{-1}): time-altitude cross-sections in the (**a**) winter (75°S, 140°W), and (**b**) summer (75°N, 140°W) hemispheres. Latitude-longitude distributions at ~210 km at two representative universal times: (**c**) 04:00UT and (**d**) 16:00UT. *Red* and *blue shadings* represent heating and cooling, respectively. The *white contour lines* mark the temperature (adopted from [*Yiğit and Medvedev*, 2009, Fig. 2])

25.7 Summary and Future Perspectives

Implementation of the gravity wave (GW) parameterization by *Yiğit et al.* [2008] into a comprehensive general circulation model extending from the lower to the upper atmosphere has revealed a significant dynamical and thermal influence of small-scale GWs of lower atmospheric origin on the thermosphere. The associated momentum deposition into the mean flow at F region altitudes is comparable to ion drag. GW-induced heating rates compete with the high-latitude Joule heating, while the net thermal effect of GWs is cooling, which is 10 to 20 % of that by the molecular heat conduction. These results have wide reaching implications for the modeling and interpretation of observations in the atmosphere above the turbopause.

The approach with the parameterization described in this article provides a promising new tool to account for the unresolved GWs effects in GCMs for the whole atmosphere system. Besides the influence on the mean state, GWs are likely to play an important role in coupling transient atmospheric phenomena such as tides, planetary waves, stratospheric warmings, convective events with their responses in the thermosphere. Of interest also is the solar influence on GW penetration to the TI [*Yiğit and Medvedev*, 2010], and GW impact on the constituents. Further validation and comparative work will be performed as more observations of GW activity become available in the upper atmosphere.

References

Alexander, M. J., & Dunkerton, T. J. (1999). A spectral parameterization of mean-flow forcing due to breaking gravity waves. *Journal of the Atmospheric Sciences*, *56*, 4167–4182.

Banks, P. M., & Kockarts, G. (1973). *Aeronomy, part A*. New York: Elsevier.

Beagley, S. R., de Grandepre, J., Koshyk, J. N., McFarlane, N. A., & Shepherd, T. G. (1997). Radiative-dynamical climatology of the first generation Canadian middle atmosphere model. *Atmosphere-Ocean*, *35*, 293–331.

Becker, E. (2004). Direct heating rates associated with gravity wave saturation. *Journal of Atmospheric and Solar-Terrestrial Physics*, *66*, 683–696.

Boville, B. A., & Randel, W. J. (1992). Equatorial waves in a stratospheric gcm: Effects of vertical resolution. *Journal of the Atmospheric Sciences*, *49*, 785–801.

Djuth, F. T., Sulzer, M. P., Gonzales, S. A., Mathews, J. D., Elder, J. H., & Walterscheid, R. L. (2004). A continuum of gravity waves in the Arecibo thermosphere? *Journal of Geophysical Research*, *31*, L16801. doi:10.1029/2003GL019376.

Dobbin, A. L. (2005). *Modelling studies of possible coupling mechanisms between the upper and the middle atmosphere*. Ph.D. thesis, University College London.

Fomichev, V. I., Ward, W. E., Beagley, S. R., McLandress, C., McConnell, J. C., McFarlane, N. A., & Shepherd, T. G. (2002). Extended Canadian middle atmosphere model: zonal-mean climatology and physical parameterizations. *Journal of Geophysical Research*, *107*, 4087. doi:10.1029/2001JD000479.

Fritts, D. C., & Alexander, M. J. (2003). Gravity wave dynamics and effects in the middle atmosphere. *Reviews of Geophysics*, *41*(1), 1003. doi:10.1029/2001RG000106.

Fritts, D. C., & Lu, W. (1993). Spectral estimates of gravity wave energy and momentum fluxes, ii, parameterization of wave forcing and variability. *Journal of the Atmospheric Sciences*, *50*, 3695–3713.

Gossard, E. E., & Hooke, W. H. (1975). *Waves in the atmosphere*. Amsterdam: Elsevier. 243 pp.

Harris, M. J. (2001). *A new coupled middle atmosphere and thermosphere general circulation model: studies of dynamics, energetics and photochemical coupling in the middle and upper atmosphere*. Ph.D. thesis, University of London.

Hertzog, A., Boccara, G., Vincent, R. A., Vial, F., & Cocqurez, P. (2008). Estimation of gravity wave momentum flux and phase speeds from quasi-Lagrangian stratospheric balloon flights. part ii: Results from vorcore campaign in Antarctica. *Journal of the Atmospheric Sciences*, *65*, 3056–3070.

Hines, C. O. (1960). Internal gravity waves at ionospheric heights. *Canadian Journal of Physics*, *38*, 1441–1481.

Holton, J. R. (1982). The role of gravity wave induced drag and diffusion in the momentum budget of the mesosphere. *Journal of the Atmospheric Sciences*, *39*, 791–799.

Hunsucker, R. (1982). Atmospheric gravity waves generated in the high-latitude ionosphere: a review. *Reviews of Geophysics*, *20*, 293–315.

Klostermeyer, J. (1972). Influence of viscosity, thermal conduction, and ion drag on the propagation of atmospheric gravity waves in the thermosphere. *Zeitschrift für Geophysik*, *38*, 881–890.

Lindzen, R. S. (1981). Turbulence and stress owing to gravity waves and tidal breakdown. *Journal of Geophysical Research*, *86*, 9707–9714.

Liu, H.-L., Foster, B. T., Hagan, M. E., McInerney, J. M., Maute, A., Qian, L., Richmond, A. D., Roble, R. G., Solomon, S. C., Garcia, R. R., Kinnison, D., Marsh, D. R., Smith, A. K.,

Richter, J., Sassi, F., & Oberheide, J. (2010). Thermosphere extension of the whole atmosphere community climate model. *Journal of Geophysical Research, 115*, A12302. doi:10.1029/2010JA015586.
Livneh, D. J., Seker, I., Djuth, F. T., & Mathews, J. D. (2007). Continuous quasiperiodic thermospheric waves over Arecibo. *Journal of Geophysical Research, 112*, A07313. doi:10.1029/2006JA012225.
Manzini, E., Giorgetta, M. A., Esch, M., Kornblueh, L., & Roeckner, E. (2006). The influence of sea surface temperatures on the northern winter stratosphere: ensemble simulations with the MAECHAM5 model. *Journal of Climate, 19*, 3863–3881.
Matsuno, T. (1982). A quasi one-dimensional model of the middle atmosphere circulation interacting with internal gravity waves. *Journal of the Meteorological Society of Japan, 60*, 215–226.
Medvedev, A. S., & Klaassen, G. P. (1995). Vertical evolution of gravity wave spectra and the parameterization of associated wave drag. *Journal of Geophysical Research, 100*, 25841–25853.
Medvedev, A. S., & Klaassen, G. P. (2000). Parameterization of gravity wave momentum deposition based on nonlinear wave interactions: Basic formulation and sensitivity tests. *Journal of Atmospheric and Solar-Terrestrial Physics, 62*, 1015–1033.
Medvedev, A. S., & Klaassen, G. P. (2003). Thermal effects of saturating gravity waves in the atmosphere. *Journal of Geophysical Research, 108*(D2), 4040. doi:10.1029/2002JD002504.
Medvedev, A. S., Klaassen, G. P., & Beagley, S. R. (1998). On the role of an anisotropic gravity wave spectrum in maintaining the circulation of the middle atmosphere. *Geophysical Research Letters, 25*, 509–512.
Munro (1950). Traveling disturbances in the ionosphere. *Proceedings of the Royal Society of London. Series A, Mathematical and Physical Sciences, 202*(1069), 208–223.
Oliver, W. L., Otsuka, Y., Sato, M., Takami, T., & Fukao, S. (1997). A climatology of f region gravity wave propagation over the middle and upper atmosphere radar. *Journal of Geophysical Research, 102*, 14499–14512.
Richmond, A. D., Ridley, E. C., & Roble, R. G. (1992). A thermosphere/ionosphere general circulation model with coupled electrodynamics. *Geophysical Research Letters, 19*, 601–604.
Roble, R. G., Ridley, E. C., & Richmond, A. D. (1988). A coupled thermosphere/ionosphere general circulation model. *Geophysical Research Letters, 15*(12), 1325–1328.
Schmidt, H., Brasseur, G. P., Charron, M., Manzini, E., Giorgetta, M. A., Diehl, T., Fomichev, V. I., Kinnison, D., Marsh, D., & Walters, S. (2006). The HAMMONIA chemistry climate model: sensitivity of the mesopause region to the 11-year solar cycle and CO_2 doubling. *Journal of Climate, 19*, 3903–3931.
Vadas, S. L., & Fritts, D. C. (2005). Thermospheric responses to gravity waves: influences of increasing viscosity and thermal diffusivity. *Journal of Geophysical Research, 110*, D15103. doi:10.1029/2004JD005574.
Warner, C. D., & McIntyre, M. E. (2001). An ultrasimple spectral parameterization for nonorographic gravity waves. *Journal of the Atmospheric Sciences, 58*, 1837–1857.
Weinstock, J. (1982). Nonlinear theory of gravity waves: momentum deposition, generalized Rayleigh friction, and diffusion. *Journal of the Atmospheric Sciences, 39*, 1698–1710.
Yiğit, E. (2009). *Modelling atmospheric vertical coupling: role of gravity wave dissipation in the upper atmosphere*. Ph.D. thesis, University College London Doctoral Thesis.
Yiğit, E., & Medvedev, A. S. (2009). Heating and cooling of the thermosphere by internal gravity waves. *Geophysical Research Letters, 36*, L14807. doi:10.1029/2009GL038507.
Yiğit, E., & Medvedev, A. S. (2010). Internal gravity waves in the thermosphere during low and high solar activity: simulation study. *Journal of Geophysical Research, 115*, A00G02. doi:10.1029/2009JA015106.
Yiğit, E., Aylward, A. D., & Medvedev, A. S. (2008). Parameterization of the effects of vertically propagating gravity waves for thermosphere general circulation models: sensitivity study. *Journal of Geophysical Research, 113*, D19106. doi:10.1029/2008JD010135.
Yiğit, E., Medvedev, A. S., Aylward, A. D., Hartogh, P., & Harris, M. J. (2009). Modeling the effects of gravity wave momentum deposition on the general circulation above the turbopause. *Journal of Geophysical Research, 114*, D07101. doi:10.1029/2008JD011132.

Chapter 26
The Geospace Response to Nonmigrating Tides

Kathrin Häusler, Jens Oberheide, Hermann Lühr, and Ralf Koppmann

Abstract During the past five years, significant progress has been made in elucidating and delineating the geospace response to nonmigrating tides from the lower atmosphere. Satellite missions providing continuously and globally distributed measurements of the atmospheric parameters revealed astonishing findings not anticipated before. Special emphasis is put on the eastward propagating diurnal tide with zonal wavenumber 3 (DE3) which manifests itself not only in the neutral atmosphere but also in the ionosphere. The DE3 tide can be traced from its origin in the troposphere to its maximum in the mesosphere, lower thermosphere (MLT) region up to an altitude of 400 km. Thereby Hough Mode Extension (HME) modeling aids to bridge the data gap between satellite measurements performed in the MLT region and upper thermosphere.

K. Häusler (✉) · H. Lühr
GFZ German Research Centre for Geosciences, Telegrafenberg, 14473 Potsdam, Germany

H. Lühr
e-mail: hluehr@gfz-potsdam.de

Present address:
K. Häusler
High Altitude Observatory, National Center for Atmospheric Research, PO Box 3000, Boulder, CO, 80307-3000, USA
e-mail: kathrin@ucar.edu

J. Oberheide
Department of Physics and Astronomy, Clemson University, 102B Kinard Laboratory, Clemson, SC 29634-0978, USA
e-mail: joberhe@clemson.edu

R. Koppmann
Physics Department, University of Wuppertal, Gaußstraße 20, 42119 Wuppertal, Germany
e-mail: koppmann@uni-wuppertal.de

F.-J. Lübken (ed.), *Climate and Weather of the Sun-Earth System (CAWSES)*, Springer Atmospheric Sciences,
DOI 10.1007/978-94-007-4348-9_26,

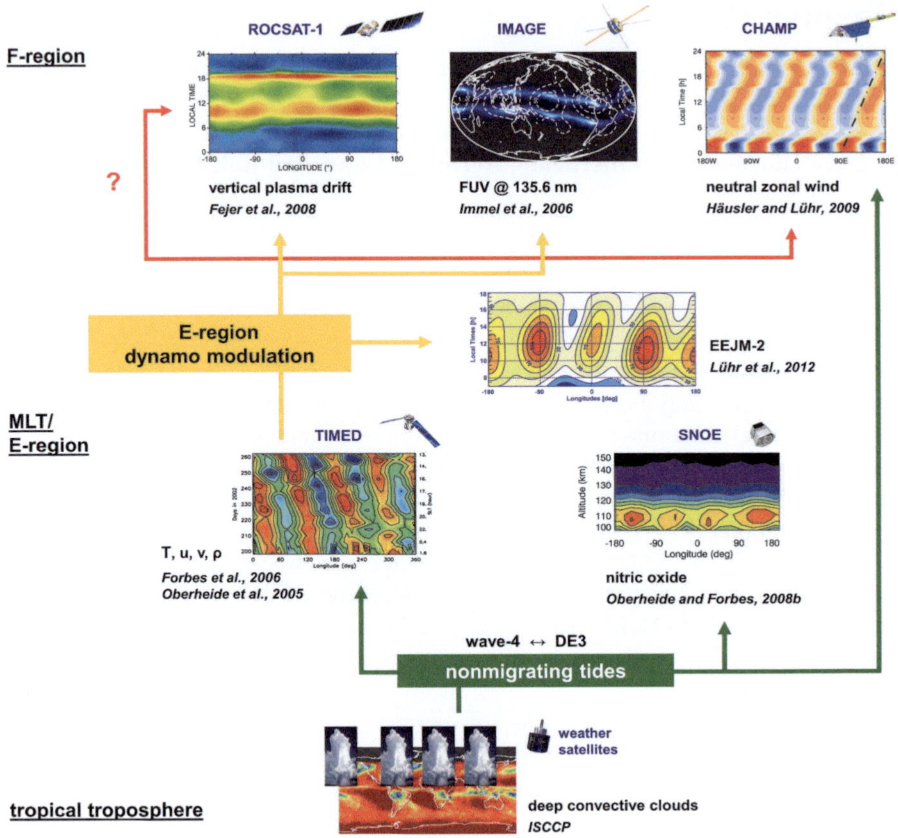

Fig. 26.1 Coupling from the troposphere to the ionosphere/thermosphere through nonmigrating tides

26.1 Introduction

Numerous observations and model studies made during the past five years have unequivocally revealed that geospace owes a considerable amount of its longitudinal, local time, seasonal/latitudinal and day-to-day variability to solar atmospheric tides from the lower atmosphere. An unexpected and new realization was the discovery that tropospheric weather is important for the "space weather" of the ionosphere-thermosphere-mesosphere (ITM) system, with tides being the main coupling mechanism (Fig. 26.1). Understanding the geospace response to lower atmospheric tides is an emphasis area of SCOSTEP's CAWSES-II program 2009–2013 and remains one of the challenges in aeronomy for the next decade, both scientifically and due to its importance for reliable space weather predictions. The German CAWSES program has significantly contributed to the current state-of-the-art by delineating the tidal morphology in the mesosphere, lower thermosphere (MLT) and upper thermosphere from remote-sensing and in-situ satellite observations, and by making the connection between tropospheric sources and the response of Earth's neutral and ionized upper atmosphere using empirical and first-principles modeling. This Chapter specifically addresses the progress made, by emphasizing the importance of non-Sun-synchronous (nonmigrating) tides.

Atmospheric tides are global-scale waves in temperature, winds and density that are forced by the periodic absorption of solar radiation in the troposphere and stratosphere. Growing exponentially when propagating upward, tides often dominate the thermal and dynamical structure of the MLT and E-region where they usually reach their maximum amplitudes. Dissipation then leads to net energy and momentum deposition in various forms. About 10^{16} Joule of energy propagate up daily from the lower atmosphere in the form of gravity waves (GW), large-scale planetary waves (PW) and tides [*Jarvis*, 2001] with tides being a major contributor. Tidal wave coupling from below is therefore a fundamental mechanism of redistributing solar energy back into geospace.

Leading tidal frequencies are 24 h (diurnal) and 12 h (semidiurnal), due to their solar origin. The parts of the tidal spectrum that follow the apparent westward propagation of the Sun in a Sun-synchronous manner are called migrating tides, and the non-Sun-synchronous parts are called nonmigrating tides. Migrating tides are predominantly forced by solar IR and UV radiation absorption in tropospheric water vapor and stratospheric ozone, and solar EUV absorption in the thermosphere. Nonmigrating tides have two major sources: non-linear wave-wave interaction processes in the strato-/mesosphere [*Hagan and Roble*, 2001; *Lieberman et al.*, 2004], and latent heating due to large-scale deep convection in the tropical troposphere [*Hagan and Forbes*, 2002, 2003]. Deep convection largely depends on land-sea differences and sea-surface temperatures. The periodic absorption of solar radiation at the surface thus transforms to a longitudinal structure in cloud droplet formation (heat release) at roughly the same local time of the day that acts as an efficient forcing mechanism for a number of nonmigrating tides, i.e., for the prominent diurnal eastward propagating tide of zonal wavenumber 3 (DE3). Here, a tidal component is denoted by a letter/number code. DWs or DEs is a westward or eastward propagating diurnal tide, respectively, with zonal wavenumber s. For semidiurnal tides, "D" is replaced by "S", and D0, S0 are zonally symmetric oscillations, respectively. There are some indications for nonmigrating tidal forcing in the thermosphere [*Oberheide et al.*, 2011b] but not much is currently known about it, due to the lack of suitable global data in the 120–400 km "thermospheric gap" between satellite remote-sensing and in-situ observations.

MLT and E-region tides on a global scale are elucidated from wind and temperature observations made by the SABER (Sounding of the Atmosphere using Broadband Emission Radiometry) and TIDI (TIMED Doppler Interferometer) instruments on board the TIMED (Thermosphere, Ionosphere, Mesosphere Energetics and Dynamics) satellite. Section 26.2 overviews these tidal diagnostics, discusses the seasonal and inter-annual variation of the tides, and puts them into the context of the CAWSES tidal campaigns. A methodology, called Hough Mode Extension (HME) modeling is introduced, that allows one to test the self-consistency of the TIDI and SABER observations, and to extend the MLT tides from TIMED into the upper thermosphere through empirical modeling.

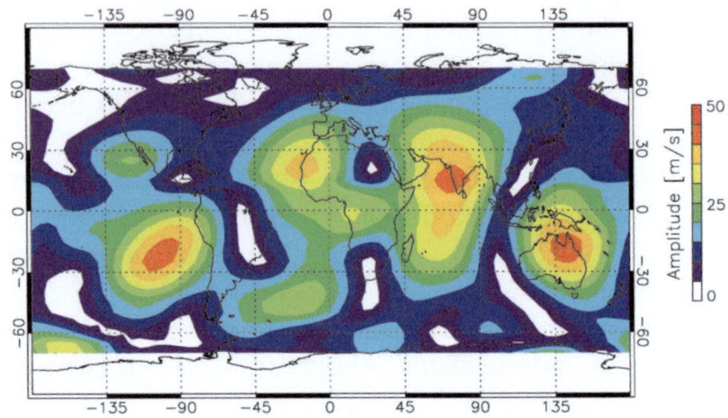

Fig. 26.2 Map of reconstructed TIDI diurnal tidal amplitudes in the zonal wind, for climatological 2002–2008 August and 100 km altitude. Migrating and nonmigrating tides are superposed

The HME-based upward extension then allows one to make the connection with in-situ tidal diagnostics from CHAMP (Challenging Minisatellite Payload), and to study how tides from the lower atmosphere impact the aeronomy of the IT system, including trace constituents, neutral density and electrodynamical coupling through tidal modulation of E-region dynamo winds (Sect. 26.3).

The German satellite CHAMP has contributed significantly to the understanding of upper thermospheric dynamics. Section 26.4 gives an overview of the important findings concerning nonmigrating tides of tropospheric origin at CHAMP altitude. The review includes the tidal spectra in equatorial latitudes (Sect. 26.4.2) as well as the comparison between CHAMP and TIME-GCM tidal diagnostics (Sect. 26.4.3). Specific attention is given to DE3 in Sect. 26.4.4 including its inter-annual variability and solar flux dependence.

Ionospheric implications of tidal wave coupling from below are emphasized in Sect. 26.5 including the equatorial electrojet (EEJ), vertical plasma drift, and crest-to-trough (CTR) ratio of the equatorial ionization anomaly (EIA). A summary is rounding off this chapter in Sect. 26.6.

26.2 Tides in MLT and E-Region Winds from TIDI

Tidal winds are one of the most prominent motions in the MLT and E-region. Amplitudes (Fig. 26.2) on the order of the time-averaged zonal wind, dominate the meridional wind, and exhibit variations on time scales ranging from days to years. Satellite-borne tidal winds in the MLT region were first diagnosed in UARS (Upper Atmosphere Research Satellite) data that provided invaluable information about the morphology of the migrating tides [*Hays et al.*, 1994; *McLandress et al.*, 1996]. UARS also provided the first opportunity to diagnose nonmigrating tidal winds [*Forbes et al.*, 2003]. However, the specifics of the UARS wind instruments limited nonmigrating tidal analysis to a single altitude of 95 km. The full appreciation of the importance of nonmigrating tides in general, and zonal winds in particular, is mainly a result of TIDI/TIMED observations since these data allowed for the first time to derive the tidal wave spectrum over a range of MLT altitudes and over an extended period of time.

26.2.1 Data and Analysis

The TIMED Doppler Interferometer (TIDI) [*Killeen et al.*, 2006] is the wind instrument on board the TIMED satellite that was launched in 2001. TIDI measures zonal and meridional MLT winds from the Doppler shift of green line emissions. Diurnal and semidiurnal nonmigrating tides are derived from 80–105 km and between ±75° latitude using the National Center for Atmospheric Research (NCAR) data version v0307a [*Wu et al.*, 2008] and a two-dimensional Fourier method [*Oberheide et al.*, 2005, 2006, 2007]. Amplitudes and phases are first computed as 60-day running means and are then averaged into monthly mean bins, from March 2002 to December 2010. TIDI tidal diagnostics covers 8 diurnal components (DW5, DW4, DW3, DW2, D0, DE1, DE2, DE3) and 9 semidiurnal components (SW6, SW5, SW4, SW3, SW1, S0, SE1, SE2, SE3). Migrating tides cannot be reliably recovered due to instrumental and aliasing issues but come from SABER temperature tidal diagnostics and their HME modeling, as overviewed in Sect. 26.2.4. Amplitude accuracy is on the order of 10 %, and amplitude and phase precisions are on the order of 1 m/s and 1–2 hours, respectively. See *Oberheide et al.* [2006, 2007] for details of the error analysis and how the errors vary for individual components. TIDI tides poleward of ±60° are subject to increasing planetary wave aliasing and should be used with due care there. Figure 26.2 exemplifies the tidal variability in the MLT region as seen by TIDI and highlights the importance of nonmigrating tides since the latter cause the observed longitudinal amplitude variation.

26.2.2 DE3 Seasonal and Inter-annual Variations

Figure 26.3 illustrates the TIDI tidal diagnostics for the DE3 zonal wind component at 105 km altitude, for the 2002–2009 climatological case and individual years. See *Oberheide et al.* [2006, 2007] for the full set of diurnal and semidiurnal tidal components from TIDI. DE3 is of special interest for the space weather of the ITM because of (i) its magnitude in the MLT and (ii) its importance for the E-region dynamo and the overlying ionospheric F-region structure, as discussed in Sect. 26.3. It should also be noticed that DE3 reflects meteorological impacts on geospace because it is largely forced by latent heat release in large-scale deep convective systems in the tropical troposphere [*Hagan and Forbes*, 2002].

The seasonal DE3 variation is largely annual, with a broad equatorial maximum in August/September that changes to two weaker peaks around ±20° during boreal winter. This is accompanied by a dramatic phase change from symmetric with respect to the equator in summer to antisymmetric in winter (Fig. 26.4). *Oberheide and Forbes* [2008a] provide a detailed discussion of this behavior: it basically reflects the relative strength of the two dominating tidal eigenmodes in summer and winter which in turn is the result of tropospheric heating variations [*Zhang et al.*, 2010]. Wind filtering effects are comparatively unimportant.

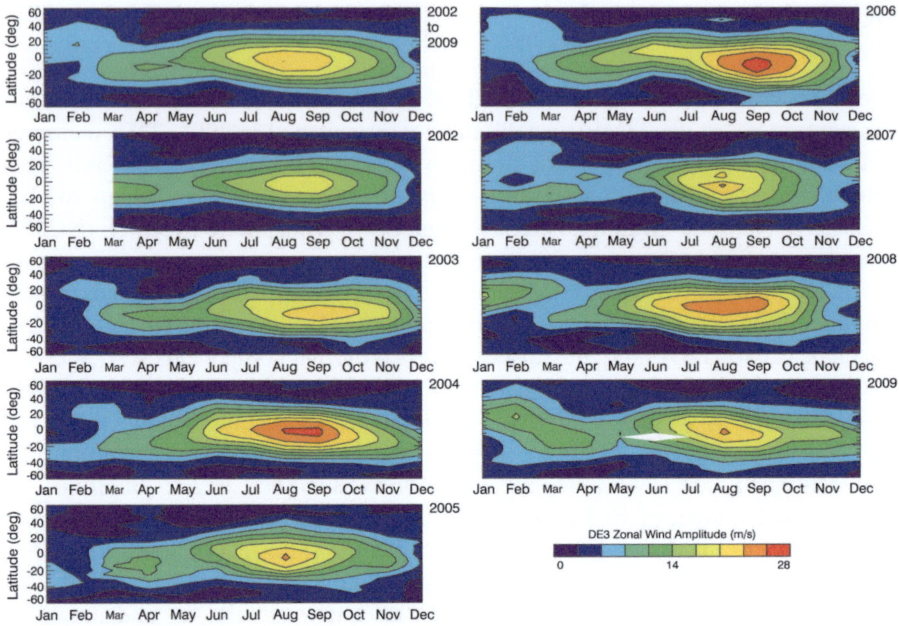

Fig. 26.3 TIDI DE3 zonal wind amplitudes at 105 km. The 2002–2009 climatology and individual years are shown. White areas indicate data gaps

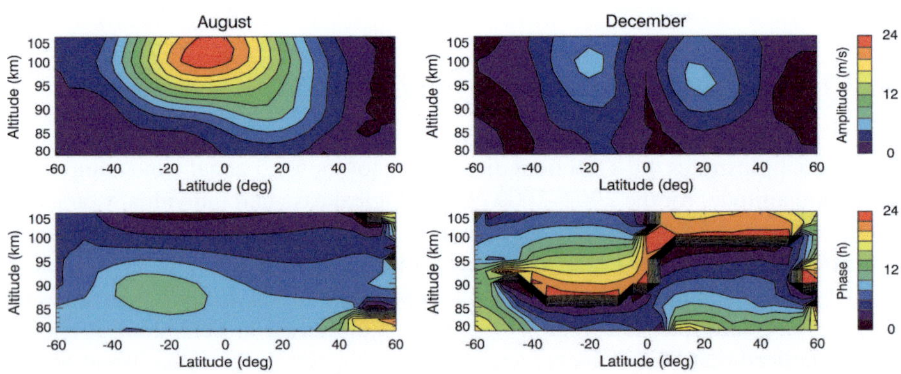

Fig. 26.4 TIDI DE3 zonal wind amplitudes (*top*) and phases (*bottom*) for the climatological cases August (*left*) and December (*right*). Phases are given in local time of maximum

A cursory inspection of Fig. 26.3 already reveals a quasi-2-year pattern in the equatorial DE3 zonal winds during summer with relative maxima in 2004, 2006, and 2008, and relative minima in 2003, 2005, 2007, and 2009. It is less perceivable "by eye" in 2002 but a more detailed analysis involving time/altitude cross-sections at the equator shows that 2002 is also a "maximum" year but with a peak altitude of about 98 km, compared to 102–105 km during the other years [*Oberheide et al.*, 2009]. Statistical analysis of the underlying symmetric and antisymmetric tidal eigenmodes provides further insight into the DE3 inter-annual amplitude modulation. Significant periods (99.9 % confidence) occur at 6, 12, and 24.3 months for the symmetric mode and thus for the equatorial summer maximum. The 2-year signal (15–20 % amplitude modulation), however, is absent in the antisymmetric mode and thus in the $\pm 20°$ latitude winter maxima.

The reason(s) for the low-latitude 2-year modulation are still under debate. Its period is suggestive of a quasi-biennial-oscillation (QBO) effect and this argument has been made for migrating tides [*Forbes et al.*, 2008; *Xu et al.*, 2009]. The mean 2002–2008 QBO period of 24.7 months compares well with the 24.3 months in the DE3 with the latter being larger in the westerly phase of the QBO. However, it is not clear where and how the QBO modulates the DE3. For example, QBO background wind variations extend well into the MLT where they are out-of-phase with the stratospheric QBO [e.g., *Burrage et al.*, 1996]. A mesospheric QBO influence on the DE3 would indeed be more plausible because easterly mean zonal winds Doppler shift the DE3 frequency toward higher values and such higher-frequency waves are harder to dissipate than a lower-frequency wave of comparable wavelength [*Ekanayake et al.*, 1997]. Additional and/or alternative QBO effects may also include stratospheric filtering during upward propagation, i.e., vertical wavelength distortions due to mean winds and hence an altered dissipation. These and other wave-wave and wave-mean flow interactions are poorly understood and an important objective of SCOSTEP's CAWSES-II program 2009–2013 (www.cawses.org).

The arguments made above and the more detailed discussion in *Oberheide et al.* [2009] support a QBO interpretation, presumably because of enhanced (reduced) tidal dissipation during the westerly (easterly) phase of the mesospheric QBO due to frequency Doppler shift. The DE3, however, will also react to tropospheric forcing variations such as the El Niño/Southern Oscillation (ENSO) since the latter changes the locations of tropical deep convective systems. *Lieberman et al.* [2007] found a largely negative correlation between the Southern Oscillation Index (SOI) and latent heat release over the central and eastern Pacific. A statistical analysis of 2002–2008 SOI data yields a 26.8 month period at the 99 % confidence level [*Oberheide et al.*, 2009]. This is not far from the 24.3 month period in the TIDI DE3 zonal winds and certainly an issue worth to be pursued. Further insight is expected to come from an ongoing investigation of the strong 2009–2010 El Niño and the 2010–2011 La Niña.

26.2.3 CAWSES Tidal Campaigns and Validation

TIDI tides have been extensively validated as part of the CAWSES tidal campaigns [*Ward et al.*, 2010] and significantly contributed to their scientific accomplishments. The tidal campaigns were organized as a project within SCOSTEP's CAWSES-I program around four observation campaigns in September–October 2005, March–April 2007, June–August 2007 and December–January 2008. Another one, now part of CAWSES-II, is scheduled for August–September 2011.

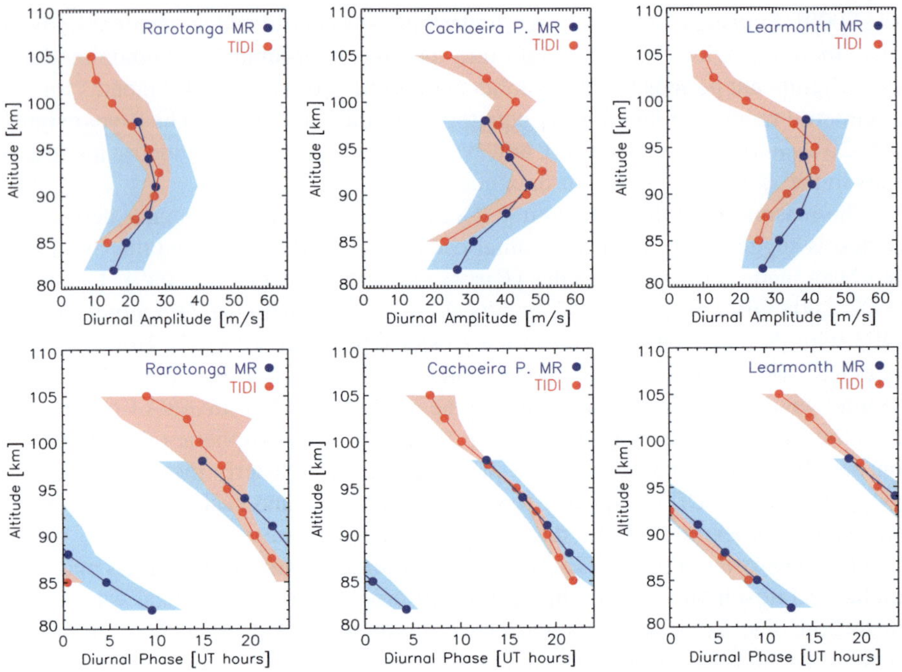

Fig. 26.5 Height profiles (September–October 2005, first campaign) of the diurnal (*top*) amplitude and (*bottom*) phase of the meridional wind for TIDI and three meteor radar stations (Rarotonga: 21°S, 160°W; Cachoeira Paulista: 23°S, 45°W; Learmonth: 22°S, 114°E). *Shading* indicates standard deviation. Original from *Ward et al.* [2010]. Reproduced/modified by permission of American Geophysical Union

Figure 26.5 shows an example for the validation of TIDI diurnal tides during the first campaign. Sixty day vector averages of three ground-based meteor radars located at different locations near the 20°S latitude circle are compared to reconstructed tidal fields from TIDI that are generated by superposing the diagnosed diurnal tidal components. It should be noted that the migrating tidal winds in the reconstruction are not from TIDI observations, for reasons outlined in Sect. 26.2.1, but from SABER temperature tides and tidal theory (see Sect. 26.2.4). This is not a problem here since the tidal amplitude variations along longitude are due to non-migrating tides alone. The top (bottom) panels in Fig. 26.5 show the amplitudes (phases) for the three locations with the solid line showing the data and the shaded regions the standard deviation within the 60-day time window. TIDI meridional wind amplitudes at 90 km at the three stations vary significantly, from 25 m/s (Rarotonga) to 50 m/s (Cachoeira Paulista) and 32 m/s (Learmonth). The meteor radars agree with these values within the experimental error. Furthermore, the agreement extends throughout the profiles both in amplitude and phase.

These and other TIDI/radar and TIDI/lidar comparisons [*Ward et al.*, 2010] show for the first time that the tidal signatures from ground-based radar and satellite optical observations are consistent with each other. This progress was made possible due to the now available nonmigrating tidal definitions over a range of MLT altitudes. It appears that many of the differences between the various sources of information on tides are now resolved. Systematic tidal investigations can thus proceed in a more coherent manner, i.e., by using combined ground-based and satellite tidal diagnostics.

26.2.4 HME Modeling and Internal Consistency with SABER Tidal Temperatures

A further source for tidal diagnostics in the MLT region is the SABER instrument on TIMED. SABER measures temperature and selected trace constituents from 20–120 km from infrared emissions [*Russell et al.*, 1999]. Migrating and nonmigrating temperature tides are analyzed in a similar manner as TIDI and provide the same monthly mean tidal components with an amplitude error of 1–2 K [*Forbes et al.*, 2006, 2008]. It is now possible to explore the internal consistency of the TIDI wind tides and the SABER temperature tides within theoretical expectations.

The approach is called Hough Mode Extension (HME) modeling and can be thought of as constraining a tidal model with observations. HMEs represent the numerical solution (pole-to-pole, 0–400 km altitude) to the linearized dynamical equations of the atmosphere taking into account dissipative effects above the tidal forcing region and height dependent background temperature. Going back to *Lindzen et al.* [1977] and *Forbes and Hagan* [1982], an HME is an extension of a classical Hough mode that represents the solution to Laplace's tidal equation in an isothermal atmosphere without mean winds and dissipation. Each HME is a self-consistent latitude vs. height set of amplitudes and phases for the perturbation fields in temperature, zonal, meridional and vertical winds, and density (T, u, v, w, ρ). HMEs are computed following *Svoboda et al.* [2005] using a simplified version of the GSWM linear tidal model [*Hagan and Forbes*, 2002], which only maintains the tidal dissipation scheme. Model background winds are set to zero and the standard model temperature and density profiles are replaced by globally averaged profiles for different solar flux conditions. Each HME is then computed using an arbitrarily calibrated, simple tropospheric heat source with a latitudinal structure equivalent to that of a classical Hough Mode. See *Oberheide and Forbes* [2008a] and *Oberheide et al.* [2009, 2011a, 2011b] for examples of the full latitude vs. height HME structure.

The HMEs for each tidal component are then least-squares fitted to the observed tides in the MLT region, as detailed in *Oberheide et al.* [2011b], and tidal amplitudes and phases are reconstructed as function of latitude, altitude, and month. This approach accounts for mean wind effects although the latter are set to zero in the HME computation. *Svoboda et al.* [2005] show that the distortion of tidal structures due to mean winds can be viewed as mode coupling, that is, the excitation of higher-order modes which combine together to approximate the distortion. HME fit errors due to uncertainties in the observed tidal fields and due to the specifics of the methodology

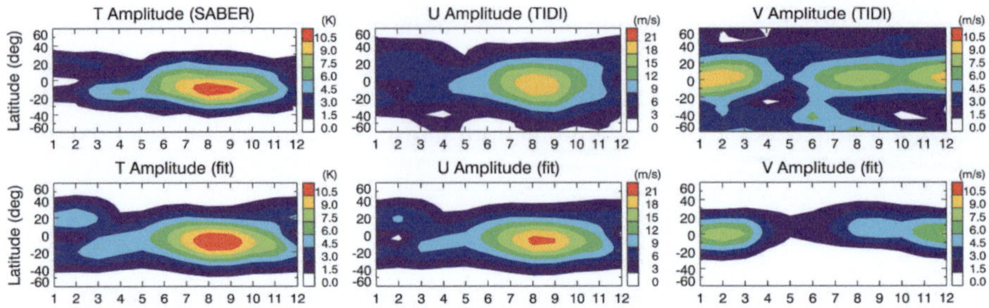

Fig. 26.6 (*Top*) Observed 2002–2006 average DE3 amplitudes in temperature (*left*), zonal (*middle*) and meridional wind (*right*) from SABER and TIDI at 100 km. (*Bottom*) Amplitudes from the HME approach. Adopted from *Oberheide and Forbes* [2008a]. Reproduced/modified by permission of American Geophysical Union

are on the order of 1 m/s for horizontal winds and 1 K for temperature amplitudes. Corresponding phase errors are 1–2 hours [*Oberheide et al.*, 2011b].

Figure 26.6 exemplifies the feasibility and strengths of the HME approach, by comparing observed seasonal variation of the DE3 tidal component at 100 km (2002–2006 climatology) and the result when fitting HMEs to the observed DE3 tides from SABER and TIDI. The HME fits agree within one color level (about the systematic error of the data) with the observations and as such establish the self-consistency of the TIDI and SABER tides within the framework of the tidal theory. The same level of agreement is found for individual years [*Oberheide et al.*, 2009] and for most other diurnal and semidiurnal tides [*Oberheide et al.*, 2009, 2011a].

26.3 Tidal Upward Propagation into the Ionosphere-Thermosphere (IT) System

TIMED tidal diagnostics is limited to altitudes below 105 km (winds) and 120 km (temperature). A major challenge in delineating and understanding the tidal wave spectrum at higher altitudes is the lack of global observations in the "thermospheric gap" between 120 km and 400 km where in-situ tidal diagnostic, i.e., from CHAMP (cf. Sect. 26.4), can be made. To date, the only approach to overcome this shortcoming and to obtain at least observation-based tides in the thermosphere are HME fits to TIMED tides observed in the MLT. This also allows one to interpret longitudinal variations in other parameters important to the aeronomy of the IT system, such as nitric oxide.

26.3.1 Nonmigrating Tides in Thermospheric Nitric Oxide

Nitric oxide (NO) acts as the upper atmosphere's natural thermostat through its 5.3 μm IR emissions [*Mlynczak et al.*, 2003]. Any NO longitudinal and local time

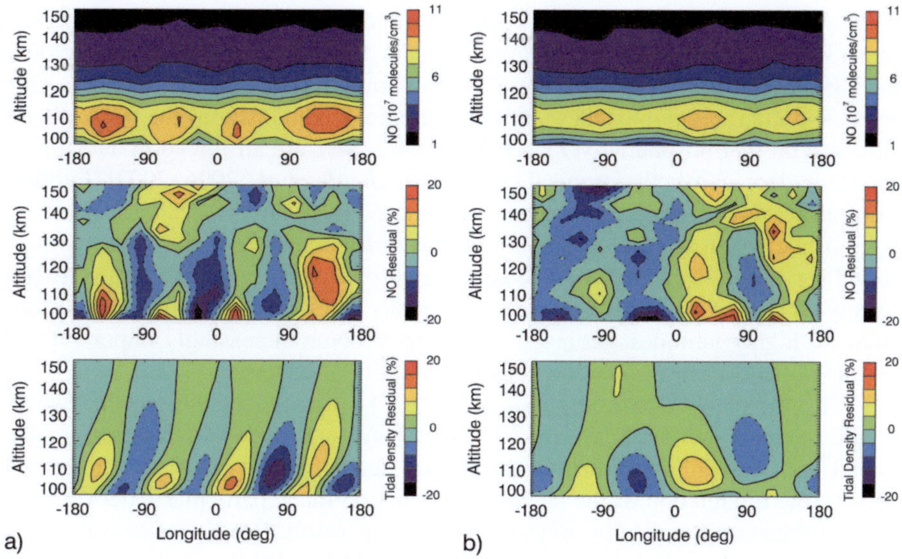

Fig. 26.7 Monthly-averaged (*top*) NO observations from SNOE, (*middle*) NO residuals from SNOE, and (*bottom*) tidal density residuals from HME fits to TIMED at the equator for (**a**) September and (**b**) December. Local solar time is 10.5 h. Adopted from *Oberheide and Forbes* [2008b]. Reproduced/modified by permission of American Geophysical Union

variability imposed by tides will thus modulate energy loss rates and hence the thermospheric energy budget. The first step in identifying the amount of NO tidal variability and its causes is to study the NO number density such as measured by the SNOE (Student Nitric Oxide Explorer) satellite. SNOE [*Barth et al.*, 2003] was in a Sun-synchronous orbit and its NO measurements were limited to daylight such that the mean local solar time of the NO data collected between March 1998 and September 2000 is 10.5 hour. Monthly averaged NO from SNOE at the equator show a clear 4-peaked longitudinal variation between March and October, and a 3-peaked longitudinal variation in December and January [*Oberheide and Forbes*, 2008b]. This is most pronounced in September and December (Fig. 26.7, top panels). The middle panels show the same data but as percentage residuals, that is, after subtraction of and division by the zonal mean. Peak-to-peak variations can approach 40 % within 40°–60° longitude intervals.

The bottom panels of Fig. 26.7 show the corresponding tidal density residuals for 10.5 hour LST. They are reconstructed from the full set of diurnal and semidiurnal components that have been inferred by fitting HMEs to the TIDI tidal winds and the SABER tidal temperatures. Note that the removal of the zonal means also removes the migrating tides because the latter are observed as zonally symmetric features in a constant LST reference frame [*Oberheide et al.*, 2003]. The tidal density residuals closely resemble the NO residuals in the 100–135 km range. The discrepancy above 135 km comes not unexpected because of altitude dependent NO production processes. More importantly, the longitudinal structure in the low-latitude NO residuals between 100–135 km is clearly a result of nonmigrating tidal density variations with DE3 (observed as a "wave-4") dominating between March and October and DE2, SE1 (observed as a "wave-3") dominating in December and January. This report and interpretation of these striking variations demonstrates the tropospheric influence on thermospheric NO since DE3, DE2, SE1 are all forced by latent heat release in deep convective tropical clouds [*Oberheide et al.*, 2006, 2007]. Furthermore, it is a first order effect in understanding NO variability because it is on the order of the equatorial NO density variations as function of solar variability as reported by *Barth et al.* [2003]. The same level of agreement between SNOE NO density residuals and HME-based neutral density tides is found between ±30° latitude. At higher latitudes, geomagnetic effects become increasingly important and they do not allow one to unambiguously identify tidal effects.

First principles modeling by *Ren et al.* [2011] recently reproduced the "wave-4" and "wave-3" longitudinal variations in NO and neutral density in a quantitative way. These simulations show that the neutral density effect is mainly transmitted through the modulation of the chemical production rates, particularly between excited nitrogen and molecular oxygen. The SNOE results proof that nonmigrating tides extend well into the thermosphere and that HMEs provide realistic estimates of their magnitude and phasing. It is thus possible to further follow the HME approach to study the upward propagation of the tides throughout the whole thermosphere up to 400 km altitude where the HME predictions can then be validated using CHAMP in-situ tidal diagnostics.

26.3.2 Inter-annual and Solar Cycle Variability

Science challenge questions in the IT system, that is, in the 120–400 km "thermospheric gap" include: (1) How large is the tidal variability in the upper thermosphere due to direct upward propagation from the tropospheric weather sources? (2) How large is the inter-annual and solar cycle variability of the tides and what causes them? TIMED-based HME fits provide first answers to these questions in that they provide the magnitude of the variability and allow one to study the physical reasons for the latter, i.e., in terms of zonal mean flow effects and wave dissipation. The following discussion focuses again on "wave-4" type longitudinal variations due to the DE3 and thus on tropospheric weather influences on the upper thermosphere.

Figure 26.8 is a time-series of DE3 HME fits to TIMED tidal wind and temperature diagnostics. Wave dissipation sets in above ~110 km and amplitudes relax to a constant value at altitudes above ~200 km when molecular diffusion becomes dominant. Nonmigrating tides in the MLT region do not show a statistically significant solar cycle dependence [*Oberheide et al.*, 2009]. The apparent amplitude increase above ~120 km with decreasing solar activity is a dissipative effect that is mainly due to the inverse dependence of vertical diffusion of heat and momentum on the background density. This is accounted for in Fig. 26.8 since the base HMEs in this case have been computed as function of solar F10.7 cm radio flux. Solar min/max

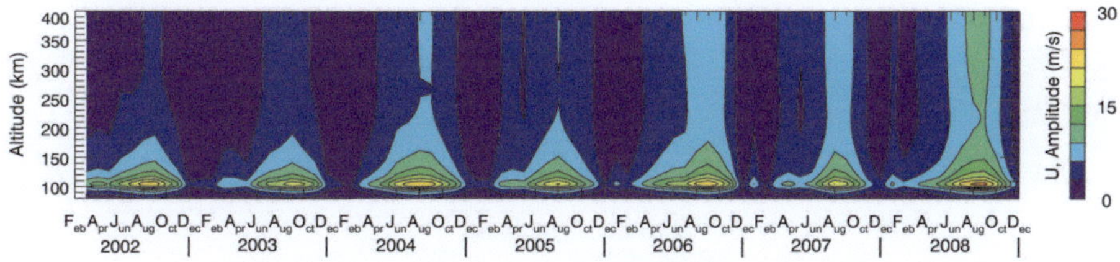

Fig. 26.8 Time series of HME-based DE3 zonal wind amplitudes at the equator. Adopted from *Oberheide et al.* [2009]. Reproduced/modified by permission of American Geophysical Union

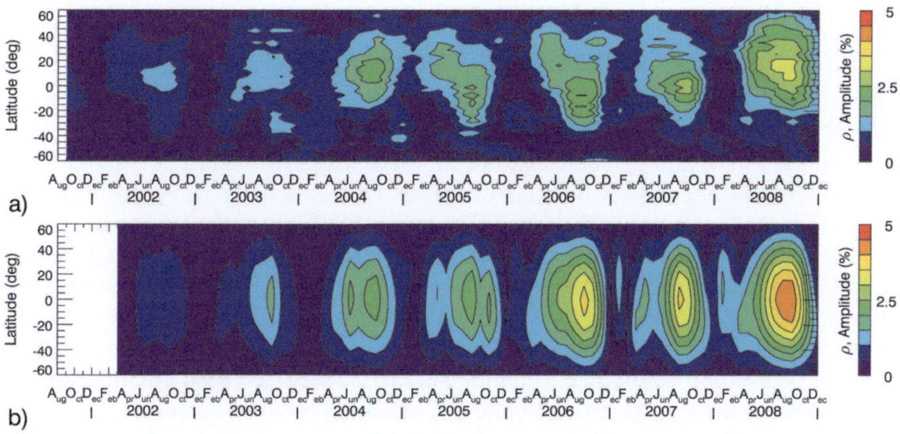

Fig. 26.9 (**a**) CHAMP DE3 relative density amplitudes normalized to 390 km. (**b**) Corresponding HME fit to TIMED tides in the MLT. Same altitude and color code

amplitude differences in the upper thermosphere are on the order of a factor of 3 in the zonal wind, 60 % in temperature and a factor of 5 in density [*Oberheide et al.*, 2009]. The amplitude error in Fig. 26.8 is ∼1 m/s and as such well below the observed solar cycle variability. Superposed to the latter is the 2-year oscillation from the MLT (Sect. 26.2.2) that is presumably caused by the QBO.

That these thermospheric extensions of MLT observations over 300 km altitude can be trusted is demonstrated in Fig. 26.9. The upper panel shows CHAMP DE3 relative density amplitudes normalized to 390 km as function of latitude and time. The lower panel is the HME prediction in the same color code which compares exceptionally well with the observation in terms of magnitude, and seasonal and solar cycle variation. Relative amplitude differences are on the order of 10 % with phase differences (not shown) of 1–2 hour [*Oberheide et al.*, 2011a]. A similar level of agreement is obtained when comparing HME zonal winds to CHAMP zonal wind tides (see Fig. 26.14) and HME temperatures to combined CHAMP/GRACE (Gravity Recovery And Climate Experiment) DE3 exosphere temperature diagnostics [*Forbes et al.*, 2009].

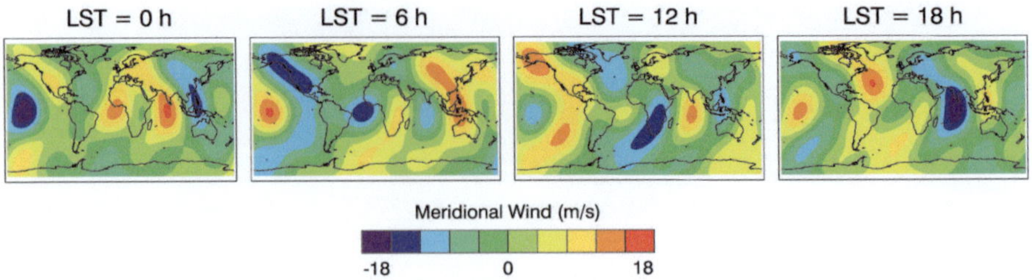

Fig. 26.10 Meridional tidal wind perturbations in September due to upward propagating nonmigrating diurnal and semidiurnal tides at 400 km as function of longitude, latitude and local time. Migrating tides are not included

26.3.3 Meridional Tidal Winds in the F-Region

Similar diagnostics for other diurnal and semidiurnal tides provide additional insight into the morphology and causes of IT variability. An important component is the semidiurnal SE2 tide, which also compares exceptionally well with CHAMP densities [*Oberheide et al.*, 2011a]. SE2 and its MLT and thermosphere structure are considerably more complex than DE3 since it is a result of tropospheric forcing and the non-linear interaction between the migrating diurnal tide DW1 and the DE3 [*Hagan et al.*, 2009; *Oberheide et al.*, 2011a]. SE2 is particularly interesting due to its meridional wind structure in the upper thermosphere and the resulting implications for neutral-ion coupling at F-region heights. A well-accepted tidal/ionosphere coupling mechanism is E-region dynamo modulation by nonmigrating tides due to the DE3 [*Immel et al.*, 2006; *Hagan et al.*, 2007; *Jin et al.*, 2008]. This produces F-region plasma density variations of "wave-4" type when observed in a constant local time frame [*Sagawa et al.*, 2005; *Lin et al.*, 2007]. SE2 (also observed as a "wave-4") in turn does not have an E-region zonal wind structure suitable for E-region dynamo modulation but exhibits upper thermospheric trans-equatorial wind speeds in excess of 10 m/s [*Oberheide et al.*, 2011a]. A comparison with the SAMI2 ionospheric model [*Huba*, 2000] suggests this will considerably contribute to the "wave-4" variability in the F-region plasma since 20 m/s are sufficient to produce a "wave-4" structure that is as large as the one due to E-region wind dynamo modulation [*England et al.*, 2010]. Superposed diurnal and semidiurnal tidal winds from HMEs produce even larger meridional winds at low latitudes. This is shown in Fig. 26.10 which combines 12 diurnal and semidiurnal nonmigrating tides (migrating tides excluded) at 400 km, based on the 2002–2008 climatological HME fits from *Oberheide et al.* [2011b].

26.4 Tides in Low-Latitude Thermospheric Zonal Winds from CHAMP

The German mini-satellite CHAMP was in orbit for ten years, two months, and four days. Its sensitive accelerometer at the spacecraft's center of gravity allowed to derive the zonal wind globally at about 400 km altitude. An introduction to the mission and an illustration is given in Chap. 11 (see Fig. 11.1). Unexpected findings of the CHAMP upper thermospheric zonal wind concerning longitudinal variability and nonmigrating tidal signatures have been published by *Häusler et al.* [2007, 2010], *Häusler and Lühr* [2009], and are summarized in *Häusler and Lühr* [2011]. Subsequently, some of these findings are echoed and new observations highlighted.

26.4.1 Data and Analysis

The zonal wind is calculated from CHAMP accelerometer measurements. The initial approach assumed that the drag force is aligned with the spacecraft velocity vector in an air rest frame [e.g., *Liu et al.*, 2006; *Häusler et al.*, 2007]. Now an iterative algorithm is used, which allows for independent drag coefficients of the various surfaces, to determine the crosstrack wind from the CHAMP accelerometer measurements [*Doornbos et al.*, 2010]. The difference between the two approaches is small (few percent) as long as relative variations are considered. Only crosstrack winds are deducible from CHAMP and due to the near polar orbit they correspond to zonal winds in equatorial and mid-latitude regions. Upper thermospheric zonal winds are available every 10 s starting from June 2001 until the beginning of September 2010. The CHAMP satellite precessed through one hour of local time every 11 days, thus having taken its daily measurements at a quasi constant local time which is crucial for the analysis. Combining the ascending (northward bound) and descending (southward bound) orbital arcs, a 24 h local time coverage is obtained after 131 days.

For the research, zonal wind residuals are used that are free of the zonal mean. The removed zonal mean comprises 15 consecutive orbits centered around the orbit of interest. These 15 consecutive orbits provide just an almost equal coverage of all longitudes. After calculating the wind residuals, they are sorted according to season, month, or time (daily means), respectively, and separated into local time hours and 24 overlapping longitude bins of 30° width each. This binning introduces a damping of the tidal amplitude which has been accounted for in the final results as described in *Häusler and Lühr* [2009]. Regarding the prominent wave-4 structure, the diurnal and semidiurnal tides contributing to the wave-4 are underestimated by the applied approach by 20 % and 26 %, respectively.

The observed wavenumbers in the longitudinal variation of the zonal wind are calculated by performing a Fourier transform on the data while a two-dimensional (2-D) Fourier transform yields the diurnal and semidiurnal spectra embedded in the zonal wind at CHAMP altitude [*Häusler and Lühr*, 2009]. Here, a 24 h local time coverage is a mandatory requirement to apply a 2-D Fourier transform. This is satisfied for each month when combining five years of CHAMP data.

Due to the large CHAMP data set available, the applied method to deduce the tidal signatures in the zonal wind renders small formal uncertainties between 0.1 m/s and 0.2 m/s with largest values occurring around June solstice. This effect can be

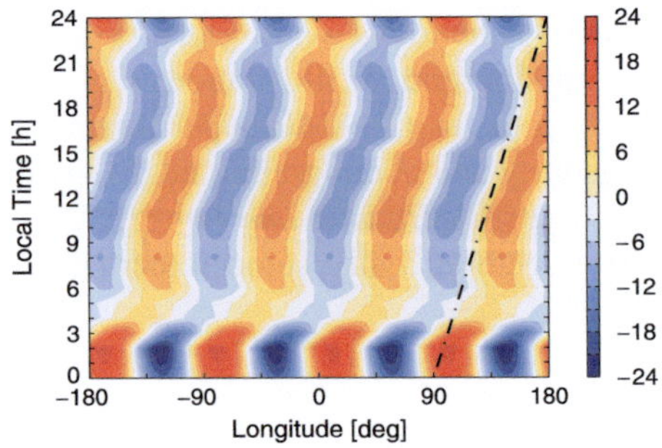

Fig. 26.11 Longitude vs. local time distribution of the observed wave-4 structure in the zonal wind [m/s] for August. The *black guiding line* is indicating a phase propagation of 90° to the east within 24 h LT. Modified version of Fig. 2 in *Häusler and Lühr* [2009]

related to the reduced thermospheric density during the middle of the year which causes a larger scatter of wind data. The achieved precision of a single wind velocity reading derived from CHAMP measurements amounts to about 20 m/s as stated in the appendix of *Liu et al.* [2006].

26.4.2 Tidal Spectra

When investigating the longitudinal variation of the CHAMP zonal wind, a four-peaked longitudinal structure dominating the spectrum during some local times was found [*Häusler et al.*, 2007]. The evolution of the wave-4 structure over the course of a day is depicted in Fig. 26.11 for the month of August. Clearly visible is a shift of 90° to the east within 24 h LT of the wave-4 pattern. This distinct phase shift is attributable to DE3 [*Häusler and Lühr*, 2009]. While the months of July–September exhibit a very strong wave-4 signature with the DE3 denoting phase shift, the signal is reduced or even vanishes in the remaining months. Applying a 2-D Fourier transform on the data, DE3 was identified as the leading source for the observed wave-4 structure in the CHAMP zonal wind [*Häusler and Lühr*, 2009].

For illustration, Fig. 26.12 displays the diurnal tidal spectrum of the zonal wind observed along the magnetic and along the geographic equator during the years 2002–2005. Owing to the fact that the zonal mean was removed before applying the Fourier transform, the migrating tide (DW1) is filtered out and hence does not appear in the figure. Inspecting the diurnal tidal spectra in Fig. 26.12, overall larger amplitudes are observed along the magnetic equator (left panel). The strongest components are DW2 and D0 peaking in December. These tides are probably generated in-situ via the nonlinear interaction of SPW1 and DW1. The DE3 tide exhibits a maximum in July–September and a minor maximum in March/April explaining the observed annual variation of the identified wave-4 pattern. A puzzling result is revealed for the annual behavior of the DE2 and DE3 tides within both coordinate systems. Though the generation mechanism is the same for both tides, DE3 amplitudes maximize in the magnetic reference frame, while DE2 amplitudes are slightly larger in the geographic reference frame.

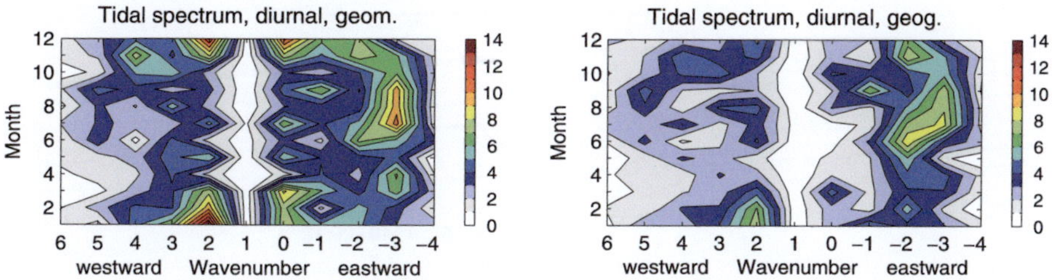

Fig. 26.12 Diurnal tidal spectra of the zonal wind [m/s] along the magnetic equator (*left panel*) and the geographic equator (*right panel*) observed by CHAMP during the years 2002–2005. Positive (negative) wavenumbers correspond to westward (eastward) propagating tides. Modified version of Fig. 4 in *Häusler and Lühr* [2009]

26.4.3 CHAMP/TIME-GCM Tidal Comparison

Besides the HME modeling approach, there exists a variety of numerical atmospheric models which also include atmospheric tides. One of them is the thermosphere-ionosphere-mesosphere-electrodynamics general circulation model (TIME-GCM) developed at the National Center for Atmospheric Research (NCAR). TIME-GCM is a three-dimensional, time-dependent global grid point model that calculates the dynamics and electrodynamics as well as neutral gas heating, photoionization and the compositional structure of the middle and upper atmosphere and ionosphere. For detailed information on the model the reader is referred to *Roble and Ridley* [1994], *Roble* [1995, 1996] and references therein. *Hagan et al.* [2007, 2009] showed that DE3 is capable of propagating well into the upper thermosphere. Having identified DE3 as the main source for the observed longitudinal wave-4 pattern in CHAMP zonal wind, a direct comparison of tidal signals in zonal wind at upper thermospheric altitudes observed in situ by CHAMP and obtained from TIME-GCM simulations was carried out. The goal was to quantify the performance of TIME-GCM regarding nonmigrating tides in the upper thermosphere. For the comparison, the model was run for the 15th day of March, July, September, and December until it settled to a diurnally reproducible state. In the beginning, the solar flux F10.7 value was set to 75 sfu (solar flux unit, sfu $= 10^{-22}\ \mathrm{m}^{-2}\,\mathrm{Hz}^{-1}$) reflecting solar minimum conditions. Important to mention is the fact that TIME-GCM inherently calculates the atmospheric tides that are excited by the absorption of UV and EUV radiation in the middle and upper atmosphere. Yet it cannot account for tides excited in the troposphere and lower stratosphere due to the fact that the model does not extend to the ground [*Hagan et al.*, 2007]. However, this absence can be compensated by perturbing the TIME-GCM lower boundary with the results of the global scale wave model (GSWM) [*Hagan and Roble*, 2001] which can account for tides driven outside the model domain [*Hagan and Forbes*, 2002, 2003]. Here, the focus is set on DE3 and DE2 only since they are of tropospheric origin. More details of the comparison can be found in *Häusler et al.* [2010].

Figure 26.13 shows the intra-annual variation of the DE3 and the DE2 tide as observed by CHAMP and predicted by TIME-GCM. As for CHAMP, the displayed

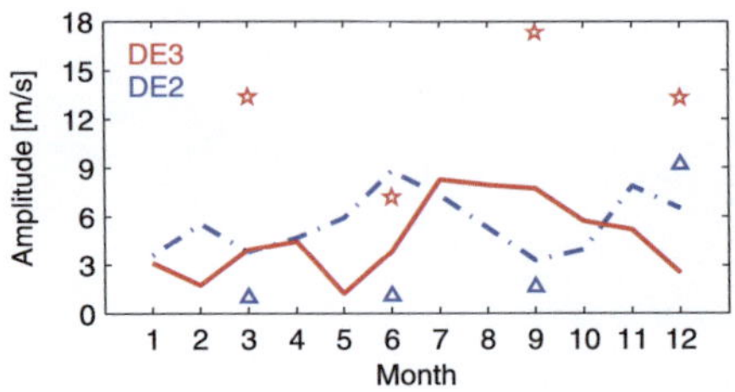

Fig. 26.13 Comparison between the CHAMP DE3 (*solid*) and DE2 (*dashed*) tidal amplitudes with the TIME-GCM DE3 (*pentagram*) and DE2 (*triangle*) tidal amplitudes obtained from the March, June, September, and December model runs under solar quiet conditions (F10.7 = 75 sfu). With kind permission from Springer Science+Business Media: modified version of Fig. 25.5 in *Häusler and Lühr* [2011]

intra-annual variation of DE2 and DE3 amplitudes corresponds to the ones shown in Fig. 26.12, right panel. It is visible that TIME-GCM exhibits a shortcoming in simulating the observed intra-annual variation for both tides. The model overestimates the strength of DE3 at CHAMP altitude in all simulated months with the largest deviation occurring in December when the model amplitude is five times larger than the observed one. The opposite is discernible for the DE2 tide. Mostly the modeled DE2 exhibits amplitudes less than 1.6 m/s and only in December a significant contribution is detectable.

The discrepancies for DE2 between the model and observations can be attributed to insufficient lower boundary forcing. The GSWM DE2 amplitudes reported by *Hagan and Forbes* [2002] are also comparatively weak. Inaccuracies in the model dissipation schemes could be a factor that leads to the high DE3 amplitudes in TIME-GCM [*Hagan et al.*, 2009]. In order to overcome these discrepancies, the TIME-GCM lower boundary conditions are adjusted with new GSWM forcing. The new forcing includes updated and more realistic background winds and temperatures. These are based on SABER observations, along with radiative and latent heating rates based on ISCCP (International Satellite Cloud Climatology Project) and TRMM (Tropical Rainfall Measuring Mission) data [*Zhang et al.*, 2010]. According to *Zhang et al.* [2010], the implemented updates produce significant improvements for the tidal temperature in the MLT region. The effects on the upper thermospheric results still need to be explored. Nevertheless, good agreement was found between TIME-GCM and CHAMP with respect to solar flux dependence of the various tides which is one of the topics of the next section.

26.4.4 Inter-annual and Solar Cycle Variability of DE3

TIME-GCM simulations show that the amplitudes of DW2 and D0 increase with increasing solar flux in June, while the DE3 amplitude increases for decreasing solar flux conditions. The predicted dependence on solar flux was also confirmed by the CHAMP observations [*Häusler et al.*, 2010]. Here the inter-annual and solar

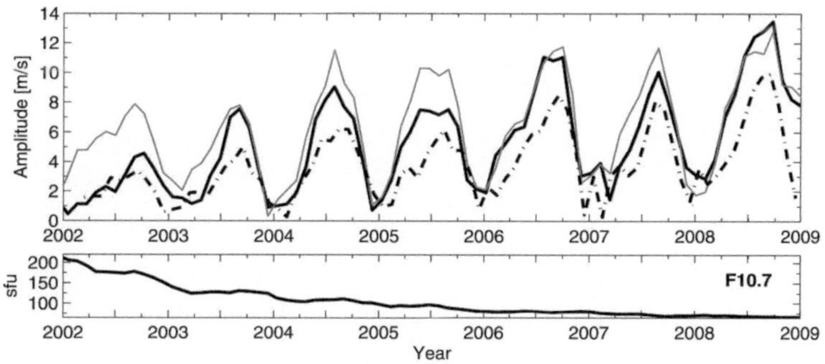

Fig. 26.14 (*Top*) DE3 zonal wind amplitudes from CHAMP along the geographic equator (*solid black line*) and the magnetic equator (*thin gray line*) and from the HME analysis at 390 km altitude (*dash-dot line*) averaged between 5°N and 5°S. (*Bottom*) trend of the F10.7 index for the time period 2002–2008

cycle variability of DE3 shown in Fig. 26.14 for the time period 2002–2008 shall be explored in more detail. Clear to see is the distinct annual variation of DE3 with peak amplitudes around the months of July/August/September. But on top of it, the same quasi-2-year pattern is observed that was already identified in the zonal wind at 105 km (cf. Sect. 26.2.2). Whatever might be the cause for the 2-year modulation in the MLT, its signal also manifests itself in the upper thermosphere depicting another example for the strong coupling between the various atmospheric layers.

Further shown in Fig. 26.14 are the results for the HME zonal wind predictions extended to 390 km altitude (cf. Sect. 26.2.4 and Fig. 26.8). Considering the good agreement between CHAMP DE3 zonal wind amplitudes and HME analysis, *Oberheide et al.* [2009] conclude that within the limits of uncertainties the DE3 tidal winds at CHAMP altitude are fully attributable to troposphere forcing. Likewise TIME-GCM supports direct DE3 tidal upward propagation from the troposphere to the upper thermosphere [*Hagan et al.*, 2007, 2009]. Yet an open issue is the question why stronger DE3 amplitudes are observed along the magnetic equator compared to the geographic equator [*Häusler and Lühr*, 2009]. Examining Fig. 26.14 a little closer, it is found that the difference between the two equators is vanishing with time and finally in 2008 stronger amplitudes are observed along the geographic equator. Probably an additional electrodynamical coupling between ions and neutrals which depends on solar flux level has to be considered to explain this surprising result.

The bottom panel of Fig. 26.14 shows the magnitude of the solar flux within the 7 years of CHAMP measurements that coincide with the declining phase of solar cycle 23. With every year, the DE3 amplitudes are increasing steadily from about 4 m/s in 2002 to roughly 14 m/s in 2008. Conversely, the DE3 amplitudes in the MLT region show no solar cycle dependence (cf. Fig. 26.16). Reduced dissipation in the lower thermosphere is believed to be responsible for the increasing DE3 amplitudes with decreasing solar flux (cf. Sect. 26.3.2).

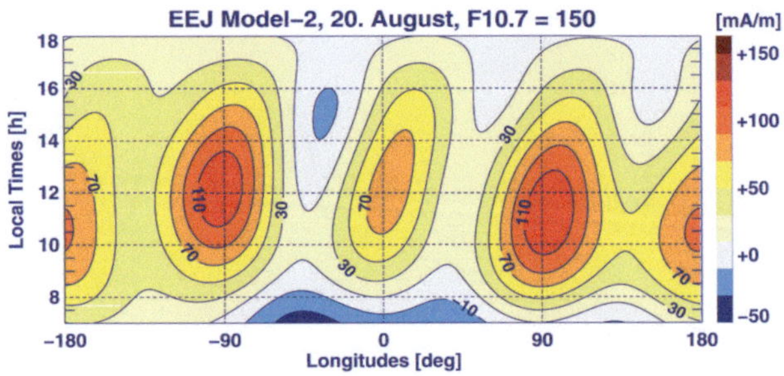

Fig. 26.15 Local time dependence of the longitudinal variation of the EEJ peak current density. Modified version of Fig. 2 in *Lühr et al.* [2012]. Reprinted with permission from Elsevier

26.5 Ionospheric Implications of Tidal Wave Coupling from Below

The much discussed wave-4 structure also shows footprints in ionospheric quantities such as the electron density [e.g., *Lin et al.*, 2007], ionospheric emissions [*Sagawa et al.*, 2005; *Immel et al.*, 2006] or equatorial electrojet (EEJ) [*England et al.*, 2006] just to name a few. As already mentioned in Sect. 26.3.3, the E-region wind dynamo with its generated electric fields is a well-accepted tidal/ionosphere coupling mechanism. However, it is quite challenging to make systematic electric field measurements directly in the E-region. An indirect way of probing the electric field near the magnetic equator is measuring the intensity of the equatorial electrojet. The equatorial electrojet, itself, is an intense current in the E-region confined to a narrow band along the magnetic equator on the dayside. It is primarily driven by the zonal polarization electric field and zonal wind in the E-region [*Heelis*, 2004] and it vanishes at night due to the ceasing E-region conductivity during the dark hours. With the help of the empirical equatorial electrojet model (EEJM) developed by *Alken and Maus* [2007], it is now possible for the first time to visualize the longitudinal variation of the EEJ for various seasons, local times, and solar flux intensities on a global scale. The EEJM is a climatological model based on more than 95 000 EEJ passes coming from the satellites Ørsted, CHAMP, and SAC-C.

Using the EEJM-1, *Lühr et al.* [2008] analyzed the longitudinal EEJ variation and, not surprisingly, found a four-peaked longitudinal structure as well. Figure 26.15 shows the local time dependence of the longitudinal variation of the EEJ as deduced from the EEJM-2 for a day in August. The typical wave-4 pattern is clearly visible with an eastward tilt of the phase front matching that of the DE3. EEJM-2 is an updated model version which also includes the influence of the moon phase as well as the seasonal difference between vernal and autumnal equinoxes. Both are missing in EEJM-1. Because the EEJ is confined to daylight hours, it is not possible to apply a 2-D Fourier transform to extract the single tidal components. Instead *Lühr et al.* [2008] used a numerical fitting procedure to quantify the amplitudes and phases of the tidal components.

Revisiting the subject with the improved EEJM-2 model, *Lühr et al.* [2012] report a similar annual variation of the DE3 amplitude in EEJ as it is observed in the zonal wind at 105 km altitude by TIDI, providing a strong argument for the direct modulation of the EEJ intensity by the zonal wind. In the month of August, the

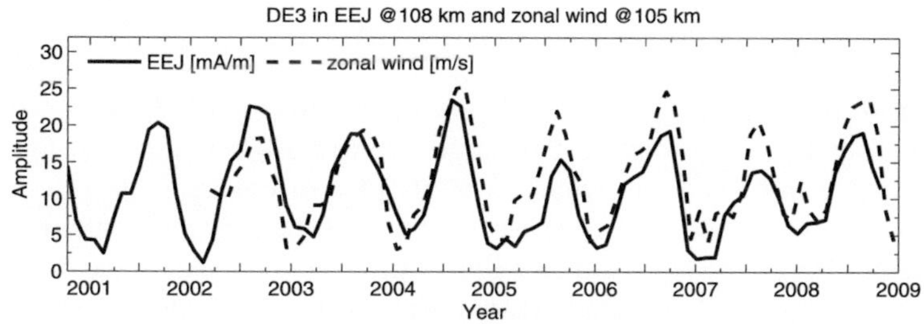

Fig. 26.16 Inter-annual variation of the DE3 amplitude in the equatorial electrojet (*solid line*) as observed by CHAMP and in the zonal wind (*dashed line*) as observed by TIDI

EEJ DE3 amplitude reaches 50 % of the background current density. Figure 26.16 displays the inter-annual variation of the DE3 tide derived from CHAMP EEJ data compared to the TIDI zonal wind. Again, the match between the two quantities is remarkable. Even the two year peak amplitude modulation is discernible in the EEJ data. As reported in Sect. 26.3.2, there is no significant solar cycle dependence for the DE3 in the zonal wind. Yet, looking at Fig. 26.16 a slight decrease in DE3 amplitude with decreasing solar flux might be detected for the EEJ which could be attributed to reduced ionospheric conductivity during solar minimum.

The vertical plasma drift [*Fejer et al.*, 2008] carries the DE3 signal to higher altitudes and hence it is not unexpected to find a wave-4 structure in the equatorial ionization anomaly (EIA). *Lühr et al.* [2012] found a modulation of 20 % for the vertical plasma drift by DE3 and 25 % for the equatorial ionization anomaly. Thereby the EIA is characterized by the crest-to-trough ratio of the electron density profile along the CHAMP orbit which gives a single number representing the strength of the EIA [*Lühr et al.*, 2007]. The close relation between the vertical plasma drift and the crest-to-trough ratio is also confirmed by the very similar variations of DE2 and DE3 tides in the two quantities. Interestingly, the phases of CTR are systematically delayed by 1–2 hours with respect to those of the plasma drift. This is consistent with the time it takes to lift the plasma to CHAMP altitudes [*Stolle et al.*, 2008]. As expected, tidal signals of the EEJ and vertical plasma drift exhibit about the same phase values [*Lühr et al.*, 2012]. This is in accordance to a common dynamo process in the ionospheric E-region.

Proper DE2 and DE3 tidal signals in ionospheric quantities disappear after sunset. This is evident in the EEJ, plasma drift and also F-region dynamo current [*Park et al.*, 2011]. It is regarded as an additional evidence for an upward propagation of DE2 and DE3 tidal signal which is coupled into the ionospheric plasma at E-region heights. After sunset the E-region quickly decays and the coupling ceases. Waves excited in situ in the ionosphere do not seem to cause the observed tidal components.

Longitudinal structures of the equatorial ionization anomaly seem to behave slightly differently. Wavenumber 4 patterns can be observed in the pre-midnight sector [*Immel et al.*, 2006]. Closer inspection, however, reveals that their characteristics after sunset differ from that of the DE3 tide, both in phase propagation and in annual variation [e.g., *Lühr et al.*, 2007; *Liu et al.*, 2009]. So far there is no conclusive explanation for the night time EIA longitudinal structure.

26.6 Summary

Within the framework of the German priority program CAWSES, contributions were made to the understanding that large-scale weather systems in the tropics do significantly influence the dynamics aloft. This discovery is an essential new realization and important for the "space weather" of the ionosphere-thermosphere-mesosphere (ITM) system. Figure 26.1 illustrates the knowledge of the tropospheric tidal coupling into the ITM system. Latent heat release due to deep convective processes in the tropics excites tidal waves propagating upward into the mesosphere, lower thermosphere (MLT) region reaching maximum amplitudes there. The predominance of wave-4 topography and land-sea differences in the tropics favors the DE3 tide which is dominating by far the tidal spectrum in the MLT over large parts of the year. In this region, the presence of DE3 is observed in various quantities such as temperature, wind, or nitric oxide. Because the measurements are taken by satellites in a quasi constant local time frame, DE3 manifests itself as a longitudinal wave-4 pattern as shown in Fig. 26.1 (see also Fig. 26.11). The DE3 tidal wind modulation of the E-region electric fields is the governing process behind the tidal coupling into the ionosphere (yellow arrows). The generated electric fields modulate in turn the vertical plasma drifts that finally control the EIA structure as illustrated by the 135.6 nm airglow brightness measurements. The vertical plasma drift at 600 km altitude derived from ROCSAT-1 measurements still exhibits a prominent longitudinal wave-4 pattern as does the equatorial electrojet (EEJ) in the E-region. The longitudinal distribution of the wave-4 peaks is similarly present in location, local time, and seasons suggesting a close connection between the vertical plasma drift and the equatorial electrojet. The crest-to-trough ratio of the EIA tracks closely the tidal variations of the vertical plasma drift but everything appears 1–2 hours later. Ionospheric signatures of DE3 tides are in general limited to daylight hours, supporting the fact that the coupling between neutrals and ions takes place primarily in the E-region. Nevertheless, it was also shown that SE2 in the meridional wind contributes to the neutral-ion coupling at F-region heights. Thus the observed "wave-4" pattern cannot be attributed to the DE3 alone though DE3 remains the leading contributer. The DE3 tidal signal is coupled directly into the upper thermospheric zonal wind as indicated by the long green arrow. An additional connection, if any, between the ionospheric wave-4 pattern in the F-region and the neutral zonal wind at this altitude still has to be resolved and is thus marked with a red question mark in Fig. 26.1. In order to bridge the gap between the observations at 100 km altitude and 400 km altitude, an empirical fit model based on Hough Mode Extensions was developed.

Acknowledgements KH and HL thank E. Doornbos for recalibrating the CHAMP accelerometer data. KH thanks M. E. Hagan and A. Maute for the help with TIME-GCM. JO thanks J. M. Forbes for providing HMEs, and X. Zhang and S. L. Bruinsma for providing CHAMP neutral density tidal diagnostics. KH was supported by DFG CAWSES grants LU 446/9, LU 446/10 whose results are presented in Sects. 26.4 and 26.5. JO was supported by DFG CAWSES grants OB 299/2-1, OB 299/2-2, OB 299/2-3 whose results are presented in Sects. 26.2 and 26.3. RK was partly supported by DFG grant OB 299/2-3.

References

Alken, P., & Maus, S. (2007). Spatio-temporal characterization of the equatorial electrojet from CHAMP, Ørsted, and SAC-C satellite magnetic measurements. *Journal of Geophysical Research, 112*, A09305. doi:10.1029/2007JA012524.

Barth, C. A., Mankoff, K. D., Bailey, S. M., & Solomon, S. C. (2003). Global observations of nitric oxide in the thermosphere. *Journal of Geophysical Research, 108*(A1), 1027. doi:10.1029/2002JA009458.

Burrage, M., Vincent, R. A., Mayr, H. G., Skinner, W. R., Arnold, N. F., & Hays, P. B. (1996). Long-term variability in the equatorial mesosphere and lower thermosphere zonal wind. *Journal of Geophysical Research, 101*(D8), 12847–12854. doi:10.1029/96JD00575.

Doornbos, E., van den IJssel, J., Lühr, H., Förster, M., & Koppenwallner, G. (2010). Neutral density and crosswind determination from arbitrarily oriented multiaxis accelerometers on satellites. *Journal of Spacecraft and Rockets, 47*(4), 580–589. doi:10.2514/1.48114.

Ekanayke, E. M. P., Aso, T., & Miyahara, S. (1997). Background wind effect on propagation of nonmigrating diurnal tides in the middle atmosphere. *Journal of Atmospheric and Solar-Terrestrial Physics, 59*(4), 401–429. doi:10.1016/S1364-6826(96)00012-0.

England, S. L., Maus, S., Immel, T. J., & Mende, S. B. (2006). Longitude variation of the E-region electric fields caused by atmospheric tides. *Geophysical Research Letters, 33*, L21105. doi:10.1029/2006GL027465.

England, S. L., Immel, T. J., Huba, J. D., Hagan, M. E., Maute, A., & DeMajistre, R. (2010). Modeling of multiple effects of atmospheric tides on the ionosphere: an examination of possible coupling mechanisms responsible for the longitudinal structure of equatorial ionosphere. *Journal of Geophysical Research, 115*, A05308. doi:10.1029/2009JA014894.

Fejer, B. G., Jensen, J. W., & Su, S.-Y. (2008). Quiet time equatorial F region vertical plasma drift model derived from ROCSAT-1 observations. *Journal of Geophysical Research, 113*, A05304. doi:10.1029/2007JA012801.

Forbes, J. M., & Hagan, M. E. (1982). Thermospheric extensions of the classical expansion functions for semidiurnal tides. *Journal of Geophysical Research, 87*(A7), 5253–5259. doi:10.1029/JA087iA07p05253.

Forbes, J. M., Zhang, X., Talaat, E. R., & Ward, W. (2003). Nonmigrating diurnal tides in the thermosphere. *Journal of Geophysical Research, 108*, 1033. doi:10.1029/2002JA009262.

Forbes, J. M., Russell, J., Miyahara, S., Zhang, X., Palo, S., Mlynczak, M., Mertens, C. J., & Hagan, M. E. (2006). Troposphere-thermosphere tidal coupling as measured by the SABER instrument on TIMED during July–September 2002. *Journal of Geophysical Research, 111*, A10S06. doi:10.1029/2005JA011492.

Forbes, J. M., Zhang, X., Palo, S., Russell, J., Mertens, M., & Mlynczak, C. J. (2008). Tidal variability in the ionospheric dynamo region. *Journal of Geophysical Research, 113*, A02310. doi:10.1029/2007JA012737.

Forbes, J. M., Bruinsma, S. L., Zhang, X., & Oberheide, J. (2009). Surface-exosphere coupling due to thermal tides. *Geophysical Research Letters, 36*, L15812. doi:10.1029/2009GL038748.

Hagan, M. E., & Forbes, J. M. (2002). Migrating and nonmigrating diurnal tides in the middle and upper atmosphere excited by tropospheric latent heat release. *Journal of Geophysical Research, 107*(D24), 4754. doi:10.1029/2001JD001236.

Hagan, M. E., & Forbes, J. M. (2003). Migrating and nonmigrating semidiurnal tides in the upper atmosphere excited by tropospheric latent heat release. *Journal of Geophysical Research, 108*(A2), 1062. doi:10.1029/2002JA009466.

Hagan, M. E., & Roble, R. G. (2001). Modeling the diurnal tidal variability with the National Center for Atmospheric Research thermosphere-ionosphere-mesosphere-electrodynamics general circulation model. *Journal of Geophysical Research, 106*(A11), 24869–24882. doi:10.1029/2001JA000057.

Hagan, M. E., Maute, A., Roble, R. G., Richmond, A. D., Immel, T. J., & England, S. L. (2007). Connections between deep tropical clouds and the Earth's ionosphere. *Geophysical Research Letters, 34*, L20109. doi:10.1029/2007GL030142.

Hagan, M. E., Maute, A., & Roble, R. G. (2009). Tropospheric tidal effects on the middle and upper atmosphere. *Journal of Geophysical Research, 114*, A01302. doi:10.1029/2008JA013637.

Häusler, K., & Lühr, H. (2009). Nonmigrating tidal signals in the upper thermospheric zonal wind at equatorial latitudes as observed by CHAMP. *Annales Geophysicae, 27*, 2643–2652. doi:10.5194/angeo-27-2643-2009.

Häusler, K., & Lühr, H. (2011). Longitudinal variations of the thermospheric zonal wind induced by nonmigrating tides as observed by CHAMP. In M. A. Abdu, D. Pancheva & A. Bhattacharyya (Eds.), *IAGA special sopron book series: Vol. 2. Aeronomy of the Earth's atmosphere and ionosphere*. Berlin: Springer. doi:10.1007/978-94-007-0326-1_25.

Häusler, K., Lühr, H., Rentz, S., & Köhler, W. (2007). A statistical analysis of longitudinal dependences of upper thermospheric zonal winds at dip equator latitudes derived from CHAMP. *Journal of Atmospheric and Solar-Terrestrial Physics, 69*, 1419–1430. doi:10.1016/j.jastp.2007.04.004.

Häusler, K., Lühr, H., Hagan, M. E., Maute, A., & Roble, R. G. (2010). Comparison of CHAMP and TIME-GCM nonmigrating tidal signals in the thermospheric zonal wind. *Journal of Geophysical Research, 115*, D00I08. doi:10.1029/2009JD012394.

Hays, P. B., Wu, D. L., Burrage, M. D., Gell, D. A., Grassl, H. J., Lieberman, R. S., Marshall, A. R., Morton, Y. T., Ortland, D. A., & Skinner, W. R. (1994). Observations of the diurnal tide form space. *Journal of the Atmospheric Sciences, 51*, 3077–3093. doi:10.1175/1520-0469(1994)051<3077:OOTDTF>2.0.CO;2.

Heelis, R. A. (2004). Electrodynamics in the low and middle latitude ionosphere: a tutorial. *Journal of Atmospheric and Solar-Terrestrial Physics, 66*, 825–838. doi:10.1016/j.jastp.2004.01.034.

Huba, J. D., Joyce, G., & Fedder, J. A. (2000). Sami2 is Another model of the ionosphere (SAMI2): a new low-latitude ionosphere model. *Journal of Geophysical Research, 105*(A10), 23035–23053. doi:10.1029/2000JA000035.

Immel, T. J., Sagawa, E., England, S. L., Henderson, S. B., Hagan, M. E., Mende, S. B., Frey, H. U., Swenson, C. M., & Paxton, L. J. (2006). Control of equatorial ionospheric morphology by atmospheric tides. *Geophysical Research Letters, 33*, L15108. doi:10.1029/2006GL026161.

Jarvis, M. J. (2001). Bridging the atmospheric divide. *Science, 293*(5538), 2218–2219. doi:10.1126/science.1064467.

Jin, H., Miyoshi, Y., Fujiwara, H., & Shinagawa, H. (2008). Electrodynamics of the formation of ionospheric wave number 4 longitudinal structure. *Journal of Geophysical Research, 113*, A09307. doi:10.1029/2008JA013301.

Killeen, T. L., Wu, Q., Solomon, S. C., Ortland, D. A., Skinner, W. R., Niciejewski, R. J., & Gell, D. A. (2006). TIMED Doppler interferometer: overview and recent results. *Journal of Geophysical Research, 111*, A10S01. doi:10.1029/2005JA011484.

Lieberman, R. S., Oberheide, J., Hagan, M. E., Remsberg, E. E., & Gordley, L. L. (2004). Variability of diurnal tides and planetary waves during November 1978–May 1979. *Journal of Atmospheric and Solar-Terrestrial Physics, 66*, 517–528. doi:10.1016/j.jastp.2004.01.006.

Lieberman, R. S., Riggin, D. M., Ortland, D. A., Nesbitt, S. W., & Vincent, R. A. (2007). Variability of mesospheric diurnal tides and tropospheric diurnal heating during 1997–1998. *Journal of Geophysical Research, 112*, D20110. doi:10.1029/2007JD008578.

Lin, C. H., Wang, W., Hagan, M. E., Hsiao, C. C., Immel, T. J., Hsu, M. L., Liu, J. Y., Paxton, L. J., Fang, T. W., & Liu, C. H. (2007). Plausible effect of atmospheric tides on the equatorial ionosphere observed by the FORMOSAT-3/COSMIC: three-dimensional electron density structures. *Geophysical Research Letters, 34*, L11112. doi:10.1029/2007GL029265.

Lindzen, R. S., Hong, S. S., & Forbes, J. M. (1977). *Semidiurnal Hough mode extensions in the thermosphere and their application* (Memo. Rep. 3442). Nav. Res. Lab., Washington.

Liu, H., Lühr, H., Watanabe, S., Köhler, W., Henize, V., & Visser, P. (2006). Zonal winds in the equatorial upper thermosphere: decomposing the solar flux, geomagnetic activity, and seasonal dependencies. *Journal of Geophysical Research, 111*, A07307. doi:10.1029/2005JA011415.

Liu, H., Yamamoto, M., & Lühr, H. (2009). Wave-4 pattern of the equatorial mass density anomaly: a thermospheric signature of tropical deep convection. *Geophysical Research Letters, 36*, L18104. doi:10.1029/2009GL039865.

Lühr, H., Häusler, K., & Stolle, C. (2007). Longitudinal variation of F region electron density and thermospheric zonal wind caused by atmospheric tides. *Geophysical Research Letters*, *34*, L16102. doi:10.1029/2007GL030639.

Lühr, H., Rother, M., Häusler, K., Alken, P., & Maus, S. (2008). The influence of nonmigrating tides on the longitudinal variation of the equatorial electrojet. *Journal of Geophysical Research*, *113*, A08313. doi:10.1029/2008JA013064.

Lühr, H., Rother, M., Häusler, K., Fejer, B., & Alken, P. (2012). Direct comparison of non-migrating tidal signatures in the electrojet, vertical plasma drift and equatorial ionization anomaly. *Journal of Atmospheric and Solar-Terrestrial Physics*, *75–76*, 31–43. doi:10.1016/j.jastp.2011.07.009.

McLandress, C., Shepherd, G. G., & Solheim, B. H. (1996). Satellite observations of thermospheric tides: results from the Wind Imaging Interferometer on UARS. *Journal of Geophysical Research*, *101*(D2), 4093–4114. doi:10.1029/95JD03359.

Mlynczak, M., Martin-Torres, F. J., Russell, J., Beaumont, K., Jacobson, S., Kozyra, J., Lopez-Puertas, M., Funke, B., Mertens, C., Gordley, L., Picard, R., Winick, J., Wintersteiner, P., & Paxton, L. (2003). The natural thermostat of nitric oxide emission at 5.3 μm in the thermosphere observed during the solar storms of April 2002. *Geophysical Research Letters*, *30*(21), 2100. doi:10.1029/2003GL017693.

Oberheide, J., & Forbes, J. M. (2008a). Tidal propagation of deep tropical cloud signatures into the thermosphere from TIMED observations. *Geophysical Research Letters*, *35*, L04816. doi:10.1029/2007GL032397.

Oberheide, J., & Forbes, J. M. (2008b). Thermospheric nitric oxide variability induced by nonmigrating tides. *Geophysical Research Letters*, *35*, L16814. doi:10.1029/2008GL034825.

Oberheide, J., Hagan, M. E., & Roble, R. G. (2003). Tidal signatures and aliasing in temperature data from slowly precessing satellites. *Journal of Geophysical Research*, *108*(A2), 1055. doi:10.1029/2002JA009585.

Oberheide, J., Wu, Q., Ortland, D. A., Killeen, T. L., Hagan, M. E., Roble, R. G., Niciejewski, R. J., & Skinner, W. R. (2005). Non-migrating diurnal tides as measured by the TIMED Doppler interferometer: preliminary results. *Advances in Space Research*, *35*, 1911–1917. doi:10.1016/j.asr.2005.01.063.

Oberheide, J., Wu, Q., Killeen, T. L., Hagan, M. E., & Roble, R. G. (2006). Diurnal nonmigrating tides from TIMED Doppler Interferometer wind data: monthly climatologies and seasonal variations. *Journal of Geophysical Research*, *111*, A10S03. doi:10.1029/2005JA011491.

Oberheide, J., Wu, Q., Killeen, T. L., Hagan, M. E., & Roble, R. G. (2007). A climatology of nonmigrating semidiurnal tides from TIMED Doppler Interferometer (TIDI) wind data. *Journal of Atmospheric and Solar-Terrestrial Physics*, *69*, 2203–2218. doi:10.1016/j.jastp.2007.05.010.

Oberheide, J., Forbes, J. M., Häusler, K., Wu, Q., & Bruinsma, S. L. (2009). Tropospheric tides from 80 to 400 km: propagation, inter-annual variability, and solar cycle effects. *Journal of Geophysical Research*, *114*, D00I05. doi:10.1029/2009JD012388.

Oberheide, J., Forbes, J. M., Zhang, X., & Bruinsma, S. L. (2011a). Wave-driven variability in the ionosphere-thermosphere-mesosphere system from TIMED observations: what contributes to the "wave 4"? *Journal of Geophysical Research*, *116*, A01306. doi:10.1029/2010JA015911.

Oberheide, J., Forbes, J. M., Zhang, X., & Bruinsma, S. L. (2011b). Climatology of upward propagating diurnal and semidiurnal tides in the thermosphere. *Journal of Geophysical Research*, *116*, A11306. doi:10.1029/2011JA016784.

Park, J., Lühr, H., & Min, K. W. (2011). Climatology of the inter-hemispheric field-aligned current system in the equatorial ionosphere as observed by CHAMP. *Annales Geophysicae*, *29*, 573–582. doi:10.5194/angeo-29-573-2011.

Ren, Z., Wan, W., Liu, L., & Xiong, J. (2011). Simulated longitudinal variations in the lower thermospheric nitric oxide induced by nonmigrating tides. *Journal of Geophysical Research*, *116*, A04301. doi:10.1029/2010JA016131.

Roble, R. G. (1995). Energetics of the mesosphere and thermosphere. In R. M. Johnson & T. L. Killeen (Eds.), *Geophys. monogr. ser.: Vol. 87. The upper mesosphere and lower thermosphere: a review of experiment and theory* (pp. 1–21). Washington: AGU.

Roble, R. G. (1996). The NCAR thermosphere-ionosphere-mesosphere-electrodynamics general circulation model (TIME-GCM). In R. W. Schunk (Ed.), *STEP handbook on ionospheric models* (pp. 281–288). Logan: Utah State University.

Roble, R. G., & Ridley, E. C. (1994). A thermosphere-ionosphere-mesosphere-electrodynamics general circulation model (time-GCM): equinox solar cycle minimum simulations (30–500 km). *Geophysical Research Letters*, *21*, 417–420. doi:10.1029/93GL03391.

Russell, J. M., Mlynczak, M. G., Gordley, L. L., Transock, J., & Espin, R. (1999). An overview of the SABER experiment and preliminary calibration results. *Proceedings - SPIE*, *3756*, 277–288. doi:10.1117/12.366382.

Sagawa, E., Immel, T. J., Frey, H. U., & Mende, S. B. (2005). Longitudinal structure of the equatorial anomaly observed by IMAGE/FUV. *Journal of Geophysical Research*, *110*, A11302. doi:10.1029/2004JA010848.

Stolle, C., Manoj, C., Lühr, H., Maus, S., & Alken, P. (2008). Estimating the day time Equatorial Ionization Anomaly strength from electric field. *Journal of Geophysical Research*, *113*, A09310. doi:10.1029/2007JA012781.

Svoboda, A., Forbes, J. M., & Miyahara, S. (2005). A space-based climatology of diurnal MLT tidal winds, temperatures and densities from UARS wind measurements. *Journal of Atmospheric and Solar-Terrestrial Physics*, *67*(16), 1533–1543. doi:10.1016/j.jastp.2005.08.018.

Ward, W. E., Oberheide, J., Goncharenko, L. P., Nakamura, T., Hoffmann, P., Singer, W., Chang, L. C., Du, J., Wang, D.-Y., Batista, P., Clemesha, B., Manson, A. H., Meek, C., Riggin, D. M., She, C.-Y., Tsuda, T., & Yuan, T. (2010). On the consistency of model and ground-based and satellite observations of tidal signatures: initial results from the CAWSES tidal campaigns. *Journal of Geophysical Research*, *115*, D07107. doi:10.1029/2009JD012593.

Wu, Q., Ortland, D. A., Killeen, T. L., Roble, R. G., Hagan, M. E., Liu, H.-L., Solomon, S. C., Xu, J., Skinner, W. R., & Niciejewski, R. J. (2008). Global distribution and interannual variations of mesospheric and lower thermospheric neutral wind diurnal tide: 2. Nonmigrating tide. *Journal of Geophysical Research*, *113*, A05309. doi:10.1029/2007JA012543.

Xu, J., Smith, A. K., Liu, H.-L., Yuan, W., Wu, Q., Jiang, G., Mlynczak, M. G., Russell III, J. M., & Franke, S. J. S. J. (2009). Seasonal and quasi-biennial variations in the migrating diurnal tide observed by Thermosphere, Ionosphere, Mesosphere, Energetics and Dynamics (TIMED). *Journal of Geophysical Research*, *114*, D13107. doi:10.1029/2008JD011298.

Zhang, X., Forbes, J. M., & Hagan, M. E. (2010). Longitudinal variation of tides in the MLT region: 2. Relative effects of solar radiative and latent heating. *Journal of Geophysical Research*, *115*, A06317. doi:10.1029/2009JA014898.

Chapter 27
Solar Diurnal Tides in the Middle Atmosphere: Interactions with the Zonal-Mean Flow, Planetary Waves and Gravity Waves

Ulrich Achatz, Fabian Senf, and Norbert Grieger

Abstract The dynamics of solar tides is investigated with regard to variations of the background atmosphere, including planetary waves (PW), and to the interaction with gravity waves (GW). (1) Using a linear model with a clear cause-effect relationship, it is shown that planetary waves play an important role in tidal dynamics, most importantly by inducing non-migrating tidal components from a migrating thermal forcing. (2) Ray-tracing simulations are used to analyze the GW force on the large-scale flow including the solar tides. In comparison to classic GW parameterizations, the inclusion of time-dependence and horizontal refraction leads to a significant decrease of the GW drag.

27.1 Introduction

The diurnal cycle of solar heating represents a forcing of the atmosphere at the diurnal period (24 h) and its higher harmonics. Corresponding large-scale waves, i.e. solar tides, are emitted, with signatures in all dynamic fields, including wind and temperature, propagating upwards into the mesosphere/lower thermosphere (MLT) region [*Chapman and Lindzen*, 1970]. Due to the enormous density decrease tidal amplitudes grow significantly so that they represent a major component of atmospheric variability in the MLT. Tides are very sensitive to the propagation conditions they encounter on their way from the forcing region. Hence, if the mean atmospheric circulation is modified by solar variability this could leave a significant footprint on the tides encountered in the MLT. Moreover, many processes, discussed in this book, by which solar variability could have an impact on the tropopause region and below,

U. Achatz (✉)
Institute for Atmospheric and Environmental Sciences, Goethe University, Frankfurt (Main), Germany
e-mail: achatz@iau.uni-frankfurt.de

F. Senf · N. Grieger
Leibniz-Institute of Atmospheric Physics, Schlossstr. 6, 18225 Kühlungsborn, Germany

F. Senf
e-mail: senf@iap-kborn.de

F.-J. Lübken (ed.), *Climate and Weather of the Sun-Earth System (CAWSES)*, Springer Atmospheric Sciences,
DOI 10.1007/978-94-007-4348-9_27,

rely on dynamic coupling mechanisms between the lower atmosphere and higher regions where the influence of a varying sun acts primarily. It seems important to understand these well. Solar tides, however, are controlled just by such mechanisms so that research on tides might be helpful not only for the detection whether solar variability can influence the lower layers of the atmosphere, but also for a better understanding how this might happen. We here take the chance to summarize recent work within this framework [*Achatz et al.*, 2008, 2010; *Senf and Achatz*, 2011]. Therein the focus mainly is on how much various processes and mechanisms contribute to the tidal signature in the MLT.

One of these is the diurnal cycle of atmospheric heating. Two of the main heating processes subject to a significant diurnal cycle, and contributing to solar tides in the MLT, are the direct absorption of solar radiation by ozone in the stratosphere and by water vapor in the troposphere. This is supplemented by the diurnal cycle of tropospheric latent heat release and convection, mostly over tropical land masses. In propagating from the regions of these source processes to the MLT solar tides encounter propagation conditions which can have a strong impact on them. Here one discriminates between the impact of the zonal mean of the atmosphere on tides, which cannot change their zonal wave number, and the interaction between tides and planetary waves, which can do so. In general, the spatial and time dependence of the signature of the solar tides in any dynamic variable X is given by

$$X(\lambda,\phi,z,t) = \sum_{n=1}^{\infty}\Big\{A_{n,0}(\phi,z)\cos\big(n\Omega t - \Phi^{e}_{n,0}\big) + \sum_{s=1}^{\infty}\big[A^{e}_{n,s}(\phi,z)\cos\big(n\Omega t - s\lambda - \Phi^{e}_{n,s}\big) + A^{w}_{n,s}(\phi,z)\cos\big(n\Omega t + s\lambda - \Phi^{w}_{n,s}\big)\big]\Big\} \quad (27.1)$$

Here λ and ϕ denote the geographic longitude and latitude, respectively. The altitude is given by z, and t is the universal time. The rotation rate of the earth is $\Omega = 2\pi/24$ h. The temporal subharmonics corresponding to $n = 1, 2, 3$ are the diurnal, semidiurnal, and terdiurnal tide, respectively. Each is decomposed into a zonally symmetric part, with zonal wave number $s = 0$, and east- and westward traveling components at zonal wave numbers $s > 0$ with amplitudes $A^{e}_{n,s}$ and $A^{w}_{n,s}$, and phases $\Phi^{e}_{n,s}$ and $\Phi^{w}_{n,s}$, respectively, both depending both on latitude and altitude. For conciseness, a westward or eastward traveling diurnal component at wave number s will be called DWs or DEs, respectively. The name for the corresponding zonally symmetric component is DS0. Since the apparent movement of the sun around the globe is westward, a leading tidal component of each temporal subharmonic is the westward traveling one at zonal wave number $s = n$, called the migrating tide. Its horizontal propagation is synchronous with that of the sun. Since also the tidal forcing by the diurnal cycle of solar heating can be decomposed into migrating and non-migrating components, the question arises which tidal components eventually dominate when the thus forced waves propagate into the MLT. Consider the interaction between tides and large-scale stationary planetary waves in terms of wave-wave interactions, governed by the triad condition $s_2 = s_1 \pm m$, with $s_{1,2}$ being zonal wave numbers of the tide and its forcing, and m the zonal wave number of a planetary wave. Clearly, a planetary wave can lead to the emergence of a non-migrating tide in the MLT from a migrating forcing further below, while the zonal-mean part of the atmosphere, with zonal wave number $m = 0$ cannot do so. Note that these are mechanisms already captured by a linear model where the dynamics of solar tides is represented after a linearization about an atmospheric reference state containing both zonal mean and stationary planetary waves.

While already the above leads to interesting questions, i.e. how much does each tidal forcing contribute to the total tidal signal, and how do interactions between tides and zonal mean or planetary waves influence what finally emerges as tide in the MLT, this is by far not all. Upward propagating small-scale gravity waves (GWs) transport a significant amount of momentum and energy from the lower to the middle atmosphere [*Fritts and Alexander*, 2003]. Again due to the density decrease the GW amplitude rises in the course of their propagation from the troposphere to the MLT so that they eventually become unstable and break. Because solar tides, mainly their horizontal winds, induce changes in the GW propagation and properties, e.g. in GW vertical wave number, the wave breaking itself and the resulting momentum deposition [*Lindzen*, 1981] are periodically modulated. This way results a diurnal variation of the GW force on the large-scale flow which can have a feedback on the tides. This effect of gravity waves on thermal tides has not been well understood so far [*Ortland and Alexander*, 2006]. Many linear models for the description of solar tides incorporate it by simple Rayleigh friction, which most likely is a crude over-simplification, needing either verification or replacement by better approaches.

These issues are addressed in the work reported here. We sketch in Sect. 27.2 some steps and results in the research on the impact of zonal-mean and planetary-wave background variations on solar tides, give an account of new findings on interaction between tides and GWs in Sect. 27.3, and finally summarize in Sect. 27.4. All of the work reported here exclusively addresses the diurnal tide.

27.2 The Seasonal Cycle of the Diurnal Solar Tide in Its Interaction with Zonal-Mean Variations and Planetary Waves

The effect of the interaction between tidal forcing, migrating and non-migrating, and zonal-mean atmosphere or planetary waves on the various diurnal tidal components has been an original focus of research [*Hagan and Roble*, 2001; *Grieger et al.*, 2004, e.g.]. Its modulation by the seasonal cycle represents an important test of our current understanding of tidal dynamics, and therefore has more recently attracted considerable attention. The MLT amplitude of the migrating diurnal tide exhibits a strong semiannual variation with maxima at equinox and minima at solstice. *McLandress* [2002a, 2002b] explains this in terms of the impact of the seasonally varying zonal-mean wind on the tidal propagation conditions. The seasonal cycle of the leading non-migrating tides (DE3, DS0, and DW2) in the MLT has been addressed by *Oberheide et al.* [2005, 2006]. They follow a similar strategy to the one from *Hagan and Roble* [2001] in using a linear model (GSWM) and a general circulation model (TIME-GCM) for simulating the seasonal cycle. The latter has a lower boundary at 30 km altitude, where a migrating diurnal tide was prescribed, as obtained from the linear model, which again has a zonally symmetric background. The TIME-GCM is then integrated with prescribed planetary wave activity at the lower boundary. To some degree the two models complement each other: The linear GSWM captures the direct effect from the non-migrating forcing in a zonally-symmetric background while the TIME-GCM, without non-migrating forcing, can only produce non-migrating tides by an interaction between migrating tides and planetary waves. The linear model is reported to reproduce the seasonal cycle of DE3. This component therefore seems to result directly from a tide being forced by non-migrating heating, due to latent heat release, and its propagation through the zonal-mean background atmosphere. An open question is which part both factors play in comparison in determining the seasonal cycle of DE3. In the same studies both the linear GSWM and the nonlinear TIME-GCM yield seasonally varying components DS0 and DW2. Therefore, the excitation of DS0 and DW2 seem to be controlled by two processes, the interaction of migrating forcing and planetary waves, and the direct non-migrating forcing in interaction with a varying zonal mean in the background atmosphere. However, the respective role of variations in the zonal-mean background and the non-migrating forcing is not clarified within GSWM. Moreover, the TIME-GCM integrations are fully nonlinear; corresponding feedbacks are not excluded. It is therefore not possible to simply add the GSWM result to the one from the GCM so as to obtain the complete tidal signal. A more conclusive picture could arise from a linear model with a background atmosphere incorporating the most important stationary planetary waves. Such analysis have been done by *Achatz et al.* [2008].

Their model uses the primitive equations, linearized about a time-independent, but fully three-dimensional, background atmosphere, with a spectral discretization in the horizontal (T14) and 60 hybrid-coordinate layers in the vertical between the ground and about 140 km altitude. Sub-grid-scale processes are parameterized crudely by vertically dependent Rayleigh friction, Newtonian cooling and horizontal diffusion. The model is forced by diurnal oscillations in the heat sources, taken, just as the background atmosphere, from the monthly climatology of a state-of-the-art GCM HAMMONIA [*Schmidt et al.*, 2006] ranging from the ground far into the thermosphere. The heat sources comprise, among others, the absorption of solar radiation, including the ultraviolet and extreme ultraviolet wavelength regime, long-wave radiation, and heating by latent heat release and convection. Instead of a brute forward integration, the equations including the forcing are Fourier analyzed in time, so that for each tidal period a system of linear equations is obtained, which are solved iteratively by a preconditioned conjugate gradient solver. The result is the complete three-dimensional tidal structure, with all migrating and non-migrating components. Note that the background state is three-dimensional so that the effect of stationary planetary waves is included.

The linear model is used for analyzing the seasonal behavior of the tides in the GCM, which it reproduces reasonably well. We here focus on the three most relevant non-migrating tidal components. Perhaps least surprising is the important role played by condensation and convection in the forcing of *DE3*. An analysis of the seasonal cycle near the mesopause shows that the amplitude maximum between November and February is mostly due favorable propagation conditions given then by the zonal mean. *McLandress* [2002b] argues that the decisive factor in the modulation of the migrating tide by the zonal mean is the seasonal dependence of the zonal-mean vorticity in the background atmosphere. This might as well be the case for DE3. An erroneous maximum in HAMMONIA, as compared to observations, in August is prevented in the linear model by a counteracting effect due to the seasonal cycle in the forcing. The interplay between the zonal-mean background and the forcing thus seems to be essential for explaining the complete seasonal cycle.

Around the time of its maximum (April–June) *DS0* in the linear model is mostly excited by the direct non-migrating forcing by the absorption of short-wave solar radiation and by condensational heating. Since the planetary waves are weak during this time it is no surprise that their effect is not so important then. Indeed, it is found that most of the seasonal cycle can be understood as an effect of variations in the zonal-mean background. However, this even holds between December and February when the planetary waves are strong. Closer analysis shows that the total signal is very similar to the direct non-migrating input from condensation and convection. Both, the effects from the direct non-migrating forcing by the absorption of solar short-wave radiation by tropospheric water vapor and the modulation of the migrating forcing by the planetary waves are also strong, but cancel each other. It thus seems that destructive interference effects such as here might also be an essential factor of the planetary wave effect on non-migrating tides.

An example where the planetary waves actually enhance the amplitude of a non-migrating tide is *DW2*, as illustrated in Fig. 27.1. This tidal component is driven to a large proportion by the non-migrating forcing due to condensation and convection. The seasonal cycle can be explained to the largest part by the seasonal variations of the zonal-mean propagation conditions of this directly forced non-migrating tide. One also has, however, a quantitatively important impact from the planetary wave modulation of the migrating tide forced in the troposphere. This holds both for the total signal as such and for the simulated seasonal cycle. In conclusion, planetary waves do seem to be a factor to be taken into account in the dynamics of solar tides.

27.3 Interaction Between GWs and Solar Tides

In previous efforts of tidal modeling [also see *Ortland and Alexander*, 2006, and reference therein], the interaction between tides and GW parameterizations has been

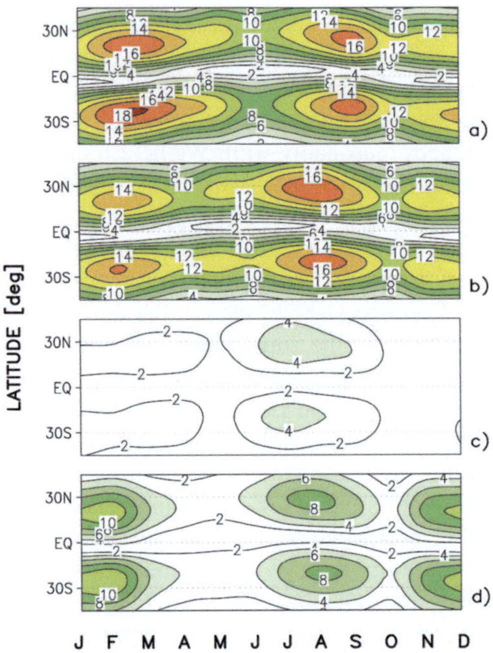

Fig. 27.1 From the linear model, the seasonal cycle in the amplitudes of the non-migrating component DW2 of the diurnal tide in the meridional wind at 95 km altitude (**a**), the part due to variations of the background atmosphere (**b**), the part due to the seasonally varying part of the diurnal heat sources (**c**), and the corresponding contribution from the seasonal cycle of the stationary planetary waves (**d**). Units are m/s. Taken from *Achatz, U., N. Grieger, and H. Schmidt, Mechanisms controlling the diurnal solar tide: Analysis using a GCM and a linear model, J. Geophys. Res., 113, A08303, 2008*. Copyright 2008 American Geophysical Union. Modified by permission of American Geophysical Union

investigated under strong assumptions. Conventional GW parameterizations work in vertical columns which are assumed to be independent from each other, ignoring horizontal inhomogeneities in the large-scale flow [*McLandress*, 1998]. Furthermore, time-dependence of the large-scale background (BG) conditions is neglected. It is supposed that GW fields just see a quasi-stationary mean flow and adjust instantaneously to its changes. In reality, however, GWs exhibit horizontal propagation and they are refracted at horizontal inhomogeneities of the BG. Time dependence of the latter changes the GW frequency. Furthermore, GWs propagate with a finite group velocity. Hence, if the time scale of the BG is short enough to get comparable to or smaller than the GW propagation time scale, significant deviations from the assumption of instantaneous adjustment appear. This might most likely be the case for solar tides in their effect on GW propagation.

In *Senf and Achatz* [2011] the effects of GW propagation and dissipation in realistic tidal fields are investigated with the help of global ray-tracing simulations, thus extending the simplified calculations by *Eckermann and Marks* [1996]. In ray tracing [*Achatz et al.*, 2010, e.g.], a locally monochromatic gravity wave field is propagated through a slowly changing environment. The GW field, or rather its local wave numbers and amplitudes are followed along characteristics, the so-called rays, determined from the local group velocity [*Hasha et al.*, 2008, e.g.]. The time-dependence of the BG wind, in our case the effect of the diurnal tide, induces a modulation of GW observed frequency along the ray. The horizontal gradients in the BG conditions lead to changes in the horizontal GW wave numbers. Following *Grimshaw* [1975], the GW amplitude is predicted from the conservation of wave action, supplemented by a damping rate estimated via the highly simplified saturation theory [*Lindzen*, 1981]. Once a GW amplitude grows beyond the threshold at which it induces local overturning of isentropes it is forced back to this convective instability threshold. Additionally, in the MLT region molecular viscosity and thermal diffusivity become more important and are included into the damping process. Our new global ray-tracer RAPAGI (RAy PArameterization of Gravity-wave Impacts) solves the ray-tracing equations on the sphere, using a small and highly idealized GW ensemble at source spectrum at 20 km altitude. Each of the 14 individual and independent GW components is integrated forward separately. It has been shown by *Becker and Schmitz* [2003] that the mean residual circulation of the middle atmosphere is well reproduced in a large-scale GCM when this GW ensemble is used in a classic parameterization according to *Lindzen* [1981]. We note, however, that, as that mostly resulted from tuning, this GW ensemble is just one of many possibilities. Therefore, the simple GW ensemble is viewed as a reasonably-motivated toy configuration, while the analysis of even better chosen scenarios is left for the future.

The superposition of monthly means and tidal fields from the HAMMONIA model are chosen as background for three different off-line experiments, named "full", "noREF" and "TS", with decreasing complexity. The "full" experiment refers to a full ray-tracing simulation without any approximations for horizontal and time dependence. Thus, changes in frequency and all wave-number components appear and are induced by mean flow changes. The geographical distribution of the GW fields is altered as well. "noREF" (no refraction) is a simplified ray-tracing experiment in which neither horizontal refraction nor horizontal propagation are allowed. Only the vertical ray propagation is taken into account. Nonetheless, the rays have a finite group velocity and feel the transience of the BG wind. The horizontal wave-number components are constant along each ray, but frequency and vertical wave number vary to compensate temporal and vertical changes in the BG conditions, respectively. The third experiment is denoted by "TS" (time slicing) and equivalent to a classic single-column and steady-state parameterization. Only vertical variations are taken into account. The tidal phase was fixed in 3 h steps to sample the diurnal cycle, and finally the results from the different tidal phases were combined to a daily cycle. With the three experiments, effects of frequency modulation and the refraction of horizontal wave vector can be extracted. Differences between "TS" and "noREF" are attributed to the first, whereas differences between "noREF" and "full" to the latter. As the simpler simulations "TS" and "noREF" are obtained by successively simplifying the "full" one, a consistent comparison of the results is possible while keeping implementation aspects the same.

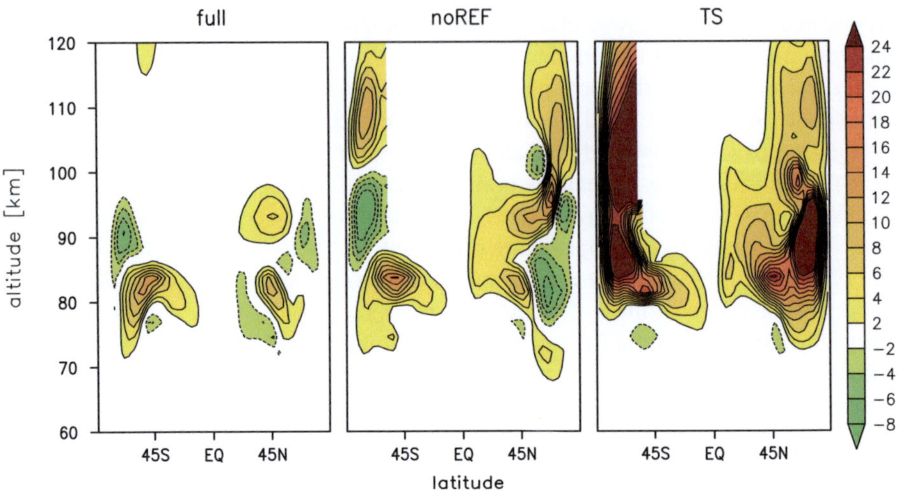

Fig. 27.2 The zonally averaged real part of the equivalent Rayleigh friction coefficient for the three different experiments: "full" (*left*), "noREF" (*middle*) and "TS" (*right*). The shading interval is $2 \times 10^{-6}\ \mathrm{s}^{-1}$

Since GW fields in the MLT region are periodically modulated by tidal winds, they produce a periodic force acting back on the diurnal tides. The relevant forcing of the mean flow, in our case temporally averaged flow plus diurnal tides, is given by the divergence of the pseudo-momentum fluxes from all 14 GW ensemble members. The diurnal amplitude of the zonal force can be analyzed most conveniently on the basis of equivalent Rayleigh friction coefficients (ERFs). These have been introduced in the context of GW-tidal interaction by e.g. *Miyahara and Forbes* [1991] and further discussed e.g. by *McLandress* [2002a]. With the help of ERFs, the effects of GWs can be incorporated into a linear tidal model. Effectively, the tidal component of the GW force is approximated by $f_\lambda \approx -\gamma_R u_T - \frac{\gamma_I}{\Omega}\partial_t u_T$, where u_T is the tidal wind, and γ_R and γ_I are the real and imaginary part of the ERFs, respectively. Positive real parts of the ERFs indicate regions of tidal damping and vice versa. The imaginary part of ERF acts on the tidal phase structure. A reduction of tidal vertical wave length is a very robust result in previous investigations, whereas the GW effect on tidal amplitudes is controversial [*Ortland and Alexander*, 2006, and references therein]. The real parts of ERFs are shown in Fig. 27.2 for the three simulations. For the reference simulation "TS" in Fig. 27.2(c), large positive peaks up to $60 \times 10^{-6}\ \mathrm{s}^{-1}$ occur. The damping of tidal amplitudes is a quite typical result of Lindzen saturation parameterization. Others show qualitatively different effects on tides [e.g. *Ortland and Alexander*, 2006]. For the "noREF" experiment, in Fig. 27.2(b), the magnitude of γ_R is reduced. As analyzed in *Senf and Achatz* [2011] this results from an avoidance of critical levels, due to the time dependence of the tidal fields. The latitude-altitude structure is wave-like with a vertical wave length comparable to the tidal wave length. In Fig. 27.2(a), the magnitude of the ERF is further decreased. The influence of γ_R is drastically lowered at high latitudes and in the thermosphere. This is mostly due to the meridional refraction of the GW horizontal wave number by the mean wind gradients and the corresponding horizontal GW propagation [*Senf and Achatz*, 2011]. The bottom line of this is that frequency modulation and horizontal refraction by the tidal background significantly reduces the GW forcing in the MLT.

27.4 Conclusions

The dynamics of solar tides is still a challenging field. We have here reported progress on the linear modeling of these forced waves. This approach has the advantage that cause-effect relationships can be established which would remain hidden in an all-nonlinear approach. We are able to trace back characteristics of the seasonal cycle of important components of the diurnal solar tide to the impact of either migrating and non-migrating diurnal heating, or the impact of planetary waves on the tidal propagation. The latter can lead to a non-migrating signal in the mesosphere, caused by migrating forcing in the troposphere. An important problem remains the GW-tidal interaction. So far we are able to show that it is incorrectly described on the basis of single-column and steady-state GW parameterization approaches. The time dependence of the solar tides, and the spatial gradients set by the background atmosphere tend to reduce the GW impact much below what has been assumed so far. The feedback on the tides, however, remains to be examined.

Acknowledgements The authors thank Erich Becker for inspiring discussions and Hauke Schmidt from MPI Hamburg for providing the set of HAMMONIA data. U.A. thanks Deutsche Forschungsgemeinschaft for partial support through the MetStröm Priority Research Program (SPP 1276), and through Grant Ac 71/4-1. U.A. and F.S. thank Deutsche Forschungsgemeinschaft for partial support through the CAWSES Priority Research Program (SPP 1176), and through Grant Ac 71/2-1.

References

Achatz, U., Grieger, N., & Schmidt, H. (2008). Mechanisms controlling the diurnal solar tide: analysis using a GCM and a linear model. *Journal of Geophysical Research*, *113*, A08303. doi:10.1029/2007JA012967.

Achatz, U., Klein, R., & Senf, F. (2010). Gravity waves, scale asymptotics, and the pseudo-incompressible equations. *Journal of Fluid Mechanics*, *663*, 120–147.

Becker, E., & Schmitz, G. (2003). Climatological effects of orography and land-sea heating contrasts on the gravity wave-driven circulation of the mesosphere. *Journal of the Atmospheric Sciences*, *60*, 103–118. doi:10.1175/1520-0469(2003)060.

Chapman, S., & Lindzen, R. S. (1970). *Atmospheric tides*. Dordrecht: Reidel.

Eckermann, S. D., & Marks, C. J. (1996). An idealized ray model of gravity wave-tidal interactions. *Journal of Geophysical Research*, *101*, 21195–21212.

Fritts, D. C., & Alexander, M. J. (2003). Gravity wave dynamics and effects in the middle atmosphere. *Reviews of Geophysics*, *41*, 1003. doi:10.1029/2001RG000106.

Grieger, N., Schmitz, G., & Achatz, U. (2004). The dependence of the nonmigrating diurnal tide in the mesosphere and lower thermosphere on stationary planetary waves. *Journal of Atmospheric and Solar-Terrestrial Physics*, *66*, 733–754. doi:10.1016/j.jastp.2004.01.022.

Grimshaw, R. H. J. (1975). Nonlinear internal gravity waves in a rotating fluid. *Journal of Fluid Mechanics*, *71*, 497–512. doi:10.1017/S0022112075002704.

Hagan, M. E., & Roble, R. G. (2001). Modelling diurnal tidal variability with the NCAR TIME-GCM. *Journal of Geophysical Research*, *106*, 24869–24882.

Hasha, A., Bühler, O., & Scinocca, J. (2008). Gravity wave refraction by three-dimensionally varying winds and the global transport of angular momentum. *Journal of the Atmospheric Sciences*, *65*, 2892–2906. doi:10.1175/2007JAS2561.1.

Lindzen, R. S. (1981). Turbulence and stress owing to gravity wave and tidal breakdown. *Journal of Geophysical Research*, *86*, 9707–9714. doi:10.1029/JC086iC10p09707.

McLandress, C. (1998). On the importance of gravity waves in the middle atmosphere and their parameterization in general circulation models. *Journal of Atmospheric and Solar-Terrestrial Physics*, *60*, 1357–1383.

McLandress, C. (2002a). The seasonal variation of the propagating diurnal tide in the mesosphere and lower thermosphere. Part I: The role of gravity waves and planetary waves. *Journal of the Atmospheric Sciences*, *59*, 893–906.

McLandress, C. (2002b). The seasonal variation of the propagating diurnal tide in the mesosphere and lower thermosphere. Part II: The role of tidal heating and zonal mean winds. *Journal of the Atmospheric Sciences*, *59*, 907–922.

Miyahara, S., & Forbes, J. (1991). Interactions between gravity waves and the diurnal tide in the mesosphere and lower thermosphere. *Journal of the Meteorological Society of Japan*, *69*(5), 523–531.

Oberheide, J., Wu, Q., Ortland, D. A., Killeen, T. L., Hagan, M. E., Roble, R. G., Niciejewski, R. J., & Skinner, W. R. (2005). Non-migrating diurnal tides as measured by the TIMED Doppler interferometer: preliminary results. *Advances in Space Research*, *35*, 1911–1917. doi:10.1016/j.asr.2005.01.063.

Oberheide, J., Wu, Q., Killeen, T. L., Hagan, M. E., & Roble, R. G. (2006). Diurnal nonmigrating tides from TIMED Doppler interferometer wind data: monthly climatologies and seasonal variations. *Journal of Geophysical Research*, *111*, A10S03. doi:10.1029/2005JA011491.

Ortland, D. A., & Alexander, M. J. (2006). Gravity wave influence on the global structure of the diurnal tide in the mesosphere and lower thermosphere. *Journal of Geophysical Research*, *111*, A10S10.

Schmidt, H., Brasseur, G. P., Charron, M., Manzini, E., Giorgetta, M. A., Diehl, T., Fomichev, V. I., Kinnison, D., Marsh, D., & Walters, S. (2006). The HAMMONIA chemistry climate model: sensitivity of the mesopause region to the 11-year solar cycle and CO2 doubling. *Journal of Climate*, *19*, 3903–3931. doi:10.1175/JCLI3829.1.

Senf, F., & Achatz, U. (2011). On the impact of middle-atmosphere thermal tides on the propagation and dissipation of gravity waves. *Journal of Geophysical Research*, *116*, D24110. doi:10.1029/2011JD015794.

Chapter 28
Short Period Dynamics in the Mesosphere: Morphology, Trends, and the General Circulation

Dirk Offermann and Ralf Koppmann

Abstract Measured data of wind and temperature changes (waves) in the middle atmosphere are obtained from several ground stations in Middle Europe (wind radar, OH temperature, intensity) and from the CRISTA, SABER, and MLS satellite instruments. Emphasis of the data interpretation is on the dynamics of the mesosphere with long-term changes and solar cycle influences. A comparative analysis of global wave activity shows that gravity waves play a leading role in most part of the year, at latitudes 20°N–70°N and altitudes 70–100 km. Traveling planetary waves are in second place. Very short period oscillations are studied down to 3 min period, i.e. to well below the Brunt-Väisälä period (5 min). They are interpreted in terms of gravity wave activity and dissipation. This applies as well to pronounced seasonal variations (with maxima near equinox) as to long-term trends. Similar seasonal variations are also found in turbulent eddy coefficients K_{zz} of a General Circulation Model (WACCM 3.5) and in the turbopause altitude. Short period gravity wave activity/dissipation is compared to changes of the zonal wind speed. A long-term increase of the wind (0.5 m/s/year) and of gravity wave intensity (1.5 %/year in 16 years at 87 km altitude) is found. This is related to an increase of the summer duration that is observed since 1988 at a rate of 1.2 days/year at 87 km. Corresponding changes are found in the middle stratosphere. A long-term change is also seen in very short period planetary waves, i.e. the Quasi-Two Day Waves. These waves have been analyzed since 1988 (at 87 km altitude) and exhibit threefold maxima in summer. They show a considerable long-term change of the stability structure of the mesosphere. These different dynamical results suggest an important long-term change in the general circulation.

28.1 Introduction

Short period/small scale structures are important for mesosphere/lower thermosphere (MLT) dynamics and circulation [e.g. *Holton*, 1983; *Garcia and Solomon*,

---

D. Offermann (✉) · R. Koppmann
Faculty of Mathematics and Natural Sciences, Physics Department, University of Wuppertal, Gauss Strasse 20, 42119 Wuppertal, Germany
e-mail: offerm@uni-wuppertal.de

F.-J. Lübken (ed.), *Climate and Weather of the Sun-Earth System (CAWSES)*, Springer Atmospheric Sciences,
DOI 10.1007/978-94-007-4348-9_28,

1985]. They especially are seen as planetary waves (PW), gravity waves (GW), and "ripples". Wave instability or interaction with wind shears leads to convective or Kelvin-Helmholtz instabilities, "ripples", and small scale turbulence [e.g. *Lindzen*, 1981; *Garcia*, 1991; *Fritts et al.*, 2009; *Hecht et al.*, 2007].

Examples of short period planetary waves are the Quasi-Two Day Waves (QTDW) [*Muller*, 1972; *Baumgaertner et al.*, 2008]. They are indicative of mesospheric instability and hence important for the study of mesospheric dynamics.

Gravity waves have been intensively studied since the work of *Hines* [1960], and it appears sometimes as if Pandora's box has been opened then [e.g. *Fritts and Alexander*, 2003; *Preusse et al.*, 2009]. Short period GW are especially difficult to measure [e.g. *Snively et al.*, 2010].

Ripples have been measured many times by ground based air glow imagers. Their time periods may be well below the Brunt-Väisälä period (5 min), and a maximum of occurrence near 3 minutes has been observed [*Hecht et al.*, 2007].

The morphology of mesospheric dynamics, i.e. its latitudinal/longitudinal, altitude, and seasonal variations can be determined from measurements of ground stations (winds, temperatures, intensities) and satellites (temperature, winds, trace gases). Ground stations as wind radars, OH temperature and lidar stations are especially suitable to determine fast time variations and in part vertical profiles. Measurements of satellite instruments like CRISTA, SABER, or MLS give global latitude/longitude and altitude distributions. For an overview of these various measurements see *Offermann et al.* [2006, 2007, 2009, 2010, 2011a, 2011b, and the references therein].

The importance of short period structures for the general circulation of the atmosphere raises the question as to the stability of their distributions and occurrence frequencies. Many of the short period features are found to change with time, indeed. Beyond seasonal variations that are treated as morphology features here, they show changes during the solar cycle and also decadal or long-term trends. Not too much is known about this. Satellite data are not suitable for such analyses yet as the data series presently are too short. Hence most of the long-term knowledge rests on the continuous measurements of a number of ground stations [e.g. *Offermann et al.*, 2010; 2011a, 2011b, and references therein]. Many of the data used here are from the OH stations at Wuppertal (51°N, 7°E) and Hohenpeissenberg (48°N, 11°E) where two similar OH spectrometers (GRIPS) are operated.

This chapter is organized as follows: In Sect. 28.2 the morphology of various waves is discussed including some aspects of their dissipation. Short period GW breaking with ripple and turbulence formation are analyzed. The distribution of QTDW during summer and the underlying stability of the mesosphere are studied. In Sect. 28.3 long-term trends including solar cycle influences are discussed. This applies to temperatures, zonal winds, and short period GW. In Sect. 28.4 the different dynamical results are compared to each other and are discussed with respect to atmospheric circulation and its long-term changes.

28.2 Morphology

28.2.1 Middle Atmosphere Waves

Major parameters to be analyzed in the middle atmosphere (stratosphere, mesosphere, and lower thermosphere) are temperature and winds. They can be determined by ground based stations or by satellites as mentioned above. The major temperature variation in the mesosphere is the seasonal variation, an example of which is shown in Fig. 28.1. A three-parameter fit has been drawn to the measurements here, with an annual, semi-annual, and ter-annual component. The differences of the measured data and the fit curve (residues) are given in the bottom of the picture. They can be used for wave estimation.

Many different types of waves are superimposed on these seasonal variations. The most important ones are gravity waves (GW) with periods up to several hours, tides with periods of one day or fractions of it, traveling planetary waves (TPW) and quasi-stationary planetary waves (SPW) with periods of days to weeks. These waves are very variable as concerns their amplitudes and periods, changing with altitude, latitude, and possibly longitude. Therefore climatologies of them are not easily available.

The residues shown in Fig. 28.1 are nightly values and can therefore be used for the analysis of PW with periods longer than 2 days according to the Nyquist theorem. A convenient measure of this wave activity is the standard deviation σ_M of these residues from their (monthly) mean value. This- or similar standard deviations-are frequently used in the literature and in the present paper as wave proxies.

Atmospheric oscillations with periods shorter than the Brunt-Väisälä period (5 min) are believed to be mostly "ripples" which result from non-linear behavior/interaction of (gravity) waves and wind shears. They are discussed below together with the GW in Sect. 28.2.2.

A climatology of the four most important wave types has been attempted by *Offermann et al.* [2009] mostly on the basis of SABER temperatures. They find the following wave ranking: At extra-tropical latitudes GW mostly are the leading wave type. Traveling PW are in second place, as they were inferred from the total variance of SABER temperatures. Stationary PW and tides are of lesser importance. An overview of the altitude/latitude distributions of these waves is given for a selected altitude range in Fig. 28.2. The variances of the four different wave types (values of single curves squared) add up to the total variance (red curve squared). These results must be considered preliminary as they are from a limited number of SABER data (up to 4 years) in an earlier data version (V1.06).

A remark appears appropriate concerning the airglow intensities used to derive the OH temperatures: these intensities show very complicated seasonal, latitudinal, diurnal, and other variations, too (e.g. *Liu et al.* [2008]). These are, however, not necessarily the same as the simultaneous temperature variations. This is because the temperatures are derived from the intensity ratios of two or three airglow lines, but not from the intensities themselves. A good example for this is a pronounced maximum found for the line intensities used for Fig. 28.1 that occurs right in the

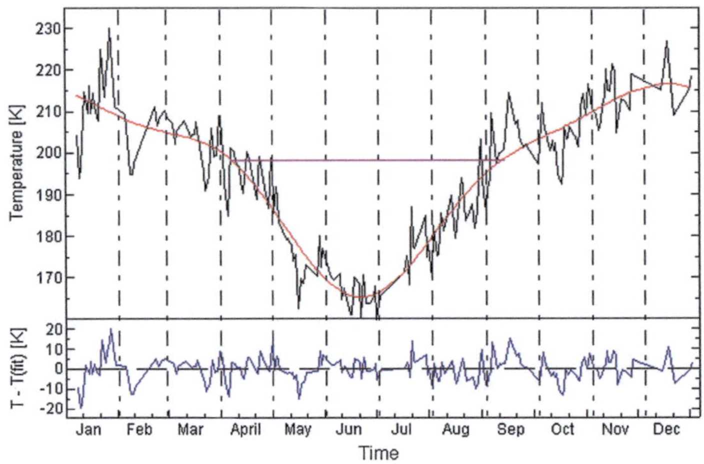

Fig. 28.1 Seasonal variation of OH temperatures at Wuppertal (51°N, 7°E) in year 2005. Nightly mean values are shown. *Solid red curve* is a three component harmonic fit to the measured data. *Horizontal red line* indicates the Equivalent Summer Duration (ESD, see Sect. 28.3.1 below). Differences of the measured data and the fit curve (residues) are given in the *bottom* of the picture [*Offermann et al.*, 2010]

middle of summer (not shown here; see also *Espy and Stegman* [2002]; *Liu et al.* [2008]). At this time the temperature has its lowest values in the year at all!

28.2.2 Gravity Waves and "Ripples"

Gravity waves on the one hand are among the most important waves in the mesosphere/lower thermosphere. On the other hand they are extremely manyfold. Restricting the analysis to a limited ensemble of GW therefore might be helpful. A corresponding study of GW limited to short and very short periods was performed by *Offermann et al.* [2011b] on the basis of OH temperatures measured at Wuppertal (51°N, 7°E). The time resolution of these measurements is 1.3 minutes. Fourier analyses were therefore possible down to periods of 3 minutes. The noise at these high frequencies is large. The following precaution was therefore taken to suppress it: The OH temperatures are derived from the relative intensities of three P-band rotational lines near 1.5 μm wavelength. An oscillation in the temperature Fourier spectrum was considered as valid only if corresponding spectra of all three line intensities showed oscillations as well (at the same frequency). A detailed justification of this method is given in *Offermann et al.* [2011b].

Examples of such spectra are shown in Fig. 28.3. Mean spectra in three-monthly intervals of year 1997 are given. The picture shows sizeable amplitudes even at the shortest periods (3 min). These mostly are not GW as GW in their majority have periods longer than the Brunt-Väisälä-Period (5 minutes). They are rather interpreted

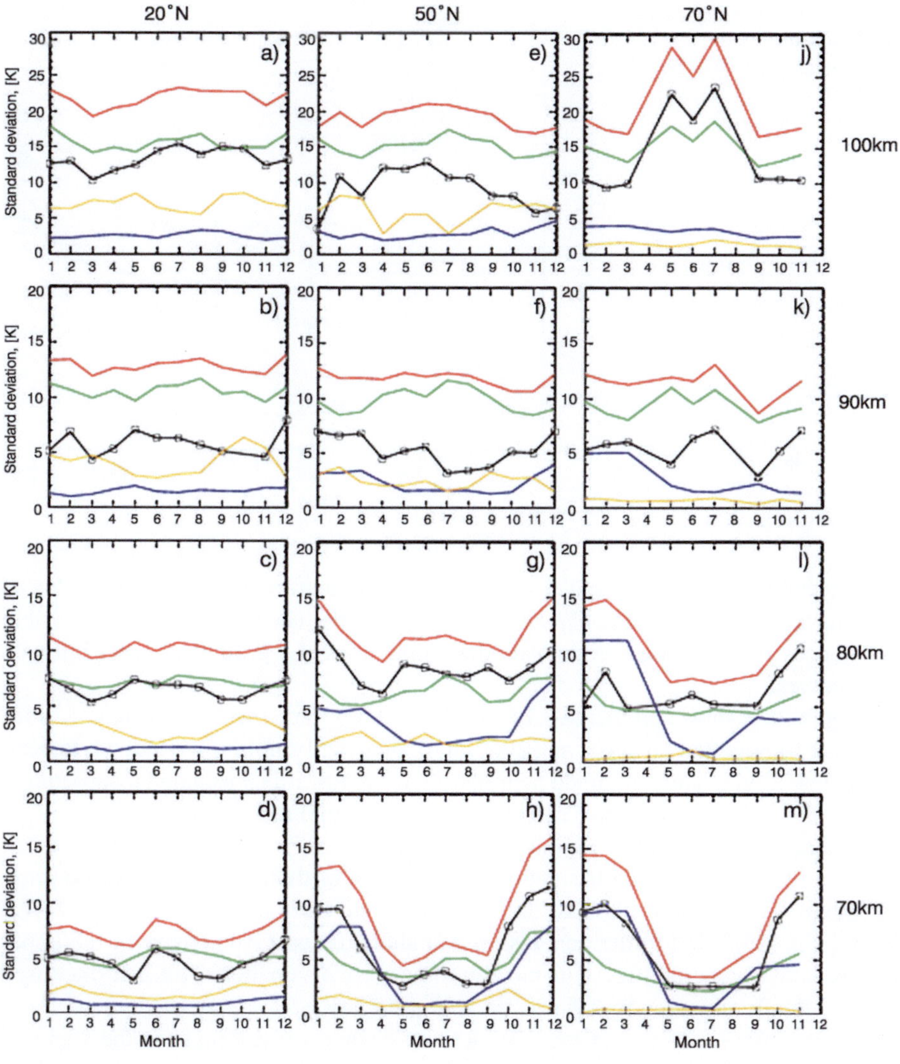

Fig. 28.2 Wave ranking (seasonal variations) at latitudes of 20°N, 50°N, and 70°N and altitudes of 70 km, 80 km, 90 km, and 100 km. The four waves are: gravity waves (*green*), traveling planetary waves (*black squares*), quasi-stationary planetary waves (*blue*), tides (*safran*). Top curves are measured SABER fluctuations (*red*). Note the expanded scale in pictures **a**, **e**, **j**: It runs from 0 to 30 K. In the other figures it runs from 0 to 20 K

as "ripples" that are frequently seen in the pictures measured by nightglow imagers [e.g. *Taylor et al.*, 2007; *Hecht et al.*, 2007]. Ripples are believed to indicate convective instabilities or Kelvin-Helmhotz instabilities that result from GW breaking. Spectra as shown in Fig. 28.3 are therefore indicative of GW and/or their dissipation.

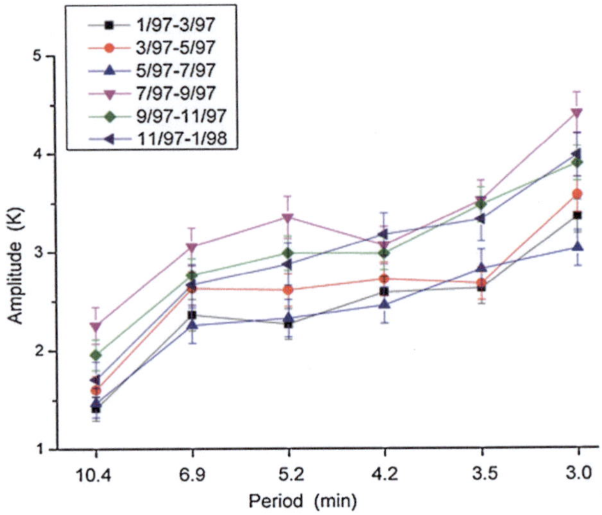

Fig. 28.3 Mean temperature oscillation spectra in three-monthly intervals in 1997

Detectability of short wavelength structures by instruments with large fields of view have been discussed in *Offermann et al.* [2011b]. The same applies to possible differences of the wave frequencies measured from the ground and the intrinsic frequencies.

Figure 28.3 shows an increase of oscillation amplitudes with frequency. It is interesting to note that *Hecht et al.* [2007] have analyzed ripple spectra down to periods below 2 min. They obtained a similar increase towards 3 min and a subsequent decrease, i.e. the maximum of ripple activity was found near 3 min.

Especially, these 3 min oscillations have been used in a detailed analysis as indication of wave breaking/dissipation. Comparison of them with short period gravity waves (up to 5 hour periods) and with turbulent eddy coefficients (Sect. 28.2.3 below, Fig. 28.5) led to the conclusion that short-period GW activity in the upper mesosphere is generally accompanied by some wave dissipation. This is further used below. Details are given by *Offermann et al.* [2011b].

A convenient measure of the GW activity during a given night is the standard deviation σ_N from the mean temperature in that night. This is a similar approach as the monthly standard deviation σ_M used for PW above. In order to compare σ_N with the short period spectra in Fig. 28.3 a measure σ_a for the wave activity in Fig. 28.3 is calculated. (It is essentially the square root of the sum of the squared amplitudes of the six oscillations shown in Fig. 28.3. For details see *Offermann et al.* [2011b]). The two parameters σ_a and σ_N are found to be highly correlated. The correlation coefficient for all nights in 1997 is $r = 0.91$. This is important because σ_N contains GW with longer periods up to five hours. This is the average length of the nightly measurement window at Wuppertal. The nightly σ_N values of course contain also fluctuations other than gravity waves. The high correlation shows, however, that GW yield a large fraction of it, and hence the parameter is quite suitable for the relative comparisons given below.

The short-period σ_a values from Wuppertal OH measurements can be compared with the SABER data in Fig. 28.2f. An even better comparison is obtained from Fig. 6 in *Offermann et al.* [2009] that is closer in altitude (88 km) to the OH measurements. The annual mean of these seasonal GW variations is 8.8 K. The data are from 2002–2005 (SABER V1.06). Long-term development of standard deviations σ_a of the Fourier spectra in Fig. 28.3 are shown in Fig. 28.11 and discussed below. Their mean value for the SABER time interval is 9.4 K. This agreement is very interesting and might be partly fortuitous as the measurement techniques are quite different (limb vs. approximately zenith looking). If the short period oscillations are linked to short horizontal wavelengths a limb viewing instrument may have some difficulties seeing them.

Gravity waves measured by SABER as well as by the OH ground based instruments show seasonal variations. These are well correlated as is discussed below. These results and the close correlations mentioned above suggest further studies, especially as the very short period GW and ripples appear to represent a substantial fraction of the total GW inventory.

Gravity wave activity varies with altitude, latitude, and season as shown in Fig. 28.2. The data taken at one OH station (as Wuppertal) allow GW analyses with high time resolution at a fixed altitude and latitude. Standard deviations σ_N are analyzed for this purpose, and three types of time variations are found. Short-term changes on a daily basis are very strong. If they are filtered out by smoothing a seasonal variation and a long-term trend emerge. The trend data are discussed below (Sect. 28.3.3).

The seasonal variation is characterized by two peaks that occur near equinox. The height of these peaks is itself quite variable, and in some years there appears to be only one peak at all. A similar picture is obtained at the OH station of Hohenpeissenberg (48°N, 11°E).

A clearer picture is obtained if the average of a number of years is calculated. This is shown in Fig. 28.4 that gives mean Wuppertal data of years 1994–2009. The picture clearly shows the double peak structure. The spring peak is consistently smaller than the autumn peak. Circulation turnaround in the stratosphere and mesosphere is indicated by vertical dashed lines (black and red, respectively). For details see *Offermann et al.* [2011b]. It is obvious that the two peaks—taken as GW representations here—occur before turnaround. Assuming the picture proposed here that GW activity goes along with wave dissipation this result suggests that GW might contribute to the turnaround near equinox.

The seasonal variation of GW activity has been determined many times by various stations/experiments [see *Offermann et al.*, 2011b]. They generally find one or two peaks in the seasonal distribution. These occur, however, not at equinox but mostly around solstices and thus appear contradictory to the OH results. On the other hand the satellite results shown in Fig. 28.2f exhibit a double peak structure quite similar to Fig. 28.4. The correlation of these data with Fig. 28.4 have a coefficient of 0.53 at 93 % significance [see *Offermann et al.*, 2011b]. Figure 28.2 also shows similar double peak structures at higher latitudes and altitudes (Fig. 28.2j, k). Furthermore, it is seen from Fig. 28.2 that GW seasonal variation can strongly

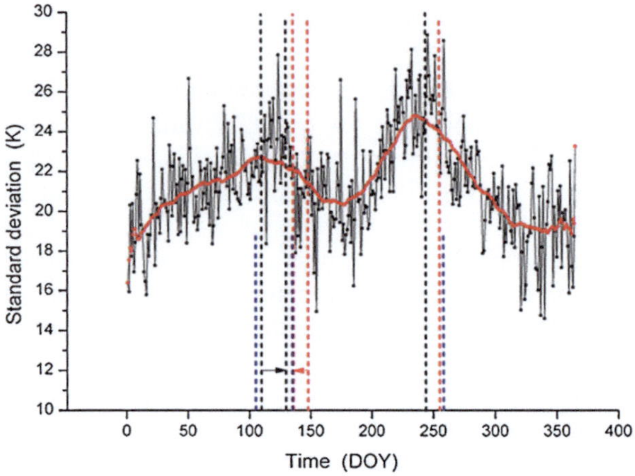

Fig. 28.4 Mean temperature standard deviations σ_N for the years 1994–2009 at Wuppertal. The seasonal variation is shown by the red curve (50 point running mean). *Vertical dashed lines* indicate circulation turnaround in the stratosphere (*black*) and mesosphere (*red*). *Black arrow* indicates a trend of turnaround in the stratosphere in the time interval 1988–2008. *Red arrow* indicates a trend of turnaround in the mesosphere in the time interval 1993–2008. *Blue dashed vertical* lines indicate maxima of eddy coefficients taken from Fig. 28.5

vary with altitude and latitude. For instance at middle latitudes and altitudes 70–80 km (Fig. 28.2g, h) peaks are seen to occur at solstice, indeed. Hence, many of the apparent observational discrepancies may be due to differences of the measurement altitudes and latitudes. Another reason may be that many of the results were obtained from radar wind measurements. These measure essentially the kinetic energy of GW, whereas temperature measurements (at OH stations) determine the GW potential energy. It was recently shown by *Geller and Gong* [2010] that different energies have sensitivities to different GW frequencies. Gravity waves with different frequencies, however, can have quite different seasonal variations, as was for instance shown by *Dowdy et al.* [2007].

28.2.3 Turbulence

The working hypothesis followed here is that GW activity is associated with dissipation which would also mean turbulence production. If this is true Fig. 28.4 suggests some seasonal variation of turbulence. A consistent climatology (altitude, season, latitude) of measured turbulent eddy coefficients is difficult to obtain. Figure 28.5 therefore shows computational values obtained from the general circulation model WACCM 3.5 on a monthly basis [for details see *Offermann et al.*, 2011b]. The picture shows a seasonal variation of turbulence, indeed. Two pronounced peaks of eddy coefficients are seen around the equinoxes in April, May, and September (DOYs 105, 135, and 258). These days are indicated in Fig. 28.4 as dashed (blue)

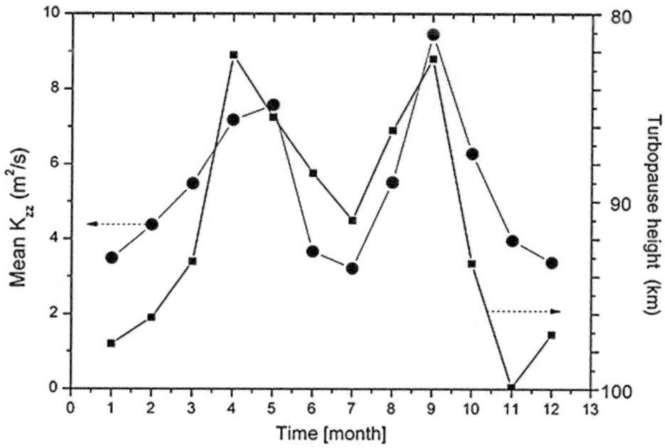

Fig. 28.5 Turbulent eddy coefficients K_{zz} (*dots*) as derived from the WACCM 3.5 model. Nineteen years of free model runs have been averaged, and resulting monthly means are shown for 85 km altitude. The turbopause height [*Offermann et al.*, 2007] is given for comparison (*squares*). Please note that its ordinate has been inverted (see text)

vertical lines. They are situated near the GW peaks as expected. The correlation of the monthly eddy coefficients with monthly σ_N values calculated from Fig. 28.4 is close. The correlation coefficient is 0.71 at 99 % significance [see *Offermann et al.*, 2011b].

In Fig. 28.5 the seasonal variation of the turbopause height is shown for comparison [*Offermann et al.*, 2011b]. The definition of this "wave turbopause" can be seen from Fig. 28.6 which is taken from *Offermann et al.* [2006]. The picture shows two altitude regimes with small (below) and large (above) increases of wave amplitudes (represented by their standard deviation). The dotted lines indicate increases expected for an atmosphere without damping. The picture thus shows strong damping below the turbopause, and much reduced damping above it. Damping is attributed to wave breaking/turbulence production which explains the designation of the intersection/diamond. The altitude at which the turbopause occurs can be quite variable. This has consequences for the wave amplitudes observed at some fixed higher altitudes, e.g. 100 km in Fig. 28.6. If the diamond moves downward along the flat blue line the steep blue line follows (in a first approximation). Thus the wave amplitude at 100 km (intersection of the steep blue line with the 100 km level) increases. Hence an anticorrelation of the height of the turbopause and the wave amplitude at higher altitude (higher than the turbopause) is expected. This is the reason why the corresponding scale in Fig. 28.5 has been inverted. The inverted curve of the turbopause indeed closely correlates with that of the eddy coefficients and thus with the GW activity shown in Fig. 28.4. The corresponding maxima occur closely together.

The concept of wave turbopause thus works quite well for GW. It is, however, not limited to GW but has been used for mesospheric wave activity in general [*Offermann et al.*, 2007]. It is found a convenient tool to study the turbulent state of the mesosphere which turns out to be very variable. If the turbopause curve in Fig. 28.5 is re-inverted it exhibits maxima in summer and winter, and minima at equinox. This is a typical seasonal variation at middle latitudes higher than the subtropics. At even higher latitudes (at and beyond 60°N) the picture is quite different. The summer maximum completely disappears and a deep summer minimum and a high winter maximum remains. These seasonal variations and the corresponding latitudinal

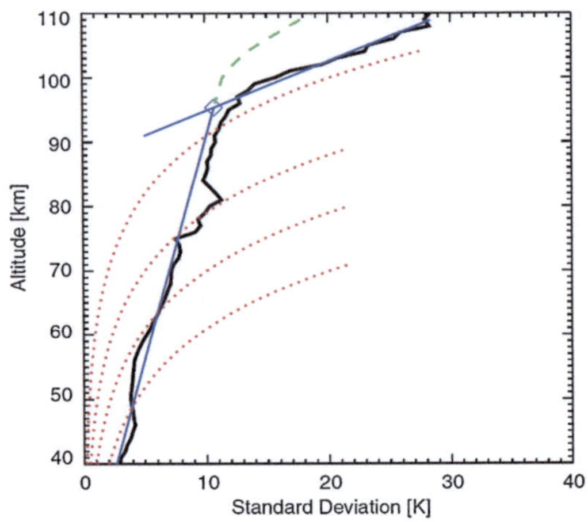

Fig. 28.6 Temperature standard deviations vs altitude at 40°N in August 1997 (CRISTA-2 data). *Black profile* shows two regimes with different gradients. These are approximated by *two blue fit lines*. The intersection point of the two (diamond) is the "wave turbopause" (for details see text)

changes have been studied in detail in CRISTA and SABER data and have been found in a similar way in the HAMMONIA GCM [*Offermann et al.*, 2006, 2007, 2009]. These analyses have shown that the altitude of the wave turbopause is mostly determined by the relative intensities of GW and PW. It is furthermore found that the turbopause occurs in a layer about 8 km thick with an additional 3 km mixing layer on top [*Offermann et al.*, 2009].

28.2.4 Short Period Planetary Waves

Short period PW can be determined from the nightly measurements of an OH station down to periods of two days according to the Nyquist theorem. Typical examples are the Quasi-Two Day Waves (QTDW) that are pronounced in summer and in winter in the mesosphere. They have been analyzed for instance by means of the combined measurements of the stations at Wuppertal and Hohenpeissenberg. Combination of these two stations about 500 km apart is allowed because of the large horizontal scales of these planetary waves (wave numbers 2, 3, and 4). It offers the advantage of greatly improved weather statistics [*Offermann et al.*, 2011a].

Typical QTDW have periods near 2.1 days, and the summer values show a maximum somewhat after the solstices. The combined data have therefore been Fourier analyzed (FFT) and a period of 2.3 days has been studied in the summers 2004–2009. (The Hohenpeissenberg instrument began operations in 2003 only.) Temperature residues as shown in Fig. 28.1 have been used. Wuppertal data in the interval 1988–1993 showed good weather statistics and were also used. The amplitudes of the QTDW are found to be very variable. Hence the data from a number of years need to be averaged to obtain identifiable structures. One major result is shown in Fig. 28.7. It yields the mid-summer maximum as expected. Not expected, however, were two additional maxima found some 45–60 days before and after the center

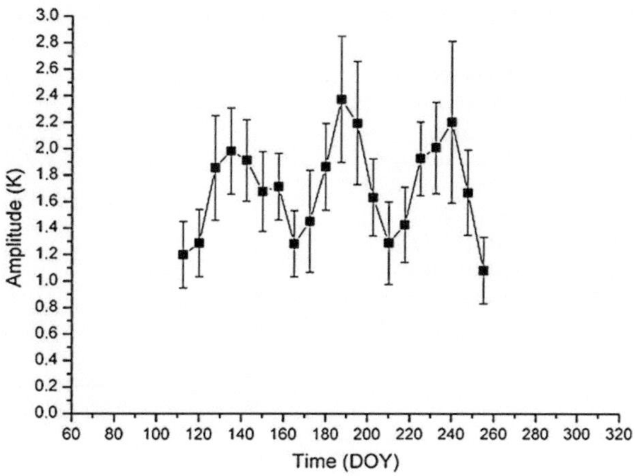

Fig. 28.7 Quasi Two-Day Wave (QTDW) amplitudes during the course of summer. Period is 2.3 days. Means of years 1988–1993 and 2004–2009 are shown. Time resolution is 7.5 days (DOY = Day Of the Year)

peak. The error bars in Fig. 28.7 are large. The triple structure is nevertheless statistically significant as the error bars of maxima and adjacent minima do not overlap.

The triplet structure in Fig. 28.7 appears uncommon and peculiar. Two steps have therefore been taken to check it. Additional temperature data from a different measurement technique have been analyzed for a triplet structure. Furthermore the mesosphere has been searched for structures suitable to excite QTDW in a way that a triplet is formed.

Additional data beyond those of the two OH stations were taken from MLS measurements on the AURA satellite. They were analyzed for waves with 2.3 day period and zonal wave numbers 2, 3, and 4. An approximate triple structure is found, indeed, which is mostly due to wave mode W3. Westward traveling QTDW of zonal wave number 3 (W3) have been cited in the literature as a typical QTDW. For an overview see *Offermann et al.* [2011a]. Zonal winds in the upper mesosphere are found to bc favorable for wave excitation in the time interval covered in Fig. 28.7.

QTDW are thought to originate from wind shears that lead to baroclinic and/or barotropic instabilities in the upper mesosphere. As an instability criterion the meridional derivative of the zonal mean, quasi-geostrophic potential vorticity Q should become negative [*Salby and Callaghan*, 2001; *Riggin et al.*, 2006]. This gradient is given by

$$\frac{\mathrm{d}Q}{\mathrm{d}y} = \beta - \frac{1}{a^2}\frac{\mathrm{d}}{\mathrm{d}\Phi}\left(\frac{1}{\cos\Phi}\right)\frac{\mathrm{d}}{\mathrm{d}\Phi}(u\cos\Phi) - f^2\frac{\mathrm{d}}{\mathrm{d}p}\left(\frac{1}{\sigma}\frac{\mathrm{d}u}{\mathrm{d}p}\right) \tag{28.1}$$

where $\beta = 2\Omega\cos\Phi/a$, $\Phi =$ latitude, $\Omega =$ earth angular velocity, $a =$ earth radius; f is the Coriolis parameter; $\sigma = -(RT/p)\mathrm{d}(\ln\Theta)/\mathrm{d}p$ is a static stability parameter with $R =$ gas constant, $\Theta =$ potential temperature.

The instability parameter $-Q_y$ (i.e. $-\mathrm{d}Q/\mathrm{d}y$) was calculated for the summer of 2005 with quasi-geostrophic zonal winds derived from MLS measurements. The values of $-Q_y$ were compared with the QTDW amplitudes (from the OH data), and a relatively close correlation was found. The correlation coefficient is $r = 0.57$ at

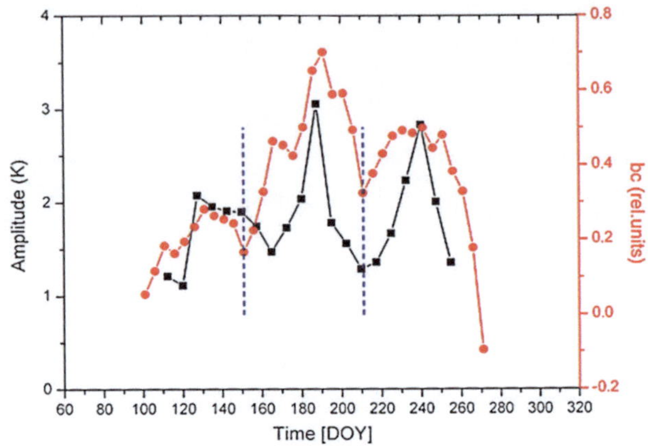

Fig. 28.8 Mean values of Quasi-Two Day Wave amplitudes and baroclinic instability parameter bc. QTDW data are from 2004–2009 (*black squares, left ordinate*). Mean *bc* values are from 2003–2008 (74 km altitude, *red dots, right ordinate*). *Dashed vertical lines* (*blue*) denote *bc* minima at DOY 151 and DOY 211

98 % significance. A better correlation might not be expected as a high instability value alone is not a sufficient condition for wave excitation.

The last term in Eq. (28.1) is the baroclinic instability. It can be approximated by an expression bc that is only dependent on the zonal wind field, i.e. its vertical gradient and curvature [for details see *Offermann et al.*, 2011a]. Parameter *bc* may thus be used as a proxy of total instability. Suitable wind data are available from a wind radar station (Juliusruh, 55°N, 13°E) that is relatively near to Wuppertal. The *bc* data obtained compare well with the $-Q_y$ values on the one hand, and with QTDW amplitudes on the other. Mean values of several years are given in Fig. 28.8. They show this agreement for mean values from six-year time intervals.

It is interesting to see in Fig. 28.8 that there is an indication of a triple structure in the bc values as well. This suggests that the QTDW triplet has its foundation in the stability structure (wind structure) of the atmosphere. It is also worth mentioning that the triplet is thus found in three data sets that are quite different: the OH local temperature measurements, the satellite temperature measurements of MLS, and the bc values from local radar wind measurements.

A detailed analysis shows that the QTDW originate from the interplay of zonal mean winds and the mesosphere stability structure [*Offermann et al.*, 2011a]. Hence QTDW can be used as indicators of atmospheric stability. They are used to study long-term changes of the stability situation in Sect. 28.4.3.

28.3 Solar Cycle and Trends

28.3.1 Temperature

Analyses of solar cycle influences and long-term trends require long data series. Satellite series are generally not suitable as they are too short as yet. Long-term data series are available from a number of ground stations that measure mesospheric winds or OH temperatures. For the long-term evaluation it is important that the data series is homogeneous, i.e. that data have been taken and treated in the same manner during the whole time interval. It should also be as complete as possible, i.e. without major data gaps.

The Wuppertal OH temperature data have been treated in a standardized manner since 1987. The OH layer is situated at about 87 km altitude. At this height the temperature shows a strong seasonal variation, which is about the strongest variation at this altitude at all. The standardized procedure models this variation by a three-component harmonic analysis that yields an annual mean temperature value, the amplitudes of annual, semi-annual, and ter-annual oscillations and the corresponding phases as mentioned above (see Fig. 28.1). The special form of the seasonal variation in this picture (high temperatures in winter, low values in summer) is a consequence of the general circulation. If changes of this circulation shall be studied analysis of the seasonal form in Fig. 28.1 is a convenient means. Hence a corresponding shape parameter ESD (Equivalent Summer Duration) has been introduced in Fig. 28.1 that measures the width of the summer valley. This is the time interval when the temperature is below 198 K. It is indicated as a horizontal red bar. The parameter is discussed in detail in *Offermann et al.* [2010] (their Sect. 5.2). Changes of ESD are discussed in Sect. 28.4.2.

The Wuppertal OH temperatures are homogeneous and fairly complete. They have been validated by comparison especially with SABER temperatures (V1.07). There appears to be a warm bias of about 3 K which is, however, unimportant for analysis of the relative variations discussed here. This data set and its accuracies have been described by *Offermann et al.* [2010]. It covers about two solar cycles (#22, #23), and is thus quite suitable for long-term analyses.

The data analysis shows that temperature variations due to solar influences and due to long-term changes are of similar magnitude. Hence special care must be taken to disentangle the two effects. Simultaneous analysis of temperature trends and temperature sensitivity to the solar fluxes shows a strong dependence of the results on the length of the time interval used. Figure 28.9a for instance shows that the solar flux sensitivity of temperature can vary by a factor 5 depending on the time window selected (with the length of one solar cycle)! Figure 28.9b shows the sensitivity dependence on the window length. Two important conclusions are drawn: (1) To give useful results the data window must be much longer than one solar cycle. (2) In each trend and solar cycle analysis published the time interval used needs to be stated.

For the Wuppertal analysis cited the window 1988–2008 was used and yielded a long-term temperature trend of −0.23 K/year and a solar flux sensitivity of 0.035 K/SFU (solar flux unit). Details for the variations of the amplitudes and phases of the three seasonal components are given in *Offermann et al.* [2010].

The high density of the Wuppertal data series allows trend analyses at a time resolution better than one year. For instance long-term data can be grouped in monthly bins, and these intervals be treated separately. A corresponding monthly analysis is shown in Fig. 28.10. It is a bi-variate analysis for long-term trends and solar flux influences. The picture shows the long-term trends. There are surprisingly large changes during the course of the year. Furthermore, the changes from month to

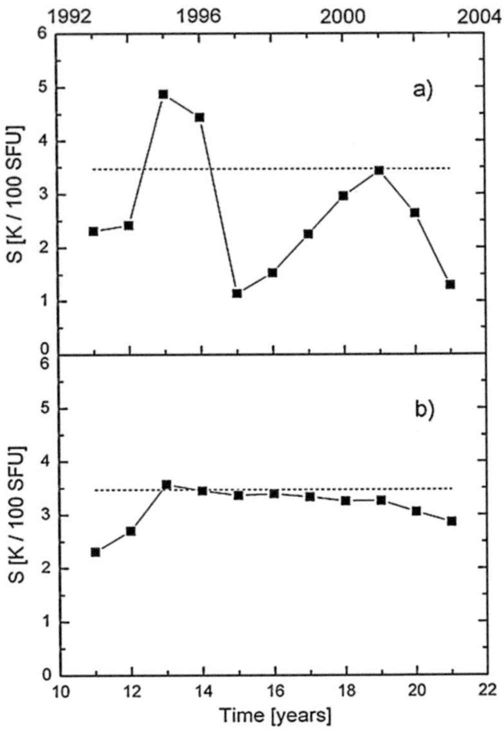

Fig. 28.9 (**a**) Temperature sensitivity *S* to solar flux density F10.7 cm in 11 year time windows shifted through the data interval 1988–2008 in steps of 1 year. *Abscissa* (*above*) gives the year of window center and runs from 1992 to 2004.
(**b**) Temperature sensitivity *S* in time windows of lengths increasing from 11 to 21 years. *Abscissa* (*below*) runs from 10 to 22 years. *Dashed horizontal lines* are for 3.5 K/100 SFU (see text)

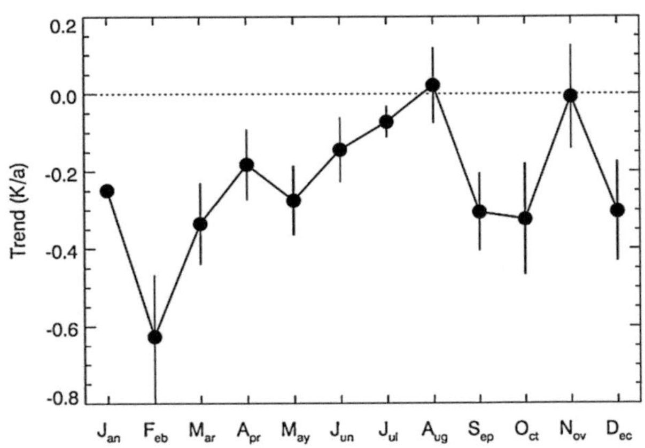

Fig. 28.10 Monthly long-term temperature trends (K/year) corrected for solar flux influences, 1987–2008. Annual mean is −0.23 K/year

month can be abrupt and very substantial. The trend distribution in Fig. 28.10 is fairly asymmetric with respect to the summer/winter or spring/autumn seasons. This leads to a long-term deformation of the seasonal curves as that given in Fig. 28.1. A corresponding change of the shape parameter ESD results. This is further discussed in Sect. 28.4.2.

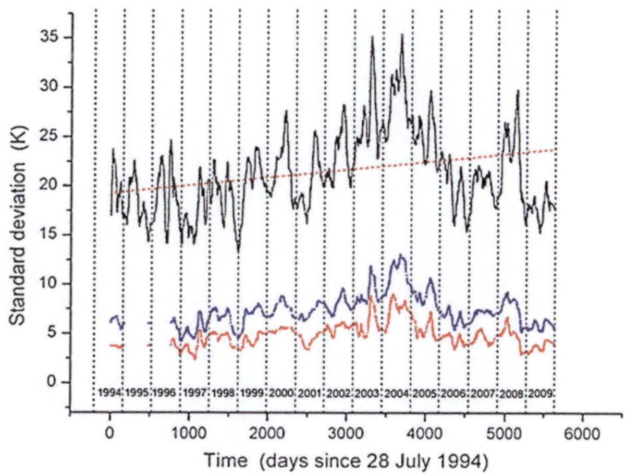

Fig. 28.11 Temperature standard deviations σ_N (*black*) for the time interval 1994–2009 at Wuppertal. For comparison σ_a (*blue*) and 3-minutes amplitudes (*red*) are given (see text). A 50 pts smoothing has been applied (for details see text)

28.3.2 Zonal Wind

Zonal winds used in the present study are mostly radar winds measured at the station of Juliusruh (55°N, 13°E). Measurement technique and data analysis have been described by *Keuer et al.* [2007]. These authors give a detailed analysis of zonal and meridional winds in the mesosphere including long-term trends and solar flux influences. They obtain a long-term increase of zonal wind speed of 0.52 m/s per year at 87.5 km altitude (their Fig. 15, solar influences removed). This will be used in Sect. 28.4.2 below.

28.3.3 Gravity Waves

Short period gravity waves are represented by standard deviations σ_N from nightly mean OH temperatures. The long-term development of σ_N since 1994 is shown in Fig. 28.11. For comparison very short period oscillations are shown as represented by their standard deviation σ_a and the amplitudes of the three-minutes oscillations (see Sect. 28.2.2). The σ_N data are quite variable. If a solar cycle variation is contained in them it should become visible at the solar maximum around years 2000/2001 and the solar minima at about 1996 and 2008/2009. Nothing of this kind is detectable, however, in the strong σ_N variability. The data rather show a trend-like increase until 2004 and a subsequent decrease. Second largest variations are the seasonal variations discussed above (Sect. 28.2.2). A minor intra-decadal variation is also found [*Offermann et al.*, 2011b]. Its time scale appears to be on the order of four years. The amplitude is about 2 K (not shown here).

A linear fit to the whole data set is shown in Fig. 28.11 as a dashed (red) line. It shows a moderate, but significant positive gradient of 0.29 ± 0.02 K/year.

The two other curves in Fig. 28.11 for the short period oscillations behave similarly. They show a close correlation with σ_N. The correlation coefficients are

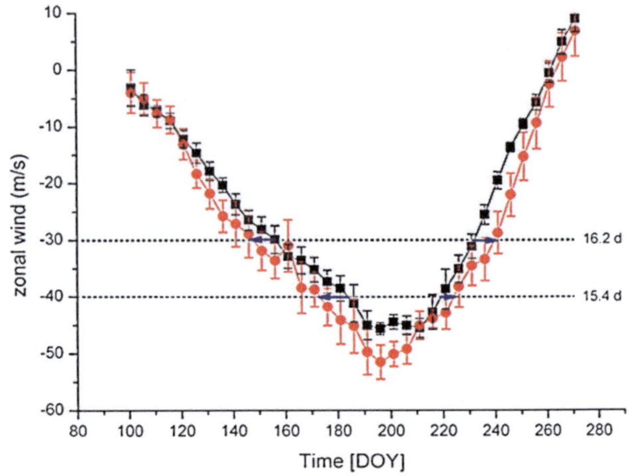

Fig. 28.12 Mean zonal winds at 55°N, 73 km in time intervals 1990–1995 (*black squares*) and 1997–2002 (*red dots*). Errors of the mean are also given. *Dashed horizontal lines* indicate phase speeds of Quasi-Two Day Waves W4 (above) and W3 (below). For *blue horizontal arrows* and *numbers* given see text

$r = 0.94$ for σ_a (blue curve) and $r = 0.88$ for the 3 min amplitudes (red curve). From this very close correlation it is concluded that σ_N is representative of the GW activity and GW dissipation [*Offermann et al.*, 2011b]. The σ_N increase is about 1.5 %/year. As GW dissipation is one of the major forcings in the mesosphere the question is whether corresponding consequences can be seen in measured data. This is discussed in Sect. 28.4.

28.4 Circulation Changes

Many of the experimental results described above appear to be indicative of or related to changes of the general circulation. The reason common to them are possibly the gravity wave trends detected. This is discussed here.

28.4.1 Zonal Wind

It is generally believed that GW dissipation in the upper mesosphere forces the zonal wind. The increase of the GW proxy σ_N in Fig. 28.11 suggests a 1.5 % increase of GW activity and dissipation at 87 km altitude. This and the above mentioned (Sect. 28.3.2) increase in zonal wind speed of 0.52 m/s per year at 87.5 km are thus in agreement with the general picture. An increase in zonal wind may certainly be considered as part of a change in general circulation. If however the details are studied two results are found that do not quite fit the picture.

1. At altitudes below 80 km the zonal wind trend in summer is opposite to that at the higher altitudes [*Keuer et al.*, 2007]. This is also seen from Fig. 28.12 which gives zonal winds at Juliusruh (73 km altitude) in two different time intervals. At

a given day in summer the mean zonal wind in the second time interval 1997–2002 is lower (more negative) than in the earlier interval 1990–1995 [*Offermann et al.*, 2011a]. This has consequences for the Quasi-Two Day Waves discussed below (Sect. 28.4.3).

2. The general belief concerning zonal wind forcing in the mesosphere assumes a special seasonal GW variation. Maximum GW activity is expected around the solstices [e.g. *Fritts and Alexander*, 2003, their Fig. 25]. This is not the case for the OH data in Fig. 28.4 and the satellite data in Fig. 28.2. The maxima rather occur at wind turn around, i.e. nearer to equinox. The same is seen for the eddy coefficients obtained from the WACCM general circulation model (Fig. 28.5).

The double peak structure of the GW seen in Fig. 28.2 occurs in a rather limited altitude and latitude regime. At 50°N it is found at about 85–95 km. At 70°N it extends somewhat higher up. Above and below this area at 50°N the maxima tend to occur around the solstices as expected (Fig. 28.2). In consequence the altitude regime around the mesopause at middle and higher latitude appears to play a special role that is more complicated than the basic picture. It should also be noted that the double peak structure is not a deep modulation of the seasonal GW curves but a rather moderate change as well in the OH temperature data as in the satellite data.

28.4.2 Summer Duration

The seasonal variation of temperature near the mesopause (Fig. 28.1) is strongly influenced by the general circulation (ascending air in summer and descending air in winter). If a change in general circulation occurs one would expect a change of that seasonal curve. The width of the summer valley, i.e. the Equivalent Summer Duration (ESD) has been used to monitor possible changes. Such a change is seen, indeed. Since 1988 an ESD increase is observed with a rate of 1.2 days per year [*Offermann et al.*, 2010]. The analysis has taken into account corrections for solar flux influences.

The cumulative change in 21 years can be made visible by means of the monthly trends given in Fig. 28.10. This is shown in Fig. 28.13, where the temperature data of year 1988 are given and have been decreased by the monthly trends in 21 years. The resulting (red) curve shows a wider valley which can be seen from the dashed horizontal line at 198 K temperature. *Offermann et al.* [2010] have discussed in detail that only part of this change is due to the general temperature decrease mentioned above, and large part must be due to dynamics/circulation effects. These authors have also analyzed changes in summer duration (SD) in the stratosphere. The SD parameter was determined as the time difference between autumn and spring reversal of the zonal wind. They found a trend, indeed, that was of similar magnitude as in the mesosphere. The sign is, however, opposite, i.e. the summer duration in the stratosphere decreases.

A mesospheric change of summer length cannot be derived from zonal wind reversal because corresponding data are available for spring only, not for autumn

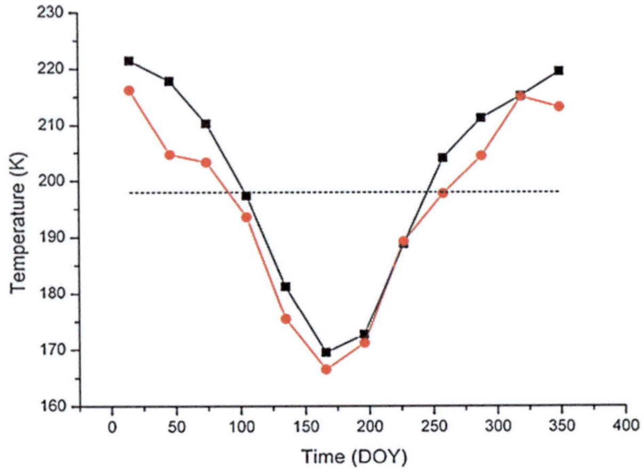

Fig. 28.13 Seasonal temperature variation as taken from a harmonic fit to measured data of year 1988 (*black squares*). Decreased temperatures after 21 years are shown as *red dots* ("trend results"), calculated by means of the monthly trends in Fig. 28.10

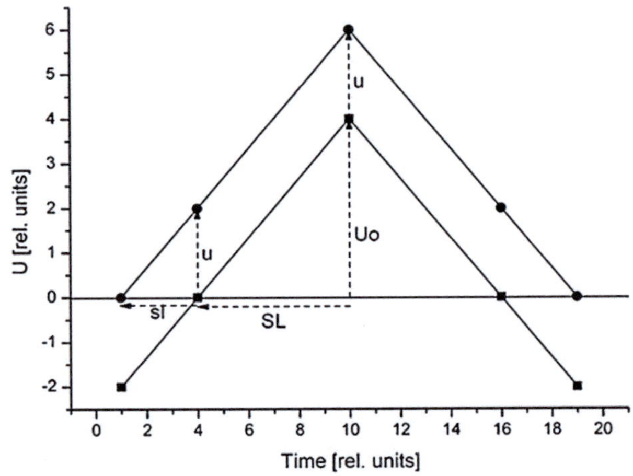

Fig. 28.14 Sketch of zonal wind variation in summer (87.5 km, see text). Summer length is $2 * SL$. Increase of summer length is by $2 * sl$ and results from an increase of zonal wind U by u

[*Offermann et al.*, 2011b]. It can, however, be related to the zonal wind trends in a qualitative manner in the following way: The zonal wind U increases from spring turnaround ($U = 0$) towards the middle of summer, and subsequently decreases towards autumn turnaround ($U = 0$). This is sketched in a very simplified manner in Fig. 28.14 by assuming linear changes (87.5 km). The summer length is $2 * SL$. The parameter u denotes the increase of the wind speed (for instance in one year). It is assumed here for simplicity that u is the same at all times of the summer. This leads to a shift sl of the turnaround point ($U = 0$) to earlier times in the first half of summer. From the geometry of Fig. 28.14 it follows that $U_0/SL = u/sl$. Mid summer maximum wind speed U_0 is about 12 m/s, half of the summer length is $SL = 62$ days (data for 87.5 km from *Offermann et al.* [2011a, their Fig. 6]). The increase rate u as derived above from *Keuer et al.* [2007] is $u = 0.52$ (m/s)/year. The resulting increase of the summer length is $sl = 2.7$ days/year (shift of spring turnaround towards earlier times), which is half of the total increase.

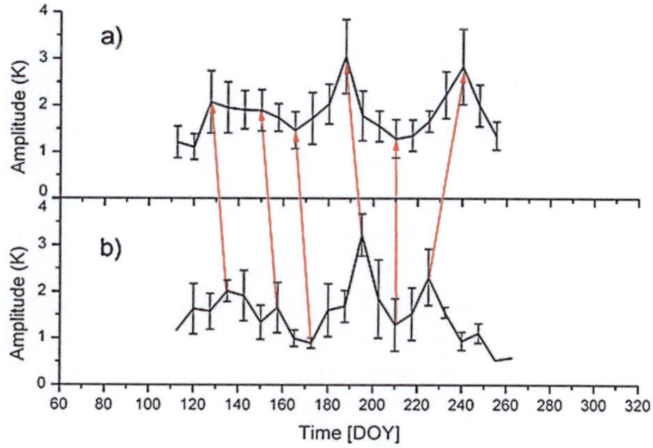

Fig. 28.15 QTDW amplitudes during the course of summer in the time intervals: (**a**) 2004–2009, Combi data; (**b**) 1988–1993, Wuppertal. Time resolution is 7.5 days. For the *red arrows* see text

A more detailed analysis is given by *Offermann et al.* [2011b]. They find that these results are an exaggeration, but that basically the sign and the order of magnitude of the trend are correct. Hence the measured ESD increase is in principle related to the GW increase.

28.4.3 Quasi-Two Day Waves

Quasi-Two Day Waves (QTDW) are expected and observed if the mesosphere is dynamically unstable (Eq. (28.1)) and the zonal wind has a suitable speed, i.e. it matches the QTDW phase speed. Either aspect is found to play a role in the present data (see above).

If a complex phenomenon like a change in general circulation is studied trends of simple parameters as temperature or wind speed are sometimes not too helpful because they need careful quantitative analyses. It is easier to use characteristic "earmarks" if they are available. The ESD parameter used above rests on such earmarks, as it is determined from the circulation turnarounds or from the characteristic spring and autumn points (198 K) of the seasonal temperature variation. Other earmarks are the three peaks of the QTDW triplet (Fig. 28.7). Their distance in time defines a temporal scale that is easy to study. As an example Fig. 28.15 shows QTDW triplets from two different time intervals 15.5 years apart. Corresponding maxima and minima in the two curves have been connected by red arrows. These arrows are not parallel but diverge instead. This shows that the time scale has changed. The distance between the first and the last triplet peaks is 112.5 days in the upper panel and 90 days in the lower. Hence the time distance between the earmarks has increased by 22.5 days in 15.5 years [*Offermann et al.*, 2011a]. This is a rate of 1.5 days per year. (Note the ESD increase of 1.2 days per year in Sect. 28.4.2, which is a similar summer structure.) The detailed reasons for these changes appear to be twofold as mentioned.

1. The zonal wind field has changed, as is shown in Fig. 28.12. The QTDW mode is mostly W3 with a phase speed of about −40 m/s. The days when this speed is met by the (measured) zonal wind is favorable for QTDW excitation. The blue horizontal arrows in Fig. 28.12 indicate that these favorable days have shifted to earlier times in the first half of summer, and to later times in its second half. This occurred over a time distance of seven years and yielded an accumulated time shift (sum of the two blue arrows) of 15.4 days as indicated in Fig. 28.12 for the W3 phase speed. This is the right direction and the right order of magnitude for the changes in Fig. 28.15. For details see *Offermann et al.* [2011a].
2. Mesospheric instability can be derived from satellite determination of temperature and quasi-geostrophic winds according to Eq. (28.1) [see *Offermann et al.*, 2011a]. Long-term analysis is, however, not yet possible as the satellite data series presently available are too short. Suitable long-term series are available from single station (radar) wind measurements as Juliusruh. The measured zonal wind vertical structure can be used to calculate the baroclinic instability part of Eq. (28.1) (last term of the equation). The corresponding parameter *bc* can be used as an instability proxy as discussed above (Sect. 28.2.4). If there are long-term trends of temperatures and zonal winds corresponding changes in bc are expected. From the changes seen in Fig. 28.15 one would preferentially expect changes in the early and late summer wings, respectively, of the curve shown in Fig. 28.8, i.e. *bc* should increase in absolute values. An increase was found, indeed, by *Offermann et al.* [2011a] especially in the early part of summer (DOY 100–140). The baroclinic instability increased by about a factor 1.6 (±6 %) in seven years (using the intervals 1990–1995 and 1997–2002 for comparison). This can be compared to the σ_N data in the same time interval (1993–2000). The latter increase by a factor 1.5 (±13 %) in this time interval. This is an interesting coincidence that suggests further studies.
   The late summer branch of the curve in Fig. 28.8 does not show a change in this time interval. There are, however, indications that there might have been an increase at somewhat later times. These data could, however, not be compared to the earlier ones because of technical reconstructions of the radar instrument. Further analyses of radar data are therefore needed.

It has been noted in the literature that gravity wave increases should lead to increased atmospheric instability in the mesosphere [*Hecht*, 2004; *Horinouchi et al.*, 2002]. It remains to be determined whether/how the more local events of GW breaking and GW interaction with the horizontal winds are related in detail to the large scale barotropic/baroclinic instabilities given in Eq. (28.1).

28.5 Summary and Conclusions

Ground stations for OH temperature and radar wind measurements yield long-term data series. These are used to look for trends in mesospheric dynamics. A long-term increase of short period gravity waves and their dissipation is found in OH temperature data at 87 km altitude. Wuppertal data (51°N. 7°E) are analyzed since 1988. The increase rate is 1.5 %/year.

Long-term zonal wind measurements were taken at Juliusruh (55°N, 13°E). Published data show an increase of zonal wind speed at a rate of 0.52 m/s/year at 87.5 km altitude. Temperature and wind data cited have been corrected for solar flux influences which are appreciable.

The seasonal temperature variation in the mesosphere is used to estimate the length of summer. The Equivalent Summer Duration ESD is strongly influenced by the general circulation, and hence an ESD change should indicate a corresponding change in circulation. An increase of ESD is observed at a rate of 1.2 days/year since 1988. Similar values, but of opposite sign, are observed in the middle stratosphere. The latter are obtained from the time difference between circulation turnaround in spring and autumn. Corresponding values for the mesosphere can only be estimated because the wind data at mesopause altitude do not reach far enough back. The estimated values indicate similar results, though. They use the increase rate of the zonal wind.

Another important structure in the mesosphere is the summer distribution of the Quasi-Two Day Waves (QTDW). This distribution shows three characteristic maxima. They are earmarks well suitable to monitor long-term changes of summer time scales, similarly as do the ESD values. Changes of these scales are observed, indeed, with increasing distance between the first and the last QTDW maximum in summer at a rate of 1.5 days/year. QTDWs occur in the summer mesosphere if the zonal wind compares favorably with the phase speed of the QTDW, and if the atmosphere is baroclinically/barotropically unstable. Either feature was found in the long-term data analyzed.

The long-term changes of zonal wind, ESD, and QTDWs fit the idea of a changing general circulation. A common reason may be the long-term increase in GW. The origin of the GW trend for its part may be in the lower atmosphere, possibly a change of wind filtering in the stratosphere. This suggests a coupling to stratospheric circulation changes. This is nothing but speculation at present, but deserves future detailed studies. The trends in GW, zonal wind, ESD, and QTDW can all be further followed in the coming years and will show whether the picture presented here will substantiate.

Acknowledgements We thank D. Marsh (ACD, NCAR, Boulder, USA) for the WACCM 3.5 eddy coefficients. Thanks for many helpful discussions go to P. Hoffmann (IAP), Ch. Kalicinsky, P. Knieling, W. Steinbrecht (DWD), and J. Wintel. Some part of this work was funded by DFG (SPP 1176) under grant No. OF 12/3-1.

References

Baumgaertner, A., McDonald, A., Hibbins, R., Fritts, D., Murphy, D., & Vincent, R. (2008). Short-period planetary waves in the antarctic middle atmosphere. *Journal of Atmospheric and Solar-Terrestrial Physics*, *70*, 1336–1350. doi:10.1016/j.jastp.2008.04.007.

Dowdy, A. J., Vincent, R. A., Tsutsumi, M., Igarashi, K., Murayama, Y., Singer, W., & Murphy, D. J. (2007). Polar mesosphere and lower thermosphere dynamics: 1. Mean winds and gravity wave climatologies. *Journal of Geophysical Research, 112*(D17), D17104. doi:10.1029/2006JD008126.

Espy, P. J., & Stegman, J. (2002). Trends and variability of mesospheric temperature at high-latitudes. *Physics and Chemistry of the Earth, 27*, 543–553.

Fritts, D. C., & Alexander, M. J. (2003). Gravity wave dynamics and effects in the middle atmosphere. *Reviews of Geophysics, 41*(1), 1–64. doi:10.1029/2001RG000106.

Fritts, D. C., & Wang, W. J. (2009). Gravity wave-fine structure interactions: a reservoir of small-scale and large scale turbulence energy. *Geophysical Research Letters, 36*(19), L19805. doi:10.1029/2009GL039501.

Garcia, R., & Solomon, S. (1985). The effect of breaking gravity waves on the dynamics and chemical composition of the mesosphere and lower thermosphere. *Journal of Geophysical Research, 90*(D2), 3850–3868.

Garcia, R. R. (1991). Parameterization of planetary wave breaking in the middle atmosphere. *Journal of the Atmospheric Sciences, 48*(11), 1405–1419.

Geller, M. A., & Gong, J. (2010). Gravity wave kinetic, potential, and vertical fluctuation energies as indicators of different frequency gravity waves. *Journal of Geophysical Research, 115*, D11111. doi:10.1029/2009JD012266.

Hecht, J. H. (2004). Instability layers and airglow imaging. *Reviews of Geophysics, 42*, RG1001.

Hecht, J. H., Liu, A. Z., Walterscheid, R. L., Franke, S. J., Rudy, R. J., Taylor, M. J., & Pautet, P.-D. (2007). Characteristics of short-period wavelike features near 87 km altitude from airglow and lidar observations over Maui. *Journal of Geophysical Research, 112*, D16101. doi:10.1029/2006JD008148.

Hines, C. O. (1960). Internal atmospheric gravity waves at ionospheric heights. *Canadian Journal of Physics, 38*, 1441–1481.

Holton, J. R. (1983). The influence of gravity wave breaking on the general circulation of the middle atmosphere. *Journal of the Atmospheric Sciences, 40*, 2497–2507.

Horinouchi, T., Nakamura, T., & Kosaka, J. (2002). Connectively generated mesoscale gravity waves simulated throughout the middle atmosphere. *Geophysical Research Letters, 29*, 21. doi:10.1029/2002GL016069.

Keuer, D., Hoffmann, P., Singer, W., & Bremer, J. (2007). Long-term variations of the mesospheric wind field at mid-latitudes. *Annales Geophysicae, 25*(8), 1779–1790.

Lindzen, R. S. (1981). Turbulence and stress owing to gravity wave and tidal breakdown. *Journal of Geophysical Research, 86*, 9707–9714.

Liu, G., Shepherd, G. G., & Roble, R. G. (2008). Seasonal variations of the nighttime $O(^1S)$ and OH airglow emission rates at mid-to-high latitudes in the context of the large-scale circulation. *Journal of Geophysical Research, 113*, A06302. doi:10.1029/2007JA012854.

Muller, H. G. (1972). Long period wind oscillations. *Philosophical Transactions of the Royal Society of London, Series A: Mathematical and Physical Sciences, 272*, 585–598.

Offermann, D., Jarisch, M., Oberheide, J., Gusev, O., Wohltmann, I., Russell III, J. M., & Mlynczak, M. G. (2006). Global wave activity from upper stratosphere to lower thermosphere: a new turbopause concept. *Journal of Atmospheric and Solar-Terrestrial Physics, 68*, 1709–1729.

Offermann, D., Jarisch, M., Schmidt, H., Oberheide, J., Grossmann, K.-U., Gusev, O., Russell III, J. M., & Mlynczak, M. G. (2007). The "wave turbopause". *Journal of Atmospheric and Solar-Terrestrial Physics, 69*, 2139–2158. doi:10.1026/j.jastp.2007.05.012.

Offermann, D., Gusev, O., Donner, M., Forbes, J. M., Hagan, M., Mlynczak, M. G., Oberheide, J., Preusse, P., Schmidt, H., & Russell III, J. M. (2009). Relative intensities of middle atmosphere waves. *Journal of Geophysical Research, 114*, D06110. doi:10.1029/2008JD010662. Copyright 2009 American Geophysical Union. Reproduced by permission of American Geophysical Union.

Offermann, D., Hoffmann, P., Knieling, P., Koppmann, R., Oberheide, J., & Steinbrecht, W. (2010). Long-term trends and solar cycle variations of mesospheric temperature and dynamics. *Journal of Geophysical Research*, *115*, D18127. doi:10.1029/2009JD013363. Copyright 2010 American Geophysical Union. Reproduced by permission of American Geophysical Union.

Offermann, D., Hoffmann, P., Knieling, P., Koppmann, R., Oberheide, J., Riggin, D. M., Tunbridge, V. M., & Steinbrecht, W. (2011a). Quasi-two day waves in the summer mesosphere: triple structure of amplitudes and long-term development. *Journal of Geophysical Research*, *116*, D00P02. doi:10.1029/2010JD015051. Copyright 2011 American Geophysical Union. Reproduced by permission of American Geophysical Union.

Offermann, D., Wintel, J., Kalicinsky, C., Knieling, P., Koppmann, R., & Steinbrecht, W. (2011b). Long-term development of short period gravity waves in middle Europe. *Journal of Geophysical Research*, *116*, D00P07. doi:10.1029/2010JD015544. Copyright 2011 American Geophysical Union. Reproduced by permission of American Geophysical Union.

Preusse, P., Eckermann, S. D., Ern, M., Oberheide, J., Picard, R. H., Roble, R. G., Riese, M., Russell III, J. M., & Mlynczak, M. G. (2009). Global ray tracing simulations of the SABER gravity wave climatology. *Journal of Geophysical Research*, *114*, D08126. doi:10.1029/2008JD011214.

Riggin, D. M., Liu, H. L., Lieberman, R. S., Roble, R. R., Russell III, J. M., Mertens, C. J. H., Mlynczak, M. G., Pancheva, D., Franke, S. J., Murayama, Y., Manson, A. H., Meek, C. E., & Vincent, R. A. (2006). Observations of the 5-day wave in the mesosphere and lower thermosphere. *Journal of Atmospheric and Solar-Terrestrial Physics*, *68*, 323–339.

Salby, M. L., & Callaghan, P. F. (2001). Seasonal amplification of the 2-day wave: relationship between normal mode and instability. *Journal of the Atmospheric Sciences*, *58*, 1858–1869.

Snively, J., Pasko, V. P., & Taylor, M. J. (2010). OH and OI airglow modulation by ducted short-period gravity waves: effects of trapping altitude. *Journal of Geophysical Research*, *115*, A11311. doi:10.1029/2009JA015236.

Taylor, M. J., Pendleton, W., Pautet, P.-D., Zhao, Y., Olsen, C., Surendra Babu, H. K., Medeiros, A. F., & Takahashi, H. (2007). Recent progress in mesospheric gravity wave studies using nightglow imaging system. *Revista Brasileira de Geofisica*, *25*(suppl. 2). doi:10.1590/S0102-261X2007000600007.

Chapter 29
Solar Effects on Chemistry and Climate Including Ocean Interactions

Ulrike Langematz, Anne Kubin, Christoph Brühl, Andreas J.G. Baumgaertner, Ulrich Cubasch, and Thomas Spangehl

Abstract In the Project on Solar Effects on Chemistry and Climate Including Ocean Interactions (ProSECCO) fundamental questions of the impact of solar variability on Earth have been investigated with improved climate system models and observations. On the decadal time scale, the atmospheric signature of the 11-year Schwabe cycle and the underlying mechanisms have been studied. This included the impact of variations in UV radiation and particle precipitation on stratospheric chemistry and ozone, as well as on the solar signal in the troposphere and on climate. On the centennial to millennium time scale, effects of solar variability on climate of different pre-industrial periods, focusing on the Maunder Minimum and the mid-Holocene, have been addressed.

29.1 Introduction

Variations in solar activity represent a natural external forcing of Earth's atmosphere which affects all height ranges from the upper atmosphere (ionosphere,

U. Langematz (✉) · A. Kubin · U. Cubasch · T. Spangehl
Institut für Meteorologie, Freie Universität Berlin, Carl-Heinrich-Becker-Weg 6-10, 12165 Berlin, Germany
e-mail: ulrike.langematz@met.fu-berlin.de

C. Brühl · A.J.G. Baumgaertner
Max-Planck-Institut für Chemie, Hahn-Meitner-Weg 1, 55128 Mainz, Germany

C. Brühl
e-mail: christoph.bruehl@mpic.de

Present address:
A.J.G. Baumgaertner
Deutsches Zentrum für Luft-und Raumfahrt (DLR), Project Management Agency, Heinrich-Konen-Str. 1, 53227 Bonn, Germany
e-mail: andreas.baumgaertner@dlr.de

Present address:
T. Spangehl
Deutscher Wetterdienst, Frankfurter Str. 135, 63067 Offenbach am Main, Germany
e-mail: thomas.spangehl@dwd.de

F.-J. Lübken (ed.), *Climate and Weather of the Sun-Earth System (CAWSES)*, Springer Atmospheric Sciences,
DOI 10.1007/978-94-007-4348-9_29,

thermosphere, >80 km) and the middle atmosphere (mesosphere, stratosphere, 10–80 km) down to the lower atmosphere (troposphere, 0–10 km) and the surface. Observations have shown clear responses to the 11-year solar Schwabe cycle in middle atmosphere temperature, stratospheric ozone concentration, or tropospheric climate variability patterns. The mechanisms which lead to the solar signal in the atmosphere are only partly understood. Solar activity variations affect the atmosphere directly by variations in in-coming electromagnetic radiation and energetic particles. While total solar irradiance (TSI) at the top of the atmosphere varies about 0.1 % during the 11-year solar cycle, variations at ultraviolet (UV) wavelengths which are responsible for ozone photo-chemistry are of the order of 5–8 % [*Lean*, 1997] and reach 100 % in the Lyman-α line which influences photo-dissociation of water vapour and carbon dioxide in the upper mesosphere. Variations of energetic particle precipitation with the solar cycle, like solar protons or low and high energy electrons lead to an activation of chemically active substances, such as nitrous oxides (NO_X), and hence to a variation of composition with the solar cycle. While the direct changes in electromagnetic radiation and charged particles can explain the response of the upper atmosphere, an impact of these variations on the solar signal in the lower stratosphere and troposphere can only occur by a transfer from above by indirect mechanisms. *Kodera and Kuroda* [2002] explained the poleward downward propagation of upper stratospheric westerly wind anomalies during solar maximum with an iterative interaction of planetary wave dissipation and the zonal mean background circulation in winter. This dynamical feedback is influenced itself by the phase of the tropical Quasi-Biennial-Oscillation (QBO) of the zonal wind in the lower stratosphere. An overview of recent observations and the current understanding of the underlying mechanisms is given in *Gray et al.* [2010].

Solar activity also varies on time scales beyond the 11-year cycle. Sunspots can be used to reconstruct solar activity back to the beginning of the 17th century. Furthermore, the cosmogenic isotopes/nuclides ^{10}Be and ^{14}C are used as proxies to reconstruct solar activity for the complete Holocene, the current geological epoch which began about 10,000 years ago (cf. Chap. 2). Periodic variations in solar activity have been documented for the Holocene record such as the Gleissberg and Suess (de Vries) cycles which show distinct spectral peaks at 87 and 208 years [e.g., *Wanner et al.*, 2008, and references therein] and millennial time scales such as the 2300 year Hallstatt cycle [*Wanner et al.*, 2008; *Gray et al.*, 2010]. The Holocene represents the current interglacial following the Last Glacial Maximum (LGM) around 20,000 before present (hereafter referred to as BP). The transition phases between glacials and interglacials are characterised by the build-up or decay of continental ice sheets that can be explained by variations of Earth's orbit around the Sun and related modifications of the latitudinal and seasonal distribution of Earth's insolation [*Kaspar et al.*, 2007, and references therein]. The retreat of the ice sheets subsequent to the LGM including continental freshwater outflow events considerably affected the climate until the early- to mid-Holocene (major sea level changes only ceased after 6000 BP). Still, the period from the mid-Holocene onwards underlies climate variations which have been linked to migrations, societal collapses, wars [e.g., *Wanner et al.*, 2008, and references therein]. The time span from the mid- to late Holocene is characterised by a southward shift of the northern hemisphere (NH) summertime innertropical convergence zone (ITCZ), a weakening of the African and Indian Monsoon systems, an increase in the El Niño-Southern Oscillation (ENSO) variability and a shift of the Arctic Oscillation (AO)/North Atlantic Oscillation (NAO) towards its negative phase. Model studies support the role of the orbital forcing for these long term shifts [e.g., *Wanner et al.*, 2008].

Much attention has been drawn to the climate of the last Millennium. It is characterised by the transition from the Medieval Warm Period (MWP, around 900–1350 AD) to the Little Ice Age (LIA, around 1500–1850 AD) and the subsequent modern warming [*Solomon et al.*, 2007]. Model based studies indicate that the solar forcing is important to capture the spatial response patterns [*Mann et al.*, 2009]. Here the realistic representation of stratospheric forcing and coupling mechanisms might be of importance [*Shindell et al.*, 2001], besides the dynamic response of the ocean [*Cubasch et al.*, 1997].

The goal of ProSECCO was to contribute to the investigation of the solar impact on Earth's atmosphere from the decadal to the millennium time scales by applying sophisticated climate system models and observations. The studies have addressed open questions concerning the decadal solar signal in stratospheric ozone and temperature and the role of the so-called 'top-down' stratospheric solar forcing for the tropospheric weather and climate. The assessment of the solar impact during the Little Ice Age and the mid-Holocene allows us to derive conclusions of the atmospheric response to solar forcing for different solar forcing amplitudes and under different climate states.

29.2 Models for Solar Variability Studies

Early studies of the 11-year stratospheric solar signal employing 3-dimensional general circulation models (GCMs) with prescribed TSI or UV changes for either the maximum or minimum phase of the 11-year solar cycle and externally calculated solar induced ozone anomalies uniformly reproduced a significant annual mean temperature response in the upper stratosphere of $\sim$1 K [*Matthes et al.*, 2003, and references therein]. However, large discrepancies between the models were obvious at high latitudes in winter which were attributed to missing dynamical feedback in the models [e.g., *Kodera et al.*, 2003]. On the centennial time scale, early simulations with coupled atmosphere-ocean models of the solar signal during the Maunder Minimum were able to explain features of climate reconstructions by solar irradiance anomalies. However, these studies either included ocean interactions neglecting the upper atmosphere [e.g., *Cubasch et al.*, 1997] or included stratospheric processes neglecting interactive ocean responses [e.g., *Langematz et al.*, 2005]. With the development of chemistry-climate models (CCMs) that involve complex modules to calculate changes in atmospheric composition, and of stratosphere resolving climate models, improved tools became available to derive the solar signal in the atmosphere and to understand the underlying mechanisms. In this study we employ two state-of-the-art numerical models that complement each other to better address the research questions of our project at the different time scales of solar impact.

On the decadal time scale simulations were performed with the ECHAM-MESSy Atmospheric Chemistry (EMAC) model [*Jöckel et al.*, 2005, 2006; *Roeckner et al.*, 2006]. EMAC is a CCM that resolves the troposphere and middle atmosphere up to an altitude of about 80 km. It includes an interactive chemistry module [*Sander et al.*, 2005] to account for the effects of solar irradiance variations on stratospheric ozone. In addition, EMAC employs a short-wave (SW) radiation scheme with improved spectral resolution that was specifically designed for solar variability studies [FUBRad, *Nissen et al.*, 2007]. To consider the effects of solar cycle variations in auroral electron precipitation and of solar proton events (SPEs) on the NO_x and ozone concentrations, EMAC was extended by new modules [*Baumgaertner et al.*, 2009, 2010]. For a more detailed description of this work it is referred to Chap. 17 of this book. Transient simulations covering the period 1960–2000 (2005) performed according to the recommendations of the Stratospheric Processes and their Role in Climate (SPARC) Chemistry-Climate-Model Validation (CCMVal) initiative for changes in greenhouse gas (GHG) concentrations, ozone depleting substances, sea surface temperatures and sea-ice distributions, as well as aerosols [REF-B1 simulations, *SPARC*, 2010] were analysed. For the first time, the solar signal could be examined based on daily spectral irradiance data, prescribed at the top of the atmosphere (TOA) (cf. http://www.geo.fu-berlin.de/en/met/ag/strat/forschung/SOLARIS/index.html) hence allowing to derive the transient solar signal throughout the past four decades and to derive uncertainties in the attribution of atmospheric change caused by observed solar variability. Two REF-B1 simulations were available, performed at MPI for Chemistry in T42L90 resolution, corresponding to a horizontal resolution of $2.8° \times 2.8°$ and 90 layers with internally resolved Quasi-Biennial Oscillation [QBO, *Giorgetta et al.*, 2006], and at FUB in T42L39 resolution, i.e. with reduced vertical resolution of 39 layers and nudged QBO (hereafter called EMAC-FUB). The solar signal was extracted from the transient simulations by applying different analysis methods, like multiple linear regression analysis [adapted from *Bodeker et al.*, 1998] and empirical mode decomposition (EMD).

For our studies of solar effects on the centennial to millennium time scale a fully coupled atmosphere-ocean general circulation model (AO-GCM) including a detailed representation of the stratosphere was used [EGMAM, *Huebener et al.*, 2007; *Spangehl et al.*, 2010]. EGMAM (ECHO-G with Middle Atmosphere Model) has already been utilised to analyse variations within the climate system [e.g., *Huebener et al.*, 2007; *Spangehl et al.*, 2010; *Bal et al.*, 2011]. The atmospheric component incorporates a T30 horizontal and L39 vertical resolution with the top level located at 0.01 hPa (80 km). For a better representation of solar induced UV/visible variations in EGMAM the FUBRad radiation code was used [*Nissen et al.*, 2007]. The model does not generate a QBO resulting in predominantly weak easterly winds in the tropical lower stratosphere. The coupled ocean model has a horizontal resolution of 2.8° with equator refinement and 21 levels in the vertical covering the deep ocean. The coupled model was driven by time dependent solar irradiance representing solar variability and volcanic activity. Moreover, time dependent GHG concentrations were prescribed. Three transient simulations have been performed covering the period from 1630 to 2000 AD using slightly different solar forcings. For EGMAM-1 fixed climatological ozone concentrations were used while for EGMAM-2 and EGMAM-3 time dependent ozone concentrations representing the effect of solar variability on stratospheric ozone were prescribed. EGMAM-1 and EGMAM-2 used a relative MM to PD TSI increase of about 0.3 % [*Spangehl et al.*, 2010], contrary to a 0.1 % increase in EGMAM-3. EGMAM-3 is additionally driven by spectrally high resolved solar irradiance fluxes (reconstructed solar forcing data for EGMAM-3 provided by N. Krivova, cf. Chap. 2). Detailed information on the model version used for EGMAM-1 and EGMAM-2 is given by *Spangehl* [2010]. The model version used for EGMAM-3 was already applied by *Bal et al.* [2011] for idealised solar cycle studies. To analyse the effect of changes in solar irradiance on the millennial time-scale two simulations were performed in accordance with the Paleoclimate Modelling Intercomparison Project Phase-II (PMIP2, http://pmip2.lsce.ipsl.fr/). The first simulation represents the mid-Holocene (6000 BP). The simulation is compared with a second simulation which represents a pre-industrial climate (1750 AD). These simulations mainly differ in the setting of the orbital parameters resulting in an enhanced seasonal cycle for the mid-Holocene [e.g., the NH summer insolation at 60°N is increased by about 25 W/m^2 [*Kirsch*, 2010]]. As solar activity is assumed to be the same in both simulations the dominating effect is the change of the orbital parameters. Table 29.1 gives an overview of the model systems and the simulations used in the project.

29.3 Solar Influence on Climate on the Decadal Time Scale

29.3.1 The Solar Signal in the Stratosphere

29.3.1.1 The Impact of Solar Decadal Irradiance Variations

While the TSI, i.e. the spectrally integrated solar irradiance at the top of Earth's atmosphere, varies only by about 0.1 % over the 11-year solar cycle, larger variations occur in the UV part of the spectrum, reaching several percent in the ozone absorption bands that generate the major part of SW heating in the stratosphere [*Lean*, 1997]. However, because of the historical focus of numerical global modelling on the troposphere where absorption of solar UV radiation by ozone plays only a minor role, SW radiation codes in climate or chemistry-climate models quite often employ broad-band parameterisations using TSI to derive solar heating rates in one or two SW absorption bands, mostly neglecting the UV spectral region. A comparison of the performance of SW radiation codes within the SPARC CCMVal initiative showed that the observed solar temperature signal in the stratosphere can only be reproduced in models that allow for the effects of spectral variations between solar

Table 29.1 Models and simulations used in this study

Simulation	Model	Period	Horizontal resolution	Vert. res./ Top height	Specifics	References
EMAC	CCM (based on ECHAM5)	1960–2000 (REF-B1)	T42 2.8° × 2.8°	L90/ 0.01 hPa	high vertical resolution interactive chemistry internal QBO (weakly nudged) FUBRad radiation scheme no coupled ocean	Jöckel et al. 2005 Jöckel et al. 2006 Sander et al. 2005 Nissen et al. 2007
EMAC-FUB	EMAC CCM (based on ECHAM5)	1960–2005 (REF-B1) 1960–2100 (SCN-B2d)	T42 2.8° × 2.8°	L39/ 0.01 hPa	interactive chemistry nudged QBO FUBRad radiation scheme no coupled ocean	like EMAC
EGMAM-1	AO-GCM EGMAM (based on ECHAM4/ HOPE-G)	1630–2000	T30 3.75° × 3.75°	L39/ 0.01 hPa	no chemistry, no QBO with coupled ocean fixed climatological ozone	Huebener et al. 2007 Spangehl et al. 2010
EGMAM-2	like EGMAM-1	1630–2000	T30 3.75° × 3.75°	L39 0.01 hPa	no chemistry, no QBO with coupled ocean prescribed time-varying ozone	like EGMAM-1

Table 29.1 *(Continued)*

Simulation	Model	Period	Horizontal resolution	Vert. res./ Top height	Specifics	References
EGMAM-3	like EGMAM-1	1630–2000	T30 3.75° × 3.75°	L39 0.01 hPa	no chemistry, no QBO with coupled ocean prescribed time-varying ozone FUBRad radiation scheme	like EGMAM-1
Holocene	like EGMAM-1	6000 BP	T30 3.75° × 3.75°	L39 0.01 hPa	no chemistry, no QBO with coupled ocean fixed climatological ozone	like EGMAM-1
Pre-Industrial	like EGMAM-1	1750 AD	T30 3.75° × 3.75°	0.01 hPa	no chemistry, no QBO with coupled ocean fixed climatological ozone	like EGMAM-1

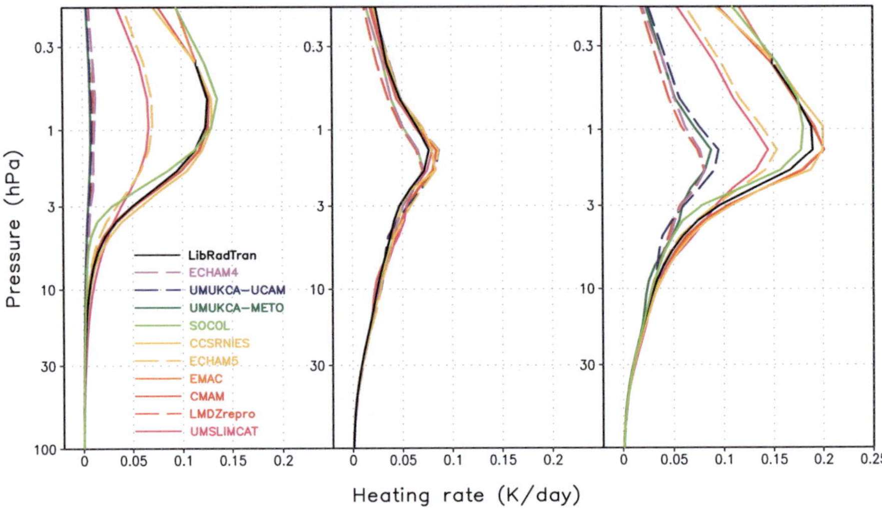

Fig. 29.1 Global mean, short-wave heating rate differences between minimum and maximum of the 11-year solar cycle in January (K/day), calculated off-line in CCM radiation schemes and one reference line-by-line model. *Left*: radiative response to prescribed solar irradiance change, *middle*: radiative response to solar induced ozone change, *right*: total radiative response. Figure taken from *SPARC* [2010] and *Forster et al.* [2011]

minimum and maximum [*SPARC*, 2010; *Forster et al.*, 2011]. As shown in Fig. 29.1, the EMAC-FUBRad SW radiation code agrees well with reference calculations of the SW heating rate differences for prescribed spectral UV variations (left), solar induced ozone variations and the total variations between the minimum and maximum phases of the 11-year solar cycle.

A different aspect influencing the solar signal in the atmosphere is discussed since different spectral irradiance data sets became available for use in radiation models [e.g., *Haigh et al.*, 2010; *Cahalan et al.*, 2010]. *Oberländer et al.* [2012] find in off-line calculations with the FUBRad SW radiation code for solar minimum conditions lower solar heating rates when using updated spectral information from Solar-Stellar Irradiance Comparison Experiment (SOLSTICE) and Solar Spectral irradiance (SOLSPEC) measurements [NRLSSI data, based on *Lean*, 2000] than when using newer data derived from Solar and Heliospheric Observatory (SOHO) Michelson Doppler Imager (MDI) and KPNSO (Kitt Peak National Solar Observatory) full-disk magnetograms [*Krivova et al.*, 2009] or Scanning Imaging Absorption Spectrometer for Atmospheric Chartography (SCIAMACHY) measurements [*Pagaran et al.*, 2009]. Consequently, using the *Krivova et al.* [2009] data in the EMAC-FUB CCM leads to significantly higher temperatures by up to 2 K in the upper summer stratosphere at solar minimum. Heating rate differences between solar minimum and maximum vary for the three irradiance data sets by up to 40 %. However, this discrepancy has a minor effect on temperature as the simulated temperature differences between solar minimum and maximum are quantitatively similar for the spectral irradiance data sets. In contrast, when using the Spectral Irradiance

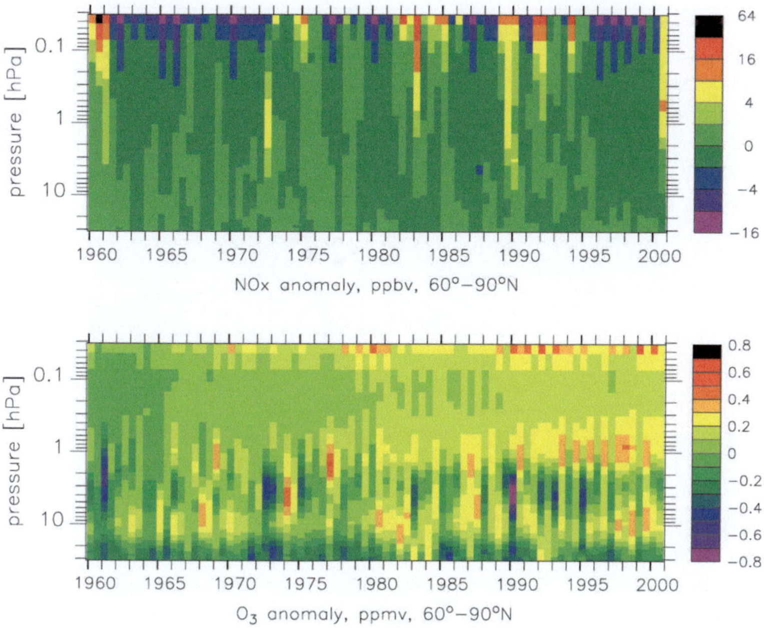

Fig. 29.2 Semiannual (winter + spring and summer + fall) NO_x and ozone anomalies related to particles calculated by EMAC, climatological seasonal cycle and ozone trend due to CFCs eliminated

Monitor (SIM) data [*Harder et al.*, 2009], changes in SW heating from 2004 to 2007, i.e. about half the complete solar cycle, are up to six times stronger than for the NRLSSI data. This leads to a significant solar cycle effect in the summer upper stratosphere and mesosphere temperature by more than 1 K compared to the NRLSSI data, implying a potentially stronger effect of solar variability on the atmosphere than previously assumed.

29.3.1.2 The Influence of Precipitating Particles on Stratospheric Ozone

Besides the UV changes associated with the 11-year solar cycle the rate of particles precipitating into Earth's atmosphere such as low energy electrons (LEE) or protons varies with the level of solar activity. The effects of precipitating particles on the composition of the middle atmosphere have been studied with EMAC using the parameterisations for the influence of LEE and solar proton events (SPE). The production of NO_x from precipitating LEE in the thermosphere has been shown in observations to be correlated with geomagnetic activity (expressed by the A_p-index observed since 1932) [*Randall et al.*, 2007]. These NO_x enhancements are transported downward by the Brewer-Dobson circulation to the lower mesosphere and stratosphere during polar winter where NO_x can catalytically destroy ozone [*Randall et al.*, 2007; *Baumgaertner et al.*, 2009]. Figure 29.2 shows the time series of semiannual anomalies of NO_x and ozone for the period 1960–2000 from a simulation

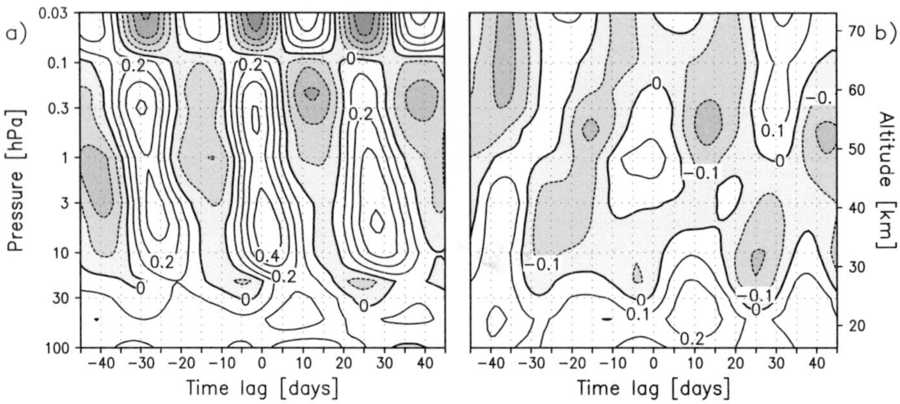

Fig. 29.3 Cross-correlation of tropical (20°S to 20°N) ozone (**a**) and temperature (**b**) with the solar flux at 205 nm. Contour interval is 0.1. Adaption of Fig. 4 from *Kubin et al.* [2011]

with EMAC averaged over 60°N–90°N with respect to the long-term mean seasonal cycle. For ozone the trend due to CFCs has been eliminated for clarity. As expected, high NO_x corresponds to reduced ozone. Prominent NO_x enhancements and ozone reduction are related to strong solar proton events (e.g. 1972, 1989, and 2000) [*Jackman et al.*, 1990; *Baumgaertner et al.*, 2010] which produce NO_x and HO_x over a wide altitude range locally and which are more likely in the years following solar maximum. The large SPE in October 2003 has been studied in more detail by *Baumgaertner et al.* [2010], including an evaluation with ENVISAT/MIPAS satellite observations. The largest LEE effects are in spring 1961, 1983 and 1992. Further details on the different particles, their energy and the induced production of ozone depleting radicals are given in Chap. 17.

29.3.1.3 The Ozone Response to 27-Day Solar Irradiance Variability

During maximum phases of the 11-year solar cycle there is a pronounced short-term variability of solar UV irradiance on the solar rotational time scale of about 27 days. This is due to a relatively uniform facular brightening which is modulated by sunspot darkening during the transit of spots. The variable UV irradiance affects the abundance of stratospheric ozone, the short-wave heating and, consequently, temperature in the middle atmosphere. Via the reaction rates of the most important ozone destruction reactions which increase with increasing temperature there arises a feedback on the ozone response [*Brasseur et al.*, 1987].

Making use for the first time of a daily irradiance data set it could be shown that these processes are well represented in EMAC-FUB. In a sensitivity study with two idealised simulations, one with prescribed daily spectral irradiances, hence explicitly resolving the 27-day variability, and a control simulation with a fixed intermediate spectral irradiance [*Kubin et al.*, 2011] the effect of daily resolved irradiances on ozone and temperature was examined. Consistent changes were prescribed to the photolysis module of EMAC-FUB. Figure 29.3a shows the cross-correlation of tropical ozone with the solar irradiance at 205 nm. The correlation maximises between 5 and 3 hPa and shows a second peak in the lower mesosphere. The shift from negative time lags at high altitudes to positive lags at lower levels reflects the temperature feedback on the ozone response (Fig. 29.3b). Higher up in the mesosphere an increasing production of odd hydrogen by increased photolysis of water vapour leads to ozone loss which overcompensates the enhanced ozone production. Tropical ozone is most sensitive to UV changes in the upper stratosphere at about 3 hPa (44 km). The magnitude of about 0.37 % per 1 % change at 205 nm is in good agreement with observational results from e.g., *Hood and Zhou* [1998]. The correlation and the ozone sensitivity in the upper stratosphere increase when a sub-period of particularly strong 27-day variability is analysed and they degrade during periods of weak daily irradiance variability. Our results suggest that ignoring daily solar flux variations on the 27-day time scale in transient CCM simulations does not lead to a significant degradation of the time mean ozone response in the stratosphere, while in the tropical mesosphere significant errors of up to 3 % may occur [*Kubin et al.*, 2011].

29.3.1.4 The 11-Year Solar Signal in Ozone and Temperature

An Empirical Mode Decomposition (EMD) analysis of tropical total column ozone as simulated by EMAC from 1960 to 2000 with six intrinsic mode functions (IMFs, see Fig. 29.4) yields a signal with a period of approx. 11 years, therefore possibly of solar origin, of the order of 4 DU (IMF5). This is similar to the effects of man-made halocarbons (measured as equivalent effective stratospheric chlorine, EESC), and of the QBO (IMF3). As a guide to the eye, the 11-year cycle and the QBO (at 10 hPa) are overlayed in green and red, respectively. IMF4 has a period of around four years and a closer examination shows that it is potentially related to the ENSO phenomenon (see inlet ENSO index). The results are rather similar to standard multiple regression methods. EMD is a signal processing technique for analysis of non-stationary and non-linear time series. Previous studies have shown its ability to identify amplitude and frequency modulations [e.g., *McDonald et al.*, 2007]. One of its strengths is the detection of natural variations in the atmosphere without the need for a priori information on the form of these variations. The method can be also applied to compute correlation coefficients between the F10.7 cm solar radio flux and IMF5 for vertical distributions of zonally averaged ozone. A large positive correlation is found in the tropical middle and upper stratosphere and in the lower mesosphere at mid-latitudes (not shown). In the high latitude middle stratosphere there is a negative correlation, presumably related to precipitating particles. There are, however, also regions where the attribution is not unique.

The signal of the 11-year solar cycle in the vertically resolved tropical ozone mixing ratio is strongest in the middle and upper stratosphere as shown in Fig. 29.5. These results from a 41-year simulation with EMAC are in agreement with the observationally based analysis by *Steinbrecht et al.* [2006] where a multiple regression technique was used to separate the effects.

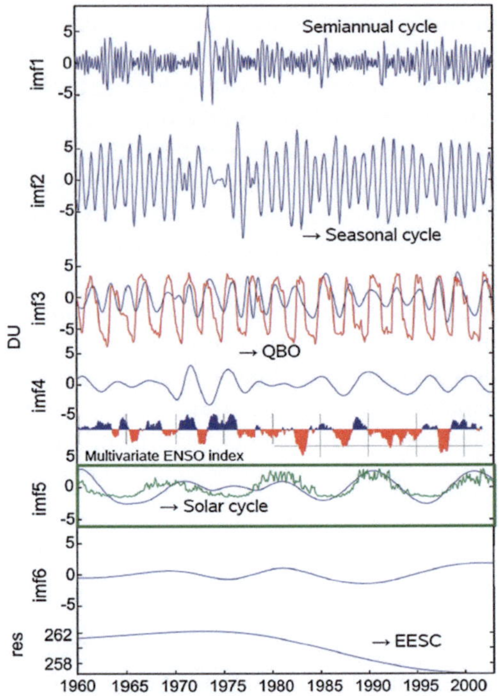

Fig. 29.4 Empirical mode decomposition of total column ozone in the tropics (20°S to 20°N) in DU, *blue curves*: intrinsic mode functions, *red*: QBO at 10 hPa, *green*: F10.7 cm solar radio flux, IMF5 is potentially related to the 11-year solar cycle

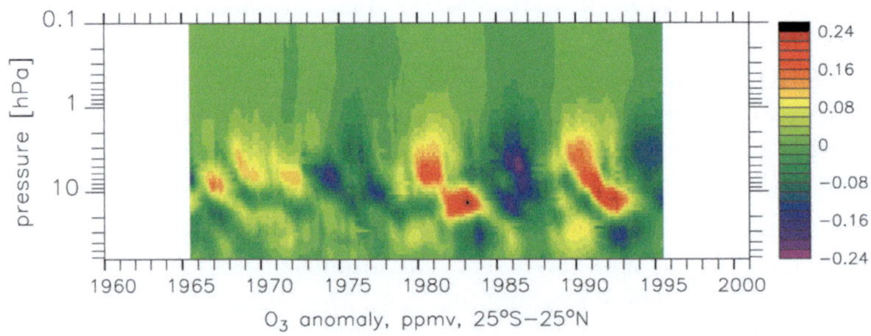

Fig. 29.5 Tropical ozone anomaly due to the 11-year solar cycle as simulated by EMAC. Difference between the results of running means over 2 and 11 years

The global picture of the ozone response to 11-year solar cycle irradiance changes is depicted on the left hand side of Fig. 29.6 as obtained from a 46-year transient simulation with EMAC-FUB. Ozone reacts to the imposed solar irradiance with a positive anomaly throughout most of the stratosphere. The spatial distribution with a maximum of about 1.5 to 2 % at both the southern and northern hemisphere mid-latitudes at about 5 hPa agrees with observations derived from SAGE I and SAGE II [*Randel and Wu*, 2007] as well as SBUV/2 and UARS HALOE data [*Soukharev and Hood*, 2006]. However, the model does not simulate the observed secondary maximum of ozone in the tropical lower stratosphere. The mesospheric

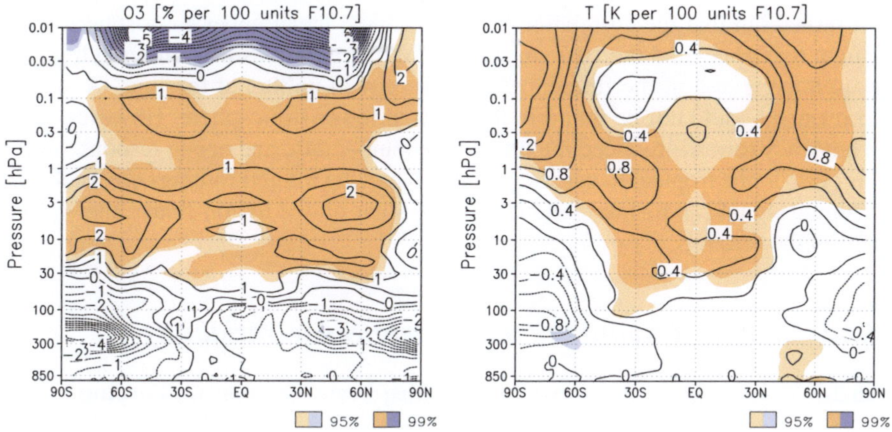

Fig. 29.6 The annual mean zonal mean 11-year solar signal in ozone (*left*) in % per 100 units F10.7 cm flux and temperature (*right*) in K per 100 units F10.7 cm flux as simulated by EMAC-FUB. Contour interval is 0.5 % and 0.2 K. Light (heavy) shading denotes statistical significance at the 95 % (99 %)-level

response is of opposite sign, i.e. the ozone volume mixing ratio decreases as solar activity increases. This is due to the enhanced abundance of OH as a consequence of increased water vapour photolysis at solar maximum.

The annual mean 11-year solar signal in zonal mean temperature (Fig. 29.6, right) shows a significant warming in the tropical and subtropical stratosphere. The peak values of about 0.8 K per 100 units F10.7 solar flux are found at the subtropical to mid-latitude stratopause. The annual mean temperature response at high northern latitudes is dominated by the winter signal with lower temperatures in the lower stratosphere and higher values in the upper stratosphere and lower mesosphere at solar maximum. The simulated solar signal agrees well in shape and strength with SSU-satellite measurements of the solar cycles between the years 1979 and 2009 [*Randel et al.*, 2009]. Discrepancies exist, however, with the solar signal derived from ERA-40 reanalysis data [*Frame and Gray*, 2010] in the tropics, where ERA-40 data show a stronger signal at the stratopause and a secondary maximum in the lower stratosphere. The differences in the observed solar signal point at uncertainties and possible caveats in the observational platforms and reanalysis procedure, an issue which is still unsolved.

It is currently a matter of debate as to what causes the appearance of secondary maxima in the 11-year solar cycle responses of ozone and temperature in the tropical lower stratosphere. Such secondary maxima are seen in observational data [*Frame and Gray*, 2010]. However, models have difficulties to reproduce these features. It is assumed that (non-)linear interactions and aliasing effects between the 11-year solar irradiance variability, the QBO and the sea surface temperatures contribute to the observed signal [*Marsh and Garcia*, 2007; *Smith and Matthes*, 2008]. To further investigate this a transient simulation with band-pass filtered QBO and SSTs, hence, linearly independent forcings was performed within the framework of the Solar Influence for SPARC (SOLARIS) initiative.

It turned out that the correlation between annual mean upper stratospheric temperature (or ozone) and the 10.7 cm solar radio flux improves when filtered forcings are used compared to the reference simulation with unfiltered forcings. A similar improvement could not be found in the tropical lower stratosphere. However, there is a tendency towards reduced maxima in the temperature and ozone response to the solar cycle in the tropical lower stratosphere when the filtered forcings are used. The remaining signal can be interpreted as purely solar or due to non-linear interactions between the forcings.

29.3.2 Climate Response to the 11-Year Solar Variability

29.3.2.1 Tropospheric Response to a Stratospheric Momentum Forcing

While the downward propagation of the 11-year solar signal from the upper to the lower stratosphere can be explained by dynamical interactions (see Sect. 29.3.2.2) the processes leading to a tropospheric response to solar variability are less clear. To investigate the potential implications of a solar induced disturbance in the lower stratosphere on the troposphere a pilot study with sensitivity simulations using an idealised stratospheric momentum forcing according to *Thuburn and Craig* [2000] has been performed with EMAC-FUB. The momentum forcing produces a stronger stratospheric zonal wind, which modifies the propagation conditions for planetary waves which in turn affects the residual mean meridional circulation (MMC). In consequence, anomalous downwelling in the tropical lower stratosphere leads to an adiabatic warming and anomalous upwelling at high latitudes to adiabatic cooling which creates a negative equator-to-pole temperature gradient, a situation as it occurs during 11-year solar maximum solstice seasons. The tropospheric jet on the winter hemisphere is shifted polewards in response to this lower stratospheric temperature anomaly, a behaviour that was also found in NCEP-NCAR data [*Haigh et al.*, 2005]. A similar but weaker signal is seen in the summer hemisphere where no stratospheric wind forcing was applied. Apart from the jet shift there is a stratospheric influence on the tropical tropopause which is located at lower levels when the forcing is active. The Hadley circulation is weakened in the forced simulations, as is shown for July in Fig. 29.7, which leads to reduced convective precipitation in the tropics. These results from a comprehensive chemistry-climate model confirm the possibility of a purely dynamically induced signal in the troposphere owing to stratospheric changes during maxima of the 11-year solar cycle, as shown in simplified GCM studies of *Simpson et al.* [2009] and *Haigh et al.* [2005]. The tropospheric effectivity of this coupling mechanism was found to be qualitatively unchanged but sensitive to the vertical resolution of the model, particularly in the upper troposphere/lower stratosphere.

29.3.2.2 The 11-Year Solar Signal in the Troposphere

The radiatively and photochemically induced stratospheric 11-year solar signal is characterised by an altered ozone abundance, modified SW heating rates and

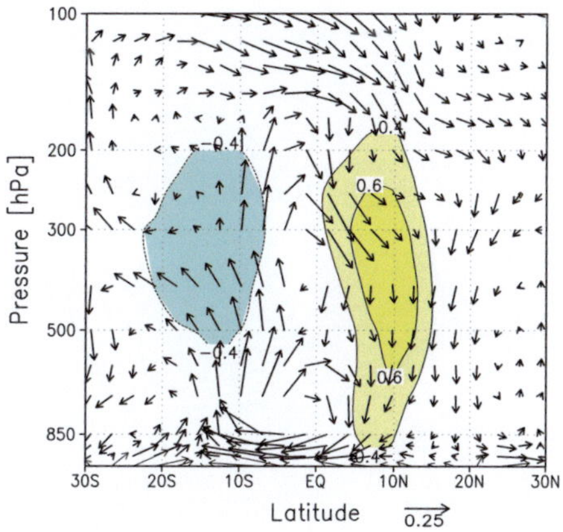

Fig. 29.7 Tropical meridional circulation in July. *Arrows*: difference between the forced and the control experiment, *Colour* and *contours*: vertical velocity from the control experiment in mm/s, *yellow colours* denote upward motion, *blue* downward motion

temperature which creates an anomalous meridional temperature gradient. Via the thermal wind relationship the zonal wind is influenced which implies a change in the propagation conditions for planetary waves. As these waves drive the MMC a solar influence on the summer hemisphere upwelling and winter hemisphere downwelling can be explained. The MMC changes may even influence the upwelling and the meridional circulation in the troposphere. The predominantly stratospheric (top-down) pathway of 11-year solar cycle influence was proposed by *Kodera and Kuroda* [2002] and is most effective during the solstice seasons.

In a transient simulation with EMAC-FUB driven by observed forcings it was possible to show that this mechanism was active during the period 1960 to 2005. The 11-year solar signal was extracted with an advanced multiple linear regression method [*SPARC*, 2010, Chapter 8] as opposed to composite analyses employed in earlier perpetual solar maximum and minimum GCM studies [e.g., *Matthes et al.*, 2006]. The leading modes of natural atmospheric variability such as the Northern and Southern Annular Mode (NAM, SAM) are susceptible to an 11-year solar influence as was shown in earlier studies [*Langematz et al.*, 2008; *Kuroda*, 2007; *Kuroda and Yamazaki*, 2010]. Strong and statistically significant positive NAM anomalies are, e.g., found in early and mid northern winter (December and January) in the lower stratosphere at 30 hPa (not shown). By February the stratospheric anomaly has decayed but a significant positive NAM pattern with positive geopotential height anomalies over the Atlantic-European as well as the Asian-Pacific sector at mid-latitudes and negative geopotential height deviations of about 40 m near the pole is evident in the troposphere at 700 hPa (Fig. 29.8, left), consistent with observational results of *Kuroda* [2007].

The geopotential height changes are associated with a poleward shift of the tropospheric jet stream and with a significant temperature increase at 850 hPa over the European and Asian continent of about 1 to 1.5 K between solar minimum and maximum (Fig. 29.8, right). A similar downward transfer from the stratosphere to the troposphere within one month was seen in southern hemisphere winter and spring. In the troposphere a negative geopotential height anomaly near the Antarctic peninsula flanked by higher than average heights to the east and to the west was detected in the

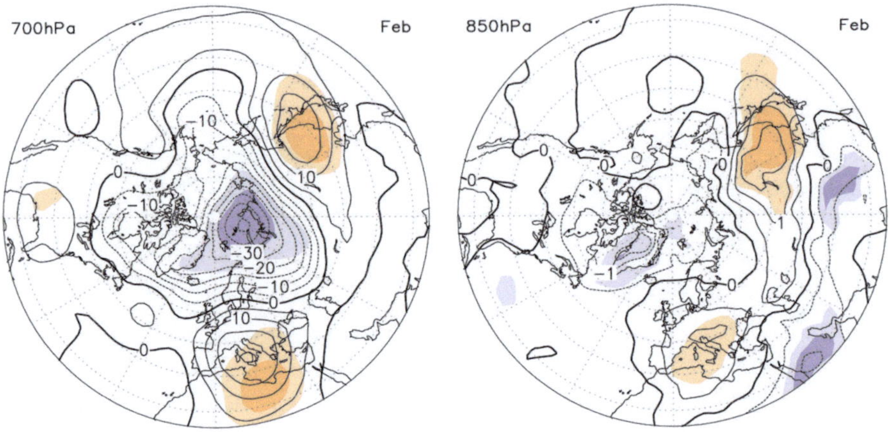

Fig. 29.8 The February monthly mean 11-year solar signal in geopotential height in m per 100 units F10.7 cm solar radio flux from 20°N to 90°N at 700 hPa (*left*), and in temperature in K per 100 units F10.7 cm solar radio flux at 850 hPa (*right*) from a transient simulation of the period 1960 to 2005. *Light* (*heavy*) *shading* denotes statistical significance at the 95 % (99 %)-level

CCM simulation in October which is comparable to reanalysis results of *Kuroda and Kodera* [2002]. These patterns are found in transient simulations with a realistically evolving 11-year solar cycle as well as in equilibrium simulations with fixed perpetual solar cycle phases and prescribed solar induced ozone changes (not shown).

A tropospheric signal has also been detected in the northern hemisphere summer season, in particular, in the Indian summer monsoon. In accordance with the observational analysis of *Kodera* [2004] statistically significant correlations between the 10.7 cm solar radio flux and the upward vertical velocity at 300 hPa over the southern Arabian peninsula and the equatorial Indian Ocean were found in a transient simulation of the past half century (not shown). In these regions stronger upwelling occurs at solar maximum which yields higher convective precipitation rates. However, the observed signal could only be retrieved in a simulation with observed SSTs prescribed at the lower boundary. Similar transient simulations from 1960 to the present with EMAC-FUB using modelled SSTs or SSTs fixed at 1960-values repeating each year could not reproduce the observed anomaly. This suggests that atmosphere-ocean interaction processes play an important role for the 11-year solar signal in the Indian summer monsoon.

29.3.2.3 The 11-Year Solar Signal in a Changing Climate

An important prerequisite for a successful simulation of the solar signal with a CCM is a realistic background climatology. As this climatology is itself subject to the anthropogenic climate change it is possible that the 11-year solar signal will appear in a different way in a future warmer climate. This is investigated in a simulation with the EMAC-FUB CCM covering the period 1960 to 2100. The model was forced with

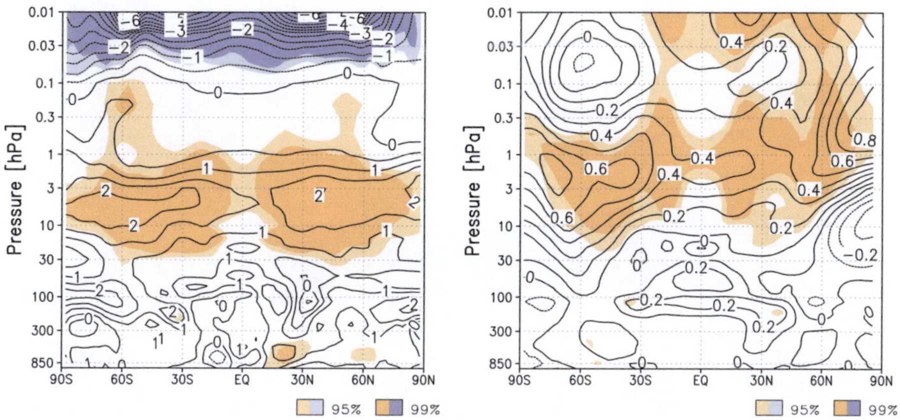

Fig. 29.9 The annual mean zonal mean 11-year solar signal in the 2nd half of the 21st century in ozone (*left*) in % per 100 units F10.7 cm flux and temperature (*right*) in K per 100 units F10.7 cm flux as simulated by EMAC-FUB. Contour interval is 0.5 % and 0.1 K. *Light* (*heavy*) *shading* denotes statistical significance at the 95 % (99 %)-level

prescribed modelled sea surface temperatures and sea ice, projected abundances of greenhouse gases according to the SRES A1B scenario and ozone depleting substances following the adjusted WMO A1 scenario. Spectral solar irradiance data from the solar cycles 20 to 23 were repeated to prescribe solar variability in the future. The solar signal is extracted by means of a multiple linear regression approach. A comparison of the annual mean solar signal in ozone and temperature in the 20th and in the second half of the 21st century when the CO_2 concentration has more than doubled compared to 1960 values reveals a similar spatial pattern as in the past. The ozone response to an increase in solar irradiance peaks at mid-latitudes in the upper stratosphere, see Fig. 29.9, left. The strongest temperature increase is found in the stratopause region (Fig. 29.9, right). The area with statistically significant changes has shrunk from the past to the future period. This points towards an increased variability in the atmosphere in a warming climate. An uncertainty is, however, the specification of the spectral irradiances in the future since the solar activity will not necessarily remain on the present relatively high level [*Abreu et al.*, 2008].

29.4 Solar Influence on Climate on the Centennial to Millennial Time Scale

The role of external drivers (e.g., solar variability, volcanoes, GHG concentrations, changes of the Earth's orbit around the Sun) for long term climate variations is investigated by means of climate model simulations for selected episodes of the past. The focus is on (i) the role of solar variability for the period from the early 17th century (Maunder Minimum) to present and (ii) the role of the orbital forcing for the mid-Holocene.

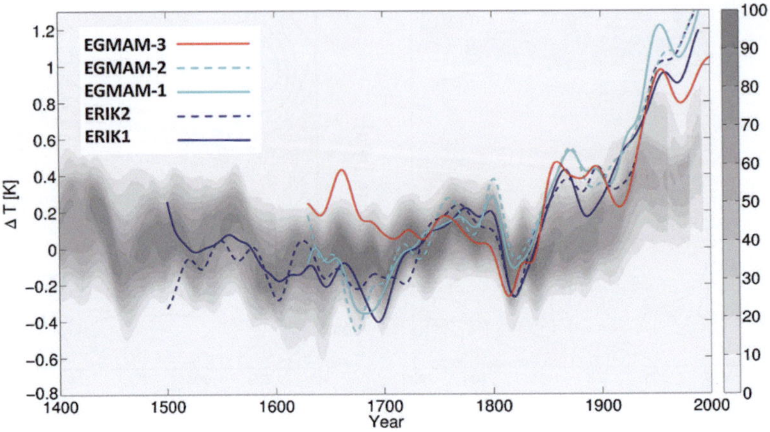

Fig. 29.10 Annual Northern Hemisphere mean 2 m temperature anomaly (K) for simulations with EGMAM-FUBRad (EGMAM-3 time-dependent ozone and high resolved fluxes for UV/vis, *red/solid line*), EGMAM (EGMAM-2 time-dependent ozone and low resolved short wave fluxes, *cyan/dashed line*; EGMAM-1 fixed ozone and low resolved short wave fluxes, *cyan/solid line*) and two simulations with ECHO-G (ERIK1, *dark blue/solid line*; ERIK2, *dark blue/dashed line*, data for ECHO-G provided by E. Zorita, GKSS). *Grey shading* represents overlap of NH temperature reconstructions from IPCC AR4. Data from reconstructions and simulations with ECHO-G are expressed as anomalies from their 1500–1899 means. Simulations with EGMAM-FUBRad and EGMAM are centred to 1716–1790 mean value from the centred (1500–1899) ECHO-G simulations. All time series are smoothed with a 30 year low-pass filter. The figure is modified after *Spangehl et al.* [2010] (forcings for EGMAM and ECHO-G (GHG, volcanic and solar) are shown in the same paper). EGMAM-3 (EGMAM-FUBRad) is additionally shown and uses a weaker TSI forcing than the other simulations

29.4.1 Maunder Minimum to Present

The Maunder Minimum (MM, 1645–1715 AD), which falls within the Little Ice Age, is defined by the nearly complete absence of sunspots and thereby represents a period with exceptional low solar activity lasting over several decades (cf. Chap. 2). While recent studies focusing on the entire millennium [e.g., *Mann et al.*, 2009; *Jungclaus et al.*, 2010], do not resolve seasons due to the lack of information from proxy-data, the period from the MM onwards provides a case study for a more detailed investigation of the role of external drivers such as solar variability, volcanic eruptions and GHG concentrations for climate [*Spangehl and Raible*, 2008, and references therein].

29.4.1.1 Northern Hemisphere Near Surface Temperature

The simulated annual northern hemisphere (NH) mean near surface temperature is shown in Fig. 29.10. EGMAM-1 and EGMAM-2 (see Sect. 29.2 for details of the experiments) are close to simulations with the tropospheric version of the model (ECHO-G) indicating that the inclusion of a detailed stratosphere does only play a moderate role for the simulated climate on the hemispheric scale when using similar forcings [*Spangehl et al.*, 2010]. Larger differences are seen for EGMAM-3. This simulation fails to capture the temperature increase during the 18th century. Still, the simulation is within the range of the proxy based temperature reconstructions with respect to the simulated pre-industrial climate from the beginning of the Late Maunder Minimum (LMM, 1675–1715 AD) onwards. *Jungclaus et al.* [2010] use an Earth system model for ensemble simulations driven by realistic estimates of a number of relevant forcings (solar, volcanic, GHG, and aerosol forcing as well as land cover changes). They show that a small solar TSI forcing together with the internal variability of their coupled system might be sufficient to reproduce general features of the climate of the last millennium and thereby highlight the role of an interactively coupled carbon cycle. Internal variability together with the small TSI forcing used in EGMAM-3 might contribute to the weaker temperature increase from LMM onwards in EGMAM-3. It should be noted that the inclusion of the new radiation code was associated with a minor decrease of the incoming solar irradiance at the surface which resulted in a slightly colder mean climate and a moderate increase of the North Atlantic overturning circulation. Though no further increase of the North Atlantic overturning circulation was recognised when starting the transient simulation, a slight drift towards a colder mean climate might be left. All simulations show a clear cooling for the Dalton Minimum (DM, 1790–1830 AD) and a comparably strong temperature increase from the onset of the industrialisation onwards. The volcanic forcing is necessary to simulate the cooling during DM while the absence of the sulfate aerosol forcing explains the overestimation of the subsequent temperature increase [*Spangehl et al.*, 2010, and references therein].

29.4.1.2 Thermal Structure

As discussed by *Spangehl et al.* [2010] EGMAM-1 and EGMAM-2 are close to transient simulations with ECHO-G when regarding the simulated near surface climate under present-day conditions. *Huebener et al.* [2007] show on the basis of ensemble simulations with EGMAM and ECHO-G, that the inclusion of the high resolved stratosphere mainly improves the simulated climate in the upper troposphere and lower stratosphere. All three transient simulations presented here show a similar global scale vertical temperature structure for the troposphere and lower stratosphere (Fig. 29.11, left). While EGMAM-1 shows a slight underestimation of the temperature around the stratopause, the use of a new ozone background climatology in EGMAM-2 and EGMAM-3 leads to better agreement with observations. Note that EGMAM-3 shows clearly lower temperatures in the higher mesosphere close to the observations [*Randel et al.*, 2004], which is related to the fact that ozone concentrations do not vary with height in EGMAM-1 and EGMAM-2 at levels above 0.3 hPa. To assess the response to the solar forcing the Late Maunder Minimum is compared with a subsequent pre-industrial period during the 18th century (1716–1790 AD, hereafter referred to as PI) which is characterised by higher

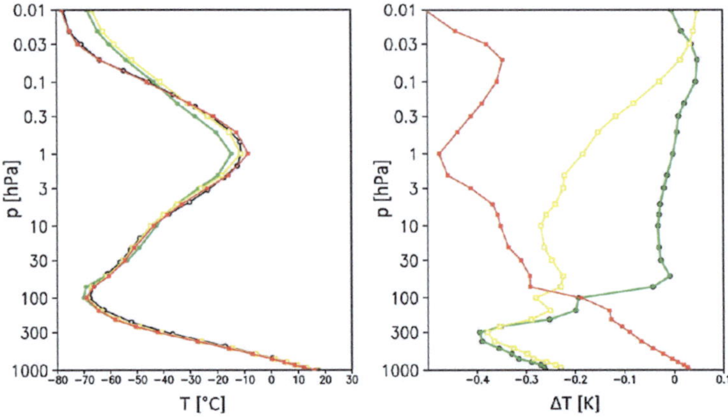

Fig. 29.11 Vertical structure of simulated annual mean temperature. (*left*) Present day (1992–1997) 80°S–80°N mean [°C]. (*right*) LMM (1675–1715) minus PI (1716–1790) global mean anomaly [K]. Colours indicate the different simulations: EGMAM-3 (*red*), EGMAM-2 (*yellow*) and EGMAM-1 (*green*). Observations are additionally shown [*black curve*, *left*, after *Randel et al.*, 2004]. Note that the *black curve* disappears when simulations are plotted on top

solar irradiance while the other external forcings show comparably small variations [*Spangehl et al.*, 2010].

All three simulations show a global cooling during LMM in the upper troposphere which reaches about 0.4 K when using the strong TSI forcing thereby exceeding the lower TSI estimate (EGMAM-3) by a factor of more than 3 (Fig. 29.11, right). In the lower stratosphere the simulations employing ozone forcing (EGMAM-2 and EGMAM-3) both show cooling during LMM which is about 25 percent larger when taking into account the additional direct radiative effect due to variations in the UV/visible spectral range (EGMAM-3). A maximum cooling of nearly 0.5 K is found around the stratopause in EGMAM-3 which exceeds the simulated cooling in EGMAM-2 by a factor of more than 2.5. Note that the strong cooling at 0.01 hPa in EGMAM-3 is due to oxygen absorption at the Lyman-α wavelength [*Nissen et al.*, 2007]. Hardly any global mean temperature changes in the stratosphere and mesosphere are seen when using fixed climatological ozone together with the low resolved short-wave radiation scheme (EGMAM-1).

29.4.1.3 Dynamic Response: Multi-decadal Time Scale

The solar induced heating in the stratosphere and upper troposphere is related to changes in the temperature and wind structure. As shown by *Bal et al.* [2011] EGMAM-3 reproduces important features of the observed response to the 11-year solar cycle. This response includes a strengthening of the NH stratospheric polar vortex during phases of high solar activity and the associated downward propagation of the solar signal from the stratosphere down to the troposphere during the NH winter season (note that the model does not generate a QBO and does therefore simulate weak easterly winds in the tropical lower stratosphere). As shown by *Spangehl et al.* [2010] the simulation including the strong solar TSI forcing together with prescribed ozone (EGMAM-2) shows statistically enhanced multi-decadal vortex variability together with a weaker vortex during the LMM and DM as well as a general strengthening from the LMM to the end of the 20th century (i.e. positive long-term trend) which is consistent with the proposed solar forcing mechanisms [e.g., *Haigh*, 1999; *Kodera and Kuroda*, 2002]. The suggested link between solar forcing and stratospheric polar vortex strength is also expressed by statistically significant enhanced wavelet-coherence and an in-phase relationship for periods around 50 years during most parts of the 18th and the first half of the 19th century (Fig. 29.12c). This signal is even larger in the simulation including the strongest stratospheric solar forcing which also shows enhanced coherence for periods larger than hundred years accompanied by a large solar lead time/nearly anti-phase relationship (EGMAM-3, cf. Fig. 29.12e). Note that the multi-decadal signal is restricted to the Dalton Minimum and to periods below 20-years indicating only a weak relationship in the simulation which neglects the stratospheric solar forcing (EGMAM-1, cf. Fig. 29.12a). *Schimanke et al.* [2011] find a multi-decadal variability in the number of sudden stratospheric warmings with a period around 52 years when analysing a pre-industrial control simulation (i.e. constant solar forcing) performed with the same model. While their study highlights the role of internally generated variability within the coupled atmosphere-ocean system, their results do not imply an in-phase relationship with the solar forcing.

All simulations show evidence for an in-phase relationship between the winter mean values of the strength of the stratospheric polar vortex and the Arctic Oscillation index in the lower troposphere on the decadal to multi-decadal time scale (Fig. 29.12b, d, f). Still, this relationship is not robust in time. This explains the simulated regional climate variability during the pre-industrial era to some extent [*Spangehl et al.* 2010]. However, a clear and consistent AO/NAO like response to the solar forcing during the pre-industrial era as suggested by *Shindell et al.* [2001] is not simulated (not shown).

29.4.1.4 Dynamic Response: Centennial Time Scale

Further evidence for a solar impact on the strength of the stratospheric polar vortex is seen on the centennial and longer time scale. The simulation employing stratospheric solar ozone and UV forcing (EGMAM-3) does not show a clear weakening of the vortex when regarding the complete Maunder Minimum and Dalton Minimum. Still, the simulation shows a strengthening of the stratospheric polar vortex during the 18th century which is also seen in the applied solar forcing (Fig. 29.13a). Both simulations employing stronger stratospheric solar forcing (EGMAM-2 and EGMAM-3) show a clear strengthening of the stratospheric polar vortex during the 18th century which is accompanied by an increase in solar activity (Fig. 29.13b, c).

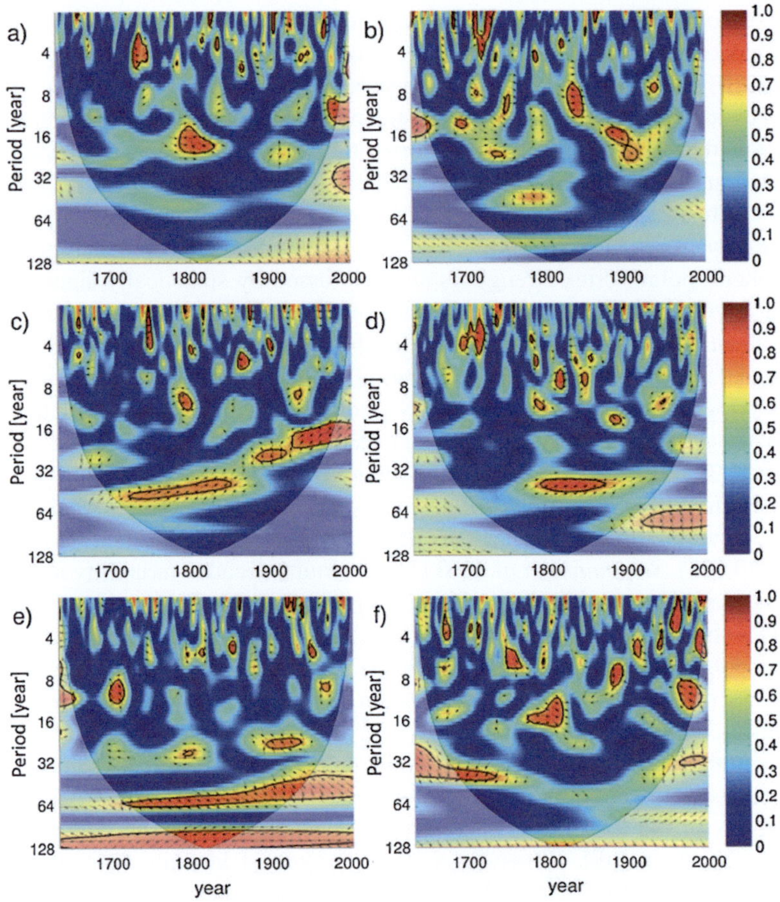

Fig. 29.12 Squared wavelet coherence after *Grinsted et al.* [2004] between (*left*) solar forcing and strength of stratospheric polar winter vortex (calculated as winter (DJF) mean of zonal mean zonal wind at 10 hPa, 60°N) and (*right*) strength of stratospheric polar winter vortex and Arctic Oscillation index (calculated as PC1 of the winter (DJF) mean series of geopotential at 700 hPa, 20°N–90°N). (**a**) and (**b**) for EGMAM-1, (**c**) and (**d**) for EGMAM-2 and (**e**) and (**f**) for EGMAM-3. *Arrows* indicate the relative phase relationship between the series (e.g. pointing right: in-phase; left: anti-phase; down: solar forcing (stratospheric vortex) leading stratospheric vortex (AO 700 hPa). Statistically significant areas are indicated by *thick black curves*. *Oval curve* indicates the cone of influence (i.e. values outside this curve are not reliable)

This increase in vortex strength exceeds the range of variability analysed in a pre-industrial control simulation by a factor of 3 which shows that internal variability is most likely not sufficient to generate such large variations in vortex strength. Note that EGMAM-3 shows a somewhat lagged response which might be related to the applied spin-up procedure.

A less consistent picture emerges for the 20th century. As shown by *Huebener et al.* [2007] for future projections according to the SRES-A2 scenario the increase

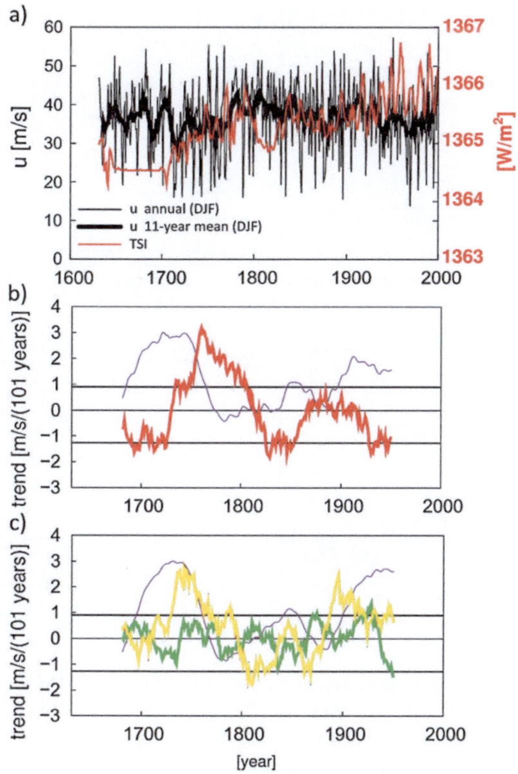

Fig. 29.13 Stratospheric polar vortex calculated as the winter (DJF) mean of zonal mean zonal wind at 10 hPa, 60°N. Annual (*thin black curve*) and 11-year running mean (*thick black curve*) of vortex strength [m/s] are shown for (**a**) the simulation including high resolved short-wave radiation scheme (EGMAM-3). *Red curve* additionally shows the used solar TSI forcing [W/m^2]. Centennial trends are shown for (**b**) EGMAM-3 and (**c**) EGMAM-2 (*orange*) and EGMAM-1 (*green*) (shown as centred 101 year linear trend [m/s/(101 years)]). The black bars in (**b**) and (**c**) give the respective maximum/minimum values of a 600 year control simulation under constant pre-industrial conditions and the *thin purple curves* show additionally the respective centennial trends of the TSI forcings used to drive the transient simulations. Note that TSI forcings used in (**b**) and (**c**) differ and that the *purple curves* are scaled to show uniform maximum values

in GHG concentrations results in enhanced tropospheric wave forcing and a related weakening of the stratospheric polar winter vortex when using the same model (EGMAM). *Spangehl et al.* [2010] emphasise that the effect is relevant for the historical simulations. A similar stratospheric effect might explain the weakening of the stratospheric vortex during the 20th century in EGMAM-3 (cf. Fig. 29.13b). Note that EGMAM-2 shows a strengthening of the stratospheric vortex during the 20th century (Fig. 29.13c) indicating that also differences in the applied solar forcing might play a role besides GHG forcing and internal variability.

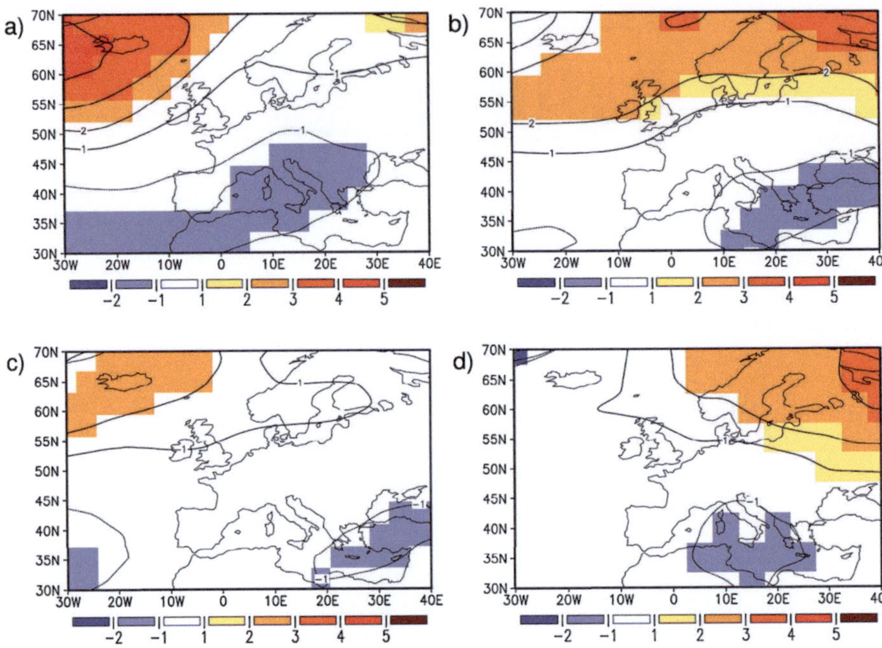

Fig. 29.14 Long-term winter mean difference of MSLP (hPa) for LMM (1675–1715) minus PD (1950–1999) for (**a**) EGMAM-3, (**b**) EGMAM-2, (**c**) EGMAM-1, and (**d**) reconstruction after *Luterbacher et al.* [2002]; *coloured regions* are statistically significant above 95th percentile value after a Student's t test (Welch's correction is applied to account for non-uniform variances, (**b**)–(**d**) modified after *Spangehl et al.* 2010)

29.4.1.5 North Atlantic/Europe

Following *Spangehl et al.* [2010] the largest impact on the tropospheric circulation is seen when comparing the Late Maunder Minimum with a present-day (PD) climate, as the solar forcing reaches a maximum together with the GHG forcing. Here, all simulations show a weakening of the zonal flow over the eastern North Atlantic/European sector for the Maunder Minimum (Fig. 29.14a–c). The NAO like change pattern can be explained by the anthropogenic GHG increase and the higher concentrations during PD. Moreover, *Scaife et al.* [2011] show on the basis of idealised simulations that the GHG induced changes in winter mean MSLP over the eastern North Atlantic are affected by troposphere/stratosphere coupling processes via a modification of the synoptic scale activity ('storm tracks'). The effect (namely a southward shift of the stratospheric jet and a related change in vertical wind shear and upper tropospheric baroclinicity) is found in the historical simulations presented here (results not shown) and explains MSLP anomalies around the British Isles [*Spangehl et al.*, 2010].

There is clear evidence for a solar impact. While EGMAM-1 shows hardly any MSLP signal over Europe the simulations employing the more realistic stratospheric solar forcing show distinct changes of MSLP and pressure gradient in this area

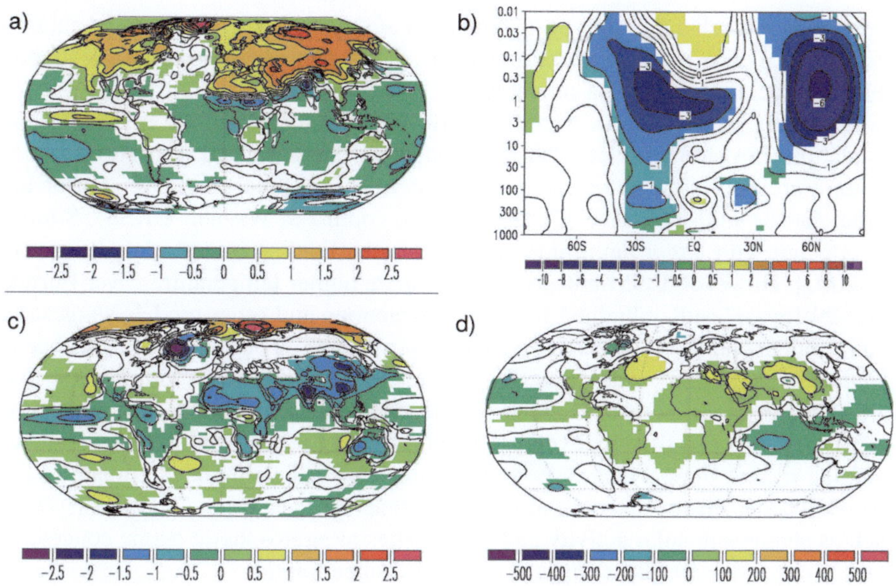

Fig. 29.15 Simulated long term mean (100 years) difference of mid-Holocene (6000 BP) minus pre-industrial state (1750) for (**a**) summer (JJA) near surface temperature [K], (**b**) winter (DJF) zonal mean zonal wind [m/s], (**c**) winter (DJF) near surface temperature [K] and (**d**) winter (DJF) MSLP [hPa]

(Fig. 29.14a, b), which is in good agreement with the reconstruction (Fig. 29.14d). In EGMAM-2 the stronger MSLP change signal over Europe is consistent with the strengthening of the stratospheric polar vortex from the LMM to PD and the 'polar-route' of the solar influence via the stratosphere [*Kodera and Kuroda*, 2002; *Gray et al.*, 2010; *Woollings et al.*, 2010]. EGMAM-3 does not show such a clear change in the strength of the stratospheric polar vortex (cf. Sect. 29.4.1.4). Here, the comparably strong MSLP signal can be explained by the lower TSI forcing which results in a weaker Arctic surface temperature response and thereby favours the strengthening of the NAO from the LMM to PD. Note that this effect is confirmed by a data comparison (results not shown) of the standard (E1) and the strong (E2) solar forcing ensemble experiments described by *Jungclaus et al.* [2010].

29.4.2 Mid-Holocene

The vertically extended AO-GCM (EGMAM) reproduces the main features of the mid-Holocene climate including the strong continental NH summer warming and cooling in the African and Indian Monsoon regions (Fig. 29.15a) which is associated with changes in the Monsoon circulation [*Kirsch*, 2010]. The main features for the winter season are a tropical/subtropical cooling and a warming in the north polar region which is related to the strong reduction of sea ice in late summer and autumn (Fig. 29.15c). The lower temperatures over the western and central North Atlantic can partly be explained by a moderate strengthening of the NAO (Fig. 29.15c, d) and a slightly weaker North Atlantic overturning circulation in the mid-Holocene simulation (not shown). Note that the changes in the simulated near surface temperature are very similar to results obtained with the tropospheric version of the model for the mid-Holocene (ECHO-G, pers. comm. S. Wagner, GKSS). Similar patterns but with larger amplitudes emerge for the Eemian interglacial [*Kaspar et al.*, 2007]. However, EGMAM does not simulate higher temperatures over Europe (Fig. 29.15c). Results obtained from reconstructions and other models suggest an even stronger change of the NAO, i.e. a more positive NAO during the mid-Holocene [*Wanner et al.*, 2008]. The vertically extended model responds with a more disturbed stratospheric vortex for the mid-Holocene which counteracts the strengthening of the NAO and thereby explains the comparably weak NAO response. Therefore, the results show a clear stratospheric impact for the simulated climate of the mid-Holocene. However, results also indicate that other factors should be considered such as changes in solar activity and sensitivities to stratospheric ozone concentrations to cover the complete range of the related uncertainties.

29.5 Conclusions and Outlook

Different aspects of solar variability on decadal and centennial to millennial time scales were analysed with a CCM and an AO-GCM, respectively. It was shown that an assessment of the decadal solar signal in atmospheric models requires the inclusion of spectral irradiance information—in contrast to formerly used TSI—as well as SW radiation schemes with sufficient spectral resolution. With newly developed parameterisations the influence of precipitating particles on the composition of the middle atmosphere has successfully been simulated. Based on these new developments the decadal solar signal was analysed for the past four decades in realistic transient simulations with the EMAC CCM, a stratosphere resolving state of the art chemistry-climate model that includes an interactive chemical model as well as a high resolution SW radiation scheme, hence providing optimal conditions to derive solar signals in the atmosphere. The focus of our analyses was on the photochemically and radiatively induced 11-year solar signal in ozone, temperature, geopotential height and wind speed in the stratosphere and troposphere. These results contributed to the SPARC Chemistry Climate Model Validation (CCMVal) initiative and helped to derive multi-model estimates of the solar induced contributions to climate variability and climate change. It could be shown that the 'top-down effect' of decadal stratospheric UV variations contributed at least partially to the tropospheric signal, which is most prominent in the late solstice seasons with positive Annular Mode anomalies in the lower troposphere.

For the first time the influence of the solar forcing on the European winter climate during the Maunder Minimum and the subsequent centuries is demonstrated by comparing transient climate simulations performed with a stratosphere resolving AO-GCM with proxy-based climate reconstructions. Simulations of the long-term changes in solar irradiance focused on circulation changes during the Maunder Minimum and the mid-Holocene. The temporal evolution of hemispheric scale temperature from the Maunder Minimum to present day is well captured by the stratosphere resolving AO-GCM. During the phase of extremely low solar activity lower temperatures were found in the stratosphere and upper troposphere as well as a weaker polar vortex and a negative NAO anomaly in surface pressure. These results are in line with the decadal solar signal which favours a strong polar vortex and an opposite tropospheric anomaly at high solar activity. By additional mid-Holocene simulations we show that the centennial scale orbital forcing results in dynamical changes in the stratosphere which in turn affect surface climate. Our results suggest the use of stratosphere resolving AO-GCMs in ongoing modelling activities such as PMIP (Paleo Model Intercomparison Project) and CMIP (Coupled Model Intercomparison Project).

Future studies will be conducted to investigate the interactions of the 11-year solar signal with the QBO and ENSO within the framework of the SPARC SOLARIS initiative. The open question whether the new measurements of spectral irradiance variations throughout the 11-year solar cycle will be confirmed in future and how this will alter our understanding of the atmospheric response to solar variability will further be pursued. The role of long-term changes in solar activity and QBO interactions will be analysed for the Holocene.

Acknowledgements Model simulations with EMAC-FUB were performed at an SGI-Altix supercomputer of HLRN Konrad-Zuse-Zentrum Berlin and on an IBM power-6 at ECMWF in Reading, UK with computing time provided by DWD. The EMAC (L90) simulations were performed on the IBM power-6 computer at Rechenzentrum Garching of Max-Planck-Society. Simulations with EGMAM were performed on a NEC-SX6 supercomputer at DKRZ, Hamburg. Data of simulations with ECHO-G were provided by E. Zorita (Helmholtz-Zentrum Geesthacht, formerly GKSS).

References

Abreu, J., Beer, J., Steinhilber, F., Tobias, S., & Weiss, N. (2008). For how long will the current grand maximum of solar activity persist? *Geophysical Research Letters*, *35*, L20109.

Bal, S., Schimanke, S., Spangehl, T., & Cubasch, U. (2011). On the robustness of the solar cycle signal in the pacific region. *GRL*, *38*. doi:10.1029/2011GL047964.

Baumgaertner, A., Jöckel, P., & Brühl, C. (2009). Energetic particle precipitation in ECHAM5/MESSy1 Part 1: downward transport of upper atmospheric NO_x produced by low energy electrons. *Atmospheric Chemistry and Physics*, *9*, 2729–2740.

Baumgaertner, A., Jöckel, P., Riede, H., Stiller, G., & Funke, B. (2010). Energetic particle precipitation in ECHAM5/MESSy1 Part 2: solar proton events. *Atmospheric Chemistry and Physics*, *10*, 7285–7302.

Bodeker, G., Boyd, I., & Matthews, W. (1998). Trends and variability in vertical ozone and temperature profiles measured by ozonesondes at Lauder, New Zealand: 1986–1996. *Journal of Geophysical Research*, *103*(D22), 28661–28681.

Brasseur, G., De Rudder, A., Keating, G., & Pitts, M. (1987). Response of middle atmosphere to short-term solar ultraviolet variations: 2. Theory. *Journal of Geophysical Research*, *92*(D1), 903–914.

Cahalan, R., Wen, G., Harder, J., & Pilewskie, P. (2010). Temperature response to spectral solar variability on decadal time scales. *Geophysical Research Letters*, *37*. doi:10.1029/2009GL041898.

Cubasch, U., Voss, R., Hegerl, G. C., Waszkewitz, J., & Crowley, T. J. (1997). Simulation of the influence of solar radiation variations on the global climate with an ocean-atmosphere general circulation model. *Climate Dynamics*, *13*(11), 757–767.

Forster, P., Fomichev, V., Rozanov, E., Cagnazzo, C., Jonsson, A., Langematz, U., Fomin, B., Iacono, M., Mayer, B., Mlawer, E., Myhre, G., Portmann, R., Akiyoshi, H., Falaleeva, V., Gillett, N., Karpechko, A., Li, J., Lemennais, P., Morgenstern, O., Oberländer, S., Sigmond, M., & Shibata, K. (2011). Evaluation of radiation scheme performance within chemistry-climate models. *Journal of Geophysical Research*, D10302. doi:10.1029/2010JD015361.

Frame, T. H. A., & Gray, L. J. (2010). The 11-year solar cycle in ERA-40 data: an update to 2008. *Journal of Climate*, *23*, 2213–2222.

Giorgetta, M., Manzini, E., Roeckner, E., Esch, M., & Bengtsson, L. (2006). Climatology and forcing of the quasi-biennial oscillation in the MAECHAM5 model. *Journal of Climate*, *19*, 3882–3901.

Gray, L., Beer, J., Geller, M., Haigh, J., Lockwood, M., Matthes, K., Cubasch, U., Fleitmann, D., Harrison, G., Hood, L., Luterbacher, J., Meehl, G., Shindell, D., van Geel, B., & White, W. (2010). Solar influences on climate. *Review of Geophysics*, *48*(RG4001).

Grinsted, A., Moore, J. C., & Jevrejeva, S. (2004). Application of the cross wavelet transform and wavelet coherence to geophysical time series. *Nonlinear Processes in Geophysics*, *11*(5/6), 561–566.

Haigh, J. (1999). A GCM study of climate change in response to the 11-year solar cycle. *Quarterly Journal of the Royal Meteorological Society*, *125*, 871–892.

Haigh, J., Blackburn, M., & Day, R. (2005). The response of tropospheric circulation to perturbations in lower-stratospheric temperature. *Journal of Climate*, *18*, 3672–3691.

Haigh, J., Winning, A., Toumi, R., & Harder, J. W. (2010). An influence of solar spectral variations on radiative forcing of climate. *Nature*, *467*, 696–699.

Harder, J., Fontenla, J., Pilewskie, P., Richard, E., & Woods, T. (2009). Trends in solar spectral irradiance variability in the visible and infrared. *Geophysical Research Letters*, *36*, L07801.

Hood, L., & Zhou, S. (1998). Stratospheric effects of 27-day solar ultraviolet variations: an analysis of UARS MLS ozone and temperature data. *Journal of Geophysical Research*, *103*(D3), 3629–3638.

Huebener, H., Cubasch, U., Langematz, U., Spangehl, T., Niehörster, F., Fast, I., & Kunze, M. (2007). Ensemble climate simulations using a fully coupled ocean-troposphere-stratosphere general circulation model. *Philosophical Transactions of the Royal Society of London. A*, *365*(1857), 2089–2101. doi:10.1098/rsta.2007.2078.

Jackman, C., Douglass, A., Rood, R., & McPeters, R. (1990). Effect of solar proton events on the middle atmosphere during the past two solar cycles as computed using a two-dimensional model. *Journal of Geophysical Research*, *95*, 7417–7428.

Jöckel, P., Sander, R., Kerkweg, A., Tost, H., & Lelieveld, J. (2005). Technical note: the modular earth submodel system (MESSy)—a new approach towards earth system modeling. *Atmospheric Chemistry and Physics*, *5*, 433–444.

Jöckel, P., Tost, H., Pozzer, A., Brühl, C., Buchholz, J., Ganzeveld, L., Hoor, P., Kerkweg, A., Lawrence, M. G., Sander, R., Steil, B., Stiller, G., Tanarhte, M., Taraborrelli, D., van Aardenne, J., & Lelieveld, J. (2006). The atmospheric chemistry general circulation model ECHAM5/MESSy1: consistent simulation of ozone from the surface to the mesosphere. *Atmospheric Chemistry and Physics*, *6*, 5067–5104.

Jungclaus, J. H., Lorenz, S. J., Timmreck, C., Reick, C. H., Brovkin, V., Six, K., Segschneider, J., Giorgetta, M. A., Crowley, T. J., Pongratz, J., Krivova, N. A., Vieira, L. E., Solanki, S. K., Klocke, D., Botzet, M., Esch, M., Gayler, V., Haak, H., Raddatz, T. J., Roeckner, E., Schnur, R., Widmann, H., Claussen, M., Stevens, B., & Marotzke, J. (2010). Climate and carbon-cycle variability over the last millennium. *Climate Past*, *6*(5), 723–737.

Kaspar, F., Spangehl, T., & Cubasch, U. (2007). Northern hemisphere winter storm tracks of the Eemian interglacial and the last glacial inception. *Climate Past*, *3*(2), 181–192. doi:10.5194/cp-3-181-2007.

Kirsch, A. (2010). Untersuchung des indischen Monsuns im Mittleren Holozän und im Maunder Minimum im Vergleich zum heutigen Klima mit dem gekoppelten Modell EGMAM. Diploma thesis, Institut für Meteorologie: Freie Universität Berlin.

Kodera, K. (2004). Solar influence on the Indian ocean monsoon through dynamical processes. *Geophysical Research Letters*, *31*, L24209.

Kodera, K., & Kuroda, Y. (2002). Dynamical response to the solar cycle. *Journal of Geophysical Research*, *107*(D24), 4749.

Kodera, K., Matthes, K., Shibata, K., Langematz, U., & Kuroda, Y. (2003). Solar impact on the lower mesospheric subtropical jet: a comparative study with general circulation model simulations. *Geophysical Research Letters*, *30*, 6.

Krivova, N., Solanki, S., Wenzler, T., & Podlipnik, B. (2009). Reconstruction of solar UV irradiance since 1974. *Journal of Geophysical Research*, *114*, D00I04.

Kubin, A., Langematz, U., & Brühl, C. (2011). Chemistry climate model simulations of the effect of the 27 day solar rotational cycle on ozone, *Journal of Geophysical Research*, D15301, doi:10.1029/2011JD015665.

Kuroda, Y. (2007). Effect of QBO and ENSO on the solar cycle modulation of winter North Atlantic oscillation. *Journal of Meteorological Society of Japan*, *85*(6), 889–898.

Kuroda, Y., & Kodera, K. (2002). Effect of solar activity on the polar-night jet oscillation in the northern and southern hemisphere winter. *Journal of Meteorological Society of Japan*, *80*(4B), 973–984.

Kuroda, Y., & Yamazaki, K. (2010). Influence of the solar cycle and QBO modulation on the southern annular mode. *Geophysical Research Letters*, *37*, L12703.

Langematz, U., Claussnitzer, A., Matthes, K., & Kunze, M. (2005). The climate during the Maunder Minimum: a simulation with the Freie Universität Berlin climate middle atmosphere model (FUB-CMAM). *Journal of Atmospheric and Solar-Terrestrial Physics*, *67*, 55–69.

Langematz, U., Brönnimann, S., Graf, H.-F., & Kapala, A. (2008). Die Stratosphäre, Vulkanismus und die NAO/AO. *Promet*, *34*(3), 122–129.

Lean, J. (1997). The sun's variable radiation and its relevance for earth. *Annual Review of Astronomy and Astrophysics*, *35*, 33–67.

Lean, J. L. (2000). Evolution of the sun's spectral irradiance since the Maunder Minimum. *Geophysical Research Letters*, *27*(16), 2425–2428.

Luterbacher, J., Xoplaki, E., Dietrich, D., Rickli, R., Jacobeit, J., Beck, C., Gyalistras, D., Schmutz, C., & Wanner, H. (2002). Reconstruction of sea level pressure fields over the Eastern North Atlantic and Europe back to 1500. *Climate Dynamics*, *18*(7), 545–561.

Mann, M. E., Zhang, Z., Rutherford, S., Bradley, R. S., Hughes, M. K., Shindell, D., Ammann, C., Faluvegi, G., & Ni, F. (2009). Global signatures and dynamical origins of the little ice age and medieval climate anomaly. *Science*, *326*(5957), 1256–1260.

Marsh, D., & Garcia, R. (2007). Attribution of decadal variability in lower-stratospheric tropical ozone. *Geophysical Research Letters*, *34*, L21807.

Matthes, K., Kodera, K., Haigh, J., Shindell, D. T., Shibata, K., Langematz, U., Rozanov, E., & Kuroda, Y. (2003). GRIPS solar experiments intercomparison project: initial results. *Papers in Meteorology and Geophysics*, *54*, 71–90.

Matthes, K., Kuroda, Y., Kodera, K., & Langematz, U. (2006). Transfer of the solar signal from the stratosphere to the troposphere: northern winter. *Journal of Geophysical Research*, *111*, D06108.

McDonald, A. J., Baumgaertner, A. J. G., Fraser, G. J., George, S. E., & Marsh, S. (2007). Empirical mode decomposition of the atmospheric wave field. *Annales Geophysicae*, *25*, 375–384.

Nissen, K., Matthes, K., Langematz, U., & Mayer, B. (2007). Towards a better representation of the solar cycle in general circulation models. *Atmospheric Chemistry and Physics*, *7*, 5391–5400.

Oberländer, S., Langematz, U., Matthes, K., Kunze, M., Kubin, A., Harder, J., Krivova, N., Solanki, S., Pagaran, J., & Weber, M. (2012). The influence of spectral solar irradiance data on stratospheric heating rates during the 11 year solar cycle. *Geophysical Research Letters*, *39*, L01801.

Pagaran, J., Weber, M., & Burrows, J. (2009). Solar variability from 240 to 1750 nm in terms of faculae brightening and sunspot darkening from SCIAMACHY. *Astrophysical Journal*, *700*, 1884–1895.

Randall, C., Harvey, V., Singleton, C., Bailey, S., Bernath, P., Codrescu, M., Nakajima, H., & Russell, J. (2007). Energetic particle precipitation effects on the Southern hemisphere stratosphere in 1992–2005. *Journal of Geophysical Research*, *112*, D08308.

Randel, W., & Wu, F. (2007). A stratospheric ozone profile data set for 1979–2005: variability, trends, and comparisons with column ozone data. *Journal of Geophysical Research*, *112*, D06313.

Randel, W., Udelhofen, P., Fleming, E., Geller, M., Gelman, M., Hamilton, K., Karoly, D., Ortland, D., Pawson, S., Swinbank, R., Wu, F., Baldwin, M., Chanin, M.-L., Keckhut, P., Labitzke, K., Remsberg, E., Simmons, A., & Wu, D. (2004). The SPARC intercomparison of middle-atmosphere climatologies. *Journal of Climate*, *17*(5), 986–1003.

Randel, W., Shine, K., Austin, J., Barnett, J., Claud, C., Gillet, N., Keckhut, P., Langematz, U., Lin, R., Long, C., Mears, C., Miller, A., Nash, J., Seidel, D., Thompson, D., Wu, F., & Yoden, S. (2009). An update of observed stratospheric temperature trends. *Journal of Geophysical Research*, *114*, D02107.

Roeckner, E., Brokopf, R., Esch, M., Giorgetta, M., Hagemann, S., Kornblueh, L., Manzini, E., Schlese, U., & Schulzweida, U. (2006). Sensitivity of simulated climate to horizontal and vertical resolution in the ECHAM5 atmosphere model. *Journal of Climate*, *19*, 3771–3791.

Sander, R., Kerkweg, A., Jöckel, P., & Lelieveld, J. (2005). Technical note: the new comprehensive atmospheric chemistry module MECCA. *Atmospheric Chemistry and Physics*, *5*, 445–450.

Scaife, A. A., Spangehl, T., Fereday, D. R., Cubasch, U., Langematz, U., Akiyoshi, H., Bekki, S., Braesicke, P., Butchart, N., Chipperfield, M. P., Gettelman, A., Hardiman, S. C., Michou, M., Rozanov, E., & Shepherd, T. G. (2011). Climate change projections and stratosphere-troposphere interaction. *Climate Dynamics*. doi:10.1007/s00382-011-1080-7.

Schimanke, S., Körper, J., Spangehl, T., & Cubasch, U. (2011). Multi-decadal variability of sudden stratospheric warmings in an AOGCM. *Geophysical Research Letters*, *38*(1), L01801. doi:10.1029/2010GL045756.

Shindell, D. T., Schmidt, G. A., Mann, M. E., Rind, D., & Waple, A. (2001). Solar forcing of regional climate change during the maunder minimum. *Science*, *294*(5549), 2149–2152.

Simpson, I., Blackburn, M., & Haigh, J. (2009). The role of eddies in driving the tropospheric response to stratospheric heating perturbations. *Journal of the Atmospheric Sciences*, *66*(5), 1347–1365.

Smith, A., & Matthes, K. (2008). Decadal-scale periodicities in the stratosphere associated with the solar cycle and the QBO. *Journal of Geophysical Research*, *113*, D05311.

Solomon, S., Qin, D., Manning, M., Chen, Z., Marquis, M., Averyt, K. B., Tignor, M., & Miller, H. L. (Eds.) (2007). *Climate change 2007: the physical science basis. Contribution of working group I to the fourth assessment report of the intergovernmental panel on climate change*. Cambridge: Cambridge University Press.

Soukharev, B., & Hood, L. (2006). Solar cycle variation of stratospheric ozone: multiple regression analysis of long-term satellite data sets and comparisons with models. *Journal of Geophysical Research*, *111*, D20314.

Spangehl, T., & Raible, C. C. (2008). Variationen der NAO auf Basis von langen Zeitreihen, Datenrekonstruktionen und Simulationen der letzten 500 Jahre. *Promet*, *34*(3), 101–107.

Spangehl, T., Cubasch, U., Raible, C. C., Schimanke, S., Körper, J., & Hofer, D. (2010). Transient climate simulations from the Maunder Minimum to present day: role of the stratosphere. *Journal of Geophysical Research*, *115*, D00I10. doi:10.1029/2009JD012358.

SPARC (2010). *SPARC report on the evaluation of chemistry-climate models* (Tech. Rep. WMO/TD-No. 1526). SPARC Report, No. 5, WCRP-132.

Steinbrecht, W., Claude, H., Schönenborn, F., McDermid, I. S., Leblanc, T., Godin, S., Song, T., Swart, D., Meijer, Y. J., Bodeker, G. E., Connor, B. J., Kämpfer, N., Hocke, K., Calisesi, Y., Schneider, N., de la Noë, J., Parrish, A., Boyd, I. S., Brühl, C., Steil, B., Giorgetta, M.,

Manzini, E., Thomason, L., Zawodny, J., McCormick, M., Russell, J., Bhartia, P., Stolarski, R., & Hollandsworth-Frith, S. (2006). Long-term evolution of upper stratospheric ozone at selected stations of the network for the detection of stratospheric change. *Journal of Geophysical Research*, *111*, D10308.

Thuburn, J., & Craig, G. (2000). Stratospheric influence on tropopause height: the radiative constraint. *Journal of the Atmospheric Sciences*, *57*, 17–28.

Wanner, H., Beer, J., Bütikofer, J., Crowley, T. J., Cubasch, U., Flückiger, J., Goosse, H., Grosjean, M., Joos, F., Kaplan, J. O., Küttel, M., Müller, S. A., Prentice, I. C., Solomina, O., Stocker, T. F., Tarasov, P., Wagner, M., & Widmann, M. (2008). Mid- to late Holocene climate change: an overview. *Quaternary Science Reviews*, *27*(19–20), 1791–1828. doi:10.1029/2010GL045756.

Woollings, T., Lockwood, M., Masato, G., Bell, C., & Gray, L. (2010). Enhanced signature of solar variability in Eurasian winter climate. *Geophysical Research Letters*, *37*(20), L20805.

Chapter 30
Interannual Variability and Trends in the Stratosphere

Karin Labitzke and Markus Kunze

Abstract Large effects of solar variability related to the 11-year sunspot cycle (SSC) are seen in the stratosphere throughout the year, but they are clearly larger if the data are grouped according to the phase of the QBO (quasi-biennial oscillation). New results based on an extended, 70 year long data set, fully confirm earlier findings. By means of teleconnections, the dynamical interaction between the Arctic and the Tropics in the stratosphere and in the troposphere is investigated for the whole data set and compared with the anomalies of single events.

30.1 Introduction

The standard deviations (scatter about the mean) of the monthly mean 30-hPa temperatures averaged along latitude circles in Fig. 30.1 serve as a measure of the interannual variability. As pointed out by *Labitzke and Van Loon* [1994], this variability is small in the tropics and in the summer hemisphere, but increases with latitude and season. It is largest in the polar regions in winter and much larger in the Arctic than in the Antarctic winter. The larger Arctic variability is related to the frequent occurrence of Major Midwinter Warmings which are disturbing and sometimes destroying the otherwise quiet stratospheric polar vortex, see discussion in Sect. 30.5.

In the southern hemisphere the largest variability can be observed near the South Pole during spring, which can be attributed to the final warmings during the transition to the stratospheric summer circulation. The band of increased standard deviation around the equator is a result of the QBO induced secondary meridional circulation [*Reed*, 1964]. The data available today for Fig. 30.1 ($n = 43$) are almost two times more than originally available [*Labitzke and van Loon*, 1999, their Fig. 2.11].

The focus in this chapter is mainly on the northern hemisphere during winter and spring. The stratospheric variability during this time period may be attributed mainly to planetary scale waves, generated in the troposphere which can propagate

K. Labitzke (✉) · M. Kunze
Institut für Meteorologie, Freie Universität Berlin, Carl-Heinrich-Becker-Weg 6-10, 12165 Berlin, Germany
e-mail: Karin.Labitzke@met.fu-berlin.de

F.-J. Lübken (ed.), *Climate and Weather of the Sun-Earth System (CAWSES)*, Springer Atmospheric Sciences,
DOI 10.1007/978-94-007-4348-9_30,

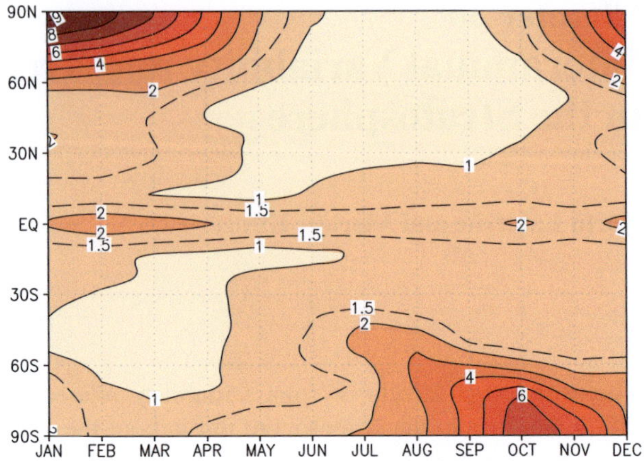

Fig. 30.1 The global distribution of standard deviations (K) of the 30-hPa monthly mean temperatures for the period 1968–2010 [based on NCEP/NCAR re-analyses; update of Fig. 2.11 in *Labitzke and van Loon*, 1999]

upward in the prevailing westerly flow. Tropospheric oscillations, like the Southern Oscillation (SO) or the North Atlantic Oscillation, modulate the planetary scale waves and thus may have an influence on stratospheric variability [*van Loon et al.*, 1982; *Perlwitz and Graf*, 1995]. Here, we want to give an update on the influence of the 11-year sunspot cycle (SSC) on the stratosphere and its modulation by the quasi-biennial oscillation (QBO) [*Labitzke*, 1987; *Labitzke and van Loon*, 1988].

The large variability during the northern and southern hemisphere winter and spring makes it difficult to detect significant temperature trends in the stratosphere, where trends can change sign, dependent on the time period selected (cf. Sect. 30.4). During the dynamically least disturbed summer seasons however significant negative trends can be derived [*Randel et al.*, 2009].

30.2 Data and Methods

In earlier studies concerned with the influence of the 11-year SSC on the stratosphere and its modulation by the QBO, the available data were analyses of the lower and middle stratosphere based on radiosondes [*Labitzke*, 1987; *Labitzke and van Loon*, 1988]. With the availability of the NCEP/NCAR re-analyses from 1948 till 2010 [*Kalnay et al.*, 1996], it is now possible to analyse the solar influence for a much longer time period. In this study we use the temperature and geopotential height data of the NCEP/NCAR re-analyses. A shorter time period from 1968 to 2010 is used for all analyses extending into the northern hemisphere tropics and into the southern hemisphere, because re-analyses of the stratosphere are less reliable during the early years because of the lack of radiosonde stations, especially over the southern hemisphere, the lack of high reaching balloons in the early years, and the scarce satellite information before 1979. More recently the time period was expanded back to 1942 for the geopotential height data at the North Pole [*Labitzke et al.*, 2006] by using the reconstructions of *Brönnimann et al.* [2006].

The QBO is an oscillation in the atmosphere which is best observed with radiosonde data in the stratosphere above the equator, where the zonal winds change with time between winds from the east and from the west. The period of the QBO varies in time and space, with an average value near 28 months at all levels, see reviews by *Naujokat* [1986] and *Baldwin et al.* [2001].

The QBO appears to modulate the solar signal in the stratosphere, and is in turn modulated by the sun [e.g. *Soukharev and Hood*, 2001; *Labitzke*, 2005], although there is some controversy on this latter point [e.g. *Fischer and Tung*, 2008; *Hamilton*, 2002]. Therefore, it is necessary to stratify the data into years for which the equatorial QBO in the lower stratosphere [at about 45 hPa, e.g. *Holton and Tan*, 1980] was in its westwind or eastwind phase. To define the QBO phase the monthly mean QBO data set derived from single radiosonde soundings at near equatorial stations (Canton Island (closed 1967), Gan/Maledive Islands (closed 1975), and Singapore) covering the period from 1953 to 2010 [*Naujokat*, 1986; *Labitzke and Collaborators*, 2002] are used. The average of the monthly mean zonal winds at 50 and 40 hPa defines the QBO phase of the respective month. There was no threshold used, once easterly winds are observed the QBO phase is defined as east and vice versa.

Monthly mean values of the 10.7 cm solar flux are used as a proxy for variations through the SSC. The flux values are expressed in solar flux units (1 s.f.u. = $10^{-22}\,\mathrm{W\,m^{-2}\,Hz^{-1}}$). This is an objectively measured radio wave intensity, highly and positively correlated with the 11-year SSC and particularly with the UV part of the solar spectrum [*Hood*, 2003].

The differences between solar minima (about 70 units) and solar maxima (about 200 units) are calculated by using a simple linear regression $y = a + bx$, where x in this case is the 10.7 cm solar flux and b is the slope. This slope is used here, multiplied by 130 units, in order to get the differences of stratospheric parameters between solar minima and maxima [*van Loon and Labitzke*, 2000; *Labitzke*, 2003].

30.3 Solar Influence on Stratospheric Temperatures over the Tropics and Subtropics During the Northern Summer

As pointed out in the beginning (Fig.30.1), the inter-annual variability is very weak over the subtropics throughout the year. However, it turned out that the stratosphere is influenced by the 11-year SSC and by the phase of the QBO [*van Loon and Labitzke*, 2000]. In the following, we are giving a few examples which show this relationship. We are mainly showing data from July, as this is a summer month over the Northern Hemisphere when the atmosphere is dynamically least disturbed.

The time series of the 10.7 cm solar flux (dashed lines) correlate moderately well with the 30-hPa temperatures in July at a grid point near New Delhi, if the data of all years are considered, uppermost panel, Fig. 30.2. But the correlation improves to values above $r = 0.85$, if only July in the east phase of the QBO is used, (middle panel), while there is a weak correlation if only the data during the west phase

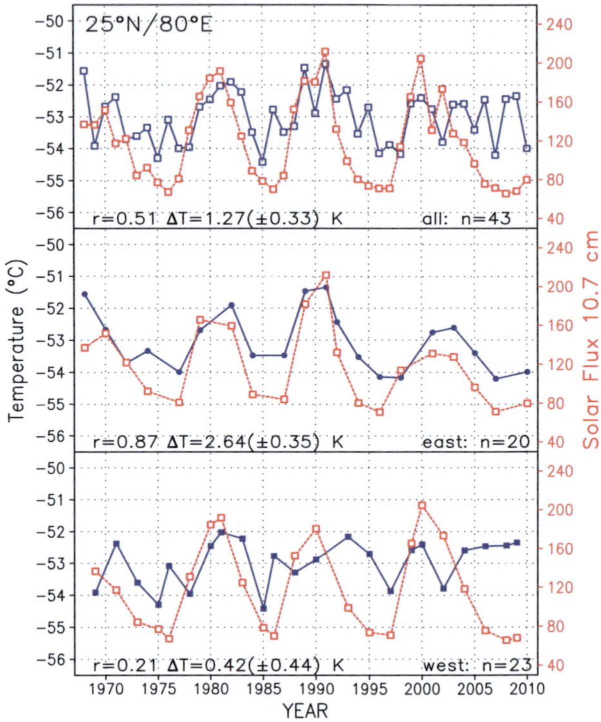

Fig. 30.2 Time series of monthly mean values of the 10.7 cm solar flux (*red lines*) in solar flux units (1 s.f.u. = 10^{-22} W m^{-2} Hz^{-1}) and detrended 30-hPa temperatures in July at the grid point 25°N/80°E (near New Delhi). The *upper panel* shows all years ($n = 43$), the *middle panel* only years in the east phase of the QBO ($n = 20$), and the *lower panel* only years in the west phase of the QBO ($n = 23$). r: correlation coefficients; ΔT: temperature difference (K) between solar maxima and minima (data: NCEP/NCAR re-analyses, 1968–2010)

of the QBO are considered. This very clear result is robust since about 1968, i.e., since we have reliable data for the tropics and subtropics. With ΔT we indicate the temperature difference between solar maxima and solar minima. In the east phase the difference is 2.6 K, and this is more than 2 standard deviations.

Figure 30.3 shows very clearly the influence of the 11-year SSC on the middle stratosphere, with correlations reaching 0.87 in July, when the sunlight is strongest over the tropics and the temperature differences between solar maximum and solar minimum are largest, more than 2 standard deviations (Fig. 30.1). Until October the sun moves towards the southern sub-tropics, the intensity of the sunlight and the correlations weaken, as well as the temperature differences.

30.4 Temperature Trends in the Stratosphere During the Arctic Winters

"It is almost impossible to determine a trend in the Arctic in winter even with a sample of 40 years because the interannual variability is so large" [*Labitzke and van Loon*, 1994]. Today, with the NCEP/NCAR re-analyses, we have 64 years of data (1948–2011) for heights and temperatures and 6 more years for heights (1942–1947), based on the reconstructions of *Brönnimann et al.* [2006], [see *Labitzke et al.*, 2006].

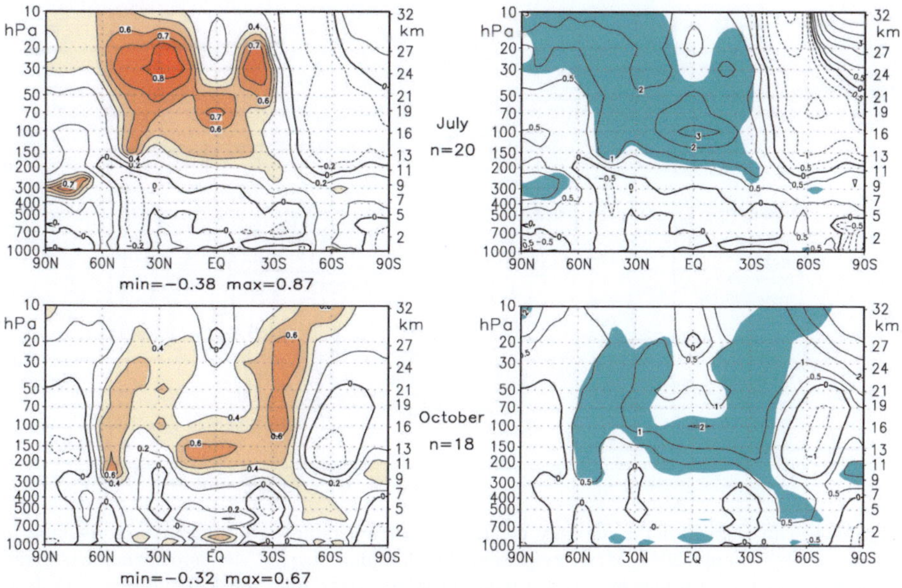

Fig. 30.3 *Left*: Vertical meridional sections of the correlations between the 10.7 cm solar flux and the detrended zonal mean temperatures in July (*upper panel*) and October (*lower panel*), only during the east phase of the QBO; *shaded* for emphasis where the correlations are above 0.4. *Right*: The respective temperature differences (K) between solar maxima and minima, shaded where the corresponding correlations on the left hand side are above 0.4 (data: NCEP/NCAR re-analyses, 1968–2010)

In Figs. 30.4 and 30.5, the time series of the 30-hPa monthly mean temperatures over the North Pole in February and March, are shown as examples of the large inter-annual variability during the Arctic winter. And we indicate the possibility of different trends, with the trend for the whole period (1948–2011, $n = 64$) being 0.1 K/decade in March, with clearly no significance. And we stress, that it is difficult to determine a significant trend in the Arctic, at least with the data available so far [cf. *Labitzke and Kunze*, 2005].

As shown in Figs. 30.4 and 30.5, the North Pole temperatures during February and March of 2011 were among the lowest of the whole time series, and it is of interest to see the extent of this anomaly. This is best done with the maps of the monthly mean deviations from the long-term mean 1968–2010 ($n = 43$) which are centred over the Arctic: Over the polar region the temperature anomalies are as large as two standard deviations, Fig. 30.6. The stable and cold polar vortex persisted in March 2011, leading to an even larger negative anomaly in geopotential height and temperature going beyond two standard deviations. With a major final warming, occurring at the beginning of April 2011, the cold conditions over the polar region ended. The still very cold vortex was shifted towards central Siberia, where the negative geopotential height and temperature anomaly exceed three standard deviations (s. Fig. 30.6), extending over an area from Europe to central Siberia.

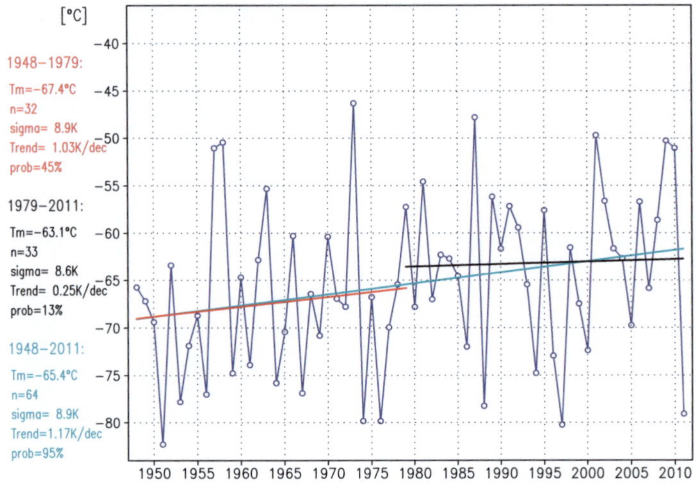

Fig. 30.4 Time series of the monthly mean 30-hPa temperatures (°C) over the North Pole in February, 1948 through 2011. Mean temperature (Tm), sample size (*n*), standard deviation (sigma), linear trend, and significance level of the trend (prob) are given for three different time periods as indicated. The trend for all 64 years is 1.17 K/dec (data: NCEP/NCAR re-analyses)

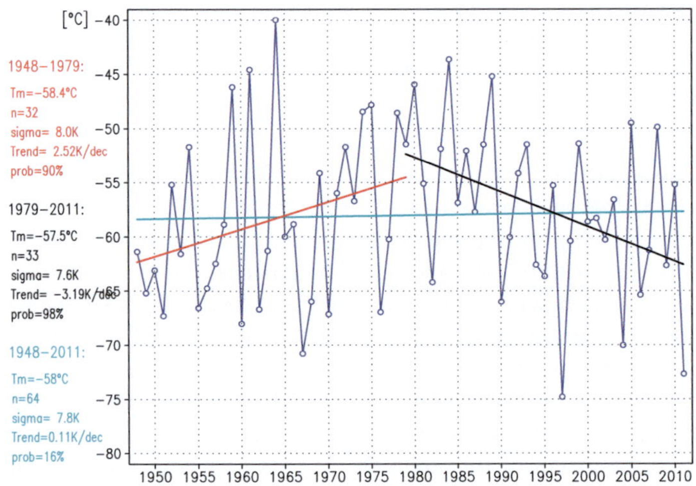

Fig. 30.5 As Fig. 30.4 for the time series of the 30-hPa monthly mean temperatures at the North Pole in March [adapted from Fig. 2.12 in *Labitzke and van Loon*, 1999]; data: NCEP/NCAR re-analyses

30.5 Arctic Variability Connected to the SSC and the QBO

As indicated already several times in this study, large dynamical anomalies may occur during the course of the winter, often leading from a very cold polar vortex

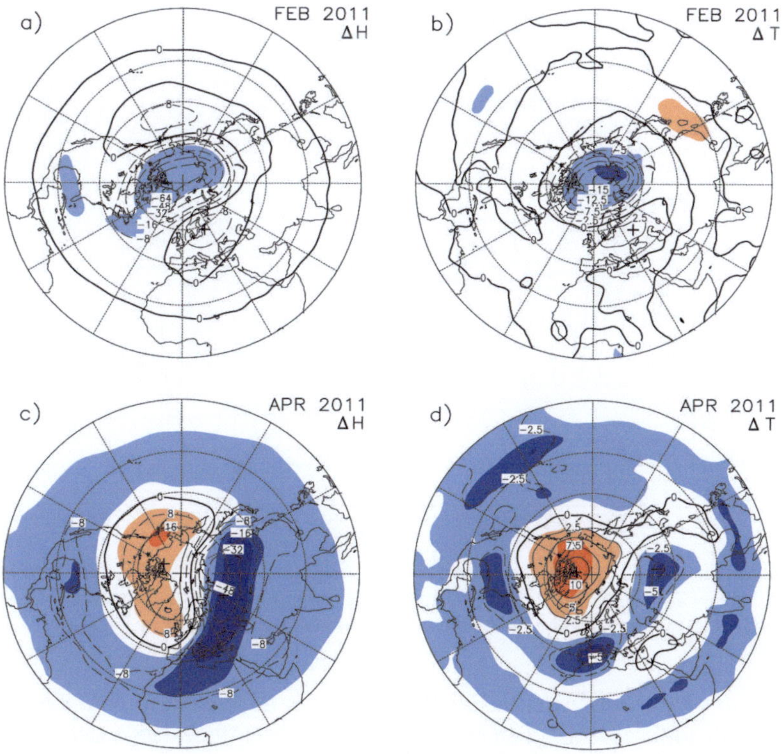

Fig. 30.6 30-hPa monthly mean fields of deviation from the long-term mean (1968–2010, $n = 43$) over the northern hemisphere (Eq—90°N) for February 2011 and April 2011; (**a**) and (**c**) geopotential height in gpdam; (**b**) and (**d**) temperature in K. In *shaded areas* the deviations are larger than 1, 2 or 3 (*blue*/*red*) standard deviations (data: NCEP/NCAR re-analyses)

to the breakdown of the vortex and intensive warming of the stratospheric Arctic. With additional data of up to 70 years this variability has been studied with the aim of understanding and possibly forecasting the type of the coming winter, in the stratosphere and also in the troposphere. Today, there is general agreement that the variability of the stratospheric circulation during the Arctic winters is influenced by different forcing mechanisms: by the tropospheric planetary waves, which penetrate into the stratosphere, by the QBO and the SO in the tropics, and by the 11-year sunspot cycle which interacts with the QBO and very likely with the SO.

In Sect. 30.4 we discussed temperature trends over the Arctic, and we are now turning to the 30-Pa heights, which are an integral from the troposphere to the stratosphere of the temperatures measured below. Figure 30.7 shows the monthly mean 30-hPa heights over the North Pole for February, from 1942 till 2011, i.e. 70 years. The very large inter-annual variability is outstanding and no simple connection can be made. However, *Labitzke* [1987] and *Labitzke and van Loon* [1988] showed a clear connection between the SSC and the 30-hPa heights, but only if the data were grouped based on the phase of the QBO.

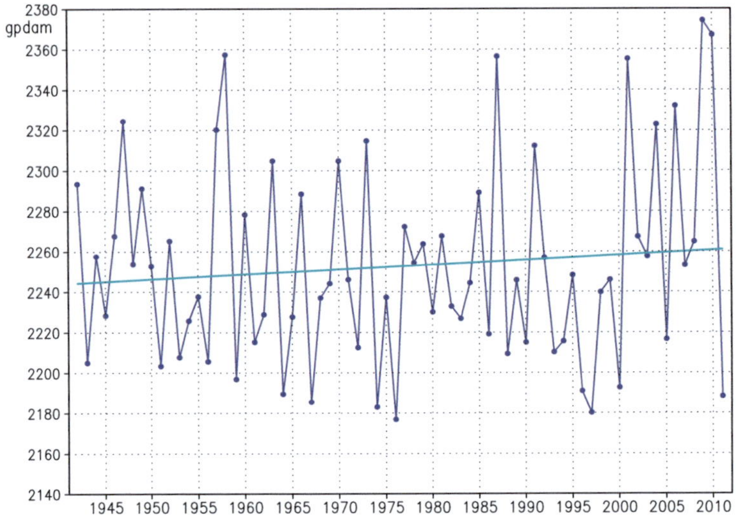

Fig. 30.7 Time series of monthly mean 30-hPa heights over the North Pole in February; 1942–2011; $n = 70$ years; linear trend: 2.4 gpdam/dec; significance level of the trend: 59 % (data: NCEP/NCAR re-analyses and reconstructions (1942–1947))

The two groups are given in Fig. 30.8, and it is of interest to note that the first paper on this topic was based on only 32 years of data, [1956–1987, *Labitzke*, 1987] while we have 70 years available today. And it is obvious that the early results were stable during the following 40 years, with only one clear outlier in the west wind group, i.e. the winter 2008/09 [*Labitzke and Kunze*, 2009].

30.6 Summary

We have shown some examples of temperature and geopotential height time series at the North Pole to illustrate the large interannual variability of the Northern hemisphere polar stratosphere during winter and spring, which does not allow the detection of significant trends during these seasons. A part of the variability can be attributed to the 11-year SSC, when the data is grouped according to the phase of the QBO.

30.6.1 Conclusion for the Summer

During the summer the detection of the solar signal is relatively straight forward: over the tropics and subtropics the temperatures of the middle stratosphere are highly and positively correlated with the SSC during the east phase of the QBO,

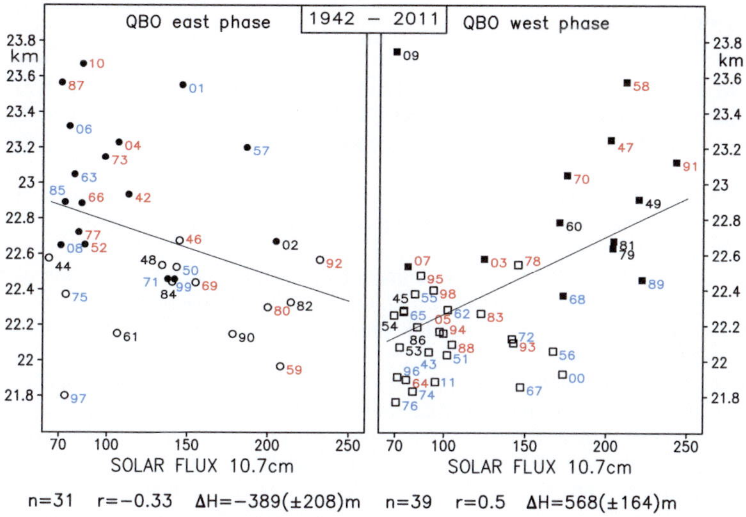

Fig. 30.8 The values of the monthly mean 30-hPa heights (geopot. km) are plotted against the 10.7 cm solar flux, grouped according to the QBO phases with the west wind group on the right hand side and the east wind group on the *left*. Sample size (n), correlation coefficient (r), and the height difference from solar minimum to solar maximum (i.e. 130 s.f.u.) (ΔH) are indicated for each group. *Filled symbols* indicate that during this winter month a "Major Midwinter Warming" took place, connected with the breakdown of the polar vortex. The numbers in the plots indicate the respective years, with years during the warm/cold phase of the SO marked *red/blue* (data: NCEP/NCAR re-analyses and reconstructions (1942–1947))

with practically no solar signal during the west phase of the QBO, over the whole globe. In the east phase the warming between solar maxima and solar minima is on the order of 2 standard deviations, which is highly significant.

30.6.2 Conclusion for the Winter

The QBO and SSC explain a large portion of the variance of the winter monthly values of Arctic geopotential heights. The behaviour of the Arctic stratosphere is strongly related to the QBO phase and the SSC, yet other factors such as internal variability also play an important role. The Arctic 2008/2009 winter is an excellent example of how these other factors, such as internal variability, can sometimes overwhelm the strong statistical relationships involving the QBO and the SSC. This emphasises the probabilistic nature of forecasting Arctic stratosphere behaviour, and we must keep in mind that correlations do not prove causality.

Acknowledgements We thank the members of the Stratospheric Research Group, FUB for professional support. The 10.7 cm solar flux data are from the World Data Center A, Boulder, Colorado. NCEP/NCAR re-analysis data are provided by the NOAA/OAR/ESRL PSD, Boulder, Colorado, USA, from their Web site at http://www.cdc.noaa.gov/.

References

Baldwin, M. P., Gray, L. J., Dunkerton, T. J., Hamilton, K., Haynes, P. H., Randel, W. J., Holton, J. R., Alexander, M. J., Hirota, I., Horinouchi, T., Jones, D. B. A., Kinnersley, J. S., Marquardt, C., Sato, K., & Takahashi, M. (2001). The quasi-biennial oscillation. *Reviews of Geophysics*, *39*(2), 179–229.

Brönnimann, S., Ewen, T., Griesser, T., & Jenne, R. (2006). Multidecadal signal of solar variability in the upper troposphere during the 20th century. *Space Science Reviews*, *125*, 305–317. doi:10.1007/s11214-006-9065-2.

Fischer, P., & Tung, K. K. (2008). A reexamination of the QBO period modulation by the solar cycle. *Journal of Geophysical Research*, *113*. doi:10.1029/2007JD008983.

Hamilton, K. (2002). On the quasi-decadal modulation of the stratospheric QBO period. *Journal of Climate*, *15*, 2562–2565.

Holton, J. R., & Tan, H. (1980). The influence of the equatorial quasi-biennial oscillation on the global circulation at 50 mb. *Journal of the Atmospheric Sciences*, *37*, 2200–2208.

Hood, L. L. (2003). Thermal response of the tropical tropopause region to solar ultraviolet variations. *Geophysical Research Letters*, *30*(23), 2215. doi:10.1029/2003GL018364.

Kalnay, E., Kanamitsu, M., Kistler, R., Collin, W., Deaven, D., Gandin, L., Iredell, M., Saha, S., White, G., Woollen, J., Zhu, Y., Leetmaa, A., Reynolds, B., Chelliah, M., Ebisuzaki, W., Higgins, W., Janowiak, J., Mo, K. C., Ropelewski, C., Wang, J., Jenne, R., & Joseph, D. (1996). The NCEP/NCAR 40-year reanalysis project. *Bulletin of the American Meteorological Society*, *77*, 437–471.

Labitzke, K. (1987). Sunspots, the QBO, and the stratospheric temperature in the north polar region. *Geophysical Research Letters*, *14*, 535–537.

Labitzke, K. (2003). The global signal of the 11-year solar cycle in the atmosphere: when do we need the QBO? *Meteorologische Zeitschrift*, *12*, 209–216.

Labitzke, K. (2005). On the solar cycle—QBO relationship: a summary. *Journal of Atmospheric and Solar-Terrestrial Physics*, *67*, 45–54.

Labitzke, K., & Collaborators (2002). The Berlin stratospheric data series. CD from Meteorological Institute, Free University Berlin.

Labitzke, K., & Kunze, M. (2005). Stratospheric temperatures over the arctic: comparison of three data sets. *Meteorologische Zeitschrift*, *14*, 65–74. doi:10.1127/0941-2948/2005/0014-0065.

Labitzke, K., & Kunze, M. (2009). On the remarkable arctic winter in 2008/09. *Journal of Geophysical Research*, *114*, D00I02. doi:10.1029/2009JD012273.

Labitzke, K., & van Loon, H. (1988). Associations between the 11-year solar cycle, the QBO and the atmosphere. Part I: the troposphere and stratosphere in the northern hemisphere winter. *Journal of Atmospheric and Terrestrial Physics*, *50*, 197–206.

Labitzke, K., & van Loon, H. (1994). Connection between the troposphere and stratosphere on a decadal scale. *Tellus*, *46A*, 275–286.

Labitzke, K., & van Loon, H. (1999). *The stratosphere*. Berlin: Springer.

Labitzke, K., Kunze, M., & Brönnimann, S. (2006). Sunspots, the QBO, and the stratosphere in the north polar region—20 years later. *Meteorologische Zeitschrift*, *15*, 355–363. doi:10.1127/0941-2948/2006/0136.

Naujokat, B. (1986). An update of the observed quasi-biennial oscillation of the stratospheric winds over the tropics. *Journal of the Atmospheric Sciences*, *43*, 1873–1877.

Perlwitz, J., & Graf, H.-F. (1995). The statistical connection between tropospheric and stratospheric circulation of the northern hemisphere in winter. *Journal of Climate*, *8*(10), 2281–2295.

Randel, W., Smith, K., Austin, J., Barnett, J., Claud, C., Gillett, N., Keckhut, P., Langematz, U., Lin, R., Long, C., Mearsm, C., Miller, A., Nash, J., Seidel, D., Thompson, D., Wu, F., & Yoden, S. (2009). An update of stratospheric temperature trends. *Journal of Geophysical Research*, *114*, D02107. doi:10.1029/2005JD006744.

Reed, R. J. (1964). A tentative model of the 26-month oscillation in tropical latitudes. *Quarterly Journal of the Royal Meteorological Society*, *90*(386), 441–466. doi:10.1002/qj.49709038607.

Soukharev, B., & Hood, L. L. (2001). Possible solar modulation of the equatorial quasi-biennial oscillation: additional statistical evidence. *Journal of Geophysical Research*, *106*, 14855–14868.
van Loon, H., & Labitzke, K. (2000). The influence of the 11-year solar cycle on the stratosphere below 30 km: a review. *Space Science Reviews*, *94*(1–2), 259–278. doi:10.1023/A:1026731625713.
van Loon, H., Zerefos, C., & Repais, C. (1982). The southern oscillation in the stratosphere. *Monthly Weather Review*, *110*, 225–229. Notes and correspondence.

Chapter 31
The Atmospheric Response to Solar Variability: Simulations with a General Circulation and Chemistry Model for the Entire Atmosphere

Hauke Schmidt, Jens Kieser, Stergios Misios, and Aleksandr N. Gruzdev

Abstract The coupled general circulation and chemistry model HAMMONIA and the MPI-ESM, consisting of the MAECHAM5 atmospheric GCM and the ocean model MPIOM, have been applied in a multitude of setups to study the response of the earth system to the variable forcing from the sun. This paper motivates the use of complex entire atmosphere models for the study of solar-terrestrial relations, and presents numerical results concerning solar rotational forcing, the response of the atmosphere-ocean system up to the lower thermosphere to 11-year forcing, and the response to particle precipitation. An issue analyzed in more detail is the so-called "secondary" response maximum in equatorial lower stratospheric ozone and temperature. Comparing numerical experiments with a variety of simulation setups it is argued that solar signals in this atmospheric region are easily obscured by variability stemming in particular from ENSO. However, simulations with solar variability as the only variable forcing suggest that indeed, the secondary maximum is of solar origin.

31.1 Introduction

One of the key issues in climate research today is to separate natural and anthropogenic causes of the observed changes. Variability of solar electromagnetic radiation but also particles emitted by the sun contribute substantially to the observed natural variability in the middle and upper atmosphere. And also for the troposphere, correlations between solar variability and for example the temperature are documented [see e.g. *Gray et al.*, 2010]. The mechanisms producing the atmospheric response to the inconstant sun are still not well understood, despite many decades of research. This is partly related to the relative shortness of many observational

H. Schmidt (✉) · J. Kieser · S. Misios
Max Planck Institute for Meteorology, Bundesstr. 53, 20146 Hamburg, Germany
e-mail: hauke.schmidt@zmaw.de

A.N. Gruzdev
A. M. Obukhov Institute of Atmospheric Physics, Russian Academy of Sciences, Pyzhevsky per. 3, 119017 Moscow, Russia

F.-J. Lübken (ed.), *Climate and Weather of the Sun-Earth System (CAWSES)*, Springer Atmospheric Sciences,
DOI 10.1007/978-94-007-4348-9_31,

records with respect to the length of e.g. the solar irradiance variation in the 11-year sunspot cycle. Additionally, solar signals have to be identified in a background atmosphere that is characterized by partly much larger variability caused internally or by other types of forcing. Numerical modeling is therefore a crucial tool to test hypotheses concerning possible mechanisms related to the occurrence of solar signals in the Earth system as a whole but in particular in the atmosphere. This paper describes a large set of numerical experiments with two complex models, the Hamburg Model of the Neutral and Ionized Atmosphere (HAMMONIA) which is an atmospheric general circulation and chemistry model covering the altitude range from the thermosphere to the surface, and the Max Planck Institute-Earth System Model (MPI-ESM), an atmosphere-ocean general circulation model that extends well into the mesosphere in the version used here. Different types of experiments have been performed: such where the only applied forcing originates from the sun in order to identify the "pure" solar signal, and such where a realistic inclusion of all important forcing types was included in order to test to what extent it is possible to reproduce observations.

The following two subsections provide background on solar forcing and its signals in the atmosphere, and on the current status of entire atmosphere modeling. Section 31.2 contains a description of the models used and the numerical experiments performed in this study. Section 31.3 provides results concerning the atmospheric response to the 27-day solar rotational variability. Section 31.4 describes simulated responses to the 11-year solar cycle variability in the stratosphere, troposphere, and ocean. Effects of energetic precipitating particles are dealt with in Sect. 31.5.

The aim of this paper is to provide an overview of results concerning the effect of solar variability on the Earth system obtained with numerical models belonging to the ECHAM family in general, and the HAMMONIA model in particular. Many results of the experiments are only summarized. This is the case, e.g., for the solar rotational effect, the high latitude stratospheric response, and the response of the equatorial oceans where the reader is referred to earlier publications for details. One issue addressed in more detail is the stratospheric equatorial response, and in particular the secondary response maximum in ozone and temperature in the lower equatorial stratosphere. The multitude of experiments performed offers the opportunity to assess the question to what extent this response is purely of solar origin or related to interaction or aliasing with other sources of variability like the QBO and the El Nino-Southern oscillation (ENSO).

31.1.1 Solar Forcing and Atmospheric Signals

For a comprehensive overview on observational and modeling results concerning the influence of solar variability the reader is referred to a review paper recently published by *Gray et al.* [2010]. This and the following section can only provide a brief overview of solar signals and model prerequisites to simulate these signals.

The response of the atmosphere to solar variability can be characterized by several "regimes":

1.) The large variability in the extreme ultraviolet (EUV) and x-ray parts of the solar spectrum leads to dramatic signals in the thermosphere [hundreds of Kelvin in temperature, large changes in density and in the concentration of neutral and ion species; see e.g. *Qian et al.*, 2006; *Emmert et al.*, 2008]. However, the influence of these signals on the lower levels of the atmosphere may be negligible.
2.) The relatively limited variability in the Schumann-Runge continuum and bands as well as in the Herzberg continuum of the spectrum leads to relatively small signals in the concentration of key species like ozone and in the temperature of the middle atmosphere [see e.g. *Chandra and McPeters*, 1994; *Soukharev and Hood*, 2006; *Randel et al.*, 2009]. Dynamical feedbacks associated with changes in heating rates and hence in zonal winds and planetary wave propagation [early studies in this field were performed e.g. by *Geller and Alpert*, 1980; *Haigh*, 1996] are still not fully understood, and should be addressed through coupled radiative chemical-dynamical three-dimensional models with an explicit formulation of key dynamical processes such as the quasi-biennial oscillation (QBO) of equatorial stratospheric winds. *Labitzke and van Loon* [1988] showed that temperature and geopotential height in the stratosphere are correlated with the 11-year solar cycle, in particular when the studied seasons are grouped according to the phase of the QBO.
3.) The absence of significant solar variability for wavelengths beyond 300 nm protects the troposphere from large direct solar influences. Using statistical methods, different authors [*Lean and Rind*, 2008; *Camp and Tung*, 2007] estimated the solar signal of the 11-year cycle at the Earth's surface to be not larger than 0.1 K in the global mean. However, stratospheric signals may dynamically propagate downward and regionally influence the troposphere. *Kodera and Kuroda* [2002], for instance, suggested that stratospheric changes in the zonal mean winds, and hence in the vertical propagation of planetary waves could modify the large-scale meridional circulation in winter with influences on the tropospheric dynamics and specifically on the modes of variability including the Arctic Oscillation. Again, this possible effect can only be addressed by an advanced general circulation model that is capable of simulating the coupling between the troposphere and stratosphere, including the downward propagation of the modes of variability [*Baldwin and Dunkerton*, 1999]. Another suggested pathway of downward coupling works via a temperature signal in the lower-most equatorial stratosphere [*Haigh et al.*, 2005, see also Sect. 31.4]. The dynamics of the atmosphere, at least up to the mesosphere [*Schmidt et al.*, 2006], are also influenced by changes in the sea surface temperature (SST). Since SSTs appear to be responding to solar variability at long time scales [e.g. *Reid et al.*, 1991], a full model simulation addressing these issues should include coupling to an ocean model.
4.) Episodic injection of energetic particles (as well as the penetration of cosmic galactic rays) affects all layers of the atmosphere, depending on the initial energy of the incoming particles. Particle precipitation may significantly influence the chemical composition down to the stratosphere [e.g. *Funke et al.*, 2011]. Besides the direct influence of precipitating solar protons, also indirect effects

of magnetospheric electrons and solar irradiance that influence thermospheric and mesospheric nitrogen and subsequently stratospheric ozone are discussed [e.g. *Randall et al.*, 2007]. Recent studies suggest that this influence may be of similar importance as that of solar irradiance variability [*Marsh et al.*, 2007]. In order to study the full range of solar influences, a numerical model should hence also include processes related to particle precipitation.

31.1.2 3D Modeling of the Entire Atmosphere

As described above, due to the complexity of processes and the large altitude range involved in producing the atmospheric response to solar variability, an ideal numerical model should cover physics and chemistry in the entire atmosphere, represent complex dynamical features like the QBO realistically, and be coupled to a comprehensive ocean model. Such models are still not existing, however, large progress has been made.

Until recently, most general circulation models (GCM) extended typically from the surface to approximately 2–10 hPa with the stratospheric layers being considered as a buffer between the tropopause and the top of the model. Some GCMs have been extended to approximately 80 km altitude [e.g. *Fels et al.*, 1980; *Boville*, 1995; *Manzini et al.*, 1997] and used for solar studies [e.g. *Haigh et al.*, 1996; *Shindell et al.*, 1999; *Matthes et al.*, 2004]. Prompted by the increasing knowledge on the importance of vertical coupling processes for climate and the availability of faster computers, several major climate research centers have decided to use this type of vertically extended models coupled to ocean models for climate simulations in the CMIP5 [Climate Modeling Intercomparison Project 5, *Taylor et al.*, 2009] context that will form the basis for the next assessment report (AR5, to be published in 2013).

Middle atmosphere GCMs have also been coupled to detailed chemical schemes [e.g. *Rasch et al.*, 1995; *de Grandpré et al.*, 2000; *Steil et al.*, 2003]. Such models have been used in the CCMVAL (Chemistry Climate Model Evaluation) intercomparisons [*Eyring et al.*, 2006, 2010]. These simulations focused on stratospheric ozone but have also been analyzed extensively with respect to solar signals [e.g. *Austin et al.*, 2008]. Only in the last decade, a few GCMs with coupled chemistry which extend from the surface even to the thermosphere have been developed. This is the case of the Extended Canadian Middle Atmosphere Model [CMAM, e.g. *Fomichev et al.*, 2002], of the Whole Atmosphere Community Climate Model [WACCM, e.g. *Marsh et al.*, 2007] developed at the National Center for Atmospheric Research (NCAR), and of HAMMONIA [*Schmidt et al.*, 2006], developed at the Max Planck Institute for Meteorology. Because of the large vertical extension and the chemistry coupling (including lower thermospheric ion chemistry) these models are particularly well adapted to assess the impact of solar variability on the atmosphere. However, due to their large computational cost, they can not be run over very long time-scales and their coupling to ocean models is still in its infancy.

31.2 Model Description and Simulation Setup

31.2.1 ECHAM5, MAECHAM5, and HAMMONIA

ECHAM5 [*Roeckner et al.*, 2003, 2006] is a state-of-the art general circulation model with a spectral dynamical core that includes comprehensive descriptions of the energy budget, the water cycle, land surface exchanges, and other physical processes. In its standard configuration it extends from the surface up to 10 hPa. It has been applied in numerous climate and atmospheric process studies. In its version coupled to the Max Planck Institute Ocean Model [*Marsland et al.*, 2003, MPIOM], for instance, it has participated in the Climate Model Intercomparison Project 3 (CMIP3) of which the results have been used as a basis for the fourth assessment report (AR4) of the Intergovernmental Panel for Climate Change [*Solomon et al.*, 2007].

The middle atmosphere version MAECHAM5 [*Manzini et al.*, 2006] has its upper lid at 0.01 hPa ($\sim$80 km). In a few of the simulations analyzed here, MAECHAM5 has been used in versions coupled and uncoupled to MPIOM. Most results of this study, however, have been obtained with the Hamburg Model of the Neutral and Ionized Atmosphere (HAMMONIA) that is a vertical extension of MAECHAM5 up to $1.7*10^{-7}$ hPa ($\sim$250 km). The model physics have been enhanced by including processes relevant for the mesosphere and the lower thermosphere: solar heating at ultraviolet and extreme ultraviolet wavelengths, a radiative scheme considering non-LTE (non-Local Thermodynamic Equilibrium) effects, vertical molecular diffusion and conduction, and a simple parameterization of electromagnetic forces in the thermosphere (ion drag and Lorenz forces). Furthermore, HAMMONIA is interactively coupled to the MOZART3 chemical module [*Kinnison et al.*, 2007] that is used here in a version with 48 chemical compounds. A more detailed description of HAMMONIA is given by *Schmidt et al.* [2006]. For the simulation of precipitating particle effects as presented in Chap. 13 five positive ions and several reactions representing ion chemistry have been included into the model.

Both atmospheric models (MAECHAM5 and HAMMONIA) have been used with the relatively coarse horizontal resolution of T31, i.e. a triangular truncation at wavenumber 31, but most simulations use a high vertical resolution (90 and 119 layers, respectively) that allows for an internal generation of the QBO [see *Giorgetta et al.*, 2006].

31.2.2 The Simulations

Table 31.1 contains brief details of the simulations presented in this study. More detailed information can be found in the indicated references. As explained above, most simulations have been performed with vertical resolutions allowing for a free QBO simulation. Exceptions are simulation H27, because the QBO is not considered important in the context of 27-day variability, and simulations Hlo-max/min

Table 31.1 Numerical simulations

Name	Model[a]	No. of years	Solar forcing	Reference
H27	HAMMONIA (L67)	5	idealized 27-day variation	*Gruzdev et al.* [2009]
Hlo-max, Hlo-min	HAMMONIA (L67)	35, 35	permanent[c] solar maximum, minimum	*Schmidt et al.* [2006]
Hhi-max, Hhi-min	HAMMONIA (L119)	35, 35	permanent solar maximum, minimum	*Schmidt et al.* [2010]
Hhi-trans-free	HAMMONIA (L119)	2*47[b]	transient[c] (1960–2006) solar, GHG, volcanoes, and SSTs	*Chiodo and Schmidt* [2012]
Hhi-trans-nudg	HAMMONIA (L119)	2*47[b]	like Hhi-trans-free but nudged QBO	*Chiodo and Schmidt* [2012]
Mc-strans	MAECHAM5-MPIOM (L90-L40)	11*52[b]	transient (1955–2006) solar, all other forcings constant	*Misios and Schmidt* [2012]
Ma-strans	MAECHAM5 (L90)	9*49[b]	transient (1958–2006) solar, all other forcings constant	*Misios and Schmidt* [2012]
H-epp-pa	HAMMONIA (L67)	5*1[b]	time dependent precipitation of protons and α-particles	*Kieser* [2011]
H-epp-pea	HAMMONIA (L67)	5*1[b]	like H-epp-pa but additionally with precipitating electrons	*Kieser* [2011]

[a]The number of vertical model layers is given in brackets

[b]These simulations have been performed as ensembles with 2, 11, 9, and 5 members, respectively. Ensemble members are started from differing initial conditions

[c]"Permanent" means that solar forcing is kept constant over the simulated time period, "transient" means varying with time

which allow an estimation of the role of the QBO when compared to Hhi-max/min. Hhi-trans-nudg is weakly nudged towards the observed QBO in order to assess (in comparison to Hhi-trans-free) the possible influence of a specific phase relation between the QBO and the solar forcing.

In all simulations information on solar spectral variability has been used as provided by *Lean* [2000] with a resolution of 1 nm. In the case of HAMMONIA, these data have been rebinned to more than 200 intervals in the near infrared, visible, and UV, while the EUV is parameterized following *Richards et al.* [1994]. In the case of MAECHAM5 the spectral resolution of solar radiation is much coarser with only 6 bands between 4500 and 170 nm [see *Cagnazzo et al.*, 2007]. Because of the missing interactive chemistry, the effect of solar variability on ozone is represented in MAECHAM5 by adding to the background ozone climatology an anomaly (taken from HAMMONIA simulations) scaled with the 10.7 cm solar radio flux (F10.7).

Simulations Hhi-trans-free/nudg have been performed according to the specifications for the CCMVAL2-RefB1 experiment *Eyring et al.* [2010], i.e. using boundary conditions in terms of sea surface temperatures (SST), greenhouse gas (GHG) concentrations, heating and cooling effects of volcanic aerosols and solar irradiance as observed for the period 1960 to 2006. Contrary, in the simulations with MAECHAM5, only solar irradiance is varying while all other forcings are fixed to values characteristic for present-day conditions.

In the case of the Mc-strans simulations, SSTs (and sea ice) have been calculated interactively using the ocean model MPI-OM. In all other cases monthly averaged values are prescribed. These stem from the HadISST1 data set recommended for the CCMVAL2 simulations (Hhi-trans), climatological values averaged over the AMIP2 period 1979–1996 (Hlo/hi-max/min, H27), and from a control run without solar forcing performed with the coupled model also used for Mc-trans (Ma-trans).

The analysis of 11-year solar signals was performed either by subtracting time averages of two simulations (Hlo/hi-max/min) or by a multiple linear regression (MLR) analysis, where the simulated time dependent temperature $T(t)$ is regressed to a number of proxies:

$$T(t) = T_{ave} + \beta_{trend} * t + \beta_{solar} * F10.7(t) + \beta_{enso} * Nino3.4(t) \tag{31.1}$$

$$+ \beta_{qbo} * u_{10\mathrm{hPa}}(t) + \beta_{qbo,orth} * u_{28\mathrm{hPa}}(t) + \beta_{volc} * AOD_{volc}(t) + R(t), \tag{31.2}$$

where β denotes the regression coefficients, T_{ave} is the average temperature, and $R(t)$ the residual. The following proxies are used: $F10.7(t)$: 10.7 cm solar radio flux; $Nino3.4(t)$: Nino3.4 index; $u_{10/28\mathrm{hPa}}(t)$: equatorial zonal average zonal winds at 10 and 28 hPa which are assumed to be almost orthogonal and should sufficiently describe the QBO [see e.g. *Lee and Smith*, 2003]; $AOD_{volc}(t)$: globally averaged optical depth of volcanic aerosol. Depending on the model simulations, a reduced number of proxies is applied. A trend term, e.g. is only needed in the "realistic" transient simulations Hhi-trans, an ENSO-proxy is not needed in simulations with climatological average SSTs (Ma-strans). The MLR is applied to deseasonalized monthly average time series. Solar signals are usually normalized to a forcing of 100 sfu (units of the 10.7 cm solar radio flux). Typical solar cycles show peak-to-peak amplitudes of about 130 sfu.

31.3 The Atmospheric Response to Solar Rotational Variability

The variation of solar radiation with a period of approximately 27 days is related to the sun's rotation and the inhomogeneous distribution of magnetic field structures and, thereby, inhomogeneous distribution of active areas on its surface. This variation is interesting to study for at least two reasons: a) Due to its shortness, it is much easier to analyze a large number of cycles and get proper statistics than from the 11-year solar cycle of which single satellite instruments usually do not cover much more than one cycle. b) As the 27-day variability is large during maximum phases of the 11-year cycle and negligible during minimum phases, it can not be excluded that 11-year signals are partly related to 27-day forcing.

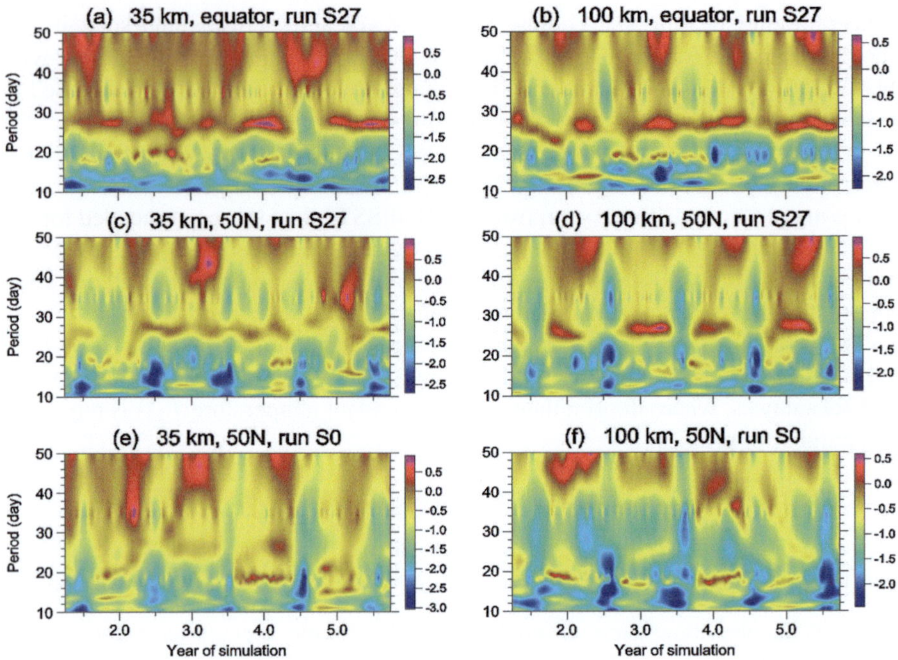

Fig. 31.1 Logarithm of the power spectral density of the ozone mixing ratio as a function of time at altitudes of (**a**, **c**, **e**) 35 km and (**b**, **d**, **f**) 100 km. Panels (**a–d**) are from the simulation including a 27-day solar variability, panels (**e–f**) from a reference simulation with solar irradiance constant in time. Panels (**a–b**) show results for the equator, panels (**c–f**) for 50N. This figure is taken from *Gruzdev et al.* [2009]

Since the 1980s, many authors have studied 27-day signals in ozone and temperature fields in the middle atmosphere from both observations and numerical simulations. An overview of existing work is given by *Gruzdev et al.* [2009]. Remaining open questions concern in particular possible effects on atmospheric waves, the dependence of signals on atmospheric background conditions, signals in the upper mesosphere and lower thermosphere (where few observations exist), and signals in chemical species other than ozone. To study some of these issues, *Gruzdev et al.* [2009] have performed HAMMONIA simulations with an idealized 27-day sinusoidal cycle of constant but realistic amplitude (H27 in Table 31.1), and compared it to a reference simulation without 27-day forcing.

Figure 31.1 presents time series of power spectral densities for periods from 10 to 50 days calculated from the ozone mixing ratio from different simulations and for several altitudes. An interesting result of the simulations is the intermittency of the 27-day signal (despite a continuous forcing) as shown in Fig. 31.1 (a–d) for the stratosphere and the mesopause region both at tropical and northern mid-latitudes. This means that the signal is influenced by the interannual variability of the background dynamical state of the atmosphere. However, a simple causal dependency (like a clear seasonal cycle) could not be identified except for the mesopause where the signal is more prominent in winter than in summer. Another intriguing feature is the presence of strong wintertime oscillations with periods between 16 and 20 days in the simulations without 27-day forcing (Fig. 31.1, e–f) which seem to be damped when 27-day forcing is applied. This means that either the forcing affects the background atmospheric states and thereby the generation or propagation conditions for planetary waves, or the forcing interacts non-linearly with such intrinsic atmospheric oscillations. Further results from these simulations concern a non-linearity in the response of both ozone and temperature, the daytime dependence of the ozone signal in the mesopause region and potential signals in other observable trace gases like nitrogen oxides and the hydroxyl radical [*Gruzdev et al.*, 2009].

31.4 The Atmospheric Response to 11-Year Solar Cycle Forcing

The amplitude of solar 11-year variability depends on the wavelength. The strongest variability is observed at very short wavelengths. This implies that the direct radiative 11-year solar signals should be largest at high atmospheric altitudes where the shorter wavelengths are absorbed. Accordingly, in the thermosphere, temperature signals of several 100 K are simulated. Figure 31.2 shows signals in zonal mean temperature and ozone from the lower thermosphere down to the surface calculated as the difference from simulations Hhi-max and Hhi-min. In the mesopause region, simulated signals per 100 sfu are of about 2 K and 5 % in temperature and ozone, respectively, up to 0.7 K and 2 % in the upper stratosphere, and very small (mostly insignificant) in the troposphere. These values are in general in the range suggested by observations (see e.g. *Beig et al.* [2012] for mesopause temperature and ozone, *Soukharev and Hood* [2006] for stratospheric ozone, and *Randel et al.* [2009] for stratospheric temperature) and other model studies and mainly explained by the increased absorption of UV radiation during solar maximum. The presented signal of diurnal average ozone in the mesopause region may be misleading as a strong diurnal cycle exists in absolute ozone mixing ratios as well as locally in the solar signal [see *Beig et al.*, 2012]. The small band of negative ozone response close to 0.01 hPa is related to a positive signal in the hydroxyl radical OH caused by stronger photolysis of water vapor during solar maximum. A more detailed discussion of these issues is given by *Schmidt et al.* [2006].

The latitudinal dependence of the signal in the troposphere and stratosphere is expected to be influenced by dynamical responses to the radiative forcing. This is less well established from observations. The ozone responses estimated from three different satellite instruments that operated over different periods as analyzed by *Soukharev and Hood* [2006] show fairly different latitudinal structures. The same is true for temperature responses deduced from different re-analysis datasets [compare e.g. *Frame and Gray*, 2010; *Labitzke*, 2005]. It is also unclear if a part of the analyzed latitudinal structure may be related to the difficulty in disentangling solar and QBO signals [see *Smith and Matthes*, 2008].

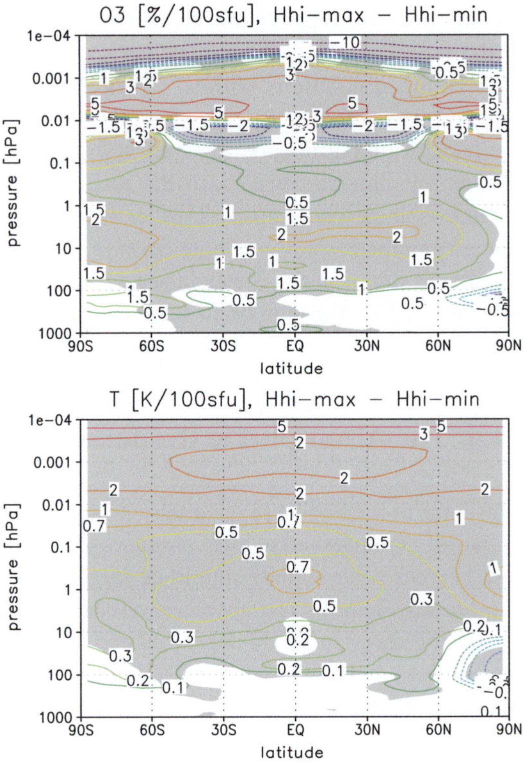

Fig. 31.2 Solar cycle responses of the annual mean zonal average ozone mixing ratio (*top*, in %) and temperature (*bottom*, in K) normalized to 100 sfu from HAMMONIA simulations (Hhi-max—Hhi-min). *Gray shading* indicates statistical significance of the signal larger than 95 % calculated with Student's T-test. These figures have been published as b/w versions by *Schmidt et al.* [2010]

The increased gradient of solar heating during solar maximum between the sunlit low to middle latitudes and the polar night region should lead to an acceleration of the stratospheric polar night jet. According to observational analyses and a mechanism proposed by *Kodera and Kuroda* [2002] this positive anomaly in the jet should propagate poleward and downward during the winter and finally influence the tropospheric circulation. *Labitzke* [1987] and *Labitzke and von Loon* [1988] have suggested first that the high latitude wintertime stratospheric solar signal interacts non-linearly with the QBO. Simulations of these high latitude effects are, so far, inconclusive. Some simulations [e.g. *Matthes et al.*, 2004] reproduce the mechanism proposed by *Kodera and Kuroda* [2002] albeit in general with relatively small signals. *Schmidt et al.* [2010] have simulated an interaction of QBO and solar cycle albeit also with weak amplitudes and with different timing than observed. Figure 31.2 shows significant response minima in temperature and ozone in the northern polar lower stratosphere. These signals are dominated by the response in northern winter and are related to an anomalous upwelling. This is consistent with the anomalous tropical downwelling as part of a weaker Brewer-Dobson circulation that is discussed in the following section. However, the variability of the northern winter is very high and solar signals in this region have to be interpreted with caution as indicated by large ensemble simulations (e.g. Mc-strans) where signals may differ strongly between ensemble members. For a further discussion of high latitude

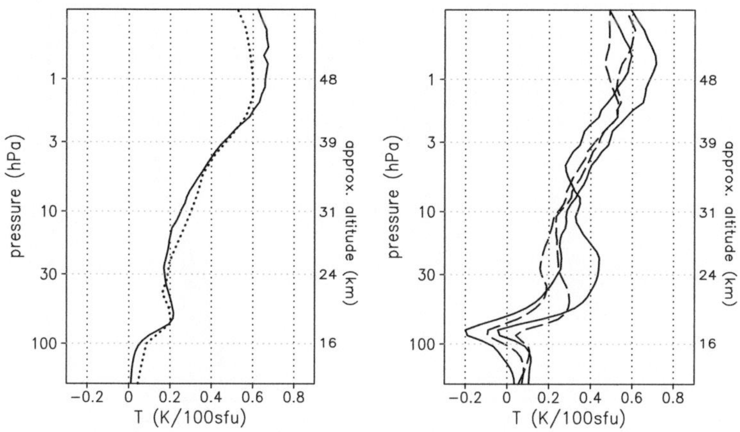

Fig. 31.3 Solar cycle responses of equatorial (25S–25N) annual mean temperature normalized to 100 sfu from HAMMONIA simulations. *Left*: *dotted*: Hlo-max—Hlo-min (simulations without QBO); *solid*: Hhi-max—Hhi-min (simulations with QBO). *Right*: response calculated with MLR from two realizations with internally produced QBO (Hhi-trans-free; *solid*) and two realizations with "nudged" QBO (Hhi-trans-nudg; *dashed*)

responses the reader is referred to *Schmidt et al.* [2010]. In the following we will concentrate on the solar signals in the equatorial stratosphere and in the equatorial Pacific.

31.4.1 The Response in the Equatorial Stratosphere

Besides the response maximum in temperature and ozone in the upper stratosphere, Fig. 31.2 also indicates secondary maxima in the lower-most equatorial stratosphere. This is more evident in a vertical profile of the temperature signals averaged over 25S to 25N as shown in Fig. 31.3 (left). The secondary response maximum would appear more pronounced directly at the equator (and in ozone) as presented by *Schmidt et al.* [2010] but we discuss here the tropical average in order to compare to other published results. Some observations and reanalysis data confirm the existence of a secondary maximum [see e.g. *Soukharev and Hood*, 2006; *Labitzke*, 2005; *Frame and Gray*, 2010]. In other observations the signal is unclear but may be obscured by poor vertical resolution. *Kodera and Kuroda* [2002] have suggested that the occurrence of the secondary maximum is related to reduced equatorial upwelling caused by changed wave-mean flow interactions in the extratropics. In most previous simulations with permanent solar maximum and minimum conditions the signal could not be reproduced. In transient simulations with prescribed observed SSTs of the period 1960 to 2006, however, a number of models have reproduced a secondary maximum [*Austin et al.*, 2008; *Eyring et al.*, 2010], albeit with quite different amplitudes and peak altitudes.

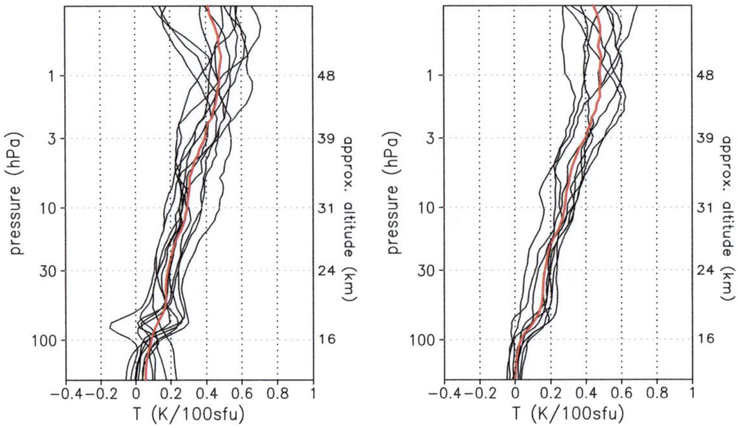

Fig. 31.4 Solar cycle responses of equatorial (25S–25N) annual mean temperature normalized to 100 sfu from MAECHAM simulations calculated with MLR. *Left*: 11 realizations of the MAECHAM5 simulation coupled to MPIOM (Mc-strans) and their ensemble mean (*red*). *Right*: 9 realizations of the atmosphere-only simulation (Ma-strans) and their ensemble mean (*red*)

As can be seen in Fig. 31.3 (left), HAMMONIA reproduces such a secondary maximum also with prescribed climatological average SSTs, and independent of the vertical resolution, i.e. independent of the existence of a QBO. HAMMONIA has now also been applied using the settings of the CCMVAL RefB1 scenario where other models have also produced a secondary maximum [see Fig. 8.11 of *Eyring et al.*, 2010]. We have performed four HAMMONIA simulations, two with an internally produced QBO and two with exactly the same model configuration but weak "nudging" to observational data in the equatorial stratosphere to force the QBO to the observed phasing. All four simulations (see Fig. 31.3, right) are characterized by similar vertical structures with response minima close to 80 hPa. This structure seems not to depend strongly on the specific QBO phasing. At least in the "nudged" simulations a weak local response maximum can be identified at around 50 hPa. These vertical profiles are very different to those from simulations Hhi/lo-min/max but quite similar to those of some models of the CCMVAL comparison [Fig. 8.11 *Eyring et al.*, 2010]. It has been discussed if the lower stratospheric response maximum in observations is partly caused by an aliasing of solar and ENSO signals [*Marsh and Garcia*, 2007; *Hood et al.*, 2010], but over the period 1960–2006 the correlation of ENSO and solar indices is very weak and the vertical profiles analyzed from our simulations change only weakly when the ENSO index is excluded from the MLR (not shown). However, the differences between the left and right panels of Fig. 31.3 suggest a strong dependence of the simulated solar signal on the choice of boundary conditions and provokes the questions if the signal analyzed by the MLR for the transient simulations is indeed purely of solar origin. The most likely boundary condition of influence is the ENSO for which HAMMONIA and MAECHAM5 simulate a very strong response peak in this region. According to *Calvo et al.* [2010], this response (that occurs also in observations and other models) is related to changes in upwelling caused by anomalous orographic gravity wave forcing during ENSO events.

To further analyze this issue we have performed two ensembles of simulations (9 and 11 members) with the MAECHAM5 model with and without ocean coupling (Ma-strans and Mc-strans, respectively). The solar signals analyzed by MLR for many of the coupled ensemble members (Fig. 31.4) show strong (negative or positive) response maxima in the lower equatorial stratosphere. The members of the uncoupled ensemble show much less variability in this region. Hence, the reason for the peaks in the coupled simulations is either the incapability of the MLR to properly disentangle solar and ENSO signals or an interaction of the two forcings.

It should be noted that the secondary maximum in the solar response is suggested to depend on the QBO, being strong only in the east phase. Simulations by *Matthes et al.* [2010] and to a lesser extent also our Hhi-trans-nudg/free simulations [not shown, see *Chiodo and Schmidt*, 2012] confirm this.

31.4.2 Responses of Troposphere and Ocean

The secondary temperature response maximum in the lower equatorial stratosphere may also be instrumental in the downward propagation of the solar signal to the troposphere. According to sensitivity simulations with prescribed temperature anomalies by *Haigh et al.* [2005] and *Haigh and Blackburn* [2006] this leads to changes in the stability of the tropical tropopause region, changes of tropospheric eddy momentum fluxes and a poleward shift of the subtropical jets. A similar solar response pattern is deduced from reanalysis data by *Frame and Gray* [2010]. Our simulations Hhi-min/max, that produce lower stratospheric equatorial temperature anomalies (see above) also show a similar shift of tropospheric zonal wind patterns [see *Schmidt et al.*, 2010].

In simulations with prescribed SSTs, however, the solar signal near the surface may be damped as SSTs can not react, and possible feedbacks resulting from a change in SSTs would be suppressed. Hence, to study these effects we performed the simulations Mc-strans with the coupled atmosphere ocean model MAECHAM5/MPIOM. In the remaining part of this section we concentrate on the solar response of the equatorial Pacific as analyzed in detail by *Misios and Schmidt* [2012].

There is an ongoing debate concerning the solar signal in this region. According to an analysis of observed SSTs by *van Loon et al.* [2007], the equatorial Pacific shows a negative temperature response during peak years of solar maximum. *Meehl et al.* [2009] attributed a similar response in a model simulation to a combination of a direct radiative effect and a dynamical signal propagating downward from the stratosphere. Another analysis of SST observations by *Roy and Haigh* [2010], however, identifies a weak El-Nino like (i.e. warming) signal during solar maximum.

The ensemble mean tropical SST anomalies of our Mc-strans simulations are filtered by the Multichannel Singular Spectrum Analysis [MSSA, *Ghil et al.*, 2002].

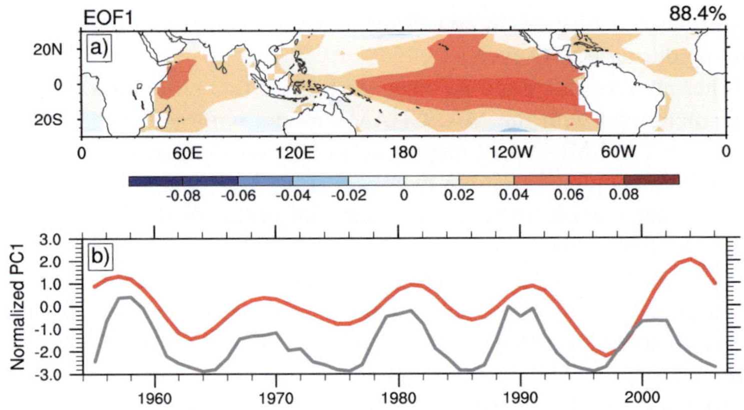

Fig. 31.5 *Top*: Leading Empirical Orthogonal Function [EOF1] of the simulated tropical annual SST anomalies filtered with the Multichannel Singular Spectrum Analysis. EOF1 explains 88.4 % of the filtered variance. To present EOF1 in meaningful units it is expressed in terms of the regression coefficients of the filtered SSTs onto the first principal component (PC1) resulting from the EOF analysis. Units are K per standard deviation. *Bottom*: Time series of the PC1 (normalized to one standard deviation) over the simulated period 1955–2006 (*red*). The annual 10.7 cm radio fluxes (*gray*) are superimposed for reference

The 7th and 8th oscillatory mode (4.2 % and 4.1 % of the total variance) exhibit spectral peaks at around 10.9 years which may be associated with the solar cycle. The first six oscillatory modes describe ENSO and other intra-decadal frequencies. The leading Empirical Orthogonal Function of the SST anomalies reconstructed from the modes with 10.9-year periods is displayed in Fig. 31.5. A weak (but clear) temperature response in phase with the solar forcing can be identified. The response in the equatorial Pacific is stronger than in other equatorial oceans. *Misios and Schmidt* [2012] explain this amplification by a synergetic effect of a water-vapor feedback and ocean dynamics.

31.5 The Atmospheric Response to Energetic Particle Precipitation

The precipitation of energetic particles can influence chemistry and dynamics of the atmosphere from the surface down to the stratosphere. Strongest signals are identified in the polar cap and polar oval where most particles enter the atmosphere. Important classes of particles are solar protons that occasionally (after large solar eruptions) have strong influences down to the stratosphere, and magnetospheric electrons of which the flux depends on geomagnetic activity and thereby indirectly also on solar activity. More information on effects of particle precipitation on the atmosphere can be found e.g. in studies by *Marsh et al.* [2007], *Randall et al.* [2007], *Funke et al.* [2011], *Kieser* [2011] and references therein. In this volume, particle effects are discussed in more detail for example in Chaps. 13, 15, 16, and 17.

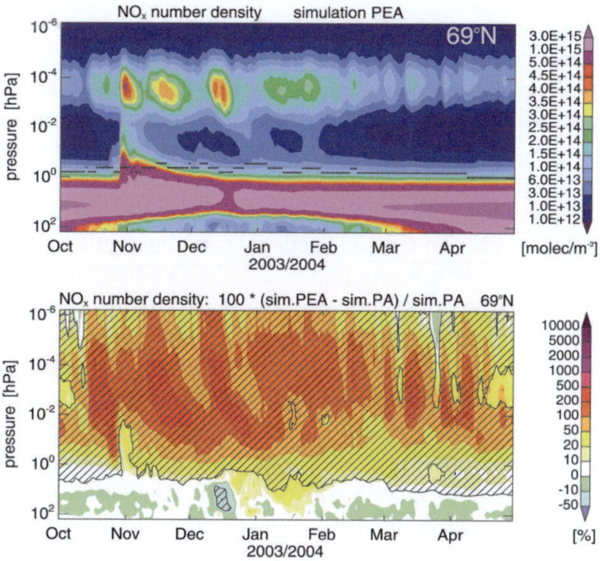

Fig. 31.6 *Top*: Nitrogen oxide (NO_x) number density [1/cm^3] as simulated with HAMMONIA in simulation H-epp-pea (including effects from protons, electrons and α-particles) at 69N geomagnetic latitude. *Bottom*: Difference [%] in NO_x number densities between simulations H-epp-pea and H-epp-pa, i.e. increase of NO_x resulting from electrons compared to a simulation including only protons and α-particles. Hatching indicates statistical significance above 95 %

Hammonia ensemble simulations including ion chemistry and ionization rates calculated with the AIMOS model (see Chap. 13) have been performed to study the relative effects of different particle classes during the winter 2003/2004 that was characterized by a strong solar proton event around 1 November (known as "Halloween event"), and strong downward transport of mesospheric air masses in late winter. Figure 31.6 shows the importance of precipitating energetic electrons for the NO_x concentration near the northern polar oval from the thermosphere down to the upper stratosphere. Compared to a simulation neglecting electron effects, NO_x concentrations are increased up to a factor of 5. The relative effect is smallest during the Halloween event because of the large contribution of protons. However, studies of proton events that neglect electrons would considerably underestimate the background NO_x concentration and thereby overestimate the proton effect.

Further results from these numerical experiments are presented by *Kieser* [2011], *Wissing et al.* [2011], and in the model intercomparison study by *Funke et al.* [2011].

31.6 Conclusions

The coupled general circulation and chemistry model HAMMONIA and the MPI-ESM consisting of the MAECHAM5 atmospheric GCM and the ocean model MPIOM have been applied in a multitude of setups to study the response of the earth system to the variable forcing from the sun. Results have been obtained concerning solar rotational forcing, the response of mesosphere, stratosphere, troposphere, and ocean to 11-year forcing, and the response to particle precipitation. An issue analyzed in more detail is the secondary response maximum in the equatorial lower stratosphere that will be discussed in the following as an example for possible benefits from and shortcomings of extensive modeling studies like presented here.

The time slice simulations with permanent solar maximum and minimum conditions, respectively, show that such a secondary maximum may be produced as a pure solar signal. However, analyzed signals from different transient simulations with a full set of forcings indicate a fairly different vertical profile of the solar response. One may conclude that the solar signal can either not be analyzed properly with the chosen (MLR) method or is the result of interaction with other forcings. A crucial role is likely played by ENSO which produces a signal of large amplitude in the region of the secondary solar maximum. The employed HAMMONIA model has of course deficiencies, e.g. it has only been applied at a very coarse resolution (T31), the effects of gravity waves have to be parameterized and no dependence of the sources on the actual meteorology is implemented, and the fixed SSTs would dampen solar signals. To overcome the latter deficit (and to allow for large ensemble simulations) we have additionally employed a coupled atmosphere-ocean model where the atmosphere, however, is not interactively coupled to a chemistry module. The ensemble simulations indicate a large variability (with strong positive and negative responses in single ensemble members) in the region of the assumed secondary solar response maximum. A comparison with the rather similar responses of the four Hhi-trans-free/nudg simulations (with prescribed SST time series) suggest that the analyzed solar response may appear very different with different underlying SSTs. It may be very difficult to analyze the solar signal in this region from a time series covering only five observed solar cycles. It can however not be excluded that the model simulations overestimate the problem of the contamination (or interaction) with ENSO signals as the ENSO variability produced by the MPI-ESM is known to be too regular and too strong.

Although model results are not always conclusive, for further progress in solar-terrestrial research it seems inevitable to continue applying further improved complex models in long (or large ensemble) simulations. One possible uncertainty is also provided by the assumptions on the spectral solar variability. Recent observations with the SIM instrument onboard the SORCE satellite suggest a much stronger UV variability than previously assumed and out of phase variability in a part of the visible spectrum [*Harder et al.*, 2009]. If these observations are confirmed, all the model simulations performed here (and in other studies) will be of very limited use for the explanation of the actual solar signals.

Acknowledgements This work was funded by the *Deutsche Forschungsgemeinschaft* (DFG) under grants SCHM2158/1-(1-3). Most of the numerical simulations have been performed on the computers and with the support of the *German Climate Computing Centre* (DKRZ). The authors would like to thank Guy P. Brasseur and Marco A. Giorgetta for many valuable discussions during the course of the project, and Heinz-Jürgen Punge for help with the MLR.

References

Austin, J., Tourpali, K., Rozanov, E., Akiyoshi, H., Bekki, S., Bodeker, G., Bruehl, C., Butchart, N., Chipperfield, M., Deushi, M., Fomichev, V. I., Giorgetta, M. A., Gray, L., Kodera, K., Lott,

F., Manzini, E., Marsh, D., Matthes, K., Nagashima, T., Shibata, K., Stolarski, R. S., Struthers, H., & Tian, W. (2008). Coupled chemistry climate model simulations of the solar cycle in ozone and temperature. *Journal of Geophysical Research, 113*(D11). doi:10.1029/2007JD009391.

Baldwin, M. P., & Dunkerton, T. J. (1999). Propagation of the Arctic Oscillation from the stratosphere to the troposphere. *Journal of Geophysical Research, 104*, 30937–30946. doi:10.1029/1999JD900445.

Beig, G., Fadnavis, S., Schmidt, H., & Brasseuer, G. P. (2012). Inter-comparison of 11-year solar cycle response in mesospheric ozone and temperature obtained by HALOE satellite data and HAMMONIA model. *Journal of Geophysical Research, 117*, D00P10. doi:10.1029/2011JD015697.

Boville, B. A. (1995). Middle atmosphere version of the CCM2 (MACCM2): annual cycle and interannual variability. *Journal of Geophysical Research, 100*, 9017–9039.

Cagnazzo, C., Manzini, E., Giorgetta, M. A., Forster, P. M. D. F., & Morcrette, J. J. (2007). Impact of an improved shortwave radiation scheme in the MAECHAM5 General Circulation Model. *Atmospheric Chemistry & Physics, 7*, 2503–2515.

Calvo, N., Garcia, R. R., Randel, W. J., & Marsh, D. R. (2010). Dynamical mechanism for the increase in tropical upwelling in the lowermost tropical stratosphere during warm ENSO events. *Journal of the Atmospheric Sciences, 67*, 2331–2340. doi:10.1175/2010JAS3433.1.

Camp, C. D., & Tung, K. K. (2007). Surface warming by the solar cycle as revealed by the composite mean difference projection. *Geophysical Research Letters, 34*, 14703–14706. doi:10.1029/2007GL030207.

Chandra, S., & McPeters, R. D. (1994). The solar cycle variation of ozone in the stratosphere inferred from Nimbus 7 and NOAA 11 satellites. *Journal of Geophysical Research, 99*, 20665. doi:10.1029/94JD02010.

Chiodo, G., & Schmidt, H. (2012). 11-yr solar-cycle effects in transient hammonia simulations. In preparation.

de Grandpré, J., Beagly, S. R., Fomichev, V. I., Griffioen, E., McConnell, J. C., & Medvedev, A. S. (2000). Ozone climatology using interactive chemistry: results from the Canadian middle atmosphere model. *Journal of Geophysical Research, 105*, 26475–26491.

Emmert, J. T., Picone, J. M., & Meier, R. R. (2008). Thermospheric global average density trends, 1967–2007, derived from orbits of 5000 near-earth objects. *Geophysical Research Letters, 35*, L05101. doi:10.1029/2007GL032809.

Eyring, V., Butchart, N., Waugh, D. W., Akiyoshi, H., Austin, J., Bekki, S., Bodeker, G. E., Boville, B. A., Brühl, C., Chipperfield, M. P., Cordero, E., Dameris, M., Deushi, M., Fioletov, V. E., Frith, S. M., Garcia, R. R., Gettelman, A., Giorgetta, M. A., Grewe, V., Jourdain, L., Kinnison, D. E., Mancini, E., Manzini, E., Marchand, M., Marsh, D. R., Nagashima, T., Newman, P. A., Nielsen, J. E., Pawson, S., Pitari, G., Plummer, D. A., Rozanov, E., Schraner, M., Shepherd, T. G., Shibata, K., Stolarski, R. S., Struthers, H., Tian, W., & Yoshiki, M. (2006). Assessment of temperature, trace species, and ozone in chemistry-climate model simulations of the recent past. *Journal of Geophysical Research, 111*(D10), D22308. doi:10.1029/2006JD007327.

Eyring, V., Shepherd, T. G., & Waugh, D. W. (Eds.) (2010). *SPARC report on the evaluation of chemistry-climate models*.

Fels, S. B., Mahlman, J. D., Schwarzkopf, M. D., & Sinclair, R. W. (1980). Stratospheric sensitivity to perturbations in ozone and carbon dioxide: radiative and dynamical response. *Journal of the Atmospheric Sciences, 37*, 2265–2297.

Fomichev, V. I., Ward, W. E., Beagley, S. R., McLandress, C., McConnell, J. C., McFarlane, N. A., & Shepherd, T. G. (2002). The extended Canadian middle atmosphere model: zonal mean climatology and physical parameterizations. *Journal of Geophysical Research, 107*. doi:10.1029/2001JD000479.

Frame, T. H. A., & Gray, L. J. (2010). The 11-year solar cycle in ERA-40 data: an update to 2008. *Journal of Climate, 23*, 2213–2222. doi:10.1175/2009JCLI3150.1.

Funke, B., Baumgaertner, A., Calisto, M., Egorova, T., Jackman, C. H., Kieser, J., Krivolutsky, A., López-Puertas, M., Marsh, D. R., Reddmann, T., Rozanov, E., Salmi, S.-M., Sinnhuber, M., Stiller, G. P., Verronen, P. T., Versick, S., von Clarmann, T., Vyushkova, T. Y., Wieters, N., & Wissing, J. M. (2011). Composition changes after the "Halloween" solar proton event: the High-Energy Particle Precipitation in the Atmosphere (HEPPA) model versus MIPAS data intercomparison study. *Atmospheric Chemistry & Physics Discussions*, *11*, 9407–9514. doi:10.5194/acpd-11-9407-2011.

Geller, M. A., & Alpert, J. C. (1980). Planetary wave coupling between the troposphere and the middle atmosphere as a possible Sun-weather mechanism. *Journal of the Atmospheric Sciences*, *37*, 1197–1215. doi:10.1175/1520-0469(1980)037<1197:PWCBTT>2.0.CO;2.

Ghil, M., Allen, M. R., Dettinger, M. D., Ide, K., Kondrashov, D., Mann, M. E., Robertson, A. W., Saunders, A., Tian, Y., Varadi, F., & Yiou, P. (2002). Advanced spectral methods for climatic time series. *Reviews of Geophysics*, *40*, 1003.

Giorgetta, M. A., Manzini, E., Roeckner, E., Esch, M., & Bengtsson, L. (2006). Climatology and forcing of the quasi-biennial oscillation in the MAECHAM5 model. *Journal of Climate*, *19*, 3882–3901 doi:10.1175/JCLI3830.1.

Gray, L. J., Beer, J., Geller, M., Haigh, J. D., Lockwood, M., Matthes, K., Cubasch, U., Fleitmann, D., Harrison, G., Hood, L., Luterbacher, J., Meehl, G. A., Shindell, D., van Geel, B., & White, W. (2010). Solar influences on climate. *Reviews of Geophysics*, *48*, RG4001.

Gruzdev, A. N., Schmidt, H., & Brasseur, G. P. (2009). The effect of the solar rotational irradiance variation on the middle and upper atmosphere calculated by a three-dimensional chemistry-climate model. *Atmospheric Chemistry and Physics*, *9*, 595–614.

Haigh, J., Blackburn, M., & Day, R. (2005). The response of tropospheric circulation to perturbations in lower-stratospheric temperature. *Journal of Climate*, *18*, 3672–3685.

Haigh, J. D. (1996). The impact of solar variability on climate. *Science*, *272*, 981–984. doi:10.1126/science.272.5264.981.

Haigh, J. D., & Blackburn, M. (2006). Solar influences on dynamical coupling between the stratosphere and troposphere. *Space Science Reviews*, *125*, 331–344.

Harder, J. W., Fontenla, J. M., Pilewskie, P., Richard, E. C., & Woods, T. N. (2009). Trends in solar spectral irradiance variability in the visible and infrared. *Geophysical Research Letters*, *36*, L07801. doi:10.1029/2008GL036797.

Hood, L. L., Soukharev, B. E., & McCormack, J. P. (2010). Decadal variability of the tropical stratosphere: secondary influence of the El Niño-Southern Oscillation. *Journal of Geophysical Research*, *115*(D14), D11113. doi:10.1029/2009JD012291.

Kieser, J. (2011). *The influence of precipitating energetic particles on the entire atmosphere—Simulations with HAMMONIA*. Ph.D. thesis, Univ. of Hamburg.

Kinnison, D. E., Brasseur, G. P., Walters, S., Garcia, R. R., Marsh, D. R., Sassi, F., Harvey, V. L., Randall, C. E., Emmons, L., Lamarque, J. F., Hess, P., Orlando, J. J., Tie, X. X., Randel, W., Pan, L. L., Gettelman, A., Granier, C., Diehl, T., Niemeier, U., & Simmons, A. J. (2007). Sensitivity of chemical tracers to meteorological parameters in the MOZART-3 chemical transport model. *Journal of Geophysical Research*, *112*(D20). doi:10.1029/2006JD007879.

Kodera, K., & Kuroda, Y. (2002). Dynamical response to the solar cycle. *Journal of Geophysical Research*, *107*. doi:10.1029/2002JD002224.

Labitzke, K. (1987). Sunspots, the QBO, and the stratospheric temperature in the north polar region. *Geophysical Research Letters*, *14*, 535–537. doi:10.1029/GL014i005p00535.

Labitzke, K. (2005). On the solar cycle-QBO relationship: a summary. *Journal of Atmospheric and Solar-Terrestrial Physics*, *67*(1–2), 45–54.

Labitzke, K., & van Loon, H. (1988). Associations between the 11-year solar cycle, the QBO and the atmosphere, part I: the troposphere and the stratosphere in the northern hemisphere winter. *Journal of Atmospheric and Solar-Terrestrial Physics*, *64*, 203–210.

Lean, J. (2000). Evolution of the Sun's spectral irradiance since the Maunder Minimum. *Geophysical Research Letters*, *27*, 2425–2428.

Lean, J., & Rind, D. H. (2008). How natural and anthropogenic influences alter global and regional surface temperatures: 1889 to 2006. *Geophysical Research Letters*, *35*, L18701.

Lee, H., & Smith, A. K. (2003). Simulation of the combined effects of solar cycle, quasi-biennial oscillation, and volcanic forcing on stratospheric ozone changes in recent decades. *Journal of Geophysical Research*, *108*. doi:10.1029/2001JD001503.

Manzini, E., McFarlane, N. A., & McLandress, C. (1997). Impact of the Doppler spread parameterization on the simulation of the middle atmosphere circulation using the MA/ECHAM4 general circulation model. *Journal of Geophysical Research*, *102*, 25751–25762.

Manzini, E., Giorgetta, M. A., Esch, M., Kornblueh, L., & Roeckner, E. (2006). The influence of sea surface temperatures on the northern winter stratosphere: ensemble simulations with the MAECHAM5 model. *Journal of Climate*, *19*(16), 3863–3881.

Marsh, D. R., & Garcia, R. R. (2007). Attribution of decadal variability in lower-stratospheric tropical ozone. *Geophysical Research Letters*, *34*. doi:10.1029/2007GL030935.

Marsh, D. R., Garcia, R. R., Kinnison, D. E., Boville, B. A., Sassi, F., Solomon, S. C., & Matthes, K. (2007). Modeling the whole atmosphere response to solar cycle changes in radiative and geomagnetic forcing. *Journal of Geophysical Research*, *112*(D23). doi:10.1029/2006JD008306.

Marsland, S. J., Haak, H., Jungclaus, J. H., Latif, M., & Röske, F. (2003). The Max-Planck-Institute global ocean/sea ice model with orthogonal curvilinear coordinates. *Ocean Modelling*, *5*, 91–127.

Matthes, K., Langematz, U., Gray, L. J., Kodera, K., & Labitzke, K. (2004). Improved 11-year solar signal in the Freie Universität Berlin Climate Middle Atmosphere Model (FUB-CMAM). *Journal of Geophysical Research*, *109*. doi:10.1029/2003JD004012.

Matthes, K., Marsh, D. R., Garcia, R. R., Kinnison, D. E., Sassi, F., & Walters, S. (2010). Role of the QBO in modulating the influence of the 11 year solar cycle on the atmosphere using constant forcings. *Journal of Geophysical Research*, *115*(D14), D18110. doi:10.1029/2009JD013020.

Meehl, G. A., Arblaster, J. M., Matthes, K., Sassi, F., & van Loon, H. (2009). Amplifying the pacific climate system response to a small 11-year solar cycle forcing. *Science*, *325*, 1114–1118. doi:10.1126/science.1172872.

Misios, S., & Schmidt, H. (2012, in press). Mechanisms involved in the amplification of the 11-yr solar cycle signal in the tropical Pacific Ocean. *Journal of Climate*. doi:10.1175/JCLI-D-11-00261.1.

Qian, L., Roble, R. G., Solomon, S. C., & Kane, T. J. (2006). Calculated and observed climate change in the thermosphere, and a prediction for solar cycle 24. *Geophysical Research Letters*, *33*, L23705. doi:10.1029/2006GL027185.

Randall, C. E., Harvey, V. L., Singleton, C. S., Bailey, S. M., Bernath, P. F., Codrescu, M., N. H. & Russell, J. M. (2007). Energetic particle precipitation effects on the southern hemisphere stratosphere in 1992–2005. *Journal of Geophysical Research*, *112*.

Randel, W. J., Shine, K. P., Austin, J., Barnett, J., Claud, C., Gillett, N. P., Keckhut, P., Langematz, U., Lin, R., Long, C., Mears, C., Miller, A., Nash, J., Seidel, D. J., Thompson, D. W. J., Wu, F., & Yoden, S. (2009). An update of observed stratospheric temperature trends. *Journal of Geophysical Research*, *114*(D13), D02107. doi:10.1029/2008JD010421.

Rasch, P. J., Boville, B. A., & Brasseur, G. P. (1995). A three-dimensional general circulation model with coupled chemistry for the middle atmosphere. *Journal of Geophysical Research*, *100*, 9041–9071.

Reid, G. C., Solomon, S., & Garcia, R. R. (1991). Response of the middle atmosphere to the solar proton events of August–December, 1989. *Geophysical Research Letters*, *18*, 1019–1022. doi:10.1029/91GL01049.

Richards, P. G., Fennelly, J. A., & Torr, D. G. (1994). A solar EUV flux model for aeronomic calculations. *Journal of Geophysical Research*, *99*, 8981–8992 (Correction, JGR, 99, 13283, 1994).

Roeckner, E., Bäuml, G., Bonaventura, L., Brokopf, R., Esch, M., Giorgetta, M., Hagemann, S., Kirchner, I., Kornblueh, L., Manzini, E., Rhodin, A., Schlese, U., Schulzweida, U., & Tompkins, A. (2003). *The atmospheric general circulation model ECHAM 5. Part I: model description* (Tech. Rep. 349). MPI for Meteorology, Hamburg, Germany.

Roeckner, E., Brokopf, R., Esch, M., Giorgetta, M., Hagemann, S., Kornblueh, L., Manzini, E., Schlese, U., & Schulzweida, U. (2006). Sensitivity of simulated climate to horizontal and vertical resolution in the ECHAM5 atmosphere model. *Journal of Climate*, *19*, 3771–3791.

Roy, I., & Haigh, J. D. (2010). Solar cycle signals in sea level pressure and sea surface temperature. *Atmospheric Chemistry & Physics*, *10*, 3147–3153.

Schmidt, H., Brasseur, G. P., & Giorgetta, M. A. (2010). Solar cycle signal in a general circulation and chemistry model with internally generated quasi-biennial oscillation. *Journal of Geophysical Research*, *115*(D14), D00I14. doi:10.1029/2009JD012542.

Schmidt, H., Brasseur, G. P., Charron, M., Manzini, E., Giorgetta, M. A., Diehl, T., Fomichev, V. I., Kinnison, D., Marsh, D., & Walters, S. (2006). The HAMMONIA chemistry climate model: sensitivity of the mesopause region to the 11-year solar cycle and CO2 doubling. *Journal of Climate*, *19*(16), 3903–3931.

Shindell, D., Rind, D., Balachandran, N., Lean, J., & Lonergan, J. (1999). Solar cycle variability, ozone, and climate. *Science*, *284*, 305–308.

Smith, A. K., & Matthes, K. (2008). Decadal-scale periodicities in the stratosphere associated with the solar cycle and the QBO. *Journal of Geophysical Research*, *113*(D12), D05311. doi:10.1029/2007JD009051.

Solomon, S., Qin, D., Manning, M., Chen, Z., Marquis, M., Averyt, K. B., Tignor, M., & Miller, H. L. (Eds.) (2007). *Climate change 2007: the physical science basis. Contribution of working group I to the fourth assessment report of the intergovernmental panel on climate change*. Cambridge: Cambridge University Press.

Soukharev, B. E., & Hood, L. L. (2006). Solar cycle variation of stratospheric ozone: multiple regression analysis of long-term satellite data sets and comparisons with models. *Journal of Geophysical Research*, *111*. doi:10.1029/2006JD007107.

Steil, B., Brühl, C., Manzini, E., Crutzen, P. J., Lelieveld, J., Rasch, P. J., Roeckner, E., & Krüger, K. (2003). A new interactive chemistry-climate model: 1. Present-day climatology and interannual variability of the middle atmosphere using the model and 9 years of HALOE/UARS data. *Journal of Geophysical Research*, *108*(D9), 4290. doi:10.1029/2002JD002971.

Taylor, K. E., Stouffer, R. J., & Meehl, G. A. (2009). A summary of the CMIP5 experiment design. http://cmip-pcmdi.llnl.gov/cmip5/docs/Taylor CMIP5 design.pdf.

van Loon, H., Meehl, G. A., & Shea, D. J. (2007). Coupled air-sea response to solar forcing in the Pacific region during northern winter. *Journal of Geophysical Research*, *112*(D2). doi:10.1029/2006JD007378.

Wissing, M., Kallenrode, M.-B., Kieser, J., Schmidt, H., Rietveld, M. T., Stroemme, A., & Erickson, P. T. (2011). Atmospheric ionization module osnabrück (aimos) 3: comparison of electron density simulations by aimos/hammonia and incoherent scatter radar measurements. *Journal of Geophysical Research*, *116*, A08305. doi:10.1029/2010JA016300.

Chapter 32
Long-Term Behaviour of Stratospheric Transport and Mean Age as Observed from Balloon and Satellite Platforms

Gabriele Stiller, Andreas Engel, Harald Bönisch, Norbert Glatthor, Florian Haenel, Andrea Linden, Tanja Möbius, and Thomas von Clarmann

Abstract Tracer observations from balloon-borne in-situ measurements and satellite observations have been transferred into mean age of stratospheric air. A 30-year time series of mean age of air from balloon observations for the period 1975 to 2005 has been generated. This time series indicated a trend of the stratospheric mean age for Northern midlatitudes which was positive or consistent to zero within its error bars, in apparent contradiction to results from Chemistry-Climate models. Satellite observations from the MIPAS instrument onboard of Envisat provided, for the first time ever, a global view of the stratospheric mean age, covering the period from 2002 to 2010, and the altitude range from 10 to 40 km. Analysis of MIPAS observations confirmed the positive trend of age of air for Northern midlatitudes, and provided a vertically resolved picture.

32.1 Introduction

The mean age of stratospheric air is a fundamental transport parameter, which summarises the ability of the atmosphere to transport tropospheric air to a certain location in the stratosphere [*Hall and Plumb*, 1994]. Mean age is therefore a "diagnostic of stratospheric transport". It was *Kida* [1983] who first realised that age has a statistical nature, i.e. that air parcels (macroscopic units, consisting of many molecules) do not maintain their integrity during the transport in the stratosphere. Due to mixing processes, a stratospheric air parcel therefore has a mean age and not a single

G. Stiller (✉) · N. Glatthor · F. Haenel · A. Linden · T. von Clarmann
Karlsruhe Institute of Technology, Institute for Meteorology and Climate Research, Hermann-von-Helmholtz-Platz 1, 76344 Eggenstein-Leopoldshafen, Germany
e-mail: gabriele.stiller@kit.edu

A. Engel · H. Bönisch · T. Möbius
Experimental Atmospheric Research Institute for Atmospheric and Environmental Sciences, Goethe-University Frankfurt, Altenhöferallee 1, 60438 Frankfurt am Main, Germany

A. Engel
e-mail: an.engel@iau.uni-frankfurt.de

F.-J. Lübken (ed.), *Climate and Weather of the Sun-Earth System (CAWSES)*, Springer Atmospheric Sciences,
DOI 10.1007/978-94-007-4348-9_32,

well defined transport time. As shown by *Li and Waugh* [1999] mean age is sensitive to the strength of the advective stream function (meridional circulation) and to the strength of horizontal and vertical mixing. The strongest effect results from a faster meridional circulation, which results in lower mean ages.

The increase of greenhouse gas abundances in the atmosphere is associated with an increased radiative forcing, leading to a warming of the troposphere and a cooling of the stratosphere [*IPCC*, 2007]. A secondary effect of increasing levels of greenhouse gases is a possible change in the stratospheric circulation [*Butchart et al.*, 2006] with substantial feed-backs on chlorofluorocarbon lifetimes [*Butchart and Scaife*, 2001], ozone [*Sheperd*, 2008], and the climate system [*Baldwin et al.*, 2007]. Model calculations [*Waugh and Hall*, 2002] have shown that the mean age of air in the stratosphere is a good indicator of the strength of the meridional circulation [*Li and Waugh*, 1999] and that mean age is expected to decrease [*Austin et al.*, 2007; *Austin and Li*, 2006; *Garcia and Randel*, 2008]. An increase in the rate of up-welling in the tropical lower stratosphere is predicted by all atmospheric general circulation models [*Butchart et al.*, 2006] and is consistent with the observed long term temperature decrease in the tropical tropopause region [*Thompson and Solomon*, 2005]. The models indicate that a change in the mean age of air has occurred over the past 40 years, with a decrease of mean age mainly occurring after 1975, coupled to an increase in mean tropical upwelling. For instance, a decrease in mean age from about 4.5 years to about 4 years for northern high latitudes (60–90°N) at 35 hPa is derived from the model runs.

Based on meteorological data, positive anomalies in tropical upwelling have also been derived for the period from 2001 to 2004 when compared to the long term mean [*Randel et al.*, 2006]. If this increased upwelling also leads to an increase in the overall residual circulation of the stratosphere (Brewer-Dobson circulation), then shorter transport times and shorter residence times of some greenhouse gases and ozone-depleting substances (ODS) are expected, as these are mainly photolysed at altitudes above 20 km in tropics, leading to decreased atmospheric lifetimes. An increased upwelling which is restricted to the lowest part of the tropical stratosphere, however, would have a much smaller feed-back on the lifetimes of ODSs.

The studies above indicate that a change in the mean circulation may occur, which could lead to changes in mean age and may have a strong impact on the overall composition of the stratosphere. The mean age may be the most suitable tracer to detect such changes in atmospheric coupling. As the expected changes in mean age are rather small (on the order of 10 to 20 % [*Austin et al.*, 2006]), they are difficult to detect from observations, which are influenced by atmospheric variability and by systematic errors in the derivation of mean age.

Mean age can be derived from observations of tracers which increase with time in the atmosphere and show neither sinks nor sources in the middle atmosphere. The two tracers that have been used most widely to derive mean age are CO_2 and SF_6. Both of these tracers have somewhat different characteristics: CO_2 has a seasonal cycle in the troposphere which can propagate into the stratosphere and makes the determination of mean age values below 2 years ambiguous. It also has a stratospheric source due to the oxidation of methane, which can however be corrected for if methane is measured simultaneously. Furthermore, the growth rate of CO_2 shows considerable interannual variability, as it is strongly coupled to the biosphere. Such interannual variability may lead to systematic uncertainties in the dating of the air. SF_6 on the other hand increases monotonically and with small interannual variability in the troposphere but has a sink in the mesosphere, which can cause an artefact when dating old air [*Waugh and Hall*, 2002; *Strunk et al.*, 2000; *Reddmann et al.*, 2001; *Engel et al.*, 2002, 2006; *Ray et al.*, 2002; *Plumb et al.*, 2002].

Mesospheric SF_6 is influenced by chemical breakdown, both due to short wavelength photolysis and to electron attachment processes (see *Reddmann et al.* [2001]). Free electrons in the middle atmosphere are produced from photoionization processes or by precipitating energetic particles [*Brasseur and Solomon*, 1986]. As short wavelength radiation below 200 nm shows a strong variation with solar activity, both of these processes vary with the solar cycle. It is therefore probable that the mesospheric sink of SF_6 may vary with the solar cycle. On the other hand, real changes in mean age may occur under the influence of the solar cycle, too, as stratospheric transport is also modulated by the solar cycle.

While the total solar irradiance changes only by about 0.1 % over a solar cycle, much larger variation (4–8 %) are found in the UV flux between 200 and 250 nm [*Lean et al.*, 1997]. At even shorter wavelengths, the solar cycle related changes become even stronger. For instance, Lyman-α radiation varies with a factor of about 2 over a solar cycle. Although the variability in total irradiance is small, some significant effects on stratospheric circulation and composition are observed, particularly during high latitude winter (see *Haigh et al.* [2004]). *Kodera and Kuroda* [2002] noted that it is expected that the Brewer-Dobson circulation should be weakened during solar maximum conditions. As noted by *Labitzke* [1987], polar stratospheric dynamics show variations with the solar cycle and with the Quasi Biannual Oscillation (QBO). This could lead to a difference between mean age of stratospheric air during solar minimum and solar maximum conditions.

Several experimental studies have provided data on the distribution of mean age derived from observations of very long lived tracers (mainly CO_2 and SF_6) in the stratosphere (e.g. *Schmidt and Khedim* [1991]; *Volk et al.* [1997]; *Boering et al.* [1996]; *Strunk et al.* [2000]; *Andrews et al.* [2001]; *Engel et al.* [2002]). However, only the studies by *Schmidt and Khedim* [1991] and by *Andrews et al.* [2001] investigated the temporal evolution of mean age. The earlier study by *Schmidt and Khedim* [1991] showed much more variability than the later observations by *Andrews et al.* [2001]. This can only be partly explained by the better precision of the later observations. The mean values derived in both studies showed good agreement. Observations made by *Engel et al.* [2003] after 1990 also showed much lower scatter in the observed mean age. This might suggest that mean age has shown larger variabilities before 1990 than during later years.

Mean age studies based on our samples have been conducted with both SF_6 [*Strunk et al.*, 2000; *Engel et al.*, 2002, 2006] and CO_2 [*Schmidt and Khedim*, 1991; *Strunk et al.*, 2000; *Engel et al.*, 2002, 2006] data. *Strunk et al.* [2000] and *Engel et al.* [2002] used a combination of SF_6 and CO_2 observations to investigate whether both age tracers gave comparable results. Their conclusion was that under certain conditions mesospheric loss of SF_6 can lead to an overestimation of mean age [*Strunk et al.*, 2000; *Engel et al.*, 2002, 2006] which is in agreement with model calculations [*Reddmann et al.*, 2001]. Systematic differences increase with mean age. For air parcels older than 5 years, SF_6 dating leads to an overestimation of mean age [*Engel et al.*, 2002]. On the other hand, the interannual variability in growth rates and the seasonal cycle in CO_2 also introduce some systematic errors in CO_2-dating, making neither of these two age tracers an ideal tracer.

An ideal tracer for the determination of mean age would have no sinks or sources in the middle atmosphere and increase linearly with time. Possible tracers include perfluorocarbons and highly fluorinated fluorochlorocarbons. However, the non-availability of good tropospheric reference data and of precise and accurate stratospheric observations has so far prevented the use of these tracers for mean age determination.

With the launch of the Envisat satellite and the MIPAS infra-red limb emission spectrometer on board, for the first time the opportunity was provided to observe SF_6 with global coverage from space. The first global distributions of SF_6 from satellite observations have been derived from MIPAS data by *Burgess et al.* [2004, 2006]; however, the global distributions derived by these authors were biased low, and, hence, analysis of age of air distributions was not attempted.

The aim of the work described in this paper is to characterise stratospheric transport time scales and the influence of solar variability and long term change on these time scales, using observations of the age tracers CO_2 and SF_6 in the stratosphere. Both balloon-borne in-situ observations and global satellite data from the MIPAS/Envisat instrument have been used. The coupling between the mesosphere and the stratosphere was investigated by the analysis of variations in the tracer observations. Existing data sets, in combination with reanalyses of archived stratospheric whole air samples and observations performed with a new light weight balloon-borne cryogenic whole air sampler were used to generate a 30-year time series from in-situ observations, covering a period starting in 1975 up to today. Further, the potential of MIPAS satellite SF_6 data for the derivation of mean age was analysed. In co-operation with several modelling groups the data were used for model validation. In particular, the following questions were addressed:

- What is the temporal evolution and variability of mean age of air over decadal time scales?
- Can changes of age of air be attributed to solar cycle influence via the mesospheric chemistry of SF_6, and does this lead to an artefact in SF_6 based mean age determination?
- Or does solar variability influence stratospheric transport to the extent that this leads to changes in the mean age?
- How often is mesospheric air transported into the polar vortices in significant amounts and which dynamic processes drive this effect?
- What other tracers can be used to derive mean age of air?

32.2 Methods and Tools

32.2.1 Balloon-Borne In-situ Measurements

Two ways to improve observational knowledge on age of air from in-situ measurements have been pursued: First, the series of balloon-borne measurements has been extended, for which a new light-weight instrument was developed, and second, the already available measurements of the last 30 years have been reanalysed.

Measurements of stratospheric trace gases have been performed for more than two decades using the technique of stratospheric whole air sampling [*Schmidt et al.*, 1991; *Schmidt and Khedim*, 1991; *Engel et al.*, 1998, 2002, 2006; *Strunk et al.*, 2000] by the University of Frankfurt research group. The advantage of balloon-borne observations is that they can reach altitudes up to 35 km, thus allowing to study the middle stratosphere, in contrast to research aircrafts which are limited to an altitude of about 20 km. This series of measurements has been extended: A newly developed cryogenic whole air sampler was flown for the first time on October 15, 2006. The instrument has 26 sample bottles, which are held in a custom built Dewar made from glass fibre reinforced plastic. Each sample bottle is sealed with a glass cap, which is broken during the flight by releasing a small hammer, so that air can enter the sample bottle. The ambient air entering the sample bottle condenses on the cold surfaces (27 K, liquid Neon), so that the sample bottles act as very efficient cryo-pumps. After a predefined time, the sample is closed, by firing a small pyro-actuator. By this means, large amounts of air can be collected in a very short time. At the same time, this system works without using any polymers for sealing, which tend to be a contamination factor due to out-gassing. The instrument is able to keep cold for about 5–7 hours after filling. After this time, the instrument should gradually heat up, making the cryo-sampling of ambient air impossible. We have analysed/reanalysed a large suite of archived stratospheric whole air samples for their SF_6 and CFC-12 mixing ratios. The old samples had been analysed for a number of species but not for SF_6. A new method which allowed precise and well calibrated measurements at the very low mixing ratios present in these samples has been developed at University Frankfurt and applied to the old archieved samples.

Many samples have also been analysed for N_2O and CFC-11. These include samples from the MPAE whole air sampler (e.g. *Fabian et al.* [1979]) and NCAR whole air samples (e.g. *Ehhalt et al.* [1975]) archived at the University of Miami. In total, 79 samples of the MPAE whole air samplers taken between 1987 and 1999, at low, mid and high latitudes and 115 samples of the NCAR sampler taken from tropical to high latitudes have been analysed. The analytical precision for SF_6 has been improved considerably for this effort, mainly by improving the analytical procedure and by more sophisticated data processing. The analytical precision of a single measurement of SF_6 is better than 0.7 %.

Based on assumptions about the growth rates, the analytical errors and the possible differences in absolute calibration scales, we estimate that a long term change on the order of 0.5 years over the entire time range covered should be detectable in our data set for both age tracers.

Most of the data included in the consolidated SF_6 data set are from our own analysis, as the focus is on the altitude range above 20 km. The OMS data included in the data set are publicly available, as well as a large suite of ER-2 data. All these data are from the NOAA/ESRL (former NOAA/CMDL) laboratory. Unfortunately, the absolute calibration scale of this laboratory has been modified over the last years. NOAA/ESRL laboratory planned to reprocess the data and to produce a data set on a consistent calibration scale (Bradley Hall, NOAA/ESRL, private communication, 2006). In order to be able to link our data to this scale, a primary standard was provided by NOAA/ESRL to our institute. An intercomparison between the two scales (Maiss and Levin vs. NOAA/ESRL) was possible, and the difference in calibration scales was below 1 %.

32.2.2 Satellite Data

MIPAS radiance data have been analysed and global distributions of about 30 different trace species have been derived by KIT/IMK [*von Clarmann et al.*, 2010a]. The approach is research-oriented in contrast to those applied by operational processing centres like ESA or DLR and focuses on the derivation of trace gas distributions of minor constituents. Global SF_6 distributions have been shown to be retrievable from MIPAS data before by *Burgess et al.* [2004].

The general approach of SF_6 retrieval from MIPAS data has been described in detail in *Stiller et al.* [2008]. In short, the spectroscopic signature of SF_6 around 948 cm^{-1} is used to retrieve SF_6 mixing ratio profiles from MIPAS radiance profiles by applying a non-linear regularised least-squares-fit approach. Interfering signatures of other species like CO_2 or H_2O are jointly fitted. The retrieved SF_6 profiles have a precision on the order of 0.5 pptv (10 %) and a vertical resolution of 4–6 km and cover the altitude range from about 10 km (or above cloud top height) to 40 km.

As described in the appendix of *Stiller et al.* [2008], the MIPAS radiance data of version 4.61 from the period 2002 to 2004 which were used for the retrievals were affected by an artefact due to erroneous radiance calibration. The artefact showed up as oscillations in the radiance baseline for certain calibration episodes. The oscillations, however, where below the precision specification for the MIPAS instrument, and affected difficult-to-retrieve species like SF_6 (and HO_2NO_2 and others) only. Since rapid correction by re-processing of the radiance spectra by ESA could not be expected, a correction method was developed to remove biases caused by the artefact from SF_6 distributions. Re-retrieval of SF_6 from a new, corrected version of MIPAS radiance spectra will ultimately solve the calibration problem.

32.3 Balloon-Borne In-situ 30-Year Time Series of Northern Midlatitude Mean Age of Stratospheric Air

In-situ stratospheric observations from balloon-borne instrumentation for the two mean age tracers SF_6 and CO_2 have been compiled from 27 high altitude balloon

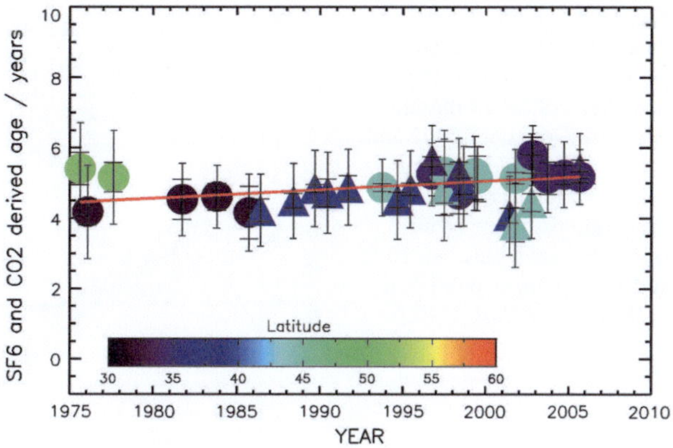

Fig. 32.1 Long term trend of mean age derived from observations during 27 balloon flights. The derived long term trend is positive, but the level of uncertainty does not put very strong constraints on model predictions yet. Further observations are necessary to give stronger constraints (figure from *Engel et al.* [2009])

flights up to 35 km altitude (SF_6 data from one flight are available up to 43 km altitude) from 1975 to 2005 at Northern Hemisphere midlatitudes between 32°N and 51°N [*Engel et al.*, 2009]. Many whole air samples from these flights are archived and have recently been reanalysed for SF_6, while other measurements were made in-situ during the balloon flights. Care was taken that only observations made with sufficient altitude coverage and quality were used which can be linked to reference tropospheric time. For CO_2, data of sufficient quality are available for the period from 1986 through 2005. For SF_6, the selected data span the years from 1975 to 2005, with a gap between the years 1985 and 1994. Note that most profile observations are from the May to October period, when stratospheric variability in the Northern Hemisphere is expected to be lower than during the winter period. The main result of this work is that, in contrast to the model prediction, the observations do not show a significant decrease in mean age. Of particular importance for this study was the question of the uncertainty in this estimate. A detailed error estimation has been performed (see details in *Engel et al.* [2009]) which includes errors in the observations, uncertainties in the age-spectra, uncertainty in the tropospheric reference time series and their representativeness, uncertainties in the seasonal variation of air mass transport into the stratosphere and also the uncertainty associated with the sparse sampling of the balloon profiles and the latitude range covered by the observations. Taking all these sources of uncertainty into account, the measurements show an increase in mean age of $+0.24 \pm 0.22$ years per decade ($1 - \sigma$ uncertainty level; see Fig. 32.1). The published model predictions of change in mean age are between -0.25 and -0.05 years per decade [*Waugh*, 2009]. While the upper limit of these predictions is inconsistent with our observations at a confidence level of 95 %, the lower end of the predictions (-0.05 years per decade) can not be falsified by our data on the 90 % confidence level. In order to put a tighter constraint on the range of model predictions of changes in mean age which is compatible with the observations, a longer time series is needed.

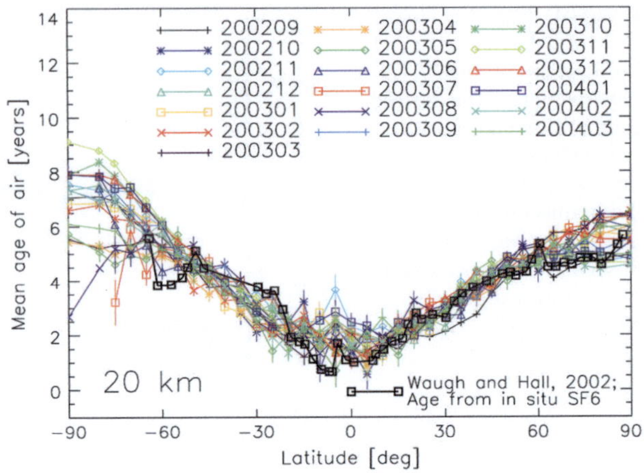

Fig. 32.2 Monthly zonal mean distributions of age of air at 20 km altitudes derived from MIPAS for 19 months between September 2002 and March 2004 (*coloured lines*), compared to aircraft observations of age of air from SF_6 taken between 1992 and 1997 (data as published in *Waugh and Hall* [2002])

32.4 Global MIPAS 8-Year Time Series of Mean Age of Stratospheric Air

32.4.1 Full Resolution Measurement Phase 2002 to 2004

The first global data set of the mean age of stratospheric air derived from SF_6 has been derived from MIPAS/Envisat data for the period September 2002 to March 2004. This work has been published by *Stiller et al.* [2008]. The retrieval approach has been developed and fine-tuned, taking into account, in particular, non-LTE effects on the emission of CO_2 lines in immediate neighbourhood of the SF_6 signature. Based on the retrieval approach developed, monthly global mean distributions of SF_6 for the altitude range of ~5 km to ~35 km and the period September 2002 to March 2004 could be derived, with a precision of better than 5 % for the 5° latitudinal zonal mean values. Our approach resulted in SF_6 distributions well consistent to in-situ balloon observations of SF_6 (see Fig. 4 of *Stiller et al.* [2008]).

The tropospheric SF_6 increase was in excellent agreement with ground-based observations by NOAA/ESRL/GMD and was consistent with a linear increase of 0.230 ± 0.008 pptv/year. The zonal mean distribution of mean age of air at 20 km for all months covered agrees well with previous aircraft measurements (see Fig. 32.2).

32.4.2 Comparison to Models

The global distributions of mean age of stratospheric air derived from MIPAS SF_6 observations have been compared to age calculations with the KASIMA model by *Stiller et al.* [2008]. The overall global agreement was very good. The MIPAS time series of mean age of air at all potential temperature levels above 550 K revealed very high apparent ages during polar winters, indicating subsidence of SF_6-poor

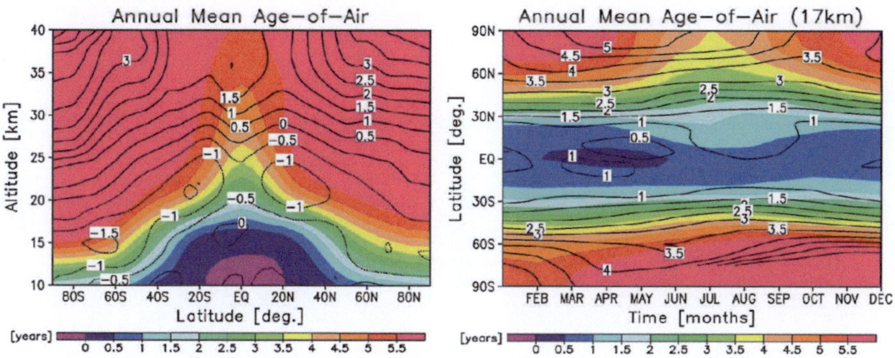

Fig. 32.3 *Left*: A five years average of zonal mean age of air from the UM (*colour shades*), and differences to the MIPAS age observations (isolines). Positive numbers indicate older air in observations. *Right*: Annual cycle of modelled zonal means of mean age of air (5 years average) at 17 km altitude, and observed zonal means of age of air for 2003 (isolines) (figures from *Braesicke et al.* [2008a, 2008b])

mesospheric air. However, if in KASIMA the mesospheric sink of SF_6 was taken into account, the very high apparent ages could be reproduced, while the real age was on the order of 6 to 7 years. The atmospheric lifetime of SF_6 was assessed at about 4500 years by this comparison [*Stiller et al.*, 2008; *Reddmann et al.*, 2001]. This confirms the relevance of the mesospheric sink in case of age determination from SF_6 and offers the opportunity to further quantify this sink.

Besides the comparison of global age of air distributions derived from MIPAS and the KASIMA model, MIPAS global mean age distributions have been compared to the Met Office Unified Model (UM) developed at University of Cambridge for purposes of model validation and for analysis of transport of air through the tropopause. *Braesicke et al.* [2008a] have compared zonal means of age of air as produced by the Met Office Unified Model (UM) with MIPAS zonal means of age of air distributions and found reasonable agreement, with the model producing somewhat higher ages above the tropical tropopause (appr. 0.5 yr) and younger air in the polar and midlatitude mid-stratosphere (see Fig. 32.3). In a second study, *Braesicke et al.* [2008b] compared the longitudinal distribution of mean age of air just above the tropical tropopause between model and observations, and found, in both data sets, indications for air younger than average during summer at the location of the Asian monsoon anticyclone (not shown).

32.4.3 Reduced Spectral Resolution Mode of MIPAS Since 2005

The time series of SF_6 derived from MIPAS has been extended up to January 2010 so far to cover also the so-called reduced-resolution mode of MIPAS from January 2005 on (see Fig. 32.4). MIPAS resumed operation in January 2005 in this mode after an instrument failure in March 2004. The spectral resolution was decreased

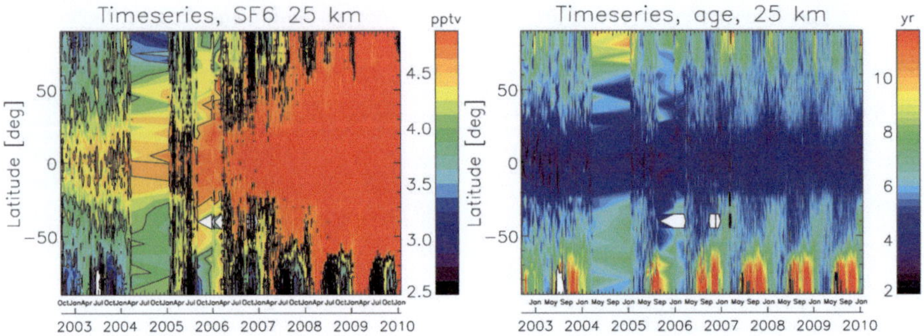

Fig. 32.4 *Left*: Latitude-time cross section of SF_6 at 25 km for September 2002 to January 2010 derived from MIPAS measurements (*left*); *right*: similar, but for age of air as derived from SF_6. From April to December 2004 no observations from MIPAS were available. In January 2005 MIPAS resumed operations in the reduced-resolution operation mode

to 0.0625 cm^{-1} from 0.025 cm^{-1} before the instrument failure, while the sampling along-track and in the vertical was improved. This resulted in better vertical and horizontal resolution of the retrievals. The reduced spectral resolution led to broader and stronger interfering spectral signatures, so that it had not been clear from the beginning if the SF_6 signature could be separated well enough from the interfering strong CO_2 lines. However, test retrievals demonstrated that the spectral resolution was still good enough to perform significant retrievals. The data set of SF_6 presented in Fig. 32.4 has been retrieved with retrieval settings adapted from the high-resolution data version V3O_SF6_6 as published in *Stiller et al.* [2008]. Radiance baseline oscillations due to an incorrect radiance calibration procedure which had been a serious difficulty to be overcome for retrieving a reliable data set from full-resolution data (see *Stiller et al.* [2008]) were fortunately not present in the spectral version of the reduced resolution data set. However, comparison of MIPAS tropical tropospheric daily means to in situ ground-based tropical data and the combined flask/in situ global means of NOAA ESDL [*Hall et al.*, 2011] revealed a bias between the full-resolution and the reduced-resolution data set of 0.111 pptv (reduced-resolution data being lower). The complete data set has been corrected for this bias, however, it cannot be excluded that a latitude/altitude dependent residual bias remained. The time series shows a continuous increase of SF_6 as expected from the tropospheric trend. Episodes of very low SF_6 hinting towards mesospheric intrusions during Arctic and Antarctic winters appear regularly, in particular during the Antarctic winters.

32.4.4 *Impact of Mesospheric SF_6-depleted Air on Age-of-Air Determination at Low and Middle Latitudes*

In each polar winter mesospheric air intrudes into the stratospheric polar vortex. During final warming events southern polar vortices are vertically divided at

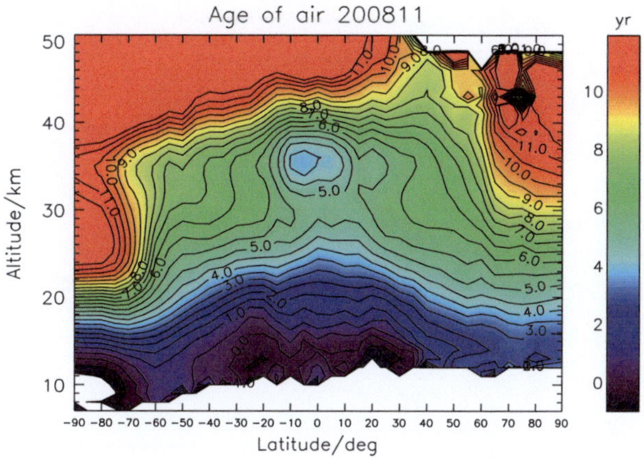

Fig. 32.5 Southern hemispheric vortex break-down as seen from a zonal mean distribution of age of air for November 2008; young air from lower latitudes enters polar latitudes around 30–35 km and divides the vortex air into a part below the entry altitude which will finally be mixed into the hemispheric stratospheric air, and a part above the entry altitude which will be pushed back into the mesosphere

typically 35 km altitude in an upper and a lower part by midlatitudinal air penetrating into polar regions (cf. Fig. 32.5). The upper part of the vortex remnant ascends back into the mesosphere while its mixing barriers—discernable as regions of still strong age gradients—remain intact. The lower vortex remnant is mixed with midlatitudinal air, thus contains air irreversibly subsided from the mesosphere into the stratosphere, affecting the mean age in a sense that through diluted polar ex-mesospheric air the apparent age is higher than the true one. Similar behaviour is observed for northern polar vortices, except that the subsidence of mesospheric air is less pronounced and that the vertical splitting of the vortex is not typically visible in latitudinal means (which certainly does not exclude that it may be present in a longitudinally resolved representation). The entire vortex air seems to be mixed into the midlatitudes. The effect of too large apparent age of air is also to be considered here. We have assessed the over-aging of non-tropical (polar non-vortex and midlatitudinal) air by mesospheric contamination from estimates of the mean excess age of polar vortex air and the amount of air mixed with midlatitudinal air after the final warming for the winters covered by the MIPAS measurements. The estimated excess age ranges from 0.003 years for the Northern hemispheric winter 2002–2003 to 0.21 years for the Southern hemispheric winter 2009. Due to high variability of the vortex size and the lowest altitude reached by subsiding air, the over-aging of midlatitude air varies strongly from winter to winter, and no significant trend in excess age was found.

32.4.5 Analysis of the Seasonal, Interannual and Decadal-Scale Variability Within the Global MIPAS Data

The global altitude-resolved age of air time series as shown in Fig 32.4 has been analysed for midlatitudes (30–40°N) and compared to the 30-years time series derived from air-borne in-situ observations (see Fig. 32.6). The monthly averages

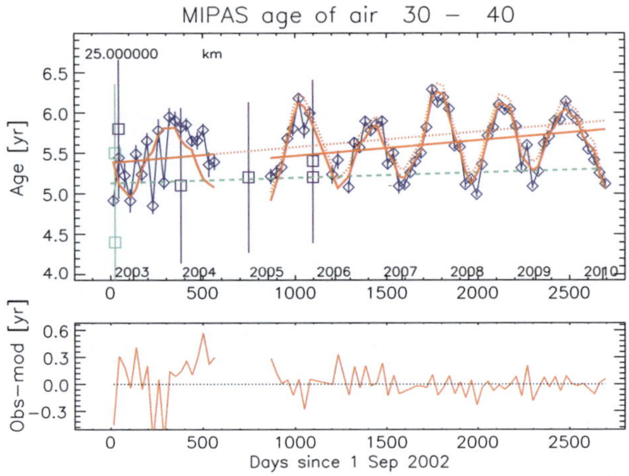

Fig. 32.6 Time series of age of air at 25 km altitude for 30 to 40°N from MIPAS data (*dark blue diamonds*: monthly averages together with their standard errors), and a fit including a linear trend, seasonal, QBO, and semi-annual variations (*solid orange curve*), compared to the data points from balloon observations and the trend derived by *Engel et al.* [2009] (*green* and *violet squares* and *green dashed line*). The *solid orange line* is the linear term of the regression fit. The *dotted orange line* and *dotted orange curve* give the linear term and the full regression fit without the bias between the two MIPAS data sets considered. The *lower panel* provides the residual between MIPAS data points and the *solid orange curve*

of mean age have been fitted by a model including a constant value (axis intercept), a linear trend, two QBO proxies from normalised Singapore winds at 30 and 50 hPa, and sinusoidal variations with periods of 12 months (seasonal cycle), 6 months (semi-annual variation), as well as further higher harmonics to account for a shape deviating from the pure sinusoidal shape. A potential residual altitude/latitude-dependent bias between the full-resolution and reduced-resolution data set was accounted for by applying the trend analysis method for correlated data of *von Clarmann et al.* [2010b]. This model reproduced the temporal evolution in a satisfactory manner. The data points of *Engel et al.* [2009] and MIPAS data are consistent within their errors; the linear regression lines, however, show an offset of about 0.3 years. While, in general, a bias between the balloon-borne and satellite-borne data cannot be excluded, it must be kept in mind, that the regression line of the *Engel et al.* [2009] data set represents the period 1975 to 2005, while the regression line from the MIPAS data set represents the period 2002 to 2010. The linear term provided a trend of 0.70 ± 0.08 years per decade (orange solid line) which is—considering the 2σ error bars—only slightly larger than the trend derived by *Engel et al.* [2009] (green dashed line). However, in contrast to *Engel et al.* [2009], this trend is significantly distinct from zero and not consistent to model results. Due to its vertical resolution, MIPAS provides also a picture on the variation of the trend with altitude (Fig. 32.7). In Fig. 32.7 the trend for Northern midlatitudes, this time 30–50°N, is shown; it is always positive and largest at the lowest altitude at about 17 km; there are two ranges to be separated—below 20 km and above 23 km with

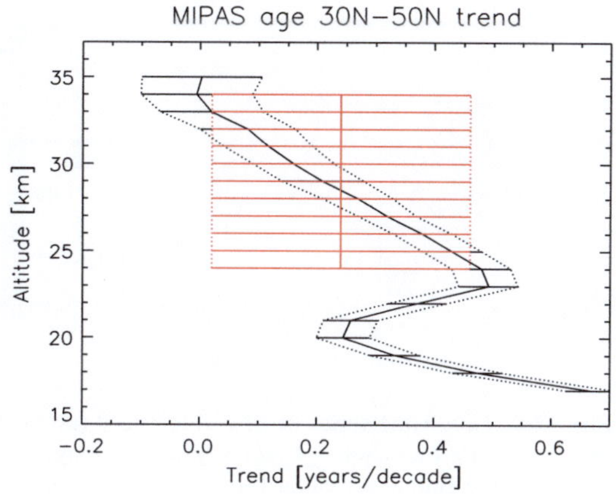

Fig. 32.7 Vertical profile of the midlatitudinal age of air trend, this time for 30–50°N from MIPAS, compared to the trend derived by *Engel et al.* [2009] (in *red*)

a transition zone of 20 to 23 km in between—which coincide well with the shallow and the deep branch of the Brewer-Dobson circulation. In both ranges the trends decrease with altitude; above 30 km the trend is no longer distinct from zero within the $1-\sigma$ uncertainty. It is to be noted that due to the MIPAS measurement period this trend describes the declining phase of the solar cycle from maximum to minimum only, so that any impact of the solar cycle on the stratospheric circulation cannot be separated. A longer time series would be necessary for this.

32.5 Impact of the Solar Cycle on Age of Air Observations

Since SF_6 is destroyed in the mesosphere by electron capture processes, the mesospheric loss should in principle depend on the availability of electrons in the mesosphere, which in turn depends on energetic particle precipitation. We expected to observe signals of mesospheric SF_6 destruction in subsided air originating from the mesosphere. Therefore we analysed time series of SF_6 anomalies (with respect to October 4, 2003 as reference day) in correlation with CO time series for the Arctic winter 2003/2004, in particular during the Solar Proton Event (SPE) in Oct/Nov 2003, and during the episode of strong subsidence in late winter 2004 (see Fig. 32.8, left). CO is in principle an excellent tracer for subsided mesospheric air (see Fig. 32.8, right). For the altitudes covered by MIPAS SF_6 data, however, i.e. below 1000 K, the signal of downward transport in the CO data is not very clear. Comparing the SF_6 time series with CO time series below 1000 K, we can mainly identify the effect of the mid-winter major warming when CO-rich, SF_6-rich air was transported into lowest altitudes (see around day 85 for SF_6 and around Jan 1, 2004 for CO, respectively). Direct effects of the SPE event or the strong subsidence at higher altitudes in Jan-Mar 2004 do not reach the lower stratosphere, according to the CO time series. The SF_6 enhancements during the SPE (days 25–31) and the decrease afterwards (until day 80) are not yet understood and contradict expectations.

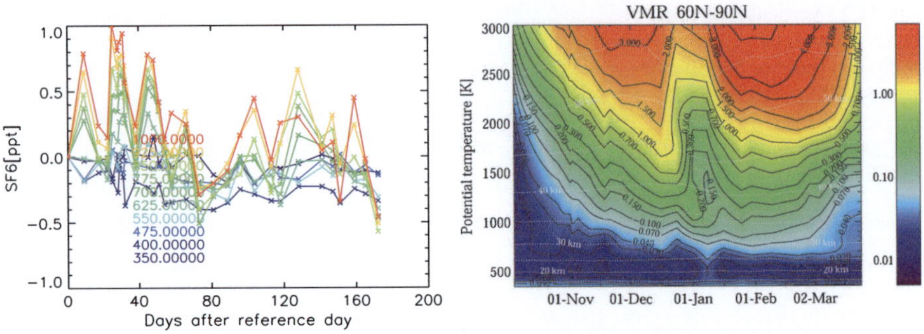

Fig. 32.8 *Left*: Time series of SF_6 differences (averaged over 60° to 90°N equivalent latitudes) relative to SF_6 on October 4, 2003 (reference day) for different potential temperature levels as indicated; the Halloween Solar Proton Event was on days 25 to 31 (October 29 to November 4); the major warming was between day 80 and 103; the strong subsidence in late winter was after day 103. *Right*: CO time series (averaged for 60° to 90°N equivalent latitudes) from Oct 1, 2003 to March 25, 2004

We concluded therefore that the SF_6 data derived so far are of insufficient quality since they do not cover the upper stratosphere, and thus do not allow to derive conclusive results on the dependence of mesospheric loss processes on the solar activity. Further improvements of the SF_6 retrievals to provide age of air of the upper stratosphere are under way.

32.6 Alternative Age of Air Tracers

32.6.1 Balloon-Borne In-situ Measurements

Due to the problems associated with the use of SF_6 and CO_2 as age tracers, the suitability of other very long-lived tracers for the determination of mean age was investigated. Possible other tracers should have no (or negligible) chemical sinks in the middle atmosphere and a well documented long term steady increase, which in an ideal case should be linear over a period of about 20 years. Possible tracers which may fulfil these criteria include perfluorocarbons and highly fluorinated chlorofluorocarbons like e.g. CF_4, C_2ClF_5 (CFC-115) and C_2F_6 (CFC-116). In particular, CF_4 is considered as another potential long-lived tracer for estimating the age of stratospheric air, with a tropospheric increase of about 1 % (or 1 pptv) per year (known from ground-based in-situ observations), a tropospheric burden of about 80 pptv, and an atmospheric lifetime of about 50.000 years. A gaschromatography/mass spectrometry (GC/MS) system capable of measuring these compounds has been established at the University of Frankfurt. The new GC/MS system was set up with a cryogenic preconcentration unit which is based on the use of a chromatographic packing material (HayeSep D) held at −70 °C as trapping material in combination with a special chromatographic column (GasPro) which allows for a good

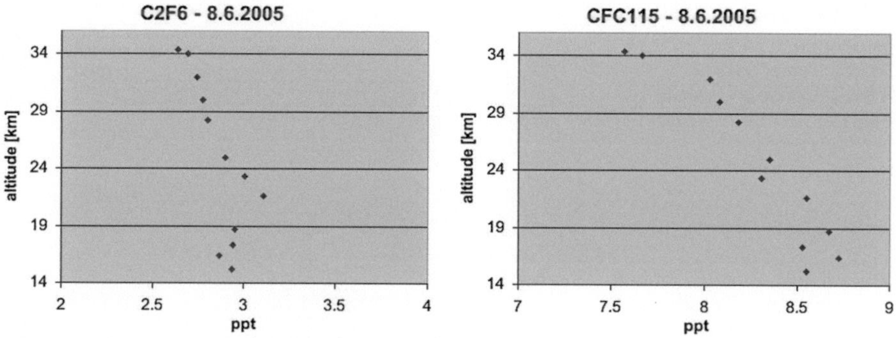

Fig. 32.9 Profiles of C_2F_6 (CFC-116) and C_2ClF_5 (CFC-115) obtained from the balloon flight on June 8, 2005

separation of low boiling halocarbons. With this method both C_2F_6 and CFC-115 (C_2F_5Cl) could be separated well from possible interfering peaks, and in addition even CF_4 (which has an even higher volatility) could be measured. A test of the reproducibility for these low boiling compounds is still pending, as is a calibration of our standard gases. Other species, however, show reproducibilities in a similar set-up of better than 0.3 %.

Vertical profiles of some of the tracers have been measured on the samples collected during two balloon flights from Teresina, Brazil (5°S) on June 8 and June 25, 2005. The analytical system was not optimal at the time of the measurements. Figure 32.9 shows the profiles of C_2F_6 (CFC-116) and C_2ClF_5 (CFC-115) obtained from this balloon flight.

32.6.2 Global Satellite Observations

CF_4 is another tracer which can be measured by infrared remote sensing. Due to its long atmospheric lifetime and, thus, almost constant vertical profile, it can provide information of the stratospheric mean age particularly at altitudes where SF_6 failed so far to provide reliable data, i.e. above appr. 40 km. CF_4 has been demonstrated to be available from infrared remote sensing instruments by *Zander et al.* [1996] and *von Clarmann et al.* [1995]. Recently *Rinsland et al.* [2006] have shown that the derivation of a stratospheric trend of CF_4 is possible from satellite infrared observations. Figure 32.10 shows the distribution of CF_4 as derived from MIPAS for September 2003 (from 9 days analysed). The distribution in the upper stratosphere is rather homogeneous, as expected, with a volume mixing ratio of about 65 pptv. Tropospheric volume mixing ratios are at maximum around 74 pptv. An artefact between 60°S and 80°S could be traced back to retrieval problems in the polar vortex boundary caused by strong horizontal temperature gradients; the retrieval set-up has meanwhile been improved with respect to this point. The retrieval error due to measurement noise for a single profile is about 3 % between 20 and 45 km, increasing to

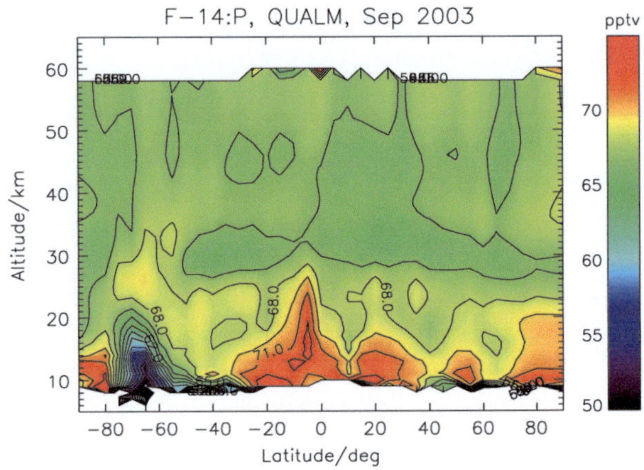

Fig. 32.10 Zonal mean distribution of CF_4 (CFC-14) derived from 9 days of MIPAS data in September 2003

8–10 % above and below, i.e. moderate averaging can provide a data set sufficiently precise to derive age of air from the data. A larger data set of CF_4 is currently produced, from which the tropospheric increase and the atmospheric lifetime of CF_4, and finally the age of air above 40 km will be derived.

32.7 Summary and Conclusions

The work presented in this paper had three foci related to stratospheric circulation and the mean age of stratospheric air. First, long-term time series of mean age of stratospheric air for Northern midlatitudes have been derived from air-borne in-situ observations of CO_2 and SF_6, and from satellite-borne global SF_6 observations. For the 30-year age of air time series from in-situ observations, archived atmospheric air samples collected during the period 1975 to 2005 have been reanalysed in a self-consistent way in order to exclude any inconsistencies due to data processing. The satellite data are from MIPAS on Envisat, which is a mid-infrared limb emission sounder with global coverage. SF_6 global distributions have been derived for the lower and middle stratosphere up to about 40 km. The 30-year in-situ time series provided a trend for the 30–50°N middle stratosphere (above 30 hPa) of 0.24 ± 0.22 years per decade. The 8-years time series from MIPAS observations covering 2002 to 2010 gave a vertically resolved trend for the same altitude range between 0.44 $\pm$ 0.04 and 0.0 years per decade, which is significantly (within 1σ) positive below 30 km. Both observational results are in contradiction to model simulations (or at least to their upper limits, in case of the in-situ time series) which have consistently produced a negative trend of mean age for the last decades.

A second focus of this work was related to the impact of solar variability. We searched for observational indications that the Brewer-Dobson circulation may vary with the solar cycle as indicated by e.g. *Labitzke* [1987]. Regarding this point, it has to be distinguished between real change of the circulation and change of apparent age of air. The latter may be produced by the mesospheric sink of SF_6 by reacting on the electron densities in the mesosphere which are influenced by energetic particle precipitation which in turn varies with the solar cycle. Although the impact of the mesospheric sink on real versus apparent age of air from SF_6 could be quantified, no clear indication has been found for a systematic trend from solar maximum to solar minimum of real nor apparent age of air on the basis of the current data set. The data quality and temporal coverage of SF_6 observations from MIPAS needs to be improved in the upper stratosphere to answer this question; extension of the MIPAS age of air data set towards the next solar maximum and to the upper stratosphere is under way.

The third and last focus of the project was on the development of alternative age of air tracers which have not the shortcomings of CO_2 (seasonal variation and interannual variability in growth rates) or SF_6 (mesospheric sink). The feasibility to measure several other tracers by in-situ measurements has been investigated, and C_2F_6 and CF_4 turned out to be very promising. For infrared remote sensing observations as from MIPAS, CF_4 is also a favourable tracer and can be derived with high precision. The work towards assessment of age of air from CF_4 will be continued in future.

Acknowledgements ESA has provided MIPAS level-1b data. The authors gratefully acknowledge that sulphur hexaflouride data from the NOAA/ESRL halocarbons in situ program have been made available. This project has been supported by DFG within the CAWSES priority program (grants STI 210/4-1, EN 367/4-1, STI 210/5-2, EN 367/4-2, and STI 210/5-3).

References

Andrews, A. E., Boering, K. A., Daube, B. C., Wofsy, S. C., Loewenstein, M., Jost, H., Podolske, J. R., Webster, C. R., Herman, R. L., Scott, D. C., Flesch, G. J., Moyer, E. J., Elkins, J. W., Dutton, G. S., Hurst, D. F., Moore, F. L., Ray, E. A., Romashkin, P. A., & Strahan, S. E. (2001). Mean ages of stratospheric air derived from in situ observations of CO_2, CH_4, and N_2O. *Journal of Geophysical Research*, *106*(D23), 32295–32314. doi:10.1029/2001JD000465.

Austin, J., & Li, F. (2006). On the relationship between the strength of the Brewer-Dobson circulation and the age of stratospheric air. *Geophysical Research Letters*, *33*, L17807. doi:10.1029/2006GL026867.

Austin, J., Wilson, J., Li, F., & Vömel, H. (2007). Evolution of water vapor and age of air in coupled chemistry climate model simulations of the stratosphere. *Journal of the Atmospheric Sciences*, *64*(3), 905–921.

Baldwin, M. P., Dameris, M., & Shepherd, T. G. (2007). How will the stratosphere affect climate change? *Science*, *316*(5831), 1576–1577. doi:10.1126/science.1144303.

Boering, K. A., Wofsy, S. C., Daube, B. C., Schneider, H. R., Loewenstein, M., Podolske, J. R., & Conway, T. J. (1996). Stratospheric mean ages and transport rates from observations of carbon dioxide and nitrous oxide. *Science*, *274*(5291), 1340–1343. doi:10.1126/science.274.5291.1340.

Braesicke, P., Stiller, G., Morgenstern, O., & Pyle, J. A. (2008a). Modelled zonal asymmetries in age-of-air: To what extent are they real. *Geophysical Research Abstracts*, *10*, EGU2008-A-04791.

Braesicke, P., Stiller, G., Morgenstern, O., & Pyle, J. A. (2008b). *Zonal asymmetries in age-of-air and their relevance for transport into the subtropical lowermost stratosphere*. Poster contribution to the IVth SPARC General Assembly, Bologna, Italy, 31 August to 5 September 2008.

Brasseur, G., & Solomon, S. (1986). *Aeronomy of the middle atmosphere* (2nd ed.). *Atmospheric science library*. Dordrecht: Reidel.

Burgess, A. B., Grainger, R. G., Dudhia, A., Payne, V. H., & Jay, V. L. (2004). MIPAS measurement of sulphur hexafluoride (SF_6). *Geophysical Research Letters*, *31*, L05112. doi:10.1029/2003GL019143.

Burgess, A. B., Grainger, R. G., & Dudhia, A. (2006). Zonal mean atmospheric distribution of sulphur hexafluoride (SF_6). *Geophysical Research Letters*, *33*, L07809. doi:10.1029/2005GL025410.

Butchart, N., & Scaife, A. A. (2001). Removal of chlorofluorocarbons by increased mass exchange between the stratosphere and troposphere in a changing climate. *Nature*, *410*(6830), 799–802. doi:10.1038/35071047.

Butchart, N., Scaife, A. A., Bourqui, M., de Grandpre, J., Hare, S. H. E., Kettleborough, J., Langematz, U., Manzini, E., Sassi, F., Shibata, K., Shindell, D., & Sigmond, M. (2006). Simulations of anthropogenic change in the strength of the Brewer-Dobson circulation. *Climate Dynamics*, *27*(7–8), 727–741. doi:10.1007/s00382-006-0162-4.

Ehhalt, D. H., Heidt, L. E., Lueb, R. H., & Martell, E. A. (1975). Concentrations of CH_4, CO, CO_2, H_2, H_2O and N_2O in the upper stratosphere. *Journal of the Atmospheric Sciences*, *32*(1), 163–169.

Engel, A., Schmidt, U., & McKenna, D. (1998). Stratospheric trends of CFC–12 over the past two decades: Recent observational evidence of declining growth rates. *Geophysical Research Letters*, *25*(17), 3319–3322.

Engel, A., Strunk, M., Müller, M., Haase, H.-P., Poss, C., Levin, I., & Schmidt, U. (2002). Temporal development of total chlorine in the high-latitude stratosphere based on reference distributions of mean age derived from CO_2 and SF_6. *Journal of Geophysical Research*, *107*(D12). doi:10.1029/2001JD000584.

Engel, A., Haase, H. P., Schmidt, U., Poss, C., & Levin, I. (2003). The temporal trend of CO_2 and the mean age of air in the stratosphere derived from balloon borne whole air samples. In N. R. P. Harris, G. T. Amanatidis & J. G. Levine (Eds.), *Proc. 6th European symposium on stratospheric ozone*, 2–6 Sep 2002, Göteborg, Sweden (pp. 152–158). Brussels: European Commission. Air Pollution Research Report 79.

Engel, A., Möbius, T., Haase, H.-P., Bönisch, H., Wetter, T., Schmidt, U., Levin, I., Reddmann, T., Oelhaf, H., Wetzel, G., Grunow, K., Huret, N., & Pirre, M. (2006). Observation of mesospheric air inside the arctic stratospheric polar vortex in early 2003. *Atmospheric Chemistry and Physics*, *6*, 267–282.

Engel, A., Möbius, T., Bönisch, H., Schmidt, U., Heinz, R., Levin, I., Atlas, E., Aoki, S., Nakazawa, T., Sugawara, S., Moore, F., Hurst, D., Elkins, J., Schauffler, S., Andrews, A., & Boering, K. (2009). Age of stratospheric air unchanged within uncertainties over the past 30 years. *Nature Geoscience*, *2*(1), 28–31. doi:10.1038/ngeo388.

Fabian, P., Borchers, R., Weiler, K. H., Schmidt, U., Volz, A., Ehhalt, D. H., & Seiler, W. (1979). Simultaneously measured vertical profiles of H_2, CH_4, CO, N_2O, $CFCl_3$, and CF_2Cl_2 in the midlatitude stratosphere. *Journal of Geophysical Research*, *84*(C6), 3149–3154. doi:10.1029/JC084iC06p03149.

Garcia, R. R., & Randel, W. J. (2008). Acceleration of the Brewer-Dobson circulation due to increases in greenhouse gases. *Journal of the Atmospheric Sciences*, *65*(8), 2731–2739. doi:10.1175/2008JAS2712.1.

Haigh, J. D., Austin, J., Butchart, N., Chanin, M.-L., Crooks, S., Gray, L. J., Halenka, T., Hampson, J., Hood, L. L., Isaksen, I. S. A., Keckhut, P., Labitzke, K., Langematz, U., Matthes, K., Palmer, M., Rognerud, B., Tourpali, K., & Zerefos, C. (2004). Solar variability and climate: selected results from the SOLICE project. *SPARC Newsletter*, *23*, 19–29.

Hall, B. D., Dutton, G. S., Mondeel, D. J., Nance, J. D., Rigby, M., Butler, J. H., Moore, F. L., Hurst, D. F., & Elkins, J. W. (2011). Improving measurements of sf_6 for the study of atmospheric transport and emissions. *Atmospheric Measurement Techniques*, *4*(11), 2441–2451. doi:10.5194/amt-4-2441-2011.

Hall, T. M., & Plumb, R. A. (1994). Age as a diagnostic of stratospheric transport. *Journal of Geophysical Research, 99*(D1), 1059–1070.

IPCC (2007). Contribution of working group I to the fourth assessment report of the intergovernmental panel on climate change. In S. Solomon, D. Qin, M. Manning, Z. Chen, M. Marquis, K. B. Averyt, M. Tignor & H. L. Miller (Eds.), *Climate change 2007: the physical science basis* Cambridge: Cambridge University Press (996 pp.)

Kida, H. (1983). General circulation of air parcels and transport characteristics derived from a hemispheric GCM, Part 2, Very long-term motions of air parcels in the troposphere and stratosphere. *Journal of the Meteorological Society of Japan, 61*, 510–522.

Kodera, K., & Kuroda, Y. (2002). Dynamical response to the solar cycle. *Journal of Geophysical Research, 107*(D24), 4749. doi:10.1029/2002JD002224.

Labitzke, K. (1987). Sunspots, the QBO, and the stratospheric temperature in the north polar region. *Geophysical Research Letters, 14*(5), 535–537.

Lean, J. L., Rottman, G. R., Kyle, H. L., Woods, T. N., Hickey, J. R., & Puga, L. R. (1997). Detection and parameterization of variations in solar mid- and near-ultraviolet radiation (200–400 nm). *Journal of Geophysical Research, 102*(D25), 29939–29956. doi:10.1029/97JD02092.

Li, S., & Waugh, D. W. (1999). Sensitivity of mean age and long-lived tracers to transport parameters in a two-dimensional model. *Journal of Geophysical Research, 104*(D23), 30559–30569. doi:10.1029/1999JD900913.

Plumb, R. A., Heres, W., Neu, J. L., Mahowald, N., del Corral, J., Toon, G. C., Ray, E., Moore, F., & Andrews, A. E. (2002). Global tracer modeling during SOLVE: high-latitude descent and mixing. *Journal of Geophysical Research, 107*, 8309. doi:10.1029/2001JD001023, printed 108(D5), 2003.

Randel, W. J., Wu, F., Vömel, H., Nedoluha, G. E., & Forster, P. (2006). Decreases in stratospheric water vapor after 2001: links to changes in the tropical tropopause and the Brewer–Dobson circulation. *Journal of Geophysical Research, 111*, D12312. doi:10.1029/2005JD006744.

Ray, E. A., Moore, F. L., Elkins, J. W., Hurst, D. F., Romashkin, P. A., Dutton, G. S., & Fahey, D. W. (2002). Descent and mixing in the 1999–2000 northern polar vortex inferred from in situ tracer measurements. *Journal of Geophysical Research, 107*(D20), 8285. doi:10.1029/2001JD000961.

Reddmann, T., Ruhnke, R., & Kouker, W. (2001). Three-dimensional model simulations of SF_6 with mesospheric chemistry. *Journal of Geophysical Research, 106*(D13), 14525–14537. doi:10.1029/2000JD900700.

Rinsland, C. P., Mahieu, E., Zander, R., Nassar, R., Bernath, P., Boone, C., & Chiou, L. S. (2006). Long-term stratospheric carbon tetrafluoride (CF_4) increase inferred from 1985–2004 infrared space-based solar occultation measurements. *Geophysical Research Letters, 33*, L02808. doi:10.1029/2005GL024709.

Schmidt, U., & Khedim, A. (1991). In situ measurements of carbon dioxide in the winter Arctic vortex and at midlatitudes: an indicator of the "age" of stratospheric air. *Geophysical Research Letters, 18*(4), 763–766.

Schmidt, U., Bauer, R., Khedim, A., Klein, E., Kulessa, G., & Schiller, C. (1991). Profile observations of long-lived trace gases in the Arctic Vortex. *Geophysical Research Letters, 18*(4), 767–770.

Shepherd, T. G. (2008). Dynamics, stratospheric ozone, and climate change. *Atmosphere-Ocean, 46*(1), 117–138. doi:10.3137/ao.460106.

Stiller, G. P., v. Clarmann, T., Höpfner, M., Glatthor, N., Grabowski, U., Kellmann, S., Kleinert, A., Linden, A., Milz, M., Reddmann, T., Steck, T., Fischer, H., Funke, B., López-Puertas, M., & Engel, A. (2008). Global distribution of mean age of stratospheric air from MIPAS SF_6 measurements. *Atmospheric Chemistry and Physics, 8*, 677–695.

Strunk, M., Engel, A., Schmidt, U., Volk, C. M., Wetter, T., Levin, I., & Glatzel-Mattheier, H. (2000). CO_2 and SF_6 as stratospheric age tracers: consistency and the effect of mesospheric SF_6-loss. *Geophysical Research Letters, 27*(3), 341–344. doi:10.1029/1999GL011044.

Thompson, D. W. J., & Solomon, S. (2005). Recent stratospheric climate trends as evidenced in radiosonde data: global structure and tropospheric linkages. *Journal of Climate, 18*(22), 4785–

4795. doi:10.1175/JCLI3585.1.
Volk, C. M., Elkins, J. W., Fahey, D. W., Dutton, G. S., Gilligan, J. M., Loewenstein, M., Podolske, J. R., Chan, K. R., & Gunson, M. R. (1997). Evaluation of source gas lifetimes from stratospheric observations. *Journal of Geophysical Research*, *102*(D21), 25543–25564. doi:10.1029/97JD02215.
von Clarmann, T., Linden, A., Oelhaf, H., Fischer, H., Friedl-Vallon, F., Piesch, C., Seefeldner, M., Völker, W., Bauer, R., Engel, A., & Schmidt, U. (1995). Determination of the stratospheric organic chlorine budget in the spring arctic vortex from MIPAS B limb emission spectra and air sampling experiments. *Journal of Geophysical Research*, *100*(D7), 13979–13997.
von Clarmann, T., Stiller, G. P., Eckert, E., Glatthor, N., Grabowski, U., Haenel, F., Höpfner, M., Kellmann, S., Laeng, A., Linden, A., Lossow, S., Orphal, J., Plieninger, J., Schieferdecker, T., Versick, S., Wiegele, A., Funke, B., García-Comas, M., López-Puertas, M., & Grutter, M. (2010a). MIPAS observations of stratospheric and upper tropospheric trace gases: an overview. In *Proc. 'ESA living planet symposium': Vol. ESA SP-686*, 28 June–2 July 2010, Bergen, Norway. CD-ROM, ESA Publications Division, ESTEC, Postbus 299, 2200 AG Noordwijk, The Netherlands.
von Clarmann, T., Stiller, G., Grabowski, U., Eckert, E., & Orphal, J. (2010b). Technical note: trend estimation from irregularly sampled, correlated data. *Atmospheric Chemistry and Physics*, *10*, 6737–6747.
Waugh, D. (2009). The age of stratospheric air. *Nature Geoscience*, *2*, 14–16.
Waugh, D. W., & Hall, T. M. (2002). Age of stratospheric air: theory, observations, and models. *Reviews of Geophysics*, *40*(4), 1010. doi:10.1029/2000RG000101.
Zander, R., Solomon, S., Mahieu, E., Goldman, A., Rinsland, C. P., Gunson, M. R., Abrams, M. C., Chang, A. Y., Salawitch, R. J., Michelsen, H. A., Newchurch, M. J., & Stiller, G. P. (1996). Increase of stratospheric carbon tetrafluoride (CF_4) based on ATMOS observations from space. *Geophysical Research Letters*, *23*(17), 2353–2356.

Acronyms and Abbreviations

ACE-FTS	Atmospheric Chemistry Experiment Fourier Transform Spectrometer
AIM	Aeronomy of Ice in the Mesosphere
AIMOS	Atmospheric Ionization Model OSnabrück
AMIP2	Atmospheric Model Intercomparison Project
ALOMAR	Arctic Lidar Observatory for Middle Atmospheric Research
ARTOS	Atmospheric Response TO Solar Variability
ATLAS	
ATMOS	
AURA	Latin for breeze
BMBF	Bundesministerium für Forschung und Bildung
CAWSES	Climate And Weather of the Sun-Earth System
CCM	Chemistry Climate Model
CCN	Cloud Condensation Nuclei
CCMVAL	Chemistry Climate Model Evaluation
CHAMP	CHAllenging Minisatellite Payload
CFC	ChloroFluorCarbon
CLaMS	Chemical Lagrangian Model of the Stratosphere
CMAM	Canadian Middle Atmosphere Model
CMAT	Middle and upper ATMosphere code
CME	Coronal Mass Ejection
COSMIC	Constellation Observing System for Meteorology, Ionosphere and Climate
CPW-TEC	Climatology of Planetary Waves seen in ionospheric F-region perturbations using TEC of GPS
CR	Cosmic Rays
CRISTA	Cryogenic Infrared Spectrometers and Telescopes for the Atmosphere
CTIP	Coupled Thermosphere-Ionosphere-Plasma
CTM	Chemistry and Transport Model
DFG	Deutsche Forschungsgemeinschaft

F.-J. Lübken (ed.), *Climate and Weather of the Sun-Earth System (CAWSES)*, Springer Atmospheric Sciences,
DOI 10.1007/978-94-007-4348-9,

DLR	Deutsches Zentrum für Luft- und Raumfahrt
ECHAM5-MESSy	atmospheric chemistry general circulation model Modular Earth Submodel System
ECMWF	European Centre for Medium-Range Weather Forecasts
EEP	Energetic Electron Precipitation
EISCAT	European Incoherent SCAtter
EMAC	ECHAM5-MESSy Atmospheric Chemistry
ENVISAT	ENVironmental SATellite
ENSO	El Niño/Southern Oscillation
EOF	Empirical Orthogonal Function
EPP	Energetic particle precipitation
ERA	ECMWF reanalysis
ESA	European Space Agency
ESR	EISCAT Svalbard Radar
EUV	Extreme UltraViolet
FZJ	Forschungszentrum Jülich
GCM	General Circulation Models
GCR	Galactic Cosmic Ray
GEANT	GEometry ANd Tracking, Monte Carlo Toolkit for particle simulations (provided by the CERN)
GEC	Global Electric Circuit
GFZ	Deutsches GeoForschungsZentrum Potsdam
GHG	Greenhouse Gas
GNSS	Global Navigation Satellite System
GOES	Geostationary Operational Environmental Satellite
GOME	Global Ozone Monitoring Experiment
GOMOS	Global Ozone Monitoring by Occultation of Stars
GPS	Global Positioning System
GRACE	Gravity Recovery And Climate Experiment
GSWM	Global Scale Wave Model
GW	Gravity Wave
GW-CODE	Gravity Wave coupling processes and their decadal variation
GWD	Gravity Wave Drag
HALOE	Halogen Occultation Experiment
HAMMONIA	HAMburg Model Of the Neutral and Ionized Atmosphere
HELIOCAUSES	Energetic particle transport in the atmosphere and environment of the Earth, cosmic rays, solar energetic particles, heliospheric and atmospheric transport
HEMT	High Electron Mobility Transistor
HEPPA	High Energy Particle Precipitation in the Atmosphere
HIRDLS	High Resolution Dynamics Limb Sounder
HME	Hough Mode Extension
HYDOX	Response of Atomic Hydrogen and Oxygen to Solar Radiation Changes: Measurements and Simulations
IAP	Leibniz Institute of Atmospheric Physics in Kühlungsborn

IGRF	International Geomagnetic Reference Field
IHY	International Heliospheric Year
IMK	Institut für Meteorologie und Klimaforschung
IPCC	Intergovernmental Panel on Climate Change
IR	InfraRed
ISCCP	International Satellite Cloud Climatology Project
ISR	Incoherent Scatter Radars
ISSI	International Space Science Institute
KASIMA	Karlsruhe Simulation Model of the middle Atmosphere
KIT	Karlsruhe Institute of Technology
KMCM	Kühlungsborn Mechanistic General Circulation Model
LIMA	Leibniz-Institute Middle Atmosphere model
LTE	Local Thermodynamic Equilibrium
MAARSY	Middle Atmosphere Alomar Radar System
MAECHAM5	Middle Atmosphere version of ECHAM5
MAIONO	Middle Atmosphere IONisation by energetic particles
MANOXUVA	Middle Atmosphere NOx variations and solar UV VAriability: Examples to study mesospheric/stratospheric coupling and the impact of solar variability on stratospheric ozone
MESSy	Modular Earth Submodel System
MF	Medium Frequency
MICHAELA	Laboratory experiments on the microphysics of electrified clouds droplets
MIPAS	Michelson Interferometer for Passive Atmospheric Sounding
MLS	Microwave Limb Sounder
MLT	Mesosphere/Lower Thermosphere
MPS	Max-Planck-Institut für Sonnensystemforschung
MR	Meteor Radar
NAM	Northern Annular Mode
NAO	North Atlantic Oscillation
NASA	National Aeronautics and Space Administration
NAT	Nitric Acid Trihydrate
NCAR	National Center for Atmospheric Research
NCEP	National Centers for Environmental Prediction
NDACC	Network for Detection of Atmospheric Composition Change
NH	Northern Hemisphere
NLC	Noctilucent Cloud
NMT-MLT	Seasonal and interannual variability of nonmigrating tides in the mesosphere and lower thermosphere
NOAA	National Oceanic and Atmospheric Administration
NRLMSISE	US Naval Research Laboratory—Mass Spectrometer and Incoherent Scatter Radar
PMC	Polar Mesospheric Clouds
PMSE	Polar Mesosphere Summer Echoes
PMWE	Polar Mesospheric Winter Echoes

POAM	Polar Ozone and Aerosol Measurement
POES	Polar Orbiting Environmental Satellites
ProSECCO	Project on Solar Effects on Chemistry and Climate Including Ozean Interactions
PSC	Polar Stratospheric Clouds
PW	Planetary Waves
QBO	Quasi-Biennial Oscillation
RMR	Rayleigh-Mie-Raman
SABER	Sounding of the Atmosphere using Broadband Emission Radiometry
SACOSAT	Sun driven Atmospheric Change Observed by ground based Stations in the Arctic and Tropics
SAGACITY	SAtellite and model studies of GAlactic cosmic rays and Clouds modulated by solar activITY
SBUV/TOMS	Solar Backscatter in the Ultraviolet, Total Ozone Mapping System
SCIAMACHY	Scanning Imaging Absorption spectroMeter for Atmospheric CHartography
SCOSTEP	Scientific Committee on Solar-Terrestrial Physics
SEM	Space Environment Monitor
SH	Southern Hemisphere
SICMA	Solar variability impacts on the chemical composition of the middle atmosphere: measurements and model predictions
SME	Solar Mesosphere Explorer
SMR	Sub-Millimetre Radiometer on ODIN
SNR	Signal-to-Noise Ratio
SOLEIL	SOLar variability and trend Effects in Ice Layers
SOLIVAR	Models of SOLar total and spectral Irridiance VARiability for climate studies
SOLOZON	Solar irridiance variability on hourly to decadal scale from SCIAMACHY and its impact on middle atmospheric ozone and ozone-climate interaction
SOLSTICE	Solar-Stellar Irradiance Comparison Experiment
SORACAL	Solar Radiation Perturbations on the Coupling of Atmospheric Layers
SOTIVAR	Solar contribution to the variability of middle atmosphere solar tides in their interaction with zonal-mean-flow variations, planetary waves and gravity waves
SPARC	Stratospheric Processes And their Role in Climate
SPE	Solar Proton Event, depending on context also Solar Particle Event
SSI	Solar Spectral Irradiance
SST	Sea Surface Temperature
SSU	Stratospheric Sounding Units
SSW	Sudden Stratospheric Warming

TEC	Total Electron Content
TID	Traveling Ionospheric Disturbance
TIDI	TIMED Doppler Interferometer
TIME-GCM	Thermosphere Ionosphere Mesosphere Electrodynamics General Circulation Model
TIMED	Thermosphere, Ionosphere, Mesosphere Energetics and Dynamics
TOMS	Total Ozone Mapping System
UARS	Upper Atmosphere Research Satellite
UBIC	University of Bremen Ion Chemistry model
UHF	Ultra High Frequency
UT	Upper Troposphere or Universal Time depending on context
UTC	Coordinated Universal Time
UV	UltraViolet
VHF	Very High Frequency
WACCM	Whole-Atmosphere Community Climate Model
WMO	World Meteorological Organization

Index

F.-J. Lübken (ed.), *Climate and Weather of the Sun-Earth System (CAWSES)*, Springer Atmospheric Sciences,
DOI 10.1007/978-94-007-4348-9, © Springer Science+Business Media Dordrecht 2013

MIX
Papier aus verantwortungsvollen Quellen
Paper from responsible sources
FSC® C105338

If you have any concerns about our products,
you can contact us on
ProductSafety@springernature.com

In case Publisher is established outside the EU,
the EU authorized representative is:
Springer Nature Customer Service Center GmbH
Europaplatz 3, 69115 Heidelberg, Germany

Printed by Libri Plureos GmbH
in Hamburg, Germany